AF535080

ECB6

Progress in Biotechnology

Volume 1 New Approaches to Research on Cereal Carbohydrates (Hill and Munck, Editors)

Volume 2 Biology of Anaerobic Bacteria (Dubourguier et al., Editors)

Volume 3 Modifications and Applications of Industrial Polysaccharides (Yalpani, Editor)

Volume 4 Interbiotech '87. Enzyme Technologies (Blažej and Zemek, Editors)

Volume 5 In Vitro Immunization in Hybridoma Technology (Borrebaeck, Editor)

Volume 6 Interbiotech '89. Mathematical Modelling in Biotechnology (Blažej and Ottová, Editors)

Volume 7 Xylans and Xylanases (Visser et al., Editors)

Volume 8 Biocatalysis in Non-Conventional Media (Tramper et al., Editors)

Volume 9 ECB6: Proceedings of the 6th European Congress on Biotechnology (Alberghina et al., Editors)

Progress in Biotechnology 9

ECB6

Proceedings of the 6th European Congress on Biotechnology Florence, Italy, June 13-17, 1993

PART I

Edited by

L. Alberghina
Department of General Physiology and Biochemistry
University of Milan
Milan, Italy

L. Frontali
Department of Cellular and Developmental Biology
University of Rome
Rome, Italy

P. Sensi
c/o Lepetit Research Centre
Gerenzano (VA), Italy

ELSEVIER

Amsterdam — London — New York — Tokyo 1994

ELSEVIER SCIENCE B.V.
Sara Burgerhartstraat 25
P.O. Box 211, 1000 AE Amsterdam, The Netherlands

ISBN: 0-444-89992-8

This book is printed on acid-free paper.

Printed in The Netherlands

PREFACE

The Proceedings of the 6th European Congress on Biotechnology (ECB6) contain the text of most of the General Lectures and of the Symposia Lectures presented at the Congress held in Florence from June 13 to June 17, 1993. The 234 papers published in these volumes reflect the State of the Art of Biotechnology as it was focused upon in the Florence Congress.

A few words on the lines followed by the Scientific Committee of the ECB6 may help to view these Proceedings in the right perspective.

The Scientific Committee acted in the belief that, at the beginning of the nineties, biotechnology can be considered as having reached both scientific depth and operational capacity in many sectors of basic research (molecular cellular biology and enabling technologies) and in its applications (health-care, industrial, environmental and agricultural biotechnology).

The first aim of the Congress was to illustrate front-line research in all these different areas, paying special attention to recent developments.

In order to make the European Congress on Biotechnology a form which encouraged open and widespread debate, the members of the Scientific Committee agreed on the policy that none of its members should be either a speaker in the General Lectures or a chairperson at the Symposia and Round Tables.

Furthermore, submissions of abstracts were considered and often invited by the chairperson of each Symposium, therefore allowing for a large representativeness in the presented papers.

New emerging areas of biotechnological research and applications, not discussed in previous European Congresses on Biotechnologies, were included among the topics of the 56 Symposia and 31 Workshops of ECB6.

The economies of the industrialized nations are presently faced with two important and interwoven problems: the raising of the competitive edge of industrial production, the protection of the environment and the improvement of the quality of life.

After decades of careless use of natural resources and of pollution of the natural ecosystems, a new model of industrial development is necessary in order to protect the environment and, at the same time, to foster economic growth and thereby yield real wealth and create new jobs.

The science-based technologies such as biotechnology can play an important role in such a new economic strategy, especially if their real economic success, which is more than merely flashy financial success, is recognized and promoted. Attention was therefore paid also to lectures, round tables and special symposia covering various aspects of this great challenge.

We hope that this volume offers to the participants in ECB6 the opportunity to read lectures they were not able to attend and to study in greater detail topics they are interested in.

To those who did not attend ECB6 these Proceedings will offer the opportunity to partake of a large part of the information presented at the Congress.

Finally we wish to acknowledge the steadfast assistance of our young colleagues Donatella Capone, Concetta Compagno, Danilo Porro, Marina Vai, who showed great competence in the development of the scientific programme and/or in the editorial activities related to the preparation of these Proceedings.

L. Alberghina
L. Frontali
P. Sensi

ACKNOWLEDGEMENTS

The Organizing Committee gratefully wishes to thank the following Organizations for their contributions that have made possible the realization of the Congress and the publication of the Abstract Books and the Congress Proceedings :

Assobiotec, Associazione Nazionale per lo sviluppo delle biotecnologie - Milan, Italy
Biocine-Sclavo - Siena, Italy
Boehringer Mannheim Italia S.p.A. - Milan, Italy
Cassa di Risparmio di Firenze, Italy
CNR - Comitato per la Chimica
CNR - Comitato Nazionale per le Biotecnologie e Biologia Molecolare
CNR - Progetto Finalizzato Ingegneria Genetica
CNR - Progetto Finalizzato Ricerche Avanzate per Innovazioni nel Sistema Agricolo
EEC - Commission of the European Communities DGXI and DGXII
Crinos S.p.A. - Villaguardia (CO), Italy
ENEA - Ente Nazionale per le Nuove tecnologie, l'Energia e l'Ambiente
Farmitalia Carlo Erba S.r.l. - Milan, Italy
Gruppo Lepetit S.p.A. - Lainate (MI), Italy
Università degli Studi di Firenze, Italy.

Thanks are also due to :

Assobirra - Rome, Italy
Fidia Farmaceutici S.p.A. - Abano Terme (PD), Italy
Schering - Berlin, Germany.

Sixth European Congress on Biotechnology

Florence, June 13-17, 1993

Organised on behalf of the European Federation of Biotechnology

ORGANIZING COMMITTEE

P. SENSI - President
L. ALBERGHINA - Vice President
F. GROS - Vice President
L. FRONTALI - General Secretary
A.L. CATAPANO - Treasurer
E. CERNIA
M. DEL BORGHI
G. DI GIORGIO
C. FRONTALI
E. GALLI
M. GHEZZI
M. POLSINELLI
S. PONTREMOLI
G. RAMPONI
M. TERZI
A.TURANO
A.VALVASSORI

SCIENTIFIC COMMITTEE

L. ALBERGHINA (Italy) - President
L.FRONTALI (Italy) - General Secretary
J.ADLER-NISSEN (Denmark)
G.M.A. van BEYNUM (The Netherlands)
C.F.A. BRYCE (United Kingdom)
R. BUESS (Belgium)
R. BURDON (United Kingdom)
J.M.S. CABRAL (Portugal)
A.L. CATAPANO (Italy)
E. CERNIA (Italy)
J.A. COLE (United Kingdom)
W. CRUEGER (Germany)
M. DEL BORGHI (Italy)
G. DI GIORGIO (Italy)
J. ECONOMIDIS (EEC)
S.O. ENFORS (Sweden)
W. FROMMER (Germany)
C. FRONTALI (Italy)
E. GALLI (Italy)
M. GHEZZI (Italy)
F. GROS (France)
M.J.A. HOFMAN (Belgium)
B. KRISTIANSEN (United Kingdom)
J.L. LEBAULT (France)
H.G. LEUENBERGER (Switzerland)
K.Ch.A.M.LUYBEN (The Netherlands)
B. MACRIS (Greece)
P. MARKKANEN (Finland)
V. MARIC (Yugoslavia)
R. MAS (France)
R.P. MELLADO (Spain)
P. MONSAN (France)
A. MOSER (Austria)
M. MOTA (Portugal)
L. NYESTE (Hungary)
M. POLSINELLI (Italy)
S. PONTREMOLI (Italy)
A. PÜHLER (Germany)
G. RAMPONI (Italy)
G. ROSSI (Italy)
G. SCHMIDT-KASTNER (Germany)
K. SCHÜGERL (Germany)
P. SENSI (Italy)
R. SPIER (United Kingdom)
U. von STOCKAR (Switzerland)
I.W. SUTHERLAND (United Kingdom)
M. TERZI (Italy)
A. TURANO (Italy)
A. VALVASSORI (Italy)
H. VERACHTERT (Belgium)
W. VERSTRAETE (Belgium)

CONTENTS

GENERAL LECTURES

Structure-based design of enzymes: engineering aspartic proteinases and their inhibitors
V. Dhanaraj, P. Nugent, J.E. Pitts, J.B. Cooper and T.L. Blundell 3
Biocatalysis in organic synthesis
H.E. Schoemaker, W.H.J. Boesten, W.J.J. van den Tweel, Q.B. Broxterman and J. Kamphuis 17
Critical problems in bioprocess engineering
J. Tramper 27
Gene pools and biodiversity
W. Arber 43
Strategies of biotechnological industries - Biotechnology in the nineties: back to business
J.C. van Suijdam and J. A. Roels 51
The role of biotechnology in agro-industry and commission policy
C.L. Mangan 57
New trends in vaccines
R. Rappuoli 67
Engineered erythrocytes as carriers and bioreactors for biomedical applications
A. De Flora, U. Benatti, M. Tonetti, A. Gasparini, M. Giovine, G. L. Damonte and A. Sdraffa 73
Biotechnology: a general survey
F. Gros 87

BASIC BIOTECHNOLOGY

Molecular Cellular Biology

CHAPERONINS AND PROTEIN FOLDING

Protein folding by molecular chaperones
F.-U. Hartl 101
Refolding recombinant proteins - Protein chemistry vs process engineering
J.B. Chaudhuri 107

A protein from the hyperthermophilic archaebacterium *Sulfolobus solfataricus* which assists protein folding at high temperature
A. Guagliardi, L. Cerchia, M. Rossi and S. Bartolucci 111

In vitro complex formation between DnaK and some recombinant proteins produced in *Escherichia coli*
K. Gustavsson, A. Veide and S.O. Enfors 115

LIPASES STRUCTURE AND FUNCTION

Relationships between structures and kinetic properties of lipases
R. Verger, H. van Tilbeurgh, F. Ferrato, E. Rogalska, M. Aoubala, A. De Caro, M. Ivanova, L. de la Fournière, C. Rivière, L. Sarda and C. Cambillau 121

Candida cylindracea lipase: approaches towards the understanding of structure-performance relationships
L. Alberghina, L. Battinelli, E. Cernia, F. Fusetti, S. Longhi, M. Lotti and S. Soro 129

Towards the design of microbial lipases
R.D. Schmid 133

Stereoselective hydrolysis of triglycerides and triglyceride - analogs by animal and microbial lipases
C. Cudrey, F. Ferrato, L. Haalck, F. Paltauf, E. Rogalska, F. Spener, P. Stadler and R. Verger 137

SECRETION AND GLYCOSILATION OF HETEROLOGOUS PROTEINS

Sequences promoting or preventing excretion in the growth medium of homologous and heterologous proteins in *Saccharomyces cerevisiae*
M. Vanoni, M. Vai, A. Morrione, M. Venturini, E. Lacanà, D. Rossini, E. Martegani, L. Popolo and L. Alberghina 143

Glycosilation, folding, oligomerization and transport competence of the bean storage protein phaseolin in plant and animal cells
E. Pedrazzini, G. Giovinazzo, R. Bollini, A. Ceriotti and A. Vitale 147

Assembly and intracellular degradation of IgM subunits during B cell development
S. Guenzi, C. Fagioli, A.M. Fra, E. Pasqualetto, C. Valetti and R. Sitia 151

RECEPTOR ENGINEERING

Engineering the hepatocyte growth factor receptor to elucidate its structure and function
A. Bardelli, F. Maina, C. Ponzetto and P.M. Comoglio 157

ANTIBODY ENGINEERING

Engineering of antibody fragments produced in bacteria: solutions to the purification problem
A. Skerra, L.-O. Essen, T. Schmidt and C. Wardenberg 167

Expression, characterization and crystallization of a recombinant anti-neuraminidase single chain antibody
M. Lah, R. Malby, A. Goldstraw, J.F. White, O. Dolezal, D. Dougan, A.A. Kortt, R.A. Irving, P.M. Colman and P.J. Hudson 171

Humanizing antibodies by epitope imprinted selection (EIS)
H.R. Hoogenboom, L.S. Jespers, A. Griffiths, M. Figini, M. Baier, A. Roberts, D.J. Chiswell and G.Winter 175
Intracellular production and renaturation from inclusion bodies of a scFv fragment fused to a serum albumin binding affinity tail
U. Oberg, G. Rundstrom, H. Gronlund, M. Uhlén and P.A. Nygren 179

ANALYSIS OF DIFFERENT GENOMES OF BIOTECHNOLOGICAL INTEREST

Sequencing the yeast genome
A. Goffeau and A. Vassarotti 185
Gene isolation from *Arabidopsis*: prospects, progress and solutions for agriculture in Europe
M. Bevan, C. Dean, R. Schmidt, I. Bancroft, G. Murphy, C. Cobbett and H. Goodman 187
Ordered genomic clone libraries as high resolution mapping tools
E. Maier, J.D. Hoheisel and H. Lehrach 191
Sequencing the *Bacillus subtilis* genome
P. Glaser, A. Danchin and F. Kunst 195

GENETIC STABILITY IN ENGINEERED MICROORGANISMS

Plasmid stability in Gram-negative bacteria: molecular structure and function of the partitioning system of the broad host-range plasmid RP4
E. Grohmann, L. Eberl and H. Schwab 199
Structural and functional analysis of plasmid vectors for *Lactobacillus*
R.J. Leer, N. van Luijk, M. Posno and P.H. Pouwels 203
The *Kluyveromyces* plasmid pKD1: mechanisms ensuring stability and copy number control
M.M. Bianchi, L. Tizzani and L. Frontali 207

BIOTECHNOLOGY OF LACTIC ACID BACTERIA

Biotechnology of lactic acid bacteria: current trends in genetic research
M. Sibakov 213
Heterologous gene expression in *Lactobacillus*
M. Posno, M. Kottenhagen, B.C. Lokman, J. Jore, R.J. Leer and P.H. Pouwels 217
Biosynthesis of the antimicrobial peptide nisin: structure, organization and function of the nisin-operon genes, protein engineering of nisin Z, and characterization of precursors of nisin
O.P. Kuipers, H.S. Rollema, M.M. Beerthuyzen, J.R. van der Meer, W.M. de Vos and R.J. Siezen 221
Diacetyl production by *Lactococcus lactis:* optimilisation and metabolic engineering
J. Hugenholtz, M.J.C. Starrenburg and A.H. Weerkamp 225
Optimization of the production of enterocin 1146, a bacteriocin produced by *Enterococcus faecium*
E. Parente and A. Ricciardi 229

RECENT DEVELOPMENTS IN NITRIFICATION/DENITRIFICATION

Nitrogen removal from wastewater
J.G. Kuenen and L.A. Robertson 235
Formation and consumption of nitric oxide by nitrifying bacteria
E. Bock, R. Stuven, R. Mansch and M. Vollmer 241
Dynamic modelling of the growth of immobilized *Nitrosomonas europeae* cells
C.D. De Gooijer, A.W. Schepers, R.H. Wijffels and J. Tramper 245
Molecular genetic analysis of nitrite reduction to ammonia: a bacterial pathway which competes with denitrification
H. Hussain, H. Crooke, A. Darwin, L. Griffiths, J. Grove, K. Tyson, S. Busby, E. Battaloglu and J. Cole 249

EXTREMOPHILES IN BIOTECHNOLOGY

Extremophiles in biotechnology
M. Rossi and M. De Rosa 255
Characterization of extracellular hydrolases from marine psychrophilic bacteria
U. Kärst, M. Woehl, K. Czempinski and R.D. Schmid 263
Characterization of the xylanotic enzyme system of the thermophilic eubacterium *Thermotoga thermarum*
A. Sunna, J. Puls and A. Antranikian 267
Isolation and characterization of new plasmids from moderately halophilic eubacteria: developing of cloning vectors
A. Ventosa, R. Fernández-Castillo, C. Vargas, E. Mellado, M.T. García and J.J. Nieto 271

BIOTECHNOLOGY OF FILAMENTOUS FUNGI

Biotechnology of filamentous fungi
J.R. Kinghorn and N. Lucena 277
Strategies for the over-production of fungal proteins in *Aspergillus*
C.A.M.J.J. van den Hondel, P.J. Punt and R.F.M. van Gorcom 287
Chitin deacetylation by enzymatic means
D. Kafetzopoulos, A. Martinou, I. Tsigos, A. Christodoulidou, P. Kavelaki and V. Bouriotis 291
Recombinant *Trichoderma reesei* strains producing improved and "tailor made" xylanase profiles
R. Hodits, A. Butterweck, S.P. Goller, A. Toronnen, R.L. Mach, R. Messner, A. Harkki and C.P. Kubicek 295
The simultaneous production of amylase and lipase by the thermophilic fungus, *Thermomyces lanuginosus*
A.S. Frich, B. Jensen and J. Olsen 299

YEAST PHYSIOLOGY AND GENETICS

A new look at carbohydrate metabolism in yeast
C. Gancedo and J.M. Gancedo 305
Bioconversion of xylose to xylitol with *in situ* generation of NAD(P)H in recombinant *Saccharomyces cerevisiae*
H.N. Carlsen, J. Hallborn, M.-F. Gorwa and B. Hahn Hagerdal 313

Ethanol metabolism in the yeast *Kluyveromyces lactis*
M. Saliola, S. Bellardi, I. Marta, and C. Falcone 317
Significance of rate of cellular adaptation to ethanol and rate of change of ethanol concentration in fermentations by *Zymomonas mobilis* and yeast
T.J. Hobley, G.A. Stanley, G. Dasari and N.B. Pamment 323
Yeast non-replicative plasmids and strain construction
Z. Zgaga, P. Koren, M. Ninkovic, K. Gjuracic and M. Alacevic 327

PHYSIOLOGY AND GENETICS OF ANAEROBES

Physiology and genetics of anaerobes
R. Daniel and G. Gottschalk 333
Regulation of solvent formation in *Clostridium acetobutylicum*
R. Fischer, U. Sauer and P. Durre 337
Molecular analysis of the pullulanase-encoding gene of *Clostridium thermosulfurogenes* EM1
G. Burchhardt, K. Sahm, M. Matuschek, H. Müller, K. Haeckel, A. Spreinat, G. Antranikian, G. Steinborn and H. Bahl 341

TAXONOMICAL ASPECTS IN BIOTECHNOLOGY

Implication of bacterial taxonomy on biotechnology
K.H. Schleifer 347
The importance of correct nomenclature in biotechnology: a brief history of *Saccharomyces sensu stricto*
A. Vaughan-Martini and A. Martini 355
Directed automated solid phase sequencing of an *in vitro* amplified segment of 16S rDNA from bacteria
B. Pettersson, K.-E. Johansson and M. Uhlén 359
A general method for generating specific DNA probes by arbitrarily-primed PCR (AP-PCR)
R. Fani, G. Damiani, C. Di Serio, E. Gallori, A. Grifoni and M. Bazzicalupo 363
Protein electrophoretic patterns for high-resolution identification of micro-organisms
K. Kersters, B. Pot, D. Dewettinck, U. Torck and L. Vauterin 367

EXPRESSION OF PROTEINS IN TRANSGENIC ANIMALS

The production of recombinant proteins in transgenic animals
L.M. Houdebine 373
The application of aqueous two phase partitioning to the purification of proteins from transgenic sheep milk
P. Harris, G. Wright, A.T. Andrews, D.L. Pyle and J.A. Asenjo 381

GENE EXPRESSION IN ANIMAL CELL CULTURES

Relationships between heterologous protein synthesis and the culture environment of Chinese hamster ovary cells
A.T. Bull 387
The biosynthesis of ribosomes: translational regulation of the ribosomal protein genes
P. Pierandrei-Amaldi and B. Cardinali 395

In situ hybridization and image analysis for mRNAs evaluation on single cells
S. Barlati, N. Zoppi, A. Ghinelli and M. Colombi 399
A cancer sensitivity test based on expression of mutant p53 in cultured human cells
C. Mothersill, J. Harney and C.B. Seymour 403
Expression of exogenous genes in porcine somatic cells using recombinant retroviral and episomal vectors
P.K. Nandi and O. Grépinet 407

EXPRESSION OF HETEROLOGOUS GENES IN YEAST SYSTEMS

Biotechnology in yeasts: alternative protein production systems
J.R. Shuster 413
Expression of a synthetic acylphosphatase gene in *E. coli* and *S. cerevisiae*
A. Modesti, G. Raugei, M. Vecchi, R. Marzocchini, G. Manao, G. Camici, G. Cappugi and G. Ramponi 417
A comparison of guar α-galactosidase production by *Saccharomyces cerevisiae* SU50B and *Hansenula polymorpha* 8/2 in continuous cultures
M.L.F. Giuseppin, J.W. Almkerk, J.C. Heistek and C.T. Verrips 421
Processing and secretion of *Cyamopsis tetragonoloba* α-galactosidase by *Saccharomyces cerevisiae* wild-type and *pmr1* strains
M.M. Harmsen, R.H. Geerse, H.A. Raué and J. Maat 425

Enabling technologies

RATIONAL DESIGN OF NOVEL BIOCATALYSTS

The blind watchmaker and protein engineering
S.B. Petersen and F. Drabløs 433
Evidence of catalytic antibodies endowed with spermidine synthase-like activity
G. Pontoni, A. Borriello, M. Cartenì-Farina and V. Zappia 441
Modification of substrate specificity of haloalkane dehalogenase by site-directed mutagenesis
J.P. Schanstra, F. Pries and D.B. Janssen 445
Replacement and introduction of COOH-groups involved in catalysis, acarbose binding, and stability in *A. niger* glucoamylase by site-directed mutagenesis
T.P. Frandsen, J. Lehmbeck, C. Dupont, B. Stoffer, M.R. Sierks, M.T. Hansen and B. Svensson 449

BIOCATALYSIS IN ORGANIC SYNTHESIS: THE POTENTIALITY OF LYASES

The potential of lyases in organic synthesis
W.J.J. van den Tweel, M.J. van der Werf, S. Hartmans, H.E. Schoemaker, J. Kamphuis and J.A.M. de Bont 455
Development of transketolase as a catalyst for carbon-carbon bond formation
G.R. Hobbs, M.D. Lilly, N.J. Turner, J.M. Ward and J.M. Woodley 463

Application of microbial transformation of prochiral 1,3-dichloro-2-propanol into (R)-3-chloro-1,2-propanediol
T. Nagasawa, T. Nakamura and H. Yamada 467
D-malate and D-citramalate production with maleate hydro-lyase from *Pseudomonas pseudoalcaligenes*
M.J. van der Werf, W.J.J. van den Tweel, J. Kamphuis, S. Hartmans and J.A.M. de Bont 471

PHYSIOLOGICAL AND ENGINEERING PROBLEMS LINKED TO IMMOBILIZED CELL FERMENTATIONS

Immobilized viable cell fermentations as applied to microbial enzyme production: problems and promises
S. Linko, R. Haapala and P. Linko 477
Semi-continuous production of the enzymes by a novel solid-state fermentation system using urethan foam as carriers
S. Ozawa, K. Amada, K. Sato, T. Nagamune and I. Endo 485
Specific ATP concentrations on freely suspended and immobilized cells degrading 3,4 dichloroaniline
P. Gikas and A.G. Livingston 489
Immobilization of yeast cells in spherical porous glass particles: consequences of intraparticle convection
L. De Backer and G. Baron 493
Production of actinorhodin by immobilised and freely suspended *Streptomyces coelicolor*
K. Ozergin and F. Mavituna 497

NEW SEPARATION METHODS IN BIOTECHNOLOGY

New separation methods in biotechnology
J.A. Asenjo 503
Cross-flow electrofiltration of disrupted microbial cells
A. Brors, K.H. Kroner and W.D. Deckwer 511
Modelling of centrifugal partition chromatography
M.J. van Buel, L.A.M. van der Wielen, R. Meester, R. Verpoorte and K.C.A.M. Luyben 515
High performance membrane chromatography (HPMC), superior to conventional protein-HPLC?
O.-W. Reif, V. Nier and R. Freitag 521

AFFINITY BASED PROCESSES IN BIOSEPARATIONS

Affinity methods in bioseparations
Y.D. Clonis 527
Kinetics and equilibria of interactions between lysozyme and dye/dextran conjugates
A.G. Mayes, R. Eisenthal and J. Hubble 535
Purification of secreted proteins by immunoaffinity chromatography with cross-reactive and anti-peptide antibodies
S. Katoh and M. Terashima 539
Sequence-specific DNA affinity chromatography: application of a group specific adsorbent for the isolation of restriction endonucleases
C. Pozidis, G. Vlatakis and V. Bouriotis 543

New diffusive matrix to improve throughput in affinity chromatography applications
E. Boschetti 547

BIOSENSOR DESIGN

Report on biosensor technology in Europe
M. Mascini 553
Enviromental monitoring of pesticides using biosensors
J.-L. Marty 557

QUANTITATIVE FLOW MICROFLUORIMETRY IN BIOTECHNOLOGY

Flow cytometry in biotechnology: potential and limitations
F. Srienc 563
Flow cytometry of micro-colonies as an approach for mutant selection and industrial strain improvement
E. Sahar, R. Nir, T. Molcho and R. Lamed 571
Analysis of the respiratory activity in growing budding yeast by flow cytometry
D. Porro, C. Smeraldi, B.M. Ranzi, E. Martegani and L. Alberghina 577
A new method for measurement of poly-β-hydroxybutyrate inside *Alcaligenes eutrophus* during cultivation by flow cytometry and spectrofluorometry
T. Scheper, A. Degelau and M. Ballmaier 581
Process application of flow cytometry to the analysis of growth and physiological response in animal cells
M. Al-Rubeai and A.N. Emery 587

GENETIC AND PHYSIOLOGICAL APPROACHES FOR THE PRODUCTION OF AMINO ACIDS

Biosynthesis and secretion of amino acids in coryneform bacteria
R. Kramer, A. de Graaf, L. Eggeling, B. Eikmanns and H. Sahm 595
Advanced genetic techniques for amino acid producing corynebacteria
A. Puhler, W. Jager, F. Kassing, A. Schafer, A. Schwarzer, A. Tauch and J. Kalinowski 603
Expression of *Vitreoscilla* hemoglobin in *Corynebacterium glutamicum* increases final concentration and yield of L-lysine
F.C. Sander, R.A. Fachini, D.E. Hughes, J.L.Galazzo and J.E. Bailey 607

BIOTECHNOLOGICAL INNOVATIONS IN FOOD PRESERVATION PROCESSING AND ANALYSIS

Recent development in yeast strains for baking
T. Benítez, A.C. Codón and J.M. Gasent-Ramírez 613
Effect of HeNe and CO_2 lasers on yeast and dough
L. Bolognani, F. Casacci, N. Haegens, C. Magnani and T. Venturelli 623
Mixed culture pre-ferment of lactic and propionic acid bacteria for improved shelf-life of sour rye bread
P. Javanainen and Y.-Y. Linko 627

Mould interactions between contaminants and starter cultures used for sausage fermentation
S.J. Andersen 631
Partial characterization of pectin methylesterase inhibitor from kiwi fruit (*Actinidia chinensis*) purified by affinity chromatography
L. Camardella, G. Di Prisco, A. Giovane, L. Quagliuolo and C. Balestrieri 635

BIOINFORMATICS FOR BIOTECHNOLOGY

Bioinformatics for biotechnology
C. Saccone 641
BIOREP-permanent inventory of biotechnology research projects in the European Communities
M.L.H. Lalieu 649
Hypervariable regions in B-type nuclear lamins
A. Lisa, G. Biamonti, G. Perini, F. Weighardt and C. Matessi 653
Computer system for genes structure prediction
L. Milanesi, I.V. Ishenko, I.B. Rogozin, V. Kapitonov, N.A. Kolchanov, A.E. Kel and M.P. Ponomarenko 657

APPLIED BIOTECHNOLOGY

Health-care biotechnology

BIOTECHNOLOGICAL APPLICATIONS OF ANTISENSE OLIGONUCLEOTIDES

Biotechnological applications of antisense oligonucleotides
E. Uhlmann 667
Control of tumor cell growth: targetting hybrid transcripts in human lymphoma
S. Morelli, L. Calastretti, E. Copreni, M. Lauletta and A. Nicolin 675
New developments in the field of antisense oligonucleotides
H. Seliger, B. Krist, H. Rösch, R. Rösch, A. Rück, R. Steiner, K.C. Gupta and J.F.R. Ortigao 681
Gene therapy of AIDS: inhibition of HIV by retroviral vectors
M.A. Biasolo, C. Parolin, J. Sodroski, W.A. Haseltine and G. Palù 685
Inhibition of human immunodeficiency virus reactivation from latency by *tat* antisense RNA and a *tat* transdominant negative mutant
A. Caputo, M.P. Grossi, P.G. Balboni, R. Bozzini, P.C. Marconi, S. Zucchini, G. Barbanti-Brodano and R. Manservigi 689

NEW DEVELOPMENTS IN GENETIC DIAGNOSIS

An integrated colorimetric PCR-method for qualitative and quantitative diagnosis
J. Lundeberg, A. Hedrum, N. Seesod and M. Uhlén 697

NEW DEVELOPMENTS IN IMMUNODIAGNOSTICS

The impact of biotechnology on immunodiagnostics
A. Albertini and C. Signorini 703
Mycotoxin detecting monoclonal antibodies
A. Gyöngyösi, I. Barna Vetrò and L. Solti 709
New immunodiagnostic colloidal carbon label: quantitative computer image analysis of a human chorionic gonadotropin sol particle immunoassay
A. van Amerongen, L.B.J.M. Berendsen, A.J.M. Timmermans and J.H. Wichers 713
New immunoanalysis systems for application in biotechnology
B. Schulze, A. Schlösser, C. Middendorf, C. Schelp, T. Scheper, K. Schügerl, W. Noé, H. Hoffmann and M. Howald 717

BIOTECHNOLOGY OF BLOOD COAGULATIONS AND FIBRINOLYTIC FACTORS

Biotechnology of blood coagulation and fibrinolytic factors
W.D. Schleuning 723
Novel and safe plasminogen activators from vampire bat (*Desmodus rotundus)* salivary glands
P. Donner, P. Bringmann, T. Petri, M. Mc Caman, L. Cashion and W.D. Schleuning 727
Manufacturing of recombinant factor VIII (Kogenate) for the treatment of hemophilia A
H.D. Hörlein 731

GENETIC ENGINEERING FOR THE PRODUCTION OF ANTIBIOTICS

Organization and expression of genes for biosynthesis of ß-lactam antibiotics in filamentous fungi and *Actinomycetes*
J.F. Martìn, S. Gutiérrez, J.J.R. Coque, E. Montenegro, F.J. Fernández, J. Velasco, F. Fierro, S. Gil, A.T. Marcos, C. Arenós, A. Rodríguez, R.E. Cardoza, F.J. Llarena, F. Enguita, M.A. Lumbreras and P. Liras 737
Metabolic engineering of cephalosporin biosynthesis in *Streptomyces clavuligerus:* genetic manipulation of Lysine ε-aminotransferase (*lat*) and ACV synthetase (*pcbAB*)
L.-H. Malmberg, W.-S. Hu and D.H. Sherman 745
Characterization of a giant linear plasmid prime carrying the oxytetracycline genes in *Streptomyces rimosus*
D. Hranueli, B. Gravius, G. Biukovic, M. Redenbach and J. Cullum 749
Stimulation of erythromycin A yield by integration of a chromosomal DNA fragment including the *eryC1* gene into the chromosome of *Saccharopolyspora erythraea*
F. Hänel, G. Schumann, G. Fiedler, K. Dornberger and H. Krügel 753
Genetic studies on lincomycin production in *Streptomyces lincolnensis*
J. Spízek, Z. Vanêk and P. Tichy 757

ENZYMES AND ENZYME INHIBITORS AS THERAPEUTIC AGENTS

Enzymes and enzyme cofactors as therapeutic agents: a challenge for modern biotechnology
G. Rotilio 763

The structure and regulation of L-asparaginases from *E. coli* and *Salmonella*
I.R. Beacham, M.P. Jennings, S.P. Scott, D.M. Burns, T.M. Yeadon, M. Chao, K. Andrews and J. Lewin 771

Amine oxidases as possible antineoplastic drugs
B. Mondovì, E. Agostinelli, F. Przybytkowski, M.A. Mateescu and D.A. Averill-Bates 775

Inhibition of cathepsin B from *Schistosoma mansoni* as a possible antiparasitic approach
M.Q. Klinkert, L. Pica-Mattoccia, E. Shaw and D. Cioli 779

Synthesis of the mimics with glutathione peroxidase activity and preparation of the conjugate of mimic-IgG(Fcγ)
G.M. Luo, W.C. Ma, Z.M. Wang, Z.L. Huang and T.S. Yang 783

ANTIGEN PRESENTATION AND PROCESSING

Generation of functional dendritic and macrophage cell lines: a new tool to explore antigen presenting cell functions *in vitro*
P. Paglia, G.P. Girolomoni, F. Robbiatti, F. Granucci and P. Ricciardi-Castagnoli 791

Antigen reactivities of human γδ cells
G. De Libero and D. Rosa 795

Discovering consensus motifs for MHC class II binding peptides using M13 display epitope libraries
J. Hammer, P. Valsasnini and F. Sinigaglia 799

Peptide binding to HLA class I molecules measured by a quantitative molecular binding assay
A. Sette, J. Sidney, M.-F. del Guercio, S. Southwood, J. Ruppert, C. Dahlberg, H.M. Grey and R.T. Kubo 803

Industrial, environmental and agriculture biotechnology

- **Development of bioprocess**

IMPACTS OF GENETIC ENGINEERING ON BIOPROCESSING

Impact of genetic engineering on the production of recombinant proteins
S. Ståhl, P.-Å. Nygren and M. Uhlén 811

Recombinant protein fermentation during high cell density cultivation (HCDC) of *Escherichia coli*
D. Riesenberg, H.-D. Pohl, V. Schroeckh and W.A. Knorre 817

The application of aqueous two phase partitioning to the purification of proteins from transgenic sheep milk
P. Harris, G. Wright, A.T. Andrews, D.L. Pyle and J.A. Asenjo 823

Regulable expression systems for the optimisation of recombinant protein production in *Escherichia coli*
W. Kramer, D. Mattanovich, G. Elmecker, R. Weik, C. Lüttich, K. Bayer and H. Katinger 827
Manipulation of alginate and gellan gum pathways in *Pseudomonas aeruginosa* or *Pseudomonas elodea* by recombinant DNA technology
J.H. Leitão, A.M. Fialho and I. Sá-Correia 831

MODELLING OF BIOPROCESSES

Identification of mathematical models for biotechnological processes: recipe and application to a culture of animal cells
V. Chotteau and G. Bastin 837
Mathematical model for filamentous growth of mycelial microorganisms in lab chambers and in batch, fed-batch and continuous cultures
H. Yang 845
A structured model for the penicillin fermentation
C.R. Thomas, G.C. Paul and A. Kent 849
Evaluation of cellular parameters in bioprocesses by segregated modeling
L. Cazzador and L. Mariani 853
Non-linear models for indirect measurements and on-line estimation of kinetic parameters in lactic acid fermentation
G. Acuna, E. Latrille, C. Beal and G. Corrieu 857

ON-LINE MEASUREMENTS OF BIOPROCESSES

Optical sensor systems for bioprocess monitoring
T. Scheper 865
Flow injection bioanalysis: a convenient tool in process monitoring and control
B. Mattiasson 869

DYNAMICS IN BIOREACTORS

Dynamics in bioreactors
L.M. Harvey and B. McNeil 875
The dynamic step response of a fed-batch bakers' yeast process
M. Keulers, L. Ariaans and M.L.F. Giuseppin 879
Metabolic response of *Saccharomyces cerevisiae* in a scale-down two compartment reactor with controlled substrate oscillations
S. George, G. Larsson and S.-O. Enfors 883
Dynamics of *Saccharomyces cerevisiae* in continuous culture
S. Frandsen, J. Nielsen and J. Villadsen 887
Study of control problems in large scale multiturbine fermentors
M.N. Pons, J.P. Corriou and L. Belfares 891

INTEGRATED BIOPROCESSING

Integrated bioprocessing
K. Biedermann 897
Development of an integrated process for the production of lactic acid from whey permeate
P. Börgardts, W. Krischke, H. Chmiel and W. Trösch 905

Two-stage process for increasing cell disruption of *E. coli* for intracellular products recovery
R.G. Lutzer, C.W. Robinson and B.R. Glick 909
Recovery of extracellular human insulin like growth factor-I and II as a fusion protein from *Escherichia coli* culture broth by aqueous two-phase extraction
K. Köhler, B. Nilsson and A. Veide 913

SCALE-UP OF HETEROGENEOUS BIOPROCESSES

Scale-up of heterogeneous bioprocesses
J. Feijen, J.J.M. Hofmeester and D. Groen 919
Large scale production of a clinical grade murine monoclonal antibody
R. Ziliotto, A. Fretta, L. Piacenza and L. Garofano 927
Characterization of bioreactors using isotope techniques
A.G. Pedersen, J. Nielsen and J. Villadsen 931
Verification of integrated microbial and fluid dynamics: *Saccharomyces cerevisiae* production on 30 m³ scale
H. Noorman, S.-O. Enfors, B.H. Hjertager, G. Larsson, K. Morud, C. Trägardh and M. Törnkvist 935

MULTIPHASE BIOREACTORS

Multiphase bioreactors
M.M.R. da Fonseca, S.S. Alves, D. Mateus and I.M.L. Nunes 941
Extended pseudo-homogeneous model for three phase mass transfer with reaction
E. Nagy and A. Moser 949
Dynamics of biofilm formation on small suspended particles in airlift reactors
L. Tijhuis, H. Zwols, K.-H. van Meekeren, M.C.M. van Loosdrecht and J.J. Heijnen 953
Kinetic characterisation of aqueous-organic biphasic enzymatic catalysis for the design of multiphasic bioreactors
P.J. Cunnah and J.M. Woodley 957
Impeller-associated shear effects in fungal fermentations and implications for bioreactor design
Y. Chisti and M. Moo-Young 961

STANDARDISATION IN BIOTECHNOLOGY

Biotechnology standardisation in Europe
B.P. Ager 967
Standardised chromatographic media products by operative change control
S. Castensson 973
Validation of chromatographic processes
G. Sofer 977

• Applications

BIOTECHNOLOGICAL PRODUCTION OF FINE CHEMICALS

Trends in the search and production of novel microbial fine chemicals and biopharmaceuticals
E.J. Vandamme 985
Factors influencing the formation of 3-keto-derivatives from disaccharides by *Agrobacterium tumefaciens*
E. Stoppok, J. Walter and K. Buchholz 999
Catechols from nitroaromatics
F.D. Marvin-Sikkema, P.E.J. Groenewegen and J.A.M. de Bont 1003

Operational strategies to improve lipase production by *Candida rugosa*
M.A. Gordillo, S. López, J.L. Montesinos, N. Obradors, F. Valero, J. Lafuente and C. Solà 1007
Fully-controllable, high surface-to-volume ratio photobioreactors for the production of speciality chemicals from oxygenic phototrophs
M. Tredici, S. Biagiolini, G. Chini Zittelli, E. Montaini, F. Favilli, D. Mannelli and R. Materassi 1011

BIOTECHNOLOGICAL PRODUCTION OF ENERGETIC SOURCES

Photosynthetic bacteria for solar energy conversion
J. Miyake 1019
Biological hydrogen production via bioelectrochemical fuel cells connected with water electrolysis
K. Schenck, K. Vuorilehto, D. Sell and G. Kreysa 1027
Biological hydrogen metabolism and hydrogenases
P. Pedroni, G. Galli, G.M. Mura, C. Pratesi and G. Grandi 1031
Xylose fermentation by immobilized *Pichia stipitis* in two-layer beads
S. Kuzmanova, E. Vandeska and I. Georgievska 1035

TRANSGENIC PLANTS AND CROP PRODUCTIVITY

Genetic engineering of agronomically important legumes and cereals
P. Christou 1041
Consuming transgenic food crops: the toxicological and safety aspects of tomato expressing CryIA(b) and NPTII
H.P.J.M. Noteborn, M.E. Bienenmann-Ploum, J.H.J. van den Berg, G.A. Alink and H.A. Kuiper 1045
Expression of a TGMV anti-sense RNA in transgenic tobacco plants confers resistance to other related geminiviruses
E.R. Bejarano and C.P. Lichtenstein 1049
Appearance of extensive DNA polymorphism in transgenic rice (*Oryza sativa* L.) and in its progeny
F. Sala, S. Castiglione, P.H. Bao, W. Li, G. Wang, C. Giordani, A. Callegarin, C. Bandi, G. Damiani, K. Datta, I. Potrykus and S.K. Datta 1053

BIOFERTILIZERS, BIOPESTICIDES AND BIOLOGICAL CONTROL

Genetic engineering of siderophore receptor genes in *Pseudomonas* directed at evaluating biocontrol capabilities
J. Morris and F. O'Gara 1059

Quantitative and qualitative study of the production of broad-spectrum antifungal lipopeptides from *Bacillus subtilis* S499
P. Jacques, C. Hbid, F. Vanhentenryck, J. Destain, G. Baré, H. Razafindralambo, M. Paquot and P. Thonart 1067

A new larvicidal activity in *Bacillus sphaericus* cells
D. Klein, I. Uspensky and S. Braun 1071

Bacillus subtilis as a biopesticide: biochemical and technological aspects
P. Thonart, V. Heuzé, J. Destain, E. Meurisse, P. Jacques, C. Hbid, P. Lepoivre, H. Razafindralambo and M. Paquot 1075

Biochemical and molecular characterization of crystal protein-negative mutants of *Bacillus thuringiensis*
P. Afkhami, C. Halicigil, G. Özcengiz and N.G. Alaeddinoglu 1081

Optimization of survival of *Pseudomonas fluorescens* by manipulation of growth conditions and adaptation to stress
L.M. Noble, M. Legg, and G.M. Stephens 1085

NEW DEVELOPMENTS IN THE PRODUCTION OF ALCOHOLIC BEVERAGES

New developments in the production of alcoholic beverages - Modification of existing biochemical pathways in yeast
M.C. Kielland-Brandt 1091

Volatile compounds kinetics in mead production
M.N. Pons and S. Schutze 1097

Brewer's yeast activity in correlation with reduced glutathione (GSH) and adenine nucleotide concentrations
N. Kolb and V. Maric 1101

Population dynamics of a mixed yeast strain culture during fermentation
M. Schütz and J. Gafner 1107

Deacidification of grape musts and wines by included *Schizosaccharomyces* yeast
P. Taillandier and P. Strehaiano 1111

MARINE BIOTECHNOLOGY

Microalgae as a source of pharmacologically valuable polysaccharides
O. Pulz and E. Koehler 1117

Eicosapentaenoic acid production by the microalga *Phaeodactylum tricornutum* using continuous culture
A. Reis, L. Gouveia, V. Veloso, H.L. Fernandes, J.A. Empis and J.M. Novais 1121

An evaluation of marine biotechnology for atmosphere control in diesel-electric powered submarines
S.P. Roe 1125

A new identification technology of marine heterotrophic aerobic bacteria performed on an informatic algorithm based on a neural net elaboration of whole cellular fatty acid profiles
S. Bertone, L. Casareto, R. Berti, S. Reina, M. Giacomini, C. Ruggiero and L. Calegari 1129

AGROINDUSTRIAL WASTES UTILIZATION BY GENETICALLY ENGINEERED MICROORGANISMS

Use of recombinant *Saccharomyces cerevisiae* cells for the utilization of agroindustrial wastes
E. Martegani, C. Compagno, D. Porro, A. Tura, L. Brambilla, M. Vanoni, B.M. Ranzi and L. Alberghina 1135
Utilization of xylose with recombinant *Saccharomyces cerevisiae* harbouring genes for xylose metabolism from *Pichia stipitis*
N. Meinander, J. Hallborn, S. Keränen, H. Ojamo, M. Penttilä, M. Walfridsson and B. Hahn-Hägerdal 1143
Water treatment with planktic microalgae
O. Pulz and H. Petzold 1147
Characterization of a *Streptomyces* endoglucanase gene
G. Mastromei, E. Hanahart, T. Irdani, B. Perito and M. Polsinelli 1151
Production and purification of *Serratia marcescens* protease from whey
C. Quirós, L.A. García and M. Díaz 1155

BIOHYDROMETALLURGY

Thiobacillus ferrooxidans, a versatile mineworker
P. Bos, R. Meulenberg, J.T. Pronk, W. Hazeu and J.G. Kuenen 1161
Development of genetic systems for moderately thermophilic, mineral sulphide-oxidizing bacteria
N.P. Burton, F.E. Gibson, J.C. Murrell and P.R. Norris 1169
A comparative study of microbial leaching of metals from anaerobically digested sewage sludge with autotrophic and heterotrophic microorganisms
H. Strasser, H. Brunner and F. Schinner 1173
Semi-continuous reactor studies of a high temperature copper bioleaching process
J.C. Duarte, N. Nunes, M.J. Moura, H. Beaumont and S. Calhau 1177

GASEOUS EFFLUENTS TREATMENT USING BIOTECHNOLOGICAL APPROACHES

Biological purification of exhaust air - Applications, process technology
K. Kirchner, U. Schlachter and S. Wagner 1183
Modelling experiments on the kinetics of mixed-solvent removal from waste gas in a biofilter
M.A. Deshusses and I.J. Dunn 1191
Application of biofiltration for the deodorization of phenol gaseous emissions
M. Zilli, A. Converti, G. Delucchi, M. Del Borghi and G. Ferraiolo 1199
Membrane bioreactor for waste gas treatment: removal of pollutants with a low solubility in water
M.W. Reij and S. Hartmans 1203
Removal of poorly water-soluble compounds from waste gases using water-immiscible organic solvents
M.T. Cesario, H.H. Beeftink and J. Tramper 1207

BIOREMEDIATION, DETOXIFICATION AND METABOLISM OF ENVIRONMENTAL CHEMICALS

The potential of microorganisms for bioremediation
W. Fritsche 1213

Factors influencing biological soil decontamination
E. Bauer, C. Pennerstorfer, S. Renner, B. Slavica and R. Braun 1223

Evolution of an acrylamide-degrading amidase in *Methylophilus methylotrophus*
N.J. Silman, M.A. Carver and C.W. Jones 1227

Decontamination biosorption to concentrate metals on granular activated carbon-influence of pH and temperature, and use of this material to adsorb chloroform
J.A. Scott and A.M. Karanjkar 1231

A new model of n-alkane degradation in heavy oils by an aerobic microorganism
L. Setti, G. Lanzarini and P.G. Pifferi 1235

ECOLOGICALLY SUSTAINABLE PROCESSING TECHNOLOGY

Ecologic process engineering - A new dimension of biotechnology
M. Narodoslawsky 1241

Eco-principles for restructuring technology in direction of ecological sustainability
A. Moser 1251

Problems of practical implementation of ecologic process technologies
K. Soyez 1255

Invasiveness and coproductivity. Criteria for the assessment of different biotechnology paths
A. von Gleich and L.H. Grimme 1259

A strategy for waste minimization in biotechnological processes
A.B. Larsen and F.H. Funch 1263

SPECIAL SYMPOSIA

EDUCATION AND TRAINING IN BIOTECHNOLOGY

Training through research in the framework of the EC programmes "BRIDGE", "BIOTECH" and "HCM" (Human Capital and Mobility)
D. de Nettancourt 1271

Education and training in biotechnology: an overview of some EC initiatives
B. Cardiff and M. Claessens 1277

The role of the EFB working party on education in the European Network
J.G. Kuenen and D.J. Bennett 1281

Education: a necessary element in the development of public perception on biotechnology? The EIBE-project
V.R.A. Damen 1285

BIOSAFETY RESEARCH IN THE EEC

The use of genetically modified organisms (GMOs) in the environment: biosafety results of field tests and key scientific issues
M.P. Nuti, A. Squartini, A. Giacomini and V. Corich 1291
Safety and efficacy assessment of a vaccinia-rabies recombinant virus for oral vaccination of foxes against rabies
P.P. Pastoret, D. Boulanger and B. Brochier 1297
Reviewing plant biosafety research in the EC
P. Rüdelsheim 1301
Biosafety research: what is being done with bacteria?
J.E. Beringer, H.E. Kay and B.A. Handley 1305

SCIENCE PARKS FOR BIOTECHNOLOGICAL DEVELOPMENT

Knowledge-based development in a global society: new challenges and roles for cities-creating conditions conducive to science and innovation
R.V. Knight 1311
Biotechnology/biomedical technology parks, incubators, research centers and institutes in the United States: an overview
E. Klein 1321
The "Biopolo Milano" project
L. Alberghina and E. Cernia 1325

SUBJECT INDEX 1329

G E N E R A L
L E C T U R E S

ECB6: Proceedings of the 6th European Congress on Biotechnology
edited by L. Alberghina, L. Frontali, P. Sensi

Structure-based Design of Enzymes: Engineering Aspartic Proteinases and their Inhibitors.

V. Dhanaraj, P. Nugent, J. E. Pitts, J. B. Cooper and T. L. Blundell.

Laboratory of Molecular Biology, Department of Crystallography, Birkbeck College, University of London, Malet Street, London WC1E 7HX, UK.

INTRODUCTION

Knowledge of the three-dimensional structures of proteins and powerful molecular modelling techniques provide a powerful approach for the design of novel biomolecules with specifically engineered properties. Structure-based design is a central feature of both protein engineering and drug design, which can both be described as multidisciplinary design cycles [1] illustrated in Figure 1. Recombinant DNA, biochemical and organic chemical approaches to prepare proteins and their ligands can be combined with X-ray analysis, NMR and computational modelling techniques to determine and model three-dimensional structures. In one cycle mutant proteins with altered functions, specificities or stabilities can be designed and expressed. In a parallel cycle modified ligands are designed and synthesised with more specific binding to the targeted protein.

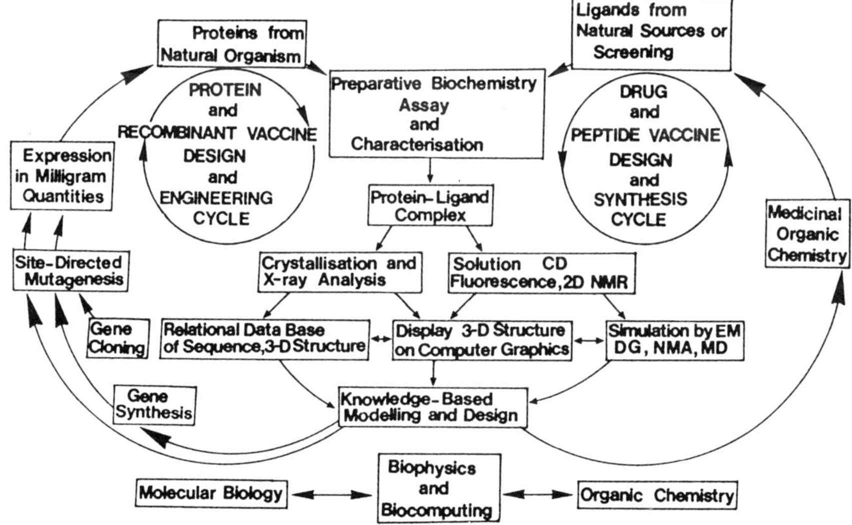

Figure 1. The multi-disciplinary cycle illustrating the design steps

Usually the determination of the three-dimensional structure of the target protein is a rate-limiting step in this process. However, quite often another protein with a common fold can be identified [2] and much can be learnt from comparative modelling using the sequence of the target protein and the three-dimensional structures of one or more related proteins [3]. Even when the structure of the target protein has been determined, the comparative study of families of homologues and their ligand complexes can be very useful in understanding specificity and stability, and in designing novel proteins and ligands. The research carried out at Birkbeck on aspartic proteinases provides good examples to illustrate these themes.

Aspartic proteinases form a family of endopeptidases belonging to a wide range of biological species with varying substrate specificities. The members of the family are of considerable commercial importance since enzymes such as chymosin and *Mucor pusillus* pepsin have been exploited by the food industry in cheese and soya processing, while renins, cathepsins and the retroviral proteinases are prime targets of the pharmaceutical industries in drug design. Several aspartic proteinases have been the subject of extensive X-ray structural analyses of both the native and inhibitor-complexed forms. These high resolution structures provide an excellent database for the development of biomolecular design principles. The structures of about 40 peptidomimetic inhibitor complexes have provided a detailed definition of the shape and geometry of the specificity pockets; this knowledge is indispensable in the design of potent inhibitors of pharmacological interest.

This article describes examples of design cycles both for protein engineering and for ligand design. For protein engineering we describe the expression and structural analysis of a chymosin mutant; this is part of a programme designed to investigate the role of loop regions in the structure and stability of aspartic proteinases. The major theme is to produce chimaeric proteins by replacing variable regions in one enzyme with the equivalent region - often of a different length and quite different sequence. For drug design we show how comparative analyses of renin-inhibitors complexed with different aspartic proteinases can be used to improve understanding of specificity. Both of these studies are truly multidisciplinary, but both are absolutely dependent on the existence of many high resolution three-dimensional structures of various homologues and their complexes with inhibitors.

PROTEIN ENGINEERING

The protein engineering programme at Birkbeck seeks to develop generic methods based on a design cycle involving biochemical preparation and characterization, determination and comparative analysis of three-dimensional structures, rule-based design, site-directed mutagenesis and expression of the mutants. The research has concentrated mainly on chymosin. The ability of chymosin to cleave specifically the Phe^{105} - Met^{106} bond in κ-caesin to initiate milk clotting makes it useful for manufacturing cheese. The objective of the protein engineering exercise is to target alterations to a number of properties including size, specificity and stability. For instance, in the case of alteration of size, the aim is to reduce the length of surface loops to allow more efficient penetration of complex substrates used in milk processing. By comparing the three-dimensional structure of chymosin B [4] with those of the other members of the aspartic proteinase family, a surface loop ($q_N r_N$) formed by residues 155-165 (pepsin numbering) was initially targeted for modification by replacement with homologous sequences from the other members of the family as well as with other modelled loops of variable chain length. The sections below describe the expression, crystallization and structural analysis of such a loop-exchange mutant where the target loop was substituted by the corresponding loop from *Rhizopus chinensis* pepsin [5].

Expression and purification of the loop-exchange mutant

The 61-mer oligonucleotide corresponding to the $q_N r_N$ loop of *Rhizopus chinensis* pepsin was introduced into chymosin B cDNA using site-directed mutagenesis and cloned into the fungal expression vector pAMH104 [6]. A *Trichoderma reesei* strain, RutC30 was transformed under acetamide selection and grown in an Applikon bioreactor under conditions resulting in autocatalytic activation to yield the 27mg/l of the loop exchange mutant [7]. The mutant was purified in four steps; ammonium sulphate precipitation, anion-exchange chromatography, gel filtration and finally affinity-binding on a **V-dL-P-F-F-V-dL** inhibitor column. Enzyme kinetics using a synthetic chromogenic peptide indicated that the activity of the mutant is similar to that of wild-type chymosin B.

Crystallization and data collection

The purified mutant was concentrated to 10mg/ml in sodium phosphate buffer at pH 5.6 and crystallised by the hanging drop vapour diffusion method using sodium chloride as the precipitant. Plate-like crystals of dimensions 0.20 X 0.15 X 0.05 mm, grown over a period of 3 to 4 weeks, were found to be isomorphous with the native crystals. X-ray intensity data were collected by the oscillation technique with a MARRESEARCH image plate detector using the Daresbury Laboratory Synchrotron Radiation Source. Although the X-ray diffraction data extended to 2.2Å resolution, the images showed splitting due to the existence of multiple crystals that were impossible to separate. The intensities were weak due to the small size of the crystals at resolutions beyond 2.5Å. Oscillation images from two differently-oriented crystals were processed using the **CCP4** software suite and merged to a unique set of reflections (R_{merge} 12.4%) extending to a resolution of 2.5Å.

Refinement and results

The atomic coordinates of native chymosin B [4], after the removal of the mutated loop, were used to obtain the initial difference Fourier maps. Interactive model - building on an **Evans & Sutherland PS300** with the graphics software **FRODO**[8], followed by least squares refinement cycles using **RESTRAIN**[9], gradually revealed the atomic positions of the residues in the altered loop. The current crystallographic agreement factor and correlation coefficient for the refined model are 0.206 and 0.917 respectively, for reflections in the resolution range 8.0-2.5 Å.

Figure 2 shows the main-chain atoms corresponding to the mutated loop; the equivalent loop in rhizopuspepsin structure [5] has been superposed for comparison.

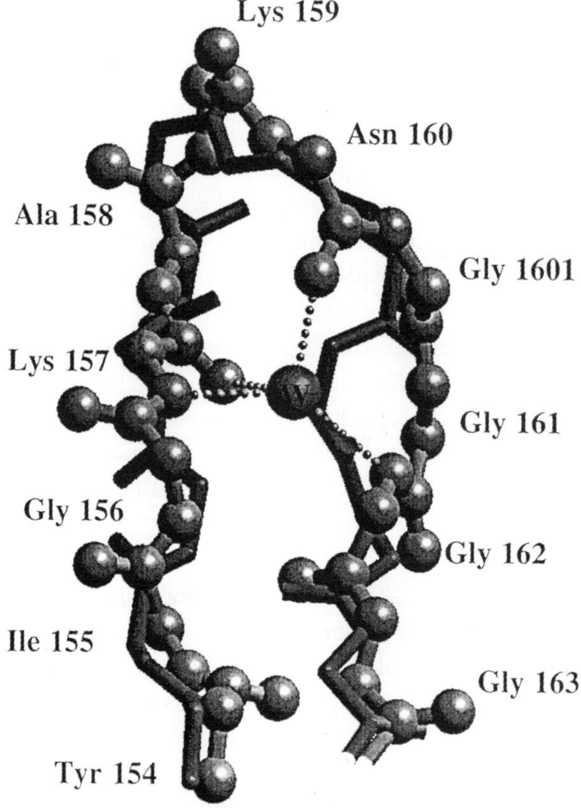

Figure 2. The main-chain atoms corresponding to the mutated loop (in ball and stick representation) and the water molecule (W) which stabilises the β-hairpin conformation of the mutant; the equivalent loop in rhizopuspepsin structure (smooth cylindrical joints) has been superposed for comparison. The pepsin numbering scheme has been used for labelling.

There are considerable differences between the two loops, most significantly at Gly 161 and Gly 162. An interesting feature of the mutant loop is the presence of a water molecule in the vicinity of the hairpin. As illustrated in Figure 2, this water molecule stabilises the beta-hairpin conformation by making hydrogen bonds with the main-chain nitrogen and carbonyl oxygen atoms of the residues in the loops. It is also interesting to note that rhizopus pepsin does not have any water molecule in an equivalent position.

INHIBITOR DESIGN

The first inhibitor complexes of aspartic proteinases to be studied by X-ray analysis were pepstatin with rhizopuspepsin [10] and a pepstatin fragment with penicillopepsin [11]. X-ray analyses of reduced bond inhibitors complexed with endothiapepsin defined interactions at P_1' to P_3' for the first time [12-14]. Crystallographic studies on endothiapepsin inhibitor complexes showed that statine acts as a dipeptide analogue in the binding cleft [15]. Studies of other complexes [14,16,17] have shown that all inhibitors have an extended beta-strand conformation, which allows the inhibitor main chain to make hydrogen bonds to each lobe of the enzyme as shown in Figure 3.

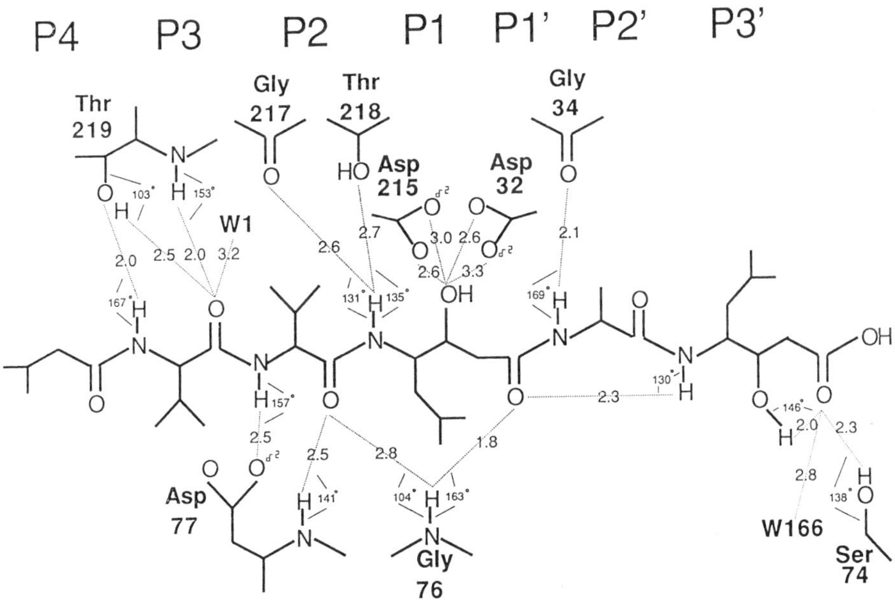

Figure 3. The arrangement of hydrogen bonds in an inhibitor complex of endothiapepsin.

The conclusions concerning specificity subsites, based on fungal enzymes complexes, have been largely confirmed from analyses of mammalian enzymes. For example, Dhanaraj *et al.*[18] describe the X-ray analyses of both recombinant renin complexed with a P_4 to P_1' norstatine-containing inhibitor CP-85,339

(Pro-Phe-SMeCys-norCStaOiPr hydrochloride)

and mouse submandibular renin complexed with a P_6 to P_4' decapeptide hydroxyethylene isostere inhibitor CH-66

(Piv-His-Pro-Phe-His-Leu-ψ[CH(OH)CH_2]-Leu-Tyr-Tyr-Ser-NH_2).

The structures of both complexes were solved by molecular replacement using the refined structure of pepsin as the search model. Least squares refinement was carried out using **RESTRAIN**[9]. The agreement factors and correlation coefficients were 0.19 & 0.91 for human renin at 2.8Å resolution and 0.18 & 0.95 for mouse renin at 1.9Å resolution.

The side chains alternate on either side of the main chain and make hydrogen bonds and van der Waals interactions with both domains in the molecule so defining subsites, S_6 to S_4'. The common hydrogen bonds to the mainchains of the inhibitors result in a similar conformation. A beta-hairpin, comprising residues 72 to 81 and known as the 'flap', interacts with the central region of the inhibitor and shields the active site from the solvent region.

Although the general topology of the two renins is quite similar to that of other aspartic proteinases, the active site cleft has a less open arrangement in renins. This is illustrated in Figure 4 (shown on the next page). It is partly due to a difference in relative position of the rigid body comprising the C-domain. For instance, there is a rotation of ~4° and a translation of ~0.1Å in the human renin complex with respect to the endothiapepsin-difluorostatone complex which makes the cleft smaller in renins. The entrance to the active site is made even narrower as a consequence of differences in the positions and composition of several well-defined loops and secondary structure elements.

Unique to the renins is a *cis*-proline, Pro 111, which caps the helix h_{N2} and contributes to the subsites S_3 and S_5. On an equivalent loop in the C-lobe (related by the intra-molecular pseudo 2-fold symmetry), there is a sequence of prolines - the Pro292 - Pro293 - Pro294 segment with Pro294 and Pro 297 in a *cis* configuration; this contributes to the subsites at S_3' and S_4'. A similar structure is also found in yeast proteinase A [19] and in cathepsin D [20,21], but it is positioned further away from the inhibitor. This poly-proline loop and the loop comprised of residues 241 - 250 lie on either side of the 'flap' formed by residues 72 - 81.

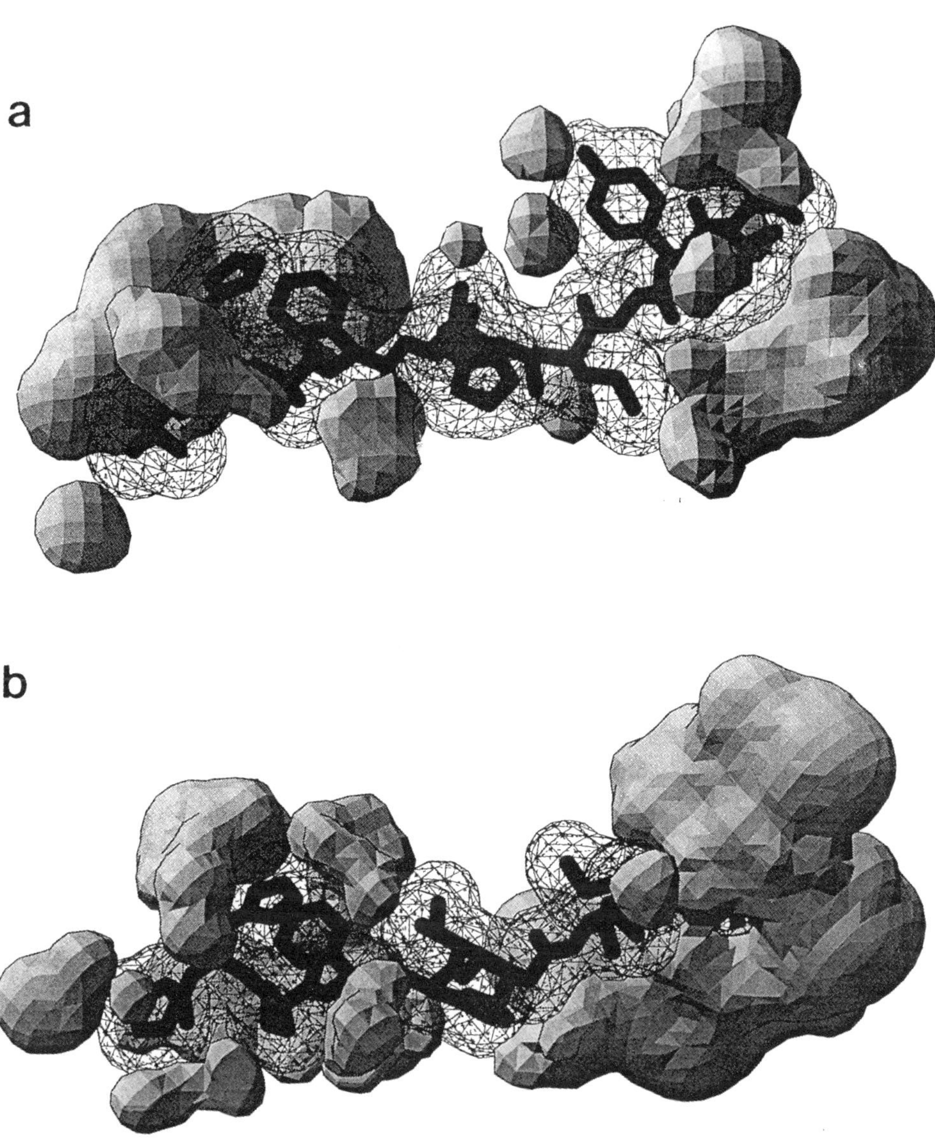

Figure 4. Schematic illustrations of the gaps between enzyme and inhibitor in the crystal structures of (a) mouse renin complexed with CH-66 and (b) endothiapepsin complexed with H-189. The figures were produced using the software **GapE**. The inhibitors (dark bonds) are enclosed by net surfaces and the gaps (where the enzyme and inhibitor are not in contact) are represented by solid surfaces.

Figure 5 shows the inhibitor binding of mouse and human renins. In the inhibitor-bound renins the cleft is covered by 'flaps' from both lobes rather than from the N-lobe alone as in other pepsin-like aspartic proteinases; this gives a superficial similarity to the dimeric, retroviral proteinases where each subunit provides an equivalent flap that closes down on top of the inhibitor.

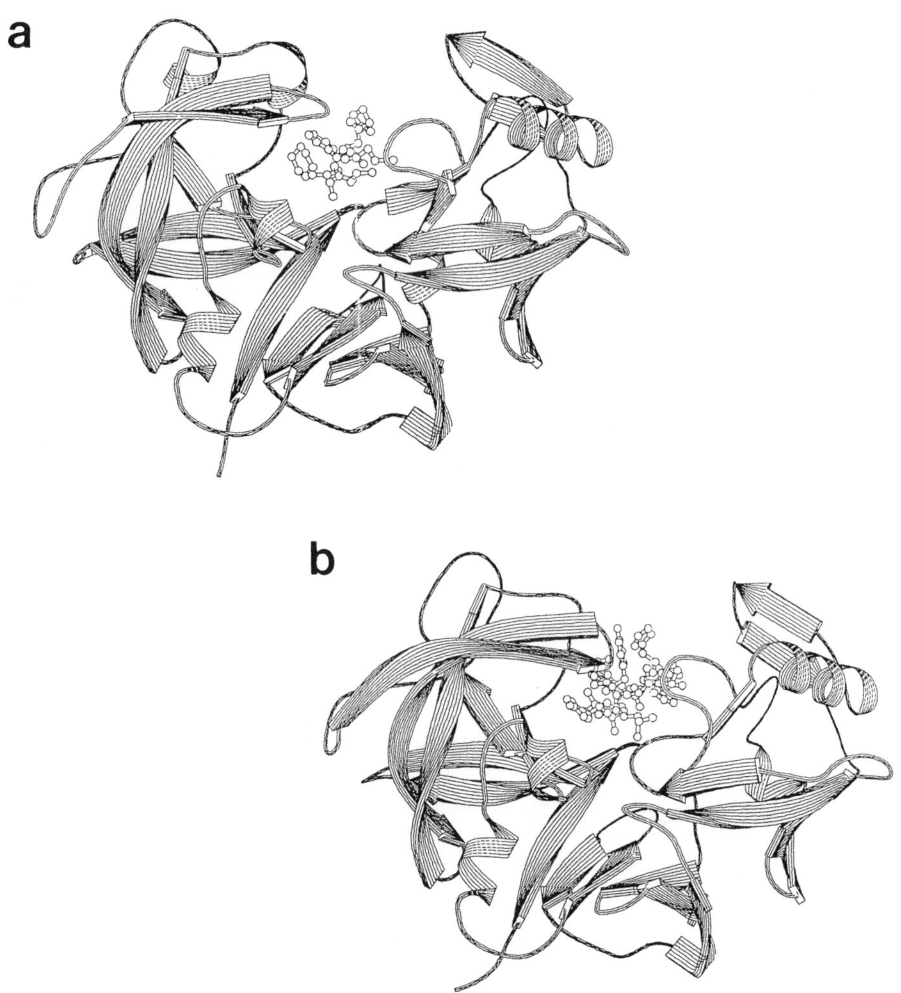

Figure 5. Schematic cartoons of (a) human and (b) mouse renin crystal structures showing the arrangement of α-helices and β-strands. The inhibitors are shown in full.

The high resolution structures of the two renin inhibitor complexes allow us to consider how the differences in the substrate specificities of aspartic proteinases are achieved. Table 1 lists the enzyme residues lining the specificity pockets in the human and mouse renin complexes.

Table. 1

Specificity pockets for the human & mouse renin complexes

Subsite	Residues in contact[a] with the inhibitor	
	Human renin complex	Mouse renin complex
S6	---	S 12, F 220
S5	---	P 111, L 114
S4	T 77, L 114	S 219, F 220, H 287
S3	Q 13, T 77, P 111, L 114, A 115, F 117, A 218, S 219	Q 13, S 77, P 111, L 114, A 115, F 117, G 217, S 218, S 219
S2	S 76, T 77, G 217, S 222, M 289	Y 75, G 76, S 77, S 218, S 222, H 287, M 289, V 300
S1	D 32, Y 75, F 112, F 117, V 120, D 215, G 217	D 32, G 34, Y 75, V 120, G 217
S1'	G 34, Y 75, S 76, L 213, D 215, I 291	G 34, Y 75, G 76, V 213, D 215, V 298
S2'	---	G 34, S 35, I 73, H 74, Q 128, V 130, T 295
S3'	---	H 74, Y 75, I 291, P 292, T 295
S4'	---	P 294, T 295

[a] Distance cut-off for contacts: 4.0Å

In modelling exercises [22,23] it was assumed that specificities derive from differences in the sizes of the residues in the specificity pockets (S_n) which complement the corresponding side chains at positions P_n in the substrate/inhibitor. However, the high resolution crystal structures of the renins show that this simple assumption only partly accounts for the steric basis of specificity. For example, the phenyl rings of Phe P_3 occupy almost identical positions in both renin inhibitor complexes. Modelling studies have predicted the specificity subsite S_3 to be larger in renins than in other aspartic proteinases due to substitution of smaller residues, Pro 111, Leu 114 and Ala 115, in place of larger ones in mammalian and fungal proteinases[23]. However, a compensatory movement (˜5Å) of the helix h_{N2} makes the pocket quite compact and complementary to the aromatic ring. Thus, the positions of an element of secondary structure differ between renin and other aspartic proteinases with a consequent important difference in the specificity pocket. In a similar way differences in the positions of secondary structural elements may also account for the specificity differences at S_2'.

The three-dimensional structures show that the situation at S_1' is more complex. Differences in the specificity subsites at S_1' for the human and mouse renins are complementary to the valine and leucine at P_1' in human and mouse angiotensinogens; residue 213 is leucine in human renin and valine in mouse renin. The S_1' pockets of chymosin, pepsin and endothiapepsin have an aromatic side-chain at residue 189 whilst the renins have amino acids with smaller side-chains (valine in human & serine in mouse renins). This would be expected to make the pocket larger in renins. However the structure of the mouse renin complex shows that the substrate moves closer to the enzyme in renins as a result of the smaller residue at 189 and the pocket is made even more compact due to changes in the position and composition of the poly-proline loop (residues 290 - 297). Thus, the specificity difference at this site arises not only from the movement of a secondary structure, in this case a loop region, but also from the substitution of an enzyme residue that allows the substrate to come closer to the body of the enzyme.

The specificity pockets are also influenced by the elaboration of loops on the periphery of the binding cleft in renins. This is most marked at P_3' and P_4'. For instance, in endothiapepsin, which has been the subject of the greatest number of studies, different conformations are adopted at P_3' and the residue at P_4' is generally disordered. In contrast these residues are clearly defined in mouse renin. The X-ray analysis of the mouse renin complex shows that the S_3'& S_4' subsites are formed by the poly-proline loop together with residues of the 'flap' and a similar situation is likely to occur in human renin. The well-defined interactions of P_3' described in the mouse renin complex explain the significant affinity when inhibitors have phenylalanine or tyrosine at P_3' as well as the importance of a P_3' residue for catalytic cleavage of a substrate by renin.

In the drug discovery processes high affinity inhibitors of enzymes can be synthesised that emulate the transition state and/or are complementary to the subsites away from the catalytic site. These studies provide a basis for understanding both of these important factors.

With respect to the design of transition state isosteres, it is notable that the catalytic site and mechanism seem to have been highly conserved even in a very divergent family of enzymes that includes the retroviral proteinases and the pepsin-like enzymes. Evidence of this is found in the fact that structures of complexes of different aspartic proteinases in the region of the catalytic aspartates and the scissile bond surrogate are virtually identical. Thus, although the structure provides a blue-print for the design of better transition state isosteres, ligands containing such groups would probably bind well to all aspartic proteinases; we can anticipate that increased affinity would be gained at the expense of specificity.

The studies of inhibitor complexes reported here show that even minor alterations in the positions of secondary structural elements can lead to major changes in the disposition of the subsites. The specificity pockets are modified by elaboration, particularly of surface loops, which can be disordered in the uncomplexed enzymes and difficult to model with precision from homologous structures. Since these define the species-specificity and determine the catalytic efficiency of the enzymes, a thorough understanding of their composition is indispensable for the synthesis of suitable inhibitors.

These data not only establish a new foundation for the rational design of enzymes and inhibitors, but also provide insights into the factors that have general value where protein structures are used to assist the design of commercially useful molecules.

Acknowledgements

The authors thank our collaborators Dr P.M.Hobart, Dr K.F. Geoghegan, Mr D.E. Danley, Dr D.J.Hoover, Prof.M.Szelke and Prof.K.Murakami for supplying enzymes and inhibitors and Dr. Roman Laskowski of University College, London for use of the software **GapE**. We thank Dr C.Frazao, Mr M.Badasso, Dr B.L. Sibanda, Dr I.J.Tickle, Dr M.Newman, Dr C.Aguilar and Dr S.P.Wood for their help in structural studies. We thank the AFRC (V.D., P.N.) and SERC for financial support.

REFERENCES

1. T.L. Blundell, G. Elliott, S.P. Gardner, T. Hubbard, S. Islam, M.S. Johnson, D. Mantafounis, P. Murray-Rust, J. Overington, J.E. Pitts, A. Šali, B.L. Sibanda, J. Singh, M.J.E. Sternberg, M.J. Sutcliffe, J.M. Thornton, P. Travers (1989) Phil. Trans. R. Soc. Lond. **B324**, 447-460.

2. T.L. Blundell & M.S. Johnson (1993) Protein Science, **2**, 877-883.

3. A. Šali, J.P. Overington, M.S. Johnson & T.L. Blundell (1990) Trends Biochem. Sci. **15**, 235-240.

4. M. Newman, M. Safro, C. Frazao, G. Khan, A. Zdanov, I.J. Tickle, T.L. Blundell & N.S. Andreeva (1991) J. Mol. Biol. **221**, 1295-1309.

5. K. Suguna, S. Sheriff, G.E. Cohen & D.R. Davies (1987) J. Mol. Biol. **196**, 877-900.

6. A. Harkki, J. Uusitalo, M. Bailey, M. Pentilä & J.K.C. Knowles (1989) Bio/Technology, **7**, 596-603.

7. M. Penttilä, H. Nevalainen, M. Rättö, E. Salminen & J.K.C. Knowles (1987) Gene, **61**, 155-164.

8. T.A. Jones (1978) J. Appl. Cryst. **11**, 268-272.

9. I. Haneef, D.S. Moss, M.J. Stanford & N. Borkakoti (1985) Acta Cryst. **A41**, 426-433.

10. R. Bott, E. Subramanian & D.R. Davies (1982) Biochem. **21**, 6956-6962.

11. M.N.G. James, A.R. Sielecki, F. Salituro, D.H. Rich & T. Hofmann (1982) Proc. Natl. Acad. Sci. USA, **79**, 6137-6142.

12. S.I. Foundling, J.B. Cooper, F.E. Watson, A. Cleasby, L.H. Pearl, B.L. Sibanda, A. Hemmings, S.P. Wood, T.L. Blundell, M.J. Valler, C.G. Norey, J. Kay, J. Boger, B.M. Dunn, B.J. Leckie, D.M. Jones, B. Atrash, A. Hallett & M. Szelke, (1987) Nature, **327**, 349-352.

13. T.L. Blundell, J.B. Cooper, S.I. Foundling, D.M. Jones, B. Atrash & M. Szelke (1987) Biochem., **26**, 5585-5590.

14. J.B. Cooper, S.I. Foundling, A. Hemmings, T.L. Blundell, D.M. Jones, A. Hallett & M. Szelke (1987) Eur. J. Biochem. **169**, 215-221.

15. J.B. Cooper, S.I. Foundling, T.L. Blundell, J. Boger, R.A. Jupp & J. Kay (1989) Biochem. **28**, 8596-8603.

16. A. Šali, B. Veerapandian, J.B. Cooper, S.I. Foundling, D.J. Hoover & T.L. Blundell (1989) EMBO J. **8**, 2179-2188.

17. B. Veerapandian, J.B. Cooper, A. Šali & T.L. Blundell (1990) J. Mol. Biol. **216**, 1017-1029.

18. V. Dhanaraj, C.G. Dealwis, C. Frazao, M. Badasso, B.L. Sibanda, I.J. Tickle, J.B. Cooper, H.P.C. Driessen, M. Newman, C. Aguilar, S.P. Wood, T.L. Blundell, P.M. Hobart, K.F. Geoghegan, M.J. Ammirati, D.E. Danley, B.A. O'Connor & D.J. Hoover (1992) Nature, **357**, 466-472.

19. C. Aguilar, M. Badasso, S.P. Wood & T.L. Blundell (1991) Unpublished results.

20. E.T. Baldwin, T.N. Bhat, S. Gulnik, M.V. Hosur, R.C. Sowder, R.E. Cachau, J. Collins, A.M. Silva & J.W. Erickson (1993) Proc. Natl. Acad. Sci. USA, **90**, 6796-6800.

21. P. Metcalf & M. Fusek (1993) EMBO J. **12**, 1293-1302.

22. T.L. Blundell, B.L. Sibanda & L.H. Pearl (1983) Nature **304**, 273-275.

23. B.L. Sibanda, T.L. Blundell, P.M. Hobart, M. Fogliano, J.S. Bindra, B.W. Dominy & J.M. Chirgwin (1984) FEBS Lett. **174**, 102-111.

ECB6: Proceedings of the 6th European Congress on Biotechnology
edited by L. Alberghina, L. Frontali, P. Sensi

BIOCATALYSIS IN ORGANIC SYNTHESIS

H.E. Schoemaker, W.H.J. Boesten, W.J.J. van den Tweel,
Q.B. Broxterman and J. Kamphuis

DSM Research, Bio-organic chemistry section, P.O. Box 18,
6160 MD Geleen, The Netherlands

1. INTRODUCTION

At the end of the seventies, expectations for the application of biotechnological methods in the chemical industry were high. This enthousiasm, however, was tempered somewhat in the eighties due to the rather slow progress in implementing the new technologies in production processes, especially in the fine chemical area. Fortunately, nowadays we notice again a steady upswing with the introduction of several large-scale processes. After a short evaluation of possible reasons for this rather long incubation time, in this paper we will point out some (scientific) key-success factors for the introduction of biotechnological processes in the fine chemical industry.

In academia, biocatalysis is now a well established tool in organic synthesis, particularly in the preparation of enantiomerically pure compounds. On a laboratory scale all classes of enzymes are applied. In the fine chemical industry, however, the application of enzymes is in general restricted to the use of either hydrolytic enzymes or lyases. Only a limited number of enzymes belonging to other classes are used. Recently, also some conversions using whole-cells have been introduced. It should be noted that in this paper the term biocatalysis is used in the sense of biotransformation, i.e. substrate and product are structurally related. *De novo* synthesis using microbial processes are excluded from this discussion.

What are the reasons for the limited application of bicatalysts other than hydrolytic enzymes or lyases up till now?
From a production point of view the major advantage in the use of hydrolytic enzymes or lyases is that these enzymes donot require cofactors which need recycling. Although for some cofactors elegant and efficient *in vitro* recycling procedures have been developed, biotransformations that require multi-enzyme systems often are most conveniently performed using whole-cells. Thus, a number of highly desired organic synthetic conversions are part of the realm of microbiology. Unfortunately, most organic chemists are not familiar with this field. This, of course, shouldnot surprise us too much, since in general organic reactions are performed in organic media and most micro-organisms prefer an aqueous environment. In addition, the present chemical industry is mostly oil-based.

Consequently, most fine chemical intermediates are produced from bulk-chemicals, using highly optimized chemical procedures. Although these processes are not always ideal, they are efficient enough to produce the desired chemicals in large volumes at relatively low prices. This fact should not be overlooked if one wants to replace a chemical process by a biotechnological procedure. In fact, high-volume markets for (fine) chemicals in most cases were only developed because there was a straightforward method to make the product in the first place.
Although the number of biotechnological applications has been rapidly increasing in recent years, there is another reason why there are still relatively few large scale applications of biotransformations currently in use in the fine chemical industry. Ironically enough, this is caused by the high level of sophistication of enzyme catalysis. With the aid of enzymes, products can be made which are not readily available, or even impossible to produce, via organic chemical techniques. Consequently, there has not been enough time, or not enough material, to develop the market for such a product.
Finally, another factor hampering the introduction of new biocatalytic processes is the cost of development of the biocatalyst. Only a very few new products can bear the development costs of a dedicated biocatalyst. For that matter, this phenomenon is not restricted to the field of biocatalysis alone. In other areas of chemistry a similar situation is observed. For example, relatively few processes involving asymmetric homogeneous catalysis to produce chiral, non-racemic compounds are in use today.

Can we indicate areas where the application of enzymes is especially apt ?
The chemical industry is mostly oil-based and in most cases the technologies used involve - sometimes rather difficult to control - oxidative or reductive conversions. Therefore, it is not surprising that the major applications of enzymes in industry are found in conversions that require a high degree of regio- or stereoselectivity, mild reaction conditions and, in general, an aqueous environment. Thus, large scale applications of enzymes are found in carbohydrate conversions, amino acid and peptide synthesis and in the processing of fats. Evidently, enzymes like 'natural' substrates.
In the future it is expected that the high level of sophistication of enzyme-catalysis will bring new opportunities, especially in the field of pharmaceuticals, agro-chemicals and in the aroma- and fragrance industry. The rapidly developing insight in the mode of action of pharmaceuticals and agro-chemicals, in combination with an increasing knowledge of physiological receptors, requires a higher level of sophistication of the future generations of agrochemicals and pharmaceuticals. In addition, we observe more and more that the required physiologically active substances are designed via a rationale approach, inspired by the target enzymes or receptors. This process eventually will give rise to lead compounds which are more 'natural-like'; presenting a major opportunity for biocatalysis, since thereby enzymes can produce 'natural-like' substances.
Although the development of a single-purpose enzyme is in most cases not cost-effective, fortunately enzymes turn out to be very versatile biocatalysts. The use of an enzyme is often not restricted to the conversion of one substrate. In most cases one single enzyme will accomodate a whole class of substrates. Also, optimization of the enzyme production by fermentation is often relatively straightforward within a family of compounds.

More important, however, is the fact that commercial enzyme preparations often catalyze a whole variety of reactions, not only in water, but also in organic media. To put it briefly, enzymes can do more than you think. Therefore, it is worthwile to investigate unusual properties of commercial enzymes. For instance, the detergent industry is a rich source of crude and cheap enzyme preparations like proteases, lipases and, in the future, peroxidases. Also, new enzymes are developed for the pulp and paper industry, like xylanases and peroxidases. Moreover, new production organisms are being studied and/or used, implying that new opportunities arise for whole-cell transformations with these or related microorganisms.
Thus, with the introduction of each new biotransformation, a variety of new possibilities become feasible. Therefore, it is to be expected that the rate by which new applications are introduced, will be increasing steadily.

We will illustrate this - admittedly rather superficial - analysis with a number of commercial examples, with special emphasis on production processes introduced in our own company. Also, we will indicate some new developments and give some pointers to opportunities in the (distant) future.

2. AMINOPEPTIDASES, AMIDASES AND ACYLASES

In the seventies a program was initiated at DSM to develop a general process for the resolution of amino acids using aminopeptidases, based on the enzymic conversion of DL-amino acid amide to form a mixture of L-amino acid and D-amino acid amide (cf scheme 1, R_1=H). Both natural and synthetic amino acids can be resolved with an aminopeptidase from *Pseudomonas putida.* The commercial development of the strain was done at NOVO-Nordisk. This example illustrates that enzymes prefer 'natural' or 'natural-like' substrates and that one enzyme can be used to resolve a whole class of compounds. Concomitantly, an integral process for the production of enantiomerically pure natural and synthetic amino acids (both in the D and L form) was developed, comprising an efficient procedure for the preparation of the amino acid amide, for the separation of the L-acid from the D-amide and for the racemisation of the unwanted isomer via a Schiff base derivative. This process, which has been extensively reviewed [1,2], is operated on a commercial scale for the production of several amino acids.
More recent developments include the resolution of C-α-disubstituted amino acids (scheme 1, R_1=alkyl) using an aminopeptidase from *Mycobacterium neoaurum* or *Ochrobactrum anthropi* [3].

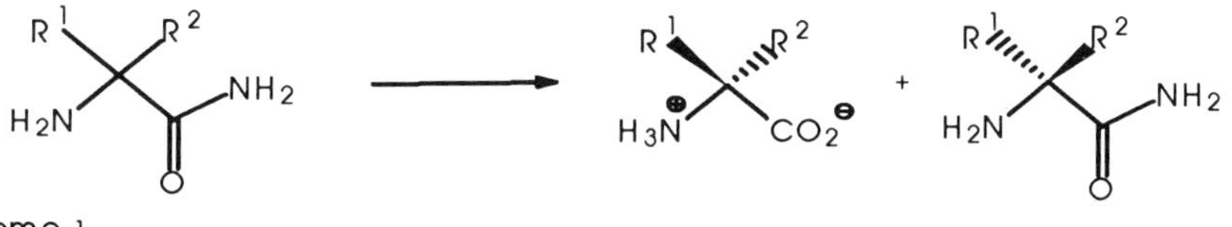

Scheme 1

In the resolution of amino acids hydrolytic enzymes like aminopeptidases and acylases [4] are being used in a rather straightforward manner. As an example of a chemical process that has been largely replaced by an enzymatic procedure- due to higher selectivity and the mild reactions conditions- the hydrolysis of penicillin G to form 6-APA using penicillin G acylase can be mentioned [see 5].

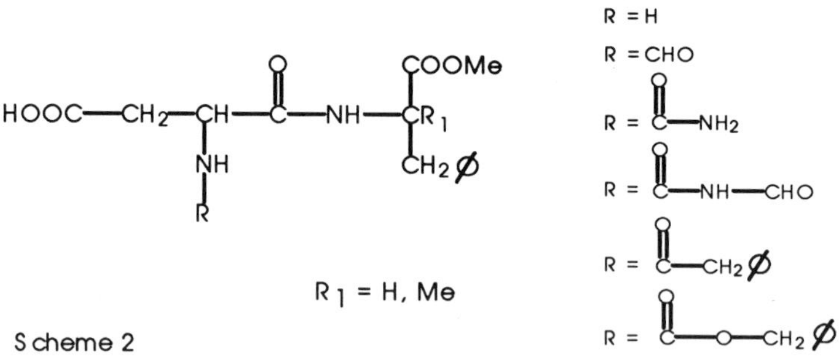

HOOC—CH2—CH(NH–R)—C(=O)—NH—CR1(COOMe)—CH2Ø

R1 = H, Me

R = H

R = CHO

R = C(=O)—NH2

R = C(=O)—NH—CHO

R = C(=O)—CH2Ø

R = C(=O)—O—CH2Ø

Scheme 2

More recently, the scope of the use of hydrolytic enzymes was dramatically enhanced by the introduction of processes in which the reverse reaction, a condensation, is used in the production process. Thus, aspartame, (see scheme 2, $R=R_1=H$) the well-known dipeptide sweetener originally developed by Searle (Nutrasweet), is produced by HSC (Holland Sweetener Company), a joint venture between TOSOH of Japan and DSM of the Netherlands, using an thermolysin catalyzed condensation between an N-protected aspartic acid derivative and phenylalanine methyl ester [6]. Moreover, in a search for alternative protective groups in the synthesis of aspartame, a new class of dipeptide sweeteners was discovered. Thus, whereas N-formyl aspartame and N-carbamoyl aspartame are not sweet, the N-formylcarbamoyl derivative of aspartame elicit a sweet taste comparable to the parent compound [7]. Moreover, also other N-formylcarbamoyl dipeptides taste sweet, even if the phenylalanine moiety is displaced by e.g. α-methyl-phenylalanine [8].

Numerous possibilities exist for the protection of aspartic acid. A rather interesting approach is the use of the phenyl-acetyl group, which can be split-off using the penicillin G acylase discussed above. The latter conversion nicely illustrates that commercial enzyme preparations can be used in other applications than the process for which the enzyme was originally developed.

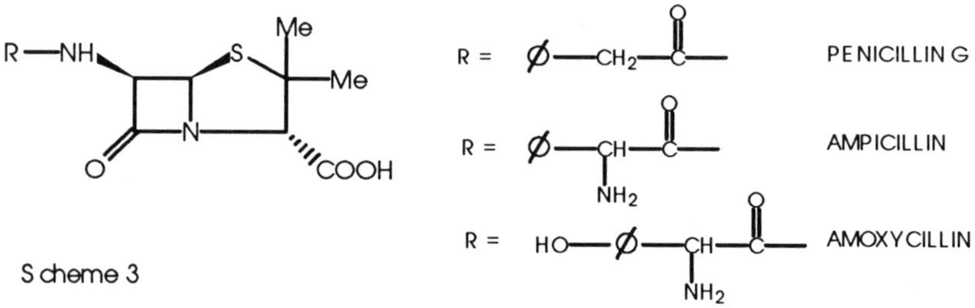

R = Ø—CH2—C(=O)— PENICILLIN G

R = Ø—CH(NH2)—C(=O)— AMPICILLIN

R = HO—Ø—CH(NH2)—C(=O)— AMOXYCILLIN

Scheme 3

Penicillin G acylase is an extremely versatile enzyme that can be used in quite a variety of applications, comprehensively reviewed by Baldaro *et al* [5]. Thus, the enzyme can not only be used for the hydrolysis of Penicillin G to form 6-APA, but also for the synthesis of semisynthetic antibiotics like ampicillin or amoxycillin using phenylglycine or p-hydroxy-phenylglycine derivatives as acyldonors. The latter examples again illustrate the versatility of relatively simple hydrolytic enzymes in organic synthesis [see scheme 3].

3. LIPASES AND ESTERASES

Both in the open and in the patent literature a wide variety of applications of lipases and esterases is described, which we will not discuss here [see 9]. We will confine ourself to a few examples of recent work from our own laboratories. Although amidases are the enzymes of choice in the preparation of amino acids (*vide supra*), and e.g. *Mycobacterium neoaurum* can be used for the synthesis of optically pure α,α-disubstituted amino acids, not all substrates are accepted by this enzyme preparation. Thus, phenylglycines with in the Cα-position an allyl- or butyl-substituent cannot be synthesized in optically pure form with this procedure. However, the desired compounds can be obtained via the enzymatic resolution [scheme 4] of the highly substituted amino acid esters using PLE [10]. Analogously, substituted mandelic acids can be synthesized using PLE from Amano [11].

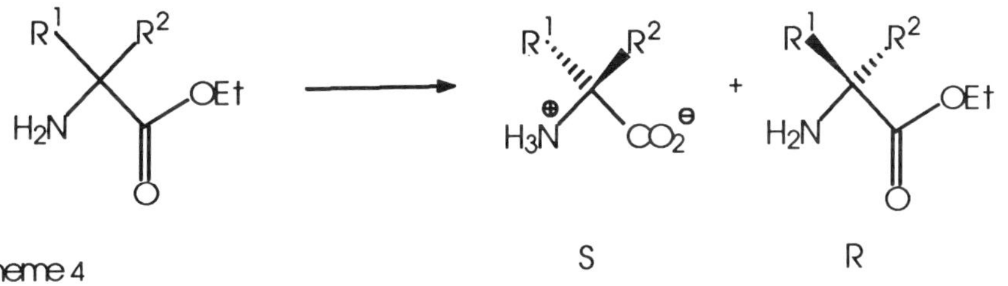

R^1 R^2 H_2N OEt O ⟶ R^1 R^2 $H_3N^{\oplus}$ $CO_2^{\ominus}$ + R^1 R^2 H_2N OEt O

S R

Scheme 4

4. LYASES

The use of lyases in the synthesis of amino acids [see 12] like L-phenylalanine from cinnamic acid (with phenylalanine ammonia lyase) or L-aspartic acid [13] from fumaric acid (with aspartase) is well known. Also hydroxy-acids like L-malic acid can be readily produced using the enzyme fumarase [14]. Recently, in a joint project with the Agricultural University of Wageningen we isolated a strain that converted maleic acid in D-malic acid via a hydratase [15]. Thus, even in this area new enzymes can still be found.

Organic chemistry is the science of carbon-carbon bonds, therefore, enzymes that catalyze C-C bond formation are especially valuable in organic synthesis. Well-known examples include the classes of oxy-nitrilases [16,17] and aldolases. The latter class of enzymes is extremely useful in complex carbohydrate synthesis [18].

5. PEROXIDASES

The reaction mechanism of peroxidase catalysis has been the subject of numerous investigations. However, the synthetic value of those enzymes is limited due to the fact that the enzymes mainly catalyze rather unspecific electron-transfer reactions. For example, the redox cycle of horse radish peroxidase (HRP) involves the two-electron oxidation of the native enzyme - containing a protoporphyrin IX-iron (III) complex as the prosthetic group - with hydrogen peroxide to form the so-called HRP-Compound I, an oxyferryl-porphyrin radical cation. After one electron reduction, the oxyferryl porphyrin species called HRP-Compound II is formed, which is further reduced to native enzyme. HRP will catalyze the oxidation of phenols to form phenoxy-radicals. Lignin peroxidase, isolated from the white-rot fungus *Phanerochaete chrysosporium* and other wood rotting fungi, has a similar redox cycle, but this enzyme will also oxidize non-phenolic substrates [19].

Peroxidase Compound I is structurally highly similar to the Compound I of cytochrome P-450. From a synthetic viewpoint the P-450 enzyme is interesting since it will catalyze stereoselective oxygen-transfer reactions. Unfortunately, P-450 is a cofactor dependent enzyme and P-450 reactions are mostly carried out via whole-cell transformations, thereby limiting the broad synthetic usefulness for organic chemists. Recently however, it was shown that peroxidases, and in particular chloroperoxidase (CPO), can also catalyze stereoselective oxygen transfer reactions [20].

Thus, thio-anisole and derivatives are converted to optically active sulfoxides, whereas a

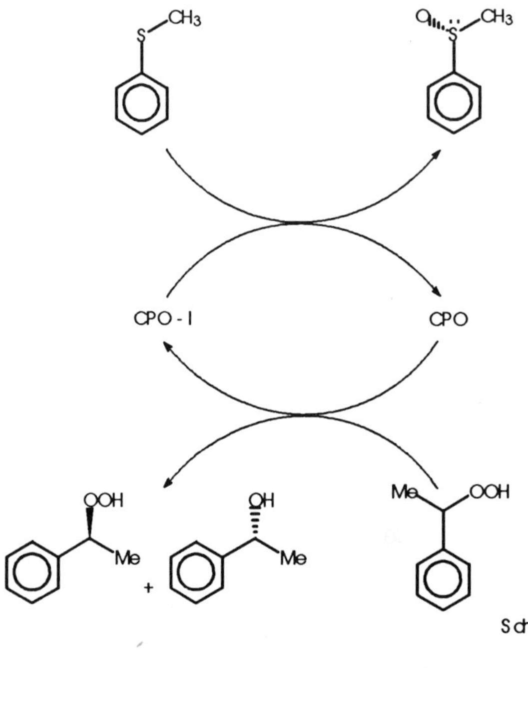

Scheme 5

benzylic hydroperoxide, as the oxygen donor for the enzyme, is converted to a mixture of an optically active alcohol and the remaining hydroperoxide [21]. Due to the developments in the detergent area and in part also in the pulp and paper industry, peroxidases will become available on a large scale in the future. Therefore, in the future more applications of peroxidases can be expected.

6. PRECURSOR FERMENTATIONS

The use of precursor fermentations in fine chemicals production is a relatively new development. Kanegafuchi reported on the synthesis of ß-hydroxy-isobutyric acid, an important intermediate in the synthesis of Captopril, starting from methyl methacrylate [22]. Lonza has [23,24] developed processes for hydroxy-nicotinate and L-carnithine [see scheme 6].

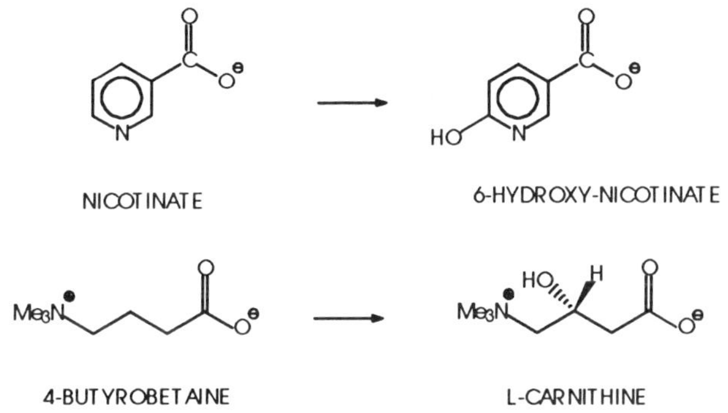

NICOTINATE → 6-HYDROXY-NICOTINATE

4-BUTYROBETAINE → L-CARNITHINE

Scheme 6

As a last example of an oxidative conversion, the formation of substituted arene cis-dihydrodiols as depicted in scheme 7 is given. Quite a number of groups have already explored the synthetic possibilities offered by these microbial conversions [see 25]. The highly reactive molecules - loss of water will lead to rearomatization - are not accessible via chemical means. These developments undoubtedly will stimulate the development of yet other examples.
In addition, it is to be expected that due to research in seemingly completely different areas, other syntheses starting from compounds with a complementary reactivity will be developed.

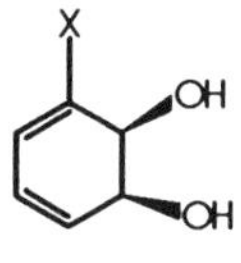

Scheme 7

To illustrate the latter point, some examples of precursor fermentations using the well-studied white-rot fungus *Phanerochaete chrysosporium* are depicted. In addition to the oxidative enzymes like lignin peroxidase and manganese peroxidase, this fungus also has a powerful reductive system. For example, an aryl alcohol dehydrogenase has been isolated and purified to homogeneity [26].

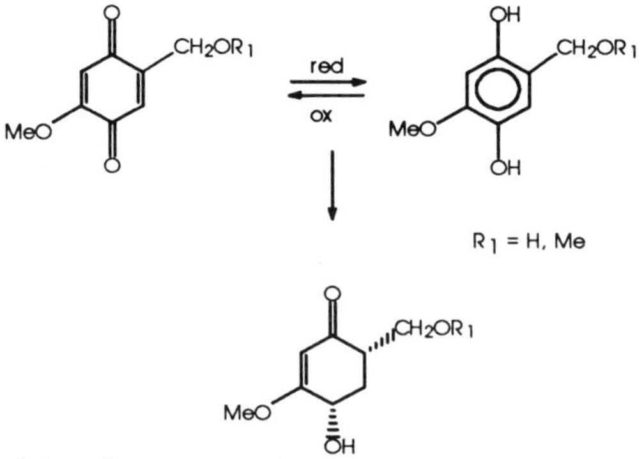

R_1 = H, Me

Scheme 8

Also, as depicted in scheme 8 quinones are reduced to form hydroquinones in whole cell cultures [27]. This reductive step is inevitably accompanied by the peroxidase catalyzed oxidation to form the quinones back. Interestingly, however, the fungus also possesses a rather unique reductive system that is capable of reducing hydroquinones to form hydroxy-substituted cyclohexenones in a very specific conversion [28]. Unfortunately, the absolute configuration of this rather labile compounds - analogous to the arene cis-dihydro-diols discussed above, loss of water will lead to aromatization - is as yet unknown. These products, which are not accessible via chemical means, possess interesting chemical reactivities, which need to be explored. Another interesting reductive conversion with *Phanerochaete chrysosporium* is depicted in scheme 9.

In a highly stereoselective fashion a keto-compound like (Z)-3-fluoro-4-phenyl-1-(p-

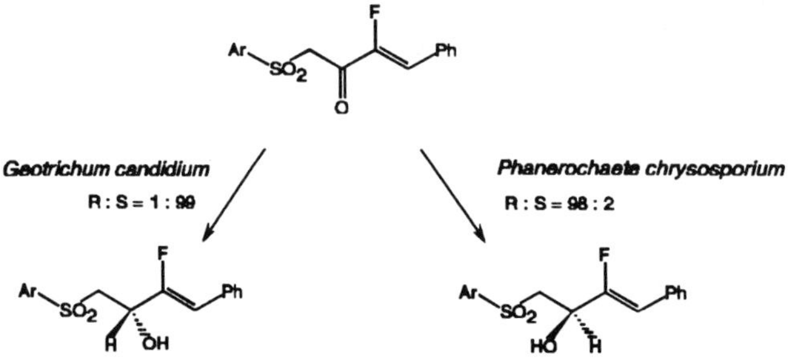

Geotrichum candidium R : S = 1 : 99

Phanerochaete chrysosporium R : S = 98 : 2

Scheme 9

tolylsulphonyl)but-3-en-2-one is converted to the corresponding enantiomerically pure (R)-alcohol [29]. It is noteworhty also, that the same starting compound yields the enantiomerically pure (S)-alcohol if *Geotrichum candidum* - a well known lipase producing microorganism - is used. These two conversions again illustrate the point made before that it is worthwhile to investigate new conversions with microorganisms that have been studied and developed for completely different fields of application.

7. CONCLUSIONS

In this paper some succesful examples of the use of enzymes in the fine chemical industry are given. Commercial applications are mostly confined to areas in which more natural-like substrates are enzymatically converted to high-added value products. Mild reaction conditions and regio- or stereo-selectivity are central in these applications. It can be expected that in the future an increasing number of biotransformations will be commercialized. The enzymes used in the present processes will be used in (seemingly) highly different applications. Also, the number of production microorganisms will be increasing steadily, thereby facilitating the use of these systems for conversions (e.g. precursor fermentations) that are completely different from the applications for which the microbial strains were originally developed. Thus, biocatalysis will have a bright future.

ACKNOWLEDGMENTS: the assistance of Mr. W. Kortenoeven in the preparation of the manuscript is gratefully acknowledged.

REFERENCES

1 H.E. Schoemaker, W.H.J. Boesten, B. Kaptein, H.F.M. Hermes, T. Sonke, Q.B. Broxterman, W.J.J. Van den Tweel and J. Kamphuis, Pure and Applied Chemistry, 64 (1992), 1171-1175
2 J. Kamphuis, E.M. Meijer, W.H.J. Boesten, T. Sonke, W.J.J. van den Tweel and H. E. Schoemaker, in: D.S. Clark and D.A. Estell (eds.) Enzyme Engineering XI, Ann. New York Acad. Sci., 672 (1992) 510-527
3 J. Kamphuis, W.H.J. Boesten, B. Kaptein, H.F.M. Hermes, T. Sonke, Q.B. Broxterman, W.J.J. Van den Tweel and H.E. Schoemaker, in: A.N. Collins, G.N. Sheldrake and J. Crosby (eds.) Chirality in Industry, John Wiley and Sons, Chichester, 1992, pp. 187-208
4 A.S. Bommarius, K. Drauz, U. Groeger and C. Wandrey, in: A.N. Collins, G.N. Sheldrake and J. Crosby (eds.) Chirality in Industry, John Wiley and Sons, Chichester, 1992, pp. 371-397
5 E. Baldaro, C. Fuganti, S. Servi, A. Tagliani and M. Terreni in: S. Servi (ed.), Microbial Reagents in Organic Synthesis, Kluwer Academic Publishers, Dordrecht 1992, pp. 175-189
6 K. Oyama, in: A.N. Collins, G.N. Sheldrake and J. Crosby (eds.) Chirality in Industry, John Wiley and Sons, Chichester, 1992, pp. 237-247
7 W.H.J. Boesten, B.H.N. Dassen, J.C.S. Kleinjans, B. van Agen, S. van der Wal, N.K. de Vries, H.E. Schoemaker and E.M. Meijer, J. Agric. Food Chem., 39 (1991) 154-158

8 S. Polinelli, Q.B. Broxterman, H.E. Schoemaker, W.H.J. Boesten, M. Crisma, G. Valle, C. Toniolo and J. Kamphuis, Bio-org. Med. Chem. Lett., 2 (1992) 453-456

9 L. Alberghina, R.D. Schmid and R. Verger (eds.) Lipases: Structure, Mechanism and Genetic Engineering, VCH Publishers, New York, 1991

10 B. Kaptein, W.H.J. Boesten, Q.B. Broxterman, P.J.H. Peters, H.E. Schoemaker and J. Kamphuis, Tetrahedron Asymmetry, in press

11 H. Moorlag, R.M. Kellogg, M. Kloosterman, B. Kaptein, J. Kamphuis and H.E. Schoemaker, J. Org. Chem., 55 (1990) 5878-5881

12 J. Crosby, in: A.N. Collins, G.N. Sheldrake and J. Crosby (eds.) Chirality in Industry, John Wiley and Sons, Chichester, 1992, pp. 1-66

13 T. Sato and T. Tosa, in: A. Tanaka, T. Tosa and T. Kobayashi (eds.) Industrial Application of Immobilized Biocatalysts, Marcel Dekker, New York, 1993, pp. 15-25

14 I. Takata and T. Tosa, in: A. Tanaka, T. Tosa and T. Kobayashi (eds.) Industrial Application of Immobilized Biocatalysts, Marcel Dekker, New York, 1993, pp. 53-67

15 M.J. Van der Werf, W.J.J. Van den Tweel and S. Hartmans, Appl. Environm. Microbiol., 58 (1992) 2854-2860

16 U. Kragl, U. Niedermeyer, M.R. Kula and C. Wandrey, in: H. Okada, A. Tanaka and H.W. Blanch (eds.), Enzyme Engineering X, Ann. New York Acad. Sci., 613 (1990) 167-175

17 C. Kruse, in: A.N. Collins, G.N. Sheldrake and J. Crosby (eds.) Chirality in Industry, John Wiley and Sons, Chichester, 1992, pp. 279-299

18 C-H. Wong, Y. Ichikawa, T. Kajimoto, K.K.-C. Liu, D.P. Dumas, Y.-C. Lin and G.C. Look, in: S. Servi (ed.), Microbial Reagents in Organic Synthesis, Kluwer Academic Publishers, Dordrecht, 1992, pp. 35-43

19 H.E. Schoemaker, Recl. Trav. Chim. Pays-Bas, 109 (1990), 255-272

20 L. Casella, M. Gullotti, R. Ghezzi, S. Poli, T. Beringhelli, S. Colonna and G. Carrea, Biochemistry, 31 (1992) 9451-9459

21 H. Fu, H. Kondo, Y. Ichikawa, G.C. Look and C-H. Wong, J. Org. Chem, 57 (1992) 7265-7270

22 T. Ohashi and J. Hasegawa, in: A.N. Collins, G.N. Sheldrake and J. Crosby (eds.) Chirality in Industry, John Wiley and Sons, Chichester, 1992, pp. 249-267

23 A. Kiener, Angew. Chem. Int. Ed. Engl., 31 (1992) 774-775

24 F.W.J.M.M. Hoeks, H.Kulla and H.-P. Meyer, J. Biotechnol, 22 (1992) 117-128

25 G.N. Sheldrake, in: A.N. Collins, G.N. Sheldrake and J. Crosby (eds.) Chirality in Industry, John Wiley and Sons, Chichester, 1992, pp. 127-166

26 A. Muheim, R. Waldner, D. Sanglard, J. Reiser, H.E. Schoemaker and M.S.A. Leisola, Eur. J. Biochem., 195 (1991) 369-375

27 D. Constam, A. Muheim, W. Zimmermann and A. Fiechter, J. Gen. Microbiol., 137 (1991) 2209-2214

28 U. Tuor, H. E. Schoemaker, M.S.A. Leisola and H.W.H. Schmidt, Appl. Microbiol. Biotechnol., 38 (1993) 674-680

29 R. Bernardi, P. Bravo, R. Cardillo, D. Ghiringhelli and G. Resnati, J. Chem. Soc. Perkin Trans. I, (1990) 579-583

ECB6: Proceedings of the 6th European Congress on Biotechnology
edited by L. Alberghina, L. Frontali, P. Sensi

CRITICAL PROBLEMS IN BIOPROCESS ENGINEERING

J. Tramper

Wageningen Agricultural University, Department of Food Science, Food and Bioprocess Engineering Group, P.O.Box 8129, 6700 EV Wageningen, The Netherlands

Bioprocess Engineering is the application-oriented integration of biological and engineering sciences in process design and development. It is, according to a recent US report [1], "the subdiscipline within biotechnology that is responsible for translating the discoveries of life sciences into practical products, processes, or systems that can serve the needs of society". As such it is similar to biochemical-engineering science, which has been defined by a group of European biochemical engineers from universities and industry, and government officials - a group pursuing incorporation of bioprocess engineering in the coming EC-Framework IV program - as "the fundamental research into all aspects of the interactions between engineering and other disciplines necessary to underpin the development of industrial-scale biological-based processes" [2].

In all definitions of bioprocess engineering, *integration* and *application* are keywords. Consequently, *communication* between bioprocess engineers, (micro)biologists, physical and material scientists, marketing and sales people, legislators, etc., plays a crucial role during all stages of process development. Ultimately, however, it is the bioprocess/biochemical engineer who integrates all necessarry inputs into total process design. Development of academic curricula that educate such bioprocess engineers is of vital importance to generate sufficient appropriate manpower for the future.

Bioreactors are the 'guts' of such processes. It is clear, however, that the bioreactor, such as the digestive system in the body, only constitutes a small part of a complex system usually consisting of numerous other vital units. In an analogous way to a biological cell, it is necessary to view the process as an integrated whole. Only by pursuing integration of disciplines and unit operations from the beginning, process research and development will become efficient and effective. Integrated process design and operation, resulting in improved process economy and reliability, and product quality, is the timely but not yet developed theme. *Money*, *time* and *manpower* for education and research are the essential ingredients for maturation.

The standard stirred-tank fermenter always has been and still is the work horse among the bioreactors, despite of the very heterogeneous nature of the hydrodynamic pattern in the tank. The recent surge of interest in bioprocess design and operation has produced a large body of research data describing novel bioreactor configurations that do not show this disadvantage. Industry is reluctant, however, to invest heavily in new systems until long-term productivity/cost benefits are proven. Clearly, this understandable industrial *conservatism* is a bottleneck in the development of alternative process schemes. Critical examination of the needs of industrial-scale operations possibly can help to refocus research to address these needs [3].

It is within the above *scope* that critical problems for the development of bioprocess engineering will be discussed in this paper; critical problems that at the same time should be read as future challenges.

1. INTRODUCTION

1.1. Role of chemical engineering in biotechnology

Basic research is supposed to supply chemical engineers with knowledge, concepts and methods required to develop and operate industrial processes in their economic and social environment [4]. In the first place chemical process industries are concerned with this research, but chemical engineering methods are extended to all kinds of other industries. Among these industries a clear distinction should be made between those dealing with large-scale production of relatively cheap commodities and those dedicated to smaller-scale production of expensive specialities. The first mainly use continuous processes with specially designed equipment and long pay-out times. The second operate batch or semi-batch processes and make use of existing multipurpose equipment.

Depending on which definition is chosen, biotechnology can be considered to be one of the oldest industrial technologies or one of the newest [5]. For the chemical engineer, however, in agreement with the distinction made above, the major distinction is in terms of scale of operation. The majority of products of the new biotechnology are extremely high-value, low-volume biologicals/biochemicals. Thus recovery (separation) processes for such products may be relatively costly and energy intensive, because of the small quantities involved, but at the same time must be highly efficient in order to minimize loss of valuable product. This is in contrast with the more traditional biotechnological processes of the food and beverage industries, or the antibiotic and medium-value pharmaceuticals manufacture and the biological waste-water treatment. Even though most of these large-scale bioprocesses are carried out (fed-)batch wise except for waste-water treatment, here the problems resemble those in the chemical or process industries. Just as the chemical industry was dominated by the chemist for decades, the bioprocess industry is still dominated by the (micro)biologist. There are consequently many areas in which improvements can be made, even to established industrial biological processes, by the application of sometimes even simple chemical-engineering concepts.

1.2. Bioprocess engineering

Bioprocess or biochemical engineering can be seen as the crucial link between process technology and biotechnology [6]. Process technology, or chemical engineering, was developed in the middle of the 20th century through the multidisciplinary concerted action of chemists and engineers. Analogously, the field of bioprocess or biochemical engineering evolves from the interaction of biologists (microbiologists, biochemists, molecular biologists, etc.) and engineers.

Bioprocess engineering, in contrast to chemical engineering, has to deal with additional complicating factors, especially biological stability and variability, sterilisation and aseptic processing, and down-stream processing of very dilute product solutions with many other components. In comparison to chemical processes, a large number of uncertainties plays a role in bioprocesses due to the biological nature of the catalyst used. Model descriptions will therefore be less accurate, harder to verify, and the predictive value will be less. On the other hand, bioprocesses are generally slower and thus more (on-line) measurement data can be made available for control purposes. However, these measurements can only be made on a limited number of parameters, because of the lack of adequate sensors. This all means that more predictive model descriptions are required in bioprocess engineering, while these are harder to get than in chemical engineering.

1.3. Research in bioprocess engineering

The research field of bioprocess engineering can be explained in terms of the combination of a bioconversion process and the preceding up-stream and subsequent down-stream processing [6]. Nowadays these processes are developed and optimized by a multi-disciplinary approach and it is more and more recognized that different disciplines should indeed contribute to find the solutions needed to come to optimal applications.

The essential and unique feature of a bioprocess is the *biocatalyst*, catalyzing the bioconversion. A biocatalyst consists of a (formulated) cell from microbial, animal or plant origin, or parts thereof. In order for the bioconversion to take place, substrate has to be brought in contact with the biocatalyst and product has to be removed from it. This takes place in the biocatalyst microenvironment, which is created by the conditions in the bioreactor. To optimize the production process, the action of the biocatalyst has to be influenced. This can be done by modification of the biocatalyst itself, for instance by genetic techniques, and/or by manipulating and controlling its microenvironment (*medium design*).

Evidently the total process contains more steps than only the bioconversion in the *bioreactor*. The pretreatment of substrate and the *down-stream processing* of product are directly involved. But also the treatment of by-products, waste streams and utilities form an integral part. In optimizing production processes it is obviously the interplay of all these steps and several other aspects such as safety, environment, labour and economy that have to be considered (*integrated process design and operation*).

The ultimate objective of bioprocess-engineering research is to provide the fundamental quantitative understanding necessary for industry to design and operate efficient, cost-effective, and safe bioprocesses [7]. This greater understanding will be attained through experimental programmes aimed at elucidating fundamental principles and collecting relevant numerical data. These must be combined with theoretical and conceptual approaches to allow accurate modelling of bioprocesses. Research must be undertaken in the three key areas of fermentation, biotransformation and down-stream processing, and pursue the important theme of an integrated approach, ultimately resulting in an improved process reproducibility and safety, and minimization of waste streams.

A more detailed, but certainly not exhaustive discussion of the various *topics* mentioned above follows below.

2. BIOCATALYSTS

"Biocatalyst" is a broad term commonly used for an industrial catalyst produced biologically. Biocatalyst is used synonymously for pure enzymes, enzyme mixtures, formulated enzyme products, crude enzymes, cell fragments, and whole cells. Subsequently, critical problems associated with enzymes and microbial, animal and plant cells are discussed.

2.1. Enzymes

Industrial enzymes are generally produced by microorganisms in large fermenter vessels. Only in a few cases enzymes are isolated from plants or animals. By mutation and selection techniques the production of the enzyme of choice can be increased to high levels, e.g. over 10% of the total protein produced by the microorganism. Often the microorganisms excrete the desired enzyme into the liquid medium in which they grow, facilitating the isolation. Large amounts of enzyme can be obtained in a relatively short time and at low cost. It must be stated though, that the development of a commercial production process for an enzyme, based on fermentation, requires a significant effort in research and development.

Catalysis by enzymes offers a number of advantages over traditional chemical catalysis. Enzymes have a high catalytic power; they increase the rates of chemical reactions by factors 10^9 to 10^{12}. Moreover, they work under mild conditions of temperature, pH and pressure, and display high specificity for the substances on which they act. It is also common knowledge now that some enzymes can perform well in organic solvents, provided these are chosen properly [8]. Despite these impressive credentials, enzymes are not without disadvantages. They may be insufficiently stable for commercial applications and, because until now they are mostly used in aqueous solutions, separating them from the reaction mixture for subsequent reuse is often difficult. Recovering the product is also difficult because both substrates and products may be present in low concentrations in the aqueous solution.

The reduction or elimination of these handicaps to the commercial exploitation of enzymes is the goal of much current research. Some of the difficulties inherent to the use of water-soluble enzymes in aqueous solutions can be overcome by the substitution of part or all of the water by a suitable organic solvent and/or by immobilizing the isolated enzymes, thus converting them to insoluble forms. Whole cells, either viable or non-viable, can also be immobilized and used for carrying out particular reactions. This is especially helpful when the reaction requires cofactors and/or the participation of several microbial enzymes. The need for cofactors of enzymes such as oxidases and reductases has largely limited industrial applications to hydrolytic enzymes. The advantages of immobilizing enzymes (or whole cells) include increased stability, reduced costs, and greater ease of enzyme separation and recovery. Furthermore, continuous processes become possible, which in turn allows a better process control and an improved productivity. All these factors are of major importance for industrial applications.

The choice of the process method can be made when the feedstock to be modified has been defined and when the specifications of the endproduct are set. Presently, the prime question is whether a chemical or a biochemical catalyst is most advantageous. Factors that play a role in this consideration are obviously costs, technical possibilities and external factors. Costs, on the one hand, that are determined by the type of reaction, or in other words performance costs, and on the other hand costs due to development of the catalyst. For bulk enzymes the contribution to the overall costs is about 2-4% of the added value of the product [9], which is comparable to those of chemical catalysts. Each application will have a number of demands with respect to production process and characteristics of endproduct. If for instance no byproducts are allowed in the endproduct or if stereoselectivity is required, enzymes can offer advantages, and in some cases there will not even be an alternative. Legal aspects, safety criteria, specific consumer demands (for example the current preference for "natural") are the external factors also determining the choice of process and catalyst.

2.2. Microbial cells [2]

Traditionally microbial fermentations have been largely optimized on the basis of empirical approaches. There is now a need for a more fundamental approach so that process improvements can be realised from a rational understanding, leading to more cost-effective processes and improved product quality.

During the last five years, improvements in the area of measurement techniques such as mass spectrometry, high-performance liquid chromatography and flow-injection analysis, have provided the necessary tools for the application of advanced control schemes in operating fermentations on a more rational basis. In addition there have been significant advances in computational fluid dynamics to describe the conditions within highly mixed gas/liquid reactors, though extensive experimental validation is still required.

A major drawback for the application of these new technologies to enhanced fermentation control is the fundamental lack of understanding of cellular kinetics on time scales relevant to the bioreactor environment. For example, there is a serious deficiency of kinetic data on substrate consumption, product formation and, in some cases, product excretion at cell concentrations and fluid dynamic conditions that pertain in real fermenters. The generation of these data and their assembly into well structured and flexible models is an absolutely fundamental component of research necessary in this field.

Specifically, promotion of inter-laboratory studies (across Europe) is essential in the following areas:

a) Experimental examination of cellular kinetics at all levels, with the aim of describing growth and product formation under realistic process conditions.
b) Development of both descriptive and mechanistic models of growth and product formation which takes account of new kinetic data, the metabolic steps involved and the dynamic bioreactor environment.
c) The further development of new, robust, 'rapid-response' sensing and analytical methods for complex marker metabolites which can be used on-line.
d) The application of advanced 2-3 phase computational fluid-dynamics models, including transport phenomena, with experimental validation, to real fermentation reactors and the explicit linkage to the full breadth of dynamic cellular responses available from the kinetic models.

The necessary research can truly be considered precompetitive with two applications of the results to industry:

1) Predictive models can be used to control the fermentation process much better than at present, to obtain highly reproducible processes with concomitant process economy and potentially higher yields. Reproducibility will also have a positive influence on down-stream processes. Furthermore, rational manipulation of fermentations will aid selectivity of product formation, by minimising byproduct formation and simplifying integration with (part of) the down-stream

processing.

2) The improved understanding of cellular kinetics may also be used to identify potential bottlenecks in cellular metabolism under realistic conditions. This will be valuable in directing genetics for metabolic engineering of new strains; the consequence will be the development of new, high-yielding strains to be used in industry in the future.

The long term consequence will be the provision of experimentally derived cellular models, which together with models at the bioreactor level and for down-stream processing can be used to simulate the complete production process. The approach outlined above does not only apply to microbial cells, but to animal and plant cells as well.

2.3. Animal cells ([10] and references cited therein)

For the rational design and scale up of bioreactors the shear sensitivity of biocatalysts should be known quantitatively. This certainly is essential for biocatalysts known to be very fragile such as animal cells. Since there is still relatively little information of this kind available, systematic experiments are usually the only way to collect quantitative data on the influence of shear on a particular cell. Although these experiments may be rather difficult in many cases, the general strategy is clear. The key parameter measured for quantification of shear effects should be closely related to the aim of the technical process under investigation, for instance, cell viability, cell growth, formation rate or quality of the desired product, and so forth. Shear sensitivity data should preferably be collected in a laboratory device that is a down-scaled version of the production system. This usually means a standard stirred-tank fermenter, despite of the severe disadvantage of the poorly defined and irregular shear levels in such stirred vessels. Data collected in a device with well defined and constant shear levels usually require an awkward translation to the practical situation. Probably the best approach is to first analyze the available shear devices and damage measuring methods for each particular cell and process aim.

The primary reason for agitating bioreactors is transfer of oxygen and maintaining homogeneity by minimizing variations of dissolved oxygen and other nutrient concentrations, cells, temperature, etc. throughout the vessel. The average liquid velocity needed to give effective homogeneity can be estimated by requiring that the cells move through the various areas of different conditions in an amount of time that is small compared to their biological response time. Even in a small vessel this generally means that the bulk liquid flow is turbulent. An analysis of shear and turbulence is thus in general essential to rationally design and scale up bioreactors meant for growth of fragile cells.

Another reason for agitation can be to substantially enhance oxygen supply by stirring so fast that dispersion of air bubbles is accomplished in case of sparging, which is unavoidable if the size of the bioreactor increases above a certain volume. A stirrer-tip speed of at least 2 m/s is needed for that, resulting in shear-stress levels that are critical for insect and other animal cells. Stirring from the point of view of substantially improving the oxygen transfer by dispersion of air bubbles in animal-cell suspensions generally is thus impossible. Various research groups have therefore directed their attention to bubble-column and air-lift type of bioreactors. It was quickly found, however, that sparging air bubbles has a damaging effect on animal cells, in particular the bursting of the bubbles at the surface. Several additives are used to minimize the damage and especially the non-ionic surfactant Pluronic has a protective effect [11]. Also the use of membrane reactors and immobilization of cells are approaches to prevent shear damage.

Presently, many studies are done to elucidate the mechanism of cell damage as result of sparging. These studies mainly concentrate on the quantification of shear and other hydrodynamic stresses during the process of bubble bursting. The picture that slowly emerges shows that locally very high shear stresses occur during the bubble burst and that the volume in which these stresses occur is of the order of magnitude as the so-called killing volume. We hypothesized this volume to be associated with each air bubble during its life time. On the basis of this hypothetical killing volume we have derived and validated in lab-scale bioreactors a very simple model describing in a quantitative way the relationship between shear sensitivity of the cells and the required oxygen supply [12]. This model is suitable for design and scale up of bubble-column and air-lift loop bioreactors. Validation of the model on pilot and technical scale remains to be done.

2.4. Plant cells ([13] and references cited therein)

The possibility of producing useful chemicals by plant-cell culture has been studied intensively for the past 30 years. However, problems associated with low product yields and culture instabilities have severely limited industrial applications of plant-cell culture so far. These problems are merely biological in nature and are not discussed any further here.

The employment of hairy-root-culture technology, developed in the last 10 years, offers new opportunities for *in vitro* production of plant secondary metabolites. In contrast to suspended-cell cultures, hairy-root cultures are characterized by relatively high biosynthetic capacity and genetic as well as biochemical stability.

Hairy roots can be obtained by genetic transformation of wounded plant tissue by a pathogenic soil bacterium, *Agrobacterium rhizogenes*. This bacterium is capable of infecting a wide range of plants, causing roots to proliferate rapidly at the infection site (hairy-root disease). A related technique, in which infection with an *Agrobacterium tumefaciens* mutant is used for the establishment of transformed and differentiating shoot cultures, makes it possible to extend the approach used with hairy-root cultures to products synthesized in green tissues.

Hairy roots can be cultivated in a simple defined medium without addition of growth hormones. The roots are characterized by a high degree of lateral branching and a profusion of root hairs. In contrast to most suspended cells, hairy roots can grow from low inocula to high final biomass densities with only a minimal lag phase being observed. In general, the growth rates of hairy-root cultures are comparable with those of suspended-cell cultures.

During fermentation, the structured nature of hairy roots leads to formation of interconnected, non-homogeneous material that is unevenly distributed throughout the fermenter. This results in a markedly different hydrodynamics compared to that of suspended-cell cultures and has thus made it necessary to investigate novel approaches to fermenter and process design.

From the technical point of view, one of the main problems in the fermentation of hairy roots is mass transfer. Mechanical agitation is not possible, since shear stress causes disorganization and callus formation with consequently lowered productivity. Also inoculation causes difficulties, because root cultures can not easily be pumped.

For the same reason, sampling of the roots during fermentation is not possible. Therefore, alternative methods for monitoring growth have to be found, e.g. by measurement of nutrient consumption or medium conductivity.

Despite the unusual properties of hairy-root cultures, they have been grown successfully in fermenters and maintained viable in a continuously growing state for extended periods. In contrast to suspended-cell cultures, cultivation of hairy-root cultures in fermenters has not affected their productivity to a great deal, even on a large scale.

The exploitation of secondary-metabolite formation of hairy-root cultures in bioreactors still requires further development. However, the production of secondary products using hairy-root cultures is apparently near commercialization in Japan, where Furuya and co-workers recently reported the large-scale production of ginseng extracts using a hairy-root culture in a 20-ton bioreactor [14].

3. MEDIUM DESIGN

Water as essential reaction medium for biocatalysts has been advocated for many years as one of the major advantages for biocatalysis. However, this so-called advantage has proven to be one of the severest limitations for broadening the scope of applications of biocatalysts. Especially when the involved reactants are poorly water soluble, use of aqueous media is a major drawback. Therefore, much effort has been devoted in the last decade to the development of biocatalysis in non-conventional media, in particular organic solvents and to a lesser extent also supercritical fluids [8, 15]. There is also a renewed interest in solid-substrate fermentations [16, 17], also called solid-state fermentation (SSF). Interesting too, but not further discussed here, are aqueous two-phase systems.

3.1. Organic solvents

The introduction of an organic solvent in the reaction system has numerous advantages. The organic solvent may improve the solubility of the reactants, thereby improving the product yield of the reaction system. The equilibrium composition of a reaction may be shifted by reduction of the water activity of the reaction mixture. This can be achieved by (partially) replacing water in the reaction system by a water-miscible organic solvent, or by introducing polymers, sugars or salts. In a two-phase system, consisting of an organic and an aqueous phase the equilibrium composition of a reaction can also be shifted in favour of the product when the product is extracted into the organic phase. High product yields may also be achieved by reduction of possible substrate or product inhibition and prevention of unwanted side-reactions such as hydrolysis of the substrate or product. When the product is extracted into the organic-solvent phase while the biocatalyst remains in the aqueous phase, product and biocatalyst recovery will be facilitated.

In spite of the many possible advantages of using organic solvents, disadvantages also exist. The biocatalyst may be denaturated or inhibited by the solvent. Introduction of an organic solvent also leads to increasing complexity of the reaction system.

In addition to partitioning, the most important requirement of the solvent is biocompatibility. A good measure for the biocompatibility of a solvent is its $\log P_{octanol}$ [18], which is defined as the logarithm of the partition coefficient of the solvent in a standard two-phase system of 1-octanol and water. Figure 1 shows this relation for the metabolic activity of tetralin-degrading cells of *Arthrobacter* and *Acinetobacter* [19]. Generally speaking, solvents having a $\log P_{octanol}$ value > 5 are biocompatible with cellular biocatalysts.

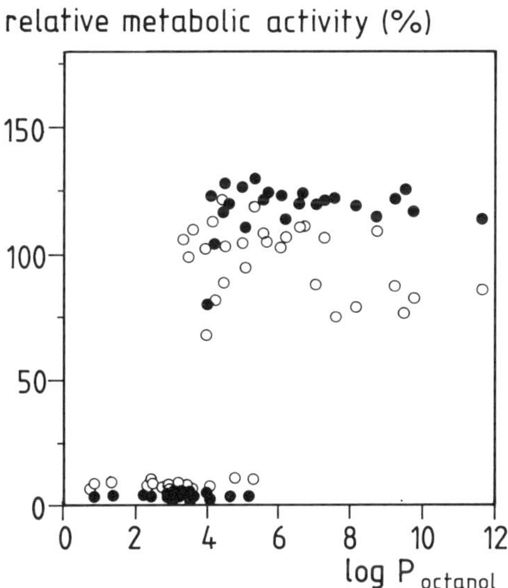

Figure 1.

Activity of tetralin-degrading cells as a function of logP [19].

Many attempts have been made to explain the empirical correlation between $\log P_{octanol}$ and the activity retention of cellular biocatalysts, but so far the mechanisms of solvent-caused toxicity are poorly understood. Bar [20] suggests that solvent toxicity in a two-liquid-phase system is caused by the presence of a second phase (phase toxicity) on the one hand, and by the solvent molecules that dissolve in the aqueous phase (molecular toxicity) on the other. Phase toxicity may result from extraction of nutrients and cell-wall components, or from limited availability of nutrients as a result of adherence of the cells to the interface, or from entrapment in an emulsion. Molecular toxicity may result from enzyme inhibition, protein denaturation and membrane modification. Obviously much more research is needed to elucidate the real mechanisms.

3.2. Solid-State Fermentation (SSF)

Solid-state (substrate) fermentation is generally defined as the growth of microorganisms on solid substrates in the absence or near absence of free-flowing water [16]. The substrate, however, must contain enough moisture that is ab(d)sorbed within the solid matrix. The moist solid substrates, which are polymeric in nature and insoluble in water, act as a source of carbon, nitrogen, minerals, water and other nutrients, and provide support for the microorganisms [17].

Solid-state fermentations have been used since antiquity in the Orient, Asia and Africa for the production of fermented foods (koji, miso, tempeh), starter inocula, mushrooms, dough fermentations, etc. In Western countries on the other hand, the SSF system has been neglected from 1940 on, while efforts were directed towards the development of submerged fermentations. However, solid-state fermentation has gained renewed interest from researchers owing to its importance in recent developments in biomass-energy conservation, in solid-waste treatment and in its application to produce secondary metabolites.

Large-scale solid-state fermentation is currently used in food processing and waste treatment, but it is still more an art than a technology. The physical and biological behaviour is poorly understood. Scale up is usually accomplished by trial and error or by maintaining geometric similarity. The former method has resulted in successful development of pilot and full-scale systems. However, extension of the results to another application will again require a trial and error procedure with a lot of expensive pilot-scale experiments. The latter method results in labour-intense tray-fermentors, which are only suitable for use in countries with cheap labour.

In contrast, recent work shows that a mechanistic description of the biokinetic and physical behaviour of solid-state fermentors is possible [21, 22]. This offers perspectives for development of highly needed, well-founded, scale-up criteria such as commonly used in the process industry. Aspects that have to be taken into consideration in this development are variations in the biomass formed, large-scale inoculum development, method of inoculation, medium sterilization, aeration, agitation, heat removal, moisture content of the solids, pH control, contamination control, heterogeneity, down-stream processing, waste management and solids handling.

Of these aspects, proper control of water activity and temperature poses the main problem. In large-scale solid-state fermentors, the metabolic heat is mainly removed by evaporation of water [23]. This results in a large moisture loss and eventually in a decrease of the water activity in the system. Even a small decrease in water activity can have a detrimental effect on the growth of microorganisms. Therefore, water addition is imperative. Most of the presently available bioreactors for solid-state fermentation are either hampered by moisture gradients from insufficient mixing, or present constructional difficulties on a large scale [24].

4. BIOREACTORS [3]

The majority of industrial fermentations are run in stirred tanks of a fairly standard design of up to 200 m^3 working volume. The impeller shaft has up to five impellers, and for pilot and production-scale fermenters, specific power inputs are in the range of 1-4 kW/m^3, the temperature range is 25-40 °C and the pH range is 4-8. Aerobic fermentations represent the most-important industrial fermentation processes, with aeration rates in the range 0.2-1 vvm (volume per volume per minute). Most industrial microorganisms thrive under these conditions, but new systems wil have to be developed for exploitation of organisms whose unusual metabolic activities represent a new resource for biotechnology. For instance, extremophiles have seen little application in industry yet due to their low growth rates, low cell yields and high shear sensitivity. Similar reasons hold for animal and plant cells and also make special reactor designs unavoidable.

The fed-batch mode of operation is the most-widely used in industrial fermentations. Fed-batch processes are particularly well suited for fermentations that exhibit substrate inhibition, catabolite repression, toxic precursors or the glucose effect. What is perhaps surprising at first sight is the relative scarcity in industry of continuous fermentation processes. Continuous operation offers several important advantages (in particular by reducing labour and capital costs, and avoiding peak demands in utility consumption), but has the requirement that the strain of the organism used is one that is stable over a large number of generations. This makes continuous fermentation less suited for the production of intermediate and secondary metabolites, where genetic drift is a problem. In addition, the technical difficulty of maintaining sterility over long periods of time means that continuous fermentations tend to be ones that do not require strict sterility, or in which the maintenance of sterility is assisted by the operating conditions, such as low a pH.

Extension of the range of continuously operated processes requires developments providing competitive advantage of the producer strain over faster-growing mutants or contaminants. This may be achieved by combination of genetic manipulation with process control. However, in view of the (increasing) dominance of the fed-batch fermentation, more research effort could be directed to the design and optimization of fed-batch control methods. Since raw materials (together with utilities) generally account for the largest part of production costs, research on efficient use of energy and raw materials is particularly relevant too.

The need for adequate oxygen transfer limits the size of aerobic fermenters to about 200 m^3. At such scales, there is little economic incentive to avoid mechanical agitation. Only when the scale of operation is much larger, alternative bioreactor designs that avoid mechanical agitation (such as the bubble-column and air-lift fermenters) offer significant economic advantage. Mechanical agitation offers an additional degree of freedom in accomodating changes in broth rheology and oxygen demand that occur not only during the course of (fed-)batch fermentations, but also as result of the application of new producer strains. In view of the pressure towards flexibility and multi-product plants, the degree of freedom provided by mechanical agitation is likely to become more, rather than less important. Control of gas-phase oxygen partial pressure can add another degree of freedom to the process scale up [25]. This can be particularly useful in case of fragile cells such as animal cells that do not survive in too intensively agitated broths.

5. DOWN-STREAM PROCESSING [2]

Down-stream processing refers to the sequence of operations in which the target substances produced in a fermentation broth or by a biotransformation are isolated and purified. The cost of down-stream processing has a critical impact on the exploitation of many biological products, amounting to as much as 90% of the production cost for a recombinant-DNA product. The critical requirements to a processing sequence are:

a) adequate resolution and purification,
b) sufficient productivity,
c) adaptability and predictability,
d) reliability and reproducibility,
e) satisfactory capital and operating costs, and,
f) safety and minimal undesirable environmental impacts.

Currently, it is almost impossible to find processes, other than very simple ones, which can combine (a) and (b). For example, a typical purification train for a macromolecule might involve seven steps, with unit yields of 80% or less; the overall yields of the target molecules are thus often 20% or less. It is also impossible to predict performance of unit operations, let alone processing trains, on new feed stocks, with the result that expensive pilot trials are needed before designs can be finalized. The result is that processing sequences are made up of individual unit operations, each of which is usually based on available equipment, so that the whole is usually suboptimal.

From these scenarios, an unsatisfactory picture emerges of processing sequences designed on the basis of empiricism. For very high-value pharmaceuticals, where regulatory considerations are paramount, the costs involved in suboptimal operation may be acceptable. However, in the exploitation of larger-volume, medium-value products requiring high purity (such as biological agrochemicals) the cost of down-stream processing remains the single most important technical barrier to commercialisation.

To tackle the problems outlined above, research is needed with the twin objectives of being able to:

1) predict (using mathematical and/or information-based models) the performance of the individual stages and by consequence the overall performance of a sequence of down-stream-processing stages, and
2) increase the range of operations available for industrial implementation.

With regard to objective 1, an adequate model would enable efficient control systems to be defined. To be able to predict quantitatively the performance of existing and new operations requires a radical rethink of existing ways of studying unit operations, and an extensive experimental program. The experimental program will have to concentrate on the molecular characteristics of the species being manipulated, measured in the separation system under investigation, rather than by unrelated techniques, which has hitherto often been the case. To carry out such characterization, advanced on-line sensors and analysis techniques will be needed. The objective of this work should be to produce a "package" of physiological, physical, chemical and thermodynamic properties for biological species in the same manner that thermodynamic packages exist for modelling chemical-engineering operations.

Coupled with the need to know molecular characteristics, there is a need for an increased knowledge on the flow and shear fields in equipment using fluid-dynamics modelling. This requires considerable work because existing modelling packages are unable to deal with the combination of non-Newtonian rheology and three-phase systems wich commonly occur in biological systems. Typical areas where such information would be of immediate use would be in examining the shear forces in decanters and centrifuges which are being used to separated viable biomass for subsequent reuse, for example in biotransformations.

With regard to objective 2, experimental work should continue on development of novel methods, together with improvement and modelling of existing operations. Current methods of separation involve electrical fields (in electrophoresis), physical/chemical forces (centrifugation), convective fluxes (chromatography), chemical potentials (affinity methods), mechanical forces (membrane separations) together with thermodynamic considerations. Most of these are adaptations from chemical or food engineering and are not necessarily optimal for biological systems. The ability to predict molecular and thermodynamic properties opens up the possibility of designing *ab initio* new operations based on novel combinations of fields and forces.

The problems of separating particles less then 1 μm in diameter including the debris from the homogenizations of bacterial and yeast cells need attention. At such sizes, centrifugal and membrane separations are ineffective and there is thus the need to increase the size of the particles by conditioning the homogenates. Another key need is the reduction of membrane fouling. Current efforts have concentrated on modifying synthetic polymeric membrane materials to alter pore sizes and to improve fluid flow. However, it is becoming increasingly clear that such methods will have to be supplemented by development of special biocompatible coatings that obviate protein denaturation at the surfaces of membranes.

6. INTEGRATED PROCESS DESIGN AND OPERATION [2, 7]

Bioprocess engineering is the crucial link between biological discoveries and real processes and products. As such it requires an understanding of both biological and engineering systems, and an integration of the demands of both in the design of optimal processes. The ultimate aim of a research program in bioprocess engineering is to provide the fundamental quantitative understanding necessary for industry to design and operate efficient, cost-effective and safe bioprocesses. This greater understanding should be attained through experimental programmes aimed at elucidating fundamental principles and obtaining relevant numerical data. These should be combined with theoretical and conceptual approaches to allow accurate

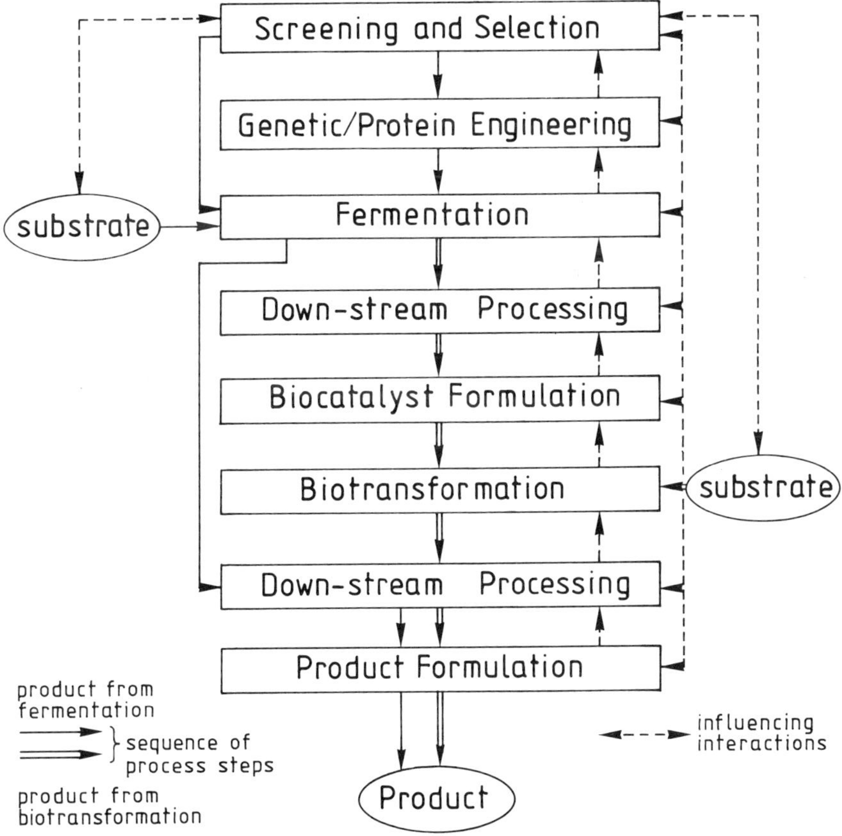

Figure 2. Schematic representation of a bioprocess.

modelling of processes which are reliable and reproducible, safe and economic to operate, environmentally acceptable, and flexible.

In order to achieve this, it is essential that effort is focused on integrated design. This means taking into account factors concerning all stages of the process from the outset and thus optimizing the design for the whole process. Figure 2 shows schematically the possible stages in a bioprocess. Changes in the parameters or properties of any one stage will have consequences for some or all others. The choice of organism and medium for example has considerable impact not only on the fermentation or biotransformation itself, but also on the down-stream processing. It is essential therefore that the requirements of the entire process are taken into account right from the start to ensure that the overall process is optimized. Optimization includes not only obtaining a high yield of the desired pure product, but doing so in an energy efficient way using the minimum possible resources and producing minimal waste. Process intensification, that is reduction in the number of steps involved and elevation of component concentrations, is therefore of key importance. The ultimate objective is to be able to devise a total approach to integrated process design and operation, resulting in process reliability and improved product quality.

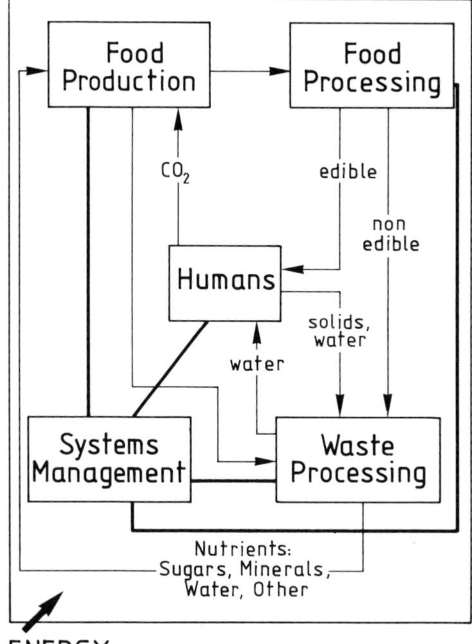

Figure 3.

Partial schematic diagram of a controlled ecological life-support system (adapted from [26]).

A splendid example illustrating that in some cases integrated bioprocessing even is indispensable, are the Controlled Ecological Life-Support Systems (CELSS). CELSS are defined as closed systems with only energy crossing the outer boundary (Figure 3, [26]). In a CELSS all raw materials are provided by recycling wastes. CELSS will be needed to provide a constant supply of food, air and water through bioregenerative means, during manned space flights of longer duration and distances from earth.

7. ACKNOWLEDGEMENTS

The author has been greatly inspired by a lecture of Prof. Villermaux in Enschede (NL) on 30 October 1992 during the opening of the Research School Process Technology. He thanks Prof. Villermaux for kindly providing him with preprints of similar lectures. The author also likes to thank his coworkers for critical reading the concept manuscript and for their many useful comments.

REFERENCES

1. Putting Biotechnology to Work: Bioprocess Engineering, Committee on Bioprocess Engineering, National Academy of Sciences, Washington, DC, 1992.

2. "Biochemical Engineering Science", Strategic Case for a Clear Focus in Framework IV, January 1992.

3. P.N. Royce, A need to refocus research in the operation of fermenters?, TIBTECH, 10 (1992) 232-234.

4. J. Villermaux, Basic chemical engineering research: Where are we going? In "The Chemical Engineer", Transactions Instit. Chem. Eng., "Research", January 1993.

5. C. Webb and B. Atkinson, The role of chemical engineering in biotechnology, Chem. Eng. J., 50 (1992) B9-B16.

6. European Biochemical Engineering Network, project proposal for EC Human Capital and Mobility Programme, Delft, 1992.

7. Proposals for a Scientific Programme on Biochemical Engineering Science in Framework IV, follow-up paper of the above-mentioned [2] strategic-case report, 1993.

8. J. Tramper, M.H.Vermuë, H.H. Beeftink and U. von Stockar (eds.), Biocatalysis in non-conventional media, Elsevier Science Publ., Amsterdam, 1992.

9. B. Poldermans, Enzymes in the food industry, European Food & Drink Review, August (1990) 5-28.

10. K. van 't Riet and J. Tramper, Basic biorector design, Marcel Dekker Inc., New York, 1991.

11. E.T. Papoutsakis, Media additives for protecting freely suspended animal cells against agitation and aeration damage, TIBTECH, 9 (1991) 316-324.

12. J. Tramper, D. Smit, J. Straatman and J.M. Vlak, Bubble-column design for groth of fragile insect cells, Bioprocess Eng., 3 (1988) 7-41.

13. L. Toivonen, Utilization of hairy root cultures for production of secondary

metabolites, Biotechnol. Progress, 9 (1993) 12-20.

14. A. Scheidegger, Plant biotechnology goes commercial in Japan, TIBTECH, 8 (1990) 197-198.

15. C. Laane, J. Tramper and M.D. Lilly M.D. (eds.) Biocatalysis in organic media, Elsevier Science Publ., Amsterdam, 1987.

16. A. Pandey, Recent Process Developments in Solid-State Fermentation, Process Biochem., 27 (1992) 109-117.

17. B.K. Lonsane, G. Saucedo-Castanedo, M. Raimbault, S. Roussos, G. Viniegra-Gonzalez, N.P. Ghildyal, M. Ramakrishna, and M.M. Krishnaiah, Scale-Up Strategies for Solid State Fermentation Systems, Process Biochem., 27 (1992) 259-273.

18. C. Laane, S. Boeren, K. Vos and C. Veeger C., Rules for optimization of biocatalysis in organic solvents, Biotechnol. Bioeng., 30 (1987) 81-87.

19. J. Tramper and M.H. Vermuë, Tailoring the medium and bioreactor for biocatalysis, Chimia, 47, April 1993.

20. R. Bar, Effect of interphase mixing on a water-organic solvent two-liquid phase microbial system: ethanol fermentation, J. Chem. Tech. Biotechnol., 43 (1988) 49-62.

21. D.A. Mitchell, D.D. Duong, P.F. Greenfield and H.W. Doelle, A semimechanistic mathematical model for growth of *Rhizopus oligosporus* in a model solid-state fermentation system, Biotechnol. Bioeng., 38 (1991) 353-362.

22. H.V.M. Hamelers, A theoretical model of composting kinetics, in Science and engineering in composting (H.A.J. Hoitink and H.M. Keener, eds.) Renaissance Publications, Worthington, USA (1993) 27-29.

23. L.M. Barstow, B.E. Dale and R.P. Tengerdy, Evaporative temperature and moisture control in solid substrate fermentation, Biotechnol. Techniques, 2 (1988) 237-242.

24. D. Ryoo, V.G. Murphy, M.N. Karim and R.P. Tengerdy, Evaporative moisture and temperature control in a rocking reactor for solid substrate fermentation, Biotechnol. Techniques, 5 (1991) 19-2.

25. L.-K. Ju and G.G. Chase, Improved scale-up strategies of bioreactors, Bioprocess Engineering, 8 (1992) 49-53.

26. P. Westgate, K. Kohlmann, R. Hendrickson and M. R. Ladisch, Bioprocessing in space, Enzyme Microb. Technol., 14 (1992) 76-79.

ECB6: Proceedings of the 6th European Congress on Biotechnology
edited by L. Alberghina, L. Frontali, P. Sensi

Gene pools and biodiversity

W. Arber

Department of Microbiology, Biozentrum, University of Basel, Klingelbergstr. 70, CH-4056 Basel, Switzerland

1. INTRODUCTION AND DEFINITIONS

Biodiversity as perceived by the human observer is a phenotypic manifestation of genetic diversity. Biological diversity can be found at various scales. At the microscale, subtle differences of specific properties lead to a genetic polymorphism in populations of members of a given species. At the other extreme, the macroscale, organisms belonging to different species may differ in a large number of biological functions and thus also considerably in their phenotypic appearance. It is the aim of this contribution to mainly reflect on the generation of diversity at the microscale. For this purpose we will specifically look at molecular mechanisms of spontaneous mutagenesis. At a longterm range and under steadily exerted selection, the same molecular processes are also thought to lead eventually to divergent forms of life by the consecutive accumulation of genetic differences.

Genetic information is known to be determined by the linear sequences of base-pairs on filamentous double-stranded DNA molecules. Part of this information is contained in open reading frames which determine the primary structure of proteins. Other parts of genetic information contain control elements for either gene expression or the physiology of nucleic acids themselves.

For the sake of simplicity, we define here as spontaneous mutation any kind of alteration of the inherited DNA sequences of an organism, independent of the mechanism of mutagenesis, provided that the mutation did not result from an intended action of an investigator. At the phenotypic level, an alteration of the DNA sequences may remain undetected. Such mutations are neutral or silent. At the other extreme, a mutation can inactivate essential life functions and thus be lethal. Other mutations may affect life less severely but still give a selective disadvantage to the mutant. Rather rarely a mutation can improve the fitness of the organism and thus provide it with a selective advantage. Since many mutations cause lethality or give a selective disadvantage, tolerable overall mutation rates must be lower than one mutation per genome per generation, at least for haploid organisms.

Gene pool has been defined as the "total genetic information possessed by the reproductive members of a population of sexually reproductive organisms" (King, 1974). Since prokaryotic microorganisms normally do not reproduce sexually, this definition cannot be strictly applied. In view of the considerable role played by horizontal gene transfer between these different microorganisms, one may look at all bacterial genes as forming one large pool of genes. Although not all genes are equally available to all organisms participating in this pool, we consider this view justifiable for our purposes.

2. EASE OF INVESTIGATIONS ON THE GENERATION OF BIO-DIVERSITY WITH PROKARYOTIC MICROORGANISMS

Organisms with haploid genomes have the advantage of rapidly manifesting the phenotype corresponding to a mutation which they had undergone. Therefore, bacteria and their viruses and plasmids are convenient objects to study spontaneous mutagenesis and the action of selection. These microorganisms have the additional advantages of relatively small genome sizes (about 5×10^6 bp for bacteria and 10^4 to 10^5 bp for viruses and plasmids), short generation times and exponential growth. Thus large populations of individuals can be obtained in short periods of time, making these microorganisms ideal objects for population genetic studies.

3. CAPACITY OF THE BIOSPHERE FOR GENETIC INFORMATION AND UNIQUENESS OF GENETIC INFORMATION

Our planet has a surface of 5×10^{14} m^2. The average volume of living cells is about 10^{-18} m^3 for bacteria and 10^{-15} m^3 for eukaryotic cells.

If we assume the biosphere as a compact surface layer of 20 cm thickness and thus of a volume of 10^{14} m^3 of living cells uniformly covering the entire planet, one can estimate the total number of living cells to be in the order of 10^{30}. Although this is an astronomically high number, it is impressive to see that upon exponential growth bacteria propagated with a generation time of 30 minutes in a very large container reach 10^{30} cells from an inoculum of one single cell within only two days!

The following simple considerations shall give a view into the remarkable uniqueness of specific DNA sequences. For this purpose we compare the number N of possible bp sequences for an entire bacterial genome of size 5×10^6 bp, that of a standard gene of 1000 bp and that of a DNA domain or motif of 75 bp length. The number of possible different bp sequences of these lengths is calculated as $N = 4^n$, if n is the number of bp composing the sequence in consideration. For the 3 examples chosen, the number N of different sequences is $10^{3'000'000}$ for the bacterial genome, 10^{600} for the standard gene and 10^{45} for the sequence domain.

Life on our planet Earth started at most about 4×10^9 years ago, i.e. 3.5×10^{13} hours. If we assume that 10^{30} living cells were steadily present since these early times and that they tolerated a mutation rate of one mutation per genome every 20 minutes, one can calculate that a total of $10^{30} \times 3 \times 3.5 \times 10^{13} = 10^{44}$ different bp sequences could have been tried out once up to now. It is surprising to see that this barely offered the chance to test for functions once for all possible domains of 75 bp length. Thus, the genetic diversity present today is an extremely small subset of all theoretically possible forms. In consequence, lost capacities are unlikely to come back rapidly.

4. BIODIVERSITY REFLECTS THE PRESENT STATE OF BIOLOGICAL EVOLUTION

Biological evolution has to be seen as a dynamic process, in which new genetic variants are steadily generated by spontaneous mutagenesis (see below). However, roughly the same proportion of variants is lost per unit of time, mainly by selection and by repair processes. The latter seriously reduce actual mutation rates, keeping these at low frequencies, tolerable for each given population. Another factor decisive for the overall genetic diversity is the size of the biosphere, roughly 10^{30} living cells as already mentioned. Thus, sampling can also be considered as a parameter limiting genetic diversity. Diversity benefits to some extent from both geographic and reproductive isolations. At the genome level, total cellular genetic microdiversity must be somewhat below 10^{30}, but its specific sequence composition is not constant; rather, it varies as a function of time. At any instance, the observed diversity reflects the state reached by biological evolution. Thus, biodiversity represents a static picture of biological evolution.

5. SEVERAL CATEGORIES OF MOLECULAR MECHANISMS CONTRIBUTE TO THE OVERALL SPONTANEOUS MUTAGENESIS

As was already mentioned, we find it convenient to define as spontaneous mutation in the broad sense any alteration occurring to preexisting nucleotide sequences. In addition to the classical categories of mutagenesis due to replication infidelities and environmental mutagens, we thus also consider DNA rearrangements and gene acquisition as causes of genetic alterations. Any of these categories of mechanisms contains several specific molecular processes. Rather than to discuss them all, we will comment here on a few representative processes.

5.1. Nucleotide substitution

Nucleotide substitution is perhaps the most classical of all events of spontaneous mutagenesis. It can result from infidelity of DNA replication or from the action of an internal or external environmental mutagen. In the former case, its origin is postulated to depend on shortliving tautomeric forms of nucleotides giving rise to unusual base pairing. Since this is an intrinsic property of matter, the resulting mutations should not be considered as mistakes. Rather, nature can be seen to profit of structural flexibilities of biological molecules to generate variation.

Nucleotide substitutions are important steps in the development of new and of altered biological functions. But we have to be aware that it often takes a great number of individual steps to reach a novel function. At each stage the state of development in progress has to withstand selective elimination.

5.2 Site-specific DNA inversion

Site-specific DNA inversion is an enzyme-mediated DNA rearrangement. The relevant enzymes, called DNA invertases, usually interact with specific sites on the DNA which are defined by their nucleotide sequences. At a pair of these crossing-over sites, the DNA becomes cut and rejoined in such a way that the result is a DNA inversion between the two crossing-over sites concerned (Glasgow et al., 1989). Since this process is repeated in the course of time, a subsequent inversion can reconstitute the original configuration of the genome. Such flip-flop systems have been shown to offer to microbial populations the possibility to display two, and in some instances even more, different phenotypes. This can represent some advantage to the organisms in changing environments.

Usually, DNA invertases do not limit their action to a single specific DNA sequence. Rather, sequences deviating from the preferred sequence can also serve for recombination, although with decreasing efficiencies for lower degrees of homology to the preferred sequence. Interestingly, even sequences widely deviating from a preferred sequence can serve at low probabilities for DNA inversion (Iida and Hiestand-Nauer, 1987). This offers a chance for a great number of DNA sites to occasionally undergo DNA inversion. Some of these events may be the source for gene and operon fusions. If such fusions happen to provide an advantage to the organism concerned, it will be selected and eventually overgrow its parent. This mechanism is therefore likely to be of great evolutionary significance (Arber, 1991). In the course of time, many different kinds of gene fusions can be tried out in large microbial populations. Genes composed of different functional domains could have their origin in this kind of DNA rearrangement processes. Thus, we consider it as an evolutionary advantage for microorganisms to encode for DNA inversion systems in their genome. In the long term, selection for the presence of such systems would then be guided by their provision to the carrier strains of a higher evolutionary flexibility serving for the development of biological functions.

5.3 Transposition

About 1 to 2% of DNA sequences of bacterial genomes represent mobile genetic elements, if one extrapolates from the situation of *E. coli* to that of other microorganisms (Birkenbihl and Vielmetter, 1989). At low rates, enzymes expressed by mobile genetic elements, transposases, mediate different kinds of DNA rearrangements. These include besides simple transposition also deletion formation, DNA inversion and replicon fusion (Iida et al., 1983). Hence transposition is an important factor in the evolution of genome structures. As with other genetic alterations, these DNA rearrangements are not directed towards specific goals. Rather, they occur to some degree randomly, but selection exerts a strong decimation among the multitude of structures resulting from rearrangement processes.

Transposition can also occur in resting cells. This is for example seen with stab cultures. These are small vials with nutrient broth-agar medium, into which bacteria are inoculated. After a primary phase of growth, the vials are tightly closed by a paraffined cork stopper. If stored at room temperature, viability of bacteria in stabs is maintained for many years. Under these conditions growth is at most very residual. Interestingly, transposition of some mobile genetic elements occurs upon storage of *E. coli* in stabs with a more or less constant rate (T. Naas, M. Blot and W. Arber, to be published). As revealed by restriction fragment length polymorphism studies, transposition affects all members of bacterial populations stored in stabs to such a degreee that after 20 to 30 years of storage all viable individuals differ in their genome sequences from the original parent, and the genome structures of many individuals are unique. Apparently, under these conditions selection is relatively weak, which allows a high genetic microdiversity to build up. These findings are of high ecological and evolutionary significance. Indeed, under natural conditions bacteria spend most of their time in a resting phase rather than replicating exponentially. Hence, also in the former condition bacteria diversify genetically.

5.4 DNA acquisition

While bacteria reproduce asexually by cell division, once in a while they take up genetic information from other cells. Different transfer mechanisms are well known to be involved in such processes. These are transformation by uptake of free DNA molecules, conjugation mediated by transferable plasmids, and transduction mediated by bacterial viruses. In the latter processes the involved plasmids and viruses play the role of natural gene vectors. Any chromosomal gene can occasionally become associated with these gene vectors, and transposition is often the cause of such association. Transposition can also bring about the integration of transferred genes into the recipient chromosome, although other DNA rearrangement processes can serve for the same purpose. Horizontal gene transfer which we call here DNA acquisition is an interesting means for bacteria to enrich their genetic potential. Indeed, a functional gene or a functional gene domain can decisively improve bacterial fitness in a single step of DNA transfer. We thus consider DNA acquisition as an important evolutionary strategy and the promoters of DNA transfer as evolutionary elements (Arber, 1993).

One of the best documentations on these processes resulted from investigations on the molecular basis of the observed increasing occurrence of antibiotic resistances of Enterobacteria. A wide, deliberate use of antibiotics in human and veterinary medicine started about 50 years ago. Already after 10 - 20 years the frequencies of resistant strains showed a medically problematic increase. Occasional acquisition of resistance genes must have always occurred in the past, but it is only under the new selective conditions in the presence of antibiotics that rare recipients of these genes could rapidly overgrow the parental populations of bacteria.

Without genetically determined factors limiting gene acquisition, a required minimal genetic stability might be put into question. Indeed, gene acquisition has to overcome a number of natural barriers. These include cell surface incompatibilities, restriction-modification systems, the requirement to get the acquired genes propagated and, last but not least, the need of compatibility of the acquired gene functions with the functional harmony of the recipient cell. Interestingly, while restriction-modification systems cut foreign invading DNA molecules into fragments, the resulting DNA fragments are recombinogenic, which offers them some chance to incorporate into the recipient genome in the short period of time before they become exonucleolytically degraded. In this process we see a strategy of DNA acquisition in small steps. This strategy also insures a better success in the selection step of the acquisition, i.e. at the level of the test for compatibility of the functions expressed by the resulting transformants, exconjugants and transductants.

5.5 Comparison of efficiencies of evolutionarily relevant developments

In view of the nondirectional nature of evolutionary developments, evolutionary efficiencies of different mutagenic processes are difficult to measure and to compare. However, one can roughly estimate such values on the basis of their observed success rates. If overall success for evolutionary progress is evaluated per single mutagenic event, DNA acquisition is likely to be much more efficient than nucleotide substitution, and development by intragenomic DNA rearrangement may appear as of intermediate efficiency. However, it is good to keep in mind that these three processes highly differ in their action specificity and that all of them, as well as many other developmental processes not specifically discussed here, have their particular roles in biological evolution. Nucleotide substitution is a source both of new biological functions and of the improvement of existing functions. But it can take a large number of individual steps to obtain a novel function. Alternatively, a bacterial cell can acquire a novel function by a single step of DNA transfer from another bacterial strain already possessing the function in question. DNA acquisition allows bacteria to share in the success of others. Finally, DNA rearrangement processes can improve available biological capacities by gene and operon fusions. Success of any of these events will of course eventually depend on selective forces exerted by the existing living conditions.

6. SPECIFIC BIOLOGICAL FUNCTIONS PROMOTE MICROBIAL EVOLUTION

The tree of microbial evolution should correctly be drawn with horizontal shunts between different branches in order to symbolically represent DNA acquisition. Evolutionary progress thus depends on a multitude of mutagenic effects acting on the entire genome, and in addition on DNA acquisition which can improve fitness at any time by adding one or a few functional elements of foreign origin to the genome. In this view, the microbial world may share a large common gene pool, although not all genes may be equally available to a given species in question. Future evolutionary development of any microbial strain may thus depend not only on its actual genetic capacities but also on any other gene function encoded by other microbial strains.

A considerable number of enzymes and organelles appear to be specifically involved in the promotion and limitation of evolutionary developments. Mutagenesis has to be kept at a tolerably low rate and limitation mechanisms can bring about a state of equilibrium between genetic variation and stability. Among the genetically determined factors involved in these processes we count DNA rearrangement enzymes such as the discussed DNA invertases and transposases, but also natural gene vectors and restriction-modification systems. In addition to these genetically encoded variation generator systems, genetic variation also depends on the intrinsic structural flexibility of organic molecules which contribute to mutagenesis. We do not consider spontaneous mutagenesis as a result of mistakes nor biological evolution as a result of accumulated errors. Rather, biological evolution depends on natural features of nucleic acids and of enzymes acting on the genome. These kinds of biological functions steadily generate genetic diversity and the resulting variants are submitted to selection. The biological functions of evolutionary relevance act at the level of populations rather than at the level of individuals. Though, per unit of time, some individuals of large populations of cells are the object of the action of these functions, and the consequences of such functions are often detrimental. Thus, concerned individuals only rarely take profit of evolutionary functions, although the entire population benefits. Interestingly, most of such accessory functions found in *E. coli* have been shown to have a different codon usage (Médigue et al., 1991) than house-keeping functions, the action of which is required in each individual cell. It is also conceptually of interest to conclude that microbial cells carry, besides their required house-keeping genes, a series of other genes which insure the evolutionary progress of the strains to which the individuals belong. Possibly this phenomenon may not only apply to bacteria but also to other organisms. Some biological functions carried out in bacteria serve both purposes at once and meet needs of individual lives and evolutionary needs at population levels. A good example may be DNA ligase. This duality is obviously trivial, since any biological activity exerted on a cell can serve for a specific purpose corresponding to its mechanistic function, as long as it remains available in a cell.

7. CONCLUSIONS

A certain degree of genetic stability is a prerequisite for the maintenance of a species, while adaptation and genetic evolution strongly depend on genetic variation in populations of living organisms. Since mutations are often lethal or reduce fitness, a relatively low, "tolerable" mutation rate may permit a species to persist over prolonged periods of time under changing living conditions. Knowledge accumulated largely from investigations in microbial genetics allows one to classify spontaneous genetic alterations into 4 categories, i.e. (1) replication infidelity, (2) effects of internal and external chemical and physical mutagens, (3) DNA rearrangements which are mostly mediated by enzymes, and (4) acquisition of genetic information from other living organisms. Relative rates of contributions of each particular mutagenic process depend both on the living conditions and on the genetic setup of a considered species. Evolutionary efficiencies of mutagenesis processes vary widely if determined per event and might be highest for the acquisition strategy. The generation and decay of genetic variation, of gene pools and of biodiversity thus depend on genetic plasticity as well as on selective forces acting on populations of living organisms. A profound understanding of molecular mechanisms of biological evolution has a double relevance for biotechnology: (1) in view of an appropriate use of specific biological functions as biotechnological resources and (2) in view of the assessment of risks which might be inherent to such use.

REFERENCES

W. Arber, J. Mol. Evol. 33 (1991)4-12.
W. Arber, Gene, in press (1993).
R.P. Birkenbihl and W. Vielmetter, Mol. Gen. Genet. 220 (1989) 147-153.
A.C. Glasgow, K.T. Hughes and M.I. Simon, in Mobile DNA, D.E. Berg and M.M. Howe (eds.), Am. Soc. Microbiol., Washington, DC, (1989) 637-659.
S. Iida, J. Meyer and W. Arber, in Mobile genetic elements, J.A. Shapiro (ed.), Academic Press, New York (1983) 159-221.
S. Iida and R. Hiestand-Nauer, Mol. Gen. Genet. 208 (1987) 464-468.
R.C. King, A dictionary of genetics, 2nd ed., Oxford University Press, (1974).
C. Médigue, T. Rouxel, P. Vigier, A. Hénaut and A. Danchin, J. Mol. Biol. 222 (1991) 851-856.

ECB6: Proceedings of the 6th European Congress on Biotechnology
edited by L. Alberghina, L. Frontali, P. Sensi

STRATEGIES OF BIOTECHNOLOGICAL INDUSTRIES
Biotechnology in the Nineties: Back to Business

Dr. J.C. van Suijdam & Dr. J.A. Roels

Gist-brocades B.V.
P.O. Box 1, 2600 MA DELFT, The Netherlands

1. BIOTECHNOLOGY: THE PAST

After many centuries of artisanial application of biotechnology, to make products like cheese, beer and wine, in the period between 1940 and 1975, breakthroughs in the understanding of the functioning of micro-organisms at the molecular level were realised, preparing the ground for developing the recombinant DNA and the hybridoma technologies.
A number of significant advances in the ability to program and manipulate the genetic information of micro-organisms, collectively labelled "genetic engineering", quickly grew to a level of refinement allowing for an efficient editing of the genetic information of micro-organisms and, later on, plants and animals.
The potential of these techniques for the creation of new processes, new products and new varieties of production organisms, cultivars and production organisms were recognised with great enthusiasm. This resulted in increasing expectations of biotechnology about innovation potential for society, including industry. In fact a situation emerged, wherein society translated technological possibilities into a near term commercial reality. Biotechnology is certainly an area of fruitful scientific debate and technological progress and a good example of merging technologies. However, as witnessed in the preceding versions of this event in the last few decades, biotechnology often was mistaken to be a business in its own right. Fortunately, it is well known, and increasingly realised - also with respect to biotechnology - that technology is only one of the factors in realising a commercial success. Other aspects of the so-called value chain may be equally or even more crucial to get "bucks" from the exploitation of "bugs".
As a positive result the great promises about the innovation potential of biotechnology provided a favourable climate for the further development of the technology base.
Firstly, there was an increase in the, partly government funded, biotechnology directed research activities in academia and institutes in the late seventies and early eighties.

Secondly, as a relatively new phenomenon in the exploitation of scientific progress, private research companies emerged. These were funded by private money, venture capital, partnership with industry, royalties and riding the stock market. These ventures did significantly contribute to the development of both the technology base and the

future business potential. However, very few of the early ventures fulfilled the expectations and are currently main players in their chosen industries.
In the same era also the traditional "biotechnology" companies, which have their roots in the "old biotechnology" reshaped their research and development efforts.
Finally, large multinational players in the pharmaceutical, the chemical and the food industries, to mention a few examples, with little or no tradition in biotechnology, but responsive to the promises and threats of the new technology, reacted in a variety of ways. Their efforts took the form of (often quite sizeable) "in-house" programmes, and/or participation in activities in other companies, including biotechnology research companies, or academia. Today, it can be observed that a number of these activities have been or are subject to a reappraisal.

No doubt it were the perceived commercial prospects, which resulted in a drastically increased level of research and development activity in biotechnology. This resulted in a quick sophistication of the technologies that comprise biotechnology, it broadened the range of technologies that were available for application. Figure 1 provides an overview of the technology portfolio of biotechnology and the position of these technologies in the so-called technology life-cycle (Ref. 1). In this life-cycle the technologies are classified according to the level in which the technology has been developed towards allowing its performance potential in application in industry or society. An embryonic technology, such as recombinant DNA technology in plants, still is in need of substantial investments in the technology perse. A mature technology, such as rDNA in bacteria is ready for application and the challenge shifts from furthering the technology base towards developing commercial products or services with sufficient financial return. It should be emphasised that figure 1 relates to the achievement of the performance potential in a technological rather than a business sense. The technology has progressed to maturity in some sectors, its translation into commercial success certainly has not.

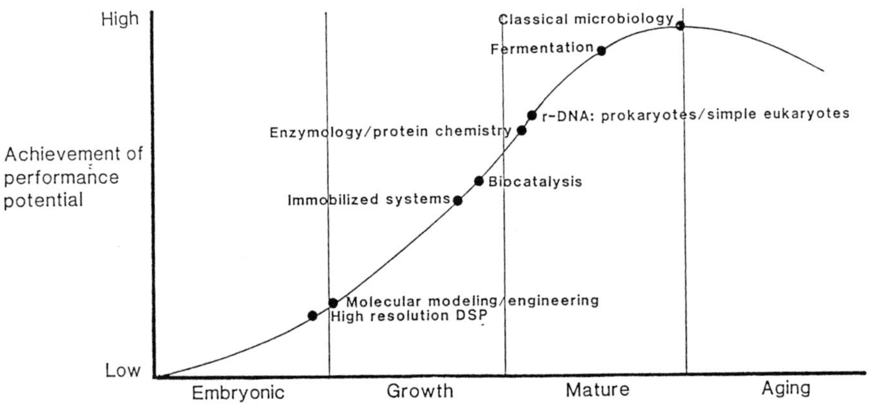

Figure 1: BIOTECHNOLOGY AND THE TECHNOLOGY LIFE CYCLE

2. BIOTECHNOLOGY: THE FUTURE

From the point of view of society and end-consumers it is a rather accepted fact that biotechnology did not live up to the expectations of the late seventies and the early eighties. There has been quite a lot of frustration on the business side and many of the early players have disappeared or, in case of the large multinationals, some turned away from biotechnology. Furthermore, the acceptance by the end-consumers was less favourable then originally was anticipated by supporters of Biotechnology and in many cases it is still one of the main barriers for rapid commercialisation, then originally was anticipated by supporters of Biotechnology.

To assess the future impact, in a business sense of biotechnology in a variety of industrial sectors, it should again be stressed that Biotechnology in itself is not a business. The commercial fruits derived from its application in industries, either already existing ones, or totally new ones that can be created based on the new technology are the products that can be turned into a business. Biotechnology did unto now, not lead to the creation of wholly new industries. It also did, until now, hardly change the structure of existing industrial sectors; in most cases strong players have largely remained strong players.

A convenient summary of this analysis has been condensed in figure 2 (Ref. 1). It provides an overview of the estimated future competitive impact of biotechnology in a number of industries and it is plotted against the maturity of the industry. The analysis reveals that the competitive impact of biotechnology is high in a number of sectors of the *pharmaceutical industry.*

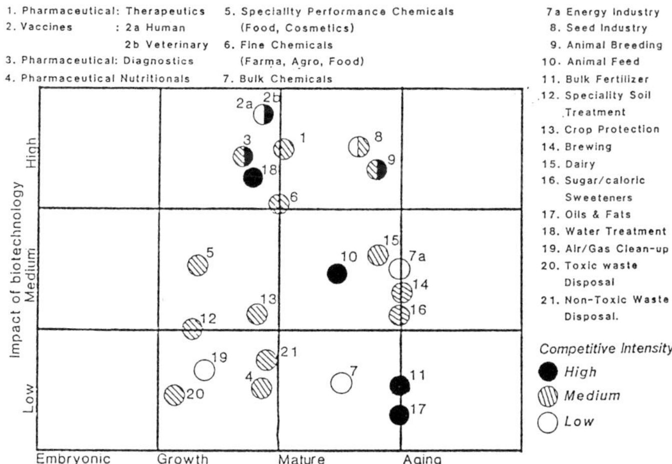

Figure 2: LIFE CYCLE CHARACTERISTICS IMPACT OF BIOTECHNOLOGY AND COMPETITIVE INTENSITY FOR A NUMBER OF INDUSTRIAL SECTORS

The pharmaceutical industry is, as an industry in the late growth to the mature stage. There is an increasing, but presently low to medium level of competitive intensity. Also the pressures on product pricing by the authorities lead to the expectation that the competitive intensity may increase in the near future. This is also apparent from the increasing consolidation in the industry. Recently large players have merged to become even larger players.

Biotechnology can be expected to have a high impact on all sectors of the pharmaceutical industry. This will, however, hardly mean a drastic change in the structure and the product range of what currently is a largely organic chemistry based industry.
To further substantiate these remarks it is useful to introduce the concept of the value chain. The so-called value chain of industrial activities is shown in figure 3. It puts technology development in its proper place, it is one of the key factors of success in bringing a product to the market place. However, other activities, such as marketing sales and distribution, may be equally or often more important to realise an innovative product. This was also very apparent as a shared view of the participants in the recent EIRMA forecasting workshop "Technology '99, phrased as: "from pure science through the market, to the law makers to the politicians" (Ref. 2). It brings the perspective that innovation goes far beyond technological progress in its own right.

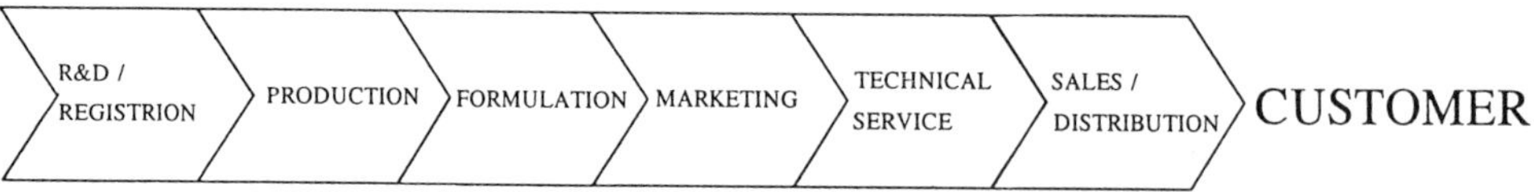

Figure 3: ANALYSIS OF THE VALUE CHAIN

Returning to the pharmaceutical industry it is concluded that, irrespective the fact that the competitive impact of the technology is high in a sizeable number of sectors and, given the basis of competition in the industry, which go far beyond the mere availability of technology, the innovation potential of biotechnology is expected to take place through its integration in the overall skill base of the industry. It is not expected that the new technology will revolutionize the industry structure. It will rather be instrumental in the further evolution in the range and the sophistication of the products of that industry.

In the *energy industry* the future impact of biotechnology is expected to be limited and significant changes in the resource base of the energy supply of society, i.e. a massive shift from fossil to renewable resources, is not expected to take place due to cost and volume considerations. The growing agricultural surpluses and the government subsidised programmes for the production of automotive fuels from agricultural resources are not expected to affect this picture in anything but a rather marginal way.

The same considerations do apply to the industry engaged in the production of *base chemicals*. Biotechnology will not change the resource base of the fossils based production of base chemicals. This expectation rests on both the integration of the base chemicals industry with the much larger energy industry and the competitive impact of biotechnology, which is rather low.
In the areas of *fine chemicals and performance chemicals* the impact of biotechnology will be larger and particularly the production of intermediates and ingredients for the pharmaceutical, the agrochemical and the food industry will be affected. Again there will be no question of a revolution but an integration of biotechnology into a broader range of skills which will assist the development of cheaper products or, perhaps more importantly, the development of products with more desirable performance characteristics. The competitive impact of biotechnology in selected sectors is estimated as varying from medium to high. The relevant technologies are generally in the late growth or the maturity phase of their life-cycle. Hence a significant penetration of biotechnological methods is expected to take place in this and the next decade.

In the *agro-alimentary industry* biotechnology is expected to make a significant impact varying by subsector. In many instances the realisation of the full potential will take us into the next century. The scene in this industry is complicated by factors of a regulatory and public acceptance nature. As experienced in the case of the recombinant chymosin product, market penetration may be significantly delayed by public acceptance problems combined by legislation by less than consistent and decisive government regulations. In this respect their is a quite large difference between Europe and the USA. In the latter country a far more realistic approach and more positive climate by government and public towards new technologies contributes to the harvesting of the fruits of biotechnology. This may have an impact which may be more important, ultimately through its effect on the speed of introduction of new products, than that of technological progress perse. In the seed and the animal breeding industries the directed manipulation of cultivars and production animals is within reach and this can revolutionise these industries. However, the technologies are still in many instances in a more early stage of the life-cycle and significant investments in the development of the technologies are still needed. The developments in these sector may in their turn affect the animal feed industry and the food industry through the raw materials for these sectors. Furthermore the food and feed industries will be gradually affected by new ingredients and new processing aids as a result of the developments in biotechnology. In the food and feed industries many innovations will follow from technologies pioneered in the pharmaceutical industry. It will, more often than not be beyond the profit potential of these sectors to develop new technologies starting from the embryonic stage in the life-cycle.

Of course, also the area of *waste management* should be mentioned. Biotechnology has a variety of impacts in this sector. This both relates to the prevention or the decrease of pollution, such as in the enzymatic cleavage of the side chain from penicillin or the addition of phytase in animal feed, and the removal or the upgrading of wastes, such as in the production of lysine from manure.

3. CONCLUSION

Biotechnology is expected to have a varying impact on a number of industrial sectors and certainly the number of applications will grow in this and the following decade. The development of the full commercial potential of biotechnology will increasingly become less related to scientific and technological progress in biotechnology. Rather the integration of the biological disciplines with other sciences and other aspects of the industries' value chain or problems of an economic, legislative or acceptance nature will become more decisive. This decade will clearly show that biotechnology is no business in its own right. We expect that the technology will be gradually become more and more integrated in the technology base of existing industries, revolutionising the industry structure will be a very rare exception rather than a rule.

REFERENCES

1. Arthur D. Little, "Biotechnologie en de Nederlandse Industrie", report to Ministry of Economic Affairs of The Netherlands, 1st December 1989.
2. Dr. P. Kramer, chairman of the EIRMA Workshop "Technology '99", Paris 11-13 October 1989.

ECB6: Proceedings of the 6th European Congress on Biotechnology
edited by L. Alberghina, L. Frontali, P. Sensi
1994 Elsevier Science B.V.

THE ROLE OF BIOTECHNOLOGY IN AGRO-INDUSTRY AND COMMISSION POLICY.

C.L.Mangan
European Commission DGXII, Research and development, Division E-2 Life Sciences and Technologies

1.INTRODUCTION:

Modern biotechnology is only about twenty years old but it has already had a major influence in various scientific domains such as medicine, environment, and agro-industry. This influence will undoubtedly grow in the future and the European Commission must devise policy which will direct it in the most useful and cost effective manner for the Community. This article will deal specifically with the application of biotechnology in agro-industrial research and the way it can be used to improve and influence the projects of EC related agro-industrial research programmes.

Because biotechnology is not a discreet entity it lends itself to different definitions, and there is sometimes confusion between new and conventional biotechnology. The latter is usually associated with the classical "natural fermentation" of agricultural raw materials to produce a wide range of different foodstuffs. New biotechnology is more of a science for the understanding of the molecular biology of plants, animals, and microbes, and the development of related tools for use in improving human and animal health, the environment, agriculture and agro-industry. With respect to agro-industrial research and biotechnology certain distinct areas can be defined. The first two are the biotechnological inputs to **Plants**, and **Animals**, which represent the primary production stage in agriculture, horticulture, and forestry and which generate the feedstocks for the other two market areas the **Food and Non-Food industries** where biotechnology plays a role in developing new and improved products and processes[1]. Food has an easily defined market area and could include produce coming from the dairy, meat/fish, and crop sectors, with mature technology already developed to produce the ever increasing wide variety of drinks and foodstuffs we find on our supermarket shelves today. Non-food on the other hand is harder to define but a broad definition could include both waste and specific biomass of new and traditional origin coming from agriculture and fisheries for the manufacture of items for many diverse markets such as energy, wood paper and pulp, fibres and composites, and bulk and speciality chemicals.

At the moment biotechnology is at a crossroads and certain factors could strongly influence the role it will play in the future of agro-industry namely;

- **Policy changes:** European agricultural production is currently moving from a situation of subsidised surplus towards products sold at world market values, as recently demonstrated by the CAP revisions and GATT negotiations.

- **Commodity imbalances**: The imbalance in the European agro-industrial area where surpluses exist in certain commodities and deficits in others, biotechnology could help redress this balance, and promote the speedy transfer of new technology to developing countries compensating for loss of exports.

- **New and Improved crops**: A willingness to introduce and scale up production of new or existing crops which are environmentally and economically sound such as transgenic and non-food crops coupled to the increase in agricultural land being set aside for growing non-conventional crops.

- **New types of food and diet and health interactions:** The need for a much better understanding between nutrition and health and the demand for specialised diets for different ages and sicknesses. The need for improved raw materials to supply this demand for specialised functional food and feed products.

- **Raw materials for chemical industry**: The need to convert surplus food or non-food raw materials from primary production, or processing by-product waste, to the chemical industry for the synthesis of speciality or commodity chemicals where niche markets for high added value environmentally sound products exist.

- **Combined Processing**: The growing need to combine traditional chemical processes with biotechnological ones in the food and non-food areas as a means to reduce pollution, and energy demand.

As industry and markets play a key role in the development of biotechnologies within the agro-industrial sector emphasis should be put on the respective industrial demands of the processing and primary production side of agriculture. The development of biotechnologies within this sector has been extensive to date and many applications already exist. Its future success depends to a large extent upon diverse developments, some outside the control of science and technology. Examples include the evolution of the European regulatory framework with sensible CAP and GATT reform, workable Environmental regulations, and realistic EC research policy. Other factors which will play a role include, economic and technical feasibility, legal and safety constraints, public acceptance, industrial and individual government policies, and intellectual property protection.

Table 1 shows the sectors in the different areas where biotechnology can play a role, and Table 2 lists for the four areas specific applications of biotechnology already commercialised.

The horizontal similarities between the different areas include, improving the health of animals, humans, and plants using diagnostics, the use of genetic technologies to modify and improve plants and animals, and the reduction of chemicals into agricultural and food productions.

Table 1
Agro-industrial sectors where biotechnology is having an influence.

Plants	Animals	Food	Non-food uses of agricultural crops
Diagnostics	Improving animal health	Rapid food safety contamination tests	Wood and energy
New genes for crops	Improving animal growth and lactation	Food enzymes	Starches and other polysaccharides
Improving plant breeding	Embryo multiplication	Biopreservation	Hydrocarbons, oils and fats
Improving plant production	Animal feeds	Novel biotechnology foods	Other plants, plant products, animal materials
Valorisation of plant biowastes	Transgenesis	Plant cell culture	Novel industrial products from transgenic plants and animals
New non-food crops	Fish	Microalgal culture	Waste treatment

It must be realised that modern biotechnology is still only a minor contributor in these industries, as new food and non-food products are coming from a wide range of different technologies. Biotechnology however has tremendous potential to provide uniform high quality products all year round, to reduce and upgrade the waste of raw material throughout the chain, and to provide custom specified crops and products such as "natural or functional foods". It is for these reasons that biotechnology is probably being hyped beyond its actual market importance at present. It should be remembered that only 10 years ago the introduction of genes into crops had still to be demonstrated. This is now routine molecular biology for certain species but much has still to be understood regarding the intricate biology of crops and animals before the application is universal. If one considers also that the world population is expected to grow by one billion this decade and with the populations of the industrial democracies continuing to decline the potential for biotechnology to increase the global food supply becomes more and more important [2].

The biotechnology products already on the market represent only a small fraction of the potential range available. At the input level for plants and animals the research being conducted is extensive but it is not yet reflected in the food and non-food markets mainly due to a lack of regulations and consumer acceptance for biotech foods, and in non-food by the competitive weakness of agricultural raw materials against oil prices. Oil prices themselves are subject to fluctuations sometimes caused by, political unrest, the percieved dwindling of global stocks, and the need to curb industrial pollution. These are some of the other driving forces behind the increased interest in non-food, particularily in the use of biomass for energy.

Table 2
Biotechnological products and processes already commercialised in the four different sectors

Plants	**Animals**	**Food**	**Non-food uses of agricultural crops**
Plant diagnostics	Animal diagnostics	New biotechnology testing methods	Conversion of biomass to energy via anaerobic fermentation to ethanol or methane gas
Microbial insecticides	New vaccines and therapeutics	Enzyme- and bio-conversion of starch to sweet products	Selective breeding and propagation techniques for trees and ornamental plants
Tissue culture techniques	*in vitro* fertilisation embryos, tissue culture techniques	Food flavours, enhancers, amino acids and other special nutrients	Plant cell culture techniques
Micropropagation techniques	Pituitary (growth) hormone to increase milk yield or growth	Fruit juice processing	Enzyme technologies for product extraction and processing (from starch, vegetable oils, etc.) and for the manufacture of agrochemicals
Genetic mapping techniques	Animal feeds and feed additives	Novel foods from fermentation, cheese enzymes, lactose-free dairy products, yeast hybrids	Expression and extraction of speciality chemicals in traditional or transgenic plants and production of high value added products

2. EC influence on biotechnology

In a European context there are two important ways that the European Commission can influence the role and apllication of biotechnology. The first is through EC framework programmes where biotechnology research is financially supported and managed through various related programmes. The second is through the introduction of environmental and social legislation. All member states have their own national biotechnology programmes but the Commission programmes, while small in comparison, have the unique advantage in tackling big and important issues on a European level through transnational co-operation and participation. The second and third framework programmes respectively, have certain sub-programmes of interest. BRIDGE and BIOTECH are broad based biotechnology programmes with a basic research approach and are primarily concerned with the understanding of specific aspects of the molecular biology of plants animals and microorganisms implemented through networked and targeted type projects.

Agricultural and agro-industrial research is catered for by the the ECLAIR (European Collaborative Linkage of Agro-industry through Research), FLAIR (Food Linked Agro-industrial Research), and AIR (Agricultural and Agro-Industry including Fisheries) research programmes. The first two are from the second framework programme and are primarily concerned with the high technology impact upon the agro-industrial food and non-food sectors, biotechnology is involved in up to 70% of the projects and is mainly applied with the objective of improving existing processes, developing new ones, reducing the costs of production and the control of the supply, and introducing new products and functions to the agro-industrial market. The current third framework AIR programme (which is jointly managed by the related research divisions of DGXII, DGVI, and DGXIV) covers the domains of agriculture, agro-industry, and fisheries, and also has a high level of biotechnological involvement particularly in the agro-industrial areas. These programmes are heavily subscribed to by European researchers mainly due to the increasing awareness of the advantages of cross border co-operation in solving common problems through the setting up of related networks, and also because member states are reducing their own internal research budgets. There is generally a broad mix of academic, public institute, and industrial participation, and the participation of small medium enterprises is strongly encouraged. The projects are on average up to five years in duration and are pre-competitive in nature in that the transfer of technology from the laboratory to the market is the main medium term objective. The vast majority of these projects are shared cost type actions in that the Commission puts up 50% of the funding over the duration of the project, concerted action projects where the commission pays for the organisation of related European networks, and demonstration projects where the Commission funds projects which intend to demonstrate the feasibility of a process to realise its market potential. Important issues tackled in these programmes on a European level include the following [3,4]:

- an understanding of the genetics of lignin, oil, fibre, and carbohydrate crops, with a view to manipulating the content and composition of the lignocellulosic ratio, and the content and quality of oils, carbohydrates, and fibres, in new and conventional crops.

- research into transgenic food and non-food plants which have conferred insect or disease resistance and increased yield, transgenic plants expressing proteins or chemicals of interest such as special unsaturated fatty acids, pharmaceuticals etc,.

- research into biosensors, diagnostics, vaccines and fertility, for animals, fish, and plant diseases, using biotechnological tools such as *in-vitro* fertilisation, DNA probes, Polymerase Chain Reaction, and monoclonal antibodies.

- genetic manipulation and utilisation of microorganisms for the improved production of conventional fermented food products and flavours, and for the production of high added value products (bioplastics, chemical commodities) from non-food sources such as surpluses and biowaste.

- development of bioprocesses and biotransformations and related equipment, machinery, and sensors in order to surpass the current limitations for scale-up set by traditional food and non-food processes.

- the use of biotechnology to develop ways to improve the quality, safety, and nutritional benefit of food such as sensors for pathogens, cloned probiotics, transgenic foods, new toxicological tests.

Table 3 Data relating to the first and second calls for the current third framework AIR programme

	AIR I	AIR 2
CALL FOR PROPOSALS	10.10.91	01.08.92
DURATION	FOUR YEARS	FOUR YEARS
NUMBER OF PROJECTS	102	204
SHARED COST PROJECTS	76	169
CONCERTED ACTIONS	19	35
DEMONSTRATION	7	none
TOTAL BUDGET	172 MECU	334 MECU
EC CONTRIBUTION	130 MECU	170 MECU
NUMBER OF PARTICIPANTS	863	1050

Table 4 A breakdown of the type of projects the AIR programme is funding.

	AIR I projects	AIR II projects
Conversion diversification extensification	8	11
Interaction of agriculture with environment	11	6
Machinery and informatics		14
Rural development	4	8
Non-food crops	3	9
Animal health and husbandry	7	17
Crop inputs and Physiology	13	28
Forestry and wood processing	14	15
Biomass for energy and non-food processing	5	23
Food processing and end use	12	35
Demonstration (non-food)	7	none
Fisheries and Aquaculture	18	34

Some of the projects from ECLAIR and FLAIR are near completion and have already yielded interesting results. Their market application in the future will depend very much on the way that the EC influences biotechnology through policy regulations and directives aimed at controlling the application of this new and technically powerful scientific discipline. The EC is an open structured institution and in this respect it is open to influence by external groups or individuals and biotechnology is a contentious issue for certain groups in Europe at the moment. A great debate is currently being held at a political and scientific level by those such as the industries who are in favour of relaxed regulations and those who are against any form of modern biotechnology from an environmental viewpoint.

Several Directorates-General in Brussels have a role in creating European biotechnology policy, DGIII (internal market), DGXI (environment), DGVI (agriculture), and DGXII (research). Three types of influence can be described[5]; **regulations** (novel foods, transgenic crops) are binding on member states and describe the ends and the means to implement the policy, **directives** (deliberate release, contained use, worker safety etc,.) are binding but member states are free to arrange implementation, and finally **recommendations** which are guides for the member states. All three contribute towards the effective political and commercial mechanism under which biotechnology can be controlled through e.g., environmental standards and risks, patent protection, harmonisation of technology, safety assessments and measures, worker protection etc,.

There are currently three directives passed at European level and these concern contained use, deliberate release, and worker safety, however few member states have actually implemented them. The deliberate release directive is quite strict, demanding environmental risk assessments at various steps along the product development and marketing phases. Other necessary legislation in the pipeline but which will also require risk assessment includes; directives on biotech patents, plant varieties, novel foods, transport, and, transgenic animals. In conclusion ideal working regulations for dealing with the administrative and legal implications of biotechnology are not yet in place at a European level and this has tended to limit the growth of biotechnology in an increasingly competitive agro-industrial market.

Biotechnology has certainly had an effect on the various political, commercial, and scientific sectors above and beyond the returns it has generated for modern society. It offers an interesting way forward and some realistic conclusions can already be drawn from developments in the agro-industrial sector which indicate useful guidelines and improvemnts for the future and the potential impact of this new technology[6];

- The competitiveness of industries using biotechnologies closely depends on the improvement of the performances of those biotechnologies already in practise. A large part of the R&D effort is thus aimed at the improvement of the returns of existing procedures (improvement of strains, of purification techniques, adaptation of available raw materials, etc.)

- Concerning the adoption of biotechnologies by new industries, biotechnologies are mostly conceived as tools to improve existing procedures and not as substitution technology.

- Products currently synthesised by biotechnologies are specialised products which represent small volumes in general. Thus in the non-food area particularly we cannot expect in the near future large quantities of new products from agricultural feedstocks. Biotechnologies thus represent only a marginal lead in the solving of the problem of European agricultural surplus. As a means to feed the ever growing global population, or as a way to introduce new funtional foods into modern diets, it has better potential at the moment.

- The history of the development of biotechnology shows the considerable weight of the regulatory framework particularly with respect to the rise in environmental concerns and how these can affect progress and hinder investment due to the increase of legislation and potential negative consumer attitudes which is especially relevant to food.

- The predicted evolutions of the CAP should favour the development of biotechnology applications in the non-food side of the agro-industrial sector, even though;
  a) the costs of European agricultural raw materials are higher than in some other world areas,
  b) it is difficult to obtain microorganisms with the same levels of productivity as chemical processes,
  c) the economies of scale for biotechnologies are limited.

In addition, when one considers that recombinant microorganisms can revert to wild type within several generations of growth, and that there are now clear indications that certain pests are becoming resistant to transgenic plants it becomes clear that in certain cases the odds are not always stacked in favour of the biotechnological route. Ways to tackle all these drawbacks are currently being explored and in many cases the critical progress has to be made at the political level. On the other hand biotechnology in its present condition offers ways to understand animal and plant mechanisms at the molecular level which were inconceivable only twenty years ago. With further rapid scientific advancements promoted and funded by European, and worldwide research, and progressive political change, it could play an important role in kickstarting a new agricultural revolution.

REFERENCES

1. Biotechnology Agriculture and Food, OECD publication Paris 1992.
2. P. Kennedy, Preparing for the Twenty-First Century, Harper Collins, 1993.
3. J. Hodgson, Europe Maastricht and Biotechnology, Biotechnology Vol 10 November 1992.
4. C.Mangan (Eds) ECLAIR Project Synopses, EC publication Luxembourg, 1993.
5. C.Mangan (Eds) ECLAIR Progress Reports, EC publication Luxembourg, 1993.
6. F. Lévêque *et al,* Non-food uses of European Agricultural Production, Monitor study, EC publication, EC Luxembourg 1993.

ECB6: Proceedings of the 6th European Congress on Biotechnology
edited by L. Alberghina, L. Frontali, P. Sensi

New trends in vaccines

R. Rappuoli

Biocine-Sclavo S.p.A., Via Fiorentina 1, 53100 Siena, Italy

SUMMARY

Vaccination, rather than chemotherapy, is the best approach to control infectious diseases.

Although several traditionally prepared vaccines have been in use for a long time, there is a strong need to improve of some of them, and to develop of new ones. Progress in vaccine research has been prompted by biotechnologies, thus setting the way for second and even third generation vaccines. The purpose of this paper is to give an overview of the possibilities which have been granted by the new technologies, by presenting the most relevant examples of newly developed vaccines.

HISTORICAL OVERVIEW

The very first idea of vaccination as prevention against infectious diseases belongs to Jenner who, as early as 1796, used cowpox to immunize humans against smallpox, a devastating disease at that time. At the beginning of the 20th century, several new vaccines were developed with either whole inactivated bacteria, viruses or semipurified bacterial toxins.

Tissue culture technology greatly improved the possibility to grow viruses thus providing a continuous supply of material to be inactivated, and opened up the way to attenuation by continued passages. In 1955 Salk developed the first inactivated polio vaccine, quickly followed by inactivated measles vaccine, mumps vaccine and rubella vaccine.

Polysaccharide vaccines were developed against invasive bacterial pathogens between 1974 and 1984.

In 1986 the first recombinant viral vaccine was developed against Hepatitis B to substitute the plasma derived one, whose availability and safety were decreasing. Today the path of cloning and expressing heterologous antigens in given microorganisms is a paved way to the formulation of safer and more efficacious vaccines.

HEPATITIS VACCINES

Several vaccines against Hepatitis A have been developed, all of which contain formalin inactivated HA virus. All these studies have been reviewed by Siegel and Lemon (1). So far, temptative recombinant HAV vaccines have been a failure due to the inability of the cloned capsidal proteins to assemble correctly to reconstitute protective epitopes. However, recently some groups may have overcome this problem and succeeded in expressing properly folded Hepatitis A empty capsids.
Hepatitis B virus is the cause of many liver disease, from chronic hepatitis to hepatocarcinoma. It is endemic in many parts of the world, and its spreading is due to the connathal transmission from chronically infected mothers to their newborns; infants will soon develop chronic hepatitis B infection. Blood transmission is also a common route of spreading infection, so that hepatitis B carriers are now estimated to be over 250,000 worldwide (2). Carriers will die from long term sequelae of chronic Hepatitis B, namely cirrhosis and hepatocellular carcinoma (3). Prevention through vaccination is the only way to fight this disease, since no effective treatment is available for chronic hepatitis B.

A first generation Hepatitis B vaccine was developed when it was discovered that in chronic carriers, a substantial quantity of viral surface antigen (HbsAg) was not incorporated into assembled virions (4). This excess HbsAg is assembled into lipoproteic spheres of 22 nanometers in diameter. The 22 nm particles were found not to be infectious, since they did not contain any nucleic acid inside. On the other hand, they did contain the major viral envelope protein and indeed they proved to be extremely immunogenic and protective.

Despite its excellent immunogenicity and safety profile, the vaccine derived from chronic carriers plasma donors was not easily accepted. Furthermore, availability was decreasing and will drop whenever the disease will be eradicated.

Once cloned and expressed in yeast (5), HbsAg showed a surprising ability to refold in such a way to form sphere particles which look like 22 nm particles under an electron microscope, and are as good immunogenic as the plasma-derived material.

Recombinant vaccine with its unlimited supplies and safety of handling and dispensing, is surely the best possible solution one could devise.
Hepatitis B recombinant vaccine is therefore the first recombinant vaccine approved for human use ever.

Another approach to novel development is the insertion of genes into "safe" viruses, which could work as carriers for unrelated antigens. Vaccinia virus and adenovirus have been manipulated for expression of HbsAg, and live virus may be administered to "infect" humans with innocuous microorganisms, which can still multiply but have lost the ability to cause disease.

At the same time, heterologous antigens are expressed, yielding an immune response which may be humoral as well as cell-mediated. Adenovirus has a further advantage in that it has already been used for humans and, moreover, could be administered orally.

For a long time, a non A non B hepatitis form could not be matched with any known microorganism. Cloning and sequencing of HCV yielded a diagnostic tool for blood supplies as well as all the information on genetic material by which one could approach the problem of a vaccine (6, 7).

HCV genome is made of a positive stranded RNA molecule of 9400 nucleotides, encoding a viral polyprotein of 3010 or 3011 aa, depending on the isolate.
Like other RNA viruses, HCV has a certain genome heterogeneity so that three basic groups of HCV can now be determined on the basis of their sequence relatedness (group I, II and III): this may end up with different abilities to interact with the host and therefore different progression of the disease.

The N terminus of the E2 envelope glycoprotein is a peculiar site of the genome: it resembles the V3 loop of HIV and is hypervariable. This could mean that it is under selective pressure, so that the important protective epitope is changed as soon as it is recognized by antibodies. Studies have also shown that the appearance of E1 antibodies rather than E2 seems to correlate with a non chronicizing development of the disease. This seems to indicate that E1 antibodies are necessary for a protection from the disease; therefore, expression of E1 in several expression systems will provide a reasonable amount of protein to test its role in protective immunity. Studies are in progress in chimps and it is a common hope that a vaccine will be available in the very near future.

ACELLULAR PERTUSSIS VACCINE

Research on acellular pertussis vaccine was mandatory, since the traditional vaccine formulation was considered less and less acceptable because of its many side effects. Traditional vaccine formulation is still in use in several countries: local and systemic side effects (redness, swelling fever) at the site of injection are associated to it, due to the fact that the vaccine is made of whole bacteria, inactivated with formol. This treatment kills the microorganisms so that they cannot infect the human host. In the final formulation several toxins are present that may be responsible for the above side effects. On top of this, other pyrogenic and toxic molecules still present in the vaccine, are thought to be the main cause of fever in a very high percentage of vaccinees (70%). Acellular formulations were devised to get a protective immunity just by providing the immunodominant molecules, aiming at very pure and defined formulation. Several molecules were tested, in order to check their potential protective role in immunization, and it was found that 69K, FHA, agglutinogens and PT were necessary for effective protection.

Studies on children recovering from the disease showed that most of the immunological responses were directed towards PT; these results proved PT to be the immunodominant molecule of *Bordetella pertussis.*

Detoxification of the molecule was necessary in order to use it in a vaccine. Several laboratories worldwide tried to detoxify the molecule with chemical treatments, which indeed denatures the toxin and reduces its toxic activity to undetectable levels. Nevertheless, it has been shown that this way of detoxification may be reversible, so that very small amounts of active toxin are found in chemically treated lots.

Genetic detoxification was then approached (8). Just by changing two aminoacids in the active site (9), a detoxified molecule was obtained, whose detoxification is irreversible.

The tertiary structure of the molecule is maintained, so that immunization yields an assay of antibodies which will recognize the native molecule when they will encounter it. Chemical detoxification seems to diminish the immunogenic potential of PT, whereas genetic detoxification does not alter the structure, therefore not altering the immunogenicity of the molecule. Substitutions of ariginin in position 9 with lysin and gentamic acid in position 129G with glycin completely abrogated the toxicity. A vaccine containing such an engineered molecule together with FHA and 69K, was tested on adult volunteers first and children afterwards; it showed to be extremely safe and immunogenic (10, 11). Children aged 2 to 18 months received this vaccine, and no appreciable side effects were shown. An efficacy study on over 13,000 children is currently being performed in Italy to check the protective efficacy of the vaccine as well as its comparative efficacy versus the traditional acellular formulation and a chemically detoxified acellular formulation. In the USA, under the supervision of the National Institute of Health (NIH), a test was carried out to control 11 different acellular formulations, from manufacturers from all over the world. The vaccine containing the genetically detoxified PT turned out to be the best in inducing neutralizing antibodies. Subsequently, the NIH decided to sponsor an efficacy trial in Italy and Sweden. This vaccine formulation will be on the market in Italy soon, and will be the second recombinant vaccine ever, the first bacterial recombinant vaccine.

HIB

Haemophilus influenzae type B is one of the predominant causes of meningitis in children. This bacterium is a normal commensal of the upper respiratory tract. Encapsulated b type microorganisms are able to reach the bloodstream, thus causing bacteremia. Polysaccharide capsule is the major virulence factor and a target for serum antibodies that mediate protection against disease.

Polysaccharides are poor immunogens, due to their inability to fit in the MHC groove. Therefore, they cannot produce a boost on repeated antigenic challenge, since they do not generate memory B cells: they are described as thymus-independent antigens. On the basis of the principle of the hapten-carrier theory of Avery and Goebel (12), several laboratories generated conjugate vaccines in which a protein of given and known immunogenicity (diphtheria toxin) is linked with polysaccharides or fractions of them, so to overcome the T independence of these antigens.

Immunization with these conjugate molecules resulted in a greatly enhanced antibody response in infants, where generally most failures of immunization occur, due to the scarcity of their B cell repertoire. Conjugate vaccines are currently available; clinical trials showed their ability to protect children from Hib infections (13, 14). These vaccines seem to stimulate a good protective antibody response.

The same approach can be taken for development of vaccines against other encapsulated bacterial, such as *N. meningitidis* or Pneumococcus.

OTHER NOVELTIES

Many other important vaccines are under clinical trials.

Recombinant vaccines have been developed against HIV and HSV, aiming at the development of both preventive and immunotherapeutic vaccines. Immunotherapy will be targeted to seropositive individuals to either halt the progression of the disease and therefore put off the onset of symptoms (HIV), or to reduce the frequency and severity of the recurrent lesions in HSV-infected individuals.

A new cholera vaccine is under study, since the traditional one does not seem to fully protect from the infection. In this case, recombinant DNA technology, has led to the development of detoxified strains, through deletions of genes coding for toxic molecules (15). The intact microorganism seems to be necessary for a protective immune response: purified antigens may not be enough to produce an effective barrier against the disease.

Helicobacter pylori is a microorganism which is most probably involved in the pathogenesis of gastric and duodenal ulcers, as well as gastric cancer (16). If a vaccine can be developed against this bacterium by immunizing with purified cytotoxic molecules, perhaps we will prevent ulcers through vaccination.

Papillomavirus is the etiologic agent for condylomas and for cervix cancer.

Viral proteins from papillomavirus have been expressed in yeast (17); when they are used for immunization of mice which have been subsequently infected with papillomavirus, they seem to prevent tumors. These studies may therefore be a first step towards anticancer vaccination.

REFERENCES

1. G. Siegel and S.M. Lemon, Virus Res., 17 (1990) 75.
2. R.P. Beasley and L.-Y Hwang, Viral hepatitis and liver disease (1984).
3. W. Szmuness, Prog. Med. Virol., 24 (1978) 40.
4. M.R. Hilleman, W.J. McAleer, E.B. Buynak and A.A. McLean, J. Infect. 7 (1983) 3.
5. P. Valenzuela, A. Medina, W.J. Rutter, G. Ammerer and B.D. Hall, Nature, 298 (1982) 347.
6. Q.-L. Choo, A.J. Weiner, L.R. Overby, G. Kuo, M. Houghton and D.W. Bradley, Br. Med. Bull., 46 (1990) 423.
7. Q.-L. Choo, K.H. Richman, J.H. Han, K. Berger, C. Lee, C. Dong, C. Gallegos, D. Coit, A. Medina-Selby, P.J. Barr, A.J. Weiner, D.W. Bradley, G. Kuo and M. Houghton, Proc. Natl. Acad. Sci. USA, 88 (1991) 2451.
8. S. Cockle, S. Loosmore, K. Radika, G. Zealey, H. Boux, K. Phillips and M. Klein, Adv. Exp. Med. Biol. 251 (1989) 209.
9. M.G. Pizza, A. Covacci, A. Bartoloni, M. Perugini, L. Nencioni, M.T. De Magistris, L. Villa, D. Nucci, R. Manetti, M. Bugnoli, F. Giovannoni, R.

Olivieri, J.T. Barbieri, H. Sato and R. Rappuoli, Science, 246 (1989) 497.
10. A. Podda, L. Nencioni, I. Marsili, S. Peppoloni, G. Volpini, D. Donati, A. Di Tommaso, M.T. De Magistris and R. Rappuoli, Vaccine, 9 (1991) 741.
11. A. Podda, E. Carapella de Luca, L. Titone, A.M. Casadei, S. Peppoloni, G. Volpini, I. Marsili, L. Nencioni and R. Rappuoli, J. Pediatr., 120 (1992) 680.
12. O.T. Avery and W.F. Goebel, J. Exp. Med., 50 (1929) 533.
13. J. Eksola, H. Peltola, A.K. Takala, H. Kayhty, M. Hakulinen, V. Karanakela ko, E. Kela, P. Rekola, P.R. Ronnberg, J.S. Samuelson, L.K. Gordon and P.H. Makela, N. Engl. J. Med., 317 (1987) 717.
14. P.H. Makela, J. Eksola, H. Peltola, A.K. Takala and H. Kayhty, Pediatrics, 85 (Suppl.) (1990) 651.
15. M.M. Levine, Proceedings of the workshop on "Cholera toxin and related enterotoxins: from disease to molecular structure" (Broadway, U.K., February 1993).
16. A. Covacci, S. Censini, M. Bugnoli, R. Petracca, D. Burroni, G. Macchia, A. Massone, E. Papini, Z. Xiang, N. Figura and R. Rappuoli, Proc. Natl. Acad. Sci. USA, (1993) in press.
17. M. Tommasino, M. Contorni and F. Cavalieri, Gene, 111 (1992) 93.

ECB6: Proceedings of the 6th European Congress on Biotechnology
edited by L. Alberghina, L. Frontali, P. Sensi

Engineered erythrocytes as carriers and bioreactors for biomedical applications*

A. De Flora, U. Benatti, M. Tonetti, A. Gasparini, M. Giovine, G.L. Damonte, A. Sdraffa

Institute of Biochemistry, University of Genoa, Viale Benedetto XV, 1 16132 Genoa Italy

1. INTRODUCTION

An emerging trend in biomedicine is represented by the withdrawal of specific cells from a patient, followed by their "in vitro" manipulations (e.g., transfection with a gene of interest) and eventually by their administration back to the same or to a compatible host where the modified cells may express new and useful functions. This general strategy underlies, for example, current attempts of somatic cell gene therapy and other advanced therapies based not necessarily upon genetic engineering, but also on alternative systems leading to modified cellular programmes, an example of which is first generation adoptive immunotherapy

Within this frame, erythrocytes have been proposed since two decades as potential carriers and bioreactors of medical interest (1). Internalization of enzyme proteins within red cells proved to be feasible by virtue of the unique properties of the erythrocyte membrane, whose elasticity and deformability allow the hypotonically induced opening of transient pores large enough to be crossed by externally placed macromolecules, before their closure in isotonic conditions (2). Alternative procedures have been described for achieving encapsulation of a number of molecular species in animal and human erythrocytes. These include, among others, transient dielectric breakdown, drug- induced (e.g., by primaquine) entrapment of chemicals and energized (e.g., ATP-driven)membrane endocytosis. However, a critical appraisal of these procedures identified those based on transient hypotonic shock followed by isotonic resealing and reanneling as the easiest, the most versatile (i.e., applicable to a wide range of chemicals) and the most conservative ones (3). The last feature is of obvious importance in view of the multiple "in vivo" applications of the variously loaded erythrocytes that require their stringent biocompatibility. Indeed, within the several versions of the technique, the procedures involving reversible hypotonic hemolysis and subsequent isotonic resealing yield a satisfactory fraction of erythrocytes showing normal biochemical, antigenic and morphological properties. Most importantly, the erythrocytes so engineered have an "in vivo" life-span strictly comparable to that of the unloaded erythrocytes, when they are re-injected in the circulatory system of an autologous or compatible organism. In addition, these red cells can be prepared on a large scale using blood banking facilities (4).

---

* This study was supported by C.N.R. Target Projects "Biotechnology and Bioinstrumentation" and "A.C.R.O.", by the "Associazione Italiana per la Ricerca sul Cancro", by the "A.I.D.S. Project" of the Italian Ministry of Health, and by the "Biotechnology Action Programme" of E.E.C.

Irrespective of the functions that are to be conferred to the engineered erythrocytes (ranging from simple carriers to true bioreactors - see below), their applications to biomedicine require a number of steps after the choice of the bioactive compound to be encapsulated therein is taken:

a) The construction of the loaded erythrocytes by means of the optimal procedure ensuring sterility and apyrogenicity, followed by an adequate evaluation of the percentages of encapsulation and of recovery of viable red cells.

b) The evaluation of the "in vitro" properties of the engineered red cells, in order to explore their biocompatibility according to a wide number of criteria (morphology, metabolic properties, rheology, antigenicity). This step should include a proper investigation of the interactions between the encapsulated compound(s) and the metabolism of erythrocytes, since the compound itself might either be steadily removed by resident enzymes or be erythrotoxic, thus not compatible with the carrier.

c) If the new function of the engineered erythrocyte is to be displayed "in vivo" at a specific site of the organism rather than in the general circulation, additional modifications of the carriers are required to ensure their targeting to selected tissues or organs . Targeting of the loaded erythrocytes is however still a critical challenge of this technology and many efforts should be directed to solve the problem, especially in those cases in which non-reticuloendothelial cells are the desired target (5).

d) The "in vivo" administration of the properly engineered red cells to the recipient organism, either intravenously or indirectly via alternative routes (e.g., intraperitoneally or within lymph nodes). This step, owing to ethical and experimental reasons, should involve at least one animal model before applications to human therapy become feasible and steps a to c must therefore be followed with the erythrocytes of the animal species under testing.

Before specific biomedical applications of the variously loaded erythrocytes are considered,a general issue concerning encapsulation of proteins should be briefly addressed. Enrichment of red cells with a specific protein yields the same result that can be achieved in nucleated cells by appropriate transfer and subsequent expression of the gene coding for that protein. As a consequence, encapsulation of an enzyme protein in erythrocytes is certainly a simple and useful tool for investigating mechanisms of intracellular regulation of enzymes. The resulting metabolic perturbation may in fact identify sites of intracellular control, much in the same way as following "de novo" synthesis of key enzyme proteins by means of proper inducers that elicit enhanced metabolic fluxes in target cells. Similar considerations, although in the opposite direction, can be made in cases of genetic enzyme deficiencies that impair specific metabolic pathways of the red cell. In these cases, best exemplified by genetic deficiency of Glucose 6-P dehydrogenase (G6PD, the rate-limiting enzyme of the pentose phosphate pathway), encapsulation of the defective or missing enzyme protein may restore metabolic competence, therefore obviating to biochemical dysfunctions. Indeed, entrapment of G6PD purified from normal human red cells within genetically G6PD- deficient erythrocytes, up to reach normal activity levels, was shown in our laboratory to correct this metabolic disorder completely (6).

Table 1 summarizes the two general performances that can be obtained by encapsulation of bioactive compounds within animal and human erythrocytes. The first one, concerning passive transport and delivery of chemicals, can be shared by other carrier systems, such as liposomes or macromolecules, although the properties of biocompatibility and biodegradability of the loaded erythrocytes make them, in principle, ideal carriers.

Table 1
Unusual functions of engineered erythrocytes

Carriers: Transport of chemicals (e.g., drugs for delivery either in circulation or at selected sites)

Bioreactors: Optimization of erythrocyte functions or display of new useful performances.
Examples:
1) Enhanced Oxygen delivery (by encapsulation of inositol hexaphosphate)
2) Detoxification (by encapsulation of heterologous enzyme proteins)
3) Vaccination (by encapsulation of antigens)
4) Drug production (by encapsulation of suitable prodrugs)

The second general (and unusual) type of performances that follow encapsulation of bioactive compounds in red cells includes optimized exchange of respiratory gases (7), display of detoxifying properties against xenobiotics (8), improved immune responses against erythrocyte-encapsulated antigens of vaccinal interest (9) and intracellular synthesis and subsequent release of drugs from properly tailored, erythrocyte-encapsulated pro-drugs (10).

2. ERYTROCYTES AS CARRIERS

Two main requirements should be met in order to obtain this general performance from properly engineered erythrocytes.
a) A reasonable encapsulation potential for the chemical compound under study, i.e. a satisfactory loading first and persistence of the internalized molecule afterwards.
b) Leakage of the encapsulated compound through various mechanisms that should however be amenable to control in order to confer flexibility to the release process. The presence in the red cell membrane of specific transport systems, e.g. for nucleobases (11), is compatible with some regulation of the kinetics of release, thus upgrading the relevant function above the possible behaviour of a passive transport system.

Table 2 lists a number of chemical compounds that have been encapsulated so far in either animal or human erythrocytes. All of these can be disseminated in circulation by the resealed erythrocytes whose viability, as mentioned, is not impaired by the current opening-resealing procedures. This is the reason why encapsulated proteins and peptides are sequestered within the red cells and cannot therefore be released throughout the circulatory system (e.g., for replacement therapies of plasma proteins). In fact, although leakage of peptides might be deliberately achieved through proper alterations of the resealing process leading to incomplete closure of the transient pores (e.g., by means of supplementation with Calcium), the resulting erythrocytes would exhibit a marked fragility producing their rapid removal from circulation (12).

Therefore, unless specific performances are pursued through the internalization of proteins and peptides within macrophages and phagocytizing cells (e.g., immunization processes or enzyme therapy of lysosomal storage diseases), protein-loaded red cells are expected to display their new functions as normally viable circulating carriers or bioreactors.

Within the list of encapsulated compounds mentioned in Table 2, three classes of

Table 2
Selected examples of chemicals encapsulated in erythrocytes for dissemination via the circulatory system

Pesticides for zooprophylactic purposes
Antibiotics
Antiviral drugs
Antineoplastic agents and cytotoxic drugs
Immunomodulators
Iron chelators in supportive transfusion therapy
Tracer compounds for diagnostic purposes
Antigens of vaccinal interest

chemicals seem to have wide biomedical applications in the future: antineoplastic and antiviral agents, because of their considerable toxicity and of potential pharmacokinetic improvement following encapsulation in erythrocytes; antigens, because of their natural, sustained delivery through loaded erythrocytes to immunopoietic organs and tissues (spleen, lymph nodes).

2.1. Antineoplastic agents.

Mention is made here to those antitumor drugs that can be encapsulated as such in either animal or human erythrocytes for a slow delivery in circulation or at selected sites. Therefore, cases such as those of encapsulated L-asparaginase or prodrugs will be considered later.

Methotrexate (MTX)

This chemotherapeutic agent was encapsulated in canine (13), murine (14) and human (15) erythrocytes. With both animal species "in vivo" studies were carried out that proved very useful for estimating toxicology and pharmacokinetics of this delivery system, while the investigations with human erythrocytes were performed "in vitro" and concerned the optimal procedures of MTX internalization, as probed by scanning and transmission electron microscopy (15). On the basis of these studies, pre-swell and electroporation methods seem to be preferable to hypotonic dialysis if prolonged delivery of MTX is required.

Experiments with the canine erythrocytes showed unequivocally that the rapid leakage of MTX from the loaded cells in circulation can be prevented upon treating the carriers with a bifunctional reagent such as glutaraldehyde (13). This treatment produces the additional advantage of targeting a large fraction of the loaded erythrocytes to the liver of dogs (16). The rationale of the "in vivo" effect seems not to be a specific reaction of the cross-linker with selected membrane proteins, but rather a rigidification of the carriers with their consequent retardation or trapping within liver sinusoids. Such explanation is supported by additional "in vivo" studies on murine erythrocytes that showed preferential sequestration of carrier erythrocytes in other organs sharing microvasculature with liver, such as spleen and lungs, according to different concentrations of the cross-linker (17).

Loading erythrocytes with MTX and subsequent exposure to glutaraldehyde allowed a targeted delivery of the drug to the liver in dogs (16). A substantial fraction of MTX entered the enterohepatic bile salt circulation. In addition, this study addressed the issue of toxicity of the target organ, that requires in principle careful consideration before clinical approaches become feasible. Although some standard tests of liver function showed mild hepatocellular necrosis as a result of liver targeting of MTX (16), toxicity was transient and should not represent a serious limiting factor for this delivery system.

Doxorubicin

Encapsulation of doxorubicin in erythrocytes, in an attempt to achieve a slow delivery system, was suggested by earlier reports indicating that the cumulative cardiotoxicity afforded by this anthracycline can be prevented or attenuated by long lasting intravenous infusion of the drug. Human erythrocytes were first loaded with doxorubicin and found not to be significantly impaired in their properties (18). When entrapment was made in murine erythrocytes, their i.v. infusion to recipient animals showed however rapid diffusion of doxorubicin from the circulating red cells, resulting in pharmacokinetic patterns apparently not dissimilar from those observed upon administering the free drug (17). The winning strategy was submitting the doxorubicin-loaded murine erythrocytes to treatment with glutaraldehyde, in the same way as previously described for MTX (see above). Like with MTX, loading the red cells with doxorubicin and incubating them with 0.1% glutaraldehyde produced two useful effects (17): 1) "in vitro", a consistently retarded efflux of the drug from the carriers; 2) "in vivo", remarkable targeting of the drug-loaded red cells to the liver. In an attempt to validate this delivery system, a murine metastatic model was used that proved to be highly reproducible in terms of time-dependent spreading of various tumor cells from the site of injection, i.e. the spleen (19). Using this model, we were able to show a better therapeutic performance of doxorubicin-loaded and glutaraldehyde-treated erythrocytes over the conventional intravenous administration of the free drug, as witnessed by an increase of the therapeutic index from 1.8 to 4.2 (19).

"In vitro" (20) and "in vivo" (21) studies on the dog confirmed the feasibility of the approach based on doxorubicin-loaded and glutaraldehyde-treated red cells. In particular, the pharmacokinetics proved to be different from and better than those observed upon conventional intravenous administration of the free drug.

Intravenous use of this slow delivery system was also made with three dogs affected by spontaneous lymphosarcomas (22). The clinical results, although demonstrating considerable myelotoxicity, were consistent with tumor remission in all cases under study (22).

So far, only one human patient, bearing massive hepatic metastases, was investigated in a systematic way using autologous drug-loaded and glutaraldehyde-treated red cells (23). This study was facilitated by locoregional infusion of the erythrocytes through a catheter implanted in the hepatic artery at the time of first resection of primary colon cancer. Preliminary scintigraphic analyses confirmed extensive and long-lasting hepatic trapping of the ^{99m}Tc -treated autologous red cells incubated with glutaraldehyde (23). Based on this promising result and on failure of three conventional treatments of chemotherapy, the patient was submitted to alternating regimens of doxorubicin administered in the hepatic artery both as free drug and in the erythrocyte-encapsulated form, respectively (the latter treatments included two glutaraldehyde concentrations). The pharmacokinetic data indicated significant decrease of plasma levels of doxorubicin when the drug was administered encapsulated compared with the values recorded following free infusion (24). Tumor growth was apparently stopped, but the patient died because of a septic shock after 5 months of treatment (his life expectancy at the onset of therapy was no more than one month).

Currently, ten additional patients affected by either primary or metastatic liver tumors are under scrutiny for a more extensive scintigraphic study on the organ distribution of autologous glutaraldehyde-treated erythrocytes administered intravenously. This route of infusion (forcefully followed, owing to lack of patients bearing a catheter in the hepatic artery) seems, however, to yield heterogeneous results in terms of hepatic targeting. Accordingly, a higher number of cases to be submitted to scintigraphic analyses is required in order to start a systematic therapy of liver tumors on those patients showing extensive targeting of their glutaraldehyde-treated red cells to the liver. In addition, careful investigations are ongoing concerning the mechanisms of myelotoxicity observed with the dogs (22).

Carboplatin

This drug was recently encapsulated in human erythrocytes up to 5 mg per ml of packed cells. In spite of a good potential of encapsulation, there seems to be little applicability of Carboplatin-loaded red cells to therapy. The main reason, rather than impairment of glutathione metabolism in the carriers, is the rapid conversion of the parent drug molecule to by-products (25). Although these partially unidentified metabolites of Carboplatin still retain some antiproliferative activity "in vitro", the process per se should be viewed as limited drug inactivation. Conversion of Carboplatin is promoted by hemoglobin and the divalent heme iron plays an important role therein as also estimated by distinctive patterns of cytotoxicity in uninduced versus induced murine erythroleukemic cells (26).

2.2. Antiviral agents.

An important and promising result has been recently achieved by Magnani et al. (27), who succeeded in encapsulating 2',3'-dideoxycytidine 5'-triphosphate (ddCTP, i.e. the pharmacologically active species of the antiretroviral agent 2',3'-dideoxycytidine) within murine and human erythrocytes. The remarkable stability of ddCTP in the loaded red cells prompted the Authors to exploit a highly specific procedure of targeting to macrophages previously developed by Turrini et al. (28). The procedure is based on $ZnCl_2$ as an agent promoting clusters of band 3 protein (the predominant transmembrane protein that functions as an anion transporter). The clusters are then stabilized by chemical cross-linking and specifically recognized by circulating autologous antibodies. Opsonization triggers complement deposition and subsequent phagocytosis by macrophages. This procedure proved to be highly efficient in targeting the ddCTP-loaded red cells to the macrophage compartment, as assessed "in vitro" by marked inhibition of HIV-1 replication in infected cells of the monocyte-macrophage lineage (27). Additionally, the efficiency of the macrophage-targeted ddCTP-loaded red cells was also validated "in vivo" using a mouse model of immunodeficiency in which lymphoadenopathy was significantly reduced (27).

These data appear to disclose remarkable potential to the erythrocyte-based technology, because, a) the pharmacologically active triphosphorylated forms of 2',3'-dideoxycytidine and of other retroviral dideoxynucleosides can hardly be formed intracellularly from the precursor drugs, in some cell types, b) infected macrophages have been recognized to play an important role in the pathogenesis of AIDS, acting as reservoirs of HIV (29). Therefore, the precision of this targeted delivery system seems to ensure high efficiency with even minute amounts of anti-HIV drugs.

2.3. Antigens.

Exploitation of erythrocytes loaded with antigens is basically suggested by the natural termination of the red cell life-span in an immunopoietic organ like the spleen. Accordingly, any heterologous protein encapsulated in red cells will reach immunocompetent cells where it will be recognized as "nonself", thereby triggering the production of specific antibodies. This "natural" fate of "non-natural" encapsulated proteins can be oriented in principle to occur either in a sustained way, if the engineered erythrocytes keep their normal viability, or alternatively in an accelerated mode, taking advantage of the manifold experimental systems available to induce premature erythrophagocytosis.

Both strategies have been followed using antigen-loaded red cells. Thus, completely viable murine erythrocytes containing three bacterial antigens (diphtheria, tetanus and pertussis toxoids) have been used to induce immunization (9). The relevant controls included each of the three antigens administered free intravenously at the same dosages and according to the same schedules, and also the antigens emulsified with complete Freund's adjuvant. Both specific and neutralizing antibodies were raised with these administration schedules. The antibody titers elicited by diphtheria and tetanus toxoids were significantly higher with the erythrocyte-encapsulated than with the free antigens, while they proved to be strictly comparable in the case of pertussis toxoid (9). Therefore, with the latter antigen the loaded red cells proved to behave as carriers and no adjuvant function was displayed, unlike the cases of diphtheria and tetanus toxoids for which the loaded red cells also acted as adjuvants. These differences may be due to distinctive patterns of antigen presentation and processing in the target macrophages. Whatever the mechanism(s), it is significant and of potential applicability that with all antigens tested the relevant antibodies were protecting against the effects of the native toxins. This finding indicates that the procedure can be reasonably exploited for antigens of vaccinal interest, mostly for those recombinant peptides and proteins that may prove to be rapidly destroyed at the site of injection .

The other approach, still with antigen-loaded murine erythrocytes but with fragilized carrier cells, seems retrospectively to have been followed by Esposito et al. (12). Indeed, presence of Calcium during the loading procedure used by these Authors is expected to interfere with closure of pores in the resealing step, thus leading to red cell fragilization, as actually observed (12). In this study, considerable antibody titers were observed using murine red cells loaded with a synthetic peptide reproducing the tandem-repeated epitope of the circumsporozoite protein of <u>Plasmodium falciparum</u>.

An additional way for constructing red cells as antigen-delivery system is the chemical coupling of proteins to the outer surface of erythrocytes by means of a biotin-avidin-biotin bridge (30). The strategy is obviously different, relying on external rather than on internalized antigens. The main advantage is that considerably less antigen (0.5 μg/mouse) is required to obtain an immune response. Moreover, evidence is available that the "in vivo" survival of the biotinylated erythrocytes in rabbits is normal (31). Problems could arise from use of poor immunogens, since in these cases the anti-avidin antibodies that are unavoidably raised may obscure the antibody production against the desired antigen (Sdraffa, A. unpublished data).

3. ERYTHROCYTES AS BIOREACTORS

There are several ways for engineered erythrocytes to behave as bioreactors. To summarize (Table 1), these patterns can include optimized performances in the peripheral oxygen delivery by encapsulation of regulators of hemoglobin function, new functions (e.g., detoxification) conferred by the entrapment of heterologous enzyme proteins and the ability to convert encapsulated prodrugs to membrane diffusable active drugs. As reported above, use of antigen-loaded red cells for vaccination purposes can be viewed as simple transport to immunocompetent cells or, with some antigens like pertussis toxoid (9), as positive immunomodulation, i.e. an adjuvant function.

3.1.Encapsulation of allosteric effectors of hemoglobin.

Entrapment in mammalian erythrocytes of the metabolically stable compound inositol hexaphosphate (IHP) produces a considerable rightward shift of the oxygen dissociation curve of hemoglobin (Hb) (7,32). Two different procedures were reported , i.e. the dimethyl sulphoxide (DMSO) osmotic pulse method (32) and the continuous flow hypotonic dialysis technique (33). The earlier "in vitro" findings suggested to move to an animal model, in view of the potential importance of this oxygen delivery system in several human diseases characterized by tissue hypoxia (myocardial and cerebral ischemia, right to left shunts, etc.) and in cases of increased tissue oxygen demand (e.g., heavy muscular exercise). Therefore piglets were carefully investigated following complete exchange of their red cells with IHP-loaded erythrocytes (34). Decrease of the oxygen affinity of the transfused, IHP-loaded red cells "in vivo" resulted in an enhanced oxygen extraction and in a decreased cardiac output, with a constant oxygen consumption (7,34). Additional studies were carried out using isolated heart perfused with IHP-loaded red cells (35). This system showed beneficial effects that were however counteracted by a cardiodepressive effect (35). Other Authors (36), still using perfused isolated rat heart reported on a toxic effect of the IHP possibly leaking from the loaded red cells. Leakage is probably the result of partial hemolysis of the engineered erythrocytes, since during the first 24 hr after transfusion of the IHP-loaded red cells in recipient piglets, a loss of 15%-20% of the transfused cells was observed (37). These more fragile engineered red cells, also demonstrated on the basis of density gradient separation (38), should be easily removed before reinfusion. However, possible adverse effects arising from a long term treatment with IHP-loaded erythrocytes as a result of IHP leaking from erythrophagocytizing macrophages, must be ruled out before clinical applications are feasible. Improved analysis of biochemical properties of IHP-loaded red cells (39) as well as precise estimation of the encapsulated IHP during storage of the engineered cells (40) and determination of purity of IHP (41) should help achieve a safe handling and use of these cells for "in vivo" studies and eventually for human and veterinary applications.

3.2.Encapsulation of enzyme proteins.

The large dimensions of the transient pores formed during the hypotonic shock step of current encapsulation procedures (42) allow an equilibration of macromolecules between the internal and external compartments of the red cell. Therefore, proteins of up to 200 kDa can be encapsulated, although with lower recoveries compared to small protein molecules. Adequate resealing will then produce persistent sequestation of the internalized protein(s). Several enzyme proteins can therefore be entrapped in red cells including both normally resident and foreign enzymes. In the former case, the loaded erythrocyte can change its metabolic properties provided that the enzyme is normally present at low activity. The best examples available are: a) the already mentioned complete correction of genetic deficiency of Glucose 6-P dehydrogenase (G6PD) activity by entrapment within the deficient red cells of G6PD purified from normal erythrocytes (6);

b) the improvement of glucose consumption through glycolysis by encapsulation of the rate-limiting enzyme hexokinase, purified from human placenta (43).

If usually absent enzyme proteins are internalized in red cells, their activity depends on availability of their substrates intracellularly and therefore on diffusion or transport of these molecules across the membrane. The performances of the bioreactor so obtained will likewise be related to the features of exit of thew product(s) from the loaded erythrocyte. This approach can be followed if abnormally high blood metabolites are present either as a result of inborn errors of metabolism and of metabolic diseases in general, or due to intoxication by xenobiotics. Properly enzyme-loaded red cells have been obtained and proposed for decreasing plasma levels of phenylalanine (44), uric acid (44,45) and glucose (46), while detoxification can be accomplished in principle in cases of blood accumulation of ethanol and acetaldehyde (47), of cyanide poisoning (48) and of quinone compounds (49).

A clinically useful exploitation of such circulating bioreactors is currently achieved by the encapsulation of L-asparaginase in human red cells from patients affected by malignant hematopoietic diseases such as acute lymphoblastic leukemia, non-Hodgkin lymphoma and lymphosarcoma (50,51). The rationale of treating these patients with L-asparaginase is the depletion of plasma asparagine, an essential amino acid that is strictly required for growth of leukemic and lymphoblastic cells. However, L-asparaginase is a bacterial protein with potent antigenicity, the intravenous administration of which can lead to serious and sometimes fatal anaphylactic reactions in the treated patients. The L-asparaginase-loaded erythrocytes proved both "in vitro" and in clinical trials "in vivo", to behave efficiently in reducing asparagine levels in plasma, with a significant improvement over the conventionally administered free enzyme (50). Moreover, no intolerance reactions were observed in any of 13 patients treated with a single administration of asparaginase-loaded erythrocytes. Kravtzoff et al. (51) have also investigated the antibody response in these patients and found no significant differences in the anti-L-asparaginase antibodies compared with subjects receiving free L-asparaginase. This pattern is therefore reminiscent of the results of immunization of mice with free versus red cell-loaded pertussis toxoid (9) and is therefore representative of those cases of antigens for which erythrocytes behave as simple carriers and not as adjuvants. A general question raised by these studies concerns the still "in vivo" undefined patterns of the immune response elicited by repeated administrations of red cell-encapsulated heterologous enzyme proteins, since this strategy, as mentioned, could be the winning one in several metabolic disorders and cases of poisoning. It seems however, from this report on single administration that the shielding erythrocytes prevent the formation of circulating immune complexes that are most likely responsible for the adverse intolerance reactions observed in some of the patients receiving the free enzyme protein (51).

3.3.Formation of drugs from encapsulated prodrugs.

The rationale of this application is based on the intracellular activity of endogenous red cell enzymes that can act upon "foreign" encapsulated drug precursors. Relevant examples include red cell encapsulation of cytosine arabinoside 5'-monophosphate (52), 5-Fluoro--2'-deoxyuridine 2'-monophosphate (FdUMP) (10), 2',3'-dideoxycytidine 5'-monophosphate (53), all three nucleotides being substrates of red cell hydrolases converting them to the corresponding pharmacologically active and membrane-releasable nucleoside forms. In addition, pro-opioid molecules (monophosphates, triphosphates and disulfates of Naloxone and Naltrexone)have been recently synthesized and encapsulated in human erythrocytes (54).

In all cases, the efficiency of the bioreactor, as measured by the output of the active drug, relies on several biochemical features including activity and substrate affinity of the bioconverting red cell enzymes and also patterns of exit of the drug across the membrane. Paradoxically, such efficiency may prove to be too high, as observed for the fast intraerythrocytic conversion of FdUMP to 5-F-2'-deoxyuridine. This may pose challenges to the development of a pharmacokinetically useful system of drug delivery. In principle, these problems can be faced by co-encapsulation of enzyme inhibitors (10) or by attempts at slowing down the process of drug extrusion through specific inhibitors of transport systems. An alternative and more general way to cope with this issue seems to be the design and synthesis of metabolically more remote precursors of the final drug. In other words, the exploitation of specific bioconverting enzymes in the erythrocyte can suggest the final production of a membrane-diffusable active drug through a multi-step pathway that starts from an encapsulated pre-prodrug (Fig. 1). Indeed, this achievement has been recently obtained with the demonstration that human erythrocytes can behave "in vitro" as bioreactors competent to the release of 5-F-2'-deoxyuridine (FUdR) in a two-step pathway involving a phosphodiesterase-catalyzed hydrolysis of an "ad hoc" synthesized dimeric fluoropyrimidine first, followed by conversion of FdUMP to FUdR (55). Ideally, the design of such encapsulatable precursors should follow extensive biochemical characterization of the relevant bioconverting enzymes, suggesting best fit-based structures to be tailored accordingly. This type of general strategy, i.e. pharmacobiochemical therapy, seems to be very promising as it is based on programmed patterns of drug biodisposition, much in the same way as pro-hormones and pro-enzymes are synthesized and then processed to yield their biologically active species.

4. CONCLUSIONS

Several lines of evidence converge to the conclusion that engineered erythrocytes can play an important role in human and veterinary medicine, mostly as advanced therapeutic devices, but with potential advantages also concerning diagnosis (56,57) and prevention (i.e., vaccination). Like in most areas of biotechnology, a critical factor is an appropriate gearing between basic research and clinical applications. In addition, once more in biotechnology, targeted transfer of the experimental data to exploitation requires interdisci-

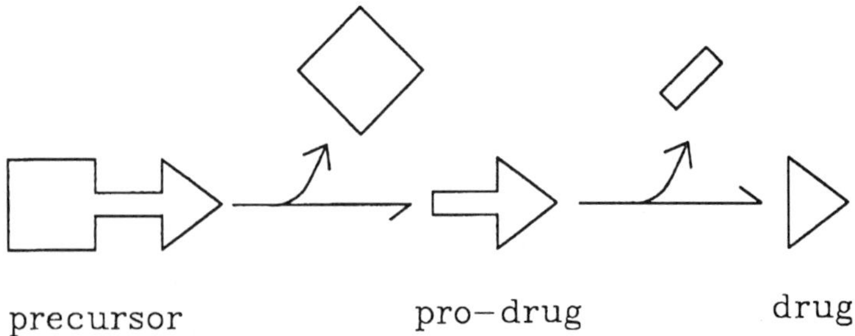

Figure 1. Scheme of conversion of an encapsulated pre-prodrug to its corresponding pharmacologically active form by specific intraerythrocytic enzymes.

plinary approaches due to involvement of multiple research skills and industrial interests (e.g., companies devoted to drug development, to vaccines, to blood derivatives, to transfusion systems). Careful attention to these challenges is required before this versatile and in principle easy technology can solve the many clinical problems considered in this paper.

REFERENCES

1. G.M. Ihler, R.H. Green & F.W. Schnure, Proc. Natl. Acad. Sci. USA, 70 (1973) 2663.
2. J.R. DeLoach, J. Appl. Biochem. 5 (1983)149.
3. J.R. DeLoach & U. Sprandel (eds.), Red Blood Cells as Carriers for Drugs, Biblthca Haemat. vol. 51, Karger, Basel, 1985.
4. C. Ropars, M. Chassaigne, M.C. Villéreal,G. Avenard, C. Hurel & C. Nicolau, in J.R. DeLoach & U. Sprandel (eds.), Biblthca Haemat. vol. 51, p. 84, Karger, Basel, 1985.
5. M. Tonetti & A. De Flora, J. Clin. Pharmacokinetics, in press.
6. A. Morelli, U. Benatti, F. Salamino, B. Sparatore, M. Michetti, E. Melloni, S. Pontremoli & A. De Flora, Arch. Biochem. Biophys. 197 (1979) 543.
7. C. Nicolau, B.P. Teisseire, C. Ropars, M.O. Vallez & R.A. Hérigault, in DeLoach & U. Sprandel (eds.), Biblthca Haemat. vol. 51, p. 92, Karger, Basel, 1985.
8. J.L. Way, P. Leung, L. Ray & C. Sanders, in DeLoach & U. Sprandel (eds.), Biblthca Haemat. vol. 51, p. 75, Karger, Basel, 1985.
9. C. Polvani, A. Gasparini, U. Benatti, A. De Flora, S. Silvestri, G. Volpini & L. Nencioni, Biotechnol. Appl. Biochem. 14 (1991) 347.
10. A. De Flora, E. Zocchi, L. Guida, C. Polvani & U. Benatti, Proc. Natl. Acad. Sci. USA 85 (1988) 3145.
11. P.G. Plagemann, R.M. Wohlueter & J. Erbe, J. Biol. Chem. 257 (1982) 12069.
12. F. Esposito, A.M. Santroni, A. Habluetzel, V. Grellani, A. Pessi, S. Lombardi, G. Falcioni & M. Brunori, Int. J. Parasitol. 20 (1990) 1109.
13. J.R. DeLoach & C. Barton, Am. J. Vet. Res. 43 (1982) 2210.
14. C.A. Kruse, C.L. Freehauf, K.R. Patel & J.D. Baldeschwidler, Biotechnol. Appl. Biochem. 9 (1987) 123.
15. C.A. Kruse & G.W. Mierau, in R. Green & J.R. DeLoach (eds.), Advances in Biosciences vol. 81, p. 139, Pergamon Press, Oxford, 1991.
16. J.R. DeLoach, C.H. Tagner & C. Barton, Res. Exp. Med. 183 (1983) 167.
17. E. Zocchi, M. Tonetti, C. Polvani, L. Guida, U. Benatti & A. De Flora, Biotechnol. Appl. Biochem. 10 (1988) 555.
18. A. De Flora, U. Benatti, L. Guida & E. Zocchi, Proc. Natl. Acad. Sci. USA 83 (1986) 7029.
19. E. Zocchi, M. Tonetti, L. Guida, C. Polvani, U. Benatti & A. De Flora, Proc. Natl. Acad. Sci. USA 86 (1989) 2040.
20. M. Tonetti, B. Astroff, W. Satterfield, A. De Flora, U. Benatti & J.R. DeLoach, Biotechnol. Appl. Biochem. 12 (1990) 621.
21. M. Tonetti, B. Astroff, W. Satterfield, A. De Flora, U. Benatti & J.R. DeLoach, Am. J. Vet. Res. 52 (1991) 1630.
22. C.M. Matherne, W.C. Satterfield, A. Gasparini, A.B. Astroff, R.D. Schmidt, R.D. Rowe & J.R. DeLoach, submitted.
23. M. Tonetti, C. Polvani, E. Zocchi, L. Guida, U. Benatti, P. Biassoni, F. Romei, A. Guglielmi, C. Aschele, A. Sobrero & A. De Flora , Eur. J. Cancer 27 (1991) 947.

24. M. Tonetti, C. Polvani, E. Zocchi, L. Guida, U. Benatti, P. Biassoni, F. Romei, A. Guglielmi, C. Aschele, A. Sobrero & A. De Flora, in M. Magnani & J.R. DeLoach (eds.) Advances in Experimental Medicine and Biology vol. 326, p. 307, Plenum Press, New York, 1992.
25. M. Tonetti, A. Gasparini, M. Giovine, D. Bini, A. Mazzuccotelli, F. De Paz, U. Benatti & A. De Flora, Biotechnol. Appl. Biochem. 15 (1992) 267.
26. M. Tonetti, A. Gasparini, M. Giovine, U. Benatti & A. De Flora, Biochem. Biophys. Res. Commun. 182 (1992) 6477.
27. M. Magnani, L. Rossi, G. Brandi, G.F. Schiavano, M. Montroni & G. Piedimonte, Proc. Natl. Acad. Sci. USA 89 (1992) 6477.
28. F. Turrini, P. Arese, J. Yuan & P.S. Low, J. Biol. Chem. 266 (1991) 23611.
29. S. Gartner, P. Markovitz, D.M. Markovitz, M.H. Kaplan, R.C. Gallo & M. Popovic, Science 233 (1986) 215.
30. M. Magnani, L. Chiarantini, E. Vittoria, U. Mancini, L. Rossi & A. Fazi, Biotechnol. Appl. Biochem. 16 (1992) 215.
31. V. Russo, R. Barker-Gear, R. Gates & R. Franco, in M. Magnani & J.R. DeLoach (eds.) Advances in Experimental Medicine and Biology vol. 326, p. 101, Plenum Press, New York, 1992.
32. R.S. Franco, M. Weiner, K. Wagner & O.J. Martello, Life Sci. 32 (1983) 2763.
33. M. C. Villéreal, C. Bailleul, M. Chassaigne & C. Ropars, in C. Ropars, M. Chassaigne & C. Nicolau (eds.), Advances in Biosciences vol. 67, p. 55, Pergamon Press, Oxford, 1987.
34. B. Teisseire, C. Ropars, M.O. Vallez, R. Hérigault & C. Nicolau, J. Appl. Physiol. 58 (1985) 1810.
35. J.F. Baron, E. Vicaut, O. Stucker, M.C. Villéreal, C. Ropars, B. Teisseire & M. Duvelleroy, in C. Ropars, M. Chassaigne & C. Nicolau (eds.), Advances in Biosciences vol. 67, p. 73, Pergamon Press, Oxford, 1987.
36. M. Samaja, R. Matterlini, M. Tarantola, M. Beretta, F. Rossi, L. Sabbioneda, M. Porcellati & A. Zanella, in R. Green & J.R. DeLoach (eds.), Advances in Biosciences vol. 81, p. 181, Pergamon Press, Oxford, 1991.
37. C. Ropars, B. Teisseire, G. Avenard, M. Chassaigne, C. Hurel, R. Ginot & C. Nicolau, Ann. N.Y. Acad. Sci. 445 (1985) 477.
38. G. Bourget, L. Boucher & C. Ropars, in M. Magnani & J.R. DeLoach (eds.) Advances in Experimental Medicine and Biology vol. 326, p. 27, Plenum Press, New York, 1992.
39. A. Zanella, L. Sabbioneda, F. Rossi, V. Russo, G. Aguggini, G. Fiorelli, S. Villa, A. Brovelli, M.A. Castellana, C. Seppi, C. Barlassina & G. Sirchia, in R. Green & J.R. DeLoach (eds.), Advances in Biosciences vol. 81, p. 219, Pergamon Press, Oxford, 1991.
40. S. Villa, F. Rossi, P.A. Biondi, V. Russo, T. Crimella, G. Fiorelli & A. Zanella, in M. Magnani & J.R. DeLoach (eds.) Advances in Experimental Medicine and Biology vol. 326, p. 41, Plenum Press, New York, 1992.
41. T. Crimella, S. Villa, F. Rossi, G. Fiorelli & A. Zanella, in M. Magnani & J.R. DeLoach (eds.) Advances in Experimental Medicine and Biology vol. 326, p. 51, Plenum Press, New York, 1992.
42. J.F. Hoffman, in M. Magnani & J.R. DeLoach (eds.) Advances in Experimental Medicine and Biology vol. 326, p. 1, Plenum Press, New York, 1992.
43. M. Magnani, L. Rossi, L. Cucchiarini, G. Fornaini, U. Benatti, L. Guida, E. Zocchi &

A. De Flora, Biochim. Biophys. Acta 972 (1988) 1.
44. U. Sprandel, in J.R. DeLoach & U. Sprandel (eds.), Biblthca Haematol. vol. 51, p. 7, Karger, Basel, 1985.
45. M. Magnani, U. Mancini, M. Bianchi & A. Fazi, in M. Magnani & J.R. DeLoach (eds.) Advances in Experimental Medicine and Biology vol. 326, p. 189, Plenum Press, New York, 1992.
46. L. Rossi, M. Bianchi, A. Fraternale & M. Magnani, in M. Magnani & J.R. DeLoach (eds.) Advances in Experimental Medicine and Biology vol. 326, p. 183, Plenum Press, New York, 1992.
47. P. Ninfali, L. Rossi, L. Baronciani, C. Ropars & M. Magnani, in M. Magnani & J.R. DeLoach (eds.) Advances in Experimental Medicine and Biology vol. 326, p. 165, Plenum Press, New York, 1992.
48. J. Way, E. Cannon, P. Leung, R. Davis, D. McGuinn, C. Yao, C. Chui & A. Naqi, in R. Green & J.R. DeLoach (eds.), Advances in Biosciences vol. 81, p. 207, Pergamon Press, Oxford, 1991.
49. U. Benatti, L. Guida, A. De Flora & P. Hochstein, Chemica Scripta 27A (1987) 169.
50. R. Kravtzoff, I. Desbois, M. Chassaigne, J.P. Muh, J.P. Lamagnere, P. Colombat & C. Ropars, in R. Green & J.R. DeLoach (eds.), Advances in Biosciences vol. 81, p. 127, Pergamon Press, Oxford, 1991.
51. R. Kravtzoff, I. Desbois, C. Daniel, P. Colombat, J.P. Lamagnere, M. Chassaigne & C. Ropars, in M. Magnani & J.R. DeLoach (eds.) Advances in Experimental Medicine and Biology vol. 326, p. 175, Plenum Press, New York, 1992.
52. J.R. DeLoach, J. Appl. Biochem. 5 (1983) 149.
53. M. Magnani, M. Bianchi, L. Rossi & V. Stocchi, Biochem. Biophys. Res. Commun. 164 (1989) 446.
54. S. Nöel-Hocquet, S. Jabburi, S. Lazar, JC. Maunier, G. Guillarmet & C. Ropars, in M. Magnani & J.R. DeLoach (eds.) Advances in Experimental Medicine and Biology vol. 326, p. 215, Plenum Press, New York, 1992.
55. M. Mazzei, T. Grandi, A. Balbi, A. De Flora, U. Benatti, A. Gasparini & G.L. Damonte, ECB 6, 6th European Congress on Biotechnology, Florence June 13-17, 1993 (abstract), in press.
56. A. De Flora, in H. Koprowsky, S. Ferrone & A. Albertini (eds.), The Impact of Biotechnology in Diagnostics, p. 223, Elsevier Science Publisher, Amsterdam, 1985.
57. R. Kravtzoff, E. Urvoase, C. Chambon & C. Ropars, in M. Magnani & J.R. DeLoach (eds.) Advances in Experimental Medicine and Biology vol. 326, p. 347, Plenum Press, New York, 1992.

ECB6: Proceedings of the 6th European Congress on Biotechnology
edited by L. Alberghina, L. Frontali, P. Sensi
1994 Elsevier Science B.V.

Biotechnology : A general survey

François Gros

Professor at Collège de France and Pasteur Institute, France

Before trying to set the scene with regard to some past and future developments of biotechnology and since it is my great privilege to give this closing lecture, I would like to express my deepest appreciation and admiration to the President of this Symposium, Professor Sensi and his collaborators. They have achieved an excellent job by conceptualizing and realizing a very exciting and lively program. They have made it so attractive that the number of participants has largely overpassed expectations. The task and responsibility involved in a Symposium like this one are very demanding; it takes months of hard thinking and preparatory approaches. As a vice President, my own work has been close to nil, but as President of the next ECB7, I am shared between enthusiasm and fear, a fear of not doing as well as our Italian colleagues ! Dr. Penasse will take the floor after me to give you a general overview of the 1995 program. But, if I may anticipate, I shall be very happy, as President of the next meeting to welcome you all to Nice, on the French Riviera, in 1995.

I also wish to say how grateful we all are to the present organizing committee, the sponsoring associations as well as the local and federal authorities of Italy.

Biotechnology, during the past decades, has oscillated in people's mind between a phase of great excitement and a stage of disillusion. This disillusion was mostly based upon the fact that many of the original predictions concerning the entrance onto the market of genetically-engineered or immunologically-derived products have not been met, at least as rapidly as what was initially thought.

Fortunately so, I think we can reasonably advance, at present, that to a period often characterized by irrealistic dreams has succeeded a period of concrete achievements. A wealth of new ideas, new techniques and new projects is merging out in biological sciences, and realism does not preclude conceptual excitement which is the motor for new discoveries. At the same time a new challenge is offered to modern biotechnology : this is to be able to meet more closely the public concern about the possible threats on environment and even human rights. I shall come back to this point at the end.

Let me begin by analyzing some of the general features of biotechnology, over the last 2 decades. I shall first do so from a scientific standpoint.

If one tries to think about some of the most significant technologies (by giving emphasis to those which are genetically-derived), in terms of the impacts they have had to improve services or create new products, the following can be said :

i) Genetic engineering, has "penetrated," all the disciplines in biology as well as in medicine, and agriculture.

ii) Ambitious programs of genome analysis and sequencing, particularly with respect to the human species, but also directed towards animal, plants or lower eukaryotes, have been launched. This has operated as a major trempolin, for the study of human genetic diseases and is laying ground for many applications in other fields (including agriculture). A new bridge has been established between biology and computer sciences.

iii) The 1980 decade has been characterized by the possibility to create a large number of transgenic organisms in the animal and vegetal kingdoms. Also, the first breakthroughs in human gene therapy have been reported and this is likely to represent a major turn in contemporary medicine.

iv) Certain, established "dogmas" in basic biology, such as the compulsory proteinaceous nature of biocatalysts, have been largely shaken following the discovery that ribonucleic acids can display catalytic activities. Although it is too early to forecast the role that ribozymes will play from a practical stand point their discovery should most probably forge new weapons in the hands of virologists.

APPLICATIONS : SOME GENERAL FEATURES

Another approach to look at the past and present achievements of biotechnology - aside from examining things as we have just done, from a technological or conceptual angle - is to ask whether it has given rise to "significant" applications. Yet, the term "significant" should be taken with caution since it has many possible connotations. For example, many interesting new drugs are in fact "orphan drugs". Their impact on human health is clear, eventhough they may never give rise to a sizeable industrial market !.

Also, the introduction onto the market of commercialized products, although often regarded as one major criterion to judge about the practical usefulness of biotechnologies, is by no means the only one ! In many circumstances biotechnologies have helped renewing classical procedures. The gains in yields of production, energy saving, safety, etc....might be considerable, albeit difficult to assess in precise financial terms.

PHARMACEUTICALS

Among the major molecules which have been commercialized since 1982, in the field of bio-pharmaceuticals, a dozen of biotech-derived (often called "new generation") products with recognized or potential therapeutic activity, or which are active as vaccines, have been on the market since the first clearance of a genetically-engineered substance, namely the recombinant insulin, by Eli Lilly in 1982.

Data recently published in an article by Houwink (and quoted by Luyben) (1) indicate that among the first to market introductions of new biotechnology products over the past decade, seven market launches were based on European developments and eight on developments in the USA.

The number of molecules, eventhough restricted to the pharmaceutical sector, might look rather modest, but two factors must be taken into account.

1) It takes on the average, from 10 to 15 years for a new molecule to be marketed as a significant drug. Financial inputs being very big constitute an obvious limiting factor.

2) We are in a sense just contemplating the "tip of the iceberg", and many new molecules are still "in the pipe".

For example, in 1992, by examining the types and numbers of biotechnology derived medical products still in development, the "Pharmaceutical Manufacturers Associations" could list 48 products waiting for FDA approval in the USA, which means that these products had already satisfied in most cases phase III requirements of the clinical assays. 132 biotechnology products were still in clinical trials, half of them being directed at the cancer market, as diagnosis or therapeutic tools, or to ameliorate the side effects of conventional therapies.

On Table 1 are depicted the principal categories of these products many of which are likely to become utilized in a not too distant future (2). It is interesting to note that the great majority of them are growth factors. All together 52 distinct biotechnology companies are presently involved.

Table 1
Biotechnology : Products in the U.S. Pipeline (1991)

CLOTTING FACTORS
factor IX, factor VIIa, antihemophilic factor

COLONY STIMULATING FACTORS
GM-CSF, M-CSF, rG-CSF

DISMUTASES
superoxyde-dismutase

ERYTHROPOIETIN
Er-β, Er-α

GROWTH FACTORS
EGF, FGF, IGF-1, PDGF

HUMAN GROWTH HORMONES
hGH, somatropin

INTERFERONS
IFN-γ-1b, IFN-αn^3,IFN-β,IFN-α2b,IFN-α2a

INTERLEUKINS
IL-2, IL-1-α ,IL-1β, IL-3, IL-4

MONOCLONAL ANTIBODIES

a) therapeutic
anti-EGF-R, anti-EGF, anti-IL-2-R, anti-leu, anti-LPS, anti-TNF, anti-pseudomonas, chimeric anti-CD4, chimeric anti-TNF, 7E3 Mab, HA-1A MAb, chimeric L6 anti CMV, Immunotoxin, LYM-1, Mab-L6,...etc...,

b) diagnostic
anti-fibrin, Tcm-FAb (colorectal, lung, breast, lymphoma, infectious diseases) etc.

RECOMBINANT SOLUBLE CD4S

TISSUE-PLASMINOGEN ACTIVATORS

TUMOR NECROSIS FACTORS

VACCINES
AIDS, BMY-35047, gp 120, hepatitis B, herpes, malaria, melanoma, etc....

FOOD SECTOR

Thus far, we have essentially discussed the situation with respect to the health related sector. But, in a survey of this type, it is interesting to examine "indicators" from other areas.

One such an area is related to Food Biotechnology. According to Barry Law (UK) (3) biotechnology has provided great added value to many conventional procedures : Beverage sweetening, fat and oil production, backing, dairy fermentations, biopreservation, food analysis. Moreover new genetically-modified food products have also been created. The problem here is whether their utilization will meet approval from the appropriate European Institutions and from the associations of consumers !

PLANT GENETICS - GENE REGULATORY MECHANISMS IN PLANT

Before addressing the macro-economical aspects, and to remain within the paradigm of possible application at a short or mid-range level, I would like to comment upon a last indicator which illustrates, perhaps more significantly, that what we are observing as a result of modern biotechnology only represents so far the tip of the iceberg : I wish to refer here to the domain of plant genetics.

It is quite clear that we are entering here into a new era which is the likely preambule of what could be regarded as a true revolution in the agricultural procedures.

First of all, great progress has been achieved, due to the recombinant DNA techniques, as far as the gene regulatory mechanisms in plants. I shall confine myself here to two examples.

The gene coding for the zein storage protein in maize, a protein of considerable economic value has been cloned. Its promoter has been shown to comprise upstream elements which respond very strongly to a newly identified plant specific transcription factor, called opaque 2 (4), a genetic construct including the promoter of the maize storage protein linked to a reporter gene. This construct is delivered specifically within the endosperm by using the particle-gun technique. The opaque 2 gene activity can be specifically monitored by the liberation of the chromophore within the endosperm. There are reasons to believe that a better knowledge of the protein storage mechanism in maize will prove of great agronomical interest.

Another example (5) stems from some recent work done in Prof. Schell's laboratory in Cologne. It concerns the effect of gene rol-C from *Agrobacterium Rhizogenes.* This gene encodes a protein which converts an inactive plant precursor into an active cytokine. This cytokine controls the balance between axial elongation and lateral growth. After gene rol-C forced expression in potatoes, the morphological effect observed is stricking : the plant exhibits only lateral growth. That one is dealing here with a control mechanism of general relevance is shown by the fact that a somewhat similar effect can be observed in *Lisianthus*, a completely different species. The possibility to handle this kind of genes might provide seed producers with many new possibilities.

TRANSGENIC PLANTS

Plant transgenesis is progressing very rapidly. Following the development of the particle gun technique (or so called "biolistics") in which small particles of gold or tungsten covered with a DNA solution are shot into living plant cells almost any kind of crops is amenable to transformation.

The number of field-tests for transgenic plants has considerably increased since 1986 going from 5 up to 156 in the single year 1991. This parallels the number of genes available for transformation (6).

The trait most commonly expressed in transgenic plants concerns the resistance to non specific herbicids (6). But significant research is also presently geared towards engineering pest tolerance (insects, fungi, viruses, nematodes) and, at a longer range, improving drought and cold tolerances. Two transgenic plants will soon be on the market : one is a long-conservation tomato which resists the effect of ethylene accumulation during "ripening", the other is a variety of insect-resistant cotton. Although it is too soon to make solid predictions regarding the impact that cultivation of transgenic plants might have on agriculture, and hence on overall economy, this impact might become considerable at the turn of the century. But this will greatly depend upon the public attitude regarding genetically-modified organisms. Also, transgenesis might become a new mean to improve utilization of plants for non alimentary purposes by facilitating their bioconversion.

FINANCIAL ASPECTS

Before concluding, I should like to discuss macroeconomical aspects by refering to some figures from the biotechnology market.

For example, the market for recombinant pharmaceutical products today is in a range of 2-2,5 billion dollars, which means only 1,5% of the total market in this field. But, according to various experts, this percentage should grow quite appreciably at the turn of the century. Optimists can view a figure of up to 40%! Estimates of the world market for biotechnology as a whole (that is to say including all fields of activities) give variable figures depending upon the experts from 7 billion dollars (EC Senior Advisory Group in Biotechnology) up to 16 billion according to SRI ! However unprecise these data might be they show that the overall consequence of biotechnology for the present world economy might look very modest at a first sight if it is computed from the total sale figures. But it should be kept in mind that biotechnologies have greatly renewed, and continue to renew, many conventional procedures from industry and agriculture and that they should also provide an important stimulus to various branches of instrumentation.

The comparative levels at which funding of Research and Development in biotechnology has been achieved within the main industrialized blocks throughout the world, enlightens the problem from a different angle. In this respect, data concerning the year 1987 are provided by a recent issue of "Biofutur", a French monthly review on biotechnology (7). Over the 466 billion francs spent at this time on Research and Development in biotechnology (by adding up public and private sectors) 60% originated from North America, 28% from Europe, 10% from Japan, the rest, 2% coming from Australia and other countries. Canada, as we have learned during this meeting from Professor Brossard has embarked in a very ambitious, long range, but quite realistic biotechnology program with well established strategies organized throughout 4 main networks interconnecting various centers of excellence. Hence, it could very well be that the present distribution of efforts throughout the world be changing rapidly, not only because of Canada's recent federal decision to largely invest in "biotechnology" but also because of the rapidly growing economical impact of countries from South East Asia and China, notwithstanding certain Eastern and Central European countries.

POSITION OF THE EUROPEAN COMMUNITY

How does Europe, and more specifically the European economic community (EEC), situate itself regarding the biotechnology "race"?

Table 2 provides a summary of the successive well known EEC programs BEP, BAP and BRIDGE to which one should add on ECLAIR and CLAIR which are more generally related to the Agricultural and the Agrofood sectors respectively.

Table 2
EEC Biotech Programmes

BEP	Biomolecular Engineering Programme (1982 - 1986) - 15 MIO-ECUS
BAP	Biotechnology Action Programme (1985 - 1989) - 75 MIO-ECUS
BRIDGE	Biotechnology Research for Innovation Development and Growth in Europe (1990 - 1993) - 100 MIO-ECUS

Describing Europe-based achievements in the field, would deserve a full lecture, but a nice illustration can be found in the field of yeast genetics, particularly after the coordinated yeast sequencing program initiated by the EEC.

Yeast chromosome III (Table 3) is the longest piece of DNA and the first entire chromosome yet sequenced from an organism. It comprises 182 ORF's.

Table 3
Sequence of yeast chromosome III : The first entire chromosome sequence from any organism

- 182 ORF's
- 315 357 bp
- 35 laboratories (16 months)
- Within the framework of BAP (Biotechnology action programme) of the EEC

This work has involved the collaboration between 35 laboratories in Europe. At present sequencing of chromosomes 2 and 11 is also virtually completed. This European yeast genome program should have many interesting outcomes. The data should help clarifying some of the key functions that yeast genes play in the overall physiology of this organism (which manifests many features of higher eukaryotes). In addition, yeast industry is quite active and has been largely involved in the program. The fourth framework program of EEC which will begin in January 1994 with a four years total funding of about 1.2 billion ECUS for Life Sciences will continue providing important support to biotechnology.

In conclusion, it is clear (and this has been amply discussed at the meeting) that biotechnologists will have to match new challenges. On the one hand, these challenges are related to the kind of (new) problems scientists, engineers or industrialists have to tackle in present time society. For instance, there are many signs of resurgence of infectious diseases, AIDS being, by far, and unfortunately, the most illustrative example. But the recent burst of drug resistant bacteria particularly in Europe (tuberculosis) and the gradually preoccupating threat of parasitic diseases, all of these situations show how much efforts are expected from biotechnologists world-wide. This, inasmuch as progress made in cell immunology (cytokines...etc.) is opening many novel research areas. But these challenges are also of a different nature, due to changes in the attitudes of contemporary society as a whole. While there is a greater public awareness about Research and Technology there are also many new concerns! Everyone here is conscious of it. It does not suffice to say that we will provide useful tools and attempt to fight new diseases or find original solutions to the world hunger or to unemployement. These tools or approaches have to cope with the respect of environment, they have to be quality controlled and in accordance with very precise regulations or more stringent ethical rules. Hence, the demand for biotechnology will not only be increasing from a quantitative standpoint. It might very well become different in nature from what it was until now. This constitutes indeed a difficult, but very exciting, and, in a sense, noble challenge since all the participants in this European Congress regard the progress in life sciences as a major component of modern endeavour and culture, to the benefit of mankind.

REFERENCES

1. "A world-wide perspective of Biotechnology" by Prof. Karel Ch. A.M.Luyben, in Manpower & Training Needs for Biotechnology in Europe in the 90's, page 34.
2. "Biotechnology : Products in the U.S. Pipeline" from the Pharmaceutical Manufacturers Association (Communication division, 1100 15th street, NW, Washington, DC 20005) August 1991.
3. B.Law, "The future for food-related biotechnology" in "Industrial biotechnology International", Edit. R.Greenshields, pub. Sterling Publications Limited, London (1993) pp.85-88.
4. "Application of genetic engineering and molecular biology for seed improvement", European Postgraduate training network - operated by the Carlsberg Laboratory, Department of Physiology (Prof. D.Von Wettstein) Commission of the European Communities, DGXII, Directorate Biology, Division Biotechnology, BRIDGE (1993)",18.

5. J.Schell, "Plant biotechnology" : a powerful tool to use plant resources and to improve the environmental impact of agriculture" Alexander von Humboldt Stiftung, AVH-Magazin (1993) 61, 9-16.
6. E.Chasseray & J.Duesing, "Field trials of transgenic plants : an overview", AgroFood-Industry Hi-tech (July/August 1992) 5-10.
7. J.C.Pinon, "Financements nationaux de la R & D en Biotechnologie : les années 80", Biofutur, (mai 1990) 90, 107-133.

BASIC BIOTECHNOLOGY

Molecular
Cellular Biology

CHAPERONINS AND PROTEIN FOLDING

ECB6: Proceedings of the 6th European Congress on Biotechnology
edited by L. Alberghina, L. Frontali, P. Sensi

Protein folding by molecular chaperones

Franz-Ulrich Hartl

Cellular Biochemistry & Biophysics Program, Memorial Sloan-Kettering Cancer Center, 1275 York Avenue, New York, NY 10021, USA

Protein folding is a process of fundamental biological and also considerable biotechnological importance. It has become clear over recent years that cells contain a complex machinery, molecular chaperones and folding catalysts, that mediates the efficient folding of newly-synthesized polypeptide chains in various cellular compartments.These components may become useful in the hands of biotechnologists in the future, for example in aiding the *in vitro* renaturation of proteins that have been produced as inclusion bodies.

1. THE CONCEPT OF MOLECULAR CHAPERONE FUNCTION

It should be stressed that the discovery of molecular chaperones and their functions does not challenge the basic principle that the primary sequence of a protein contains all the information necessary to specify its three-dimensional structure [1, 2]. Molecular chaperones belong to a heterogeneous group of components. Many of them are stress proteins but have essential functions under normal growth conditions. They have to be distinguished from true folding catalysts such as protein disulfide isomerase and peptidyl prolyl isomerase, which accelerate specific reactions during folding. In contrast to the folding catalysts, molecular chaperones generally recognize unfolded or partially denatured proteins [for review see 3-5]. Their function lies predominantly in the prevention of off-pathway reactions such as the unproductive inter- and intramolecular association of proteins that leads to their aggregation. The physiological importance of this can easily be rationalized: Rapid spontaneous re-folding of purified proteins *in vitro* requires the presence of a complete polypeptide or at least a folding domain. Under cellular conditions, however, proteins emerge from ribosomes as unfolded chains which are restricted from forming stable tertiary structure. Unfolded proteins or partially folded intermediates expose hydrophobic surfaces and thus have the tendency to aggregate given the high cellular concentration of folding polypeptides [6-8]. It is therefore believed that productive folding *in vivo* first depends on the prevention of incorrect folding by molecular chaperones at a stage when the polypeptide chain is still nascent, i.e. prior to its release from the ribosome.

2. DIFFERENT MOLECULAR CHAPERONES CAN ACT SEQUENTIALLY

It is useful to distinguish between the following two principles of chaperone action: The function of preventing incorrect folding during synthesis or membrane translocation of a protein is carried out by chaperones acting as *monomers or dimers* such as the constitutively expressed members of the Hsp70 family of stress proteins. Once a sufficient length of polypeptide chain is available for folding, a second principle of chaperone action can take over: *Oligomeric chaperones* , such as the members of the Hsp60/GroEL (chaperonin) family, mediate folding to the native state. These high molecular weight doublering complexes provide an environment that protects the polypeptide chain from unproductive interactions during the folding process. Hsp70s and Hsp60s both have ATPase activity and release their bound substrate protein upon ATP hydrolysis. In case of the oligomeric Hsp60s, this release event translates into an efficient folding reaction (see below).

The functional cooperation of molecular chaperones of different classes and of different specificities in a sequential folding reaction has become apparent by studying the import of proteins into mitochondria [8,9]. These organelles take up most of their proteins from the cytosol as precursors that typically contain amino-terminal tergeting sequences. Precursor proteins are translocated across the two mitochondrial membranes in a rather extended conformation. This occurs at translocation contact sites where the two mitochondrial membranes are in close proximity. Hsp70 in the mitochondrial matrix interacts with the incoming, extended polypeptide chains [9]. This interaction may resemble that of cytosolic Hsp70 with nascent polypeptide chains emerging from ribosomes. It is required for efficient transloction of proteins but seems not to be sufficient for their subsequent folding. Mitochondrial Hsp60, the homologue of *E. coli* GroEL, receives the newly-imported polypeptide chains from Hsp70 and mediates their folding to the native state in an ATP-dependent process [10-12]. In this reaction, Hsp60 is regulated by a further stress protein, Hsp10, the mitochondrial homologue of *E. coli* GroES.

Using the purified Hsp70 and Hsp60 homologous stress proteins of *E. coli*, DnaK and GroEL, we have recently reconstructed a folding reaction in which these molecular chaperones function sequentially [13]. This process required the heat-shock proteins DnaJ and GrpE in addition, which regulate the ATPase function of the Hsp70 protein DnaK [14].The overall reaction can be subdivided into the following three steps: 1. Stabilization of an aggregation-prone folding intermediate by DnaK and DnaJ. 2. GrpE-dependent transfer of this intermediate to the chaperonin GroEL. 3. ATP- and GroES-dependent folding to the native state at GroEL. The experimental approach was based on first unfolding a substrate protein, such as the 35 kDa monomeric enzyme rhodanese, in 6M guanidinium-Cl and then diluting it into buffer solution containing the various chaperone proteins. Rhodanese was chosen as the substrate for these studies because of its pronounced tendency to aggregate upon attempted refolding by dilution from denaturant *in vitro* [15]. DnaK and DnaJ efficiently prevented the aggregation of unfolded rhodanese and stabilized the protein presumably in the conformation of an early folding intermediate that lacks stable tertiary structure. DnaK/DnaJ stabilized rhodanese was not productive for folding to the native state when Mg-ATP was added but instead remained tightly bound to these chaperones. The small heat-shock protein GrpE and Mg-ATP was necessary to catalyze the release of rhodanese from DnaK/DnaJ. When the Hsp60 homologue GroEL and its co-fctor GroES were also present, this release resulted in efficient reactivation of rhodanese in a GroEL, GroES and ATP-dependent reaction (see below). The possibility of efficient transfer of a partially folded substrate protein from DnaK/DnaJ to GroEL also allowed us to test the ability of GroEL to carry out multiple rounds of rhodanese folding, a property of potential biotechnological interest. GroEL was added in substoichiometric concentrations to a solution containing an excess of rhodanese stabilized by DnaK and DnaJ. Upon addition of GrpE, efficient reactivation was observed whereby one GroEL 14-mer complex was able to carry out 4-6 rounds of rhodanese folding [13].

3. DOUBLERING CHAPERONINS MEDIATE PROTEIN FOLDING

The paradigm of molecular chaperones that actually mediate folding to the native state in an ATP-dependent reaction are the members of the Hsp60/GroEL family, the so-called "chaperonins" [16]. These proteins have been highly conserved during evolution. They are found in the bacterial cytosol (GroEL of E. coli), in the matrix space of mitochondria (mitochondrial Hsp60) and in the stromal space of chloroplasts (rubisco binding protein) [for review see 5]. The chaperonins are active as ~800 kDa complexes consisting of two stacked rings of seven 60 kDa subunits each. They have a weak ATPase activity that requires functional regulation by a co-chaperonin (GroES of E. coli; Hsp10 of mitochondria), a single ring of seven 10 kD subunits that forms a 1:1 complex with the chaperonin cylinder [17,18]. Convincing evidence that protein assembly in general might require chaperonins came from the study of *mif4* mutant yeast which harbor a temperature-sensitive mitochondrial Hsp60. In these yeast import of proteins into the organelle occurs normally but newly imported proteins fail to assemble [10, 19]. The kinetic and spatial separation of protein synthesis and folding/assembly in this experimental system served subsequently to demonstrate that the mitochondrial chaperonin supports even the folding of simple monomeric proteins, such as dihydrofolate reductase

(DHFR) [11, 20], which are well able to fold spontaneously *in vitro*. Since then it has become apparent that the basic role of the chaperonins is indeed the folding of protein monomers although a specific function in subunit association cannot be ruled out [reviewed in 5].

4. RECONSTITUTION OF CHAPERONIN FUNCTION

Important insight into the complex molecular mechanism of chaperonin action has been obtained through the reconstitution of chaperonin function *in vitro* [17, 18, 21-24]. Most of these studies have been carried out with the Hsp60 of *E. coli*, GroEL. The GroEL oligomer has been shown to bind only one or two molecules of unfolded polypeptide via as yet undefined structural elements. Most efficient binding is usually achieved by diluting unfolded polypeptide from strong denaturant into chaperonin containing buffer solution in the absence of hydrolyzable ATP and the co-factor GroES. Binding to the chaperonin prevents the aggregation of unfolded proteins such as rhodanses, and GroEL is able to mediate the efficient refolding of this protein [18, 23]. This reaction is strictly dependent on ATP-hydrolysis and GroES. In contrast, other substrate proteins, such as DHFR, are independent of GroES and ATP hydrolysis for their GroEL-mediated folding [18, 24]. These proteins are characterized by a low tendency to aggregate under optimal re-folding conditions *in vitro*. A common functional principle of chaperonin action explaining both types of reactions has yet to be established.

GroEL appears to bind its substrate protein in a loosely folded conformation resembling the "compact intermediate" or "molten globule"-like state [18] which may contain native secondary structure but lacks stable tertiary structure [7]. In a collaboration with W. Baumeister and coworkers we have recently analyzed structural and functional properties of the interaction between GroEL, its co-chaperonin GroES and substrate polypeptide. The stoichiometry of GroEL-14mer to GroES-7mer in the holo-chaperonin complex was found to be 1:1 indicating an asymmetrical structure for the functional chaperonin unit. Electron microscopic image analysis of GroEL-GroES confirmed the asymmetrical shape of the holo-chaperonin complex. We found that GroES binds to one end-surface of the GroEL cylinder. This triggers conformational changes in both the GroES adjacent and in the opposite end of the GroEL cylinder thus apparently lowering the affinity for binding of a second GroES [25].

The 14.5 X 16 nm chaperonin cylinder has a central cavity of about 6 nm in diameter. Bound substrate protein could be detected within this cavity, again by electron microscopy [25]. Based on this finding we proposed that chaperonin-mediated protein folding may occur while the substrate polypeptide remains in the central space of the GroEL cylinder protected from unproductive intercactions with other folding polypeptides [see also 5]. The central cavity of GroEL could indeed provide space for proteins of up to 90 kDa [26]. A substrate protein could make multiple contacts with the GroEL subunits of one seven-member ring and folding would be achieved by the cooperative, ATP-dependent release of these binding elements. By regulating the K^+-dependent ATPase of GroEL, GroES would coordinate this function [17, 18].

5. BIOTECHNOLOGICAL APPLICATION

Molecular chaperones are not yet used for the commercial production of proteins but several biotechnological companies are testing the possibilities these fascinating components may offer. Chaperone proteins, such as the GroEL/GroES system of *E. coli*, are quite robust and can be isolated in large amounts from overproducing strains. Homologues of GroEL from thermophilic bacteria and archaebacteria are also available that function in protein folding at high temperatures. How could chaperones possibly be useful at a technical scale? The following two possibilities are obvious: 1. Similar to the refolding experiments described above, purified chaperones may be used to increase the efficiency of refolding that is currently performed in most cases by diluting unfolded protein out of denaturant. Procedures to immobilize chaperones in a functional state may have to be developped in order to allow the recycling of chaperone protein and its convenient removal from the product. 2. Bacterial strains and yeast strains may be engineered which have elevated levels of certain chaperones thus increasing the folding efficiency of overproduced protein and reducing its propensity to aggregate in inclusion bodies. One way or the other, chaperones will probably soon be part of the standard repertoir of methods used by the biotechnologist concerned with protein folding problems.

REFERENCES

1. C.B. Anfinsen. Science 181 (1973) 223.
2. R. Jaenicke. Biochem., 30 (1991) 3147.
3. R.J. Ellis. Nature 328 (1987) 378.
4. M.-J. Gething and J. Sambrook. Nature 355 (1992) 33.
5. J.P. Hendrick and F.U. Hartl. Annu. Rev. Biochem., 62 (1993) 349.
6. T.E. Creighton. Biochem., 270 (1990) 1.
7. H. Christensen and R.H. Pain. Eur. Biophys. J., 19 (1991) 221.
8. F.-U. Hartl, J. Martin, and W. Neupert. Annu. Rev. Biophys. Biomol. Struct., 21 (1992) 293.
9. W. Neupert, F.U. Hartl, E. Craig, N. Pfanner. Cell 63 (1990) 447.
10. M.Y. Cheng, F.U. Hartl, J. Martin, R.A. Pollock, F. Kalousek, W. Neupert, E.M. Hallberg, R.L. Hallberg, and A.L. Horwich. Nature 337 (1989) 620.
11. J. Ostermann, A.L. Horwich, W. Neupert, and F.U. Hartl. Nature 341 (1989)125.

12. U. Manning-Krieg, P.E. Scherer, and G. Schatz. EMBO J., 10 (1991) 3273.
13. T. Langer, C. Lu, H. Echols, J. Flanagan, M.K. Hayer, and F.U. Hartl. Nature 356 (1992) 683.
14. C. Georgopoulos, D. Ang, K. Liberek, and M. Zylicz. In: R. Morimoto, A. Tissieres, C. Georgopoulos (eds.), Stress proteins in biology and medicine. Cold Spring Harbor Laboratory, Cold Spring Harbor NY, 1990.
15. S. Tandon and P.M. Horowitz. J. Biol. Chem., 261 (1986) 15615.
16. S.M. Hemmingsen, C. Woolford, S.M.v.d. Vies, K. Tilly, D.T. Dennis, C.P. Georgopoulos, R.W. Hendrix, and R.J. Ellis. Nature 333 (1988) 330.
17. P.V. Viitanen, T.H. Lubben, J. Reed, P.Goloubinoff, D.P. O'Keefe, J. Reed, P. Goloubinoff, and G.H. Lorimer. Biochem., 29 (1990) 5665.
18. J. Martin, T. Langer, R. Boteva, A. Schramel, A.L. Horwich, and F.U. Hartl. Nature 352 (1991) 36.
19. M.Y. Cheng, F.U. Hartl, and A.L. Horwich. Nature 348 (1990) 445.
20. J. Martin, A.L. Horwich, and F.U. Hartl. Science 258 (1992) 995.
21. P. Goloubinoff, J.T. Christeller, A.A. Gatenby, and G.H. Lorimer. Nature 342 (1989) 884.
22. J. Buchner, M. Schmidt, M. Fuchs, R. Jaenicke, R. Rudolph, F.X. Schmid, and T. Kiefhaber. Biochem., 30 (1991) 1586.
23. J.A. Mendoza, E. Rogers, G.H. Lorimer, and P.M. Horowitz. J. Biol. Chem., 266 (1991) 13044.
24. P.V. Viitanen, G.K. Donaldson, G.H. Lorimer, T.H. Lubben, and A.A. Gatenby. Biochem., 30 (1991) 9716.
25. T. Langer, G. Pfeifer, J. Martin, W. Baumeister, and F.U. Hartl. EMBO J., 11 (1992) 4757.
26. K. Braig, M. Simon, F. Furuya, J.F. Hainfield, and A.L. Horwich. Proc. Natl. Acad. Sci., 90 (1993) 3978.

ECB6: Proceedings of the 6th European Congress on Biotechnology
edited by L. Alberghina, L. Frontali, P. Sensi

Refolding Recombinant Proteins - Protein Chemistry vs Process Engineering

Julian B. Chaudhuri

School of Chemical Engineering, University of Bath, Bath, Avon BA2 7AY, UK.

1. INTRODUCTION

Advances in genetic engineering have made the production of biologically active mammalian proteins from microbial cells a routine procedure. The bacterium *E coli* has been the most common host cell for production of rDNA proteins, and is likely to remain in use for some time to come. *E coli* produces many recombinant proteins as solid, intracellular protein aggregates (or inclusion bodies), unglycosylated, but at very high expression levels. The protein within the inclusion bodies is generally inactive. There is, therefore, a need to refold the denatured protein to its native structure to regain biological activity.

Inclusion bodies are visible by light microscope and appear refractile to light. There is generally one inclusion body per bacterial cell. The mass and number of inclusion bodies is proportional to the product level. Within the inclusion body the recombinant protein may accumulate at levels of greater than 50% of the total cell protein. Other species present in small amounts may include other *E coli* proteins such as RNA polymerase, nucleic acids, and lipopolysaccharides (Kane and Hartley, 1991). The purity of the inclusion body protein may be greater than 60%. Production of proteins by an inclusion body route has the advantage that the product may be isolated in a highly purified, solid, and concentrated state by simple physical operations.

2. REFOLDING RECOMBINANT PROTEINS

2.1 Recovery of inclusion bodies

Following growth of *E coli* cells the inclusion bodies are released by high pressure homogenisation and separated from the cell debris by centrifugation. Contaminating *E coli* proteins are removed by washing the inclusion bodies with chemicals including urea, Triton X-100, and EDTA. This step removes any proteases present and can lead to increases in the yield of active protein (Babitt *et al*, 1990).

2.2 Inclusion body solubilisation

The protein aggregates are solubilised in strong denaturants such as urea (5-10 M) or guanidine hydrochloride (GuHCl) (4-7 M). These chaotropic agents disrupt the non-covalent bonds between polypeptide chains. The use of detergents, at concentrations above their critical micelle concentration (typically 0.01-2% wt), may also be used for aggregate solubilisation (Lu *et al*, 1992).

During solubilisation, the disulphide bonds are disrupted and reduced to give a polypeptide chain. This is carried out using thiol reagents such as β-mercaptoethanol or dithiothreitol at concentrations ranging from 10-500 mM. Alternatively a sulphitolysis reaction can be used to break the disulphide bonds and protect the free cysteine residues. Addition of sodium sulphite and sodium tetrathionate (a mild oxidising agent) results in the S-sulphonate form of the protein. The sulphonate groups are removed later during the refolding step by the addition of thiol reagents.

If the recombinant protein is expressed at low levels (<5%) it is sometimes necessary to purify the protein at this stage. Usually hydrophobic interaction or reversed phase chromatography methods are employed.

2.3 Refolding by denaturant removal

Refolding is initiated by removal of the denaturant to allow the molecule to fold into its native structure. This is achieved by dilution, dialysis, or diafiltration. The rate of folding is protein specific, and may occur spontaneously or take many days. Protein folding proceeds according to a preordained pathway, via the formation of intermediate protein structures. The formation of intermediates is usually fast, with the progression to the native structure being rate-limiting. During refolding the solubilised, partially folded protein molecules often aggregate. The aggregation reaction is irreversible and is second order or higher; refolding is first order. Because of this the denatured protein concentration is kept as low as possible (typically 10-100 μg/ml) to minimise aggregation and to promote refolding.

Dilution of the protein is the simplest refolding method and has been used widely: antibody fragments (Buchner and Rudolph, 1991); M-CSF (Halenbeck *et al*, 1989); and pro-urokinase (Orsini *et al*, 1991) The disadvantage of dilution is the large processing volumes involved increasing costs.

Dialysis has been used for some proteins including citrate synthase (Kelly and Price, 1992) and urokinase (Builder and Ogez, 1986). This process is mass transfer controlled and is very slow, and thus is not used in production scale processes.

Diafiltration is used for buffer exchange. This keeps the protein at a constant concentration whilst allowing reduction of the denaturant concentration. This method is very suitable for production processes. Proteins refolded this way include carbonic anhydrase (Vicik and De Bernadez-Clark, 1991) and prorennin (Builder and Ogez, 1986).

2.4 Reformation of disulphide bonds

The majority of eucaryotic proteins of interest contain disulphide bridges. The disulphide bonds must be correctly reformed to recover the native structure. In a protein with more than one disulphide bond there is the opportunity for the formation of incorrect disulphide bonds.

Disulphide bond formation occurs under oxidising conditions. The correct redox potential is maintained by the use of reducing agents and oxidised thiol compounds. Disulphide bond formation is achieved by allowing the disulphide bonds to rearrange by thiol-disulphide exchange ("oxido-shuffling"). The most common reagent used is the redox couple of reduced and oxidised glutathione.

After refolding, the protein is further purified and processed as required.

3. NOVEL TECHNIQUES FOR PROTEIN REFOLDING

The basis of commercial refolding processes have emerged from the fundamental studies on protein refolding. These processes employ similar methods and strategies as those developed for laboratory-scale. In many cases, the laboratory-scale protocols have been roughly scaled to production levels. Recently, novel approaches to refolding proteins have been reported

The use of fusion proteins removes the requirement to recover inclusion bodies and solublllse the protein. This approach has been reported by Samuelsson *et al* (1991) who have refolded insulin-like growth factor 1 at concentrations of 1-2 mg/ml. In this work, a fusion prolein conslsting of the insulin-like growth factor and the IgG binding domain of Protein A were expressed and secreted into the medium. The fusion protein was separated from any degraded fusion protein and a gel filtration step used to remove the monomeric protein from the aggregates The multimeric proteins were then diluted and refolded and then separated from the monomers. The fusion protein was cleaved with hydroxylamine and purified by HPLC

Another recently reported method utilised monoclonal antibodies (MAbs) raised to the native protein to assist the refolding process (Carlson and Yarmush, 1992). The S-protein fragment of ribonuclease A was reduced and mixed with a MAb raised to the native S-protein. The correct redox potential for refolding was provided using glutathione. Refolding was terminated and the S-protein recovered from the MAb and any misfolded protein. In the absence of the MAb the recovery of enzymatic activity was 13%, whereas it rose to 54% in the presence of antibody. It is likely that the antibody acted as a "refolding template "

In vitro refolding is subject to unwanted side reactions such as aggregation and the formatlon of incorrect disulphide bridges. During *in vivo* folding these reactions are suppressed by chaperones or helper proteins (Gething and Sambrook, 1992). Such proteins include the heat shock proteins DnaK, GroeE and the helper protein protein disulphide isomerase (PDI). Much activity has been directed at the use of these proteins for refolding. These approaches include the co-expression of chaperones with target proteins and the use of purified chaperones for *in vitro* refolding (Buchner *et al*, 1992).

4. PROCESS ENGINEERING STRATEGIES FOR PROTEIN REFOLDING

Considerable effort is being directed at biophysical aspects of refolding (*eg* pathway elucidation, equilibrium and kinetic studies). However, the process engineering of protein refolding is very much an empirical science. Although studies have been made on a variety of proteins there is no rational basis for designing a refolding process. The engineering challenge is, therefore, to produce fully active recombinant protein from inclusion bodies m large quantities, in a shorter time, and at high eoncentration.

Biochemical engineering strategies that may be of use in refolding include: the use of mathematical models to optimise the refolding process and to facilitate its scale-up; the study of process interactions and integration in chemical engineering has given rise to increases in productivity and may be of relevarice here; and the development of completely novel refolding strategies. There have been a few studies addressing process engineering requirements for refolding:

The relevance of a mathematical modelling approach has been illustrated by Vicik and De Bernardez-Clark (1991). A mathematical model was formulated which accounted for the kinetics of the refolding pathway of human carbonic anhydrase B. The model was coupled to the equations governing diafiltration. The kinetic constants in the refolding pathway are a function of the denaturant concentration. Thus, the rate at which these agents are removed will affect the rate of accumulation of intermediates and thus aggregation. The optimisation results indicated that optimal refolding occurred at the lower urea concentration and also that there is an optimum filtration rate to achieve the highest refolding yield. The results show that a low diafiltration rate is required: although leading to longer process times, this will be outweighed by increased yield

The use of polyethylene glycol (PEG) to enhance refolding yields has been reported (Cleland and Wang, 1990; Cleland *et al*, 1992). It was found that using PEG concentrations of 3 g/l in the refolding buffer the refolding rate of carbonic anhydrase B was increased threefold. This method is believed to work as the PEG interacts with the refolding intermediate preventing it from aggregating. This allowed refolding to occur without hinderance. (Protein concentration was 0.5 mg/ml). This approach has been applied to refolding the proteins carbonic anhydrase B, rec DNAse, t-PA and interferon-γ.

Because aggregation of the folding intermediates reduces overall yield refolding proteins in isolation from each other may give increases in productivity. The use of reverse micelle microenvironments to isolate and refold denatured proteins has been studied (Hagen *et al*, 1990). Reverse micelles offer an environment in which the proteins may refold in isolation from one another thereby minimising aggregation. The results reported 100% renaturation of denatured ribonuclease-A refolded in the reverse micelles. Although successful with ribonuclease-A, this method has not been investigated with other proteins especially those that readily aggregate.

REFERENCES

Babbitt, P C, West, B L, Buechter, D D, Kuntz, I D and Kenyon, G L (1990). *Bio/Technol*, **8**, 945-949.

Buchner, J and Rudolph, R (1991). *Bio/Technol*, **9**, 157-162.

Buchner, J, Brinkmann, U and Pastan (1992). *Bio/Technol*, **10**, 682-685.

Builder, S E and Ogez, J R (1986). *U S Pat 4,620,948*.

Carlson, J D and Yarmush, M L (1992). *Bio/Technol*, **10**, 86-91.

Cleland, J L and Wang D I C (1990). *Bio/Technol*, **8**, 1274-1278.

Cleland, J L, Builder, S E, Swartz, J R, Winkler, M, Chang, J Y and Wang D I C (1992). *Bio/Technol*, **10**, 1013-1019.

Gething, M and Sambrook, J (1992). *Nature*, **355**, 33-44.

Hagen, A J, Hatton, T A and Wang, D I C (1990). *Biotechnol Bioeng*, **35**, 955-965.

Halenbeck, R, Kawasaki, E, Wrin, J and Koths, K (1989). *Bio/Technol*, **7**, 710-715.

Kane, J F and Hartley, D L (1991). In *Purification and Analysis of Recombinant Proteins*, eds Seetharam, R and Sharma, S K, pp 121-146, Marcel Dekker Inc.

Kelly, S M and Price N C (1992). *Int J Biochem*, **24**, 627-630.

Lu, H S, Clogston, C L, Narhi, L O, Merewether, L A, Pearl, W R, and Boone, T C (1992). *J Biol Chem*, **267**, 8770-8777.

Orsini, G, Brandazza, A, Sarmientios, P, Molinari, A, Lansen, J and Cauet, G (1991). *Eur J Biochem*, **195**, 691-697.

Samuelsson, E, Wadensten, H, Hartmanis, M, Moks, T and Uhlen, M (1991). *Bio/Technol*, **9**, 363-366.

Vicik, S and De Bernardez-Clark, E (1991). In *Protein Refolding*, eds De Bernardez-Clark, E and Georgiou, G, ACS Symposium Series, **470**, pp 180-196.

ECB6: Proceedings of the 6th European Congress on Biotechnology
edited by L. Alberghina, L. Frontali, P. Sensi

A protein from the hyperthermophilic archaebacterium Sulfolobus solfataricus which assists protein folding at high temperature

A. Guagliardi[*], L. Cerchia[*], M. Rossi[*§] and S. Bartolucci[*]

[*]Dipartimento di Chimica Organica e Biologica, Università di Napoli
Via Mezzocannone 16, 80134 Napoli, Italy

[§]Istituto di Biochimica delle Proteine ed Enzimologia, C.N.R.
Via Marconi 10, 80127 Napoli, Italy

DBF (Disulfide Bond Forming) enzyme from the hyperthermophilic archaebacterium *Sulfolobus solfataricus* greatly enhances the refolding at 30°C of denatured and reduced RNAse A (1). Here we show that DBF affects in an ATP-dependent manner the *in vitro* refolding at 50°C of a thermostable alcohol dehydrogenase. The role of the proteins which assist protein folding at high temperature in the applied biocatalysis is discussed.

1.INTRODUCTION

At least two classes of proteins have been identified that assist protein folding and assembly (2-4, for reviews). The first class includes the catalysts that accelerate the rate-limiting steps of folding, namely protein disulphide isomerase, peptidyl prolyl *cis-trans* isomerase, and thioredoxin. The second class includes the proteins known as molecular chaperones that assist protein folding by preventing the formation of incorrect structures through the ATP-dependent interaction with unfolded, partially folded, or misfolded polypeptide chains.

An enzyme designated DBF (Disulfide Bond Forming) which greatly enhances the refolding of RNAse A has been purified from crude extracts of the hyperthermophilic archaebacterium *Sulfolobus solfataricus* strain MT-4 which grows optimally at 87°C and pH 3.0 (1). Here we show that DBF affects the *in vitro* refolding at 50°C of the alcohol dehydrogenase (ADH) from *S.solfataricus* (a dimer of 35,000 Da subunits) which does not contain disulfide bonds (5), and that this effect is ATP-dependent.

2. EXPERIMENTAL

2.1. Materials

ATP, NAD and guanidine HCl were purchased from Sigma. DBF was purified according to the published procedure (1). ADH (EC 1.1.1.1) from *S.solfataricus* (5) was kindly provided by Dr.C.A.Raia.

2.2. Assay of ADH Activity

ADH was assayed at 60°C in 25 mM barbital-HCl buffer, pH 8.0, containing 2 mM NAD and 5 mM benzyl alcohol (final volume 1 ml).

2.3. Unfolding and refolding of ADH

ADH was incubated overnight at 37°C in 2.4 M guanidine HCl (600 μg ml^{-1} of denaturation mixture). The refolding was obtained at 50°C by diluting 120-fold an aliquot from the denaturation mixture into 0.1 M sodium phosphate buffer, pH 8.0, 5 μM $ZnCl_2$, in the absence or in the presence of DBF (2 μg). After 50 min-incubation, 0.5 mM Mg-ATP was added into the mixtures. At designated time points, ADH activity was assayed on aliquots withdrawn from the mixtures. The regain of activity was reported as the relative renaturation calculated with respect to the activity of the same amount of native ADH, assumed as 100%.

3. RESULTS AND DISCUSSION

Upon dilution and incubation for 20 min at 50°C, the chemically denatured ADH spontaneously regained 25% of its activity; no further increase was obtained within the following 30 min of incubation (Fig.1).

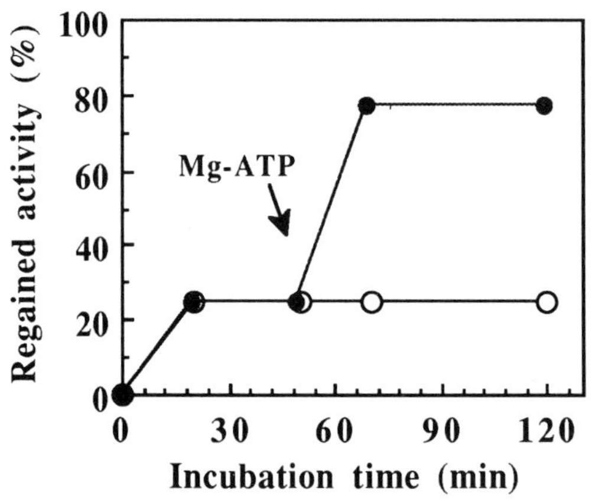

Fig. 1. Spontaneous (o) and DBF-dependent (●) refolding of ADH.

The spontaneous refolding of ADH was not influenced by the addition of bovine serum albumin into the renaturation mixture, nor did DBF have any effect on the native ADH. The presence of DBF in the renaturation mixture apparently did not influence the spontaneous refolding of ADH; however, the addition of Mg-ATP into the mixture resulted in a higher activity regain (77%) compared to that of the spontaneous refolding. The addition of Mg-ATP in the mixture without DBF (the spontaneous event) had no effect on the ADH refolding.

Because the binding and release of proteins from molecular chaperones is often associated with ATP hydrolysis, studies are under way to test whether DBF displays ATPase activity.

DBF enhances the refolding of the denatured and reduced form of RNAse A carried out at 30°C (Fig.2). The fact that ADH does not contain disulfide bonds in its structure suggests that DBF may have a general role as a "helper" protein in assisting protein folding. Strikingly, it has been recently reported that the peptidyl prolyl *cis-trans* isomerase explicates an isomerase and a chaperone activity in the refolding of carbonic anhydrase (6).

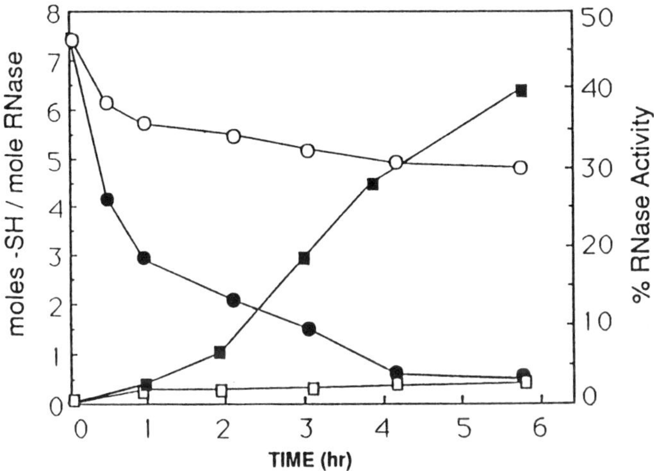

Fig. 2. Spontaneous (open symbols) and DBF-dependent (closed symbols) renaturation of reduced and denatured RNAse A: the circles refer to the disappearance of the free -SH groups, the squares to the regain of activity. The Figure was redrawn from Ref.1

Since some molecular chaperones are active as complexes of two proteins -as for the *E.coli* GroE complex consisting of GroEL and GroES- it cannot be excluded that DBF can work *in vivo* in cooperation with another molecule, and experiments are in progress to obtain more information on this possibility.

Molecular chaperones from some thermophilic microorganisms have been identified (7-10). The chaperone protein from the archaebacterium *Sulfolobus shibatae* (10) displays a high degree of aminoacid homology with TCP1, the cytoplasmic t-complex polypeptide-1 which is involved in protein folding (11,12). The archaebacterial TF-55 and the eukaryotic TCP1 have been suggested to be members of a new class of molecular chaperones (13). The study of the protein-assisted refolding at high temperature could open new prospects in the understanding of the patterns and energetics of protein folding. On the other hand, the practical exploitation of the proteins which assist protein folding at high temperature is easily foreseen. These proteins could contribute to overcame a major problem of the applied biocatalysis: the denaturation and/or inactivation of a protein following some purification manipulations, the immobilization, or the expression of the recombinant protein (14,15). The overproduction of a foreign protein can lead to the formation of inactive aggregates (inclusion bodies) from which the recovery of active material upon treatment with chaotropic agents is often discouraging.

This work was partially supported by Progetto Finalizzato Biotecnologie e Biostrumentazione, C.N.R.

REFERENCES

1. A. Guagliardi, L. Cerchia, M. De Rosa, M. Rossi and S. Bartolucci, FEBS Lett., 303 (1992) 27.
2. J.E. Rothman, Cell, 59 (1989) 591.
3. G. Fischer and F.X. Schmid, Biochemistry, 29 (1990) 2205
4. M. Gething and J. Sambrook, Nature, 355 (1992) 33.
5. S. Ammendola, C.A. Raia, C. Caruso, L. Camardella, S. D'Auria, M. De Rosa, M. Rossi *et al.* Biochemistry, 31 (1992) 12514.
6. P. Freskgard, N.Bergenhem, B. Jonsson, M. Svensson and U. Carlsson, Science, 258 (1992) 466.
7. H. Tamada, T. Ohta, T. Hamamoto, Y. Otawara-Hamamoto, M. Yanagi, H. Hiraiwa *et al.* Biochem. Biophys. Res. Commun., 179 (1991) 565.
8. H. Taguchi, J. Konishi, N.Ishii and M. Yoshida, J. Biol. Chem., 266 (1991) 22411.
9. N.M. Lissin, S.E. Sedelnikova and S.N. Ryazantsev, FEBS Lett., 311 (1992) 22.
10. J.D. Trent, E. Nimmesgern, J.S. Wall, F.U. Hartl and A.L. Horwich, Nature, 354 (1991) 490.
11. M.B. Yaffe, G.W. Farr, D. Miklos, A.L. Horwich, M.L. Sternlicht and H. Sternlicht, Nature, 358 (1992) 245.
12. V.A. Lewis, G.M. Hynes, D. Zheng, H. Saibil and K. Willison, Nature, 358 (1992) 249.
13. J. Ellis, Nature, 358 (1992) 191.
14. A. Gatenby, P.V. Viitanen and G.H. Lorimer, TIBTECH 8 (1990) 354.
15. A. Horwich, W. Neupert and F.U. Hartl, TIBTECH, 8 (1990) 126.

In vitro complex formation between DnaK and some recombinant proteins produced in *Escherichia coli*

K. Gustavsson, A. Veide and S.-O. Enfors

Department of Biochemistry and Biotechnology, Royal Institute of Technology
S-100 44 Stockholm, SWEDEN

Complex formation between the *Escherichia coli* heat-shock protein DnaK and some recombinant proteins have been studied. The kinetics of the complex formation between a recombinant staphylococcal protein A with a truncated C domain, pRIT2 protein A, and DnaK was studied *in vitro*. The half-time constant for this reaction varied between 4 and 15 minutes, depending on whether ATP was present or not. Though ATP dissociates an already formed DnaK- protein A complex, the complex formation rate was increased by ATP.

When the ratio between DnaK and pRIT2 protein A was gradually increased, a new complex was formed. This second complex consisted of more DnaK per pRIT2 protein A.

Complex formation studies with a series of proteins derived from staphylococcal protein A, the ZZT proteins, showed that binding of DnaK to the protein, as well as proteolytic sensitivity, increased when the number of a tetrapeptide unit, composed of ala-trp-trp-pro, was increased.

1. INTRODUCTION

The *E. coli* heat-shock protein DnaK is a chaperone belonging to the Hsp70 family. Under normal conditions the level of DnaK is low but it is increased under certain stress conditions. Chaperones are generally recognised as relatively unspecific helper proteins in folding an assembly of other proteins. DnaK has been ascribed several functions in the intracellular processing of proteins, but it has also been associated with proteolysis [1].

We have found that DnaK binds to different versions of truncated staphylococcal protein A when produced in *E. coli*. This was utilised to design a purification method for DnaK, based on IgG-affinity purification of the complex, followed by selective elution of the DnaK. This DnaK has been used for *in vitro* complex formation with different derivatives of staphylococcal protein A, in order to investigate the impact of target protein structure on the DnaK-binding.

2. MATERIALS AND METHODS

DnaK and pRIT 2 protein A were produced in *E. coli* RR1ΔM15 containing two plasmids, pRITcI857 and pRIT2, encoding the temperature sensitive lambda λP_R repressor and a truncated protein A, respectively [2]. The cells were cultivated in a fermentor on complex medium at 30°C at constant pH. When OD_{610} was about 10 the temperature was raised to 42°C for four hours to induce pRIT2 protein A production. The cells were harvested by centrifugation and disintegrated. Cell debris was removed by centrifugation and the supernatant was diluted and loaded on an IgG-Sepharose column. DnaK was selectively eluted from the column either by ATP and magnesium or by a pH 5.4 buffer before the pRIT2 protein A was eluted at pH 3.2. The DnaK and the pRIT2 protein A fractions were lyophilized.

Protein ZZ is an IgG-binding protein composed of two modified B domains of staphylococcal protein A. It was modified by addition of a 17 amino acids linker region (resulting in protein ZZT0) and insertion of 1-3 units of the tetrapeptide ala-trp-trp-pro into the linker region (yielding ZZT1, ZZT2 and ZZT3, respectively). Design and production of these proteins are described elsewhere [3]. Protein ZZC is obtained by hydroxylamine cleavage of the fusion protein ZZ-IGF-I [4] and it was kindly provided by Christina Kalderén (Kabi Pharmacia, Stockholm).

Analysis of the complex formation was performed with native PAGE with Pharmacia Phast system using 10-15% gradient gels. After staining with Coomassie Brilliant Blue the gels were scanned at 595 nm.

3. RESULTS AND DISCUSSION

3.1 Kinetics of DnaK- protein A complex formation.

Purified samples of DnaK and pRIT2 protein A were mixed and incubated at 37 °C for 60 minutes. Figure 1 shows a scan of a native PAGE gradient gel of the purified components and the mixture, respectively. A slowly migrating compound, representing a complex between the two molecules, is observed in the mixture of the two components.

To investigate the kinetics of this complex formation, a series of samples were mixed and incubated for different periods of times before the samples were analysed with native PAGE as described above. The peak areas were integrated and the relative peak area was plotted against time (Figure 2). DnaK has a weak ATPase activity which seems to be involved in many of DnaKs' cellular functions, but ATP is also known to dissociate DnaK-protein complexes. It was therefore interesting to see in which way ATP influences the complex formation. Thus, the complex formation was performed both with 5 mM ATP and without ATP in the buffer. The DnaK used for these experiments was produced by elution at a lower pH to minimise the ATP content in the DnaK preparation.

As can be seen in Figure 2 the complex formation is not immediate after the mixing. A maximum value of 50-60% for the complex formed is reached after about one hour when ATP is not present. The complex formation was fitted to the equation $A = A_{max}\, t/(t+k_t)$, where A is the time (t) dependent quantity of complex, A_{max} is the maximal complex formation and k_t is a half-time constant. Without ATP the half-time constant was 15 minutes. This value was reduced to 4 min by the addition of 3 mmol ATP. Both A_{max} and time constants varied between different experiments but the tendency is always the same: ATP increases the rate of complex formation but has little or no effect on the amount of complex at equilibrium.

Since the maximum relative peak area reached a value of 50-60%, not all of the initial DnaK and pRIT2 protein A formed a complex. If increasing amount of DnaK was added to a fixed amount of pRIT2 protein A, almost all pRIT2 protein A was titrated into complex. (We assume that the non constant protein A value, seen in fig 3, is caused by an increasing DnaK background with increasing distance from the migration origin. We have not been able to compensate for this background.) Figure 3 shows that, as the DnaK concentration in the sample is increased, a new complex with lower migration rate was formed.This new complex grew only slightly with increasing DnaK to protein A ratio until a critical ratio was reached. A possible explanation for this sudden increase in this second complex concentration would be that when a molar excess of DnaK is present, more than one DnaK molecule might associate with each pRIT2 protein A molecule. This would explain both the sudden increase and the lower migration rate of the second complex.

3.2 Complex formation with other proteins

To investigate complex formation with some other proteins, DnaK was incubated with a group of proteins, all derived from protein ZZ. The investigated proteins were ZZC, ZZT0, ZZT1, ZZT2 and ZZT3. No DnaK was observed in complex with ZZC and only very small amount was found in complex with ZZT0 and ZZT1. With ZZT2 and ZZT3, however, considerable complex formation was observed. The binding constants of these complexes seemed to be lower than that of the protein A complexes, since the native PAGE produced a smeared band between the ZZ-protein and the main complex band.

Other studies on the nature of the ZZT series have shown that this DnaK-binding also correlate with the proteolytic sensitivity of these proteins in *E.coli*: ZZT0 is essentially stable and ZZT1 shows only vague proteolysis while ZZT2 and ZZT3 are very susceptible to proteolysis. Thus, insertion of increasing amounts of ala-trp-trp-pro in the ZZ protein induces proteolysis as well as DnaK binding. Since insertion of ala-ile-ile-pro did not cause the same effects the conclusion is that either exposure of tryptophan itself or tryptophan induced modifications of the folding were responsible for the DnaK complex formation and the proteolysis of the ZZ proteins.

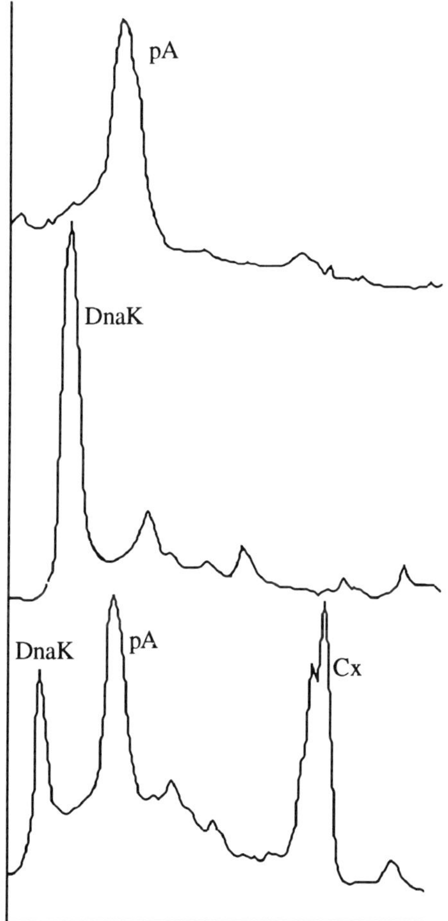

Figure 1. Native PAGE of (from top): Purified pRIT2 protein A (pA); purified DnaK and a mixture of pRIT2 protein A and DnaK.Cx= complex

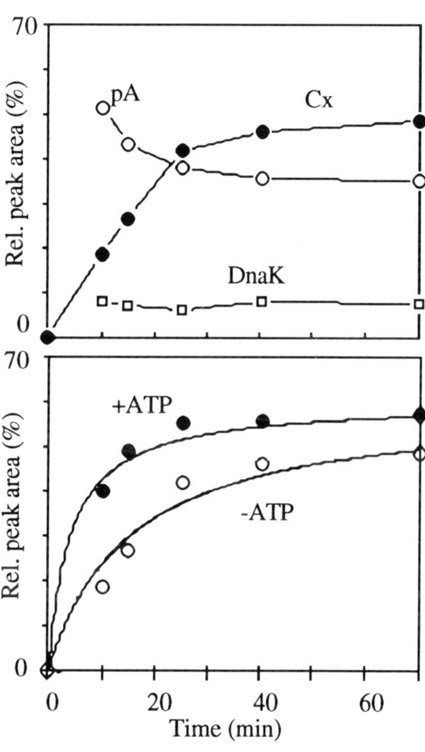

Figure 2. Upper graph shows scanning data of a mixtrure of DnaK and pRIT2 protein A as function of time. Lower graph shows the effect of 5 mM ATP on the complex formation. Solid lines represent $A=A_{max}\, t/(t+k_t)$ with A_{max}= 60% and k_t= 4 min and 15 min for experiments with and without ATP, respectively

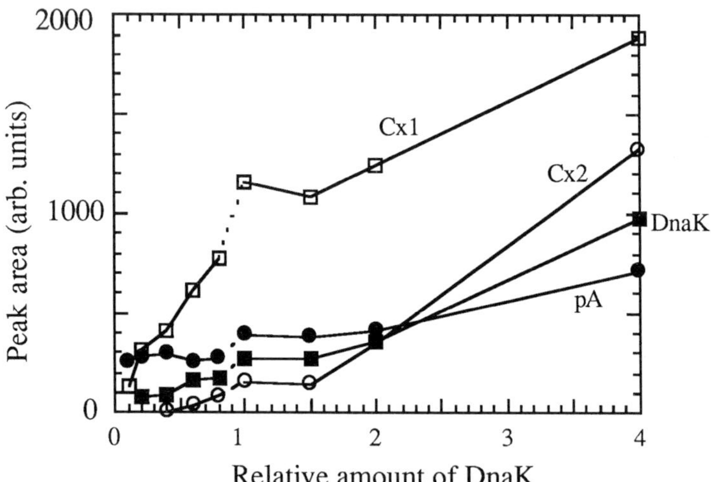

Figure 3. Scanning data of native PAGE of mixtures of DnaK and pRIT2 protein A with increasing ratio of DnaK per protein A.

4. CONCLUDING REMARKS

DnaK binds to certain proteins in the cell and this raises the question why. The fact that pRIT2 protein A as well as the ZZT2 and ZZT3 molecules all were very susceptible to proteolysis suggests that DnaK complex formation could be involved in the proteolysis. One hypothesis would be that DnaK acts as a signal for proteolysis of impermissive proteins. This is in agreement with the observation that the rate of proteolysis of a cytoplasmically located alkaline phosphatase in *E.coli* was reduced in a DnaK negative mutant Sherman [1]. Another possibility would be that DnaK recognises incorrectly folded proteins and binds to them in a refolding effort. This is in accordance with the observation that DnaK, when mixed with heat denaturated *E.coli* RNA polymerase, facilitates refolding of the protein to active enzyme [5]. Irrespectively of the biological function involved in the complex formation this phenomenon may have considerable consequences for the protein purification. In this case, when DnaK binds to a molecule that is easily purified by affinity chromatography, the complex formation could be utilised for purification of the DnaK. If complex formation of this kind between DnaK and other proteins is common during protein purification, it may cause considerable problems when non-affinity methods are used, since the physico-chemical properties of the complex may be considerably different from those utilised for separation of the protein. Such mechanisms would contribute to lower the yield in the separation step. This could also be an advantage if the complex is selectively formed with unproperly folded proteins. The phenomenon is further complicated by the fact that the complex formation kinetics is relatively slow and dependent on the presence of ATP.

5. REFERENCES

1. M.Y. Sherman and A.L. Goldberg, EMBO J., 11 (1992) 71-77.
2. H. Hellebust, M. Uhlén, and S.-O. Enfors, J. Bacteriol., 172 (1990) 5030-5034.
3. K. Köhler, *et al.*, Bio/Technology, 9 (1991) 642-646.
4. T. Moks, *et al.*, Bio/Technology,, 5 (1987) 379-382.
5. D. Skowyra, C. Georgopoulos, and M. Zylicz, Cell, 62 (1990) 939-944.

Lipases Structure and Function

ECB6: Proceedings of the 6th European Congress on Biotechnology
edited by L. Alberghina, L. Frontali, P. Sensi
1994 Elsevier Science B.V.

Relationships between structures and kinetic properties of lipases.

R. Verger[a], R. van Tilbeurgh[b], F. Ferrato[a], E. Rogalska[a], M. Aoubala[a], A. de Caro[a], M. Ivanova[a], L. de la Fournière[a], C. Rivière[a], L. Sarda[a] and C. Cambillau[b].

[a] Laboratoire de Lipolyse Enzymatique, GDR-1000 du CNRS, 31, Chemin Joseph-Aiguier, 13402 Marseille Cedex 9, France.

[b] LCCMB-CNRS, Faculté de Médecine Nord, Boulevard P. Dramard, 13916 Marseille Cedex 20, France.

BRIDGE T-project on lipases: contribution of the Marseilles' team. Apart from their general biological significance, lipolytic enzymes play an increasingly important role in biotechnology.

1. INTERFACIAL ACTIVATION OF THE LIPASE- PROCOLIPASE COMPLEX BY MIXED MICELLES REVEALED BY X-RAY CRYSTALLOGRAPHY.

The three dimensional structure of the lipase procolipase complex, cocrystallized with mixed micelles of phosphatidylcholine and bile salt, has been determined at 3 Å resolution by X-ray crystallography. Considerable changes in the secondary and tertiary structure are observed in the region around the active site (figure 1). Especially the lid, a surface helix covering the catalytic triad, adopts a totally different conformation that allows a phospholipid to bind to the active site, displacing some residues over a distance of 30 Å. The open lid is an integral part of the active site and is intensively interacting with the bound phospholipid. The open lid also interacts with procolipase and together they form the lipid-water interface binding site. This reorganisation of the lid structure provokes a second drastic conformational change in an active site loop, which in turn creates the oxyanion hole. This is an example of "induced fit" triggered by the physical properties of the substrate. Besides this, the variable conformation of the lid illustrates how the same amino acid sequence can adopt different secondary structures, depending on the local physico chemical environment.

The present crystal structure shows that the pancreatic lipase lid is a complicated structural device. Upon lipid binding it reorganizes in an ingenious way. This conformational change serves many purposes : it creates the interfacial lipid binding site, gives shape and access to the active site canyon, stabilizes the oxyanion hole loop, and finally it strengthens the interaction with procolipase. The importance of the lid in mammalian lipases has recently been illustrated by an elegant study on lipoprotein lipase. A chimeric construction consisting of the pancreatic lipase lid and the lipoprotein lipase core yielded a completely inactive enzyme. The pancreatic lipase lid grafted on a lipoprotein lipase core may not encounter the correct stabilizing interactions. It may well be that the lid regulates the different substrate specificity, interfacial behaviour and cofactor dependency of these mammalian lipases.

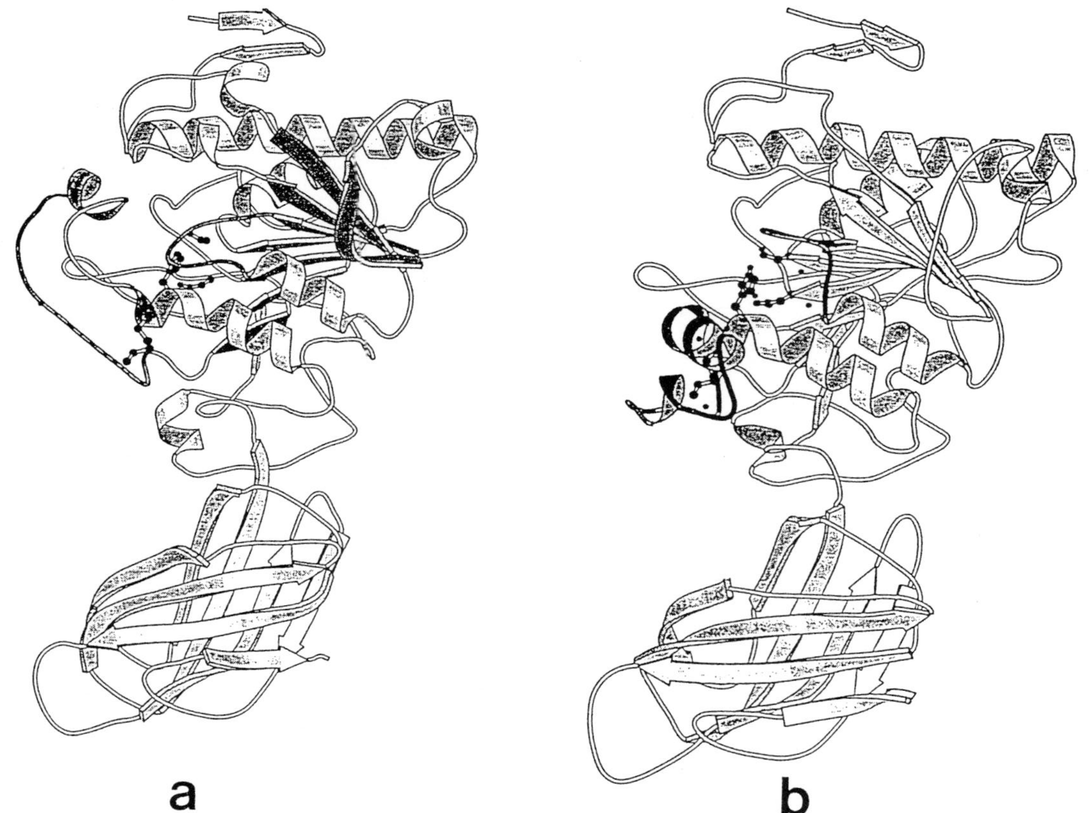

Figure 1. Comparison between the "closed" (a) and "open" (b) form of pancreatic lipase. Colipase is not represented. The catalytic triad is represented in balls and sticks. References: H. van Tilbeurgh, L. Sarda, R. Verger and C. Cambillau. Nature 359 (1992) 159-162. H. van Tilbeurgh, M.P. Egloff, C. Martinez, N. Rugani, R. Verger & C. Cambillau. Nature (1993) in the press.

2. MONOCLONCAL ANTIBODIES AGAINST HUMAN GASTRIC LIPASE.

In humans, the hydrolysis of dietary triacylglycerols begins in the stomach and is catalyzed by one major enzyme: human gastric lipase (HGL) which is located in the chief cells of fundic mucosa [1]. HGL is a glycoprotein with a molecular mass of around 50 kDa. In vitro studies have shown that HGL hydrolyses both short and long-chain triacylglycerols at comparable rates. The enzyme is highly stable and active under acidic pH conditions. The optimal pH for HGL activity is 5.4 in the case of long-chain and 6 in the case of short-chain triacylglycerols.

The amino acid sequence of HGL, deduced from the *c*DNA, consists of 379 residues and shows a conservative pentapeptide (Gly-X-Ser-X-Gly) common to all the known lipases as well as serine esterases. The essential serine within this pentapeptide is supposed to be part of the Asp/His/Ser catalytic triad to be found in esterases and lipases.

Two topographically and functionally distinct sites (a lipid or interface binding site and a catalytic site) have been postulated to exist in lipolytic enzymes, based on monolayer experiments. No information about the three dimensional structure of HGL is yet available, although several crystal types have been obtained. In order to characterize better these functional sites of HGL, an immunological approach was developed using monoclonal antibodies (MAbs) against HGL as probes for the spatial identification of these domains.

2.1. Production of MAbs against HGL. [2]

Hybridomas prepared by fusing spleen cells with myeloma cells were placed in 576 wells containing the selective medium H.A.T. and mouse peritoneal macrophages as feeder cells. Ten days after the fusion, 422 wells produced one or more clones of which 52 clones had anti-HGL activity in a direct binding ELISA test and five of them (4-3, 25-4, 35-2, 83-15, and 218-13) were selected and cloned using the limiting dilution technique. The five clones selected were then cultured for storage and production of antibodies in mice. MAbs were then purified from ascitic fluid. All these MAbs belong to the IgG1 class with a K light chain.

2.2. A new ELISA procedure to quantitate HGL in duodenal contents. [3]

The usual method for estimating enzymatic activity is based on the use of insoluble substrates present in an emulsified state. The lipolytic activity measurements in biological fluids, such as in duodenal contents, are sometimes unreliable, due to the presence of lipases of various cellular origins acting upon the same lipidic substrate. For this reason we developed a sensitive and specific double sandwich ELISA for measuring HGL in the duodenal contents, where HGL and HPL are both present. With this new procedure using purified rabbit anti-HGL polyclonal antibodies as captor antibody and biotin-labeled MAb 35-2 as detector antibody, it was possible to detect HGL concentrations as low as 1 ng/ml. Since HGL levels in duodenal contents are of the order of μg/ml the sensitivity of the ELISA test turned out to be sufficient. Moreover, the good correlation between the expected and measured concentrations of duodenal HGL (Table 1) indicates that no other duodenal compound interfered with the immunoassay. Furthermore, the HGL concentrations determined either by the ELISA or an enzymatic test, using standard pH stat titration, shows a good correlation between both assays.

Table 1
HGL detection ratio (%) of pure exogeneous HGL added to duodenal contents.
The detection ratio was calculated as follows:

$$\text{HGL detection ratio} = \frac{\text{measured HGL - endogenous HGL}}{\text{exogeneous HGL}} \times 100$$

Endogeneous HGL in duodenal contents (μg/ml)	Exogenous HGL (μg/ml)	Total measured HGL (μg/ml)	HGL detection ratio (%)
2.5	1.5	3.8	86.6
2.5	5.0	7.2	94.0
2.5	10.0	12.4	99.0
5.1	1.5	6.6	100.0
5.1	5.0	9.3	84.0
5.1	10.0	16.1	110.0

With the enzymatic assays used to measure lipase activity in biological fluids (gastric juice, duodenal contents and serum), there is a risk of overestimating lipolysis due to the presence of lipases from various tissular origins such as pancreatic lipase. The double sandwich ELISA developed in this study provides a particularly suitable means of specifically estimating HGL in biological fluids.

2.3. Immunopurification of HGL. [3]

Human gastric juice is a biological fluid which is rich in mucus and pepsinogens. An improved procedure was developed for isolating HGL by ion exchange chromatography in combination with immunoaffinity chromatography, using a column of immobilized MAb 35-2. As expected, the elution of HGL under acidic conditions (pH 2.2) from the immunoaffinity column does not alter its enzymatic activity and gives rise to very pure HGL fractions. This procedure offers the advantage of being rapid, reproducible with a better overall yield (58 %) as compared to the previously described methods using conventional chromatographic steps.

2.4. Immunoinactivation of HGL by MAbs. [2]

Since the screening of anti-HGL MAbs was performed using a native enzyme, all epitopes recognized by these antibodies are probably located on the surface of the enzyme. In order to investigate the involvement of several epitopes in the catalytic activity of HGL, we studied the effects of each MAb on the lipolytic activity of this enzyme using the following three substrates: tributyrin or soybean oil emulsions and 1,2 didecanoyl *sn*- glycerol (dicaprin) monomolecular films.

2.4.1. HGL activity on tributyrin and soybean oil emulsions.

Three MAb classes could be defined: class 1 includes MAbs 4-3, 25-4, and 35-2 (Figure.2A); class 2 includes MAb 83-15 (Figure 2B) and class 3 composed of MAb 218-13. The three MAbs of class 1 give very similar patterns, reaching a maximum inhibition level at a MAb/HGL molar ratio of 0.5. Figure 2A illustrates typical MAb 35-2 inhibition curves with tributyrin or soybean oil emulsions as HGL substrates. In the case of class 2 MAb, which contains MAb 83-15 (Figure 2B), no significant effect on HGL lipolytic activity was observed when the substrate was tributyrin, whereas on soybean oil the HGL activity was reduced by 50%. With class 3 MAb (218-13) no inhibitory effect on HGL activity was observed with either tributyrin or soybean oil emulsions.

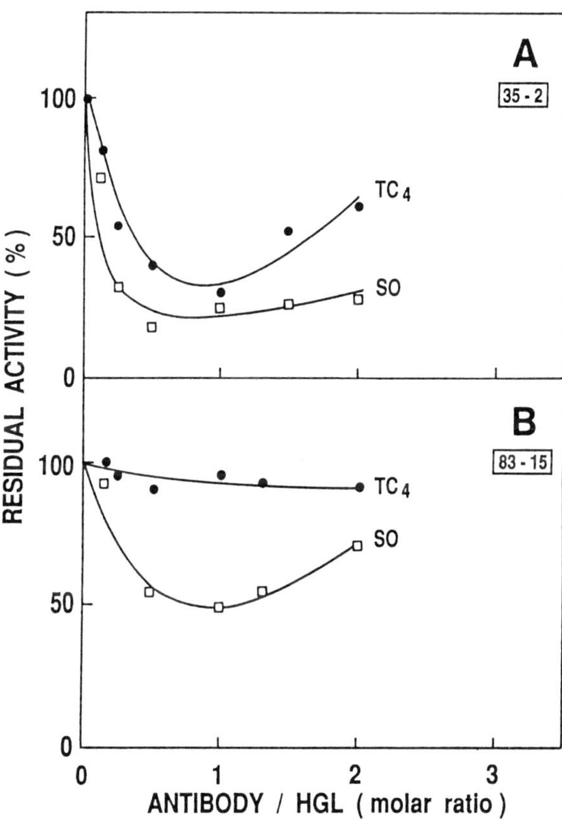

Figure 2. Effect of both MAbs 35-2 and 83-15 on HGL activity. A fixed amount of HGL was incubated with each MAb at various molar ratios. The relative residual activity of HGL was measured on either tributyrin (●) or soybean oil (❑) emulsions.

2.4.2. HGL activity on 1,2-*sn*-didecanoyl glycerol monolayers (Figure 3).

In order to optimize the inhibitory effect of various MAbs, HGL was incubated for 1 h at 37°C with each MAb at a MAb/HGL molar ratio of 0.5. We checked that the optimal inhibition of HGL activity tested on monolayers was obtained at MAbs/HGL molar ratios of 0.5 as previously observed in the case of the bulk assay systems. The lipolytic activity of HGL was measured at surface pressures ranging from 0 to 35 $mN.m^{-1}$. As shown in Figure 3, HGL does not significantly hydrolyse dicaprin films at surface pressures below 15 $mN.m^{-1}$. Above this pressure the HGL activity increases rapidly reaching a maximum value at 35 $mN.m^{-1}$. When HGL was incubated with each of the five MAbs, the previously obtained classification of the MAbs, based on bulk assay systems, was confirmed.

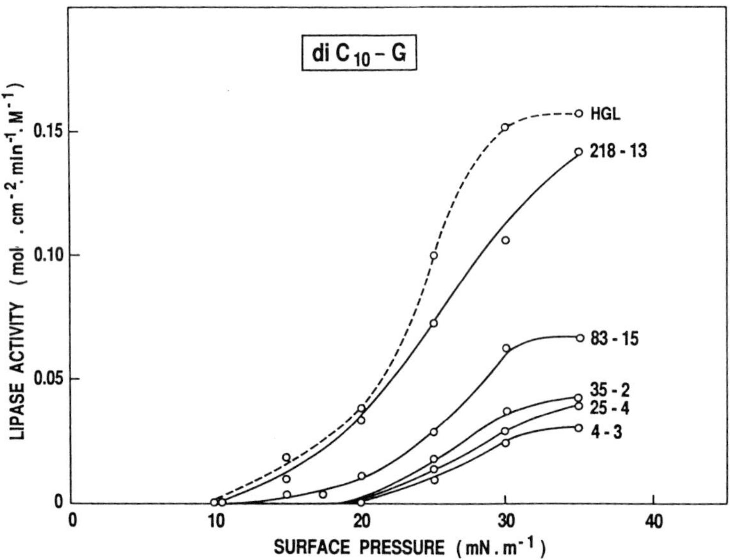

Figure 3. Effects of the five MAbs on HGL activity measured with 1,2-*sn*-didecanoylglycerol films as substrate at variable surface pressures.

2.4.3. Effect of MAbs on HGL penetration into lipid monolayers. [4]

In order to further characterize the catalytic and lipid binding domains on the HGL surface, we investigated the lipid monolayer penetration capacities of the five anti-HGL MAbs as well as their respective complexes with HGL.

Figure 4 shows the adsorption patterns at the air/water interface of HGL and the five purified anti-HGL MAbs. Each point represents the maximal surface pressure value reached after protein injection into the aqueous subphase. The HGL and MAbs adsorption behaviours are clearly very different. HGL possesses a much higher surface tensioactivity than any of the MAbs. This is a particularly favorable situation which facilitated the interpretation of the adsorption experiments performed with MAb/HGL complexes.

To identify the lipid binding domain on the HGL surface, the penetration of MAbs/HGL complexes into egg phosphatidylcholine monomolecular films was investigated (Figure 5). MAbs 4-3, 25-4, 35-2 and 83-15, unlike MAb 218-13, decreased the penetration capacity of native HGL. These results show that the inibitory MAbs 4-3, 25-4, 35-2 and 83-15 recognize epitopes overlapping the lipid binding domain of HGL, unlike the epitope reacting with MAb 218-13.

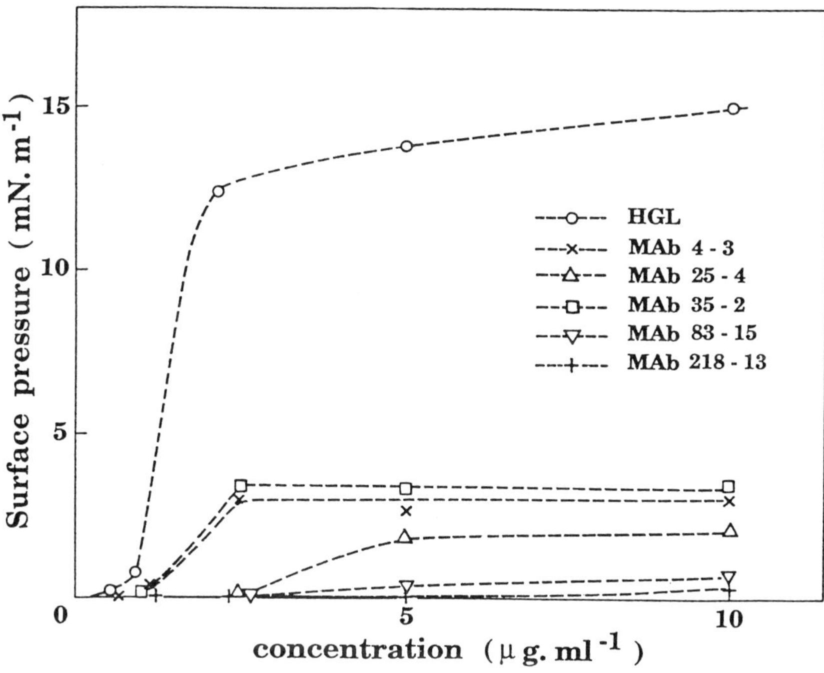

Figure 4. Variation in surface pressure of the air/water interface after injection of HGL and anti-HGL MAbs. Each point coresponds to the maximal value of the surface pressure reached after protein injection into the water subphase.

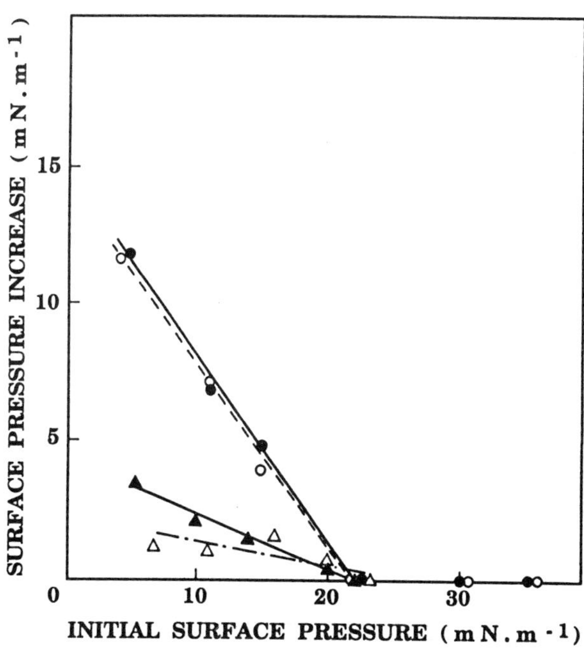

Figure 5. Maximal surface pressure increase after injections of HGL [●], MAb [△] and the MAb/HGL complexes (MAb 4-3/HGL [▲] and MAb 218-13/HGL [○]) under a monomolecular film of egg phosphatidylcholine spread at various initial surface pressures.

2.5. Epitope mapping with MAbs. [2]

To test whether the MAbs recognized overlapping epitopes on HGL, the ELISA double antibody binding test (ELISA additivity test) developed by Friguet et al. was used. The titration studies have shown that the five MAbs expressed different binding properties with HGL adsorbed on a PVC plate. Competition between couples of antibodies for the antigen was expressed by means of the additivity index. This index (expressed as an overlapping percentage) makes it possible to evaluate the simultaneous binding of two monoclonal antibodies to the antigen. The five anti-HGL MAbs were studied in all possible pairs and the values obtained ranged between 20.5 % and 75 %. From these results, it emerged that four (4-3, 25-4, 35-2 and 83-15) out of the five anti-HGL MAbs recognize closely overlapping epitopes in the same antigenic region. We have tentatively positioned the epitopes of HGL by means of a schematic spatial map (Fig 6). The additivity indexes between MAbs have been taken to be inversely related to the overlapping surfaces between pairs of epitopes. If one assumes first that an epitope has a relatively flat surface (700 $Å^2$) with protuberances and depressions formed by the amino-acid side chains of the combining antibody site and secondly that the complex formation between a MAb and its antigen does not induce any substantial conformational changes that might affect the subsequent binding of other MAbs, it is possible to calculate the distances (in Å) between the centers of the five epitopes and to build a spatial model (Fig 6). A general overview of the resulting model confirms that the various overlapping epitopes are all located in the same antigenic region of HGL.

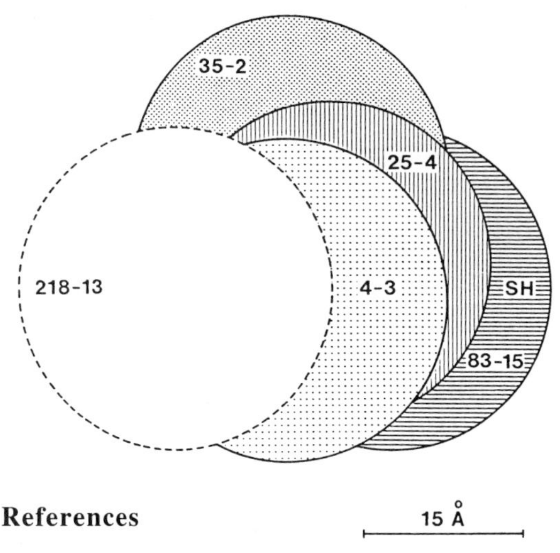

Figure 6. Spatial map of overlapping epitope recognized by the five MAbs against HGL

References

1. Y. Gargouri, H. Moreau & R. Verger. Biochim. Biophys. Acta 1006 (1989) 255.
2. M. Aoubala, C. Daniel, A. De Caro, M. Ivanova, M. Hirn, L. Sarda & R. Verger. Eur. J. Biochem. 211 (1993) 99.
3. M. Aoubala, I. Douchet, R. Laugier, M. Hirn, R. Verger & A. De Caro. Biochim. Biophys. Acta. (in the press).
4. M. Ivanova, M. Aoubala, A, De Caro, C. Daniel, J. Hirn & R. Verger. Colloïds and surfaces B: Biointerfaces. (in the press).

ECB6: Proceedings of the 6th European Congress on Biotechnology
edited by L. Alberghina, L. Frontali, P. Sensi

Candida cylindracea LIPASE: APPROACHES TOWARDS THE UNDERSTANDING OF STRUCTURE-PERFORMANCE RELATIONSHIPS

Lilia Alberghina[a], Lucia Battinelli[b], Enrico Cernia[b], Fabrizia Fusetti[a], Sonia Longhi[a], Marina Lotti[a] and Simonetta Soro[b]

[a]*Dipartimento di Fisiologia e Biochimica Generali, Sezione di Biochimica Comparata, Universita' degli Studi di Milano, Italia*
[b]*Dipartimento di Chimica, Universita' di Roma "La Sapienza", Roma, Italia*

1. INTRODUCTION

The lipase secreted by the yeast *C. cylindracea* (CCL) is described in the literature as a "non-specific lipase", referring to its ability to rapidly hydrolyse any triglycerides, regardless to the position of the fatty acid or synthetic molecule released.

Commercial preparations of CCL have been reported to be made of several protein species (1). Very recently, two enzymes (lipase A and lipase B) highly related in molecular properties but differing in carbohydrate content, hydrophobicity and substrate specificity have been purified (2). These results have suggested that *C. cylindracea* might produce several isoforms of enzyme displaying different substrate and/or reaction specificities.

This point is of relevant interest for the industrial applications of CCL and we believe that a more precise characterization of the lipase isoforms as well as the possibility to produce pure isoforms via recombinant DNA will greatly contribute to extend the range of utilization of this enzyme, even to very sophisticated processes, i.e. for the production of fine chemicals and pharmaceuticals.

As it is well known, lipase's catalytic activity is difficult to be measured in a reliable way because it is drastically influenced by the physico-chemical properties of the water/oil interface. For this reason it is important to use assays able to control these properties and to discriminate even small differences in the enzyme activities.

The work described in this paper is aimed to address this topic by a multidisciplinary approach including:

i) cloning of genes encoding the lipase isoforms and their expression in a host system;

ii) use of monolayer technique to carry out hydrolysis reactions at controlled interfacial quality;

iii) structure-performance relationship employing synthetic substrates.

This work has been supported by the EC BRIDGE Project and by the Progetto Finalizzato Biotecnologie e Biostrumentazioni of the Italian National Research Council

2. RESULTS AND DISCUSSION

2.1. Cloning and characterization of lipase genes

From the genome of *C. cylindracea* strain ATCC 14830 five different and indipendent genes (*LIP1* to *LIP5*) have been cloned by colony hybridization with lipase-specific probes (3, 4). The determination of nucleotide sequences and their translation in the corresponding amino acid sequences, allowed to predict the occurrence of mature proteins with similar molecular weight (about 57 Kda) but possibly different in their isoelectric point and degree of glycosylation. In fact, one, two or even three consensus sequences for glycosylation (Asn-X-Ser/Thr) have been alternatively recognized along the LIP sequences.

The molecular properties predicted for the lipase proteins seem to be consistent with those described for the proteins purified by other authors (2). In particular, it is notewhorty the correspondence between the N-terminal sequence determined for one of these enzymes and that predicted for the LIP3 protein and the detection of lipase proteins with different carbohydrate contents in the commercial preparations.

Gene	aa mature protein	MW (kDa)	p*I*	N-glycosylation sites N°	residues	N-terminus mature protein
LIP1	534	57.223	4.5	3	291, 314, 351	APTATLANGD
LIP2	534	57.744	4.9	1	351	APTATLANGD
LIP3	534	57.291	5.1	3	291, 314, 351	APTAKLANGD
LIP4	534	57.051	5.7	1	351	APTATLANGD
LIP5	534	56.957	5.5	3	314, 351	APTATLANGD

Protein (2)	MW (kDa)	p*I*	grade of glycosylation	N-terminus sequenced
Lipase A	60.000	5.5	8 ± 3%	APTAKLANGD
Lipase B (4 isoforms)	60.000	4.8 4.84 4.95 5.04	3.6 ± 0.8%	APTATLANGD

Comparison between the molecular features of lipases as deduced from DNA sequencing (upper row) and from purified proteins (lower row)

This correspondence between purified protein sequences and gene sequences is the first reported indication that *C. cylindracea* lipase isoforms are the products of different genes and not the result of post-translational modifications. The availability of enzymes with different specificities could confer to the yeast the possibility to grow on different nutrient sources (5).

2.2. Determination of lipase activity

In order to monitor and control, during the hydrolytic reactions, the influence of interfacial quality on the lipase activity, a monolayer technique was employed. In particular we carried out the hydrolysis of glycerol-trioctanoate monolayer catalyzed by the lipase from *C. cylindracea* at different purity degrees, in function of monolayers surface pressure.

Using two different preparations of CCL we observed an influence of monofilm surface pressure on hydrolysis rate. The profile for crude extract of CCL shows a complex behaviour probably due to molecular complexity of enzymatic preparation (Fig.1) The crude extract of CCL was purified by affinity chromatography on cross-linked polyvinylalcohol resin esterified with dodecanoic acid. Using this purified preparation we observed an optimal surface pressure value in the hydrolysis rate profile (Fig. 1).

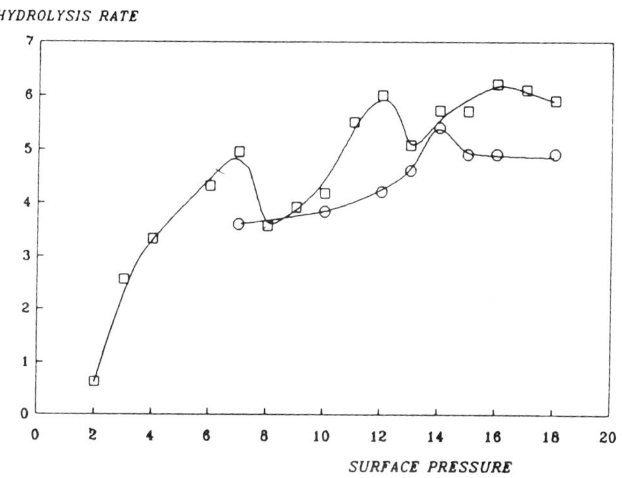

Fig.1. Glycerol-trioctanoate hydrolysis rate (n• mol/cm^{2}min*10^{13}) biocatalyzed by lipase crude extract (-□-) and purified lipase (-O-) in function of monofilm surface pressure (mN/m)

C. cylindracea lipase crude extract was then employed in the hydrolysis of esters of 2-(6-methoxy-2-naphthyl) propionic acid to study the effect of modifications of the ester-alcoholic chain on hydrolysis rates. We employed a broad range of the above given acids to determine which substrate structures are important for the hydrolytic activity of the enzyme. The results show that the esters of primary alcohol are better substrates than those containing a secondary moiety. The esters with short alcoholic chain (methyl) are hydrolysed faster than medium chain lenghts (n-butyl). The presence of an electron substituent (fluoro, chloro, bromo and iodio) at the beta carbon atom of alcoholic chain increases the hydrolysis rate (Fig.2). The initial reaction rates are strongly dependent on the steric effect of the alcoholic branched chain. The increase of activity observed for the esters with electron-substituents is in good agreement with the catalytic mechanism proposed for serine-enzyme.

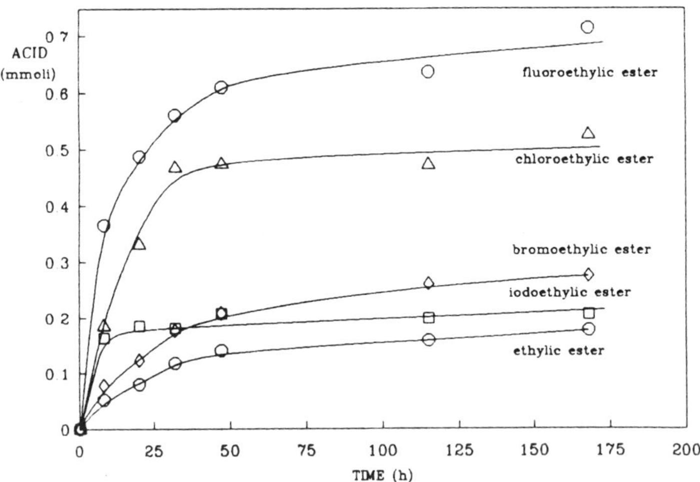

Fig.2. Relative rates of enzymatic hydrolysis of 2-(6-methoxy-2-naphtyl)propionic ethylic esters

2.3. CONCLUSIONS AND PERSPECTIVES

The availability of purified cloned lipases with different properties and of refined techniques for activity determination is the basis for a more sophisticated employement of these enzymes. In this respect, protein engineering will allow to modify the protein structure in order to clearify structure/function relationships and to modulate the enzyme activity. Lipases constitute a difficult challenge for this kind of approach, due to their extreme variability in sequence and general properties. However, the few three-dimensional structures already solved strongly suggest that all lipases would belong to the family of hydrolases (9). Should all lipases be related by a common structural fold, targeted mutagenesis would be greatly facilitated.

REFERENCES

(1) Shin-Hsiung Wu, Zhi-Wei Guo, Sih C.J. (1990). J. Am. Chem. Soc. 112: 1990-1995

(2) Rùa M.L., Dìaz-Maurino T., Fernandez V.M., Otero C. and Ballesteros A. (1993). Biochim. Biophys. Acta 1156: 181-189

(3) Longhi S., Fusetti F., Grandori R., Lotti M., Vanoni M. and Alberghina L. (1992). Biochim. Biophys. Acta 1131: 227-232

(4) Lotti M., Grandori R., Fusetti F., Longhi S., Brocca S., Tramontano A. and Alberghina L. (1993). Gene 124: 45-55

(5) Sidebottom C.M., Charton E., Dunn P.P.J., Mycoc G., Davies C., Sutton J.L., Macrae A.R. and Slabas A.R. (1991). Eur. J. Biochem. 202: 485-491

(6) Pieroni G., Gargouri Y., Sarda L. and Verger R. (1990). Adv. in Colloid and Interface Science 32:341-378

(7) Battinelli L., Borioni A., Castagnola M., Cernia E., Ortaggi G., Rabino R. (1991). In: Lipases: Structure, Function and Application to Biotransformation, Radcliffe House, University of Warwick, Coventry, U.K.

(8) Qu-Ming Gu, ching-Shih Chen, Sih C.J. (1986). Tetrahedron Letters 27(16): 1763-1766

(9) Ollis et al. (1992) Protein Engineering 5(3): 197-211

ECB6: Proceedings of the 6th European Congress on Biotechnology
edited by L. Alberghina, L. Frontali, P. Sensi
1994 Elsevier Science B.V.

TOWARDS THE DESIGN OF MICROBIAL LIPASES

Rolf D. Schmid[1]
Department of Enzyme Technology
GBF - Gesellschaft für Biotechnologische Forschung
D-3300 Braunschweig

1. Introduction

Lipases of microbial origin have found wide industrial application as detergent additives, in oleochemistry and for chemoenzymatic synthesis [1]. Over the past few years, several microbial lipases have been cloned, and progress has been made in the elucidation of the tertiary structure of this class of enzymes via x-ray analysis. Since 1991, the European Community (EC) has defined "lipase structure and mechanism of action" as one of its targets, in the framework of the BRIDGE program [2]. This project presently combines 25 laboratories in 6 member countries of the EC. As a member of this project group, we have focussed our work on the structure elucidation, the cloning and mechanistic studies of various lipases of microbial origin. This survey summarizes the results achieved in this project and pertinent work of other groups.

2. Purification and crystallization of microbial lipases

The purification of lipases to homogeneity may prove difficult due to inherent properties such as hydrophobic behaviour, with a tendency to form large associates or, in the case of lipases of eukaryotic origin, due to microheterogeneity based on glycoside moieties.
Table 1 summarizes our own studies, and those of our project group, on the purification and crystallization of various microbial lipases.

For the determination of microheterogeneity or occurence of isoenzymes, Laser-induced matrix absorption mass spectrometry (LAMA) has been advantageously used [3]. Large-scale purification was greatly assisted by in-line analysis of lipase activity using flow injection techniques [4].

3. Structure of microbial lipases

To date, the structures of 3 microbial lipases have been solved by x-ray analysis. (table 2), and of 3 microbial lipases have been suggested based on computer modelling studies.

---

[1]The work described in this article was carried out in cooperation with many researchers within and outside the GBF. In particular, the contributions of the following individuals are appreciated: Dietmar Schomburg, Hans-Jürgen Hecht, Birger Vasel, Dietmar Lang (all GBF), Masayasu Hirai (Toyota Inc.) and Kimiyasu Isobe (Amano Co.) in the structure elucidation and modelling of various lipases; Ulrich Menge, Helmut Erdmann, Marianne Kordel, Claudia Schmidt-Dannert (all GBF), Friedrich Spener, Hans-Christian Hedrich (both University of Münster) and Christine Davis (Unilever Co.) in the purification and characterization of various lipases; and Dietmar Beer (GBF) in the cloning and expression of Rhizopus lipase.

Table 1: Purification and crystallization of various microbial lipases - studies at the GBF or in collaboration with the GBF

Origin	MW [kDa]	sp.act. [U/mg]	diffract. crystals	literature
Rhizomucor miehei (Novo Co., Copenhagen)	32	1.900	+	U. Menge et al., 1989
				L. Brady et al., 1990
				B. Vasel et al., 1993
Rhizopus oryzae DSM 853	30	11.200	-	D. Beer et al. 1992
Geotrichum candidum (Amano Co., Nagoya)	62	1.050	+	U. Menge et al., 1990
				H.C. Hedrich et al., 1991
Geotrichum candidum B (Unilever Co., Colworth House)	58	2.450	+	C. Davies et al., 1992
Penicillium camambertii (Amono Co., Nagoya)	37/39	11.500	+	K. Isobe et al., 1992
Penicillium simplicissimum [a]	56	142	+	H. Sztajer et al., 1992
Pseudomonas spec. ATCC 21808	35	3.300	+	M. Kordel et al., 1991
Pseudomonas cepacia (Amano Co. , Nagoya)	34	5.800	+	U. Bornscheuer et al. [b]
Chromobacterium viscosum ATCC 6918	30	7.000	+	F. Spener et al. [b]
Staphylococcus carnosus DSM 20501	85	16	+	H. Erdmann et al., 1991
Bacillus thermocatenulatus DSM 730	ca. 16	49	-	C. Schmidt-Dannert et al. [b]
Psychrophilic prokaryote (Alfred-Wegener-Inst.)	ca. 15	n.d.	-	U. Kärst et al. [b]

[a] as classified by DSM [b] in preparation

Table 2: Structure data for microbial lipases

Origin	Resolution/ model lipase (% homology)	Literature
Humicola lanuginosa	not announced	-
Rhizomucor miehei	1.9 A	Z. Derewenda et al., 1990/1992
	2.3 A	B. Vasel et al., 1993
inhibited	3.0 A	A. M. Brzozowski et al, 1991
inhibited	2.6 A	Z. Derewenda, et al, 1992
Geotrichum candidum	1.8 A	J. D. Schrag et al., 1991 / Schlag & Cygler, 1993
Rhizopus niveus	Rhizomucor miehei (60 %)	M. Hirai et al., in preparation
Penicillium camembertii	Rhizomucor miehei (30 %)	K. Isobe et al., 1993
Candida cylindracea	Geotrichum candidum (40 %)	S. Longhi et al., 1992
Pseudomonas aeruginosa	Geotrichum candidum (10 %)	S. Ransac et al., 1992

The publication of the tertiary structures, based on x-ray analysis, of the following microbial lipases has been announced:

Candida cylindracea (M. Cygler et al., Montreal),
Penicillium camembertii (Z. Derewenda et al., Saskatoon),
Chromobacterium viscosum (D. Lang et al., Braunschweig),
Pseudomonas glumae (M. Noble et al., Oxford)

The structure of the lipase from Rhizomucor miehei has been studied in particular detail. As indicated in Fig. 1, the active site, composed of a serine-aspartate-histidine triad, is covered by a "lid" which is uncovered in the presence of triglyceride supersubstrate. While other fungal (and pancreatic/gastric) lipases have the same type of lid structure, modelling results with Pseudomonas fluorescence give no indication of the presence of a lid structure. In the absence of the structure elucidation of a suitable lipase-inhibitor complex, the function of the lid awaits further experimental data.

Fig. 1:‘Ribbon-model´of Micor miehei lipase

4. Cloning of microbial lipases

Several microbial lipases have been cloned and expressed to date (selected examples in table 3).

Table 3

Cloning and consensus sequences of the active site serine of various microbial lipases

Origin	MW [kDa]	expression system	consensus sequences of the active site serine	literature
Rhizomucor miehei	32	Aspergillus oryzae	SYKVAVTGHSLGGATALL	E. Boel et al., 1989
Rhizopuj niveus	30	Saccharomyces cerevisiae	TYKVIVTGHSLGGAQALL	W. Kugimiya et al.,1991
Rhizopuj delemar ATCC 34614	30	Escherichia coli	TYKVIVTGHSLGGAQALL	M.J. Haas et al., 1991 R.D. Joeyer et al., 1993
Rhizopus oryzae DSM 853	30	Escherichia coli	TYKVIVTGHSLGGAQALL	D. Beer et al., in prep.
Geotrichum candidum ATCC 34614	55	Escherichia coli	PDKVMIFGESAGAMSVAH	Y. Shimada et al., 1989
		E. coli, S. cerevisiae		E. Vandamme et al., 1987
Candida cylindracea ATCC 14830	57		PTKVTIFGESAGSMSVMC LIP I	Y. Kawaguchi et al., 1989
		Saccharomyces cerevisiae	PSKVTIFGESAGSMSTFV LIP II	M. Loui et al., 1993
Pseudomonas aeruginosa TE 3285	30	Escherichia coli	QPKVNLIGHSHGGPTIRY	K.E. Jaeger et al., 1990
Pseudomonas fragi IFO-12049	32	Escherichiacoli	AQRVNLIGHSQGALTARY	S. Aoyama et al., 1989
Pseudomonas glumae P1	33	Paeudomonas glumae	ATKVNLIGHSQGGLTSRY	L.G.I. Frenken et al., 1992
Pseudomonas cepacia DSM 3959	36	E.coli, B.subtilis Streptomyces lividans	ATKVNLVGHSQGGLSSRY	S. Jorgensen et al., 1991
Pseudomonas spec. ATCC 21808	35	Pseudomonas, E. coli	ATKVNLVGHSQGGLTSRY	S. Hom et al., 1990
Chromobacterium viscosum ATCC 6918	33	Escherichia coli	ATKVNLIGHSSQGGLTSRY	H. Sagai et al., 1989

5. Outlook

While much has been learned over just a few years on the mechanism of lipase catalysis, based on structural data, the evidence is still scarce and many aspects are still poorly understood. As the field widens, it may be observed that various classes of lipases exit, differing both in structure and mechanism, as this has shown already in the case of proteases. Protein design, based on the emerging structure information, may yield important clues to understand the mechanism of action of lipases on a molecular level and may also assist in the generation of lipase mutants useful in industrial catalysis.

6. Literature

[1] C.S. Chen and C.J. Sih, Angew. Chem. 1Ol: 711-724 (1989)

[2] Lipases: Structure, Mechanism and Genetic Engineering. L. Alberghina, R. D. Schmid, R. Verger, eds. VCH Publishers, Weinheim, Basel, Cambridge, New York, 1991

[3] H. C. Hedrich, K. Isobe, B. Stahl, K. Nokihara, M. Kordel, R.D. Schmid, M. Karas, F. Hillenkamp, and F. Spener, Anal. Biochem., in print

[4] G. C. Chemnitius, H. Erdmann, and R. D. Schmid, Anal. Biochem. 202: 16-24 (1992)

ECB6: Proceedings of the 6th European Congress on Biotechnology
edited by L. Alberghina, L. Frontali, P. Sensi

Stereoselective hydrolysis of triglycerides and triglyceride - analogs by animal and microbial lipases

C. Cudrey[a], F. Ferrato[a], L. Haalck[b], F. Paltauf[c], E. Rogalska[a], F. Spener[b], P. Stadler[c] and R. Verger[a]

[a] Centre de Biochimie et de Biologie Moleculaire du CNRS, Marseille, France

[b] Institut für Biochemie der Universität Münster,FRG

[c] Institut für Biochemie der Technischen Universität Graz, Austria

Lipases are ubiquitous enzymes that are essential for the utilization of triacylglycerols as nutrients for higher organisms as well as for micro-organisms. While much is known about the catalytic mechanism of lipase action which in most cases involves a serine in the catalytic center, the often observed positional and stereoselectivity of lipases is not understood at a molecular level. On the other hand it is the stereospecificity of lipases that is displayed not only with the natural substrates, but also with synthetic compounds, which makes lipases valuable tools as chiral catalysts in preparative organic chemistry. Available X-ray data do not allow a definition of substrate binding sites to an extent that would help to understand the structural basis of lipase stereoselectivity. An alternative approach to a better understanding of chiral recognition involves determination of stereoselectivity of a given lipase towards substrates of varying structure. In the present study the structure of the natural lipase substrates, triacylglycerols, was varied with respect to hydrophobicity (chain length of aliphatic moieties) and polarity in the glycerol backbone region (replacement of acylester groups by one or two alkylether groups). Polar and apolar interactions as well as steric factors are assumed to be involved in substrate recognition and binding.

Ether analogs of triacylglycerols have previously been employed in studies on lipase specificity [1-6]. The underlying idea was to replace hydrolyzable ester bonds by non-hydrolyzable ether bonds to avoid difficulties that might arise from acyl migration after partial hydrolysis or from hydrolysis at different sites of the substrate molecule at the same time. In other words, more equivocal results concerning positional and stereoselectivity of lipases were expected from experiments using these substrate analogs. When 1(3)-O-alkyl-2,3(1,2)-diacyl glycerols were used with various mammalian and human lipases [2] results were obtained that were comparable to those obtained with triacylglycerols. For example, it was confirmed that porcine pancreatic lipase did not show stereoselectivity, while lipoprotein lipase from blood serum or milk cleaved preferentially the ester in position *sn*-1. However, when 1,2(2,3)-di-O-alkyl-3(1)-acyl glycerol was used as the substrate, stereoselectivity of lipoprotein lipase was completely lost [3]. It was then concluded that the ester carbonyl in the (non-hydrolyzed) *sn*-2 position of a triacylglycerol was responsible for correct positioning of the substrate at the binding site of the enzyme. Similarly, phospholipases C and D lost stereospecificity when the acyl group in position 2 of phosphatidylcholine was replaced by an O-alkyl group [7]. These observations had two consequences: first, results obtained with ether analogs do not necessarily allow to draw conclusions as to the specificity of lipases towards the natural triglyceride substrates. Secondly, results obtained with triglyceride analogs modified in position 2 of glycerol might provide information on the mode of enzyme - substrate interaction and chiral recognition.

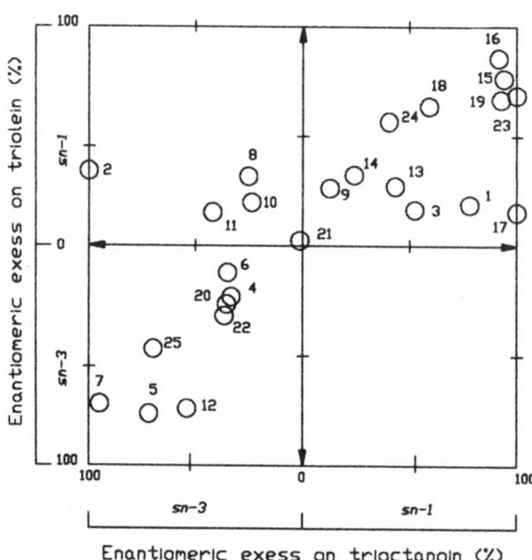

1, *Candida antartica* A lipase; 2, *Candida antartica* B lipase; 3, *Candida rugosa* lipase; 4, *Chromobactrium viscosum* lipase; 5, Dog gastric lipase; 6, Dog pancreatic lipase; 7 *Fusarium solani* cutinase; 8, *Geotrichum candidum* A lipase, 9; *Geotrichum candidum* B lipase; 10, *Geotrichum candidum* Münster lipase; 11, Guinea pig pancreatic lipase; 12, Human gastric lipase; 13, *Humicola lanuginosa* lipase; 14, Human pancreatic lipase; 15, Lipoprotein lipase (bovine milk); 16, *Mucor miehei* lipase; 17, *Pseudomonas aeruginosa* lipase; 18; *Penicillium camemberti* lipase; 19, *Pseudomonas fluorscens* lipase; 20, *Pseudomonas glumae* lipase; 21, Porcine pancreatic lipase; 22, *Penicillium simplicissimum* lipase; 23, *Pseudomonas species* lipase; 24, *Rhizopus arrhizus* lipase; 25, Rabbit gastric lipase

Figure: Lipase stereoselectivity on trioctanoin and triolein. Correlations between the stereoselectivities on the two substrates; (modified from ref. [8]).

In Marseille an extended study was undertaken to determine the stereoselectivity of 25 lipases of microbial and animal origin towards triolein and trioctanoin [8]. The steric configuration of diacylglycerols generated during partial hydrolysis was determined by reacting the diacylglycerols with R-(+)-1-phenylethylisocyanate to obtain diastereomeric carbamate mixtures that were resolved by HPLC. As can be seen in the Figure most of the lipases investigated showed preference for the *sn*-1 position. Interestingly, with four lipases, the preference switches from *sn*-3- to *sn*-1 when the acyl chain length increases from 8 to 18 carbon atoms.

Four microbial lipases differing in their positional and stereoselectivities towards triolein were chosen to test the effect of acyl for alkyl replacement on the respective selectivities (see Table). The lipases were from *C. rugosa* (CRL; low stereo- and positional specificty with triolein), *P. species* (PSL; positional, but no stereoselectivity), *R. arrhizus* (RAL; selectivity for *sn*-1) and *C. viscosum* (CVL; selectivity for *sn*-3). Results obtained with triolein and trioctanoin as the substrates essentially confirmed what had been found in Marseille; the completely different results obtained with PSL can be explained by the different origin of this lipase. With 3 of the 4 lipases chain length had little effect on stereoselectivity; only with CRL a significantly higher enantiomeric excess (ee) was found with the short-chain substrate. The presence of an ether instead of

Table: Stereoselectivities of microbial lipases tested with different triacylglycerols and triacylglycerol analogs

Lipase	CRL		CVL		PSL		RAL	
Substrate	ee (%)	*sn*	ee (%)	*sn*	ee (%)	*sn*	ee (%)	*sn*
Triolein	7.2	1	22.0	3	7.2	1	33.2	1
rac-1-O-Hexadecyl-2,3-dioleoylglycerol	n.d.		33.0	3	n.d.		28.4	1
rac-1-O-Octadecenyl-2-O-hexadecyl-3-oleoylglycerol	n.d.		31.0	1	n.d.		47.6	1
1,3-Dioleoyl-2-O-octadecenyl-*sn*-glycerol	n.d.		44.2	1	n.d.		46.0	1
1,3-Dioleoyl-2-O-hexadecyl-*sn*-glycerol	5.6	1	45.6	1	21.6	1	63.8	1
1,3-Dioleoyl-2-O-octyl-*sn*-glycerol	5.4	1	40.4	1	20.0	1	67.2	1
Trioctanoin	23.8	1	25.4	3	8.8	1	32.4	1
1,3-Dioctanoyl-2-O-hexadecyl-*sn*-glycerol	20.0	1	61.6	1	52.4	1	75.3	1
1,3-Dioctanoyl-2-O-octyl-*sn*-glycerol	27.8	1	60.8	1	54.8	1	78.4	1

CRL, *Candida rugosa* lipase; CVL, *Chromobacterium viscosum* lipase; PSL, *Pseudomonas species* lipase; RAL *Rhizopus arrhizus* lipase; all ee% data are mean values out of at least five experiments (± 4% ee).

an ester bond in one of the primary (1,3) positions of the racemic substrate, 1(3)-O-hexadecyl-2,3(1,2)-dioleoylglycerol, had little effect (compared to triolein) on the stereoselectivity of RAL and CVL. In contrast, replacement of the acylester for an alkylether in position 2 resulted in quite significant alterations of stereoselectivity of 3 of the 4 lipases tested; only CRL was insensitive to this change in substrate structure. With 1,3-dioleoyl-2-O-hexadecyl glycerol (compared to triolein), the ee increased from 7.2 to 21.6 with PSL and from 33.2 to 63.8 with RAL. The most striking and surprising effect was observed with CVL. In this case selectivity switched from *sn*-3 (ee 22.0) to *sn*-1 (ee 45.6). This change in selectivity was independent of the physical state of the substrate emulsion because the same result was obtained when a 1:1 mixture of the alkyl analog and triolein was used as the substrate. A further significant increase in stereoselectivity was observed when the acyl groups in the 1(3) positions of the 1,3-diacyl-2-O-alkylglycerol were C:8. On the other hand, variation of the chain length of the alkyl in position 2 (from 16:0 to 8:0) had no significant effect on stereoselectivity of the three lipases. Unsaturation of the alkyl chain in position 2 (9(Z)-octadecenyl, as compared to hexadecyl) had no effect with CVL and led to a decrease of ee (63.8 to 46.0) with RAL. With a diether analog (rac-1-O-octadecenyl-2-O-hexadecyl-3-oleoyl-glycerol) as the substrate, the results obtained with CVL and RAL were comparable to the results seen with the long-chain 1,3-diacyl-2-O-alkyl substrate.

These results underline the importance of the C2 position of a triradylglycerol substrate for chiral recognition by lipases with positional specificity for the primary (1,3) positions. CRL which has little positional specificity is completely insensitive to the type of bonding in position 2. From the available data one can of course only speculate on the underlying molecular mechanism. One possible explanation implies the existence, in the stereoselective lipases, of two potential binding sites that accomodate the (non-hydrolyzed) 2-acylester group of the substrate with different affinity, thus presenting either the *sn*-1 or the *sn*-3 ester preferentially to the catalytic site. Binding to one such site is obviously hindered by the bulky and more polar ester carbonyl as compared to an ether group. Therefore, presence of an ether group in position 2 enhances stereoselectivity. Other types of substrate analogs will be required to test this hypothesis.

REFERENCES

[1] H.K. Mangold and F. Paltauf (eds.), Ether Lipids. Biochemical and Biomedical Aspects, Academic Press, 1983.

[2] F. Paltauf, F. Esfandi and A. Holasek, FEBS Letters, 40 (1974) 119.

[3] F. Paltauf and E. Wagner, Biochim. Biophys. Acta, 431 (1976) 359.

[4] S. Ransac, E. Rogalska, Y. Gargouri, A.M.T.J. Deveer, F. Paltauf, G.H. de Haas and R. Verger, J. Biol. Chem., 265 (1990) 20263.

[5] J. Kötting, H. Eibl and F.-J. Fehrenbach, Chem. Phys. Lipids, 47 (1988) 117.

[6] D. Meusel, N. Weber and K.D. Mukherjee, Chem. Phys. Lipids, 61 (1992) 193.

[7] M. Bugaut, A. Kuksis and J.J. Myher, Biochim. Biophys. Acta, 853 (1985) 304.

[8] E. Rogalska, C. Cudrey, F. Ferrato and R. Verger, Chirality, 5 (1993) 24.

Secretion and Glycosilation of heterologous Proteins

ECB6: Proceedings of the 6th European Congress on Biotechnology
edited by L. Alberghina, L. Frontali, P. Sensi

Sequences Promoting Or Preventing Excretion In The Growth Medium Of Homologous And Heterologous Proteins In *Saccharomyces cerevisiae**

M. Vanoni, M. Vai, A. Morrione, M. Venturini, E. Lacanà, D.Rossini, E. Martegani, L. Popolo, L. Alberghina

Dipartimento di Fisiologia e Biochimica Generali Sezione di Biochimica Comparata Università degli Studi di Milano

1. INTRODUCTION

Most secreted proteins - except a few that follow an independent route [1] - are directed to the secretory pathway by short amino acid sequences that specifically interact with the signal recognition particle, an RNA/protein complex mediating protein translocation to the lumen of the endoplasmic reticulum (ER). Once entered in the ER, secretory proteins proceed to the cell surface unless other topogenic sequences divert them to other cell districts, such as the ER itself, the lysosome or its budding yeast counterpart, the vacuole [2].
In *Saccharomyces cerevisiae* the cell wall, mainly composed of mannoproteins and glucans [3], constitutes a barrier to secretion in the growth medium - a process that we will refer to as excretion - of proteins. Most secreted proteins, in fact, remain entrapped in the periplasmic space between cell wall itself and the plasma membrane. Although oligopeptides and small proteins are more easily released than bulkier ones, the molecular sieving effect played by the cell wall is unlikely to be the only factor involved in the choice between retention or release of secreted proteins in budding yeast. By deletion analysis and gene fusion experiments we present results supporting the notion that specific amino acid sequences may affect protein excretion in the growth medium.

2. RESULTS

We are currently working on two yeast glycoproteins: glucoamylase II (GAII) and gp115. GAII is encoded by the *STA2* gene. The primary translation product is a 778 amino acid preprotein, spanning an NH_2-terminal 32 amino acid secretion signal sequence [4]. The mature protein is a heavily glycosylated 670 KDa dimer, which is among the few proteins excreted in the medium in *Saccharomyces* spp [5]. Gp115 is a 115 KDa *N*- and *O*-glycosylated protein encoded by the *GAS1/GGP1* gene [6-7] of so far unknown function. It is anchored to the outer face of the plasma membrane by means of a glycosyl-phosphatidyl-inositol (GPI).

* This work has been partially supported by C.N.R., P.F. Ingegneria Genetica

2.1 Effect of signal sequence on β-galactosidase excretion

In a first series of experiments, the effectiveness of *Kluyveromyces lactis* killer toxin and gp115 signal sequences in promoting *Escherichia coli* β-galactosidase (β-gal) secretion was tested. β-gal was chosen as a model enzyme because, being a big cytoplasmic protein composed of four identical 116 KDa subunits [8], it has proven very difficult to secrete in both *E. coli* and *S. cerevisiae* [9,10]. Gp115 signal sequence proved unable to promote β-gal secretion, since $gp115_{1-66}$/β-gal hybrid proteins remained stuck in the ER (Table 1). On the contrary, both GAII [11] and killer toxin secretion signal sequences allowed periplasmic secretion of the enzyme (Table 1). These results confirm that secretion of large cytoplasmic enzymes is not a trivial task and that a panel of signal sequences should be tested when attempting to secrete a given heterologous protein.

Table 1
Secretion and periplasmic extraction of GAII and β-gal

Signal Sequence	Octapeptide	Protein	Synthetic medium			Rich Medium		
			Secretion[a]		PS	Secretion[a]		PS
			PS	Med	Extraction	PS	Med	Extraction
gp115	-	β-gal	-	-	NA[b]	ND[b]	ND[b]	NA[b]
GAII	+	β-gal	+	-	-	+	+	+
KT	-	β-gal	+	-	-	+	-	-
GAII	-	β-gal	+	-	-	+	-	-
GAII	+	GAII	+	+	+	+	++	+
KT	-	GAII	+	-	-	+	+/-	+/-
GAII	-	GAII	+	-	+/-	+	+/-	+/-

[a]PS= Periplasmic Space; Md= Medium
[b]NA = Not Applicable; ND = Not Done

2.2 An octapeptide promotes excretion of periplasmically secreted proteins

Although GAII itself is efficiently excreted even in synthetic medium, no excretion of GAII/β-gal hybrid proteins was detected in synthetic media, even with fusion proteins spanning as much as 451 GAII residues (data not shown). A deletion analysis was undertaken to test for the presence of sequences required for efficient excretion of GAII. By a combination of immunological and enzymatic tests, it was found that as much as 41 % of the GAII protein is dispensable for excretion in the growth medium, since carboxy-terminal deletions as long as 327 residues - and completely abolishing enzymatic activity - allow quite efficient excretion in the growth medium. Removal of preGAII residues 1-40, spanning the 32 aa-long secretion signal sequence, abolished excretion (data not shown). Addition of *K. lactis* killer toxin signal sequence restored near wild type activity and secretion in the periplasmic space. However a dramatic reduction in growth medium excretion was observed. The same reduction was obtained by using GAII own signal sequence, thus suggesting a role for GAII residues 1-8 of in growth medium excretion (Table 1).

To test whether the octapeptide could promote excretion of other proteins besides GAII, the pattern of β-gal secretion was analyzed in strains expressing fusions between β-gal and either killer toxin or GAII signal sequence. Even under conditions that are permissive for excretion of octamer-containing GAII/β-gal proteins - i.e. growth in YP medium [10] - hybrid proteins lacking the preGAII 33-40 peptide were excreted with much lower efficiency, if at all (Table 1). Since the octapeptide plays the same role when fused to unrelated protein molecules, it acts as a novel type of topogenic sequence promoting protein excretion in the growth medium.

In order to test whether ionic bonds might be involved in retention/release of periplasmic proteins, we tested whether alterations in ionic strength and/or the presence of reducing agents - which are known to loosen cell wall - might differentially affect extraction of octamer-containing *vs.* non-containing GAII and β-gal fusions. The results indicate that proteins lacking the octapeptide are poorly released under the same set of conditions that allow subtantial extraction of their octapeptide-containing counterparts. Since the octapeptide affects in parallel both *in vivo* excretion and *in vitro* extraction, the same mechanism, possibly involving ionic interactions, is likely to be responsible for both effects.

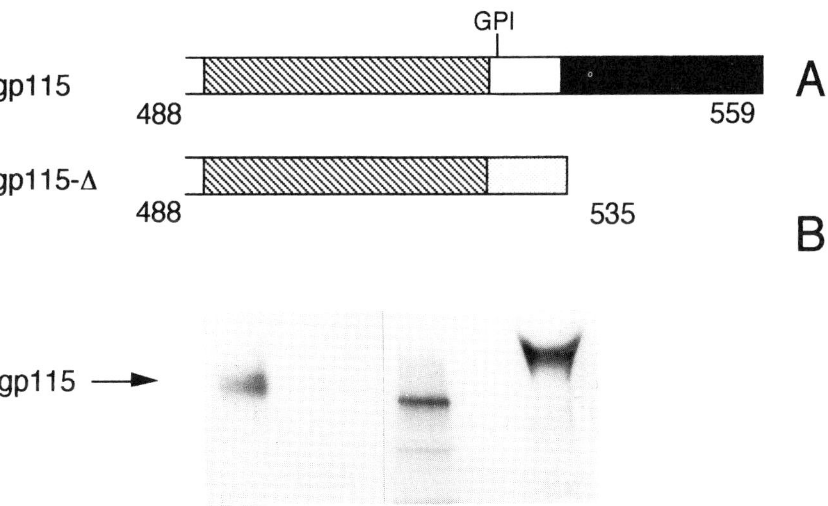

Figure 1 Structure of gp115 protein and of a deletion derivative (panel **A**). The positions of a serine-rich domain (striped box), of a hydrophobic C-terminal domain (black box) and of additional amino acids resulting from a frameshift in the deletion mutant (light grey box) are indicated. Immunoblotting with anti-gp115 antibodies (Panel **B**) of cell-bound (lanes 1 and 3) and growth medium-released (lanes 2 and 4) gp115 proteins in the wild type strain (lanes 1 and 2) and in an isogenic *ggp1* null mutant harboring a plasmid expressing C-terminal truncated gp115.

2.3 A C-terminal deletion allows secretion of gp115 in the growth medium

Addition of a GPI anchor occurs in the ER and involves the removal of a C-terminal hydrophobic sequence and the covalent attachment of the carboxyl-terminal amino acid of the mature protein to the anchor. In order to investigate the localization of a C-terminal deleted form of gp115, a truncated protein lacking the proposed anchor attachment site (Asn528) and the entire hydrophobic domain was constructed (Fig. 1A) and expressed in *ggp1* null mutants [12]. The mutant protein was unable to rescue the morphological defects brought about by *GGP1* disruption (data not shown). A fraction of the mutant protein accumulated intracellularly as a 110 KDa form , while most of the protein was hyperglycosylated (~ 140 KDa) and secreted in the growth medium (Fig. 1B), indicating that gp115 has the required features to cross the cell wall.

3. CONCLUSIONS

Our results indicate that both positive and negative acting sequences may affect excretion of proteins in *S. cerevisiae*. Together with recent results [13] which show the role of environmental factors in excretion of heterologous proteins, our data stress the importance of a global approach to protein secretion and may aid in the development of more efficient host/vector combinations and fermentation strategies, suitable for excretion of biotechnologically relevant proteins.

REFERENCES

1. Muesch, A, Hartman, E, Rohde, K, Rubartelli, A, Sitia, R and Rapoport, TA (1990) *Trends Biochem Sci* **15**, 86.
2. Rothman, JE and Orci, L (1992) *Nature* **355**, 409.
3. Ballou CE (1988) Organization of the *Saccharomyces cerevisiae* cell wall. In Varner JE (Ed) *Self assembling architecture* Liss (New York) 105-117.
4 Lambrechts, MG, Pretorius, IP, Sollitti, P and Marmur J (1991) *Gene* **100**, 95.
5. Modena D, Vanoni M, Englard S and Marmur J (1986) *Arch Biochem Biophys* **248**, 138.
6 Vai M, Gatti E, Lacanà E, Popolo L and Alberghina L (1991) *J Biol Chem* **266**, 12242.
7 Nuoffer, C, Jeno, P, Conzelmann, A and Riezman H (1991) *Mol Cell Biol* **11**, 27.
8 Kalnins, A, Otto, K and Muller-Hill, B (1983) *EMBO J* **2**, 593.
9 Moreno, F, Fowler, AV, Hall, m, Silhavy, TJ, Zabin, I and Schwartz, M. (1980) *Nature* **286**, 356.
10 Emr, SD, Schauer, I, Hansen, W, Esmon, P and schekman, R. (1984) *Mol Cell Biol* **4**, 2347.
11 Vanoni, M, Porro, D, Martegani, E and Alberghina, L. (1989) *Biochem Biophys Res Comm* **164**, 1331.
12 Popolo L, Vai M, Gatti E, Porello S, Bonfante P, Balestrini R and Alberghina L (1993) *J Bacteriol* **175**, 1879.
13 Rossini D, Porro D, Brambilla L, Venturini M, Ranzi BM, Vanoni M and Alberghina L (1992) *Yeast* **9**, 77.

ECB6: Proceedings of the 6th European Congress on Biotechnology
edited by L. Alberghina, L. Frontali, P. Sensi
1994 Elsevier Science B.V.

Glycosylation, folding, oligomerization and transport competence of the bean storage protein phaseolin in plant and animal cells.

Emanuela Pedrazzini, Giovanna Giovinazzo*, Roberto Bollini, Aldo Ceriotti and Vitale Alessandro.

Istituto Biosintesi Vegetali, Consiglio Nazionale delle Ricerche, via Bassini 15, 20133 Milano, Italy

*On leave from the Istituto di Ricerca sulle Biotecnologie Agroalimentari, Consiglio Nazionale delle Ricerche, Lecce, Italy

1. INTRODUCTION

The protein content of legume seeds is unsurpassed by the other cultivated plants and is mainly constituted by the storage proteins. These, however, are poor in sulphur aminoacids. To optimize the nutritional value of legume storage proteins by protein engineering, it is first necessary to acquire information on the molecular characteristics that allow for their efficient intracellular transport and stable accumulation into the protein storage vacuoles of cotyledonary cells. Random introduction of sulphur-rich sequences may alter such characteristics and result in protein instability (1).

Phaseolin (PHSL), the major bean storage protein, is a soluble, homotrimeric glycoprotein that is co-translationally inserted into the ER lumen and is then transported to the protein storage vacuoles (2). The three-dimensional structure of PHSL (3) indicates that the protein is composed of two structurally similar β-barrel regions and two similar α-helical regions (Figure 1, wtPHSL). In the trimers, α1 interacts with α2 of the adjacent polypeptide. Expression in *Xenopus* oocytes revealed that assembly into trimers is necessary for PHSL intracellular transport (4). We have constructed an assembly-defective PHSL mutant (Figure 1, Δ363PHSL), that lacks the α1 region. When expressed in *Xenopus* oocytes, Δ363PHSL is not competent for intracellular transport and appears to remain confined to the ER as a soluble monomer (4).

The PHSL polypeptides have two glycosylation sites, Asn252 and Asn341: in bean cotyledons, the first one is fully glycosylated, while the second one is only partially glycosylated (5, 6). Glycosylation has been suggested to play a role in the stability of PHSL, since mutated PHSL in which glycosylation sites have been destroyed accumulates at lower levels than wild-type phaseolin in transgenic tobacco (7). It is not know whether degradation occurs during intracellular transport or after deposition in the storage vacuoles. To acquire more information on the role played by the different regions of PHSL in assembly and transport competence and on the contribution of glycosylation to the structure of PHSL we have constructed new PHSL mutants. We have also developed a system for the expression of our constructs in plant cells.

2. RESULTS

To construct the AC1PHSL mutant, the nucleotide sequence coding for amino acids 178 to 220 of the α1 region were deleted and substituted with a nucleotide linker coding for a flexible glycine-rich sequence. The linker was introduced in an effort to avoid drastic effects of the deletion on the conformation of the adjacent β1 and β2 regions of the polypeptide (Figure 1, AC1). The AC2PHSL mutant was constructed by introducing the AC1PHSL mutation in Δ363PHSL (Figure 1, AC2). This mutant therefore lacks both regions thought to be involved in assembly.

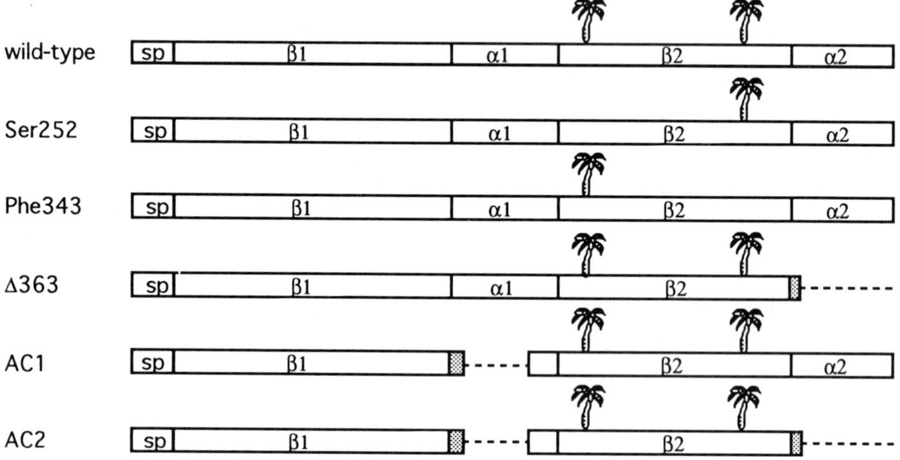

Figure 1: Phaseolin mutants. Symbols and abbreviations: sp, signal peptide; β1 and β2, β-barrel regions; α1 and α2, α-helical regions; : oligosaccharide chain. Dots: deletions. Shaded regions: new sequences in mutants.

AC1PHSL and AC2PHSL expressed in *Xenopus* oocytes failed to trimerize and were not transported intracellularly. The proteins remained monomeric and, in part, formed large aggregates suggesting that, unlike Δ363PHSL, AC1PHSL and AC2PHSL not only are assembly-defective but may also be folding-defective. We also noticed that, in AC1PHSL and AC2PHSL, glycosylation of Asn341 is more efficient than in wtPHSL and Δ363PHSL.

We have constructed two glycosylation mutants of PHSL, lacking one or the other glycosylation site. The glycosylation sites were destroyed by site-directed mutagenesis, converting Asn252 to Ser (Figure 1, Ser252 mutant) or Thr343 to Phe (Figure 1, Phe343 mutant). The conversion of Thr343 to Phe343 was chosen because soybean vicilin (a protein related to phaseolin) has a Phe residue in that position (8). We therefore think that the mutation does not affect the conformation of the polypeptide. In the soybean protein, a new glycosylation site is present six amino acids ahead of the one of PHSL, suggesting that in this region there might be an evolutive pressure for glycosylation. The first glycosylation site of PHSL is located in a region of low homology among PHSL-related proteins. Mutation of Asn252 to Ser252 was chosen because both are polar but uncharged amino acids.

Wild type PHSL (wtPHSL) and the two glycosylation mutants were expressed in *Xenopus* oocytes. Both wtPHSL and mutant PHSL showed similar kinetics of intracellular transport. Since animal cells do not have vacuoles, they secrete plant vacuolar proteins. All the constructs showed similar stability in the oocyte incubation medium. We also observed that the efficiency of glycosylation of Asn341 was increased in Ser252PHSL with respect to the wild type.

We determined that the single oligosaccharide chain (OS) in Phe343PHSL acquires full resistance to endoglucosaminidase H (endo H) digestion: this indicates that, when only Asn252 is glycosylated, the glycan became complex. The single OS in Ser252PHSL is always sensitive to endo H digestion, indicating that the OS is not accessible to the Golgi modification enzymes. It is known that when both glycans are present, neither of the two acquires resistance to endo H (6).

To test whether the results obtained by expressing the PHSL mutants in *Xenopus* oocytes are reproduced in plant cells, we are now expressing the constructs in tobacco mesophyll protoplasts. The vector for transient expression (pDHA) contains the 35S cauliflower mosaic virus promoter for constitutive expression, an alfalfa mosaic virus 5' untranslated sequence, a polycloning site and a 3' poly A+ terminator from Cauliflower mosaic virus. As a preliminary experiment, the coding region for wtPHSL was subcloned into this vector (the new plasmid is termed pDHE), and expressed in protoplasts. The protein is correctly glycosylated with one or two glycans, efficiently forms trimers and is partially degradated (Figure 2). The degradation products are part of the trimers (Figure 2). Partial degradation probably occurs in the vacuole, because degradation products are resistant to digestion with endoglycosidase H.

3. CONCLUSIONS

Deletion of a large part of the $\alpha1$ region inhibits assembly and transport competence of PHSL, indicating that this region, like $\alpha2$ (4) is essential for trimerization of PHSL. The assembly-defective mutant Δ363PHSL does not have altered pattern of co-translational glycosylation. Conversely, introduction of a mutation that appears to affect both assembly and folding results in enhanced glycosylation of Asn341. This provides *in vivo* evidence for a relationship between protein folding and glycosylation. This conclusion is also supported by the observation that efficiency of glycosylation at Asn341 is enhanced in the Ser252PHSL mutant, in which glycosylation at Asn252 cannot occur.

The pattern of formation of complex chains in the glycosylation mutants suggests that PHSL glycans are clustered in a single region of the folded polypeptide, in close contact with each other.

The stability and kinetics of intracellular transport of the Ser252PHSL and Phe343PHSL glycosylation mutants expressed in *Xenopus* oocytes are similar to the ones of wtPHSL indicating that the direct correlation between stability and glycosylation of PHSL observed in transgenic plants (7) may be the result of the increased degradation of the mutants after deposition into the protein storage vacuoles rather than to conformational alterations that impair intracellular transport. Increased degradation may result from the exposure of a protease-sensitive region of PHSL normally protected by the cluster of glycans.

We have developed a transient expression system in tobacco protoplasts, in which wtPHSL is correctly glycosylated, efficiently trimerizes and is transported intracellularly, probably to the vacuole. This system can be used to study the behaviour of the mutated PHSL polypeptides in plant cells.

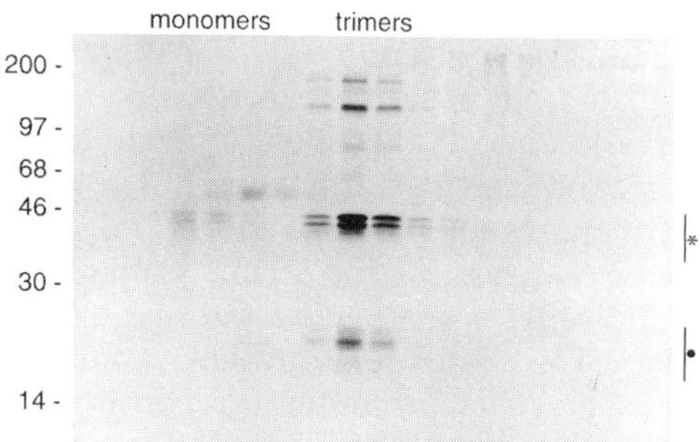

Figure 2: Expression of wtPHSL in tobacco protoplasts. Protoplasts transfected with wtPHSL and labelled for 5 h with radioactive amino acids were homogenized and the homogenate was loaded on the top of 5-25% sucrose gradient to separate monomeric and trimeric PHSL. Phaseolin polypeptides were immunoselected from each gradient fraction using α-PHSL antiserum and analyzed by SDS-PAGE and fluorography. Asterisk: intact PHSL polypeptides, with one or two oligosaccharide chains. Dot: degradation products. Positions of molecular weight markers, in kDa, on the left.

REFERENCES

1. L. M. Hoffman, et al., *Plant Mol. Biol.* **11**, 717-729 (1988).
2. R. Bollini, et al., *Physiol. Plant.* **55**, 82-92 (1982).
3. M. C. Lawrence, et al., *EMBO J.* **9**, 9-15 (1990).
4. A. Ceriotti, et al., *Eur. J. Biochem.* **202**, 959-968 (1991).
5. R. Bollini, et al., *J. Cell Biol.* **96**, 999-1007 (1983).
6. A. Sturm, et al., *J. Biol. Chem.* **262**, 13392-13403 (1987).
7. M. M. Bustos, et al., *Plant Mol. Biol.* **16**, 381-395 (1991).
8. J. J. Doyle, et al., *J. Biol. Chem.* **261**, 9228-9238 (1986).

ECB6: Proceedings of the 6th European Congress on Biotechnology
edited by L. Alberghina, L. Frontali, P. Sensi

Assembly and intracellular degradation of IgM subunits during B cell development.

Silvia Guenzi°, Claudio Fagioli°, AnnaMaria Fra°, Elena Pasqualetto°, Caterina Valetti°* and Roberto Sitia°*.

°DIBIT-HSR, 20132 Milano, Italy.

*IST, 16132 Genova, Italy.

1. INTRODUCTION

Newly synthesized proteins undergo a series of "quality control" steps which ensure that only properly folded and assembled molecules reach the final destination within the cell. In the case of membrane and secretory proteins, which are cotranslationally translocated across the endoplasmic reticulum (ER) membrane, immature molecules are retained intracellularly, largely in the ER itself. This may be achieved by formation of aggregates or specific interactions with ER resident proteins, such as BiP or protein disulfide isomerase (PDI). These interactions are likely to serve the dual role of catalysing folding and assembly, and preventing premature transport. Intracellular retention may depend on the exposure of hydrophobic patches or other 'signals' inaccessible in properly folded molecules (1,2). To maintain homeostasis, proteins which fail to attain the proper conformation must be degraded. It has been shown that the ER is equipped with a proteolytic system, independent from that of lysosomes (3).
What kind of signals allow cells to discriminate between assembled and unassembled molecules? Are these signals "general"? How is retention coupled to degradation? We have investigated the relationships between assembly, retention and degradation for a soluble oligomeric protein, IgM, whose secretion is developmentally regulated.

2. QUALITY CONTROL OF IgM SECRETION: THE ROLE OF μs Cys 575.

Surface Ig positive B cells secrete monomeric immunoglobulins, such as IgG or IgE, but not IgM; plasmacells secrete only polymeric IgM. In both lymphoid and non lymphoid cells disulphide interchange reactions involving the C terminal cysteine (Cys 575) of secretory μ heavy chains (μs) are responsible for the selective retention of IgM assembly intermediates (4,5,6).
Where and how are IgM subunits disposed of? To determine the intracellular site of degradation, pulse-chase experiments were performed in WEHI 231 B lymphoma cells (4) in the presence of NH_4Cl, an inhibitor of lysosomal proteases, or of brefeldin A (BFA), a drug which blocks transport to the Golgi apparatus (7,8) As shown in Figure 1, neither drug inhibit IgM degradation indicating that proteolysis takes place in a pre-Golgi compartment.

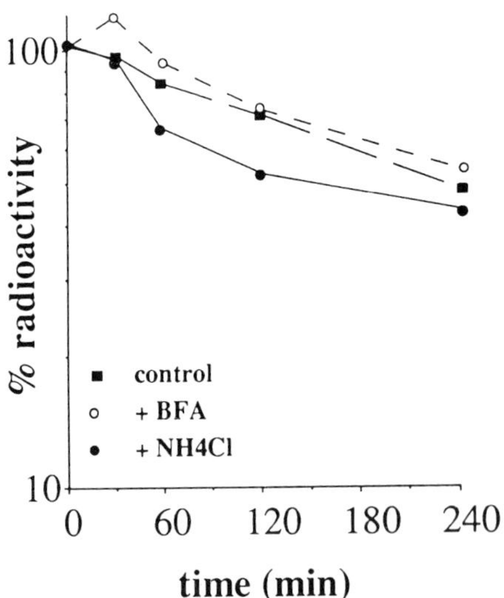

Figure 1. Degradation of IgM assembly intermediates occurs in the ER
WEHI 231 cells were pulsed for 10 minutes with ^{35}S methionine and chased for the indicated times in medium (control) or in the presence of NH_4Cl (50mM) or BFA (1 μg/ml) before lysis and immunoprecipitation with rabbit anti mouse anti-μ antibodies followed by protein A sepharose. Kinetics of degradation (expressed as the percentage of the initial μ chain-associated radioactivity) were calculated by densitometry of SDS-PAGEs.

Additional experiments were performed in myeloma cells: Figure 2 shows the kinetics of degradation of secretory μ chains (μs) expressed in the presence (J558L transfectants) or in the absence (NS0 transfectants) of light (L) chains. In both cases degradation is insensitive to BFA, showing that also in myeloma cells degradation takes place in the ER or a related subcompartment. Expression in J558L of a mutant μ chain bearing an alanine instead of Cys575 (J[μala]) determines secretion of monomeric IgM and inhibition of intracellular degradation (Figure 2a). Also in the presence of BFA, mutants lacking Cys 575 are stable (Figure 2b), indicating that ER degradation is not simply a function of the residency time in this organelle. In the absence of L, free μ chains are not secreted unless both the first constant domain (CH1) and Cys575 are removed: the presence of either element, Cys 575 or CH1, is sufficient to cause retention (4,9,10). The overall rate of degradation is faster in NSO than in J558L transfectants suggesting a stabilizing role for light chains; however, also in the absence of L, μ chains having a cysteine at position 575 are more susceptible to degradation than those with alanine (Figure 2c), irrespective of the capability of exiting the ER (Figure 2d shows the kinetics of degradation in the presence of BFA).

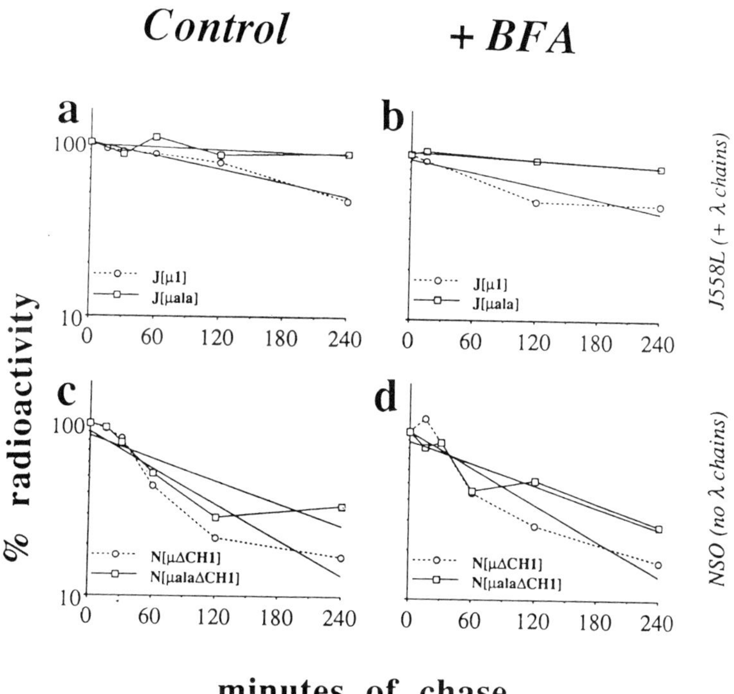

Figure 2. Cys 575 targets assembly intermediates to ER degradation.
J558L transfectants: J[μ1] and [J,μala] (panels a and b) and NSO transfectants: N[μΔCH1] and N[μalaΔCH1] (panels c and d) were pulsed for 10 minutes with ^{35}S methionine and chased for the indicated times with (panels b and d) or without BFA (1 μg/ml) (panels a and c). Immunoprecipitations, SDS-PAGE and densitometry were performed as described for WEHI 231 (see legend to Figure 1).

To investigate whether Cys 575 might work out of context, the last 20 aminoacids of μs chain (the μs tailpiece) have been added to cathepsin D (CD), a lysosomal protease. The resulting chimeric protein (CDMμtp) is retained and degraded in a pre-Golgi compartment (Fra et al. unpublished data). By contrast CDMK, a construct where CD has been extended wlth KDEL (11), is retained and stable in the ER, confirming that retention is not sufficient to cause degradation. Thus, Cys 575 plays a key role in the selective targeting of IgM subunits to ER degradation.

3. DISCUSSION

Quality - control of newly synthetized membrane and secretory proteins synthesis relies on the selective recognition and disposal of aberrant polypeptides. The 'signals' which mark conformationally immature molecules should be general and must be absent or inaccessible on properly folded and assembled proteins. It has been proposed that hydrophobic patches exposed by newborn proteins interact with chaperones in the ER, thus preventing their further transport along the exocytic pathway. We have obtained evidence that the C-terminal cysteine present in the heavy chain μ tailpiece (Cys575), can act as a retention element for unpolymerized Ig subunits. Here we show that the same element can also mediate the intracellular degradation of IgM assembly intermediates. In agreement with this, we have recently obtained evidence that, when transplanted onto the lysosomal protease cathepsin D, the μ tailpiece is sufficient to cause retention and degradation of the chimeric protein in COS transfectants only when the C-terminal cysteine is present (Fra et al. unpublished data). Thus, a model can be proposed, in which the ultimate fate of newly made proteins is determined by the competition between assembly, retention and degradation, all mediated by the same element. Assembly masks Cys575, allowing secretion of IgM polymers; when polymerization is inefficient, like in B lymphocytes, subunit breakdown prevails. On the basis of sugar analysis and immunofluorescence we propose that the above equilibria take place in a pre-Golgi compartment, probably in the ER itself. In principle, the same model could be extended to retention-degradation events which do not involve disulfide interchange reactions.

Acknowledgements
This work was supported by grants from Consiglio Nazionale Ricerche (PF BTBS, PF IG), Ministero della Sanità (AIDS Project), and Associazione Italiana Ricerca sul Cancro (AIRC). SG and AMF are recipient of fellowships from Fondazione Monte Tabor and AIRC, respectively.
The expert secretarial assistance of Cristina DeMarchi is gratefully acknowledged.

REFERENCES

1. Hurtley,S.M. and Helenius,A., Annu. Rev. Cell Biol., 5, 277-307, (1989).
2. Pelham,H.R.B., Annu. Rev. Cell Biol., 5, 1-23, (1989).
3. Bonifacino,J.S. and Lippincott-Schwartz,J., Curr.Opin.Cell Biol., 3, 592-600, (1991)
4. Sitia, R., Neuberger, M.S., Alberini,C., Bet,P., Fra,A., Valetti,C., Williams,G., and Milstein,C., Cell, 60, 781-790, (1990).
5. Alberini, C.M., P. Bet, C. Milstein, and R. Sitia, Nature, 347, 485-487, (1990).
6. Valetti, C., Grossi, C.E., Milstein, C. and Sitia, R. J. Cell Biol., 115, 983-994, (1991)
7. Misumi,Y., Miki,Y., Takatsuki,A., Tamura,G. and Ikehara,Y., J. Biol. Chem. 261, 11398-11403, (1986).
8. Lippincott-Schwarzt,J., Donaldson,J.G., Schweizer,A., Berger,E.G., Hau,H.P., Yuan,L.C. and Klausner,R.D., Cell, 60, 821-836,(1990).
9. Hendershot,L., Bole,D., Kohler,G. and Kearney, J.F., J. Cell Biol. 104, 761-767, (1987)
10. Hendershot,L.M., J. Cell Biol. 111, 829-837, (1990).
11. Pelham,H.R.B., EMBO J. 7, 913-918, (1988).

Receptor Engineering

ECB6: Proceedings of the 6th European Congress on Biotechnology
edited by L. Alberghina, L. Frontali, P. Sensi

ENGINEERING THE HEPATOCYTE GROWTH FACTOR RECEPTOR TO ELUCIDATE ITS STRUCTURE AND FUNCTION

Bardelli A., Maina F., Ponzetto C. and Comoglio P.M.

Dept. of Biomedical Sciences & Oncology, University of Torino, School of Medicine - Corso M. D'Azeglio 52 - 10126 Torino (Italy).

ABSTRACT

Hepatocyte Growth Factor/Scatter Factor (HGF) is secreted by cells of mesodermal origin and shows powerful mitogenic, motogenic and morphogenic activities on epithelial and endothelial cells. HGF/SF is produced as a single-chain biologically inactive precursor (pro-HGF/SF). Maturation into the active αβ heterodimer results from proteolytic cleavage by urokinase which acts as a pro-HGF convertase. HGF binds with high affinity to $p190^{MET}$, the transmembrane receptor encoded by the *MET* proto-oncogene. The *MET* receptor is a heterodimer of two disulfide-linked subunits (α and β). The α subunit is extracellular, while the β subunit is transmembrane and endowed with tyrosine kinase activity. The *MET* protein is the prototype of a distinct subfamily of heterodimeric tyrosine kinases, including the putative receptors *RON* and *SEA*. HGF binding triggers receptor autophosphorylation and upregulates its kinase activity. The major phosphorylation site has been mapped to Y_{1235}. Negative regulation of the receptor kinase activity occurs through distinguishable pathways involving protein kinase C activation or an increase in the intracellular Ca^{2+} concentration. Both lead to serine phosphorylation of a unique phosphopeptide of the receptor and to a decrease in its kinase activity. HGF-triggered signaling is mediated by activation of different cytoplasmic effectors, including phosphatidylinositol 3-kinase, phospholipase C-γ, Src-related tyrosine kinases and Ras-guanine nucleotide exchanger. Y_{1349} and Y_{1356} in the HGF receptor have been identified as binding sites for the p85/110 PI 3-kinase complex.

1. THE HEPATOCYTE GROWTH FACTOR AND SCATTER FACTOR

Hepatocyte Growth Factor (HGF) and Scatter Factor (SF) are identical glycoproteins secreted by cells of mesodermal origin (Gherardi and Stoker, 1990; Weidner et al., 1991). The factor has several activities on epithelial cells, including mitogenesis, dissociation of epithelial sheets, stimulation of cell motility, and promotion of matrix invasion (for a review see Gherardi and Stoker, 1991).

Analysis of the binding properties indicated the existence of two classes of binding sites in target cells with affinities one order of magnitude apart. The high affinity binding site (K_d 0.3 nM) has been identified with the receptor encoded by the *MET* proto-oncogene (Naldini et al., 1991b). The lower affinity, large capacity sites (K_d in the 10^{-9} M range) are likely to be ubiquitous heparansulfate proteoglycans.

HGF/SF is secreted as a single chain precursor (pro-HGF/SF) and is then cleaved to the mature $\alpha\beta$ heterodimer in the extracellular environment. Most of the secreted pro-HGF/SF is sequestered on the cell surface or bound to the extracellular matrix. HGF/SF functional analysis was performed using aminoacid substitution and deletion variants. Processing to the mature heterodimer is crucial for HGF/SF to induce biological responses in target cells. The primary determinant for receptor binding appears to be located within the α chain. The same domain leads to the activation of the downstream signal cascade involved in the motility response. However, the complete HGF/SF protein seems to be required for mitogenic activity (Hartmann et al., 1992; Lokker et al., 1992).

Pro-HGF/SF is correctly processed to the mature heterodimer by pure uPA (urokinase-type plasminogen activator) *in vitro*. Cleavage at a specific site is followed by conversion of the biologically inactive pro-HGF/SF into the active form, indicating that uPA acts as pro-HGF/SF convertase. Plasminogen activators also mediate the processing of pro-HGF/SF *in vivo* at the cell surface (Naldini et al., 1992).

To further elucidate its pleiotropic biological effects the HGF/SF was produced in insect cells using the baculovirus expression system. Recombinant purified HGF/SF elicits full biological activity in different bioassays when compared to wild type protein secreted by human fibroblasts.

2. THE HGF RECEPTOR

2.1 p190^{MET}, the tyrosine kinase encoded by the *MET* protooncogene.

The tyrosine kinase receptor encoded by the *MET* proto-oncogene (p190^{MET}) has unique structural features: it is a two-chain oligomer, composed of a 50 kDa (α) chain disulfide linked to a 145 kDa (β) chain in an α/β complex of 190 kDa. The α chain is exposed at the cell surface and the β chain spans the plasma membrane. The β chain contains the tyrosine kinase domain, as well as sites for tyrosine auto-phosphorylation (Giordano et al., 1988; Giordano et al., 1989a).

The *MET* protein, is the prototype of a distinct subfamily of heterodimeric tyrosine kinases, including the putative receptors *RON* and *SEA* (FIG.1). These receptors share significant sequence similarity including a canonical consensus site for proteolytic cleavage in the extracellular domain sequence of the β subunit (Ronsin et al., 1993; Huff et al., in press). In the case of $p190^{MET}$ aminoacid replacement in the proposed cleavage site inhibits the conversion to the α/β heterodimer (Mark et al., 1993). Moreover, transfection in COS cells of the cloned human *MET* cDNA leads to the expression of a correctly processed $p190^{MET}$ provided with both the α and the β chain.

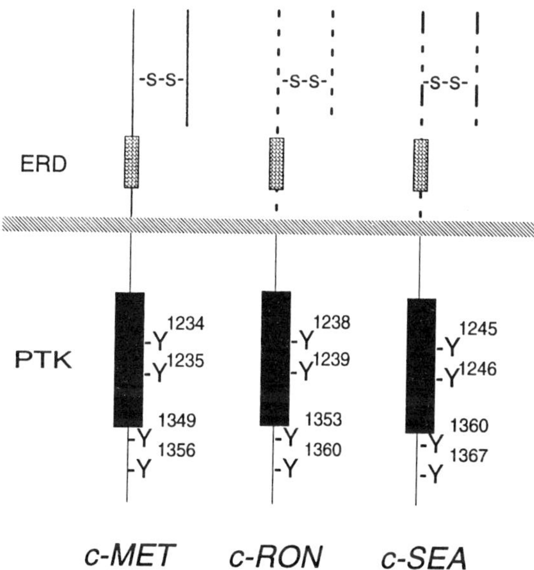

Figure 1. Schematic representation of *MET* receptor tyrosine kinase family. Conserved tyrosine residues are indicated. S-S: disulfide bridges. PTK: protein tyrosine kinase domain. ERD: extracellular related domain.

The first translational product of the c-*MET* proto-oncogene is a single chain precursor, $p170^{MET}$; its proteolysis originates the α and β chains. The mature receptor is a glycoprotein that contains N-linked oligosaccharides. Digestion with endoglycosydases resulted in a shift of the apparent molecular weight of the 170 kDa precursor to approximately 150 kDa. This is the same molecular weight of the precursor synthesized in cells in which N-linked glycosylation was inhibited by tunicamycin and matched the predicted size in aminoacids of the *Met* receptor protein. Glycosylation of the 150 kDa precursor is essential for further processing. Thus, treatment with tunicamycin prevents the correct cleavage and the appearance of the α and β chains (Giordano et al., 1989b).

By producing a panel of monoclonal antibodies reacting with the extracellular domain of the HGF receptor, two C-terminal truncated *Met* proteins were identified: a 140 kDa transmembrane form (p140^{MET}) and a 130 kDa soluble protein (p130^{MET}) which is released from the cell. The truncated forms of the HGF receptor have the same heterodimeric structure of the intact p190^{MET}, consisting of two disulfide-bonded chains.
The α chains of the truncated forms are indistinguishable from the α chain of p190^{MET}. The β chain of p140^{MET} is approximately 85 kDa (p85$^{\beta}$), while the β chain of p130^{MET} is approximately 75 kDa (p75$^{\beta}$). Both p85$^{\beta}$ and p75$^{\beta}$ are truncated at their C-termini. The transmembrane 140^{MET} originates by post-translational proteolytic cleavage of the p170^{MET} precursor. Infact, mouse cells transfected with the full length human *MET* cDNA express p190^{MET} as well as p140^{MET} and p130^{MET} (Giordano et al., 1993). The proteolytic event generating p140^{MET} occurs within the endoplasmic lumen (Crepaldi et al., submitted). The soluble p130^{MET} is likely to be generated by proteolytic cleavage at the cell surface (Prat et al., 1991a).

The p190^{MET} receptor has been proved to be the high affinity binding site for HGF/SF by: (1) heparin-insensitive, specific binding of the radioiodinated ligand to a 190 kDa protein immunopurified with antibodies directed against the product of the *MET* oncogene (p190^{MET}); (2) chemical cross-linking of the α and β chains of HGF/SF to p190^{MET}; (3) specific inhibition of the cross-linking by purified preparation of biologically active ligand (Naldini et al., 1992c).

Formal proof of the role of p190^{MET} as the sole component of the HGF receptor was obtained by reconstituting the ligand binding site in insect cells infected with a recombinant baculovirus carrying the human *MET* cDNA. A single class of binding sites was observed with nanomolar affinity (Naldini et al., 1992c). Moreover, mouse cells transfected with the full length human *MET* cDNA acquired the ability to respond to HGF/SF with increased motility and anchorage-independent growth in soft agar (Giordano et al., 1993).

The HGF receptor is widely expressed in several epithelial tissues and is often overexpressed in neoplastic cells (Iyer et al., 1990; Naldini et al., 1991a; Prat et al., 1991b; Di Renzo et al., 1991). In some tumors of the gastro-intestinal tract the *Met* tyrosine kinase is constitutively activated, either by overexpression of the amplified *MET* oncogene or by lack of cleavage of the receptor precursor, due to defective post-translational processing (Giordano et al., 1989a; Mondino et al., 1991).

2.2 Regulation of the HGF receptor kinase activity

The earliest detectable consequence of HGF stimulation is tyrosine autophosphorylation of its receptor β subunit (Naldini et al., 1991b; Bottaro et al., 1991); this powerfully stimulates the tyrosine kinase activity of p190^{MET}. The increase in phosphorylation rate is due to a several fold increase in the V_{max} of the enzyme catalyzed phosphotransfer reaction (Naldini et al., 1991a).

The major tyrosine autophosphorylation site of the p190^{MET} receptor has been identified with the tyrosine residue 1235 both *in vitro* and *in vivo* (Ferracini et al., 1991). Tyr_{1235} is part of a "three tyrosine" motif, including Tyr $_{1230}$, Tyr_{1234}, and Tyr_{1235}, conserved in several other tyrosine kinase receptors (Hanks et al.,1988).

The role of these tyrosines in the positive regulation of the HGF receptor has been studied by site-directed mutagenesis. Substitution of either Tyr_{1235} or Tyr_{1234} with Phe residues severely reduced the *in vitro* kinase activity of the mutant receptors toward exogenous substrates, without preventing their autophosphorylation. Kinetic experiments showed that the residual activity of these mutants could still be enhanced by autophosphorylation. Phosphopeptide mapping indicated that in the absence of Tyr_{1235}, the adjacent Tyr_{1234} is the vicarious phosphorylation site. Only the replacement of both Tyr_{1234} and Tyr_{1235} yielded a mutant which lost the ability to be activated by autophosphorylation (Longati et al., submitted).

The crystal structure of the catalytic core of the cAMP-dependent protein kinase has been recently described (Knigton et al., 1991). According to this model one can extrapolate the location of $Tyr_{1234-1235}$ to a loop facing the cleft between the two lobes of the catalytic core. Phosphorylation of one of these residues might induce a conformational change that renders the enzyme cleft more accessible to the substrates.

The tyrosine kinase activity of the HGF receptor is negatively modulated by protein kinase C activation, similarly to its effect on EGF and insulin receptors. The negative control by protein kinase-C seems to operate at steady-state level also in physiological conditions, i.e, in the absence of exogenous activation. In fact, depletion of protein kinase-C by prolonged treatment with TPA resulted in an increased phosphorylation of $p190^{MET}$ (Gandino et al., 1990).

The normal endogenous cellular activator of protein kinase-C is diacylglycerol which is produced transiently in the cell membrane as result of the breakdown of phosphatidyl-inositol-phosphate by phospholipase-C. This enzyme concomitantly produces inositol-1,4,5-triphosphate (IP_3), one of the soluble cytoplasmic second messengers causing the release of calcium ions from intracellular stores. Both arms of the signal transduction pathway triggered by phospholipase-C modulate the kinase activity of the HGF receptor. The increase in intra-cellular Ca^{2+} concentration is inhibitory. This negative regulation takes place independently from protein kinase-C, *via* activation of another Ca^{2+}-dependent serine kinase (Gandino et al., 1991). Notably a unique serine residue is phosphorylated both by this Ca^{2+}-dependent kinase and by protein kinase-C activation (Gandino et al., submitted). Thus two independent systems negatively modulate the HGF receptor through the same molecular mechanism.

2.3 THE HGF/SF receptor signal transduction pathways.

Hepatocyte Growth Factor/Scatter Factor (HGF/SF) elicits mitogenesis and/or motogenesis in epithelial cells. It should be noted that while its biological effects vary depending on the target cell, the HGF/SF signal is mediated by a single receptor: the tyrosine kinase encoded by the *MET* proto-oncogene. Thus it is likely that the basis for these distinct biological responses will be found in a bifurcation in the pathways lying downstream to the receptor.

A new concept has emerged in the past two years in the field of signal transduction which links tyrosine kinase receptors to downstream signalling events. A number of cytoplasmic molecules that mediate cellular responses to growth factors have been shown to interact with activated receptors. Among these are PI 3-kinase, *ras*GAP, phospholipase C-γ, *src*-related tyrosine kinases, SHC, GRB-2 and PTPase-1C (for a review see Mayer and Baltimore 1993). These molecules all contain one or more SH2 regions (*src* homology region 2, reviewed in Koch et al., 1991) which are responsible for binding to specific phosphotyrosine residues in activated receptors (Cantley et al., 1991).

Recombinant HGF receptor produced in insect cells was used to test its ability to associate with number of SH2-containing transducers. The recombinant p190^{MET} receptor associates *in vitro* with *ras*GAP, PI 3-kinase holoenzyme, PLC-γ, and p59*Fyn*, a member of the c-*src* family of kinases (FIG.2). These interactions are strictly dependent upon receptor phosphorylation (Bardelli et al., 1992).

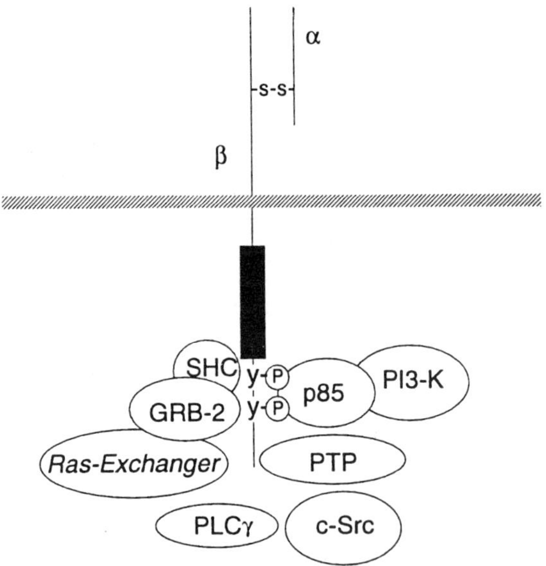

Figure 2. The HGF/SF receptor transductosome.

Addition of HGF/SF to responsive cells (A549, lung carcinoma cells) induces association of PI 3-kinase with the *Met* receptor *in vivo* (Graziani et al., 1991), and stimulates the *ras*-guanine nucleotide exchanger (Graziani et al., 1992). Moreover, we recently reported that there is a protein tyrosine phosphatase activity (PTP), wich coprecipitates with the HGF receptor in A549 cells after stimulation with HGF/SF (Villa-Moruzzi et al., 1993). These experiments show that the HGF receptor has the sequence requirements to associate with a specific subset of cytoplasmic signalling proteins containing SH2 domains.

Identification of the transducers' binding sites is a pre-requisite to develop model systems aimed at correlating different biological response to a specific signaling pathway. By a combination of two approaches, competition with synthetic phosphopeptides and association with Tyr-Phe receptor mutants, the binding sites for PI 3-kinase have been identified to Y_{1349} and Y_{1356} in the HGF receptor. Y_{1349}VHV and Y_{1356}VNV do not conform to the canonical consensus sequence YXXM for PI 3-kinase binding, and thus define YVXV as a novel recognition motif. Y_{1349} and Y_{1356} are located within the C-terminal portion of the HGF receptor and are phosphorylation sites. Interestingly these two tyrosine residues are conserved in the evolutionarly related receptors *RON* and *SEA* together with the aminoacids in the +1, +2, and +3 positions.

Stable cell lines expressing wt receptor and Tyr-Phe mutants have been estabilished to test the effects of this mutations with the HGF/SF pleiotropic biological response.

ACKNOWLEDGEMENTS

We thank A. Cignetto and E. Wright for editing the manuscript. This work was supported by grants of the Associazione Italiana Ricerche sul Cancro (AIRC), and of the Consiglio Nazionale delle Ricerche (CNR: special project ACRO).

REFERENCES

Bardelli, A., Maina, F., Gout, I., Fry. M.J., Waterfield, M.D., Comoglio, P.M., and Ponzetto, C. (1992) Oncogene 7: 1973-1978.

Bottaro, D.P., Rubin, J.S., Faletto, D.L., Chan, A.M.L., Kmiecick, T.E., Vande Woude, G.F., and Aaronson, S.A. (1991) Science 251: 802-804.

Cantley, L.C., Auger, K., Carpenter, C., Duckworth, B., Graziani, A., Kapeller, R., and Soltoff, S. (1991) Cell 64: 281-302.

Di Renzo, M.F., Narsimhan, R.P., Olivero, M., Bretti, S., Giordano, S., Medico, E., Gaglia, P., Zara, P., and Comoglio, P.M. (1991) Oncogene 6: 1997-2003.

Ferracini, R., Longati, P., Naldini, L., Vigna, E., and Comoglio, P.M. (1991) J. Biol. Chem. 266: 19558-19564.

Gandino, L., Munaron, L., Naldini, L., Magni, M., and Comoglio, P.M. (1991) J. Biol. Chem. 266: 16098-16104.

Gandino, L., M.F. Di Renzo, S. Giordano, F. Bussolino and P.M. Comoglio (1990) Oncogene 5: 721-725.

Gherardi, E., and Stoker, M. (1990) Nature 346: 228.

Giordano, S., Di Renzo, M.F., Ferracini, R., Chiado'Piat, L. and Comoglio, P.M. (1988) Mol. Cell. Biol. 8: 3510-3517.

Giordano, S., Ponzetto, C., Di Renzo, M.F., Cooper, C.S. and Comoglio, P.M. (1989a) Nature 339: 155-156.

Giordano, S., Di Renzo, M.F., Narsimhan, R., Cooper, C.S., Rosa, C., and Comoglio, P.M. (1989b) Oncogene 4: 1383-1388.

Giordano S, Zhen Z, Medico E, Gaudino G, Galimi F, Comoglio PM. (1993) Proc. Natl. Acad. Sci. USA 90: 649-653.

Graziani, A., Gramaglia, D., Cantley, L.C., and Comoglio, P.M. (1991) J. Biol. Chem. 266: 22087-22090.

Graziani, A., Gramaglia, D., Dalla Zonca, P. and Comoglio P.M. (1993) J. Biol. Chem. in press.

Hanks, S.K., Quinn, A.M. and T. Hunter (1988) Science 241: 42-52.

Hartmann, G., Naldini, L., Weidner, M., Sachs, M., Vigna, E., Comoglio, P. and Birchmeier, W. (1992) Proc. Natl. Acad. Sci. USA. 89: 11574-11578.

Huff, J.I., Jelinek, A.M., Borgman, C.A., Lansing, T.J. and Parsons, J.T. (1993) Proc. Natl. Acad. Sci. USA. in press.

Iyer, A., Kmiecik, T.E., Park, M., Daar, I., Blair, D., Dunn, J., Sutrave, P., Ihle, J.N., Bodescot, M. and Vande Woude, G. (1990) Cell Growth & Differentiation 1: 87-95.

Knighton, D. R., Zheng, J., Ten Eyck, L. F., Ashford, V. A., Xuong, N., Taylor, S., and Sowadski, J. M. (1991) Science 253: 407-413.

Koch, C.A., Anderson, D., Moran, M.F., Ellis, C., and Pawson, T., (1991) Science 252: 668-674.

Lokker, N.A., Mark, M.R., Luis, E.A., Bannet, G.L., Robbins, K.A., Baker, J.B. and Godowski, P.J. (1992) EMBO J. 11: 2503-2510.

Mark, M.R., Lokker, N.A., Zioncheck, T.F., Luis, E.A. and Godowski, P.J. (1993) J. Biol. Chem. in press.

Mayer, B.J. and Baltimore, D. (1993) Trends In Cell Biology 3: 8-13.

Naldini, L., Vigna, E., Ferracini, R., Longati, P., Gandino, L., Prat, M., and Comoglio, P.M. (1991a) Mol. Cell. Biol. 11: 1793-1803.

Naldini, L., Vigna, E., Narshiman, R.P., Gaudino, G., Zarnegar, R., Michalopoulos, G., and Comoglio, P.M. (1991b) Oncogene 6: 501-504.

Naldini, L., Weidner, M., Vigna, E., Gaudino G., Bardelli A., Ponzetto, C., Narsimhan, R., Hartmann, G., Zarnegar, R., Michalopoulos, G., Birchmeier, W., and Comoglio, P.M. (1991c) EMBO J. 10: 2867-2878.

Hartmann, G., Naldini, L., Weidner, M., Sachs, M., Vigna, E., Comoglio, P.M. and Birchmeier, W. (1992) Proc. Natl. Acad. Sci. USA 89: 11574-11578.

Prat, M., Crepaldi, T., Gandino, L., Giordano, S., Longati, P., and Comoglio, P.M. (1991a) Mol. Cell. Biol. 11: 5954-5962.

Prat, M.P., Narsimhan, R.P., Crepaldi, T., Nicotra, M.R., Natali, P.G., and Comoglio, P.M. (1991b) Int. J. Cancer 49: 323-328.

Ronsin, C., Muscatelli, F., Mattei, M.G. and Breathnach, R. (1993) Oncogene 8: 1195-1202.

Villa-Moruzzi, E., Lapi, S., Prat, M., Gaudino, G. and Comoglio, P.M. (1993) J. Biol. Chem. in press.

Weidner, K.M., Arakaki, N., Vandekerchove, J., Weingart, S., Hartmann, G., Rieder, H., Fonatsch, C., Tsubouchi, H., Hishida, T., Daikuhara, Y., and Birchmeier, W. (1991) Proc. Natl. Acad. Sci. USA 88: 7001-7005.

ANTIBODY ENGINEERING

ECB6: Proceedings of the 6th European Congress on Biotechnology
edited by L. Alberghina, L. Frontali, P. Sensi

Engineering of antibody fragments produced in bacteria: Solutions to the purification problem

A. Skerra, L.-O. Essen, T. Schmidt, and C. Wardenberg

Max-Planck-Institut f. Biophysik, Heinrich-Hoffmann-Str. 7, W-6000 Frankfurt/M. 71, Germany

1. Recombinant Ig fragments from *Escherichia coli*

The development of methods for expressing functional antibody fragments in *Escherichia coli* by secretion to the periplasm [1-3] has enabled the facile engineering of these antigen-binding Ig fragments [reviewed in 4]. Three kinds of recombinant antibody fragments are generally in use: F_v, "single chain" F_v (scF_v), and F_{ab} fragments. F_v fragments are the smallest antibody fragments that carry the whole antigen binding site. They consist of the variable domains V_H and V_L, which are associated via non-covalent forces. Unfortunately, these Ig fragments show a concentration-dependent tendency towards dissociation into the isolated domains [5,6]. Yet, the two polypeptide chains of an F_v fragment can be covalently connected by a flexible peptide linker, yielding so-called scF_v fragments [reviewed in 7]. Although scF_v fragments show improved stabilities as isolated proteins they were observed to form oligomeric structures [8] and often possess reduced antigen binding affinities [7]. In contrast to F_v fragments the larger F_{ab} fragments consist of two polypeptide chains that are covalently linked by a disulfide bond. The additional pairing of the C_H1 and C_κ (or C_λ) domains also contributes to their stable association, even in cases where the disulfide bond is absent [9]. F_{ab} fragments have previously been accessible by limited proteolytic digest of whole antibodies and a wealth of biochemical data is available for these natural Ig fragments. Therefore, even in the age of bacterial production they probably remain to be the molecules suited best for the study of antigen recognition.

2. Problems in purification

Whereas the expression of functional antibody fragments in *E. coli* has become a familiar method the development of techniques for the rapid and simple purification of recombinant Ig fragments remained largely unattended up to now. In most of the earlier, more exemplary studies on bacterially expressed antibody fragments hapten affinity columns were used for the simple affinity purification, e. g. phosphoryl choline sepharose in the case of the antibody McPC603 [1,3,9]. Nowadays, however, Ig fragments with antigen specificities of greater practical relevance have come into focus, demanding the use of antigen affinity columns for the same purpose. Antigen affinity chromatography usually makes use of rather harsh and even partially denaturing conditions for elution of the bound immunoglobulin. Given the limited conformational stabilities especially of F_v and scF_v fragments, such conditions are highly undesirable when a functional and homogeneous protein preparation is wanted. Furthermore, in many cases the antigen may not be available in the amount and purity needed and, particularly in the structure-function analysis of engineered Ig fragments, an appropriate antigen that is bound with sufficient affinity may not even exist. In order to develop methods that permit the efficient single-step purification of antibody fragments under mild conditions and without knowledge of their antigen specificities we used the concept of appending peptide sequences with defined affinity properties to their carboxy-termini. Since the carboxy-terminal end of the variable Ig domain is far away from the antigen combining site these peptide tags do not interfere with antigen binding and, consequently, they need not be removed via proteolysis for most applications.

3. Purification by metal chelate affinity chromatography

In a first approach, carboxy-terminal stretches of several consecutive histidine residues were fused to Ig fragments for the selective purification by immobilized metal affinity chromatography (IMAC). This strategy was initially applied to the scF_v fragment of the myeloma protein McPC603 which was functionally secreted to the periplasm of *E. coli* [10]. A stretch of five histidine residues was considered to be most suitable for the efficient single-step purification. Meanwhile, however, it was found that the efficiency of retardation to the metal chelate affinity column varies depending on the biochemical properties of the individual antibody fragment so that satisfactory results are usually obtained with six histidine residues.

The same purification strategy was adopted to F_{ab} fragments functionally secreted in *E. coli* [cf. 9]. In order to optimize the expression and purification conditions the variable domains of the lysozyme-binding antibody D1.3 [11] were used [Skerra & Wardenberg, unpublished]. The V_L domain was fused to the murine C_κ domain yielding the whole light chain as it is present in a natural antibody. The V_H domain was fused to the murine C_H1 domain of class IgG1. The resulting heavy chain was extended into the hinge region up to cysteine residue no. 235 that forms the link to the carboxy-terminal cysteine of the light chain. The F_{ab} fragment thus obtained possesses two carboxy-termini to both of which the oligo-histidine tail could in principle be added. In contrast to the heavy chain of an F_{ab} fragment (called F_d fragment), the light chain shows a general tendency to form a homo-dimer, also called Bence Jones protein [12,13]. In order to avoid the co-purification of the F_{ab} fragment (the hetero-dimer of F_d fragment and light chain) with the corresponding light chain homo-dimer, which is also formed during expression in *E. coli*, the hexa-histidine tail was fused to the carboxy-terminus of the heavy chain. When the periplasmic cell fraction of *E. coli* cells co-secreting the two corresponding polypeptide chains was applied to a metal chelate affinity column a single protein fraction was specifically eluted by an imidazole gradient [cf. 10]. This fraction contained the homogeneous F_{ab} fragment with the proper inter-chain disulfide bond formed. Due to the presence of the constant domains, it was possible to confirm the functionality of the F_{ab} fragment directly in an ELISA with lysozyme as the antigen and a commercial anti-mouse-Ig-enzyme conjugate for detection of the bound Ig fragment.

Contrasting to scF_v and F_{ab} fragments, an F_v fragment comprises the non-covalent assembly of two different polypeptide chains. Due to the weak association of the two variable antibody domains [5,6] and to the ability of V_L to form homo-dimers [13], the problem arose to specifically purify the hetero-dimer of V_H and V_L via IMAC. In analogy to the F_{ab} fragment it might be assumed that this could most conveniently be achieved by adding a histidine tail to the carboxy-terminus of the V_H domain, which must usually be associated to a V_L domain in order to adopt a soluble structure. This strategy was tested with a CDR-engineered F_v fragment of the antibody HyHEL-10 [14] both domains of which were co-secreted to the periplasm of *E. coli* [cf. 9]. When the periplasmic cell fraction was subjected to IMAC, however, only the V_H domain specifically bound to the column and was eluted as a homogeneous protein preparation upon application of an imidazole gradient. Not unexpectedly, the converse phenomenon was observed when the histidine tail was fused to the V_L domain instead of V_H. When, finally, the F_v fragment was expressed with histidine tails at the carboxy-termini of both of its variable domains, mixtures of V_L and V_H with varying stoichiometry were obtained by IMAC. This behaviour was attributed to the disruption of the weak non-covalent interactions between V_H and V_L under the high salt conditions (0.5 to 1 M NaCl) which were necessary to prevent non-specific binding of proteins to the charged groups of the metal chelate column by an ion exchange effect. As we found out, NaCl can be substituted by glycine betaine using otherwise identical conditions for the metal chelate chromatography. The zwitter-ionic osmolyte betaine serves at a concentration of 0.5 M equally well to prevent non-specific binding and does not interfere with metal complexation. In addition to its contribution to the ionic strength of the buffer this compound has also a generally stabilizing effect on protein structures [15]. By use of this buffer system the F_v fragment was isolated via IMAC with correct stoichiometry of V_H and V_L when both domains carried a histidine tail. Thus, it was demonstrated that the efficient single-step purification of a bacterially expressed F_v fragment can indeed be achieved by metal chelate affinity chromatography [Essen, Wardenberg, and Skerra, unpublished].

4. Purification by the *Strep tag* technique

In a second strategy towards the development of a carboxy-terminal peptide tag our aim was to establish a system that permitted purification of Ig fragments under even more physiological conditions than those described for IMAC above. For this purpose a novel affinity tag, termed *Strep tag* was derived by selection from a random peptide library [16]. The *Strep tag* consists of the nine amino acid sequence "Ala-Trp-Arg-His-Pro-Gln-Phe-Gly-Gly". Fused to the carboxy-terminus of, e. g. a variable Ig domain, this peptide displays specific binding activity towards streptavidin, a familiar biochemical reagent that is commercially available in a variety of conjugates as well as immobilized to agarose. The binding is competitive with natural ligands of streptavidin, as biotin or related compounds. Therefore, when the *Strep tag* was fused to the carboxy-terminus of the V_H domain of the D1.3 F_v fragment the bacterially secreted protein was specifically affinity-purified with streptavidin agarose by elution in the presence of low concentrations of iminobiotin under otherwise physiological buffer conditions. The intact and functional Ig fragment was isolated from either the periplasmic cell fraction or the total cell protein of *E. coli* in a single step without disruption of the native hetero-association of V_H and V_L [16]. Apart from its use for the efficient purification of functional F_v fragments, the *Strep tag* offers the additional advantage of permitting the facile and specific detection of the recombinant protein in ELISAs or on Western blots by use of a commercial conjugate of streptavidin and alkaline phosphatase [16]. Due to the absence of constant Ig domains in the case of F_v and scFv fragments, specific antisera are needed for their immuno-detection when expressed and purified according to the IMAC strategy described above. Therefore, the *Strep tag* clearly surpasses oligo-histidine fusions in this regard.

5. Conclusion

Two general methods for the efficient single-step purification of bacterially expressed scF_v, F_v, and F_{ab} fragments were developed, based on carboxy-terminal fusion peptides with defined affinity properties. These affinity tags do not interfere with the antigen binding function and obviate the need for specific antigen-affinity columns. Novel vectors were constructed that allow for the direct cloning of the Ig variable domain genes followed by subsequent expression in *E. coli* and one-step purification. In the light of the exponentially growing knowledge about antibody structure and function this technology does not only permit the rational engineering of immunoglobulin fragments derived from nature but also provides an experimental basis for attempts towards the design of artificial antibodies.

References

1. A. Skerra and A. Plückthun, Science, 240 (1988) 1038.
2. M. Better et al., Science, 240 (1988) 1041.
3. A. Plückthun and A. Skerra, Methods Enzymol., 178 (1989) 497.
4. A. Skerra, Curr. Opinion Immunol., 5 (1993) in press.
5. R. Glockshuber et al., Biochemistry, 29 (1990) 1362.
6. D.J. King et al., Biochem. J., 290 (1993) 723.
7. R.E. Bird and B.W. Walker, Trends Biotechnol., 9 (1991) 132.
8. A.D. Griffiths et al., EMBO J., 12 (1993) 725.
9. A. Skerra and A. Plückthun, Protein Eng., 4 (1991) 971.
10. A. Skerra et al., Biotechnology, 9 (1991) 273.
11. G. Boulot et al., J. Mol. Biol., 213 (1990) 617.
12. M. Klein et al., Biochemistry, 18 (1979) 1473.
13. F.J. Stevens et al., Biochemistry, 30 (1991) 6803.
14. E.A. Padlan et al., Proc. Natl. Acad. Sci. USA, 86 (1989) 5938.
15. M.M. Santoro et al., Biochemistry, 31 (1992) 5278.
16. T.G.M. Schmidt and A. Skerra, Protein Eng., 6 (1993) 109.

ECB6: Proceedings of the 6th European Congress on Biotechnology
edited by L. Alberghina, L. Frontali, P. Sensi
1994 Elsevier Science B.V.

EXPRESSION, CHARACTERIZATION AND CRYSTALLIZATION OF A RECOMBINANT ANTI-NEURAMINIDASE SINGLE CHAIN ANTIBODY

Maria Lah[1], Robyn Malby[2], Alanna Goldstraw[1], Jacinta F. White[1], Olan Dolezal[1], David Dougan[1], Alec A. Kortt[1], Robert A. Irving[1], Peter M. Colman[2] and Peter J. Hudson.[1]

CSIRO Division of Biomolecular Engineering[1] and Biomolecular Research Institute [2],
343 Royal Parade, Parkville, Vic, 3052. Australia

1. INTRODUCTION

Genetically engineered antibodies have immense potential for use in the diagnosis and treatment of human diseases. Single chain Fv (scFv) molecules have greater advantages over conventional antibodies as only those portions of the molecule involved in antigen recognition, the variable heavy and light chain (V_H and V_L) domains are linked by a flexible polypeptide. To evaluate high-level expression systems for scFv production we chose to clone the genes encoding the well characterized monoclonal antibody NC10 [an antibody that recognizes the tern influenza N9 strain of neuraminidase (NA) and whose binding surface as a Fab-NA complex has been resolved by X-ray crystallography (1, 2)] as a scFv construct into the high level secretion vector, pPOW (3, 4). The complex between the parent NC10 Fab and NA (5) is unique among the the small number of crystal structures of antibody-antigen complexes, because only four of the six complementarity-determining regions (CDRs) of the Fab make contact with antigen. Despite this, each protein has more than 700Å^2 of surface area buried in the interface.

2. CONSTRUCTION AND STRUCTURAL ANALYSIS OF A RECOMBINANT SINGLE CHAIN ANTIBODY

The V_H and V_L genes of NC10 were used to construct a synthetic gene encoding a scFv fragment in which the V_H and V_L domains were joined by a peptide linker of 15 residues' length and an octapeptide (FLAG) affinity tail was fused to the carboxyl-terminus of V_L. ScFvNC10 was expressed in *E.coli* and purified by solubilization of the bacterial membranes followed by gel filtration and anion exchange chromatography (4). ScFvNC10 was purified as a soluble molecule and binding studies using ELISA, sedimentation equilibrium and

Biacore techniques have demonstrated that the scFv had retained affinity for NA which was only two-fold lower than that of the parent Fab (5). Our expression system was successful in producing sufficient scFv for crystallization and for three-dimensional structural determination by X-ray diffraction (Fig. 1). The scFv-NA complex was crystallized and the structure solved using molecular replacement techniques and energy minimization routines. We believe this is the first successful crystallization and structural analysis of a scFv molecule in complex with target antigen.

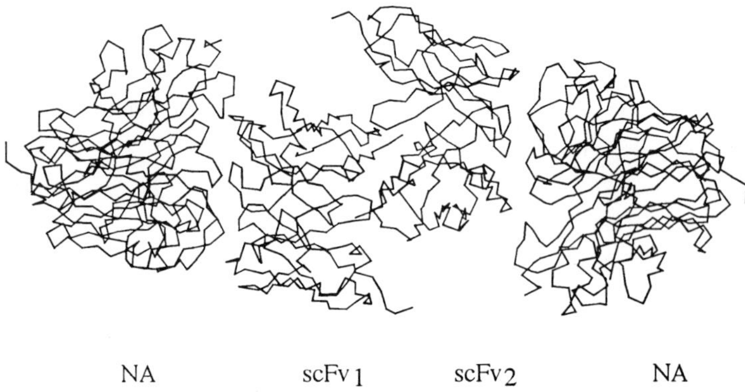

Figure 1: Cα trace showing crystal packing of two scFvNC10-NA complexes.

Preliminary comparisons of the NC10 scFv-NA structure to the NC10 Fab-NA structure indicates that the modes of attachment of Fab and scFv to NA are very similar. The scFv-NA structure was solved at a resolution of 3.0Å, and will be refined at higher resolution so that comparisons between the overall structures to determine the effect (if any) of the peptide linker and peptide affinity tail upon the binding conformation of the scFv can be assessed. Analysis of the NA-scFv structure will reveal the binding conformation of a scFv in complex with its target antigen and will provide the basis for rational protein design of scFv mutants with modified affinity and stability. We note that the scFvNC10 molecules may have crystallized as dimers with single-chain linkers forming intermolecular bonds.

3. PHAGE SURFACE PRESENTATION AND SECRETION OF A RECOMBINANT SINGLE CHAIN ANTIBODY

An alternative means of affinity maturating antibodies can be achieved by mutation and selection of antibody fragments displayed on phage surfaces using antigen affinity procedures (6). Antibody libraries based on bacteriophage display have enabled the rapid identification and selection of specific binding molecules by an enrichment and selection process. In an attempt to improve phage surface display by phagemid vectors (7), we have developed the phagemid vectors, pHFA and pHFA/SacII for high level expression of antibody that can be produced as: i) a soluble antibody fragment incorporating a stable carboxyl-terminal octapeptide (FLAG) or ii) on the surface of a bacteriophage minor coat protein, known as gene III protein (gIIIp).

The gene for the scFv NC10 was cloned into the phagemid vectors pHFA and pHFA/SacII. Recombinant clones were found to express a 30 kDa soluble fusion (scFvNC10-FLAG) protein into the supernatant and periplasm when grown in non-suppressor strains of *E.coli*, which could be detected after Western blotting using an anti-FLAG antibody. The expressed proteins were capable of binding to the NA component on influenza particles in a concentration dependent manner as determined by ELISA. When the scFvNC10 constructs were grown in suppressor strains of *E.coli* and infected with helper phage for phage surface expression of antibody encoding fragments, scFvNC10-FLAG-gIIIp fusion protein production was monitored by Western blotting and electron microscopy studies. Phage bearing the scFvNC10 fragment associate with whole influenza virus, while phage bearing no scFvNC10 inserts show no such interactions (Fig. 2).

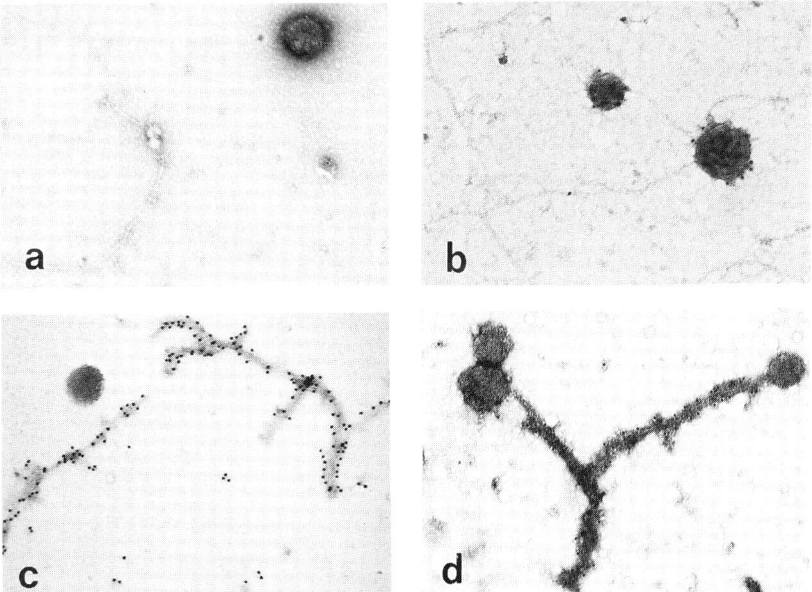

Figure 2: Immunoelectron microscopy of G70/C influenza virus (spherical in appearance) and filamentous phage (strand-like in appearance). Two different immunogold staining methods were used to analyse the interaction of the phage with the N9 NA antigen on the G70/C virus. Specific gold-labelled antibodies were used to stain either the G70/C virus

(panels a and b) or the phage (panels c and d). Phage which display the scFvNC10 (panels b and d) bind at their tips to the G70/C virus, whereas phage which do not display scFvNC10 (panels a and c) do not bind.

Binding affinities of scFvNC10 phage fusion proteins were determined by ELISA on whole influenza virus with either an anti-FLAG antibody or an anti-M13 antisera. Phage bearing scFvNC10 exhibit three-fold higher binding affinity to G70/C than phage bearing no antibody fragments.

Affinity maturation of the NC10 scFv is being assessed using the pHFA/SacII phagemid vector. The entire scFvNC10 gene is being subjected to an error prone polymerase chain reaction and recombinant clones bearing mutations are being screened and selected for by their ability to bind to NA using a two membrane system (8). Site-directed mutagenesis of residues known to be involved with antigen binding is also being used to generate recombinant clones whose affinity for NA is being determined. A detailed structural comparison of the NC10 scFv mutants generated by such mutagenesis techniques will enable a rational design to affinity maturation.

REFERENCES

1. Tulip, W.R., Varghese, J.N., Webster, R.G., Air, G.M., Laver, W.G. and Colman, P.M. (1989) *Cold Spring Harb. Symp. Quant. Biol.* **54**, 257-263.
2. Tulip, W.R., Varghese, J.N., Laver, W.G. and Colman, P.M. (1992) *J. Mol. Biol.* **227**, 122-148.
3. Power, B.E., Ivancic, N., Harley, V.R., Webster, R.G., Kortt, A.A., Irving, R.A. and Hudson P.J. (1992) *Gene* **113**, 95-99.
4. Malby, R.L., Caldwell, J.B., Gruen, L.C., Harley, V.R., Ivancic, N., Kortt, A.A., Power, B.E., Webster, R.G., Colman,P.M. and Hudson, P.J. *Proteins: Structure, Function & Genetics* (in press).
5. Colman, P.M., Tulip, W.R., Varghese, J.N., Tulloch, P.A., Baker, A.T., Laver, W.G., Air, G.M. and Webster, R.G. (1989) *Phil. Trans. R. Soc. Lond. [Biol]* **323**, 511-518.
6. Marks, J.D., Hoogenboom, R.H., Griffiths, A.D. and Winter, G. (1992) *J. Biol. Chem.* **267**, 16007-16010.
7. Hoogenboom, H.R., Griffiths, A.D., Johnson, K.S., Chiswell, D.J., Hudson, P. and Winter, G. (1991) *Nucleic Acids Res.* **19**, 4133-4137.
8. Skerra, A., Dreher, M.L. and Winter, G. (1991) *Anal. Biochem.* **196**, 151-155.

ECB6: Proceedings of the 6th European Congress on Biotechnology
edited by L. Alberghina, L. Frontali, P. Sensi

Humanizing Antibodies by Epitope Imprinted Selection (EIS)

Hennie R. Hoogenboom[a,b], Laurent S. Jespers[b], Andrew Griffiths[b], Mariangela Figini[b], Michael Baier[b], Andy Roberts[a], David J. Chiswell[a] and Greg Winter[b]

[a]Cambridge Antibody Technology Ltd., The Science Park, Melbourn, Cambridgeshire SG8 6EJ, U.K.

[b]MRC Centre for Protein Engineering, Hills Road, Cambridge CB2 2QH, U.K.

1. INTRODUCTION

The hybridoma technology has generated a huge number of interesting and very well characterised antibodies, some of which may be of use as therapeutics. Unfortunately, most of them are of rodent origin, and would evoke an immune response when used in therapy. In addition, rodent antibodies may not recruit appropriate effector functions. Human antibodies can be generated in a number of ways, with the most novel and promising techniques being the antibody gene manipulations. Human antibodies can be made directly by combinatorial cloning of human V-genes and selection of binders from phage display libaries. For example, libraries of human V-genes from unimmunised blood donors were displayed on the surface of phage fd as single-chain Fv fragments, and binding phage isolated by affinity selection. Binders with mediate affinity (association constant in the range of 10^6-10^7 M^{-1}) were isolated against a number of antigens (1,2). We also improved the affinity of one of these antibodies (with $K_d = 3.3 \times 10^{-7}$ M for the hapten phOx), by shuffling its chains with V-gene libraries and selecting higher affinity variants on antigen (upto $K_d = 1.1 \times 10^{-9}$ M)(3). Finally, we have built a germline gene antibody repertoire using V-gene germline segments, which were rearranged *in vitro* to provide a synthetic CDR3 of random sequence (4). From these artificial repertoires many different binders highly specific for haptens and proteins were identified (4 and unpublished results). Our results show that a single pot library, made without immunisation, should provide a rich source of antibody specificities, including those against pathogens or self-antigens. This technology in the future will generate many important human antibodies for therapy (5).

Alternatively, rodent antibodies can be rebuild for therapy. For example, the therapeutic efficacy of rodent antibodies can be improved by chimaerisation or humanisation (6) of their genes. We now present a method involving phage display to humanise rodent antibodies, dubbed Epitope Imprinted Selection (EIS)(7), using (1) one of the rodent antibody chains as a scaffold to imprint epitope properties onto a library of human partner chains, and (2) selection from phage display libraries to select the best binders.

2. HUMAN ANTIBODIES BY EIS

The aim of the present study is to develop a technique for rapid converting mouse antibodies into human antibodies with the same binding specificity. The idea is that the mouse antibody V-genes can serve as a mould (or scaffold) with 'antigen binding imprint', around which the human antibody can be rebuild. The combination of chain shuffling and selection of binding variants of phage display libraries was applied before, to diversify mouse (8) or affinity mature human (3) antibodies. Now we used chain shuffling to humanise an antibody (Fig. 1). One of the mouse V-genes is replaced with a library of human V-genes, and phage display used to select from this 'chimaeric' library binders directed against the original antigen. The mouse V-domain may help to 'imprint' the antigen-binding features and dock the Fab's on the original epitope. In a second shuffle, the selected human genes are reshuffled with a library of human partner V-genes.

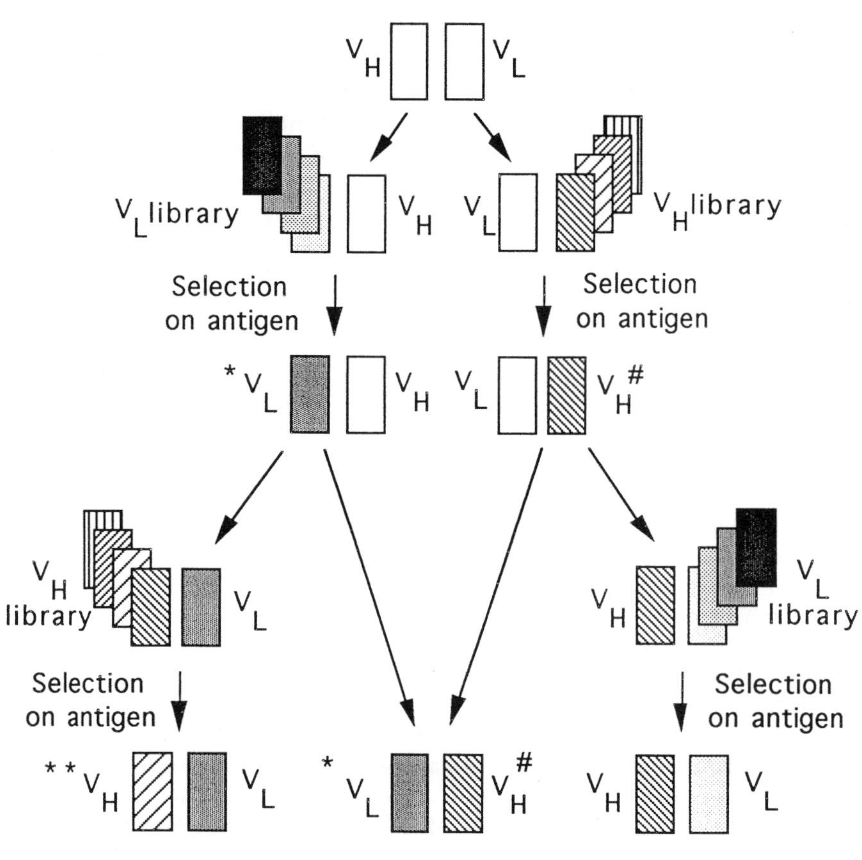

Figure 1.The principle of Epitope Imprinted Selection (EIS) (see text for details).

EIS can be envisaged using several antibody fragment formats. Here, we describe one example of EIS, using Fab-display in a 2-replicon system for the humanisation by chain shuffling. We previously described a system for the display of antibody Fab fragments on the surface of phage fd, by cloning the antibody genes into a phage(mid) vector. One domain is fused to the phage minor coat protein, while the partner domain is secreted in the periplasm. There, the two chains assemble as functional Fab fragments on the surface of nascent phage particles (9). Similarly, Fab fragments can be displayed on phage by cloning one antibody gene into fd, and using these phage to infect cells harbouring the partner chain cloned into another replicon (such as pUC119). Fab fragments with binding activities to the hapten phOx (9) and to gp120 antibodies (unpublished results) were successfully displayed in this 2-replicon system.

This Fab display system can be applied to humanise antibodies by EIS. The 'docking' chain is cloned in a pUC19-derivative for expression as a soluble chain (targeted to the periplasm), while the library of human partner chains is cloned on phage. In such 2-replicon system, the human one-chain libraries have to be made only once, and then can be used for chain shuffling experiments repeatedly. Other formats were tried as well, including an Fab format, where the murine light chain was denatured and refolded in the presence of phage displaying a repertoire of human heavy (Fd) chains, and the scFv format, in which humanisation of a mouse V-gene requires the cloning of the human V-gene libraries with the partner mouse V-gene, and after selection recloning of the human V-gene(s) with a library of human partner chains. The chain shuffling method relies on the efficient pairing of mouse V-domains with human partner V-domains. However, the association constant between natural V_H and V_L pairs is relatively low (in the range of 10^{4-6} M^{-1}), and in the Fab format, the additional association of the C_H1 and C_L regions (with the association constant in the range of 10^6 M^{-1}) may further drive the assembly of Fab fragment. Indeed, inter-species variable region sets in whole antibodies have been described before (6). Whatever the format of display will be (either Fab or scFv format), although the selection process may introduce a bias for the best fitting pairs of binders, a good association of V_H and V_L perhaps may contribute to the affinity for the antigen.

We chose a murine monoclonal antibody directed against human TNF as our target antibody for humanisation by EIS. The antibody does not inhibit the cytolytic effect of TNF (presumably because it does not inhibit the binding of TNF to its receptors), but it enhances the *in vivo* anti-cancer and anti-viral effects of the cytokine. A library of human light chains were shuffled with a mouse V_H domain, three different binding combinations selected (* in Fig. 1) and then used in a second shuffle as 'docking domains' for a library of human V_H genes. Several completely human antibodies were isolated from such 'genuine' human antibody library (** in Fig. 1). The antibodies were shown to retain specificity and to have similar affinity as the original murine antibody. Alternatively, the mouse V_L was used as docking chain for selecting human V_H partners. Such V_H domains (# in Fig. 1) can be used to find human V_L genes, or, alternatively, can be combined with human V_L domains selected with the mouse V_H domain. Indeed, binding activity was obtained by combining two independently selected

V-genes (* and # in Fig.1), pointing towards potential additivity of the EIS procedure.

3. SUMMARY

We demonstrate a new method, termed 'Epitope imprinted selection' (EIS), based on chain shuffling and selection, for humanizing animal antibodies with defined antigen binding activities. Sequences comprising part of the antigen binding binding site of a rodent antibody are combined with diverse repertoires of sequences of human antibodies that can in combination create a complete antigen binding site. This rodent-human repertoire is displayed on the surface of filamentous phage to allow selection of antigen binding phage bearing antibody fragments with partially rodent, partially human sequences. Next, the remaining rodent sequences are replaced by a repertoire of human sequences, and binding combinations selected on antigen. Using various formats of repertoire assembly and selection, we have converted three mouse antibodies into human antibodies. The sequences of the selected antibodies are entirely human and yet the antigen binding site is directed towards the same epitope of the antigen.

The EIS approach may serve to humanise antibodies more rapidly than by CDR-grafting (6), as this method requires very often a detailed knowledge of the 3-D structure of the antibody. However, the EIS method can be extended to for example antibodies repertoires obtained by phage selection from immunised rodents. Depending on which substance the immunisation was carried out with, such repertoire of V-genes with high affinity and specificity may be used in a 'repertoire' imprinted selection (RIS) to generate a range of human antibodies of high affinity and enriched for the desired specificity.

REFERENCES

1. J.D. Marks *et al.* J. Mol. Biol., 222 (1991) 581.
2. A.D. Griffiths, A.D. *et al.* EMBO J., 12 (1993) 725.
3. J.D. Marks *et al.* Bio/Techn., 10 (1992) 779.
4. H.R. Hoogenboom and G. Winter. J. Mol. Biol., 227 (1992) 381.
5. H.R. Hoogenboom *et al.* Immunol. Rev., 130 (1992) 41.
6. L. Riechmann *et al.* Nature, 332 (1988) 323.
7. H.R. Hoogenboom *et al.* PCT/GB 92/01755 (1992).
8. T.P. Clackson *et al.* Nature, 352 (1991) 624.
9. H.R. Hoogenboom *et al.* Nucl. Acids Res. 19 (1991) 4133.

ECB6: Proceedings of the 6th European Congress on Biotechnology
edited by L. Alberghina, L. Frontali, P. Sensi
1994 Elsevier Science B.V.

Intracellular production and renaturation from inclusion bodies of a scFv fragment fused to a serum albumin binding affinity tail.

Ulrika Öberg[1], Gerd Rundström[2], Hans Grönlund[2], Mathias Uhlén[1] and Per-Åke Nygren[1,*]

[1]*Royal Institute of Technology, Teknikringen 34, S-100 44 Stockholm, Sweden.* [2]*Pharmacia Diagnostics AB, S-751 82 Uppsala, Sweden.*
**Corresponding author*

1. Introduction.

Recombinant DNA technology has allowed efficient isolation and cloning of antibody encoding genes for selection of specificities [i.e. phage display libraries] and production in alternative hosts [1]. For many applications such as diagnostics, it is sufficient with smaller subfragments such as Fab, [Fab']$_2$, Fv and scFv [single chain Fv] which can be produced in secreted, soluble and active forms in *Escherichia coli* [2]. However, for large scale bacterial production, it would be beneficial to take advantage of the high-level expression levels seen for intracellular production of recombinant proteins as inclusion bodies. Such systems have shown to have the potential to yield gram amounts of recombinant products per litre culture. Here we describe a general expression system for the intracellular production of scFv fragments fused to a serum albumin binding affinity tail, allowing a general one-step affinity purification protocol, regardless of the antibody specificity. In the example discussed, a scFv fragment of a mouse α-human IgE mAb used in allergy diagnostics is produced, renatured and analysed for biological activity.

2. Expression system.

The described expression system is based on the serum albumin binding region of the streptococcal surface receptor protein G. Different fragments of this region have been subcloned, produced in *E. coli* and characterized [3]. The region is highly soluble, monomeric in structure, binds serum albumins from different species with high affinity and can be expressed at high levels in *E. coli* . Recombinant proteins produced as fusions to the tail can efficiently be purified in a one-step procedure by affinity chromatography using i.e. human serum albumin [HSA]-Sepharose. To construct an expression vector for intracellular production of fusions to the serum albumin binding tail, denoted BB, a gene fragment encoding a 25 kDa serum albumin-binding region, followed by a cloning linker region was inserted behind the *E. coli trp* promoter in a vector frame work containing a chloramphenicol resistance gene [CAT] for selection [fig 1]. The strong *trp* promoter can be induced chemically by the tryptophan analogue β-indole acrylic acid [IAA] or by simple tryptophan starvation.

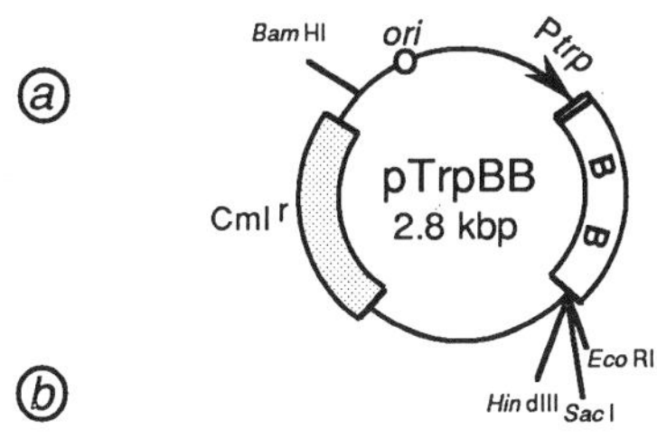

Eco RI Sac I Kpn I Bam HI Xba I Sal I Pst I Sph I Hin dIII
g aat tcg agc tcg gta ccc ggg gat cct cta gag tcg acc tgc agg cat gca agc tt

Figure 1. a: *Schematic description of the expression vector pTrpBB used for the intracellular production of the BBscFv.* ***b:*** *Restriction sites in the multilinker suitable for the insertion of gene fragments.*

3. Results.

3.1. DNA constructions

Approximately 10^6 mouse hybridoma cells producing the IgG1 [kappa] α-human IgE mouse monoclonal antibody 354 [Pharmacia Diagnostics AB] were pelleted and lysed with 1% Triton X-100 [Sigma, USA] for 5 minutes at 37°C. Messenger RNA [mRNA] was isolated from lysed hybridoma by the use of magnetic beads coupled with poly dT tails [dT_{25}-beads, Dynal AS, Norway]. Complementary DNA of the antibody encoding genes were synthesized using AMV reverse transcriptase [Pharmacia AB, Sweden] and oligonucletide primers specific for the IgG1-CH1 and kappa constant regions respectively. The genes encoding the VH-CH1 and VL-CL regions were amplified by PCR with degenerated 5´-primers and the IgG1-CH1 and kappa primers. For DNA sequencing, the PCR products were inserted into pGEM-11Zf[-] [Promega, USA] using the *Eco*RI and *Hind*III restriction sites incorporated by the PCR primers. The VH and VL sequences of five individual clones each were confirmed by solid phase DNA sequencing [4]. A scFv gene fragment, VH-linker-VL, was constructed by assembly of VH and VL genes extended by PCR with sequences encoding a 16 residue peptide linker [GGGSGGGSGGSMGGGS] and inserted into the pTrpBB expression vector using the *Sac*I and *Hin* dIII sites, resulting in pTrpBB-scFv encoding the fusion 52 kDa fusion BB-scFv [fig 2].

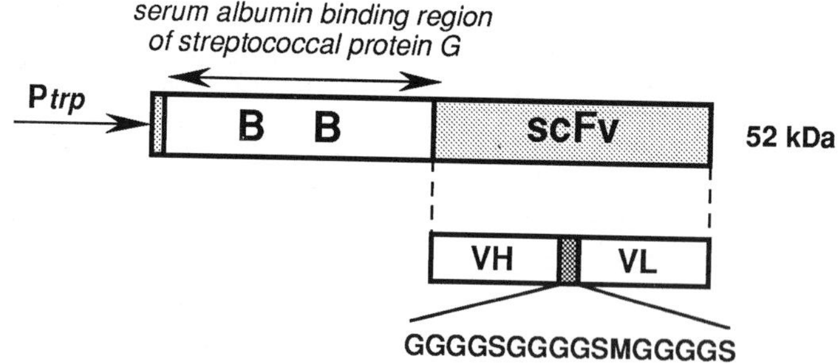

Figure 2. *Schematic description of the BB-scFv fusion. The linker sequence bridging the VH and VL regions is shown, using the one-letter code.*

3.2. Protein production and purification.

The cells from an 500 mL 30°C overnight culture of *E. coli* RR1ΔM15 [5] harboring the pTrpBBscFv construct were pelleted, resuspended in 50 mM Tris-HCl pH 7.5, 150 mM NaCl and 0.05 % Tween 20 and disrupted by sonication. SDS-PAGE analysis of total cells [fig 3, lane 2] shows that the BB-scFv product is the major protein of the total content. The cell paste was resuspended in 7 M GuaHCl, 50 mM Tris-HCl, pH 8.5, 0.1 % Tween 20 and 100 mM 2-mercaptoethanol and incubated at 37°C for 1 hour. After removal of insoluble material by centrifugation at 12 000 *g* the supernatant was diluted 2,5-fold to a final GuaHCl concentration of 3 M, and incubated for 2 h at room temperature. Reoxidation was performed by air oxidation at room temperature overnight at 0.5 M GuaHCl, 20 mM Tris-HCl pH 8.5, 10 μM ortho-phenantroline and 5 μM $CuSO_4$. Renatured BB-scFv fusion protein was subjected to affinity chromatography on HSA-Sepharose as described earlier [3] and eluted with 0.5 M HAc, pH 2.8.

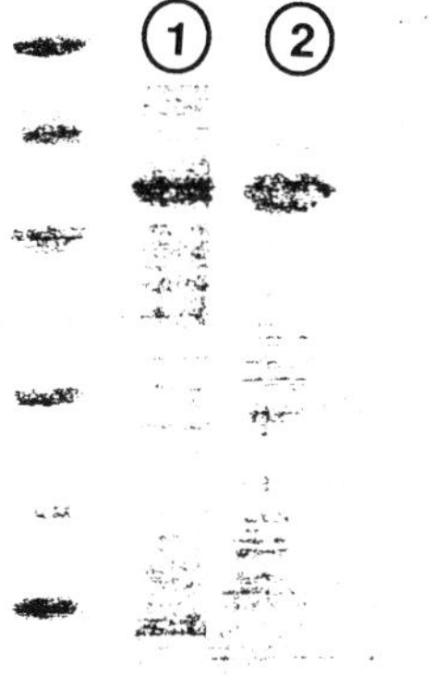

The SDS-PAGE analysis of the affinity purified proteins shows that the fusion is stable and can be purified to high homogeneity, using the one-step HSA affinity chromatography [fig 3, lane 3]. No multimeric forms of the BB-scFv fusion were present as jugded from SDS-PAGE analysis under non-reducing conditions [data not shown]. The yield using this non-optimized scheme was appr. 5 mg/litre culture. However, these amounts are sufficient for initial characterization and functional studies.

Figure 3. *12 % SDS-PAGE [reduced conditions] analysis of total cells at harvest [lane1] and HSA-affinity purified BB-scFv fusion after refolding [lane 2].*

3.3. BIAcore analysis

The biological activity of the recombinant scFv fragment was investigated using biosensor technology [BIAcore™ Pharmacia Biosensor AB, Sweden]. Human IgE [hIgE] was immobilized [1700 RU] onto a sensor chip surface via a capturing rabbit α-human lambda polyclonal sera, previously covalently coupled to the surface using EDC/NHS chemistry [Pharmacia Biosensor, Sweden]. The binding to the hIgE of 30 μl [flow 5μl/min] of recombinant BB-scFv as well as native hybridoma produced mAb 354 was analyzed at a concentration of 1100 nM for both analytes. The registered sensorgram [fig 4] shows that the recombinant BB-scFv binds to the hIgE [270 RU], indicating biological activity of the refolded scFv moiety. As expected, also the mAb 354 binds to the hIgE. The higher response [1230 RU] seen for the original mAb can partly be explained by the 3-fold higher molecular weight [150/52 kDa]. The influence of the divalent binding capacity of the mAb compared to the monovalent nature of the scFv is not fully investigated. No binding to the surface was seen for either sample when the hIgE was excluded from the assay and saturation of the hIgE with the native mAb totally blocked the binding of the BBscFv fusion, suggesting that the specificity was retained in the recombinant scFv fragment.

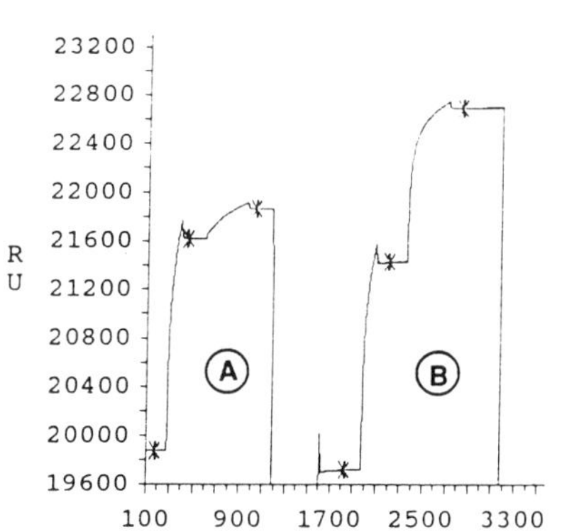

Figure 4. *Sensorgram showing the successive binding of human IgE and the BB-scFv fusion [a] or the native mAb 354 [b] to the sensor chip coated with the rabbit a-human lambda serum. The surface was regenerated with 50 mM HCl between the two runs.*

4. Discussion

The system described here offers an attractive approach to produce large amounts of active scFv fragments of different antibodies. The strategy to fuse the scFv fragment to an affinity tail offers several advantages: *i*, the purification is greatly facilitated since a general affinity chromatography scheme can be used for different antibody fragments. *ii*, the high solubility of the tail can facilitate the refolding procedure, reducing the risk of aggregation [6]. *iii*, the fusion partner can be used for directed immobilization of antibody fragments in biosensor or diagnostic applications. *iv*, by chosing a suitable tail, the *in vivo* performance can be altered for antibody fragments to be used therapeutically. It is possible that the 25 kDa serum albumin binding tail used in this study sterically can influence on the properties of the C-terminally fused scFv fragment. Therefore, vectors for N-terminal fusions to the tail are under development. Furthermore, recent data indicate that a small [5 kDa] subfragment of the BB region still displays full binding to serum albumin and can be used as an affinity tail.

References.

1. Winter, G. and Milstein, C. *Nature* **349,** 293-299 [1991]
2. Skerra, A. *Curr Opinion Immunol.* **5**, 256-262 [1993]
3. Nygren, P.-Å. *et al Eur. J. Biochem.* **193**, 143-148 [1990]
4. Hultman, T. *et al Nucl. Acids Res.* **17**, 4937-4946 [1989]
5. Rüther, U. *Nucl. Acids Res.* **10**, 5665-5772 [1982]
6. Samuelsson, E. *et al Bio/Technology* **9**, 363-366 [1991]

Analysis of different Genomes of biotechnological Interest

ECB6: Proceedings of the 6th European Congress on Biotechnology
edited by L. Alberghina, L. Frontali, P. Sensi
1994 Elsevier Science B.V.

Sequencing the yeast genome

A. Goffeau[1,2] and A. Vassarotti[1]

[1]CCE - DG XII, rue de la Loi 200, B-1049 Bruxelles

[2]Unité de Biochimie Physiologique, Université Catholique de Louvain, Place Croix du Sud 2-20, B-1348 Louvain-la-Neuve

The yeast genome contains about 13 millions of non-rDNA base pairs. This genome is very dense in information: one open reading frame longer than 100 amino acids per 1,8 thousand base pairs. It is divided in 16 chromosomes ranging from 220 thousands to 1,6 million of base pair. Ten of these chromosomes are presently being sequenced systematically in the European Community, the Usa, Japan and United Kingdom.

The European Community is funding the sequencing of six of these chromosomes within the frame work of these Biotechnology Research and Development Programmes. This work includes more than 60 laboratories in Europe which are funded at the rate of 2 ecus per final base pair sequenced. The data are assembled and analyzed by the Martinsried Institute of Protein Sequence (MIPS) located in Munich. The sequencing of each chromosome is coordinated by different scientists who are each responsible for the construction of an organized chromosomal DNA libraries, the dispatching of the DNA fragments, the quality control and the data analysis in collaboration with MIPS (chr III by S. Oliver, Umist, UK; chr XI and XV by B. Dujon, Pasteur, F; chr II by H. Feldmann, Munich, D; chr X by F. Galibert, INSERM, F; chr XIV by P. Philippsen, Basel, CH and chr VII by A. Goffeau, Luovain-la-Neuve, B). The general coordination of the project is carried out by the Université Catholique de Louvain.

---

* Some of the contributors to this Symposium could not send a written contribution. However for the sake of completeness of information the Editor decided to include the Abstracts.

At the time of writing up this abstract, a total of 1,5 million of base pairs from chr III, XI and II have been sequenced in final form by the European Network. Chromosome III, the first chromosome ever sequenced has been published in 1992. Chromosomes XI and II will be published in 1993. Today more than 800 new yeast genes have been unravelled by the EC network among which 55% turned out to be totally unknown as they show no significant sequence homology to any other gene sequenced so far. The search of the function of these unknown new EEC (esoteric, elusive but conspicious) is organized by Piotr Slonimski (Gif-sur-Yvette, F) as a pilot project funded from the EC Biotech programme.

REFERENCE

1. *Oliver and 147 authors.* The complete DNA sequence of yeast Chromosome III. Nature 357, 38-46 (1992).

ECB6: Proceedings of the 6th European Congress on Biotechnology
edited by L. Alberghina, L. Frontali, P. Sensi
1994 Elsevier Science B.V.

Gene isolation from *Arabidopsis*: prospects, progress and solutions for agriculture in Europe

M. Bevan[a], C. Dean[a], R. Schmidt[a], I. Bancroft[a], G. Murphy[a], C. Cobbett[b] and H. Goodman[c].

[a]AFRC John Innes Centre, Norwich NR4 7UJ, UK
[b]University of Melbourne, Parkville, Victoria 3042, Australia
[c]Massachusetts General Hospital, Boston MA 02114, USA

INTRODUCTION

Plant productivity has benefited in the past 75 years from large increases in the yield of storage products, enhanced disease resistance, and tolerance to more diverse environmental threats. These improvements, brought about by the application of genetic principles to plant breeding, will need to continue to be made for the foreseeable future due to pressures of population and the need to diversify and enhance the profitability of EC agriculture. Molecular genetic techniques are beginning to be successfully allied to these conventional approaches, in order to introduce well-characterised genes from other organisms into crop plants. Most of the necessary technology such as transformation methods, vectors and expression cassettes are in place, and an increasing emphasis is now being placed on identifying genes for transfer and expression in crop plants. Not suprisingly, a considerable proportion of these genes are from other plants. This goal requires effective means of identifying plant genes and ascertaining their function. Only a small proportion of plant genes can be isolated and analysed by conventional biochemical means, leaving the vast majority of genes out of reach. Map-based gene isolation, which requires only the map position of a phenotype in order to isolate the responsible gene, is a particularly promising route, but requires a relatively dense genetic map and a relatively small genome.

Arabidopsis GENOME ANALYSIS

The small crucifer *Arabidopsis thaliana* is closely related to brassicaceous plants, which include a wide variety of important crops such as oilseed rape, and has a small genome size of approximately 100Mbp, about the same as *Caenorhabditis elegans*. In addition, it has a rapid generation time of 6 weeks, and each plant produces prolific quantities of seeds (over 10,000). This has made it a favourite of geneticists for some time, resulting in a genetic map of 117 markers (Koornneef *et al*, 1983); the rate of mapping new mutations is increasing, and these include a wide variety of mutations effecting morphological, biochemical, and environmentally induced characteristics. In addition, *Arabidopsis* molecular biology is straightforward; transformation is facile, and a variety of recombinant libraries (cDNA, phage, cosmid and YAC) are available. Two fairly detailed RFLP maps totalling nearly 200 markers, from Meyerowitz and Goodman, have been frequently used in map based cloning experiments, and a RADP map of 250 markers is also available. Presently, there are over 30 map-based walking projects going on, aiming to isolate a diverse array of genes such as those specifying pathogen interactions, growth regulator action, light signal transduction and various biochemical alterations. Each of these efforts will contribute to the growing physical map of *Arabidopsis*. Nevertheless, a concerted effort is needed to ensure a timely and complete coverage of the entire genome. Principal among these efforts is that of Hauge *et al* (1991). They have assembled 17000 cosmid clones into 750 contigs using fingerprinting methods, and it is estimated that these cover 90-95% of the genome. Presently the number of contigs is being reduced using YAC hybridisation. In addition, there is an informal international collaboration to make an overlapping YAC library of the entire genome. YAC libraries with insert sizes of 150-250kb, containing about 2000-2500 clones (3 genome equivalents) are available. A "mega-YAC" library aimed at containing inserts of 500-800kb is presently being constructed as part of the BRIDGE programme in *Arabidopsis* technology development and it should be particularly useful. The small genome size of *Arabidopsis* means that with appropriate offsets, an entire library can be patched onto a small area of membrane for DNA hybridisation. Using this approach, some of the available RFLP markers on chromosomes 4 and 5 have been used to assign YACs to these chromosomes at the John Innes Centre. Similar efforts on chromosome 1 (J.Ecker at Univ. Pennsylvania) and chromosomes 2 and 3 (Massachusetts General Hospital) are also underway. Southern blot analysis of yeast and RFLP DNA verifies the colony hybridisation data and can provide information regarding possible DNA rearrangements in the YACs. Using this approach, 125 markers have generated YAC contigs covering nearly 30% of the genome (Hwang *et al*, 1991). A large proportion of chromosomes 4 and 5 are now covered, and walking experiments to interesting genes in these regions have established two contigs of nearly 2Mb each in each chromosome. In order to close the YAC map expeditiously, less labourious methods must be used. Obtaining a denser RFLP map based on recombinant inbreds and multiple probes (such as cDNAs) is a favoured strategy. Recombinant inbreds, the availability of markers such as random cDNAs, microsatellites, PCR-generated markers and the like will greatly increase the number of markers on the genome, facilitating rapid placement of YACs and eventual closure of the map. The imminent availability of large-insert YACs will hasten this process still further, as will the integration of cosmid contigs and YAC contigs.

An array of linked YAC and cosmid clones is a necessary foundation for the initiation of large-scale sequencing projects. These have been anticipated by the *Arabidopsis* Multinational Steering Committee, who had as their mission statement in 1990 "...to determine the complete sequence of the *Arabidopsis* genome by the year 2000." Presently, there are two programmes, one directed by Chris Somerville in the US and a French national programme, which have partially sequenced approximately 3500 random cDNAs. These efforts will continue in the foreseeable future and may encompass many of the 25,000-50,000 potentially expressed genes. A sister effort in Japan has already partially sequenced over 6000 rice cDNAs, providing an excellent oportunity for comparisons between dicots and monocots, the divergent groups comprising all of the flowering plants. Ever more powerful protein comparison methods applied to the burgeoning databases will mean that putative functions of encoded proteins may be able to be defined with some accuracy in the near future, thereby providing plant scientists with a rich source of genes for further study. Nevertheless, cDNA sequencing only reveals part of the genome, therefore a concerted effort to sequence the genome has just begun. This is funded in the BIOTECH programme of Framework 3, and involves a network of 18 European labs, including one industrial participant (Keygene, NL). There are two types of participant; those that sequence areas of interest to themselves, and those that contribute to medium-scale contiguous sequencing on chromosomes 1 and 4. The groups are described in Figure 1. This European effort, called ESSA (European Scientists Sequencing Arabidopsis) also covers an ongoing French national effort to sequence random cDNAs (EST sequencing). The emerging data will be sent to a central Informatics node for collation, analysis, quality control checks and release to public databases. Informatics issues in ESSA will be dealt with by MIPS, who also manage yeast sequence data emanating from EC-sponsored work. A particularly important job will be to collate emerging random cDNA sequence with genomic sequence. Because of the generally small intron size in *Arabidopsis*, the interpretation of gene structure and putative protein function will not be possible without both types of sequence. To increase the chances of obtaining cDNAs cognate to genomic sequence, cosmids used for sequencing will be used to screen a battery of cDNA libraries made from different tissues, and the isolated cDNAs wil be completely sequenced. The cDNA libaries being screened by French colleagues, made from RNA isolated from floral tissues, cultured cells, developing seeds, hypocotyls etc, will be used for cognate cDNA isolation. In the first year of the programme medium scale sequencing will produce 350kb of new sequence from contigs on chromosomes 1 and 4, while smaller scale work on different areas of the genome will provide a further 100kb. At the end of the three year programme our goals are to have sequenced a 1.5Mb contig on the top arm of chromosome 4, centered on the *fca* locus (see Figure 2), a 0.5Mb contig on the top arm of chromosome 1 centered on the *pfl* locus, 7 contiguous regions of 75kb each surrounding gene clusters, flowering time, homeotic genes and the like, and over 4000 partially and fully sequenced cDNAs. These efforts should form nuclei from which future sequencing programmes can spread. A medium-term goal is to sequence as much as possible of chromosome 4, which may be 10-15Mb long. The ESSA programme will put in place the necessary components for large-scale *Arabidopsis* sequencing for these tasks within Europe. Our readiness to continue large-scale sequencing at the end of this initial 3 year programme will be matched with the availability of extensive overlapping YAC contigs and with the availability of more efficient sequencing technologies emerging from the human genome programme. But in plants it seems that (at least with existing methods) searches for the function of these anonymous genes may lag significantly behind their discovery. Gene replacement is not yet a viable technique in higher plants, but targeting mutations caused by transposons is a possible approach. More effort needs to be put into developing this area if full advantange is to made of large-scale sequencing work.

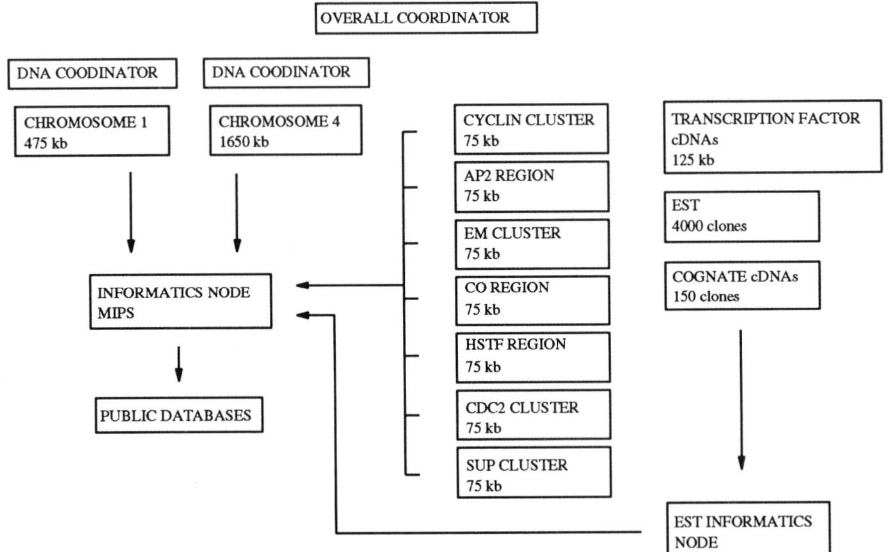

Figure 1. Structure of ESSA.

Figure 2. Region of chr. 4 to be sequenced.

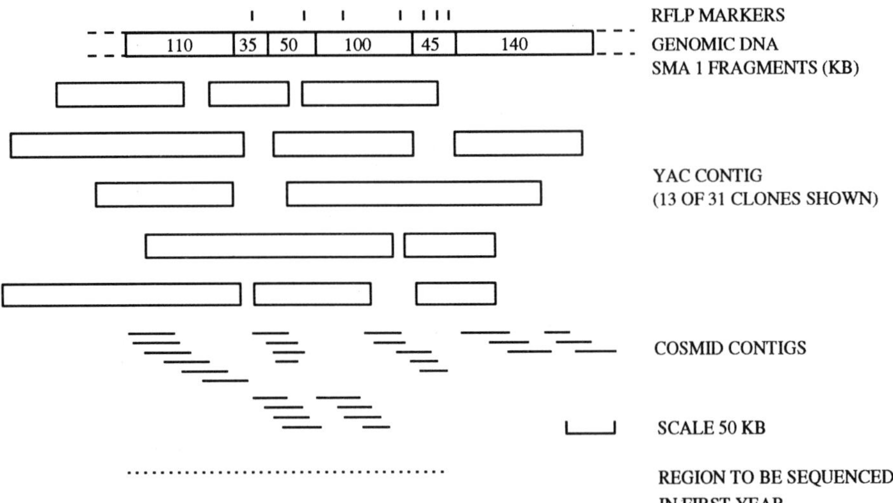

REFERENCES

Koornneef, M., van Eden, J., Hanhart, C.J., Stam, P., Braaksma, F.J. and Feenstra, W.J., J. Hered. 74 (1983) 265.

Hauge, B.M., Hanley, S., Giraudat, J., and Goodman, H.M., Molecular Biology in Plant Development, (ed Jenkins and Schuch) p 45, Company of Biologists, Cambridge, England (1991).

ECB6: Proceedings of the 6th European Congress on Biotechnology
edited by L. Alberghina, L. Frontali, P. Sensi

Ordered genomic clone libraries as high resolution mapping tools

Elmar Maier, Jörg D. Hoheisel and Hans Lehrach

Imperial Cancer Research Fund, 44 Lincoln's Inn Fields, P.O. Box 123, London WC2A 3PX, United Kingdom

In this report we describe the advances made with the help of our clone libraries of the *Schizosaccharomyces pombe* genome. These ordered libraries [1, 2] serve as high resolution mapping tools for laboratories trying to unravel the genetic information stored within the genome of the fission yeast.

1. INTRODUCTION

The genome of *S. pombe* shows all the basic characteristics of an eukaryotic genome. This allows the comparative examination of important biological functions in a relatively simple model organism, as demonstrated by the intensive studies on the cell cycle division process [3-5]. Due to this inherent biological interest and the availability of genetic [6] and physical [7] mapping information, *S. pombe* served as an excellent system to develop and test techniques for the mapping of entire genomes.

2. ORDERING *S. POMBE* CLONE LIBRARIES

Hybridisation fingerprinting on high density filters provides an efficient route to map large numbers of clones preferentially using probes detecting multiple sites in the genome [8]. However, since the genome of *S. pombe* is relatively small, the use of unique probes (Figure 1) was very efficient, permitting a direct data analysis.

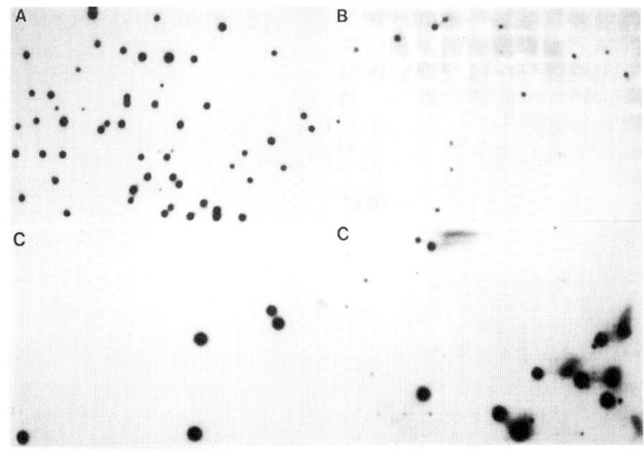

Figure 1. Autoradiographs of unique *S. pombe* gene fragment hybridisations onto nylon filters ($7*11cm^2$) containing gridded: A) 1248 clones of our YAC library (47-fold coverage) [1]; B) 3456 clones of our P1 library (17-fold coverage) [2]; C) on two filters a subset of 3072 clones (8-fold coverage) of our cosmid library (38-fold coverage) [2]; all made from *S. pombe* strain 972h$^-$.

Gridded on high-density filters, a YAC genomic library of 47-fold coverage [1], a P1 library of 17-fold coverage and a cosmid library of 8 genome equivalents [2], all made from *S. pombe* strain 972h$^-$, were ordered in parallel by hybridising genetically localised markers and anonymous clones from the three libraries. Hybridisations with clone pools and short oligonucleotides to the cosmid library were performed in order to compare the quality of the results obtained with the different probe types.

2.1. Data analysis

In the analysis, more accurate and longer clone contigs were obtained by ordering the probes first and then fitting the clones to the probe order [9]. A probe-probe comparison, based on the hybridisation frequencies of a relatively large number of clones, has a higher information content than a clone-clone comparison, based on the hybridisation data of the much smaller number of probes. The probe order was derived by algorithms using simulated-annealing producing a set of *probe tagged sites* (PTSs) [1], named in analogy to the concept of a *sequence tagged site* (STS) [10]. Another algorithm exploited a set of heuristic rules producing an almost identical order of probes.

2.2. Physical clone maps

Figure 2 shows an updated version of the physical map of the entire *S. pombe* genome [2] with the locations of 91 genetic markers already mapped to the libraries. A set of clones from each of these libraries is shown (25 YACs, 209 P1 clones, 243 cosmids). Note that clones which have only one probe in common do not necessarily overlap. Since most of the probes are cosmid clones themselves, in some cases the listed clones only overlap with either end of the relevant probe but not with each other.

The shown subset of 25 YAC clones is completely spanning the entire genome [1]. The maps of chromosomes III (3.5 Mbp) and II (4.6 Mbp) are continuous in P1 and cosmid clones, while there are three short contig breaks in chromosome I (5.7 Mbp) [2].

Restriction digests were generated as an independent confirmation of the clone order. The high-resolution clone map was aligned to both the genetic [6] and the physical [7] *Not*I map. Apart from the three most frequently occurring repeat sequences in *S. pombe* - the 10.4 kbp ribosomal DNA repeat, retrotransposable regions and centromeric sequences - six other major repeats were identified [2]. The parallel use of the YAC, P1 and cosmid libraries contributed substantially toward the resolution of the clone order on all levels.

3. GENE MAPPING WITH ORDERED CLONE LIBRARIES

Filters containing all the clones of our ordered genomic *S. pombe* YAC, P1- and cosmid libraries arrayed in a high denstity format are freely available on request (e_maier@icrf.ac.uk). Over 40 different laboratories have already used our libraries to locate their gene fragments of interest. The mapping data is stored in our relational *S. pombe* database and eventually provided to the different users. We also have a gene list of the nuclear and mitochondrial genes of the fission yeast [11, 12] which consists of a gene symbol, the physical location, a description of the product or the phenotype, the EMBL accession number (if available), and references to original publications.

Since one particular genome representation is being used as a reference system by a large number of groups, the data generated are more easily comparable. Also, apart from linking the probes to the physical clone map, and *vice versa*, we are providing other laboratories with the required genomic clones so that they can work on a more detailed (e.g., sequence) analysis of the genomic region of their interest. As the fission yeast genome is thought to comprise around 5,000 genes, much still remains to be discovered.

With this report we would like to encourage scientists working on *S. pombe* to make even further use of our ordered genomic clone libraries as high resolution mapping tools and thereby speed up the detailed analysis of the fission yeast genome.

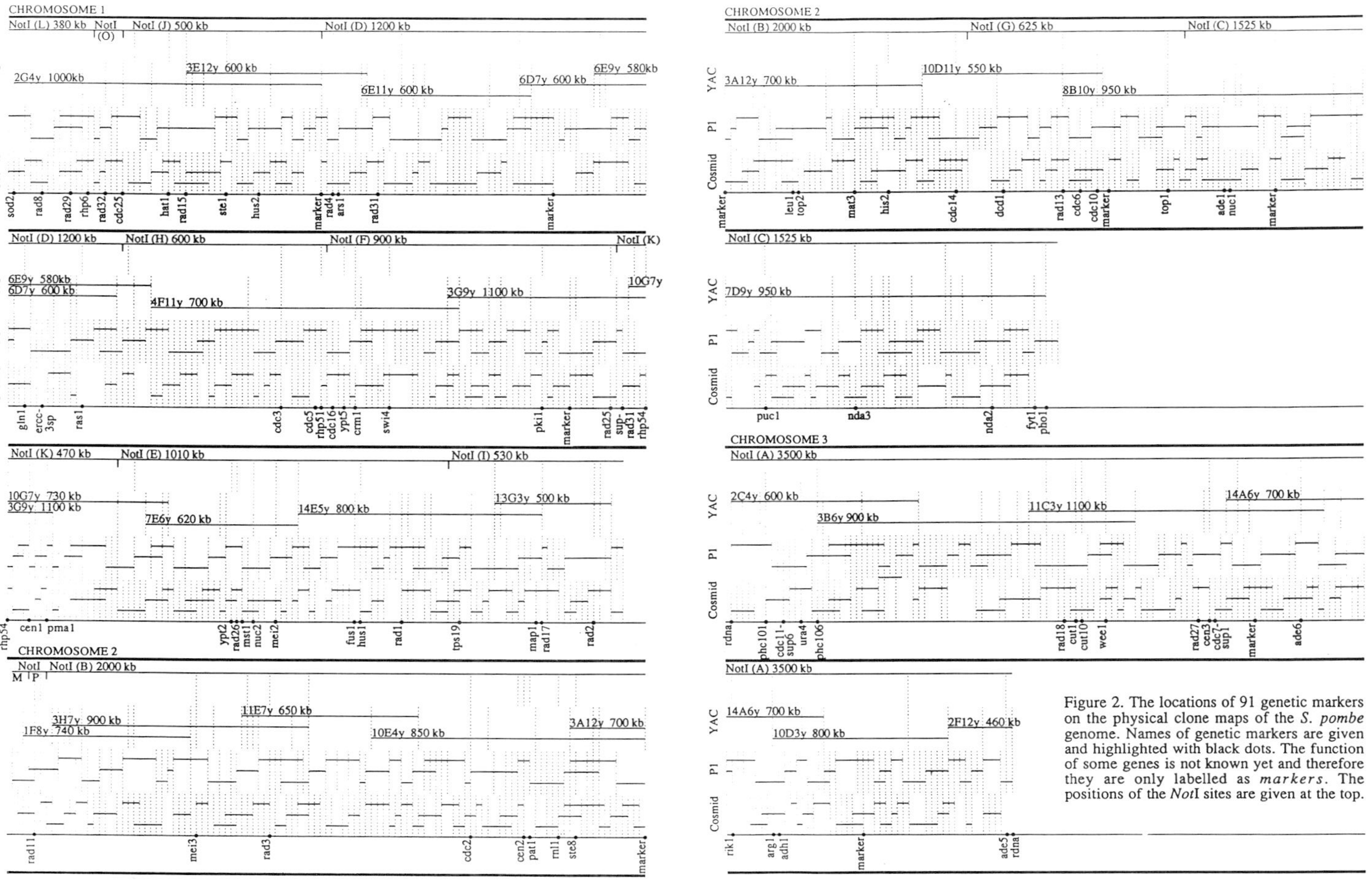

Figure 2. The locations of 91 genetic markers on the physical clone maps of the *S. pombe* genome. Names of genetic markers are given and highlighted with black dots. The function of some genes is not known yet and therefore they are only labelled as *markers*. The positions of the *Not*I sites are given at the top.

4. ACKNOWLEDGEMNTS

We thank John Armstrong, Elizabeth Bertani, Wendy Bickmore, David Bonthron, Antony Carr, Richard Egel, Peter Fantes, Susan Forsburg, Kathy Gould, Merl Hoekstra, Sergio Moreno, Daan Muris, Olaf Nielsen, Tom Patterson, Viesturs Simanis, Mitsuhiro Yanagida and Felicity Watts for providing genetic probes of *S. pombe,* and Paul Nurse for the *S. pombe* strain 972h$^-$. We also like to thank Richard Mott for helpful computing and Janet Roden for organising all the technical work. E. M. is funded by an European Community Research Fellowship.

REFERENCES

1. Maier, E., Hoheisel, J.D., McCarthy, L., Mott, R., Grigoriev, A., Monaco, T., Larin, Z. and Lehrach, H. (1992). Nature Genetics 1, 273-277.
2. Hoheisel, J.D., Maier, E., Mott, R., McCarthy, L., Grigoriev, A.V., Schalkwyk, L.C., Nizetic, D., Francis, F. and Lehrach, H. (1993). Cell 73, 109-120..
3. Booher, B. & Beach, D. (1989) Cell 57, 1009-1016.
4. Moreno, S., Hayles, J. & Nurse, P. (1989) Cell 58, 361-372.
5. Kinoshita, N., Ohkura, H. & Yanagida, M. (1990) Cell 63, 405-415.
6. Munz, P., Wolf, K., Kohli, J. and Leupold, U. (1989). In "Molecular Biology of the Fission Yeast" (A. Nasim, P. Young and B.F. Johnson, eds.), pp. 1-30, Academic Press, San Diego.
7. Fan, J.B., Chikashige, Y., Smith, C.L., Niwa, O., Yanagida, M. and Cantor, C.R. (1989). Nucleic Acids Research 17, 2801-2818.
8. Lehrach, H., Drmanac, R., Hoheisel, J., Larin, Z., Lennon, G., Monaco, A.P., Nizetic, D., Zehetner, G. and Poustka, A. (1990). In "Genome Analysis: Genetic and Physical Mapping" (K.E. Davies and S. Tilghman, eds.), pp. 39-81, Cold Spring Harbor Laboratory Press, Cold Spring Harbor, New York.
9. Mott, R., Grigoriev, A., Maier, E., Hoheisel, J. and Lehrach, H. (1993). Nucleic Acids Research, in press.
10. Olson, M.V., Hood, L., Cantor, C.R. and Botstein, D. (1989) Science 245, 1434-1435.
11. Lennon, G. and Lehrach, H. (1992). Current Genetics 21, 1-11.
12. Maier, E. and Lehrach, H. (1993). In "The Yeasts" (A.H. Rose and A. Wheals, eds.), Vol. 6, Academic Press, in press.

ECB6: Proceedings of the 6th European Congress on Biotechnology
edited by L. Alberghina, L. Frontali, P. Sensi
1994 Elsevier Science B.V.

Sequencing the *Bacillus subtilis* genome

P. Glaser[1], A. Danchin[1] and F. Kunst[2]

[1]Unité de Régulation de l'Expression Génétique
[2]Unité de Biochimie microbienne
Institut Pasteur, 25-28 rue du Dr. Roux, F-75724 Paris Cedex

Six years elapsed since the original reflection on the need for setting up an international collaborative program aimed at the sequencing of the complete *Bacillus subtilis* genome. Recently a consortium of nine European and six Japanese laboratories has started to sequence at least 25 kb of *B. subtilis* DNA every year. in the previous years several fragments of more than 25 kb have already been sequenced, including a segment of ca 100 kb. This allowed evaluation of the requirements for investigating in detail the corresponding genome segment. it now appears that the main bottleneck is not sequencing *per se*, but rather the cloning step (new techniques for chromosome walking should however provide significant improvement of this step), and, more importantly, the step consisting in interpreting the data. indeed about half of the genes differ from what is known in the data libraries. It is therefore necessary to set up new conditions for exploring unknown genes, by reverse genetics, 2-D protein gel electrophoresis, and *in silico* experiments using the most recent approaches in computer sciences.

---

* Some of the contributors to this Symposium could not send a written contribution. However for the sake of completeness of information the Editor decided to include the Abstracts.

Genetic Stability in Engineered Microorganisms

ECB6: Proceedings of the 6th European Congress on Biotechnology
edited by L. Alberghina, L. Frontali, P. Sensi

Plasmid stability in Gram-negative bacteria: Molecular structure and function of the partitioning system of the broad-host-range plasmid RP4

E.Grohmann, L.Eberl, and H.Schwab*

Institut für Biotechnologie, Arbeitsgruppe Genetik, Technische Universität Graz
Petersgasse 12, A-8010 Graz, Austria

I. INTRODUCTION

Segregational plasmid instability can seriously affect the productivity of industrial bioprocesses employing recombinant micro-organisms based on autonomously replicating plasmid vectors. Such vectors are mostly used with bacterial systems. Therefore, genetic systems that contribute to plasmid stability are of substantial importance to the development of large scale processes in bioreactors (contained use) or of bioprocesses that are open to the environment (e.g. releases).

Naturally occurring autonomously replicating bacterial plasmids are usually very stably maintained over many generations and loss of such plasmids during cell division is a rare event. This is only possible due to the action of specific genetic systems that are designed to prevent plasmid loss. Such systems can act in different ways. Some can stabilize plasmids by preventing multimer formation and thus maintaining a high copy number. Other systems act by specifically killing cells that have lost the plasmid. Finally, "true partitioning systems" have been described that actively facilitate equal partitioning of plasmid molecules to daughter cells during cell division.

In the early stages of the work for construction of plasmid vectors stabilizing functions usually have not been identified or characterized on the natural plasmids used as basic replicons. Therefore, such functions have been removed in most cases because the size of vectors was kept to a minimum and non-essential parts were deleted. Since at the laboratory scale plasmid-containing bacterial strains are routinely cultivated under (e.g. antibiotic) selection pressure for maintaining the plasmids, instability only becomes a problem in the scaling-up stage of a bioprocess or at the first stages of release trials, respectively.

2. PROPERTIES OF THE *par*-REGION OF PLASMID RP4

Plasmid RP4, which is identical to plasmid RK2, is an extremely stably maintained plasmid being able to replicate in a large number of Gram-negative bacteria. Usually no plasmid-free cells can be detected in cultures of RP4-containing bacterial strains. Since the plasmid RP4 (60 kb in size) has a rather low copy number of about four to seven copies per chromosome, a specific mechanism ensuring this high stability must be encoded.

2.1. Stabilizing ability of RP4-*par*

We have identified and cloned a gene region of plasmid RP4 that is extremely efficient in stabilizing plasmids (Saurugger et al., 1986). This system, designated as a "partitioning system" (*par*), was originally cloned as a 6 kb *Pst*I fragment located between 31 to 37 kb of the RP4 map (Lanka et al., 1983). The essential region for full Par-function could be located on a region of about 2 kb between *fiwA* and IS*8* (Gerlitz et al., 1990).
Cloning of the essential region onto different replicons revealed, that the RP4 *par* region can fully act on different replicons. Besides IncP1 replicons, pMB1 and p15A replicons have been tested and in all cases the cloning of an RP4 *par*-fragment resulted in a high stability. As with the natural plasmid RP4, no plasmid-free cells could be detected after many (100 to 200) generations of growth under non-selective conditions (Saurugger et al., 1986; Gerlitz et al., 1990; Roberts et al., 1990; Weinstein et al., 1992).
In addition, it could be demonstrated that the RP4 partitioning system can function in a broad range of Gram-negative bacterial hosts. In all of the tested cases broad-host-range IncP1 replicons were used. Table 1 shows a compilation of tested strains.

Table 1
Plasmid stabilization by RP4 *par* in Gram-negative hosts

Bacterial strain	Reference
Agrobacterium tumefaciens	Roberts et al., 1990
Alcaligenes eutrophus	Saurugger et al., 1986
Alcaligenes latus	Saurugger et al., 1986
Azotobacter chroococcum	Saurugger et al., 1986
Azotobacter vinelandii	Roberts et al., 1990
Escherichia coli C600	Saurugger et al., 1986
Escherichia coli HB101	Roberts et al., 1990; Saurugger et al., 1986
Pseudomonas aeruginosa	Roberts et al., 1990
Pseudomonas putida	Saurugger et al., 1986
Pseudomonas denitrificans	Crouzet et al., 1992
Rhizobium meliloti	Weinstein et al., 1992

2.2. Molecular structure of the RP4 *par* region

Detailed molecular analysis (Gerlitz et al., 1990; Roberts et al., 1990; Eberl et al., 1992; Davis et al., 1992) of the *par* region of RP4/RK2 revealed a rather complex system.

Five genes arranged in two divergently transcribed autoregulated operons could be identified. Fig. 1 shows a schematic drawing of the molecular organization.

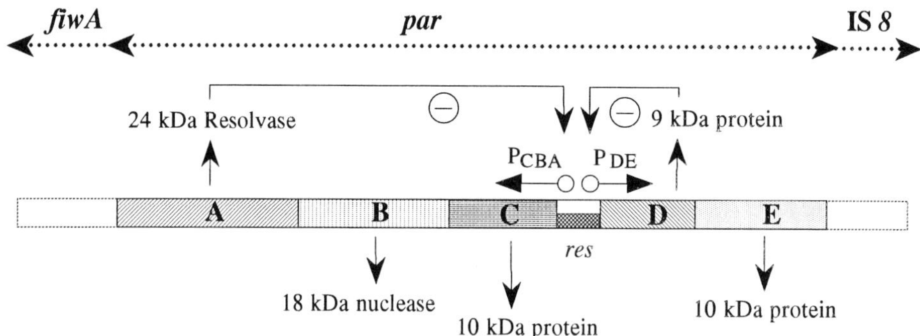

Figure 1. Molecular organization of the RP4 *par* region.

The *parA* gene encodes a resolvase acting on a specific site (*res*) within the promoter region. These two components represent a site specific recombination system that was found to be highly active *in vivo* and *in vitro*. (Eberl et al., submitted). However, it could be demonstrated that the presence of an active site specific recombination system by itself is not sufficient to stabilize plasmids (Gerlitz et al., 1990). The *parB* gene encodes for a nuclease that has homology to *Staphylococcus aureus* nuclease (Gerlitz, 1990) and to a nuclease encoded on plasmid pSa (Close and Kado, 1992). However, the role of the nuclease in the stabilization process is not yet understood. In addition, the function and the role of the other components is poorly understood at the moment. There is evidence that some of these components are needed for efficient plasmid maintenance only under certain conditions (Gerlitz et al., 1990, Roberts and Helinski, 1992; Weinstein et al., 1992).

3. EXAMPLES OF STABILIZATION WITH SYSTEMS OF COMMERCIAL INTEREST

There exist a few examples of the successfull use of the RP4 *par* region to stabilize plasmids in microorganisms of commercial interest.

Crouzet et al. (1992) reported on the stabilization of an IncP1-based plasmid carrying genes of the cobalamin biosynthesis pathway. Cloning of these genes onto an IncP1 vector causes drastic instability of the resulting plasmid in a *Pseudomonas denitrificans* strain used for the production of cobalamin. The unstabilized plasmid is rapidly lost under non-selective conditions (only about 5% plasmid-containing cells after 50 generations). Cloning of the RP4 *par* region onto such a plasmid resulted in a high stability and nearly 100% plasmid-containing cells could be still found after 90 generations of growth in non-selective medium.

In a second example a similar situation was given. Cloning of the genes for biosynthesis of poly-ß-hydroxybutyric acid (PHB) derived from *Alcaligenes eutrophus* resulted in instability of plasmids expressing these genes in *E.coli*. In this case multicopy vectors based on the pMB1 replicon (pUC-derivatives) were used. The plasmid containing the PHB-genes is very rapidly lost in batch cultures, even under selective conditions. Within 10 hours of cultivation, even in the presence of a selective pressure (addition of 100 $mg.l^{-1}$ ampicillin), the portion of plasmid-containing cells drops to nearly 60%. The instability increases when expression of the PHB genes is enhanced by putting the PHB genes under the control of the *tac* promoter (induced by IPTG). In this case only 30% of the cells contain the plasmid after 10 hours of cultivation under selective conditions. Cloning of the RP4 par region onto the plasmid containing the PHB genes results in good stability under non-selective conditions. Only in the late stages of a culture performed without selection pressure (no ampicillin added) a small population of plasmid-free cells (less than 10%) could be observed (Haigermoser et al., 1993).

A further example shows the potency of the RP4 *par* region also for applications in open systems. Weinstein et al. (1992) could demonstrate that the RP4 *par*-region is also efficient with IncP1-plasmids residing in *Rhizobium meliloti* that were growing in root nodules. This shows that the RP4 *par* region is active even under conditions of drastic changes in the metabolism of a host cell.

4. CONCLUSIONS

The RP4 *par* system is extremely efficient for stabilizing bacterial plasmids. This system acts on various replicons and is efficient in many Gram-negative bacterial hosts. Therefore, this system is of substantial interest to biotechnological applications. One focus is on large scale bioprocesses with recombinant Gram-negative bacteria to retain high productivity by stable maintenance of the recombinant plasmids without external selection pressure. A second area is the use of Gram-negative bacteria in open systems such as deliberate release. Here the use of such a stabilizing system helps to maintain the desired activity of the recombinant organism for longer periods in the target environment.

5. REFERENCES:

Close, S.M., and Kado, C.I. (1992), Mol. Microbiol., 6, 525-527.

Crouzet, J., Lévy-Schil, S., Cauchois, L., and Cameron, B. (1992), Gene, 110, 105-108.

Davis, T.L., Roberts, R.C., Helinski, D.R. (1992), Mol. Microbiol., 6, 1981-1994.

Eberl, L., Givskov, M., Schwab, H. (1991), Mol. Microbiol., 6, 1969-1979.

Gerlitz, M. (1990), Thesis, Technische Universität Graz

Gerlitz, M., Hrabak, O., Schwab, H. (1990). J.Bacteriol., 172, 6194-6203.

Haigermoser, C., Chen, G.-Q., Grohmann, E., Hrabak, O., and Schwab, H. (1993), J.Biotechnol., 28, 291-299.

Lanka, E., Lurz, R., and Fürste, J.P. (1983), Plasmid, 10, 303-307.

Roberts, R.C., Burioni, R.,Helinski, D.R. (1990). J.Bacteriol., 172, 6204-6216.

Saurugger, P.N., Hrabak, O., Schwab, H., Lafferty, R.M. (1986). J.Biotechnol., 4, 333-343.

Weinstein, M., Roberts, R.C., and Helinski, D.R. (1982), J.Bacteriol., 174, 7486-7489.

ECB6: Proceedings of the 6th European Congress on Biotechnology
edited by L. Alberghina, L. Frontali, P. Sensi

Structural and functional analysis of plasmid vectors for *Lactobacillus*

R.J. Leer, N. van Luijk, M. Posno, and P.H. Pouwels

TNO Medical Biological Laboratory, P.O.Box 5815, 2280 HV Rijswijk, The Netherlands.

1. INTRODUCTION

The economic importance of various *Lactobacillus* strains is beyond doubt. Since ancient times lactobacilli have been used in the preparation and conservation of human and animal food. *Lactobacillus* spp. have also been found to show certain health promoting properties for humans and animals. A potentially very useful property of certain *Lactobacillus* species is the capacity to evoke a mucosal as well as a systemic immune response against epitopes associated with these organisms after oral administration. To be able to fully exploit the potential of these organisms for practical application, studies have been initiated in a number of laboratories, including our own laboratory, to develop genetic tools for the introduction and stable expression of genetic material in lactobacilli.

2. DEVELOPMENT OF GENE-TRANSFER SYSTEMS FOR *LACTOBACILLUS*

Most *Lactobacillus* species harbour endogenous plasmids that are present in 1 to approximately 50 copies per bacterium. Generally, the plasmids vary from 2 to >30 kb in size, and for most of them a function is not known. Starting with small cryptic plasmids from two related species, *L.plantarum* and *L.pentosus*, we have developed shuttle vectors that can be extra-chomosomally replicated both in *E.coli* and in *Lactobacillus*. In addition, we have constructed vectors which contain a replicon for *Lactobacillus* but no *E.coli* sequences (1) as well as vectors that can be stably integrated at a specific site of the chromosome (2). Most *Lactobacillus* strains can be reproducibly transformed with these vectors using the electroporation technique. The efficiency of transformation ranges from 10^2 to 10^7 transformants per μg of DNA, depending on the strain used (1,3).

3. STRUCTURAL FEATURES OF *LACTOBACILLUS* PLASMIDS

The complete nucleotide sequence of the 1.9 kb plasmid p8014-2 from *L.plantarum* and the 2.4 kb plasmid p353-2 from *L.pentosus* has recently been established (3). The two plasmids show more than 80% similarity. The nucleotide sequence data suggests that these plasmids replicate by a mechanism of rolling circle replication. Strong evidence in favor of this hypothesis was obtained by showing that p353-2 accumulates ssDNA intermediates in the presence of rifampicin. Based on the structure of the replication protein (Rep) and its cognate plus-origin of replication, both plasmids belong to the pUB110/pC194 class of ssDNA plasmids (4). An inverted repeat was found whose sequence is strongly conserved among plasmids from different *Lactobacillus* species. This inverted repeat was shown to be involved in the conversion of single-stranded into double-stranded DNA (*ori*), as removal of part of the sequence results in accumulation of single-stranded DNA intermediates (3). The sequence of the *ori* found in *Lactobacillus* plasmids shows no resemblance to that of other ssDNA plasmids and represents a new class of minus-origins of replication, named *pal*L.

Comparison of sequence data from various plasmids indicates that all *Lactobacillus* plasmids have a modular organization similar to that of ssDNA plasmids from other Gram-positive bacteria, comprising the following elements: i) all plasmids encode a replication protein and cognate *ori*$^+$, and a *ori* ; ii) in addition, some plasmids contain a gene coding for a repressor protein, and/or a gene coding for a mobilization protein and cognate target site. These elements are found in different combinations, presumably reflecting horizontal transfer of these elements through the genus, and possibly to and from other bacterial species.

4. CONTROL OF PLASMID DNA REPLICATION

To keep plasmid copy number constant, DNA replication in bacteria is negatively controlled by mechanisms involving plasmid-encoded and host functions. Most ssDNA plasmids encode a 50-100 nt antisense or countertranscript RNA (CT-RNA), complementary to the 5' untranslated region of *rep* RNA, which acts as a repressor for DNA replication. For the pT181 family of ssDNA plasmids it has been shown that CT-RNA specifically interacts with *rep* RNA and controls transcription of the *rep* gene, by a mechanism of transcription attenuation (5). For other ssDNA plasmids, represented by pMV158, it was shown that interaction of CT-RNA with *rep* RNA negatively controls translation of *rep* RNA (6).

DNA replication of the *L.pentosus* plasmid p353-2 appears to be regulated by a similar mechanism. Nucleotide sequence analysis of the 5' untranslated region of plasmid p353-2 has revealed the presence of a putative promoter followed by a potential terminator just in front of the translation startcodon ATG, and a second promoter followed by two potential terminator sites in the strand with opposite polarity. The putative CT-RNA gene, which does not contain an open reading frame, could encode transcripts of ~75nt and 250nt in length. Northern analysis of RNA from *L.casei* containing p353-2 shows, indeed, the presence of two RNA species of ~75nt and - greatly reduced amounts -of 250nt RNA. The 5' end of CT-RNA has been mapped by primer extension analysis at a site 8nt downstream from the -10 region of the postulated promoter. The regulatory role of CT-RNA in copy control was demonstrated by removal of a nucleotide sequence entirely contained within the 5' untranslated region of *rep* RNA which includes the promoter of CT-RNA. The copy number of this deletion derivative is 5-10-fold increased. CT-RNA may exert its regulatory role by a mechanism of transcription attenuation. Wild-type p353-2 shows the presence of two transcripts specifically hybridizing to a probe derived from the 5' untranslated region: a ~190nt RNA and trace amounts of an 1100nt RNA corresponding to the length of the *rep* gene. In RNA from the deletion mutant the amount of full-length *rep* RNA is strongly increased but no 190nt RNA is found. It appears that in the presence of CT-RNA, transcription which is started at a promoter approximately 200nt 5' to the translation startcodon ATG is prematurely terminated at the transcription termination site just upstream of the startcodon ATG.

5. STRUCTURAL AND SEGREGATIONAL STABILITY OF *LACTOBACILLUS* VECTORS

For application in industrial fermentation processes, it is essential that DNA which is introduced into *Lactobacillus* bacteria remains structurally intact (structural stability) and can be maintained in the host cell in the absence of selective pressure during the fermentation process (segregational stability). Plasmid vectors remain structurally intact if the cells are cultivated in the presence of the selective agent. Structural instability was not observed after introduction into a *Lactobacillus* vector of DNA fragments derived from bacteriophage lambda varying in size from 2 to 9 kb, irrespective of whether the bacteria were cultivated in the presence or absence of selective pressure. Recombinant DNA could also be stably maintained without structural changes when DNA was introduced into the chromosome (1,3).

Most plasmid vectors are relatively rapidly lost (50 - >95% loss after 100 generations) when their host cells are cultivated in the absence of the selective agent. However, some vectors can be stably maintained for more than 200 generations in a number of *Lactobacillus* strains (1,3). Of particular interest is plasmid pLPE323 which was found to be segregationally stable in all except one *Lactobacillus* strains tested. Plasmid pLP3537 harbours the same *Lactobacillus* replicon as pLPE323 but differs from it by the presence of *E.coli* sequences. Since the two plasmids display remarkably large differences in stability, we have studied the effect of the presence in *Lactobacillus* vectors of *E.coli* sequences on segregational stability. The results show that the vectors become segregationally completely stable by removal of ~95% of the *E.coli* sequences (3). In addition, the results show that segregational instability increases if the size of the inserted DNA fragment increases.

The effect of the presence of an intact minus-origin of replication on segregational stability was investigated by comparing the stability of a derivative of pLPE323 from which part of the inverted repeat which constitutes the *ori* was deleted. The deletion derivative is segregationally unstable in two *Lactobacillus* strains, whereas the parent plasmid is fully stable. We conclude that, in agreement with results obtained for other ssDNA plasmids, that the absence of a functional *ori* is not lethal but renders the plasmid segregationally unstable.

Segregationally stable transformants have also been obtained by introduction of foreign DNA into the chromosome by homologous recombination (2). This was demonstrated by replacement of the chromosomal gene encoding the enzyme conjugated bile salt hydrolase (Cbh) by a *cbh* gene with an insertion of a chloramphenicol resistance marker, through a mechanism of double-cross over recombination. As expected, transformants carrying a mutant copy of the *cbh* gene lacked Cbh activity. The Cml^R phenotype was maintained for more than 100 generations when the bacteria were cultivated in the absence of selection pressure.

6. NOVEL VERSATILE CLONING VECTORS

We have constructed two novel cloning vectors (pLPE23M and pLPE24M) which contain a multilinker region comprising 18 unique cloning sites. Vectors pLPE23M and pLPE24M, which are derivatives of the broad-host range cloning vector pLPE323 (1), differ by the presence in the latter plasmid of an as yet unidentified mutation which increases the copy number approximately 3-5 fold. Figure 1 shows a schematic representation of plasmids pLPE23M/pLPE24M. The vectors are expected to be very useful for the cloning and expression of homologous and heterologous proteins in *Lactobacillus*.

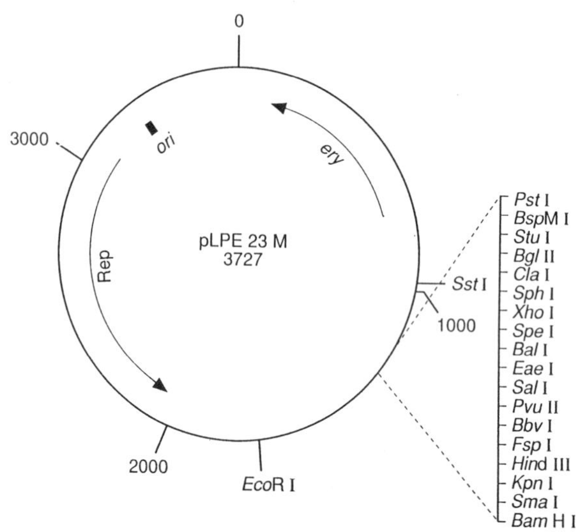

REFERENCES

1. M.Posno, R.J.Leer, N. van Luijk, M.J.F. van Giezen, P.T.H.M. Heuvelmans, B.C.Lokman and P.H.Pouwels, Appl. Environ. Microbiol. 57 (1991) 1822.
2. R.J.Leer, H.Christiaens, L.Peters, M.Posno and P.H.Pouwels, Molec. Gen. Genet. in press.
3. R.J.Leer, N. van Luijk, M.Posno and P.H.Pouwels, Molec. Gen. Genet. 234 (1992) 265.
4. A.Gruss and S.D.Ehrlich, Microbiol Reviews 53 (1989) 231.
5. R.P.Novick, S.Iordanescu, S.J.Projan, J.Kornblum and I.Edelman, 59 (1989) 395.
6. G. del Solar and M.Espinosa, Molecular Microbiology 6 (1992) 83.

ECB6: Proceedings of the 6th European Congress on Biotechnology
edited by L. Alberghina, L. Frontali, P. Sensi

The Kluyveromyces plasmid pKD1: mechanisms ensuring stability and copy number control

M.M. Bianchi, L. Tizzani and L. Frontali

Dept. of Cellular and Developmental Biology, University of Rome "La Sapienza", p.le Aldo Moro, 00185 Rome, Italy

1. INTRODUCTION

The stability of the genotype deriving from plasmids is one of the most important characteristics of transformed microrganisms. This stability is a consequence of the efficiency of the mechanisms underlying correct plasmid segregation during cell division and copy number control. In this paper we describe the replication, recombination and partitioning of pKD1, a plasmid which is stably maintained in Kluyveromyces lactis (1) and has been widely used for the construction of vectors for heterologous gene expression in this yeast (2).

2. REPLICATION OF pKD1

pKD1 is a yeast circular plasmid belonging to the 2μm family of plasmids. These plasmids (reviewed in reference 3) share the same genomic organization, their genes are equivalent in number and function, and they also possess cis-acting elements with analogous functions. Nevertheless, they have no sequence homology and their species specificity is rather narrow.

pKD1 is the smallest plasmid of the family. It is only 4757 base pairs (bp) long and its origin of replication is a short sequence which shows ARS activity both in K. lactis and in Saccharomyces cerevisiae (1,4). The origin of replication is adjacent to one of the two inverted repeats (IRs) present on pKD1 (Figure 1a).

3. PLASMID PARTITIONING

The yeast circular plasmids are actively partitioned between dividing cells. Two plasmid genes and a cis-acting sequence which in pKD1 are called, respectively, B, C and CSL (5) are involved in this mechanism. Stability is greatly affected by these elements: in fact, inactivation of the partitioning genes or deletion of the CSL results in plasmid instability (Table 1). In addition, high stability can be obtained when vectors containing only the replication origin and the CSL are used for the transformation of strains harboring the resident pKD1 (Table 1).

The sequences of the partitioning genes of the 2μm-like plasmids are completely unrelated, except for some conserved amino acids in the type B genes; also the stability loci, which show sharp structural regularities, are completely different from one another (Figure 1b).

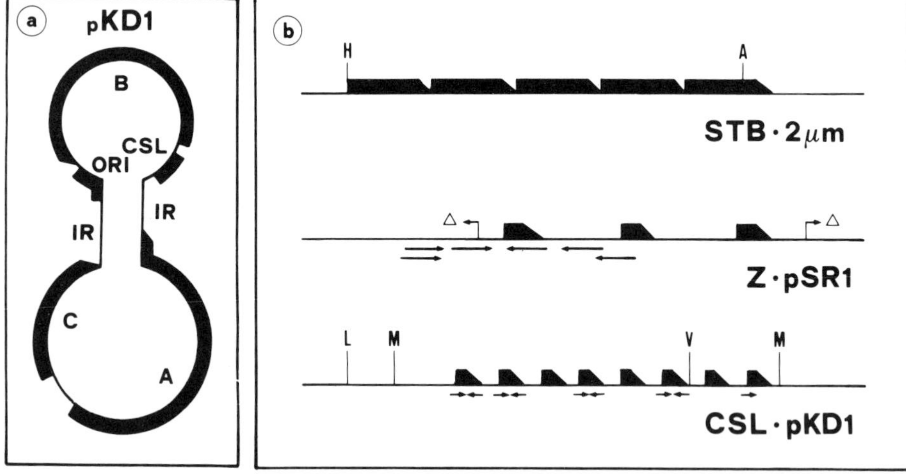

Figure 1. Figure 1a shows one of the two isomers of pKD1. The three genes A, B and C, the origin of replication, the CSL and the IRs are indicated. Figure 1b shows the structures of the three characterized cis-acting stability loci of yeast plasmids: the STB of 2μm, the Z locus of pSR1 and the CSL of pKD1. Repeated sequences are indicated by black blocks. Arrows indicate inverted repeats within the Z locus or sequences with dyad symmetry present in the CSL. Restriction sites or deletion sites (open triangles) used for functional analysis of these elements are reported: A=AvaI, H=HpaII, L=BclI, M=MaeII and V=PvuII.

4. SPECIES SPECIFICITY

Among the mechanisms that ensure the maintenance of the yeast plasmids, the ones involved in replication and in partitioning are both relevant to the determination of plasmid stability and dependent on host functions. In fact, the origin of replication must be recognized by the host replication machinery and the plasmid encoded elements involved in partitioning probably have to interact with some host nuclear factor to allow active plasmid segregation.

The species specificity of the plasmids can thus be a direct consequence, for example, of the divergence of the plasmid encoded partitioning elements, which may not be able to interact properly with the heterologous host factors.

5. STABILITY IN HETEROLOGOUS SYSTEMS

It is possible to obtain further information on the host-plasmid interactions underlying the mechanisms of plasmid maintenance from an analysis of stability in heterologous systems. In Table 1 are reported the stabilities of pKD1 based vectors in both K. lactis and S. cerevisiae. The stability values of S. cerevisiae transformants, which are much lower than the K. lactis ones, are nevertheless reproducible. They result from the high rate of plasmid loss even on selective medium which, in turn, may be due to defective replication and/or interaction with host factors. We can observe that stability in S. cerevisiae is strain dependent and independent on the presence of the endogenous 2μm. On the other hand, the dramatic decrease in stability as a result of the inactivation of genes B and C indicates that the partitioning gene products of pKD1 interact with the heterologous nuclear factors of S. cerevisiae and cannot be complemented by the REP genes of 2μm.

Table 1
Stability of pKD1-based vectors

vector		K. lactis		S. cerevisiae		
		VD1 pKD1°	MD2/1 pKD1+	AH201 cir+	AB110 cir°	FL3/1 cir+
	inactive element					
pS13	none	101.9	74.0	1.3	0.007	0.012
pKA	A	79.3	66.3	1.4	0.006	0.006
pKB	B	4.8	58.7	0.005	0.002	0.009
pKC	C	4.2	78.0	0.005	0.003	0.008
pE11	CSL	29.4	26.7	nd	nd	nd
	element present					
pKRF4	ARS	3.3	3.3	0.22	nd	nd
pACS2	ARS CSL	4.2	66.0	0.06	nd	nd
pHP1	ORI	4.3	11.0	0.011	nd	nd
pHCS1	ORI CSL	1.8	67.4	0.019	nd	nd
pU6	ORI IR	2.6	23.4	0.011	nd	nd

The stabilities are reported as the % of plasmid containing cells after a 10 generation growth on complete medium. The strains VD1 and MD2/1 are isogenic K. lactis strains, except for the presence of pKD1 in MD2/1. The S. cerevisiae strains AB110 and FL3/1 are isogenic except for the presence of 2μm in FL3/1. The structures of the vectors are described in references 5 and 6. "nd" stands for "not determined".

6. RECOMBINATION AND COPY NUMBER CONTROL

The 2μm-like plasmids present in the host cells are an equimolar mixture of two isomers resulting from site-specific recombination at the IRs. The isomers differ only in the orientation of the unique regions between the two IRs. Recombination can occur also between IRs of different molecules, and thereby gives rise to polymers.

The site-specific recombinase of pKD1 is encoded by the largest ORF of the plasmid, which is called gene A, and its target sequences are localized within the IRs (1). In the 2μm system, the plasmid recombinase and its target sequences are sufficient for site-specific recombination both in vivo and in vitro. In pKD1 the situation is different: recombination also requires the integrity of the partitioning genes B and C for complete site-specificity and directionality (6).

In a model proposed by A.B. Futcher for the 2μm (7), plasmid recombination plays a crucial role in copy number amplification: a recombination event occurring during plasmid replication leads to the synthesis of a multimeric form of the plasmid from a monomeric template, thereby increasing copy number. This model can also explain the conservation of structural characteristics among the plasmids, such as the positions of the IRs and of the replication origins.

Copy number amplification probably depends on the activity of the plasmid recombinase also in pKD1. In fact, the inactivation or overexpression of gene A lead, respectively, to copy number decreases or increases in this plasmid (6 and M.M. Bianchi; unpublished results).

7. CONCLUSIONS

The plasmid pKD1 is a useful tool for the construction of stable vectors which can be employed, for example, to express heterologous genes. High copy number of such vectors will be obtained by increasing the activity of the plasmid recombinase, while the optimal conditions for high stability can be more easily determined on the basis of the mechanisms of plasmid partitioning and host-plasmid interaction.

ACKNOWLEDGEMENTS

This work was supported by Consiglio Nazionale delle Ricerche (Progetto Finalizzato Biotecnologie e Biostrumentazioni) and by Rhône-Poulenc Rorer (contract with the University of Rome, PES0030192).

REFERENCES

1. M.M. Bianchi et al., Curr. Genet. 12 (1987) 185-192
2. R.G. Buckholz et al., Bio/Technology 9 (1991) 1067-1072
3. C. Volkert et al., Microbiol. Rev. 53 (1989) 299-317
4. X.J. Chen et al. Nucleic Acids Res. 14 (1986) 4471-4480
5. M.M. Bianchi et al., Curr. Genet. 19 (1991) 155-161
6. M.M. Bianchi, J. Bacteriol. 174 (1992) 6703-6706
7. A.B. Futcher, J. Theor. Biol. 119 (1986) 197-204

BIOTECHNOLOGY OF LACTIC ACID BACTERIA

ECB6: Proceedings of the 6th European Congress on Biotechnology
edited by L. Alberghina, L. Frontali, P. Sensi

Biotechnology of Lactic Acid Bacteria: Current Trends in Genetic Research

Mervi Sibakov

Valio Ltd, Research and Development, P.O.Box 390, 00101 Helsinki, Finland

The purpose of this paper is to discuss some of the recent achievements and trends in genetic and molecular biological studies aiming at modern biotechnological applications of lactic acid bacteria.

1. RECENT ACHIEVEMENTS

By the present time the molecular biological methods have been well established in the genus Lactococcus, towards which the most intensive research efforts have been directed, mostly because of their world wide use as starter cultures. As the result the genetic basis for the crucial process characteristics, lactose break-down and caseinolytic activity have been analyzed to the detail. Especially the analysis of disaccharide production, in particular, in lactococci has proceeded considerably during the recent years. All the sugar-specific genes, both from the lac-PTS and lacZ systems are now available (1-5) and can be used in structure-funtion studies as well as in the control and improvement of industrial fermentations utilizing lactic acid bacteria. Also in the genus Lactobacillus advances in the study of lactose metabolism have been made, but also some anomalous results have been obtained, e.g. the lack of cross-reaction or homology between the lacZ genes of seemingly closely related species (6-7). Thus the nature of lactose catabolism in some of the important thermophilic lactobacillar starters still remains to be elucidated.

The various enzymatic activities, which altogether form the complex proteolytic system of lactococci are being unraveled. Without the proper functioning of each component efficient growth in milk is prevented. The first steps in the sequence of events: the caseinolytic proteases are well-characterized, to the point of protein engineering trials, which have, indeed, been done (8-13). At present considerable effort is being put on the analysis of indispensable extra- and intracellular peptidases yielding smaller oligopeptides (14-18) and finally on the study of the specialized uptake/transport system of di- and tripeptides (15, 19). The study of the proteolytic system in lactobacilli is far behind lactococci, almost just at its beginning. However, it seems possible to use the information obtained with lactococci to speed up the analysis of this crucial system in lactobacilli.

Another area in dairy industry, where applications of modern biotechnology are more than welcome, is the problem of *bacteriophages* attacking cheese vats.

Starter cultures used to produce Cheddar- the world's most sold cheese- are especially vulnerable to phages. It is now possible to construct *phage resistant starters* based on detailed characterization of several different mechanisms by which lactococci can obtain resistance (20-21).

The last years have also seen huge increase in *nucleic acid probe* and *PCR*-technology used to identify and classify lactic acid bacteria (22-24). It has been known for a long time that conventional biochemical tests give rather vague taxonomical criteria. The new much more accurate methods are very much needed in many fields of lactic acid bacterial research and applications.

2. HIGH LIGHTING BACTERIOCINS

It is probably safe to say that *bacteriocins* are the most popular research area at present. Contrary to the genetic research in general, which is very much focussed on lactococci, the search for bacteriocins has been widened to all the representative lactic acid bacterial genera. There are extensive screening programs aimed at isolation of new bacteriocins from lactococci, lactobacilli, pediococci, leuconostocs, carnobacterium and propionibacterium (25-35). In addition to isolating new compounds there is an increased tendency to construct new starters with higher levels of bacteriocin production, to alter their specificity or to introduce bacteriocin production and immunity into previously non-producing cells.

There is one bacteriocin above others as to the information on structure, function and applications--*nisin*-- synthezised by Lactococcus lactis subsp. lactis. Several prominent groups (36-39) are studying the biosynthesis and mode of action of this post-translationally modified lantibiotic. At present seven genes of the nisin operon have been isolated and analyzed to the nucleotide sequence level from two different natural nisin variants nisA and nisZ (39-41). Although the organization of the genetic structure is about to be at hand, the physiological functions of the individual genes in the secretion, processing, maturation and immunity are still to be confirmed. Due to the intensive efforts the mechanism of nisin biosynthesis should be available in a very near future. Although most interesting, especially for the basic science, the wealth of information obtained from nisin, does not necessarily make its application in milk fermentations very easy, since most of the commonly used starters are nisin sensitive. The question of nisin resistance and immunity has not yet been solved, at least, to the level of commercial applications (42-43). There have also been attempts to use conjugation as a possibility to expand the spectrum of nisin producing starters (44), but serious instability problems have been encountered.

Until very recently nisin has represented the only lantibiotic characterized from lactic acid bacteria. Now, at least, two lanthionine containing new bacteriocins have been reported: lacticin 481 from Lactococcus lactis subsp. lactis (25) and carnocin U149 from Carnobacterium sp (33).

When applications of bacteriocins are thought of there is a clear demand for a wider target spectrum especially towards yeast and fungi. Most characterized bacteriocins or antimicrobial compounds seem to lose effect towards neutral and alkaline pH-conditions, which is also a drawback, particularly in non-dairy applications. The combined effort of screening and gene technological studies is likely to result in new and useful bacteriocin varieties.

3. AREA FOR FUTURE RESEARCH

It has become clear from the previous discussion that, where molecular biology, to the level of applications, has been established for the important mesophilic starters (lactococci), the knowledge available from thermophilic starters, (representatives of lactobacilli and Streptococcus thermophilus) lacks far behind. It is among these organisms that one encounters serious difficulties even in establishing genetic methods. For example, there is very little published information on successful gene transfers in e.g. Lactobacillus bulgaricus or L. helveticus (45-46) in spite of numerous attempts.
Streptococcus thermophilus has been somewhat easier to handle, but frequent problems with genetic variation, instability and lack of heterologous expression has been reported (47). Thermophilic starters are economically also a very important group , comprising bacteria used in yogurt, Italian and Swiss type cheeses. A concerted effort should be put forward to promote the study of their genetics.

4. REGULATIONS AND PUBLIC PERCEPTION

Progress in the genetic modification of lactic acid bacteria has now reached the stage where it is possible to apply the GMO strains constructed in laboratories in commercial applications. So far, however, there are no such applications. Although the new biotechnology or gene technology laws and regulations in different countries make it possible and acceptable to apply for commercialization of such new products or strains, individual scientists and companies are still hesitant.
There is uncertainty about consumers' attitudes towards genetically engineered food. In many cases, dairy products would represent the 'extreme' , where the final product contains live GMOs. If such applications are not readily accepted yet, there are many other possibilities to apply the results obtained through modern biotechnology even today, e.g. new analytical methods for process monitoring, determination of biocide residues or strain identification.

REFERENCES

1. P. Leong-Morgenthaler, M.C Zwahlen and H. Hottinger, J. Bacteriol.,173 (1991) 1951.
2. R.J. van Rooijen, S. van Schalkwijk and W.M. de Vos, J. Biol. Chem., 266 (1991) 7176.
3. R.J. van Rooijen, M.J. Gasson and W.M. de Vos, J. Bacteriol., 174 (1992) 2273.
4. W.M. de Vos, P. Vos, G. Simons et. al., J. Dairy Sci., 72 (1990) 3398.
5. J.S. Schroeder, C. Robert, G. Lenzen et. al., J. Gen. Microbiol.,137 (1991) 369.
6. M.A. Collins and J.K. Thompson, J. Appl. Bacteriol., 73 (1992) 237.
7. M. Sibakov, unpublished observation.

8. J. Kok, K.J. Leenhouts, A.J. Haandrikman et. al., Appl. Environ. Microbiol., 54 (1988) 231.
9. P. Vos, M. van Asseldonk, F. van Jeveren et. al., J. Bacteriol., 171 (1989) 2795.
10. J. Kok, FEMS Microbiol. Rev., 87 (1990) 15.
11. P. Vos, G. Simons, R.J. Siezen et. al., J. Biol. Chem., 264 (1989) 13579.
12. E. J. Smid, B. Poolman and W.N. Konings, Appl. Environ. Microbiol., 57 (1991) 2447.
13. P. Vos, I. J. Boerrigter, G. Buist et. al., Protein Engineering, 4 (1991) 479.
14. P.S.T. Tan and W.N. Konings, Appl. Environ. Microbiol., 56 (1990) 526.
15. E. Smid, Ph.D. Thesis. University of Groningen, The Netherlands, 1991.
16. M. Nardi, M-C. Chopin, A. Chopin et. al., Appl. Environ. Microbiol., 57 (1991) 45.
17. B. Mayo, J. Kok, K. Venema et. al., Appl. Environ. Microbiol., 57 (1991) 38.
18. M.-C. Chapot-Chartier, M. Nardi, M.-C. Chopin et.al., Appl. Environ. Microbiol., 59 (1993) 330.
19. S.Tynkkynen et. al., manuscript in preparation.
20. M.-C. Chopin and A. Chopin, Proc. 6th Internat. Symp. GIM, (1990) 467.
21. W.D. Sing and T.R. Klaenhammer, Appl. Environ. Microbiol., 59 (1993) 365.
22. M. Delley, B. Mollet and H. Hottinger, Appl. Environ. Microbiol., 56 (1990) 1967.
23. M.Salama, W. Sandine and S. Giovannoni, Appl. Environ. Microbiol., 57 (1991) 1313.
24. N. Klijn, A.H. Weerkamp and W.M. de Vos, Appl. Environ. Microbiol., 57 (1991) 3390.
25. J-C. Piard, P.M. Muriana, M.J. Desmazeaud et. al., Appl. Environ. Microbiol., 58 (1992) 279.
26. J. Nissen-Meyer, H. Holo, L.S. Håvarstein et. al., J. Bacteriol., 174 (1992) 5686
27. C.I. Mortvedt, J.Nissen-Meyer, K. Stetten et. al., Appl. Environ. Microbiol., 57 (1991) 1829.
28. P.M. Muriana and T.R. Klaenhammer, J Bacteriol., 173 (1991) 1779.
29. M.C. Joerger and T.R. Klaenhammer, J. Bacteriol., 172 (1990) 6339.
30. A.K. Bhunia, M.C. Johnson, B. Ray et.al., J. Appl. Bacteriol., 70 (1991) 25.
31. J.W. Hastings, M. Sailer, K. Johnson et.al., J. Bacteriol.,173 (1991) 7491.
32. H. Daba, S. Pandian, S. Gosselin et. al., Appl. Environ. Microbiol., 57 (1991) 3450.
33. G. Stoffels, J. Nissen-Meyer, A. Gudmundsdottir et.al., Appl. Environ. Microbiol., 58 (1992) 1417.
34. W.J. Lyon and B.A. Glatz, Appl. Environ. Microbiol., 57 (1991) 701.
35. D.A. Grinstead and S.F. Barefoot, Appl. Environ. Microbiol., 58 (1992) 215.
36. C. Kaletta and K-D Entian, J. Bacteriol., 171 (1989) 1597.
37. G. Buchman, S.H. Banerjee and J.N. Hansen, J. Biol. Chem., 263 (1988) 16260.
38. H.M. Dodd, N. Horn and M.J. Gasson, J. Gen. Microbiol., 136 (1990) 136.
39. J.R. van der Meer et.al., J. Bacteriol., (1993) in press.
40. O.P. Kuipers et. al., submitted
41. P. Saris et. al., submitted
42. A. von Wright, S. Wessels, S.Tynkkynen et.al., Appl. Environ. Microbiol., 56 (1990) 2029.
43. B. R. Froseth and L.L. McKay, Appl. Environ. Microbiol., 57 (1991) 804.
44. H.J. Tsai and W.E. Sandine, Appl. Environ. Microbiol., 53 (1987) 352.
45. C.A. Batt, Food Technol., 40 (1986) 95.
46. H. Hashiba, R. Takiguchi, K. Jyoko et. al., Biosc. Biotech. Biochem., 56 (1992) 190.
47. A. Mercenier, FEMS Microbiol. Rev., 87 (1990) 61.

ECB6: Proceedings of the 6th European Congress on Biotechnology
edited by L. Alberghina, L. Frontali, P. Sensi

Heterologous gene expression in *Lactobacillus*

M. Posno, M. Kottenhagen, B.C. Lokman, J. Jore, R.J. Leer and P.H. Pouwels.

TNO Medical Biological Laboratory, P.O. Box 5815, 2280 HV Rijswijk, The Netherlands.

1. INTRODUCTION

The economic importance of various *Lactobacillus* strains is beyond doubt. Members of this genus are known, in the first place, for their widespread use in various food and agricultural fermentation processes. Furthermore, in recent years it has become clear that some species of lactobacilli may beneficially influence the health of animals and humans, either as commensal colonizers of the gastrointestinal tract or when administered orally (1). In view of these attractive properties, it is of no surprise that strain improvement of *Lactobacillus* receives considerable attention.

The demonstration that foreign DNA could be successfully introduced into *Lactobacillus* by electroporation has opened a new area of research aimed at the application of recombinant DNA technology to these microorganisms. During the past few years gene transfer systems and suitable cloning vectors have become available for a large number of *Lactobacillus* species (2). To fully exploit the potential of genetic modification of *Lactobacillus* with regard to strain improvement, our current research focuses on the development and application of methods to efficiently express heterologous genes in *Lactobacillus*. Here we will present some results of our recent investigations on (regulation of) gene expression in *Lactobacillus*.

2. *LACTOBACILLUS* PROMOTERS

A limited number of reports have been published describing the expression in *Lactobacillus* of genes from other bacterial sources under the control of their own promoter (3). To achieve efficient expression of foreign genes, it will in many cases be desirable to use homologous expression signals instead of heterologous ones. Knowledge about gene expression and control of gene expression in *Lactobacillus*, however, is still scarce. Although the nucleotide sequence of more than fifty *Lactobacillus* genes is known to date, little information is present concerning the sequences required for transcription initiation and control of transcription.

Basically, we have followed two different approaches to isolate suitable *Lactobacillus* promoters. In the first approach, DNA fragments exhibiting promoter activity have been obtained by shotgun cloning of *Lactobacillus* chromosomal DNA fragments in front of a promoterless chloramphenicol resistance gene (promoter-probe vector). This strategy, which is fast and relatively simple, has the additional advantage that it selects for strong and most likely constitutive promoters. A number of promoter activity containing *Lactobacillus* chromosomal DNA fragments have been characterized into more detail with respect to relative promoter strength (analysis of chloramphenicol acetyl transferase activity) and promoter structure (primer extension analysis). The results showed that the structures of the promoters were quite similar (see table 1: ORFX and ORFY), although significant differences in relative promoter strength were observed. An important disadvantage of this approach, however, is that nothing is known of possible regulation mechanisms of the promoter in concern. Therefore, the majority of our research efforts have been concentrated on the second approach which deals with the analysis of upstream regions of well-characterized *Lactobacillus* genes.

We have cloned five genes involved in xylose catabolism in *L. pentosus* and have determined the transcriptional organization of the genes. Induction (by xylose) of *xylA*, which encodes xylose isomerase, and *xylB*, which codes for xylulose kinase, takes place at the level of transcription and is negatively controlled by a repressor, the product of *xylR* (4). The promoters of two of the genes, *xylR* and *xylA*, were identified by mapping of the 5' ends of the messengers by primer extension analysis and compared with promoter sequences of other *Lactobacillus* genes (table 1).

Table 1
Promoters for which transcription start sites have been mapped

Gene	Species	-35	-10	Reference
xylA	*L. pentosus*	TTTACA	TGTAAT	Unpublished
xylR	*L. pentosus*	TTGACA	CACATT	Unpublished
ORFX	*L. acidophilus*	TTGTCA	TATATT	Unpublished
ORFY	*L. acidophilus*	TTGACA	TAAAAT	Unpublished
D-*HicDH*	*L. casei*	TTCACA	TACGGT	5
fgs	*L. casei*	CTCATA	TATGAT	6
lacP	*L. bulgaricus*	TTGGTT	TATACT	7
L-*HicDH*	*L. confusus*	TGGTGA	TATAAT	8
SP-P1	*L. brevis*	TTGTAT	TATACT	9
SP-P2	*L. brevis*	TTAACA	TATACT	9
hdcA	*Lactobacillus*	TTTACT	TATAAT	10
Consensus		**TT**GACA	**TA**TAA**T**	

The nucleotide indicated is the one occurring most frequently. **Bold** capital: > 75%; capital: 50-75%. Taken, in part, from: Mercenier, Pouwels and Chassy (3).

Compilation of *Lactobacillus* promoters for which the transcription start sites have been determined (table 1) reveals the presence of sequence elements which are similar or even identical to the consensus sequence (TTGACA -17bp- TATAAT) of Gram-positive and Gram-negative promoters. A similar consensus can be derived from the nucleotide sequence of the upstream regions of *Lactobacillus* genes from which the promoter has not been mapped yet. It should be noted, however, that in these cases it is uncertain whether these elements represent functional promoters.

3. HETEROLOGOUS GENE EXPRESSION

A potentially useful property of certain *Lactobacillus* strains is the capacity to evoke a mucosal as well as a systemic immune response against epitopes associated with these microorganisms after oral administration (11). This finding and earlier observations indicating that certain *Lactobacillus* strains have adjuvant properties, may have important consequences for the development of novel oral vaccines. In this perspective, we are carrying out a research program which is aimed at the development of *Lactobacillus* as a safe (non-pathogenic) vector for the delivery of antigens in the gastrointestinal tract. To this end, methods need to be developed which allow the expression of foreign antigens in *Lactobacillus* strains of choice.

As a first step towards this goal, an expression vector was constructed on the basis of regulatory sequences of the xylose genes of *L. pentosus*. Briefly, the promoter of the *xylR* gene and the terminator of the *xylB* gene (separated by a multiple cloning site) were inserted in the *E. coli* - *Lactobacillus* shuttle-vector pLP3537 (2). The resulting vector was designated pLPCR2. Transformation of *L. casei* or *L. pentosus* with a pLPCR2-derivative in which a promoterless chloramphenicol resistance gene of *S. aureus* was cloned in the multicloning site, yielded numerous chloramphenicol resistant colonies for both strains, indicating that the *xylR* promoter in pLPCR2 is also functional in a heterologous host. Next, a gene encoding protein p72, which was inserted in the polylinker of pLPCR2, was introduced into *L. casei* by electroporation. Protein p72 consists of two copies of a foot-and-mouth-disease virus (FMDV) epitope (in tandem) which are fused with *E. coli* β-galactosidase. Synthesis of p72 could be detected in cellfree extracts of transformed cells both with antibodies specific for the FMDV epitope and antibodies specific for β-galactosidase. The level of expression, however, was relatively low (about 0.1% of total cellular protein).

To try to increase the expression level, a new vector was constructed based on a promoter that was assumed to be stronger than that of *xylR*. Using the regulatory sequences of the bile salt hydrolase gene (*cbh*) of *L. plantarum*, we had previously demonstrated that a 10-fold overproduction of a homologous protein in *L. plantarum* could be achieved (12). Therefore, this promoter was chosen for further expression studies. First, the *E. coli lacZ* gene was placed under control of the *cbh* promoter. *L. plantarum* transformants containing the resulting vector produced about 1-2% of total cellular protein, indicating that the *cbh* promoter is a suitable promoter for heterologous gene expression in *L. plantarum*. Unfortunately, however, the expression unit was not expressed in another *Lactobacillus* species, *L. casei*. Moreover, when rotavirus capsid protein VP7 was fused to the N-terminus of β-galactosidase, it appeared that the level of expression in *L. plantarum* was drastically reduced compared to β-galactosidase alone (about 0.2% of total cellular protein).

Finally, strong evidence has been obtained that in certain cases constitutive expression of a heterologous gene may be lethal to the *Lactobacillus* host. For example, we have found that a gene coding for hepatitis delta virus antigen (HDVAg) could only be introduced into *L. casei* in the case that this gene was not preceded by functional regulatory sequences. However, transformation of *L. casei* with the same expression cassette under control of a functional promoter and ribosome binding site yielded only a few transformants which could not be further cultivated. To be able to produce proteins like HDVAg, we are currently developing regulatable expression systems for intracellular and extracellular production. An attractive candidate is the promoter of the α-amylase gene of *L. amylovorus,* whose activity can be induced by for instance maltose and almost completely repressed by for instance glucose.

4. CONCLUSIONS

The results described in this paper demonstrate that expression in *Lactobacillus* of heterologous genes under the control of genuine *Lactobacillus* promoters can be achieved. The levels of expression that have been obtained vary depending on the promoters used and the nature of the gene product. This in combination with unexpected finding that the *cbh* promoter is not functional in *L. casei* clearly indicates that a substantial research effort is still required to unravel as yet unknown mechanisms of control of (heterologous) gene expression in *Lactobacillus.* Moreover, for the expression of genes whose gene product is lethal for *Lactobacillus,* regulatory expression systems need to be developed.

REFERENCES

1. L.L. McKay and K.A. Baldwin. FEMS Microbiol. Rev., 36 (1990), 3.
2. M. Posno, R.J. Leer, N. van Luijk, M.J.F. van Giezen, P.T.H.M. Heuvelmans and P.H. Pouwels. Appl. Environ. Microbiol., 57 (1991), 1822.
3. A. Mercenier, P.H. Pouwels and B.M. Chassy. In: Applied genetics of lactic acid bacteria. Eds. M. Gasson and W.M. de Vos. Chapman and Hall, London, (1993), in press.
4. B.C. Lokman, P. van Santen, J. Verdoes, J. Kruse, R.J. Leer, M. Posno and P.H. Pouwels. Mol. Gen. Genet., 230 (1991), 161.
5. H.P. Lerch, H. Blocker, H. Kalwass, J. Hoppe, H. Tsai and J. Collins. Gene, 78 (1989), 47.
6. J. Toy and A.L. Bognar. J. Biol. Chem., 265 (1990), 2492.
7. P. Leong-Morgenthaler, M.C. Zwahlen and H. Hottinger. J. Bacteriol., 173 (1991), 1951.
8. H.P. Lerch, R. Frank and J. Collins. Gene, 83 (1990), 263.
9. G. Vidgrén, I. Palva, R. Pakkanen, K. Lounatmaa and A. Palva. J. Bacteriol., 174 (1992), 7419.
10. P. Vanderslice, W. Copeland and J. Robertus. J. Biol. Chem., 261 (1986), 15186.
11. K. Gerritse, M. Posno, M.M. Schellekens, W.J.A. Boersma and E. Claassen. Res. Microbiol., 141 (1990), 955.
12. H. Christiaens, R.J. Leer, P.H. Pouwels and W. Verstraete. Appl. Environ. Microbiol., 58 (1992), 3792.

ECB6: Proceedings of the 6th European Congress on Biotechnology
edited by L. Alberghina, L. Frontali, P. Sensi

Biosynthesis of the antimicrobial peptide nisin: structure, organization and function of the nisin-operon genes, protein engineering of nisin Z, and characterization of precursors of nisin

Oscar P. Kuipers *, Harry S. Rollema, Marke M. Beerthuyzen, Jan Roelof van der Meer, Willem M. de Vos and Roland J. Siezen

Department of Biophysical Chemistry, NIZO, P.O. Box 20, 6710 BA, Ede, The Netherlands⁺.
Fax: 31-8380-50400; Tel: 31-8380-59511

1. INTRODUCTION

Nisin is produced by some strains of *Lactococcus lactis* and is used in the food industry as a preservative, since it inhibits the growth of unwanted Gram-positive bacteria. Nisin is a member of the group of antimicrobial peptides called lantibiotics (for a review, see [1]). These molecules are characterized by a low molecular weight (< 4 kD), by the presence of unsaturated amino acid side chains, such as dehydroalanine and dehydrobutyrine, and by thio-ether bridges of lanthionine residues, which are derived from the addition of sulfhydryl groups of cysteine residues to the double bonds of unsaturated residues. The primary structure of nisin Z is represented in Fig. 1.

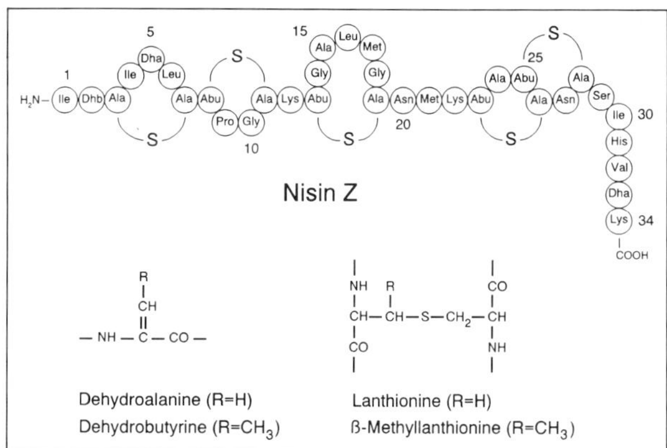

Figure 1. Primary structure of nisin Z, and structures of the uncommon amino acid residues dehydroalanine (Dha), dehydrobutyrine (Dhb), lanthionine and ß-methyllanthionine.

---

* To whom correspondence should be addressed

+ This work was supported by contract BIOT-CT91-0265 of the BRIDGE program of the European Community.

The pathway for biosynthesis of nisin has not been fully elucidated. It is assumed that post-translational modifications occur on the ribosomally synthesized precursor peptide of 57 residues, followed by secretion and processing or *vice versa*. Two natural variants of nisin have been found, *i.e.* nisin A, containing a His residue at position 27 and nisin Z, containing an Asn residue at the same position [2]. These two species display similar antimicrobial properties and their genes are widely distributed in nature [3]. Our research is directed to the overproduction of nisin in *L. lactis*, to the elucidation of the nisin-biosynthesis mechanism and to the development of altered nisins, which are tailored for specific applications.

2. ORGANIZATION OF THE NISIN GENE CLUSTER AND FUNCTION OF THE ENCODED PROTEINS

The gene cluster located downstream of the *nisA* gene on transposon Tn*5276* [4] of *Lactococcus lactis* NIZO R5 was cloned and sequenced. The gene cluster was found to contain six genes, with the order *nisBTCIPR* [5-7] (Fig. 2). The NisB and NisC proteins are likely to be involved in the catalysis of modification reactions, such as dehydration of Ser and Thr residues and of subsequent thioether-bridge formation [5,7]. The encoded NisT protein shows a high degree of sequence similarity to tranport proteins of the ABC-family, exemplified by HlyB of *E. coli* [5,7]. The *nisI* gene product has the characteristics of a lipoprotein and was shown to be involved in the development of immunity against nisin in both *L. lactis* and in EDTA-treated *E. coli* cells [5]. NisP is a subtilisin-like protease involved in the processing of a nisin precursor on the outer side of the cellular membrane [6]. Finally, NisR was shown to act as a positive regulator of expression of *nisA* and possibly also of downstream genes [6]. These genes together probably constitute the full operon required for nisin biosynthesis and immunity.

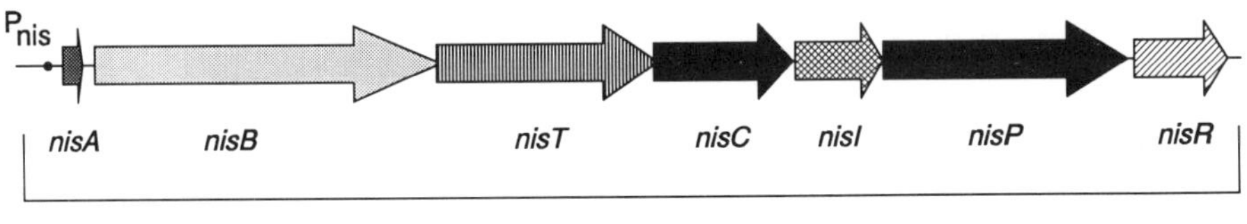

Figure 2. Organization of the nisin A gene-cluster of transposon Tn*5276* of *Lactococcus lactis* NIZO R5.

3. PROTEIN ENGINEERING OF NISIN

Expression systems for altered *nis*-genes were developed with *L. lactis* as a host. One system makes use of a nisin A producing transconjugant of *L. lactis* NIZO R5 in which mutant *nisZ* genes on plasmids are expressed under control of the *lac*-promoter [8]. Another system was developed in which a 4-bp deletion was introduced in the *nisA* gene of Tn*5276* by gene replacement [5]. This strain can be complemented by expression plasmids harboring mutant *nisZ* genes.

Several mutants of nisin Z were produced, purified on a 10-100 mg scale, and characterized. It was shown that introduction or substitution of Ser or Thr residues (S5T or M17Q/G18T) leads to mutants containing other or additional dehydrated residues [8]. These mutants showed different antimicrobial activities compared to wild-type nisin Z. Other mutations introduced in nisin Z include an M17W mutant, which was used in time-resolved fluorescence experiments, to study its interactions with micelles and vesicles of different phospholipid composition. In addition, mutants with higher solubility at pH values above 6 were produced (N27K and H31K nisin Z), which retained an antimicrobial activity similar to that of nisin Z. Protein engineering of nisin offers the possibility of studying biosynthetic substrate requirements, but also allows the development of nisins suited for specific applications.

4. PRODUCTION AND CHARACTERIZATION OF NISIN PRECURSORS

Three different precursors of nisin were isolated from *L. lactis*. The first precursor was produced by a strain deficient in NisP production [6], while the other two were obtained either by replacing the nisin leader sequence by that of the homologous lantibiotic subtilin [Kuipers et al., in preparation], or by changing the Ala residue at position -4 into an Asp residue in the leader sequence of nisin Z [van der Meer et al., in preparation]. In all three cases fully matured nisin was produced with an unmodified leader seqence still attached to it. An example of the primary structure of one of the precursors is shown in Fig. 3.

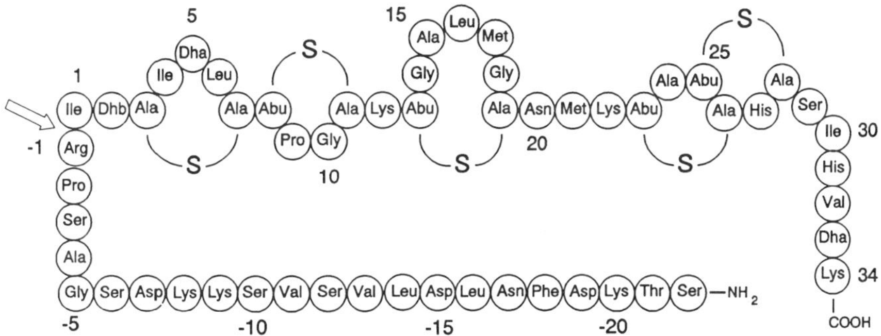

Figure 3. Primary structure of precursor nisin A.

In all three cases the N-terminal Met residue in the leader was absent, probably due to posttranslational processing, as is common in bacteria. These precursors were purified and found to have at least 200-fold lower antimicrobial activities than nisin A and nisin Z. Precursors with an Arg residue at position -1, could be activated by either trypsin or NisP to yield fully active nisin [6]. The results show that *(i)* precursor nisin is first secreted and then processed to its mature form by NisP on the outer side of the cellular membrane; *(ii)* leaders of nisin precursors are not modified, whereas the pronisin part is fully modified; *(iii)* certain mutations in the leader, or the use of a different lantibiotic leader sequence, prevents processing but not secretion; *(iv)* precursors are virtually inactive due to the attachment of the leader.

5. PROPOSED MODEL FOR NISIN BIOSYNTHESIS

Based on the results described above, we propose the following model for nisin biosynthesis and immunity (Fig. 4).

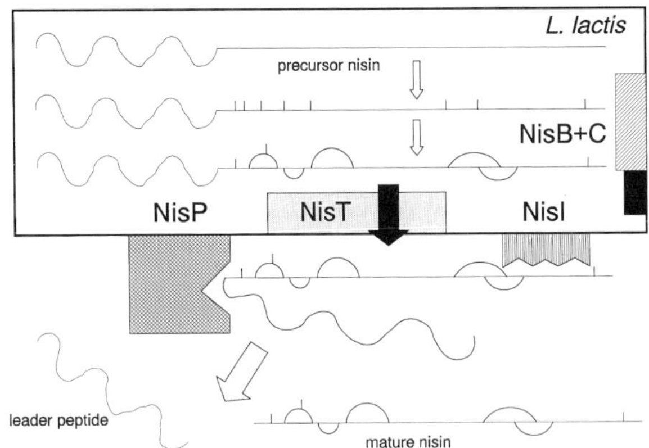

Figure 4. Proposed model for nisin biosynthesis

First, the 57 amino acid precursor nisin is ribosomally synthesized. Second, a total of eight Ser and Thr residues are dehydrated and subsequently thio-ether bonds are formed; both reactions are likely to be catalyzed by the putative membrane-bound enzymes NisB and NisC. Third, the modified precursor nisin is transported across the membrane by the putative translocator NisT. Fourth, on the outer side of the membrane the subtilisin-like enzyme NisP cleaves off the leader peptide, and releases fully matured nisin. Recent evidence suggests that NisI is a lipoprotein involved in immunity of the producer strain [5]. All biosynthetic proteins might be clustered at the membrane and synergistic action of these proteins can be envisaged.

REFERENCES

1. G. Jung, in: Nisin and novel lantibiotics, G. Jung and H.-G. Sahl, eds, ESCOM, Leiden (1991) 1-31.
2. J.W.M. Mulders, I.J. Boerrigter, H.S. Rollema, R.J. Siezen and W.M. de Vos, Eur. J. Biochem., 201 (1991) 581-584.
3. W.M. de Vos, J.W.M. Mulders, J. Hugenholtz, R.J. Siezen and O.P. Kuipers, Appl. Environm. Microbiol., 59 (1993) 213-218.
4. P.J.G. Rauch and W.M. de Vos, J. Bacteriol., 174 (1992) 1280-1287.
5. O.P. Kuipers, M.M. Beerthuyzen, R.J. Siezen and W.M. de Vos, submitted for publication.
6. J.R. van der Meer, J. Polman, M.M. Beerthuyzen, R.J. Siezen, O.P. Kuipers and W.M. de Vos, J. Bacteriol., 175 (1993), in press.
7. G. Engelke, Z. Gutowski-Eckel, M. Hammelmann and K.-D. Entian, Appl. Environm. Microbiol., 58 (1992) 3730-3743.
8. O.P. Kuipers, H.S. Rollema, W.M.G.J. Yap, H.J. Boot, R.J. Siezen and W.M. de Vos, J. Biol. Chem., 267 (1992) 24340-24346.

ECB6: Proceedings of the 6th European Congress on Biotechnology
edited by L. Alberghina, L. Frontali, P. Sensi

Diacetyl production by *Lactococcus lactis*: Optimilisation and metabolic engineering

Jeroen Hugenholtz*, Marjo J.C. Starrenburg and Anton H. Weerkamp

Department of Microbiology, Netherlands Institute for Dairy Research (NIZO), P.O. Box 20, 6710 BA, Ede, The Netherlands.

1. INTRODUCTION

The butteraroma diacetyl is produced in dairy products by the action of specific lactic acid bacteria. The citric acid in milk is converted via pyruvate to the unstable metabolic intermediate α-acetolactate. This can subsequently be converted enzymatically to acetoin or chemically converted (in the presence of oxygen) to diacetyl. To achieve optimal diacetyl production, fermentation conditions and lactic acid bacteria have to be found that favour high production of α-acetolactate from citric acid and its chemical oxidation/decarboxylation to diacetyl. Furthermore, by employing metabolic engineering efficient diacetyl production from lactose can be achieved in *Lactococcus lactis*.

2. α-ACETOLACTATE METABOLISM

Diacetyl is formed as a side-product by chemical decarboxylation of the metabolic intermediate α-acetolactate (AL), in the presence of oxygen. AL is formed biologically during the synthesis of branched-chain amino acids and during acetoin- and butanediol-formation. Two molecules of pyruvate are combined to form the C5-component in a thiamine pyrophosphate (TPP) dependent reaction (reaction 1 in Fig. 1). The enzyme catalyzing this reaction in *Lactococcus lactis*, α-acetolactate synthase, clearly has a katabolic function, and is not involved in (amino acid) biosynthesis (1). It has only appreciable activity at high concentrations of pyruvate (K_M = 50 mM), which agrees with the observed high internal concentrations of pyruvate (50 - 100 mM) and the subsequent excretion of pyruvate into the medium during acetoin formation (2, 3). AL is decarboxylated to acetoin by the enzyme α-acetolactate decarboxylase (reaction 2 in Fig. 1) and acetoin can, finally, be reduced to butanediol by the enzyme acetoin reductase (reaction 3 in Fig. 1) This is the general pathway of acetoin and butanediol formation from pyruvate (or citrate) in lactic acid bacteria. *L. lactis* subsp. *lactis* bv. *diacetylactis* strain NIZO Ru4 (4), however, - also known as strain 806 (5) and strain 425A (6) - is defective in AL decarboxylase activity (Fig. 1) and accumulates large amounts of AL during citrate metabolism (4). Dairy fermentations involving this unique strain result in high diacetyl production (7).

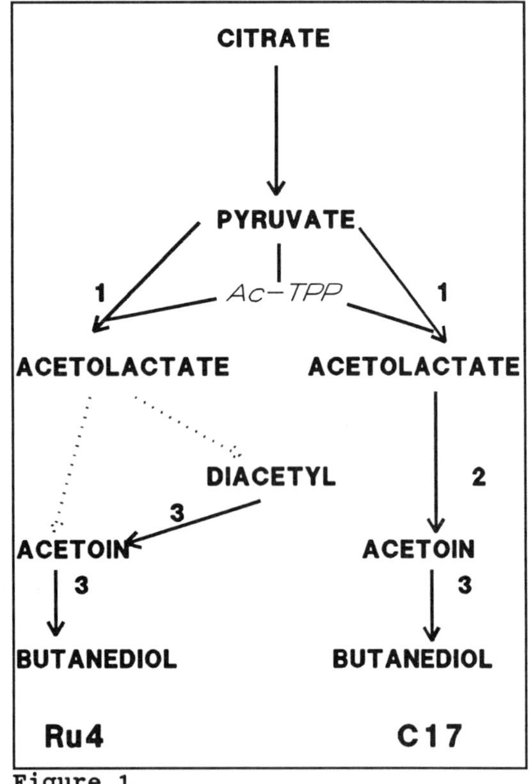

Figure 1

3. DIACETYL PRODUCTION FROM CITRATE

Diacetyl production from citrate was optimalised using *L. lactis* strain Ru4. Fermentation conditions were chosen that: 1. resulted in maximum citrate utilisation, 2. allowed rapid growth of the *L. lactis* strain, 3. resulted in efficient conversion of citrate into AL, and 4. made use of only dairy substrates. These objectives were reached by growing the *L. lactis* strain on a whey-medium with the pH controlled at 5.7 using an alkaline citrate solution (1M, pH=9.0) as neutralising agent. By setting the pH below 6.0, citrate uptake was optimal (8), while a pH above 5.5 could still support growth of *L. lactis* Ru4. By pulsed-addition of citrate, high amounts of citrate (100 mmol/l) were converted by the lactococci while the citrate concentration in the growth medium remained below the toxicity-level (10 mM). Using this fermentation strategy upto 50 mM AL could be produced in culture, even at the pilot-plant scale of 2500 liter. The produced AL could be converted for more than 50% into diacetyl by decreasing the pH after fermentation to pH 4.0 under vigorous aeration. Dairy-based fermentation fluids with upto 2000 ppm diacetyl could be obtained by this procedure.

4. DIACETYL PRODUCTION FROM SUGARS

Since AL is directly synthesised from two pyruvate molecules, it should also be possible to produce AL during sugar-metabolism. This was actually observed in lactose-limited continuous cultures of *L. lactis* strain Ru4 (4). Under these conditions the activity of lactate dehydrogenase is repressed resulting in the conversion of pyruvate into alternative endproducts such as AL. This process was optimalised for growth media based on whey. Under strong aeration in lactose-limited cultures upto 30% of the fermented lactose was converted into AL. Fermentation fluids containing upto 20 mM AL were produced. Higher production of AL by increase of the lactose-content in the medium was not succesfull. The lactococci could not be cultivated under lactose-limitation in the high lactose-containing media.

5. METABOLIC ENGINEERING

To achieve even higher AL production metabolic engineering can be employed. In this strategy fermentation technology, microbial physiology and genetic engineering is combined to direct sugar (lactose) metabolism towards AL-formation in *L. lactis*. The use of aeration and lactose-limited continuous cultures was described in the previous paragraph. For more insight in the pathway of AL production the key enzymes oxaloacetate decarboxylase (9), pyruvate dehydrogenase (1), α-acetolactate synthase (1), α-acetolactate decarboxylase and diacetyl reductase (10) have been purified and characterised from *L. lactis*. Furthermore, the genes for lactate dehydrogenase (11) and AL synthase (12) have been cloned and sequenced making it possible to specifically inactivate or overproduce these enzymes. Finally, several naturally occurring mutants

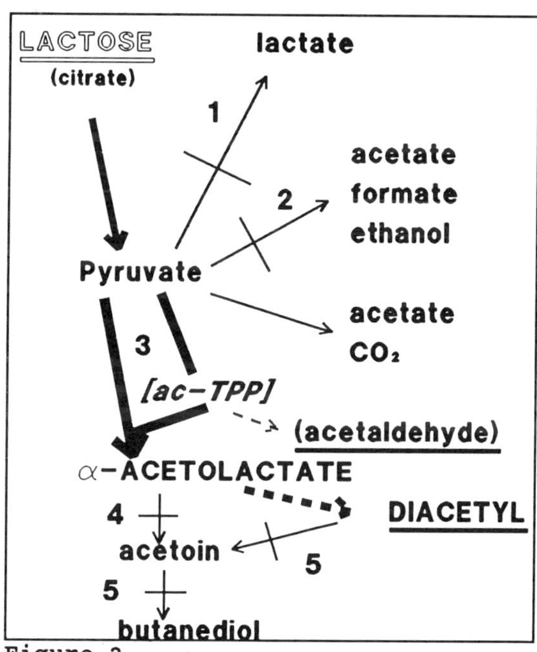

Figure 2

with altered product profiles, such as the AL decarboxylase-negative *L. lactis* Ru4, several acetoin reductase-negative lactic acid bacteria and the lactate dehydrogenase-deficient *L. lactis* KB (13), have served as basis for (genetic) engineering of diacetyl-producing lactic acid bacteria. The different strategies are summarised in Fig. 2. Strategy 1 is inactivation of lactate dehydrogenase which can be achieved both on genetic level and (partially) by fermentation under lactose-limitation. The mixed acid fermentation (formate/acetate/ethanol-formation) can be completely inhibited by aeration (Strategy 2). Production of AL can be enhanced by overproduction of AL synthase (Strategy 3), but also by deletion of the AL decarboxylase (Strategy 4) and, finally, stable diacetyl production will only be achieved in the absence of diacetyl reductase activity (Strategy 5).

Acknowledgements

This work was partly financed by EEC contract no. BIOT-CT91-0263.

REFERENCES

1. J.L. Snoep, M.J. Teixeira de Mattos, M.J.C. Starrenburg and J. Hugenholtz, J. Bacteriol, 174 (1992) 4838-4841
2. J.L. Snoep, PhD Thesis (1992), University of Amsterdam
3. M.J.C. Starrenburg and J. Hugenholtz, Appl. Environ. Microbiol., 57 (1991) 3535-3540
4. J. Hugenholtz and M.J.C. Starrenburg, Appl. Microbiol. Biotechnol., 38 (1992) 17-22
5. W.M. Verhue and F.S.B. Tjan, Appl. Environ. Microbiol., 57 (1991) 3371-3377
6. K.N. Jordan and T.M. Cogan, J. Dairy Res., 55 (1988) 227-238
7. H.A. Veringa, G. van den Berg and J. Stadhouders, Milchwissenschaft, 31 (1976) 658-662
8. S. David, PhD Thesis (1992), Agricultural University Wageningen
9. J. Hugenholtz, L. Perdon and T. Abee, Appl. Environ. Microbiol., submitted for publication
10. V.L. Crow, Appl. Environ. Microbiol., 56 (1990) 1656-1665
11. R.M. Llanos, A.J. Hillier and B.E. Davidson, J. Bacteriol., 174 (1992) 6956-6964
12. J.D. Marugg, D. Goelling, A.M. Ledeboer, U. Stahl, M.Y. Toonen, W.M. Verhue and C.T. Verrips, Appl. Environ. Microbiol., submitted for publication
13. L.L. McKay and K.A. Baldwin, J. Dairy Sci., 57 (1974) 181-186

ECB6: Proceedings of the 6th European Congress on Biotechnology
edited by L. Alberghina, L. Frontali, P. Sensi
1994 Elsevier Science B.V.

Optimization of the production of enterocin 1146, a bacteriocin produced by *Enterococcus faecium*.

E. Parente and A. Ricciardi.

Dipartimento di Biologia, Difesa e Biotecnologie Agro-Forestali, Università della Basilicata, Potenza, Italy.

1. INTRODUCTION

Bacteriocins produced by lactic acid bacteria have been the subject of a large amount of research work because of their potential to function as novel food preservatives (1). Enterocin 1146 (E1146) is a bacteriocin produced by *Enterococcus faecium* DPC1146 which was found to be active against *Listeria monocytogenes* but not against several lactic starters, thus being potentially useful in the control of this pathogen in fermented foods (2). Except for the lantibiotic nisin, studies on the effect of media ingredients and fermentation conditions on the production of bacteriocins are scarce. In a previous study, the effect of some media ingredients (tryptone, yeast extract, tween 80) and initial pH of the medium on production of E1146 and lactocin D, two bacteriocins produced by lactic acid bacteria, was evaluated (3), and media for optimal bacteriocin production were designed. In this study we evaluated the effect of some media ingredients (tryptone, acid casein hydrolysate, yeast extract, Mn^{2+}) on production and specific activity (BU/mg protein) of E1146, in order to design a medium for optimal production and recovery of the bacteriocin. Moreover, the production of E1146 was studied in a STR fermenter at three pH levels.

2. OPTIMIZATION OF CULTURE MEDIUM

A central composite design with 4 factors was set up to evaluate the effect of acid casein hydrolysate (CH, Oxoid, UK, 0-10 g/l), yeast extract (YE, Oxoid, UK, 2-10 g/l), tryptone (T, Oxoid, 0-10 g/l) and $MnSO_4 \cdot 4H_20$ (MN, 0-0.1 g/l) on E1146 production. Twenty-five different media were formulated by varying the concentration of the four factors in a basal medium containig glucose, Na-ß-glycerophosphate, $MgSO_4 \cdot 7H_20$ and Tween 80, according to the experimental design described in Parente and Hill (3). The media were inoculated with 0.01% of an overnight culture of *E. faecium* DPC1146 and incubated for 16 h at 37°C.Growth was measured as OD_{600} on culture broth diluted 1:5, while proteins (by the Biuret method) and bacteriocin activity (as BU/ml, by a photometric method, 4) were assayed on filter sterilized culture supernatants. Factor levels were coded as follows: x_1=(CH-5)/2.5; x_2=(YE-6)/2; x_3=(T-5)/2.5; x_4=(MN-

0.05)/0.025. Their effect on growth (OD_{600} of culture diluted 1:5), E1146 activity (BU/ml) and specific activity (BU/mg prot.) was modeled using a second degree polynomial. The following reduced models, which reconstructed quite well the experimental results, were obtained:

$OD=0.672-0.009x_1+0.019x_2-0.010x_4-0.014x_2^2$ $R^2=0.81$
$BU/ml=1322.5+55.05x_1+71.97x_2+145.86x_3$ $R^2=0.72$
$BU/mg\ prot.=324.6-40.46x_2-20.79x_3$ $R^2=0.77$

A contour plot of predicted activity and specific activity as a funtion of tryptone and yeast extract concentration is shown in figure 1. Although growth was not significantly influenced by T, and was best at low levels of CH and MN and intermediate (6 to 8 g/l) levels of YE, all the peptide and growth factors sources had a significant effect on E1146 activity (higher activity levels were obtained at higher levels of CH, YE and T), which in turn was not influenced by MN. CH and MN had no significant effect (at the 5% level) on specific activity which was negatively influenced by increasing concentrations of YE and T.

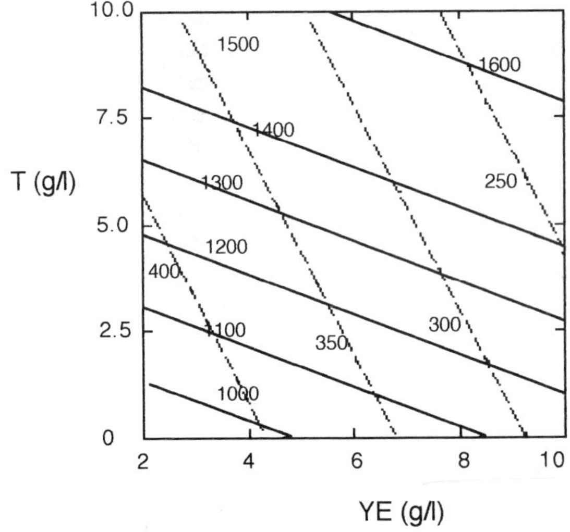

Figure 1. Contour plot of predicted bacteriocin activity (BU/ml, continuous line) and specific activity (BU/mg protein, broken line) as a function of yeast extract and tryprone concentration at 5 g/l casein hydrolysate and 0.05 g/l $MnSO_4$.

A reduced amount of contaminating peptides would be desirable for easier purification of bacteriocins: since T and YE have a higher peptide content than CH, a 3^2 factorial experiment was set up to evaluate the effect of CH (0-10 g/l) and T (5-15 g/l) at low levels of YE (2.5 g/l) and intermediate levels of MN (0.05

g/l), in the same basal medium. Factor levels were coded as follows: x_1=(CH-5)/5; x_2=(T-10)/5; and their effect on growth (OD_{600}), E1146 activity (BU/ml) and specific activity (BU/mg prot.) was modeled using a second degree polynomial.

The following reduced models, which provided an excellent reconstruction of the experimental results, were obtained:

$$OD=0.731+0.062x_2-0.010x_1x_2-0.015x_2^2 \qquad R^2=0.96$$
$$BU/ml=1417.4+169.37x_1+289.34x_2-84.10x_1^2 \qquad R^2=0.96$$
$$BU/mg\ prot.=336.9-39.40x_2-22.75x_1^2 \qquad R^2=0.84$$

The experimental results were in good agreement with the models developed in the previous experiment, thus confirming its predictive ability. A contour plot of predicted bacteriocin activity, specific activity and growth as a function of tryptone and casein hydrolysate is shown in figure 2. Maximum bacteriocin activity and specific activity are unfortunately conflicting objectives since CH (with low peptide and high free aminoacids content) can only partially substitute for T. This may be partly due to its lower N content (7.6% vs. 11.7% in tryptone). In fact an excellent correlation exists between total N content of the medium and bacteriocin activity (r=0.97). Attempts to concentrate E1146 in media with low protein content by ultrafiltration through Millipore Ultrafree PL with 10 kDa NMWL were inconclusive: bacteriocin activity was relatively unstable even at 4°C and part of the activity was lost in the permeate, and no concentration could be attained in any of the media studied.

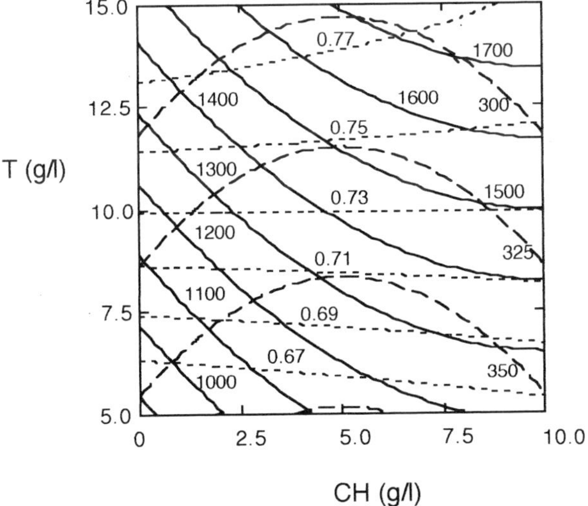

Figure 2. Contour plot of predicted bacteriocin activity (BU/ml, continuous line), specific activity (BU/mg protein, broken line, long dashes) and growth (OD_{600}, broken line short, dashes) as a function of casein hydrolysate and tryptone concentration.

3. BATCH PRODUCTION OF ENTEROCIN 1146 IN A STIRRED FERMENTER.

A medium which would provide intermediate activity and specific activity (respectively 1294 BU/ml and 331 BU mg/prot.) was selected for studies on E1146 production in a 3 l STR fermenter (Applikon, Schiedam, NL). The culture medium (glucose, 10 g/l; CH, 5 g/l; YE, 2.5 g/l; T, 7.5 g/l, KH_2PO_4, 0.5 g/l; $MgSO_4 \cdot 7H_20$, 0,25 g/l; MN, 0.05 g/l; Tween 80, 2 g/l) was inoculated (5%) with an overnight culture of *E. faecium* DPC1146 and 3 fermentations were carried out at 37°C, 300 rpm stirrer speed, controlling the pH at 6.5, 6.0 or 5.5 by addition of 6 M NaOH. Biomass was measured as cell dry weight and OD_{600}, glucose by the DNSA method, total acid produced by the amount of NaOH used to control the pH and E1146 activity by a photometric method (4). E1146 was produced throughout the logarithmic phase of growth; its production ceased during late log-phase and a decrease in activity was observed afterwards; this is probably due to the unstability of E1146 at 37°C and pH close to 7 (2); a weak proteolytic activity, mainly associated with cells (as shown by the azocasein method) may be a contributing factor. Maximum bacteriocin activity was obtained at 5 h at pH 6.5 (1516 BU/ml, 374 BU/mg prot.), at 7 h at pH 6.0 (1494 BU/ml, 368 BU/mg prot.) and at 7 h at pH 5.5 (1960 BU/ml, 483 BU/mg prot.), when OD_{600} was between 2.25 and 2.3 and glucose was between 0.7 and 1.7 g/l; prolonged incubation always resulted in decreased activity, although rate of decrease was slower at lower pH. Activity and specific activity predicted using the model developed in the previous experiment were respectively 1294 BU/ml and 331 BU mg/prot., thus showing that pH control can improve bacteriocin production.

It can be concluded that since higher activity is obtained at higher tryptone and yeast extract levels, and casein hydrolysate can only partially substitute for tryptone, increased bacteriocin production can only be obtained at the expense of specific activity. Production of E1146 can be further improved by fermentation at controlled pH. Due to the unstability of E1146, best results are obtained at lower pH (5.5) and with prompt separation and refrigeration of the culture broth.

ACKNOWLEDGEMENTS

E. faecium DPC1146 was kindly provided by Dr. T. Cogan, DPC, Moorepark, IRL. This work was funded by Ministero dell'Università e della Ricerca Scientifica e Tecnologica, Rome, Italy.

REFERENCES

1. T.R. Klaenhammer, Biochemie, 70 (1988) 337.
2. E. Parente and C. Hill, J. Food Prot., 55 (1992) 497.
3. E. Parente and C. Hill, J. Appl. Bacteriol., 73 (1992) 290.
4. E. Parente, A. Ricciardi and F. Villani. In: A. Zamorani, P. L. Manachini, V. Bottazzi and S. Coppola (eds.) "Biotechnology and molecula biology of lactic acid bacteria for the improvement of foods and feeds quality". Istituto Poligrafico e Zecca dello Stato, Rome (1993) 318.

RECENT DEVELOPMENTS
IN
NITRIFICATION/
DENITRIFICATION

ECB6: Proceedings of the 6th European Congress on Biotechnology
edited by L. Alberghina, L. Frontali, P. Sensi

Nitrogen removal from wastewater.

J.G.Kuenen & L.A.Robertson.

Department of Microbiology and Enzymology, Kluyver Laboratory for Biotechnology, Delft University of Technology, Julianalaan 67, 2628BC. Delft.

Nitrogen removal from wastewater is traditionally carried out in two phases. One phase involves an aerobic reactor employing autotrophic nitrifiers and fed with wastewater containing little or no organics, but rich in NH_4^+. The other phase is anoxic, and employs denitrifying bacteria that require the presence of a suitable electron donor. Depending on the reactor configuration, this electron donor can be organic compounds or sulphide, for example, from the wastewater, or must be added separately (e.g. methanol). The nitrifying and denitrifying phases are frequently run in separate reactors, but may also be carried out in different areas of the same tank. In spite of substantial improvements in the practise of nitrification and denitrification, there is a clear need for further research, not only to provide novel ways of removing nitrogen from wastewater, but also to improve and allow better control over existing systems. As nitrification and denitrification have been studied in greater depth over the last few years, new aspects of these phenomena have been recognized. These include heterotrophic nitrification (figure 1), aerobic denitrification, combinations of these two, anoxic ammonia oxidation (see below), and nitrogenous gas production by obligately and facultatively autotrophic nitrifiers (Robertson & Kuenen, 1992). In view of the current concern about the levels of inorganic nitrogen compounds in natural and drinking water, as well as in the atmosphere, it is essential that all aspects of the nitrification-denitrification part of the nitrogen cycle are understood.

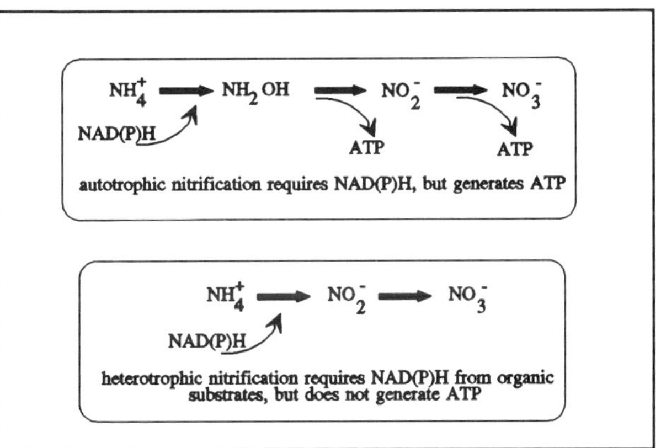

Figure 1. Autotrophic and heterotrophic nitrification.

During the 1980's it became clear that homogenous cultures of a number of bacteria were able to denitrify under fully aerobic conditions, although this property, at least in *Thiosphaera pantotropha* is easily lost under laboratory culture conditions. Extensive chemostat studies with several different species were carried out in Delft, and showed that their behaviour differed Robertson et al., 1989; van Niel et al., 1992). For example, one strain (*Thiosphaera pantotropha*) possessed a fully constitutive denitrifying system. Another (*Pseudomonas* sp.) had an inducible nitrate reductase, but the rest of the denitrifying pathway was constitutive. An *Alcaligenes* species was a constitutive denitrifier, but lacked nitrate reductases completely. Indeed, it may be the constitutive nature of the

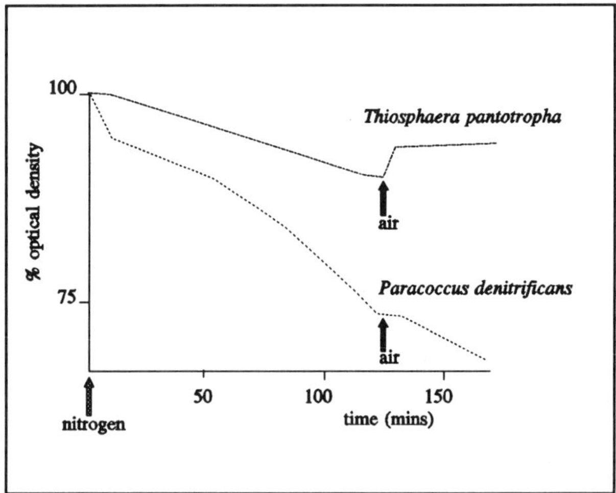

Figure 2. The response of *Tsa. pantotropha* and *Pa. denitrificans* to a sudden onset of anaerobiosis after prolonged aerobic (80% air saturation) growth (Robertson & Kuenen, 1992).

denitrifying enzymes which provide this group of bacteria with a selective advantage in situations where adaptability is needed. This is illustrated in figure 2, where the results of suddenly switching aerobic (80% air saturation) continuous cultures to anoxia are shown (Robertson & Kuenen, 1992). *Paracoccus denitrificans* is known to only denitrify in the absence of oxygen, and therefore had to induce its denitrifying enzymes. Not only did the *Pa. denitrificans* culture wash out, but the rapid accumulation of nitrite in these cultures was sufficient to inhibit oxygen uptake when the air supply was restored. The immediate onset of nitrite production might be explained by the recent observation that *Pa. denitrificans* has the periplasmic nitrate reductase associated with aerobic denitrification in *Tsa. pantotropha* (Richardson, 1993). In contrast, the *Tsa. pantotropha* culture smoothly made the transfer from aerobic to anoxic growth with little or no accumulation of intermediates, so that the return to aerobic growth was also achieved without difficulty.

One of the most universal factors observed thus far among aerobic denitrifiers is their ability to nitrify heterotrophically. Many of the strains thus far tested, reduce some or all of the oxidized nitrogen compounds produced by heterotrophic nitrification to gaseous compounds. For example, figure 3 shows $^{15,15}N_2$ production and simultaneous O_2 utilization by a vigorously aerated batch culture of *Tsa. pantotropha* supplied with $^{15}NH_4^+$ and growing on acetate (L.A. Robertson & T. Dalsgaard, unpublished results). The combined nitrification-denitrification rate was in the order of 30 nmol min $^{-1}$ mg protein^{-1} in spite of the fact that this strain could no longer denitrify aerobically to any significant extent. Some species (e.g.

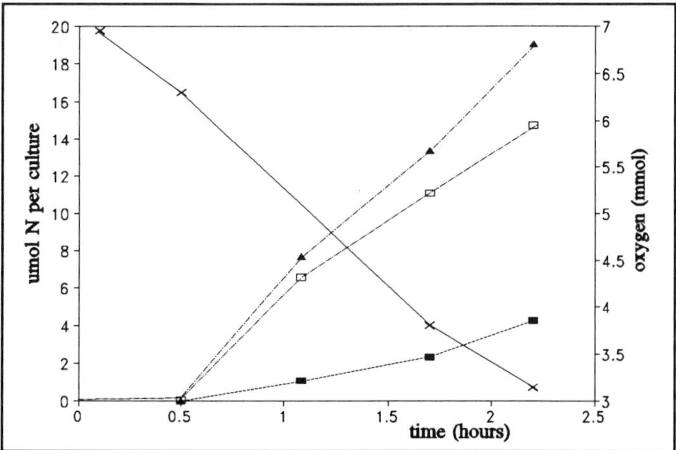

Figure 3. $^{15}N_2$ production and O_2 uptake by *Tsa. pantotropha*. ■= $^{15}N_2$; □= $^{14}N_2$; ▲=total N_2; X= O_2 (Robertson & Dalsgaard).

Tsa. pantotropha and a *Pseudomonas* species) give, at most, a transient appearance of intermediates. Others can produce sufficient nitrite to support the growth of nitrite oxidizing autotrophs. Figure 4 shows the growth of a heterotrophic nitrifier, *Alcaligenes* sp., and a nitrite-oxidizing autotroph, *Nitrobacter agilis* in batch cultures supplied with pyruvic oxime (Castignetti & Gunner, 1980). In axenic cultures, *Alcaligenes* sp. accumulated nitrite, and was eventually inhibited by it. When *Nitrobacter agilis* was added to the culture, the nitrite concentration rapidly fell to zero, and nitrate was produced.

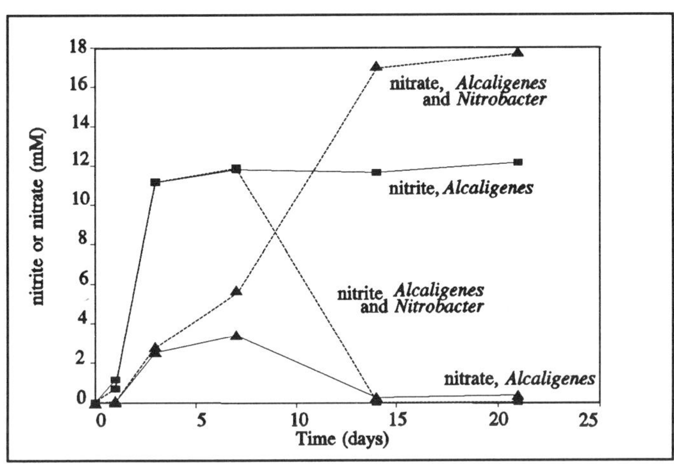

Figure 4. Products formed by *Alcaligenes* sp. alone or with *Nitrobacter agilis* on pyruvic oxime; ▲=NO_3^-; ■=NO_2^- (Castignetti & Gunner, 1980).

The N_2-producing heterotrophic nitrifiers are of obvious potential interest for single-stage nitrogen removal, even though they would have to be grown at higher sludge densities to correct for the lower conversion rates. In order to determine the outcome of various selective pressures on competition between heterotrophic and autotrophic nitrifiers, van Niel and co-workers (1993)used mixtures of *Tsa. pantotropha* and *Nitrosomonas europaea* in continuous cultures limited by acetate and ammonium. They found that heterotrophic nitrification is favoured over autotrophic nitrification at high C:N ratios (figure 5) and low dissolved oxygen concentrations (figure 6), circumstances where autotrophic nitrification proceeds poorly, if at all. A potential disadvantage was that a substantial amount of the NH_4^+ conversion at the higher C:N ratios was due to N-assimilation rather than N_2 production (figure 5).

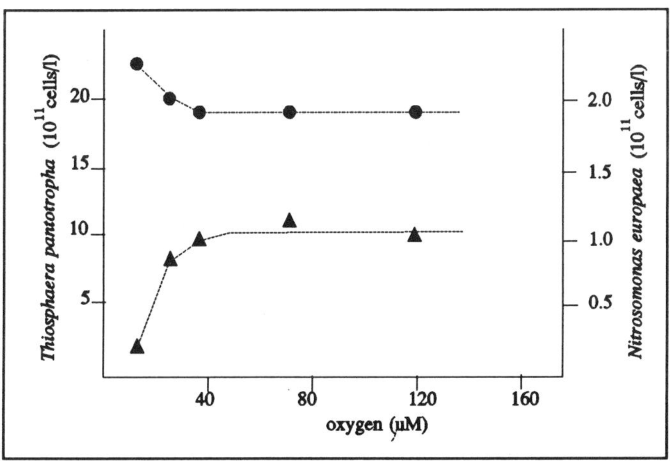

Figure 6. Behaviour of *T. pantotropha* and *N. europaea* at different O_2 levels. ▲=*N.europaea*, •=*T. pantotropha* (van Niel et al., 1993)

N_2O contributes to most atmospheric pollution problems

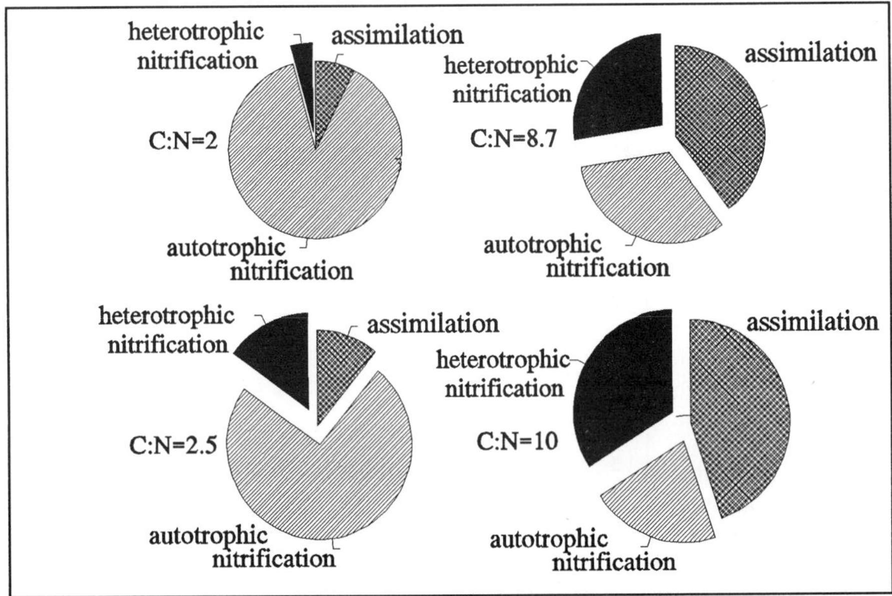

Figure 5. Fate of NH_4^+ in co-cultures of *Tsa. pantotropha* and *N. europaea* in the chemostat at different C:N ratios. (redrawn from van Niel et al, 1993).

Table 1. Atmospheric trace gases, their assumed concentrations in 1975, and the probable greenhouse effect of their concentration doubling. Major climatic changes are predicted to result from an increase in average temperature of 1°K. (from Wang et al., 1976).

Gas	Assumed concentration (ppmv)	Predicted greenhouse effect °K	Trace gas lifetime (approx)
N_2O	0.28	0.68	130 years
NH_3	$6x10^{-3}$	0.12	10 days
HNO_3		0.08	10 days
TOTAL °K		<u>0.88</u>	
SO_2	$2x10^{-3}$	0.03	10 days
CH_4	1.6	0.28	3-7 years
CO_2	330	1.26	
O_3		-1.25	

(e.g. acid rain, ozone depletion, unwanted fertilization) and is a potential product in both denitrification and nitrification. Its impact on the greenhouse effect is smaller than that of CO_2, but N_2O has an atmospheric lifetime of 130 years (table 1). N_2O and the shorter-lived NO, tend to be produced by nitrifying and denitrifying bacteria under sub-optimal conditions (e.g. pH, pO_2, loading). For example, a strain of *Alcaligenes faecalis* being studied in this laboratory has been shown to make significant quantities of $^{14,15}N_2O$ from $^{14}NH_4^+$ and $^{15}NO_2^-$, even under fully aerobic conditions. Recent estimates in the USA and the Netherlands have indicated that a substantial amount (18-25%) of NO_x in the atmosphere may come from poorly-managed wastewater treatment systems. As more nitrogen-removing wastewater treatment systems come on line, this could become a major problem if it is not controlled. Existing systems must therefore be improved, and novel systems extensively tested in order to avoid adding to the problem.

A completely unexpected anoxic ammonia oxidation with nitrate as the electron acceptor (patent EPO 88200204,1) has been shown to occur in mixed cultures being fed with effluent from a methanogenic reactor (van der Graaf et al., 1990). Theoretically, the energetics of this reaction are as favourable as when O_2 is the terminal electron acceptor:

$$NH_4^+ + NO_2^- \rightarrow N_2 + 2\ H_2O \qquad \Delta G' = -361\ kJ/mol\ NH_4^+$$
$$NH_4^+ + 0.75\ O_2 \rightarrow 0.5\ N_2 + 1.5\ H_2O + H^+ \qquad \Delta G' = -315\ kJ/mol\ NH_4^+$$

Table 2. Nitrogen balances for the anammox reactor. Data from van der Graaf et al, 1990.
(I) before anammox, (II) after anammox appeared (all concentrations as mg N per litre). v.f.a. = volatile fatty acids; form. = formed.

	Amount of NO_3^- required for			Total NO_3^- required		Measured NO_3^-
	SO_4^{2-} form.	v.f.a. use	NH_3 anammox.	without anammox	with anammox	consumption
(I)	92	18	0	110	110	100
(II)	80	18	48	98	146	150

Table 2 shows the nitrogen balances for the bioreactor before (row I) and after (row II) the appearance of the anammox phenomenon. It can be seen that before the onset of the anammox reaction, the amount of nitrate theoretically required for the oxidation of sulphide to sulphate and the metabolism of available volatile fatty acids agreed very well with the measured nitrate consumption. However, once ammonia had begun to disappear, it was necessary to include the amount of nitrate stochiometrically required to oxidize the ammonia before the theoretical and observed nitrate consumption values agreed. The biological nature of the reaction has been established by using different amounts of live and killed biomass, among other tests (van der Graaf et al., 1990). The dependence of the reaction on the presence of nitrate or nitrite, as well as the gaseous nature of the products, has been confirmed by van de Graaf et al with mass spectrometry and ^{15}N-labelled nitrogen compounds. Pulses of $^{15}NH_4^+$ fed into the reactor resulted in a large $^{14,15}N_2$ peak, indicating that anammox is a form of biological, ammonium dependent denitrification. The biomass has a very low growth rate, but despite this, conversions comparable to those achieved with purpose-built nitrification systems have been obtained (A.A. van der Graaf, unpublished data).

In conclusion, it appears that the combined aerobic (heterotrophic) nitrification-denitrification and combined anaerobic denitrification-denitrification deserve further attention as potential alternatives to the conventional aerobic nitrification and subsequent anaerobic denitrification. Furthermore, all of these processes, conventional and novel, must be better understood in order that the production of gaseous intermediates such as N_2O can be prevented.

REFERENCES.

Van Niel, E.W.J., Arts, P.A.M., Wesselink, B.J., Robertson, L.A., and Kuenen, J.G., 1993. Competition between heterotrophic and autotrophic nitrifiers for ammonia in chemostat cultures. FEMS Microbiology Ecology, 102: 109-118.

Wang, W.C., Yung, Y.L., Lacis, A.A., Mo, T., and Hansen, J.E., 1976. Greenhouse effects due to man-made perturbations of trace gases. Science, 194: 685-690.

Richardson, D.J., 1993. Roles of the membrane-bound and periplasmic nitrate reductases of *Thiosphaera pantotropha*, *Paracoccus denitrificans* and *Rhodobacter capsulatus* in aerobic and anaerobic growth. In: Abstracts of the 125th meeting of the Society for General Microbiology. Manchester.

Castignetti, D., and Gunner, H.B., 1980. Sequential nitrification by an *Alcaligenes* sp. and *Nitrobacter agilis*. Canadian Journal of Microbiology, 26: 1114-1119.

Robertson, L.A., and Kuenen, J.G., 1992. The effect of electron acceptor variations on the behaviour of *Thiosp[haera pantotropha* and *Paracoccus denitrificans* in pure and mixed cultures. FEMS Microbiology Ecology, 86: 221-228.

van der Graaf, A.A., de Bruijn, P., Robertson, L., Kuenen, J., and Mulder, A., 1990. Anoxic ammonium oxidation. In: C. Christiansen, L. Munck, and J. Villadsen (Editor), Proceeding of the 5th European Congress on Biotechnology. Munksgaard International Publisher, Copenhagen, pp. 388-391.

Robertson, L.A., Cornelisse, R., de Vos, P., Hadioetomo, R., and Kuenen, J.G., 1989. Aerobic denitrification in various aerobic denitrifiers. Antonie van leeuwenhoek, 56: 289-299.

van Niel, E.W.J., Braber, K.J., Robertson, L.A., and Kuenen, J.G., 1992. Heterotrophic nitrification and aerobic denitrification in *Alcaligenes faecalis* strain TUD. Antonie van Leeuwenhoek, 62: 231-237.

Robertson, L.A., and Kuenen, J.G., 1992. Nitrogen removal from water and waste. In: J.C. Fry, G.M. Gadd, R.A. Herbert, C.W. Jones, and I.A. Watson-Craik (Editor), Microbial Control of Pollution. Cambridge University Process, Cambridge, pp. 227-267.

ECB6: Proceedings of the 6th European Congress on Biotechnology
edited by L. Alberghina, L. Frontali, P. Sensi

Formation and Consumption of Nitric Oxide by Nitrifying Bacteria

E. Bock, R. Stüven*, R. Mansch* and M. Vollmer*

Institut für Allgemeine Botanik der Universität Hamburg, Abteilung Mikrobiologie, Ohnhorststrasse 18, W-2000 Hamburg 52, Federal Republic of Germany

1. Introduction

Nitric oxide (NO) is a radical and belongs to the trace gases of the atmosphere. NO is oxidized to NO_2 with molecular oxygen or ozone. Depending on the concentration, its half life time varies from a few seconds to hours and even to days. Nitric oxide has been shown to be a signal molecule in mammalian cells. It is well known that this molecule is produced by the activity of nitrifying bacteria. However, the mechanism of its formation and consumption is not well understood.

2. Formation

2.1. Obligately lithoautotrophic ammonia oxidizers

Hydroxylamine (NH_2OH) is an intermediate when ammonia is oxidized to nitrite by lithoautotrophic ammonia oxidizers. It was detected in low µmolar concentrations in pure cultures of *Nitrosomonas* and *Nitrosovibrio* as well as in mixed cultures together with *Nitrobacter* when formate was present.

As illustrated in Figure 1, hydroxylamine is chemically unstable and decomposes rapidly. In *Nitrosomonas* medium, the half life time of this compound is approximately 1 h. A major product of chemical decomposition is N_2O. Furthermore, nitrite is formed. As a strong reducing agent, hydroxylamine reduces nitrite to nitric oxide. Thus, the formation of N_2O and NO by ammonia oxidizers can be attributed to the decomposition of the metabolic intermediate NH_2OH and the chemodenitrification of nitrite. N_2O and NO production is enhanced in the presence of organic acids such as formate and pyruvate. This can be explained as follows: Normally, electrons needed for the ammonia monooxygenase-reaction are revealed from the subsequent oxidation of hydroxylamine. In the presence of the above mentioned organic acids, electrons from these compounds are additionally fed into the ammonia oxidation step. This leads to an imbalance in the steady state concentration of hydroxylamine. As a consequence, hydroxylamine is released into the medium [1].

---

* Results presented are part of doctoral thesis

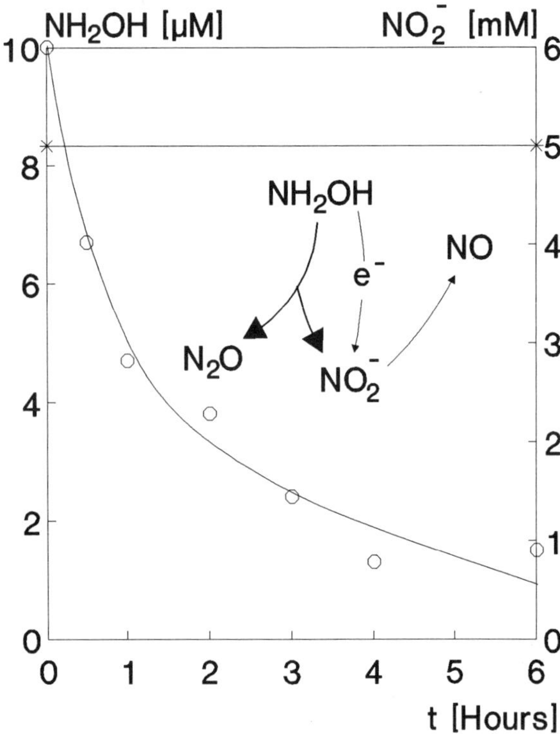

Figure 1. Decomposition of hydroxylamine (o---o) in *Nitrosomonas* medium in the presence of nitrite (∗---∗). The scheme illustrates the direct formation of N_2O and nitrite from hydroxylamine and the indirect formation of NO by reduction of nitrite.

2.2. Facultatively lithoautotrophic nitrite oxidizers

As a facultatively lithotrophic bacterium, *Nitrobacter* is able to denitrify nitrate and nitrite under anaerobic conditions. Gaseous end products are N_2O and NO [2].

3. Consumption

3.1. Inhibition of ammonia oxidation

Ammonia oxidizers of the genera *Nitrosomonas* and *Nitrosovibrio* produce nitric oxide (NO) and nitrous oxide (N_2O) not only when growing in mixotrophic media [1] but also when living endolithically in building stones [3]. In stones, nitrite accumulation was not observed. When test stones as "Ihrlersteiner Grünsandstein" were inoculated with nitrifiers and kept under high humidity, nitrite was accumulated and then oxidized to nitrate (Figure 2 A). In the presence of NO, only small amounts of nitrite and nitrate were formed (Figure 2 B). The ammonia oxidizing bacteria were inhibited by NO as it could be shown with pure cultures of *Nitrosomonas europaea*. On the other hand, the chemical formation of 0-20 µM nitrite and 1-5 µM nitrate from NO oxidation to NO_2 (sterile controle) was without influence.

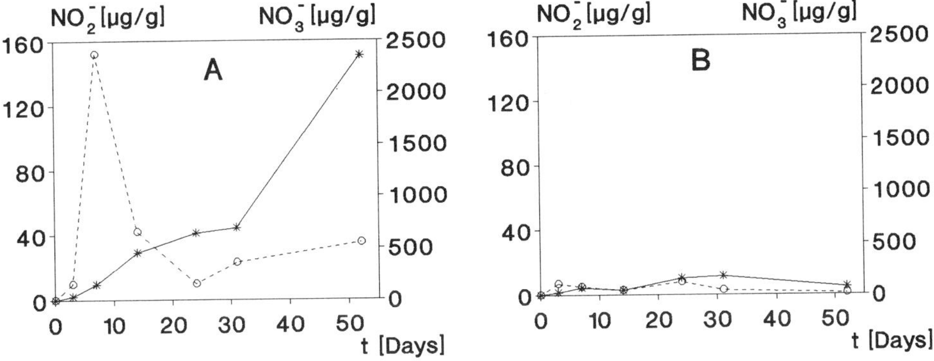

Figure 2. Formation of nitrite (O---O) and nitrate (✳---✳) on test cubes (1.5 • 1.5 cm) of "Ihrlersteiner Grünsandstein" inoculated with 1 • 10^9 cells of *Nitrosovibrio* strain K 7.1 and *Nitrobacter vulgaris* strain K 48, respectively. Incubated at 28 °C with air (**A**) and in an atmosphere of 3-5 ppmv NO (**B**) in a totally controlled gas cabinet by using a flow through system.

3.2. Stimulation of nitrite oxidation

Indications that nitric oxide could be used as an energy source for lithoautotrophic growth of *Nitrobacter* cells could not be verified. In an atmosphere of 5 ppmv NO no growth was measurable. But NO had a positive effect on the nitrite oxidizing activity of *Nitrobacter winogradskyi*. During conversion from lithotrophic to heterotrophic growth conditions the nitrite oxidoreductase activity was repressed as already shown [4]. As demonstrated in Figure 3 the repression was less extensive in the presence of 3-5 ppmv NO. After 4 weeks, the specific activity increased. This increase could be followed up to 6 months (Mansch, personal communication). The trace amounts of nitrite and nitrate which were formed chemically (see above) were not able to induce the nitrite oxidizing system. Therefore, we assume that nitric oxide is a positive effector for the induction of the nitrite oxidoreductase.

4. Discussion

The observation that nitrifiers may communicate via the trace gas nitric oxide is of ecological importance. Ammonia as well as nitrite oxidizers live on surfaces of a variety of materials e. g. soil particles or building stones. Here, they are in contact to gaseous compounds of the atmosphere but often are not in direct contact to each other (patchiness of the biotop, lack of free water). As shown for stones of historical buildings, the nitrite oxidation potentials are significantly higher than those of ammonia oxidation. High numbers of *Nitrobacter* cells can be explained by heterotrophic growth. High nitrite oxidizing activity is probably due to induction by nitric oxide.

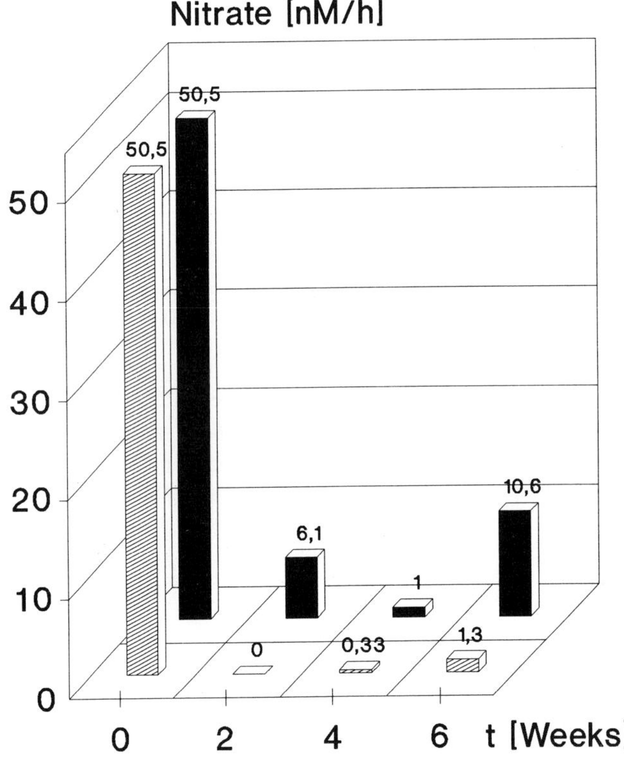

Figure 3. Repression of the nitrite oxidizing system of *Nitrobacter winogradskyi* with ■ and without ▨ NO in the presence of pyruvate (5mM) and ammonia (1mM). 2.7 • 10^7 cells were immobilized on polyurethane expanded plastic (0,1 g), which was saturated with pyruvate-ammonia medium. The cells were grown at 28 °C and at a concentration of 3-5 ppmv NO in a totally controlled gas cabinet by using a flow through system. The specific activity is expressed in nitrate formation.

References

1. R. Stüven, M. Vollmer, E. Bock, Arch. Microbiol., 158 (1992) 439.

2. A. Freitag, M. Rudert, E. Bock, FEMS Microbiol. Lett., 48 (1987) 105

3. M. Baumgärtner, F. Sameluck, E. Bock, R. Conrad, FEMS Microbiol. Ecol., 85 (1991) 95

4. W. Steinmüller and E. Bock, Arch. Microbiol., 115 (1977) 51

ECB6: Proceedings of the 6th European Congress on Biotechnology
edited by L. Alberghina, L. Frontali, P. Sensi

DYNAMIC MODELLING OF THE GROWTH OF IMMOBILIZED Nitrosomonas europaea CELLS

C.D. DE GOOIJER, A.W. SCHEPERS, R.H. WIJFFELS, and J. TRAMPER

Wageningen Agricultural University, Food and Bioprocess Engineering Group, P.O. Box 8129, 6700 EV Wageningen, The Netherlands.

1 INTRODUCTION

The use of entrapped nitrifying bacteria in wastewater treatment is gaining interest, as this can potentially lead to a more stable and well defined system[1]. The main advantage of immobilized cells is the enhancement of the overall volumetric productivity in combination with a high dilution rate.

The integration of mass transport by diffusion and simultaneous substrate consumption is described in numerous papers. Steady-state models for growing cells[2,3] can be quite satisfactory for design purposes, but may fail to describe the startup and transient states[4-6].

There are three rationales for a dynamic model: it will contribute to the understanding of the immobilized-cell process, it can be used to predict the dynamic response of the system to a change in environmental conditions, and it can be used as a tool in bioreactor design.

In previous papers[7,8] the development was described of a model that integrates simultaneous internal and external mass transport, substrate consumption, and biomass growth. The model describes the occurrence of an inhomogeneous biomass distribution with time[7,8]. It is based on non-specific bacterial kinetics and substrate transfer by diffusion, so it can be used in a general way. The model output consists of substrate and biomass concentration profiles across gel beads, and, from that, the overall substrate consumption rate by the immobilized biomass. This model has been validated with N. europaea cells immobilized in spherical particles of κ-carrageenan, with oxygen as the limiting substrate. It was shown that growth resulted in the formation of rather large colonies across which diffusion limitation might be significant.

For this reason, biomass growth has been implemented now in the form of colony growth, and diffusion limitation over the individual colonies in the gel bead is included in the model as well.

2 MODELLING

In many aerobic systems, as with nitrifying bacteria, oxygen is the limiting substrate. In the case of growing immobilized cells, substrate depletion near the centre of the biocatalyst bead will occur, whereas a high substrate concentration exists near the surface. The growth model used here[9-11] allows a negative net growth at low substrate concentrations, and a maximum growth rate at high substrate concentrations. The specific growth rate constant is described by the Monod equation. Furthermore, the well-known differential equation for simultaneous consumption and diffusion of substrate was used[7]. In order to account for external mass-transfer resistance, Fick's law was combined with the film theory[7]. The biomass concentration inside the gel beads was limited to a maximum. The set of equations was solved numerically as described before[7]. Also the observation was implemented that biomass will grow in colonies inside the gel bead[12]. In that case the biomass in the beads consists of a collection of colonies with different radii. The colony radius is determined by the radial position of the colony in the bead, and colonies at a certain radial position have identical diameters. Then, the volume of bacterial colonies and the biomass concentration can be calculated[13]. The model also accounts for diffusion limitation inside the colonies. After some time colonies reach the surface of the beads. At this point two different mechanisms were implemented. The first is based on the assumption that the colonies remain intact, and that further biomass growth is discharged into the medium. The second mechanism assumes that at the moment a colony reaches the bead surface, its entire content is washed out into the medium.

3 MODEL VALIDATION

In the transient growth of immobilized cells, 4 phases can be recognized. In phase 1 homogeneous growth occurs since diffusion is not limiting yet. Phase 2 shows inhomogeneous growth, as in the centre of the sphere substrate depletion aggravates. In phase 3, the substrate hardly reaches the core of the sphere anymore, so biomass decays, where at the outer shell the biomass concentration reaches a maximum. The biomass colonies reach the outer shell of the bead and biomass eruption takes place. As a consequence, the biomass concentration at the surface of the bead drops. In phase 4 this process of growth and decay continues and a distinct biomass film develops. Figure 1 is a model simulation of this transient process; it clearly shows that external diffusion limitation gains importance with time.

The model was validated with N. europaea cells immobilized in κ-carrageenan beads. The immobilized cells were kept in an air-lift loop reactor at a sufficiently high ammonia concentration and constant oxygen concentration, so that oxygen was the limiting substrate. The macroscopic oxygen consumption rates for a high and a low initial biomass concentration were determined in time by measuring the ammonia consumption rate. The biomass

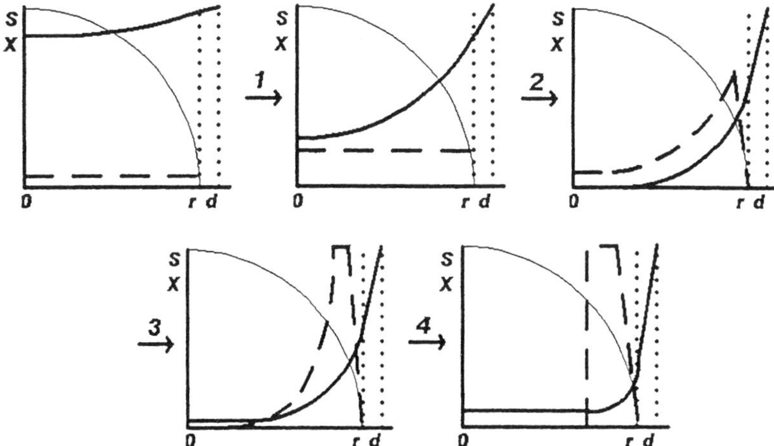

Figure 1. Typical model predictions for growth of immobilized cells. Substrate concentrations (—) and Biomass concentrations (---) versus radial position r at various times. The external stagnant layer is represented by d.

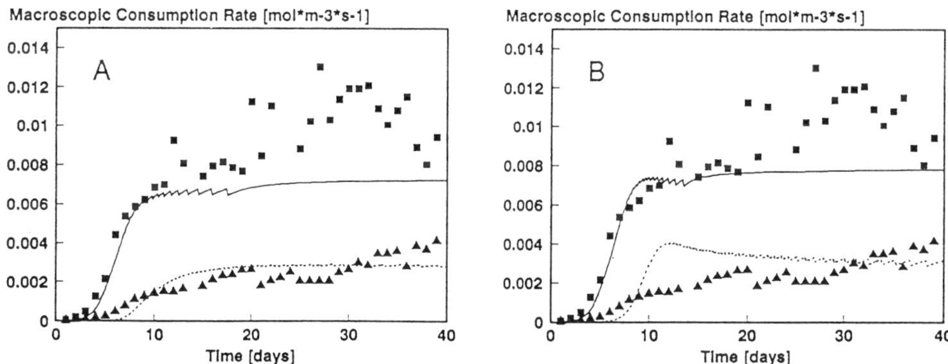

Figure 2. Macroscopic consumption rates with (A) and without (B) diffusion limitation in the colonies versus time. Lines are model predictions, markers are experimental data. Experiments were performed with an initial biomass concentration of 1.95 10^{-3} $kg.m^{-3}$ (squares) and 1.95 10^{-5} $kg.m^{-3}$ (triangles) at an oxygen bulk concentration of $1.2*10^{-1}$ $mol.m^{-3}$.

profiles were determined according to Wijffels et al[8]. The profiles were compared to model predictions. Results are shown in Figures 2 and 3. The curves predicted by the model including diffusion limitation in the colonies match the experimental results well.

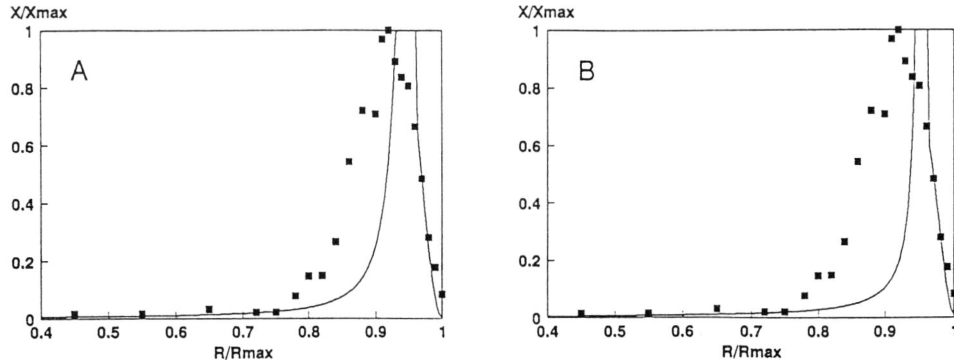

Figure 3. The dimensionless biomass concentration versus radial position. Model predictions of biomass profiles with (A) and without (B) diffusion limitation in the colonies after 35 days of cultivation (—) and experimental verification (squares).

4 CONCLUSIONS

The modeling of immobilized N. europaea cells was successfully carried out. Measured macroscopic oxygen consumption rates matched well with model predictions. Biomass profiles in the gel beads could be described satisfactorily.

REFERENCES

1. R.H. Wijffels and J. Tramper, Appl. Microb. Biotechnol. 32 (1989) 108-112.
2. H.G. Monbouquette and D.F. Ollis, Ann. NY Acad. Sc. 469 (1986) 230-244.
3. G. Andrews, Chem. Eng. J. 37 (1988) b31-b37.
4. K. Nakasaki, T. Murai and T. Akiyama, Biotechnol. Bioeng. 33 (1989) 1317-1323.
5. G.D. Sayles and D.F. Ollis, Biotechnol. Bioeng. 34 (1989) 160-170.
6. H.G. Monbouquette, G.D. Sayles and D.F. Ollis, Biotechnol. Bioeng. 35 (1990) 609-629.
7. C.D. de Gooijer, R.H. Wijffels and J. Tramper, Biotechnol. Bioeng. 38 (1991) 224-231.
8. R.H. Wijffels, C.D. de Gooijer, S. Kortekaas and J. Tramper, Biotechnol. Bioeng. 38 (1991) 232-240.
9. H.H. Beeftink, R.T.J.M. van der Heijden and J.J. Heijnen, FEMS Microbiol. Ecol. 73 (1990) 203-209.
10. S.J. Pirt, Proc. Roy. Soc. London 163 B (1965) 224-231.
11. D. Herbert, in : Recent progress in microbiology, Ed. G. Tunevall G, Almqvist & Wiksell, Stockholm (1958) 381-396.
12. P.M. Salmon, PhD Thesis, Stanford University, USA (1989).
13. C.D. de Gooijer, R.H. Wijffels and J. Tramper, in : Biofilms-Science and Technology, Eds. Melo L.F. et al., Kluwer, The Netherlands, (1992) 291-296.

ECB6: Proceedings of the 6th European Congress on Biotechnology
edited by L. Alberghina, L. Frontali, P. Sensi

Molecular genetic analysis of nitrite reduction to ammonia: a bacterial pathway which competes with denitrification.

H. Hussain, H. Crooke, A. Darwin, L. Griffiths, J. Grove, K. Tyson, S. Busby, E. Battaloglu and J. Cole.

School of Biochemistry, University of Birmingham, Birmingham B15 2TT, UK.

Many bacteria adapt readily from aerobic to anaerobic growth, but to do so must synthesize components of electron transfer pathways not found in aerobic bacteria. These alternative electron transfer pathways form the basis of waste water treatment and other processes exploited in environmental biotechnology. For example, the denitrification pathway exploited in waste water treatment requires the synthesis of a range of c-type cytochromes involved in the reduction of nitrate via nitrite, nitric oxide and nitrous oxide to N_2. As all bacterial c-type cytochromes are located either in the cytoplasmic membrane or in the periplasmic space, they must either be synthesised as mature enzymes with their associated haem groups within the cytoplasm and then be exported by some unknown mechanism to their functional location, or the individual components must be transported separately into the periplasm and, again by an undefined mechanism, be assembled into functional complexes (1,2). Any defect in either of these processes will inevitably result in loss of denitrification, nitrite reduction to ammonia or other aspects of the Biological Nitrogen Cycle of biotechnological significance.

Two independent pathways for the rapid reduction of nitrite to ammonia have been found in facultatively anaerobic bacteria. Both catalyse the dissimilation of nitrite to ammonia rather than denitrification or the assimilation of nitrate and nitrite into the pool of organic nitrogen compounds required for growth (Fig.1). The more active of these two pathways involves a soluble, NADH-dependent enzyme complex required to detoxify nitrite formed as a product of bacterial nitrate respiration. This Nir pathway has been extensively characterised (3), but appears to be restricted to the *Enterobacteriaceae*. In contrast, the alternative, electrogenic Nrf pathway (for nitrite reduction by formate) is widely distributed amongst bacterial species and involves a trans-membrane electron transfer chain terminating in the bacterial periplasm with a low redox potential cytochrome c_{552} (4). Previous studies have established that nitrite is reduced rapidly to ammonia by bacteria only during anaerobic growth. Nitrite reduction by either of the known pathways requires the

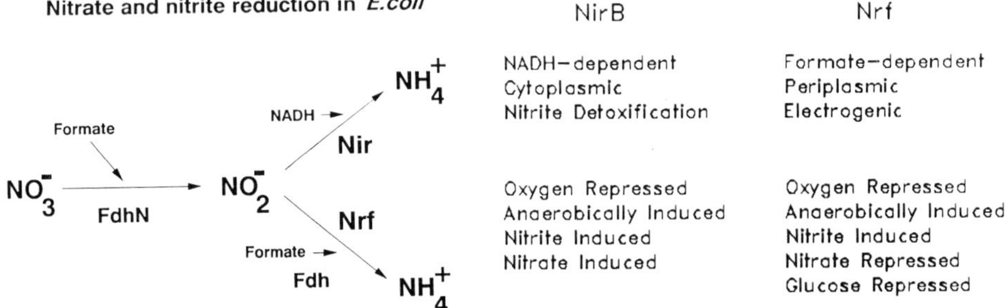

Figure 1. Nitrate reduction to NH_4^+ by fermentative bacteria.
Figure 2. Regulation and roles of the Nir and Nrf pathways.

transcription activator protein, FNR. Synthesis of components of both pathways is repressed by oxygen but induced during anaerobic growth by nitrite. Although nitrate also induces the Nir pathway, it represses cytochrome c_{552} synthesis and Nrf activity (5). Figure 2 summarises the main characteristics of these two pathways and how they are regulated genetically.

To understand how nitrate and nitrite regulate gene expression and hence promote either denitrification or nitrate and nitrite reduction to ammonia, it is essential first to establish the protein-DNA interactions involved and their specificities. In enteric bacteria, nitrate is sensed by a

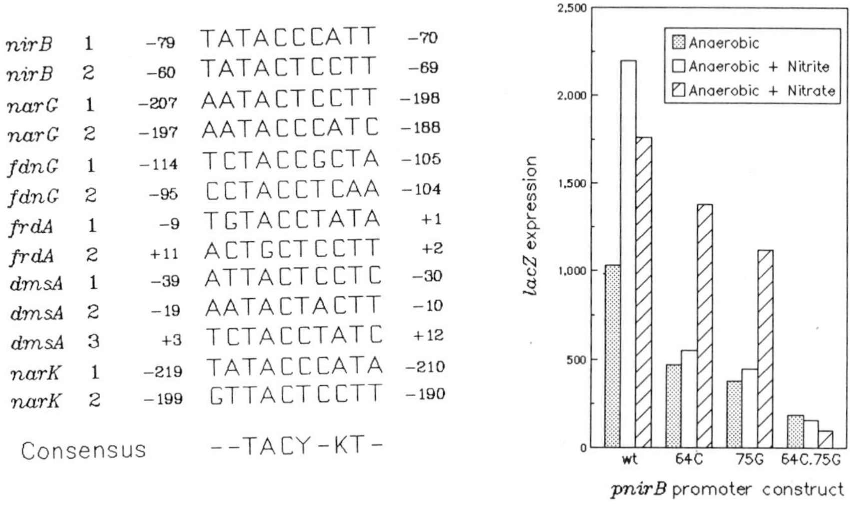

Fig 3. Similar DNA sequences at the *nir*, *nar* and other NarL dependent promoters.
Fig.4. Effect of mutations in the putative NarL binding site on the regulation of transcription from the *nir* promoter

transmembrane protein, NarX, which in the presence of nitrate phosphorylates its cognate transcription regulator, NarL. The promoters of the *nar* (nitrate reductase) and *nir* operons, both of which are activated by NarL, include two copies of similar 10-base DNA sequences which also occur at other nitrate-regulated promoters (Fig. 3). Mutagesis of one or other of these sequences at the *nir* promoter resulted in loss of nitrite induction but either sequence was sufficient for nitrate induction (Fig. 4). However, a strain unable to make NarL still responded to both nitrate and nitrite, so there must be a second sensing system in these bacteria (5; 6).

The structural gene, *nrfA*, for cytochrome c_{552} has been located at coordinate 4366 (minute 92.4) on the physical map of the *E. coli* chromosome (Fig. 5). Although there is an extensive literature which describes this class of nitrite reductase as hexahaem c-type cytochromes, the DNA sequence of *nrfA* encodes a tetrahaem *c*-type cytochrome with a predicted Mr for the unprocessed product of 53 788. Cleavage of the putative signal peptide at Ala 26 would result in a mature, periplasmic cytochrome of Mr 50 580. A cytochrome of this size was detected by staining SDS-PAGE gels for covalently bound haem. This cytochrome was partially purified by anion exchange chromatography and confirmed to be cytochrome c_{552} by difference spectroscopy. Similar cytochromes were detected in five other *E. coli* strains including strain ST 249 which was used previously to purify and characterise the protein. A plasmid with an in-phase deletion within *nrfA* directed the synthesis of a truncated haemoprotein of the predicted mass. Over 7 kb of DNA is required to complement a deletion mutant which lacks all of the *nrf* operon, so genes additional to *nrfA* are essential for the restoration of Nrf activity.

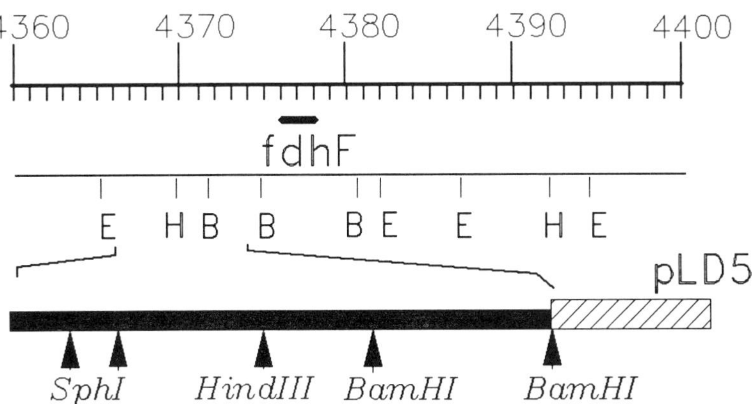

Fig.5. Location of the structural genes for the Nrf pathway in the 92 minute region of the *E. coli* chromosome.

Expression from the FNR-dependent *nrfA* promoter was repressed during aerobic growth, partially induced anaerobically in the absence of nitrite or in the presence of nitrate, but fully induced during anaerobic growth in the presence of nitrite. No nitrate repression was detected in a *narL* mutant, but nitrite induction was unaffected, indicating that the nitrite sensing mechanism is independent of the NarL protein. A putative NarL site which coincides with the consensus deduced above (Fig.3) was found in the *nrf* promoter. A point mutation in this region disrupted regulation by nitrate and nitrite; deletion of this sequence totally abolished nitrite induction.

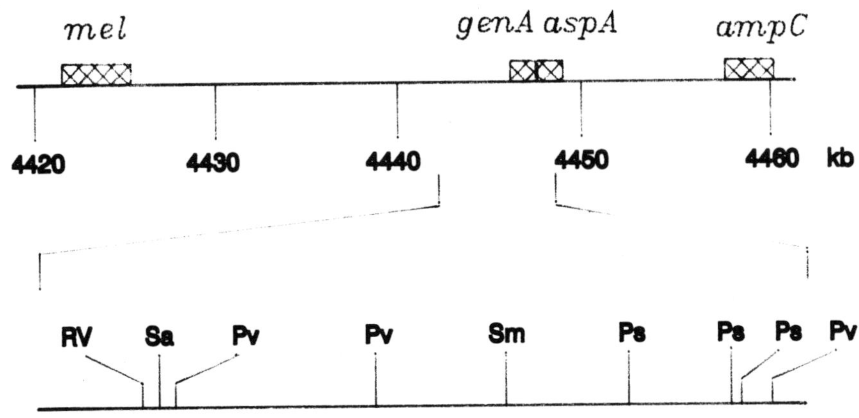

Fig. 6. Location of genes essential for Nrf activity in the 94 minute region of the *E. coli* chromosome.

Further *E. coli* mutants have been isolated which are defective in genes located in the 94 minute region of the chromosome (Fig. 6). These mutants are totally devoid of *c*-type cytochromes, are also unable to catalyse nitrite reduction by formate and are partially defective in the reduction of trimethylamine-N-oxide. All of the known defects of the mutant were corrected by a plasmid containing a 2.5 kb fragment of DNA from the 94 minute region of the *E. coli* chromosome. The sequence of the cloned DNA should reveal whether the cloned genes encode a haem lyase or proteins required to transport the cytochrome into the periplasm.

1. Page M D & Ferguson S J (1989) Mol. Microbiol. 3; 653-661.
2. Ramseier T M *et al.* (1991) J. Biol. Chem. 266; 7793-7803.
3. Harborne N R *et al.* (1992) Mol. Microbiol. 6; 2805-2813.
4. Cole J A (1989) Molecular and Genetic Aspects of Nitrate Assimilation. Oxford Science Publications. Oxford. 229-43.
5. Page L *et al.* (1990) Arch. Microbiol. 154; 349-354.
6. Tyson K L et al. (1993) Mol. Microbiol. 7; 151-157.

Extremophiles in Biotechnology

ECB6: Proceedings of the 6th European Congress on Biotechnology
edited by L. Alberghina, L. Frontali, P. Sensi

EXTREMOPHILES IN BIOTECHNOLOGY

M. Rossi[a] and M. De Rosa[b]

[a] Istituto di Biochimica delle Proteine ed Enzimologia, CNR, Via Marconi 10, 80125 Napoli, Italy and
Dipartimento di Chimica Organica e Biologica, Università di Napoli, Via Mezzocannone 16, 80134 Napoli, Italy

[b] Istituto di Biochimica delle Macromolecole, Università di Napoli, Via Costantinopoli 16, 80136 Napoli, Italy.

INTRODUCTION

There is a recent and growing interest, both academical and commercial on extremophiles (1, 2, 3, 4). The discovery of many microorganisms living optimally over 100°C, at extreme pHs, in saturated saline solutions, has prompted studies on utilization of extremophiles and their enzymes for industrial applications. Since the potential exploitation of these enzymes for practical purposes is one of the main goal in many biotechnological processes, the use of enzymes and the availability of extremophiles as a source of enzymes stable and active at high temperature and in adverse chemical environment are important developments in this area. Thermophiles, either Bacteria or Archaea, seem to be the most interesting and promising organisms for biotechnological exploitation. Table 1 shows an outline of the strategic opportunities that can be foreseen by using thermophilic microrganisms and their enzymes in industrial processes.

Table 1. Strategies in industrial uses of extremophiles.

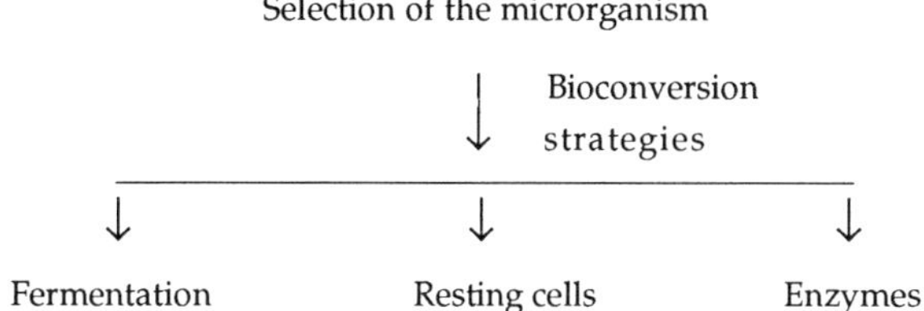

1. Fermentation at high temperature

Having chosen the microrganism, whose enzymatic patrimony is able to carry out a given bioconversion, there are several options for industrial applications. The immediate choice can be a fermentative process in extreme conditions, in which the specific transformation is directly carried out by the growing organism. This approach offers the advantage of a good bioreactor sterility, for the severe operative conditions of fermentation, and, under specific environmental conditions, reduced operative costs in comparison to conventional processes. On the other hand fermentation in extremophilic conditions requires special plants that are not available at the industrial level. Of consequence the high investment required and the strong competition of the conventional approach limit the industrial development of fermentation at high temperature only to processes in which a real advantage is achieved as is the case of ethanol production. In fact at high temperature ethanol can be continuosly removed avoiding toxic concentrations for the microrganism used.

At our knowledge, one of the most advanced project is that developed by Prof. B.S. Hartley in the frame of an European Commission Eclair project (5, 6). The target was to develop and assess the economics of a novel ethanol fermentation process at a temperature higher than 70°C, using mutants of the thermophilic bacterium, *Bacillus stearothermophilus LLD-R*, that rapidly metabolises both aerobically and anaerobically a wide range of sugars derived from hemicelluloses and pectins. These mutants cannot produce the normal anaerobic product, L-lactate: hence metabolism is directed via two alternative pathways to give ethanol yields approaching those of yeast on starch or sucrose. Work is in progress to utilize hemicellulosic feed stocks derived from steam explosion of wood or straw or from dilute acid hydrolysis of corncobs; an alternative is to derive feed stocks from sugars or fodder beat, and to make a comparison of their fermentation with that of wood or straw feed stocks. During the fermentation, for the high temperature of the process, ethanol is continuosly removed by vapor phase distillation.

The costs for ethanol production at high temperature can be further reduced by using the tons of biomass available as affluent from the process for the production of thermostable enzymes of industrial interest. Recent studies indicate that the microrganism is a good source a) of a stable glucose isomerase that has used for the production of High Fructose Syrup (HFS) at 100°C and pH 5.5-6.0 where caramelization reactions are minimal; b) of a thermostable alcohol dehydrogenase with interesting potentialities for the stereospecific and regiospecific oxidoreductions in the synthesis of fine chemicals.

2. Production of thermophilic biomasses.

Though it not very difficult to grow extremophiles in the gram scale using small glass fermentors, it is not an easy task to scale up at industrial level the production of biomasses. The high temperature and corrosive nature of the media used for growing many thermophiles represent engineering problems that cannot be solved at a reasonable cost. In Naples it has been designed and built with good results a new fully automatic 150 liter fermentor for high temperature and extreme pHs carried out in special stainless steel. But even if the fermentation could be scaled up to hundred liters, this would hardly be sufficient even for scientific use since the yield of enzymes of extremophiles is far lower than that of mesophiles. However, an increased production can be obtained by better knowledge of growth requirements, by the control of genic expression and by mutagenic tecniques.

3. Use of resting cells of thermophilic microrganisms.

The use of the resting cells of extremophilic microorganisms could represent an interesting alternative to the fermentative process. In this case the whole cell of the microorganism, no longer vital, works as a bioreactor, in which only certain enzymatic activities are able to perform biocatalysis. Resting cells permit operative conditions behind the strict limits allowed by the physiology of bacteria. In this case the bioconversion may be compatible with very high substrate concentrations and with the presence of organic solvents or tensioactive agents used to improve solubility. Another point in favour of this strategy is the low biocatalyst cost, since are not necessary the expensive steps of enzyme purification and coenzyme addition. Limits to the use of resting cells are, however, the reduced flux trough the cell wall and the risk of unwanted side-catalytic actions carried out by other enzymes inside the microorganism.

Resting cells of *Sulfolobus solfataricus* have been used for stereochemical reduction of prochiral ketones to obtain chiral ketones with optical yield near 100% (7). In this process the endogenous pool of NADH, continuosly regenerated with the coupled substrate system by using isopropanol, is the coenzyme source. The system has excellent potentialities since, by changing the organic phase, the results do not change and cells can be reutilyzed several fold. The catalitic action is reduced to 50% after seven cycles of 16 hours.

S.solfataricus resting cells have also been used for chemical modifications of several steroids such as progesterone (8). Different modification patterns were obtained by changing the reaction conditions.

4. Production and utilization of enzymes

Enzymes from extremophiles can be very similar to mesophiles except for their general stability to heat, to organic solvents, to detergents and to the commons protein denaturing agents. There is an explanation, at least in principle, of the molecular basis of these properties. In fact, it must be pointed out that the folded state of a protein is stabilized against unfolded state (conformational entropy) by weak interactions, such as hydrophobic, electrostatic and hydrogen bonding. For a typical enzyme ΔG of unfolding is of the order of 20-50 Kj/mole and derives from energy differences of the order of ~ 1000 Kj/mole. Sinçe each weak interaction can contribute ~10 Kj/mole, two or three of such bonds would be enough to change the half life of an enzyme from 1 hour at 50°C to 1 hour at 85°C.

Attempts to stabilize enzymes by generating such extra interactions have not been particularly successful, underlining the need for a better understanding of the stabilizing network of forces present, and the pathways of denaturation.

However the general stability of these enzymes should confer a number of advantages in terms of industrial applications. In fact, in addition to longer lifetime, the possibility to operate with enzyme reactors at high temperature will increase substrate solubility and diffusion rates and reduce viscosity of the medium and bacterial contamination. In the case of reactors and membrane reactors it would be easier to clean the enzyme from fowling by using, for example, urea at a certain concentration (2-4 M) and acetone.

Proteinases, cellulases, amylases, glycosidases, esterases, DNA polymerases, alcohol dehydrogenases, are the thermophilic enzymes more suitable for potential biotechnological exploitation. As shown in Table 2 proteinases from thermophiles are, with no doubts, more stable; in addition, since mesophilic proteins are the substrates, the high reaction temperature and the possible addition of detergents result in the denaturation of these proteins, which, as demonstrated, are more suscettible to proteolysis under these conditions. Consequently thermophilic proteases result to have higher specific activity than analogous mesophilic enzymes.

Cellulases, hemicelluloses and amylases are another group with several potential industrial applications since cellulose, hemicellulose and starch represent a large renewable resource from which many chemicals, including fuel and food, can be derived.

Table 2. Thermal stability of some thermophilic enzymes (adapted from Ref. 1).

Enzyme	Organism	Half Life
Proteinase	*Thermus sp.Rt41A*	13,5 h 80°C
Proteinase	*Thermococcus celer*	45' 95°C
Proteinase	*Desulfurococcus sp*	1.5 h 95°C
Proteinase	*Pyrococcus furiosus*	33 h 98°C
Endocellulase	*Clostridium thermocellum*	5 h 70°C
Endocellulase	*Acidothermus cellulolyticus*	2 h 85°C
Endocellulase	*Caldocellum saccharolyticum*	1.5 h 85°C
Exocellulase	*Thermotoga sp.*	11 h 95°C
Glycosidase	*Sulfolobus solfataricus*	2 h 95°C
Amylase	*Sulfolobus solfataricus*	> 1h 95°C
Amylase	*Desulfurococcus sp.*	> 1h 95°C
DNA polymerase	*Thermus aquaticus*	40' 95°C
DNA polymerase	*Thermococcus litoralis*	1 h 100°C
DNA polymerase	*Sulfolobus solfataricus*	30' 85°C
Esterase	*Sulfolobus acidocaldarius*	1 h 100°C
Esterase	*Bacillus acidocaldarius*	30' 85°C
Alcohol dehydrogenase	*Thermoanarobium broki*	2 h 85°C
Alcohol deydrogenase	*Sulfolobus solfataricus*	2 h 85°C

Several processes may employ thermostable cellulases, hemicellulases and amylases in the future. These include the production of liquid fuels from renewable biomass, food and fruit juice processing, production of HFS, paper manufacture and laundry detergency.

However, though enzymes from thermophiles seem well-suited for industrial applications, until now only a few of them have found application. Exceptions are the use of thermolysin in the industrial synthesis of the sweetener aspartame, and of DNA polymerase from *Termus aquaticus* in the polymerase chain reaction. This last enzyme, reasonably stable at 94°C, has permitted the automated amplification of DNA using specific oligonucleotide primers and a template DNA. The polymerase chain reaction allows the detection of one or few gene copies of viruses, and is useful to study inherited genetic disorders, and for the identification of populations and organisms in anthropological studies. Furthermore Taq polymerase has made more easy DNA sequencing procedures and this will be important in the sequencing of human genomes.

The reluctance to change existing processes and the difficulty to economically produce thermophilic enzymes on industrial scale are the keys reasons of the few applications. Same of these enzymes are still scientific curiosity and often it is a hard task to obtain the amount needed to try a crystallization.

However it should be distinguished between the production of enzymes from moderate thermophiles and that from extreme thermophiles. Since moderate thermophiles are easier to grow and give quite high biomass yield, it may be conceivable and economically reasonable to produce enzymes from these organisms. On the contrary the severe growth conditions and low biomass yield of extreme thermophiles do not allow production of enzymes other than for laboratory studies. In this case the only methodology for mass production of these enzymes lies in the cloning and overexpression of their genes in mesophilic hosts, which should be generally recognized as safe (GRAS) and characterized by high biomass yield and optimized fermentation conditions.

In our laboratory, in the attempt to set up a general methodology for the production of archaebacterial enzymes, several enzymatic activities from *Sulfolobus solfataricus*, such as a β-glycosidase (9-13), a DNA polymerase (14), an alcohol dehydrogenase (15) have been cloned and expressed either in *E.coli*, yeast and human cells. This methodology offers several advantages, also regarding the down stream processing of enzyme purification. The results can be summarized as follows:

a) Archaea utilize the universal genetic code and their genes are translated by Eukaria and Bacteria;
b) the expression product mantains its properties such as thermophilicity and thermostability;
c) thermophilic proteins are, in general, inactive at the host optimal growth temperature and, consequently, an accumulation of a higher quantity of enzyme may occur;
d) thermophilic proteins have a rigid structure at low temperature and are thus

resistant and not degraded by the host proteases;

e) the thermostability of the expressed protein makes possible its purification from mesophilic proteins by using thermal denaturation steps, which can be easily scaled up in an industrial down stream process;

f) the degree of purity needed for industrial applications of thermophilic enzymes expressed in mesophilic hosts is lower since, after a thermal denaturation step, the only remaining enzymatic activity is the thermophilic protein being host enzymes contaminants inactivated at high temperature.

We will discuss here, as example, the results obtained with the β-glycosidase isolated from *Sulfolobus solfataricus* . This enzyme was at first characterized as a β-galactosidase, because purified by using o-nitrophenil-β-D-galactopyranoside as substrate. However the homogenous enzyme was discovered to possess a broad specificity for β-D-gluco, fuco and galactosides; hydrolyzed a large number of β-linked glycoside oligomers, and had a remarkable exoglucosidase activity (13). The gene of this enzyme was cloned and expressed in *E.coli* and yeast, and its expression in mammalian cells applyed in the assays of eukariotic promoters activity (16).

The expression of the β-glycosidase in *E.coli* was achieved by inserting a DNA fragment containing the ORF and its flanking regions in the plasmid pEMBL18; the plasmid was then transferred in a β-gal$^-$*E.coli*, strain JM109. The expression of the gene was independent from the orientation of the insert in the plasmid, suggesting that its transcription was not initiated from the vector promoter but from its own regulatory sequence. The expressed enzyme was purified 20 fold with a recovery of 95% first by heating the homogenate at 75-85°C for 30-60 minutes, and then by precipitation at pH 3.5. The properties of the expressed enzyme were similar to those of the native enzyme (11).

The expression of the β-glycosidase gene in *Saccharomyces cerevisiae* (12) was obtained under the control of the galactose inducible upstream activating sequence of yeast (UAS gal) and a new heat-inducible promoter that we constructed by using the sequence isolated from the nematode *Caenorhabditis elegans* (poly-HSE), that works as a strong heat inducible activating sequence in yeast.

In the first case the enzyme was produced by the engineered yeast when galactose was added to culture medium; in the second case the expression of the enzyme was induced when the temperature was raised at 39°C. The activity progressively raised when cells were allowed to recover from the heat shock for increasing times at room temperature.

Though the cellular levels of the expressed protein are not very high, due to the different codon usage between Archaea and Eukarya, nonetheless yields of expressed enzyme per liter of fermentation medium were raughly 100 fold higher.

The same purification procedures used for *E.coli* were used for the yeast expressed β-glycosidase. Alternatively, purification was achieved by autolysis of yeast at 37°C, followed by an heat step at 85°C and precipitation at pH 3.5, to obtain the enzyme concentrated and near to homogeneity.

ENZYMATIC TRANSGLYCOSYLATION USING *Sulfolobus solfataricus*

Sulfolobus solfataricus {immobilized cells
{crude homogenate
{purified enzyme
75°C {recombinant enzyme

R-OH ================================ ≫ **R-O-β-Sugar**

ACCEPTOR **DONORS:** phenyl-β-D-galattoside
phenyl-β-D-glucoside
phenyl-β-D-fucoside
Cellobiose
Lactose

YIELDS 50-80%

An interesting result obtained by the group of Trincone and Gambacorta is the enzymatic synthesis of different β-D-glycosides using as biocatalyst various preparations of *Sulfolobus solfataricus* such as immobilized cells, crude homogenate, native and recombinant enzyme in the presence of organic solvents (17-19).

In particular the regioselectivity in the glycosylation of 2-hydroxybenzyl alcohol, salicin, 1,2-propanediol, and glycerol and the synthesis of hexyl-β-D-glucoside indicate 1) a strong preference toward primary hydroxyl group, 2) a high yield of reaction and different stereochemistry in comparison with mesophilic enzyme. By using these enzyme preparations an easy synthesis of natural glycosides may be possible.

The last point to discuss is the way in which the accumulation of basic research results on proteins from extremophiles could be applied to confer to desired proteins the unique properties of these enzymes. Since these properties cannot be attributed to a common factor, but are rather the consequence of a variety of cooperative stabilizing effects, much more structural work is needed. Until now the primary structure of only about a dozen proteins from extremophiles have been analysed, and the resolution of tridimensional structures are needed, which might allow insight into the construction principle of these proteins. Only a few extremophilic proteins have been crystallized and only one structure has been reported (10).

In a collaboration with Dr. L. Pearl of University College London we succeded in crystallizing the β-glycosidase (21). The crystals are extremely reproducible in size quality and yield, withstand X-ray exposure remarkably well and are very robust to handling and heavy soaking. In addition crystals soaked in a chromogenic substrate became coloured, indicating that enzyme is active in the crystals. Preliminary data indicate that the water content of the crystals is very low (~27%).

Since no homologous enzyme structure is available, work is in progress to determine the structure of the enzyme with multiple isomorfus replacements with heavy atoms. Once the structure will be solved, we will perform very detailed analyses of the secondary structure, side chain packing, solvent accessibility, thermal motion and other structural features. A comparison will be made with the features of mesophilic enzymes in order to understand how the *Sulfolobus* glycosidase achieves its considerable structural stability.

ACKNOWLEDGEMENTS

This work was partially supported by Progetto Finalizzato Biotecnologie e Biostrumentazione, by Progetto Finalizzato Ingegneria Genetica of the Italian National Research Council and by a grant EC Biotech Program.

REFERENCES

1. K. Peek, L.D.Ruttersmith, R.M. Daniel, H.V. Morgan and Bergquist, BFE, 9 (1992) 466.
2. T. Coolbear, R.M. Daniel and H.W. Morgan. In Advances in Chemical Engineering/Biotechnology, A. Fiechter Ed., 45 (1992) 57 - Springer Verlag.
3. D.A. Cowan. In The Archaebacteria: Biochemistry and Biotechnology. Biochemical Society Symposium n. 58 Edited by J. Danson, D.W. Hough, G.G. Lunt (1992) pp. 149-169 Portland Press London.
4. M. Rossi. In International Conference: Science and Thermophiles (1992) Reyhjavik, Iceland, Abstract p. 11
5. B.S. Hartley and G. Shama, Phil Trans. R. Soc. London A., 321 (1987) 553.
6. M. Rangarajan, B. Asboth and B.S. Hartley, Bioch. J., 285 (1992) 889.
7. A. Trincone, L. Lama, U. Lanzetti, B. Nicolaus, M. De Rosa, M. Rossi and A. Gambacorta, Biotechnology and Bioengineering, 35 (1990) 559.
8. M. De Rosa, A. Gambacorta, G. Sodano and A. Trabucco, Experientia, 37 (1981) 541.
9. M.F. Pisani, R. Rella, C. Raia, C. Rozzo, R. Nucci, A. Gambacorta, M. De Rosa and M. Rossi, Eur. J. Biochem., 187 (1990) 321.
10. M.V. Cubellis, C. Rozzo, P. Montecucchi and M. Rossi, Gene, 94 (1990) 89.
11. M. Moracci, M. De Rosa and M. Rossi. In Separation for Biotechnology, 2 (1990) 557. D.L. Pyle Ed. Elsevier Applied Science.
12. M. Moracci, A. La Volpe, J.F. Pulitzer, M. Rossi, M. Ciaramella, J. Bacteriol., 174 (1992) 873.
13. R. Nucci, M. Moracci, C. Vaccaro, N. Vespa and M. Rossi, Biotechnol App. Bioch., 17 (1993) (in press).
14. M.F. Pisani, C. De Martino and M. Rossi, Nucleic Acid Res., 20 (1992) 2711.
15. S. Ammendola, C.A. Raia, C. Caruso, L. Camardella, S. D' Auria, M. De Rosa and M. Rossi, Biochemistry, 31 (1992) 12514.
16. R. Cannio, D. De Pascale, M. Rossi and S. Bartolucci (paper submitted).
17. A. Trincone, B. Nicolaus, C. Lama, P. Morzillo, M. De Rosa and A. Gambacorta, Biotechnol. Lett., 13 (1991) 235.
18. A. Trincone, B. Nicolaus, L. Lama and A. Gambacorta, S. Chem. Soc. Perkin-Trans, 1 (1991) 2841.
19. A. Trincone, R. Improta, B. Nicolaus, L. Lama, A. Gambacorta, R. Nucci and M. Rossi, Applied Microbiology and Biotechnology (submitted).
20. M.W. Day et all., Prot. Sci., 1 (1992) 1494.
21. L.H.Pearl, A.L. Hemming, R. Nucci and M. Rossi, J. Mol. Biol., 229 (1993) 561.

ECB6: Proceedings of the 6th European Congress on Biotechnology
edited by L. Alberghina, L. Frontali, P. Sensi

Characterization of extracellular hydrolases from marine psychrophilic bacteria

U. Kärst[a], M. Woehl, K. Czempinski[b] and R. D. Schmid[a]

[a]*GBF*, Gesellschaft für Biotechnologische Forschung mbH,
Mascheroder Weg 1, D-3300 Braunschweig, Germany

[b]*IGF*, Institut für Genbiologische Forschung Berlin GmbH,
Ilmestraße 63, D-1000 Berlin 33, Germany

1. INTRODUCTION

The ecology and physiology of psychrophilic and psychrotrophic bacteria from marine environments have been studied intensively for a long time. It is only recently, however, that the enzymes formed by these extremophilic microorganisms received more attention. Psychrophiles are supposed to produce potentially useful enzymes with peculiar properties due to the adaptation to their permanently cold environment. Low temperature optima and reduced temperature stability would provide high efficiency below 30°C and facilitate thermal inactivation. These enzymatic properties are regarded as valuable for, e.g. biosensors, conservation and processing of refrigerated food and pharmaceutics, as additives to washing detergents, and for biochemical and biomedical research (*cf.* [*1* - *3*]).

2. RESULTS

We screened a collection of 18 bacterial strains isolated from different sites in permanently cold arctic and antarctic habitats for hydrolytic enzymes such as proteases, chitinases, glucanases, esterases, lipases / phospholipases, and DNA-degrading enzymes, which are regarded as useful for the purposes mentioned above, and observed plenty of these enzyme activities. Some examined strains, preferably those that displayed strong hydrolytic activities on solid media as well as fast and good growth, were grown in liquid culture, and their enzymes characterized from culture supernatants and partially purified.

Of the ten proteases characterized, eight were serine proteases, seven of which exhibited alkaline pH-optima ranging from pH 8.5 to pH 10. The temperature optima of these proteases ranged from 45°C to 65°C, except for a metalloprotease that was optimally active at 28°C. The four proteases with optima ≤ 50°C had half lives of 10 to 45 minutes at 50°C and less than five minutes at 60°C (Table 1). Two of the four proteases with temperature optima ≥ 55°C (from strains 6302 and 8955) were characterized kinetically to learn how effective they are at the optimal growth temperature of the producing strains. For both enzymes V_{max} and K_M increased almost in parallel with temperature giving an optimum curve for the specificity constant V_{max}/K_M (Fig. 1).

Table 1
Properties of the characterized proteolytic enzyme activities

Strain	class	pH optimum	pH stability[1]	temperature optimum [°C]	temperature stability 37°C, 50°C [$t_{1/2}$, min]	Activation energy [kJ/Mol]
4988	nd[2]	7.0	nd	nd	nd	nd
5064	serine	10.0	6 - 12	45	186, 10	54.3
5069	serine	9.5	4 - 10	55	stable, 46	66.9
5088	serine	6.5	4 - 12	50	stable, 63	57.6
5112	metallo	6.0	nd	28	nd	nd
6302	serine	8.0	nd	60	nd	29.8/63.7[3]
7227	serine	8.5	nd	50	nd	nd
7561	serine	8.5	6 - 12	45	220, 47	58.9
8955	serine	9.5	nd	65	nd	65.4
9121	serine	9.0	nd	60	nd	nd

[1] after 2 h incubation at the indicated pH; [2] nd, not determined; [3] nonlinear Arrhenius plot

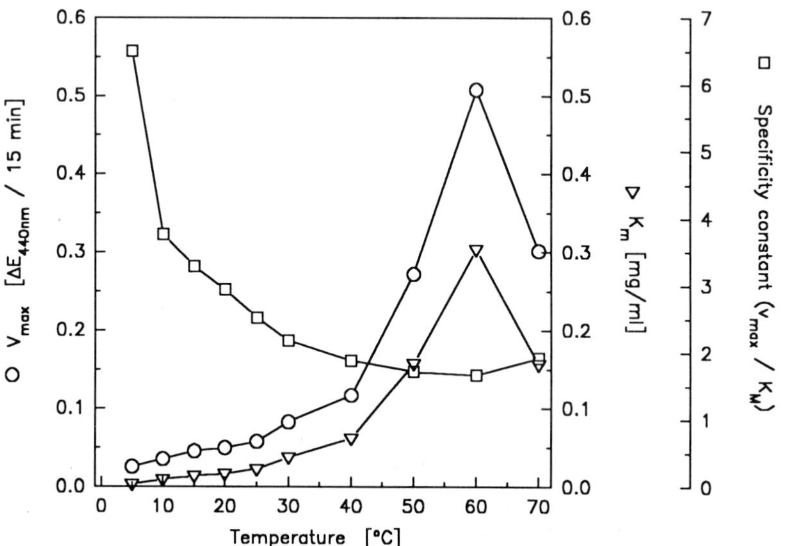

Figure 1.: Temperature dependence of K_M, V_{max}, and V_{max}/K_M of the protease of strain 6302 grown on fish peptone.

These optima of the physiological efficiency of 5°C for strain 6302 and 1°C for strain 8955 were even lower than the observed optimal growth temperatures of the cells which were 12°C and 18°C, respectively. The other enzymes retained a considerable residual activity at ≤ 30°C. The activation energies determined ranged from 18.6 to 66.6 kJ/Mol and did not correspond to the observed temperature optima. Two proteases with low temperature optima, a metalloprotease and a serine protease, were purified to homogeneity. Both are monomers with M_r of 34 kDa under denaturing conditions. The four chitinases characterized exhibited neutral pH-optima and temperature optima around 30°C, significantly lower than the proteases; they were rapidly inactivated with half-lives of 31 to 46 minutes at 37°C and less than 20 minutes at 50°C. The endoglucanase from strain 8839 exhibited a temperature optimum at 30°C and a neutral pH-optimum. The M_r of the native enzyme is 30 kDa. Three of the four examined lipases exhibited exceptionally low temperature optima ranging from 10 to 20°C, depending on the substrate; the fourth enzyme had its optimum at 33 to 42°C (Table 2). All lipases except that of strain 6302 had alkaline pH-optima around pH 8.5. The enzyme formed by strain 4988 was purified and the pI and native molecular weight determined by gradient polyacrylamide gel electrophoresis and activity staining. The pI was around pH 5 and the M_r 20 kDa.

Table 2
Properties of the examined lipases

Strain	4988	6302	7227	8899
Temperature optimum [°C]*	10 / 20	10 / 20	45 / 33	10 / 20
Temperature stability#	90	n.b.	n.b.	n.b.
pH optimum	8,5	7,3	9,0	8,5
pH stability	6 - >10	7 - >10	6 - 9	6 - >10
native M_r [kDa]	19	n.b.	n.b.	n.b.
native pI	4,8 - 5,1	n.b.	n.b.	n.b.

* Substrate dependent for pNPP / Olive oil
# remaining activity after 30 min incubation at 40°C.

The lipase activities were also examined by TLC-analysis of the degradation products obtained with triolein. The substrate specificity was broad except for three enzymes that preferentially hydrolysed the ester bonds at the 1,3-position.

3. DISCUSSION

The observed temperature optima for the enzyme activities of the isolates under consideration were in part much higher than the optimal temperatures of growth, which range from 7°C to 18°C; this is not an unexpected property, however, since many enzymes from mesophilic and thermophilic sources have been found to display temperature optima 20°C to 30°C above those of their producers' optimal growth temperatures. In spite of their partly high temperature optima, the enzymes from psychrophilic bacteria turned out to be well adapted to the low temperature environment. Either their catalytic efficiency, expressed as specificity constant V_{max}/K_M, has its optimum well below 20°C close to the ambient temperature of the organisms habitat or they retained a high residual activity at ≤ 30°C. The temperature optima of most of the observed activities were 10 to 20°C lower than those of the respective enzymes from mesophilic sources, e.g. trypsin (55°C), subtilisin (65°C), lipases from *Pseudomonas* or *Rhizopus* (37°C) and chitinase of *St. griseus* (45°C). They furthermore exhibited a reduced thermostability. Similar findings regarding the temperature optima and stabilities have also been reported for the proteases o210f 25 mainly gram-negative terrestrial isolates [*4 -7*] and an α-amylase from a psychrotrophic *Alteromonas* strain [*8*]. Four extracellular bacterial lipases from psychrotrophic *Moraxella* strains, however, displayed temperature optima around 40°C [*9*]. Thus, the enzymes of psychrophilic origin are of potential interest in biochemistry and biotechnology. Examples could be the alkaline serine proteases and lipases and the neutral endoglucanase.

REFERENCES

1. J.M. Ludlow and D.S. Clark, Crit. Rev. Biotechnol. 10, 321-345 (1991).
2. D. Scott, Food Biotechnology 2, 119-132 (1988-89)
3. H. Kobori, C.W. Sullivan, and H. Shizuya, Proc. Natl. Acad. Sci. USA 81, 6691-6695 (1984).
4. R. Margesin and F. Schinner, In: Christiansen, C. Munck, L. and Villadsen, J. (Eds.), 5th European Congress on Biotechnology Abstract Book, Munksgaard International Publisher, Copenhagen, p. 423 (1990).
5. R. Margesin and F. Schinner, FEMS Microbiol. Lett. 79, 257-262 (1991).
6. R. Margesin, N. Palma, F. Knauseder and F. Schinner, J. Biotechnol. 24, 203-206 (1992).
7. R. Margesin and F. Schinner, J. Biotechnol. 24, 203-210 (1992).
8. G. Feller, T. Lonhienne, C. Deroanne, C. Libioulle, J. Van Beeumen, and C. Gerday, J. Biol. Chem. 267, 5217-5221 (1992).
9. G. Feller, M. Thiry, J.-L. Arpigny, M. Mergeay, and C. Gerday, FEMS Microbiol. Lett. 66, 239-244 (1990).

ECB6: Proceedings of the 6th European Congress on Biotechnology
edited by L. Alberghina, L. Frontali, P. Sensi

Characterization of the xylanolytic enzyme system of the thermophilic eubacterium *Thermotoga thermarum*

A. Sunna[a], J. Puls[b] and A. Antranikian[a]

[a]Technische Universität Hamburg-Harburg, Technische Mikrobiologie/Graduiertenkolleg Biotechnologie, Denickestrasse 15, 2100 Hamburg 90, Germany

[b]Bundesforschungsanstalt für Forst-und Holzwirtschaft, Institut für Holzchemie, Leuschnerstrasse 91, 2050 Hamburg 80, Germany

1. INTRODUCTION

The polysaccharide xylan is the most abundant renewable hemicellulose and is composed of 1,4-linked ß-D-xylopyranosyl residues with substituents such as acetyl, arabinosyl and glucoronyl groups (1).The complete hydrolysis of xylan requires the action of several hydrolytic enzymes; the main enzyme being endo-1,4-ß-D-xylanase (2). Thermostable xylanases could be introduced in the pretreatment of paper pulps to remove xylan while preserving cellulose content (3). Therefore, thermostable xylanases free of cellulolytic activity are needed. Besides the thermostable xylanase from *Thermotoga sp.* (4), most of the xylanases reported so far are unstable at high temperature. In this report we present data on the characteristics of xylanolytic enzymes produced by the extreme thermophilic eubacterium *Thermotoga thermarum*. This microorganism produces an endoxylanase that is optimally active at 90°C and at a wide pH range.

2. RESULTS

2.1. Growth experiments

Growth and enzyme production was performed in a dialysis membrane fermentor anaerobically at 77°C (5). *Thermotoga thermarum* was cultivated in the presence of 0.5% (w/v) of insoluble beech xylan at an agitation speed of 500 rpm and a gassing rate of 0.5 vvm. Under these conditions the final cell concentration was 2.0 x 10^9 cells/ml (4.0 g/l wet weight) and the total xylanolytic activity was 488 U/l (Fig.1). During growth on xylan the following enzyme activities were detected: endoxylanase, ß-xylosidase and arabinofuranosidase. Under nonlimiting growth on xylan, most of these enzymes were predominantly cell associated.

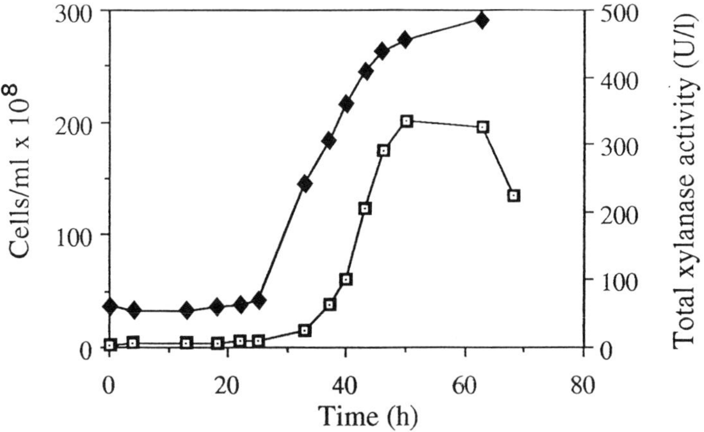

Fig. 1. Growth and xylanase production by *T. thermarum* in dialysis membrane fermentor. -□- cells/ml, -◆- total xylanolytic activity. The reactor is composed of two chambers (1 l inner chamber and a 4 l outer chamber) which are separated by a dialysis cuprophan membrane (5).

2.2. Enzyme induction

During growth of *T. thermarum* on glucose, maltose and starch very low level of xylanase activity was measured. A significant increase in the enzyme level was observed when cultivation was performed with mannan and xylan. Interestingly, enzyme induction was also achieved if the monomer xylose was used as carbon source. Hence it can be concluded that the xylanolytic enzyme system is partially constitutive. In all these experiments most of the enzyme produced was associated to the cells.

2.3. Partial characterization of the xylanolytic enzymes

The enzymes from *T. thermarum* are active over a broad temperature (40°C-105°C) and pH (4.0-9.0) range. The xylanase and ß-xylosidase produced by this microorganism have pH and temperature optima of 6.0 and 90°C, respectively (Fig.2 and 3). The arabinofuranosidase, on the other hand, showed an optimum for activity at 70°C and a pH of 6.0 (Fig. 2 and 3).
The xylanase activity was not influenced by the addition of various divalent cations. However, inhibiton of endoxylanase activity was observed with N-bromosuccinimide, p-hydroxymercuribenzoate and Pefabloc SC®. This may indicate that tryptophan, cystein, and serine residues are essential for enzymatic activity

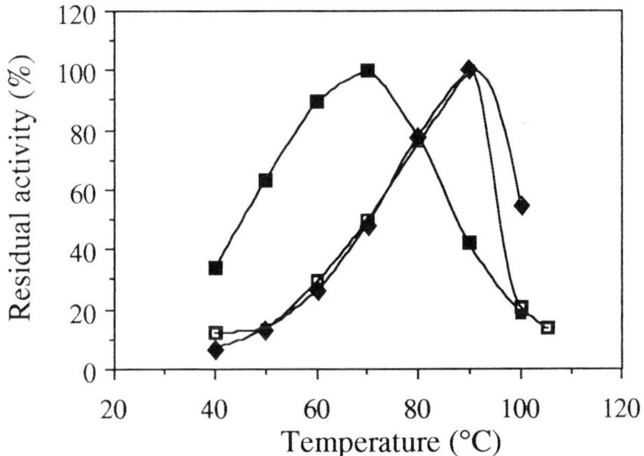

Fig. 2. The influence of temperature on the initial velocity of the enzymes produced by *T. thermarum*. -□- endoxylanase, -◆- ß-xylosidase, -■- arabinofuranosidase.

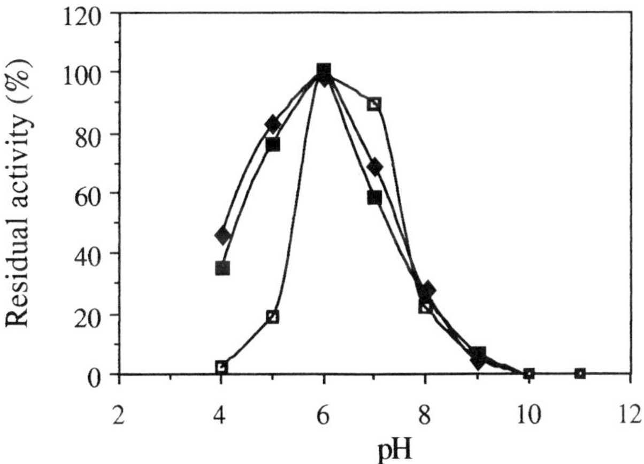

Fig. 3. The influence of pH on the activity of the enzymes produced by *T. thermarum*. -□- endoxylanase; -◆- ß-xylosidase; -■- arabinofuranosidase.

2.4. Substrate specificity

Substrate specificity experiments with different xylans indicated that commercial birchwood xylan, as well as xylan from an industrial process (Lenzing AG), were the most suitable substrates for the xylanase of *T. thermarum*. Interestingly, the xylanolytic enzyme did not show any cellulolytic activity when incubated with hydroxyethyl-cellulose or filter paper. The enzyme system was capable of hydrolyzing insoluble beech xylan, as well as partially soluble birchwood xylan, to yield mainly xylobiose, xylotriose, xylotetraose and longer xylooligosaccharides. This, hence proves the presence of a depolymerizing endo-1,4-ß-D-xylanase.

REFERENCES

1. **Wistler, R. L. A. and E. L. Richards.** 1970. Hemicelluloses. In The Carbohydrates-Chemistry and Biochemistry, 2nd edn, vol. 2A, pp. 447-469. W. Pigman and D. Horton (eds). New York: Academic Press.

2. **Biely, P.** 1985. Microbial xylanolytic systems. Trends Biotechnol. 3:286-290.

3. **Viikari, L., Sundquist and J. Kettunen.** 1991. Xylanase enzymes promote pulp bleaching. Paperi ja Puu-Paper and Timber. **73**:384-389.

4. **Simpson, H. L., U. R. Haufler and R. M. Daniel.** 1991. An extremely thermostable xylanase from the thermophilic eubacterium *Thermotoga*. Biochem. J. **277**:413-417.

5. **Märkl, H., M. Lechner and F. Götz.** 1990. A new dialysis fermentor for the production of high concentrations of extracellular enzymes. J. Ferment. Bioeng. **69**:244-249.

ECB6: Proceedings of the 6th European Congress on Biotechnology
edited by L. Alberghina, L. Frontali, P. Sensi

Isolation and characterization of new plasmids from moderately halophilic eubacteria: developing of cloning vectors

A. Ventosa, R. Fernández-Castillo, C. Vargas, E. Mellado, M.T. García and J.J. Nieto

Departamento de Microbiología y Parasitología, Facultad de Farmacia, Universidad de Sevilla, 41012 Sevilla, Spain

1. SUMMARY

Three autochthonous plasmids were isolated from the moderately halophilic eubacteria, *Chromohalobacter marismortui* ATCC 17056 (pCM1, 18.5kb), *Halomonas elongata* ATCC 33174 (pHE1, 4.2kb) and *Vibrio costicola* strain E-367 (pVC1, 2.95kb) by using the rapid alkaline lysis method. These plasmids were characterized by restriction endonuclease analysis and restriction maps were constructed. Hybridization experiments were conducted to determine the rate of homology among these plasmids and with the previously isolated pMH1. These plasmids, along with pMH1, are the only so far reported plasmids from moderate halophiles.

2. INTRODUCTION

Besides the extremely halophilic archaea (halobacteria), the moderately halophilic eubacteria constitute the most important group of microorganisms adapted to live in hypersaline habitats. Moderately halophilic bacteria are defined as those which grow optimally in media containing 0.5 to 2.5 M salt [1]. They constitute a very heterogeneous physiological group which includes a great variety of bacteria [2]. Despite extensive studies concerning their physiology and ecology have been reported [1,3,4], a scarcy information is still available on their genetics. However, besides their importance in the ecology of hypersaline environments, they are an interesting microbial group from a genetical point of view. Thus, they can grow in a wide range of salt concentrations and many of them produce interesting halophilic enzymes (nucleases, amylases, proteases, etc.) [5] or compatible solutes [6] which can be used as stabilizing agents for enzymes. Therefore, the genetic manipulation of these extremophiles for the overproduction of such products appeared as desirable. Although we have very recently demonstrated that some moderately halophilic eubacteria harbour a plasmid (pMH1) [7], the molecular biology of this microbial group remains a virtually unexplored field. In this paper we report the preliminary molecular characteristics of three new plasmids (pCM1, pHE1, and pVC1), which we have isolated from another moderately halophilic strains. They present good features for their use in the development of cloning vectors for these extremophiles.

3. MATERIALS AND METHODS

3.1. Bacterial strains and growth conditions

The three moderately halophilic strains from which the plasmids were isolated were *Halomonas elongata* ATCC 33174, *Vibrio costicola* strain E-367 and *Chromohalobacter marismortui* ATCC 17056. The strains were grown in a saline medium containing 10% total salts [8] to which 0.5% yeast extract (Difco) was added. Incubation was at 37°C in an orbital shaker (New Brunswick Sci. Co., Inc., Edison, New Jersey, USA) at 200 strokes/min. When necessary, all media were solidified by adding 2% Bacto-agar (Difco).

3.2. Isolation of DNA

Plasmid DNAs were prepared in a mini or max. scale by using the alkaline lysis method [9]. In some cases, the cells were previously washed with 0.1 % SDS in TE buffer (10 mM Tris, 1 mM EDTA, pH 8.0) to improve the procedure. Genomic DNA was isolated according to Ausubel et al. [10].

3.3. DNA characterization

Chromosomal and plasmid DNA were digested with restriction endonucleases (Boehringer) according to standard methods[11]. Restriction fragments were separated on agarose gels (Sigma) in TBE buffer (45 mM Tris-borate, 1 mM EDTA, pH 8.0). *Hin*dIII-digested lambda DNA fragments were used for the estimation of the molecular masses of the plasmid fragments. Gels were visualized after staining in 0.5 mg/ml of ethidium bromide for 10 min. and photographed under 254 nm ultraviolet light. Plasmid and genomic DNA fragments obtained after digestion were transferred

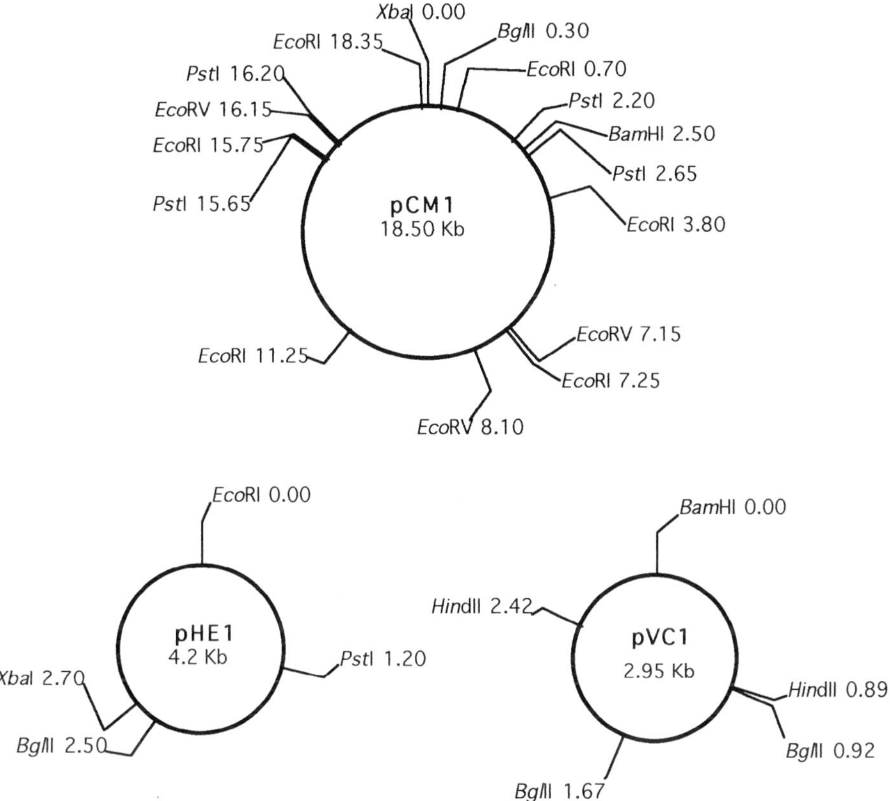

Figure 1. Restriction maps of the three plasmids, showing cleavages sites for the restriction endonucleases used.

to nylon filters (Amersham) as described by Southern [12]. Probe labelling and hybridization experiments were performed by using standard procedures [11].

4. RESULTS AND DISCUSSION

In a previous study, plasmid pMH1 was isolated from four moderately halophilic collection strains only after transformation into *Escherichia coli* strain JM109 cells. In contrast, the three plasmids that we describe herein could be easily visualized on electrophoresis gels and isolated by using the standard alkaline lysis method. These plasmids were physically characterized by using a number of restriction endonucleases and the resulting restriction maps were obtained, as shown in Fig. 1.

The biggest plasmid, pCM1, of 18.5 kb in size, shows unique cleavage sites for the enzymes *Bam*HI, *Bgl*II and *Xba*I. Moreover, pCM1 also contained more than one cleavage site for the restriction enzymes *Eco*RI, *Eco*RV and *Pst*I. The plasmid pHE1, of 4.2 kb, showed unique cleavage sites for several endonucleases (*Eco*RI, *Bgl*II, *Pst*I and *Xba*I). Finally, the smallest plasmid, pVC1, of 2.95 kb, presented only an unique cleavage site for the endonuclease *Bam*HI. Besides, *Bgl*II and *Hind*II yielded two and three, respectively, restriction sites. To determine the rate of homology amongst the three plasmids, Southern hybridization experiments were conducted. No homology among these three plasmids could be detected. Other experiments aiming to determine the existence of homology between pMH1 and these new plasmids are now in progress. Overall, plasmids pVC1 and pHE1 appeared as the most suitable for the developing of further cloning vectors on the basis of their small size and the presence of unique restriction sites for cloning purposes.

This investigation was supported by grants from the Dirección General de Investigación Científica y Técnica (PB92-0670) and from the Junta de Andalucía.

REFERENCES

1. D.J. Kushner and M. Kamekura. In Halophilic Bacteria, vol. I, pp. 109-140. F. Rodríguez-Valera (ed.). CRC Press, Boca Ratón, 1988.
2. A. Ventosa, In Halophilic Bacteria, vol. I, pp. 71-84. F. Rodríguez-Valera (ed.), CRC Press, Boca Ratón, 1988.
3. F.Rodríguez-Valera, In Halophilic bacteria, vol. I, pp. 3-30. F. Rodríguez-Valera (ed.), CRC Press, Boca Ratón, 1988.
4. B. Javor, Hypersaline Environments. Microbiology and Biogeochemistry, Springer Verlag, Berlín, 1989.
5. M. Kamekura, FEMS Microbiol. Rev., 39 (1986) 145.
6. E.A. Galinski and B.J. Tindall, In Molecular Biology & Biotechnology of Extremophiles, pp. 76-114, R.A. Herbert and R.J. Sharp (eds.), Blackie and Son, London, 1992.
7. R. Fernández-Castillo, C. Vargas, J.J. Nieto, A. Ventosa and F. Ruiz-Berraquero, J. Gen. Microbiol., 138 (1992) 1133.
8. A. Ventosa, E. Quesada, F. Rodríguez-Valera, F. Ruiz- Berraquero and A. Ramos-Cormenzana, J. Gen. Microbiol., 128 (1982) 1959.
9. G. Morelle, BRL Focus, 11 (1989) 7.
10. F.M. Ausubel, R. Brent, R.E. Kingston, D.D. Moore, J.G. Seidman, J.A. Smith and K. Struhl (eds.), Current Protocols in Molecular Biology. John Wiley and Sons, New York, 1989.
11. T. Maniatis, E. Fritsch and J. Sambrook (eds.), Molecular Cloning: A Laboratory Manual, Cold Spring Harbor Lab., Cold Spring Harbor, 1982.
12. E.M. Southern, J. Mol. Biol., 98 (1975) 503.

BIOTECHNOLOGY OF FILAMENTOUS FUNGI

ECB6: Proceedings of the 6th European Congress on Bio'echnology
edited by L. Alberghina, L. Frontali, P. Sensi

Biotechnology of Filamentous Fungi

J.R. Kinghorn and N. Lucena

Plant Sciences Laboratory, School of Biological and Medical Sciences, Sir Harold Mitchell Building, University of St. Andrews, Fife KY16 9TH, Scotland, United Kingdom

1. RANGE AND DIVERSITY OF FILAMENTOUS FUNGI

It has been estimated that there are well over one million species of filamentous fungi (compared with around 250,000 bacterial species). Such organisms range from the simple water moulds through the Zygomycetes and Ascomycetes to the complex Basidiomycetes. They exhibit tremendous diversity in habitat, morphology and size, cell type and general physiology. Many species possess exceptional metabolic capabilities which permit growth in very unfavorable nutritional and climate environments. Certain filamentous fungi produce substances which enable successfully competition with other organisms. Such features have attracted the attention of a considerable number of applied as well as fundamental research scientists.

2. FUNGAL PRODUCTS OF BIOTECHNOLOGICAL INTEREST

The filamentous fungi are probably unique in the diversity of their metabolic activities. Many of these activies are of direct biotechnological interest and a few examples are provided below (Ref.1 and reviews therein).

First, their degradative power is illustrated by the copious amounts and extensive range of extracellular enzymes produced. For instance, the current annual world market for fungal enzymes used in the food and beverage industries, is considered to be in the region of US$650M. In particular, there is the need for saccharification of starch to glucose, maltose and low molecular weight dextrins by glucoamylase and amylases. Such enzymes are initially produced by large scale fermentation of fungal species such as *Aspergillus niger* particularly but not exclusively in the Western and *Aspergillus oryzae* especially in the Eastern Hemisphere. In

Japan, beverages such as sake, shochi, as well as the condiment, soy sauce are fermented using aspergilli including *A. oryzae* and *A. sojae*. In China, the fungus *Rhizopus niveus* is traditionally involved in the production of similar beverages and food, eg. jiu, tempe and rice beer. The ability of filamentous fungi to utilise lignocellulose is another example of their considerable degradative abilities which have commercial potential. Lignocellulose, being a major plant component, is, of course, a renewable carbon and energy source. Plant material degraded by lignocellulase enzymes can be used as animal food material. The white rot lignin degrading fungi such as *Phanerochaete chysosporium* and *Sporotrichum pulverulentum* have been extensively characterised, in this regard. In addition, certain fungi such as *Trichoderma viridae* and *T. reesei* produce cellulases which degrade the many classes of cellulose molecules. Applied considerations have provided the impetus to investigate lignocellulytic and related enzymes.

Second, filamentous fungi are equipped with the ability to synthesise molecules which are often useful to man in various ways. For instances, molecules which bring about biological control of other organisms has been central to biotechnology for many decades. In particular, secondary metabolites such as ß-lactam antibiotics for the control of bacterial growth has had an enormous beneficial effect on the health and well being of humans and domestic animals. Other fungi, such as certain Deuteromycetes are useful, particularly in the development of myco insecticides against several insect pests of agricultural crops. There is a recent interest as a further example of biological control, in using filamentous fungi, such as species of *Trichoderma*, *Gliocladium* and *Talaromyces* to combat unwanted soil borne filamentous fungi.

Third, fungi can be used directly as a source of food. Mushrooms are amongst one of the oldest crop plants grown by man. Although many are edible, only a few are grown on a commercial scale. These include the Western mushrooms *Agarius biosporus* and the Oriental mushrooms *Auricularia ssp.*, *Flammulina velutipes*, *Grifola frondosa*, *Lentinus edodes*, *Lyophyllum ulmarium*, *Pholiota nameko*, *Pleurotus ssp.*, *Tremella ssp.*, *Tricholoma matsutake*, *Tuber melanosporum* and *Volvariella volvacea*. Mushrooms are very desirable as a food source, of course, since these fungi are comprised of all the necessary components for life and are socially acceptable as food. In addition, fungal extracts can be added to supplement the protein content of animal and plant protein products. Such 'mycoproteins' can be more economically desirable than, for example fish or soybean protein supplements.

3. FILAMENTOUS FUNGAL TECHNOLOGY

Here we focus on recently developed recombinant DNA technology which may be applicable to the vast range of filamentous fungi. It is pertinent to note that certain filamentous fungi such as *Aspergillus nidulans* are regarded as a model experimental fungus. Unlike most commercial fungi, *A. nidulans* is amenable to a rigorous classical genetic approaches, the techniques of which we omit here (for a review see ref.2). Second, that molecular techniques were developed, for such model fungi and the experience gained 'transfered' to industrial species. Finally, that filamentous fungi have a traditional history of industrial use (see section 2) there exists fermentations procedures and equipment which have been well documented (see ref.3) and consequently these aspects are not included here.

During the 1980's, the development of genetic engineering procedures for the filamentous fungi, in particular aspergilli, took place. There are several reasons for such attention. First, to study the regulation of gene expression was perhaps the major driving force, since a number of genes and their regulatory circuits had been identified by previous classical genetic studies especially *A.nidulans*. Second, that aspergilli have a long history in producing materials and substances beneficial to man such as food, alcoholic beverages and enzymes. Third, certain aspergilli are of medical interest such as species producing aflatoxins. Fourth, although *E. coli* and the yeast *Saccharomyces cerevisiae* were proving, in many cases, to be attractive hosts for the recombinant protein industry, filamentous fungi offered alternative systems (see section 4). We briefly focus here on fungal molecular genetic technology.

It has been over 10 years since the first report of filamentous fungal genetic transformation using *Neurospora crassa*. Since then the number of filamentous fungi that have been transformed has grown substantially, building on methods that were developed - mainly using *A. nidulans* as a model system (for reviews, refs. 4,5,6). Transformation of industrial fungi benefited from research studies of these model fungi not only in terms of methodology but also in the development of selective markers which have greatly increased in number over the past decade. These include nutritional selection markers and dominant selective markers (Table 1) of which most have been used for the transformation of filamentous fungi important in industry (for a review, see ref. 7).

Table 1
Common selective markers used in filamentous fungal transformation (*)

	Gene	Selection
Nutritional selection markers	amdS	acetamide utilisation
	argB	arginine prototrophy
	niaD	nitrate utilisation
	pyrG	uridine prototrophy
	trpC	tryptophan prototrophy
Dominant selection markers	benA	benomyl resistance
	oliC	oligomycin resistance
	G418	antibiotic resistance
	hygromycin	antibiotic resistance
	phleomycin	antibiotic resistance

(*) Reference for these selection systems can be found in reviews 4,5,6,7. It should be noted that such systems quoted in Table 1 are commonly used, although others are available. In addition, bacterial genes, which confer resistance to antibiotics (when transformed into fungi) such as G418, hygromycin or phleomycin, are usually flanked by fungal 5′ and 3′ expression sequences.

Typical filamentous fungal frequences are normally relatively low - around 10 to 50 transformants per microgram of plasmid DNA - compared with yeast of around 20,000. The difference is probably due to the fact that the incoming gene integrates into the nuclear genome, whereas the yeast incoming gene is maintained as a stable replicating plasmid. Such low frequencies in filamentous fungi make it difficult to clone genes by transformation and functional complementation of mutant strains. One way of circumventing this particular difficulty is to use cosmids, thereby increasing the chance of isolating the gene of interest. Very recently, an *A. nidulans* DNA fragment (designated AMA1) which displays replicon activity, has been isolated (8). A construct containing this fragment, yields frequences of up to 20,000 transformants and has been shown to be useful for other filamentous fungi including industrial fungi (8).

Transformation of filamentous fungi usually necessitates the removal of the cell wall and the generation of protoplasts. There is a range of lytic enzymes that can be used for this purpose, the most common of which is ß-glucuronidase which is present in the routinely used commercial preparation Novozyme 234. Usually very young mycelial cells are used as starting material and an osmoticum such as sorbitol or salts are included to prevent lysis of the protoplasts. Co-transformation of non-selected genes (and foreign genes come into this category, of course, see section 5) have been shown to occur at a satisfactory frequency in filamentous fungi. In this regard, rates of up to 90% have been demonstrated and this frequency, of course, easily allows the introduction of genes of commercial interest. However, the frequency of co-transformation is dependent on the selection system employed. Futhermore, the copy number of integrated transformed genes can vary considerably, again depending on the system used.

4.APPLICATION OF GENETIC ENGINNERING TO FUNGAL BIOTECHNOLOGY

With the development of fungal genetic transformation systems and gene isolation technology, there has been widespread interest in the application of the technology for commercial purposes. First to improve existing fungal industrial processes.

This has lead to the isolation of genes of commercial importance such as genes encoding industrial enzymes, eg amylases, cellulases, glucoamylases, ligninases, pectinases, proteases. The goals of course, are to increase production levels or to change their physical properties conferring upon them more advantagous properties such as more convenient temperature, pH optima or to extend the substrate range. Such goals have already been achieved for several systems (see ref. 1). In addition, the yield of metabolites such as antibiotics have been successfully improved by gene cloning and amplification of the biosynthetic genes involved. A further application of fungal molecular techniques is where fungal genes are transfered to another fungus - a species that can be handed better, for convenience, or grows faster, or where large scale fermentation is already 'in place'.

A second goal is generate new fungal industries. In this respect fungi may produce potentially useful enzymes or metabolites - the concentration of which may be too low for commercial exploitation. The isolation of the corresponding genes and their subsequent amplification can often lead to elevated levels commensulate with commercial production requirements. Further new fungal industries may result from the successful programming of fungi to produce foreign proteins and this aspect is discussed below.

5. EXPRESSION OF MAMMALIAN PROTEINS

The main reasons for heterologous expression studies are commercial. The availability of certain proteins such as chymosin may be erratic and so expression in a microbial host can fulfill demand. As long as the levels expressed are sufficiently high, heterologous production is therefore commercially lucrative. However, pharmaceutically useful proteins which previously were only obtained from biological samples such as serum, and these include human interferon $\alpha 2$ and tissue plasminogen activator can be profitable even if they are produced at much lower levels.

There has been interest to access the usefulness of certain filamentous fungi, particularly the *aspergilli*, as host for the production of mammalian proteins (9). The reason for such interest is four-fold. First, fungi such as aspergilli (and others) have the capacity to produce and secrete impressive amounts of their own proteins. This can be as much as 25 grams per litre or even higher. Second, as discussed above, fungi have a traditional use in the production of metabolites and enzymes. Consequently, the appopriate fermentation experience and equipment already exists (3). Third, fungi such as *A. oryzae* and *A. niger* have GRAS (Generally Regarded As Safe) status bestowed on them by the USA Food and Drug Administration and the World Health Organisation. This makes such fungi, *a priori*, acceptable hosts for the production of proteins of pharmaceutical interest. Fourth, certain proteins have been found to be difficult to express in bacteria or yeast. Finally, that may bacterial and yeast systems are encumbered by patents proved another incentive to assess filamentous fungi.

Aspergillus has been most exploited for the purpose of foreign protein expression. *A. nidulans* together with *A. niger* have been used most commonly to express mammalian proteins. Recent reviews of heterologous expression by filamentous fungi are available (10,11,12).

Before successful expression of a foreign gene by a fungal cell can take place, the foreign gene requires to be inserted between a efficient fungal transcriptional promoter system and terminator region. A number of filamentous fungal genes have been used for this purpose. A typical plasmid expression construct might be in which fungal glucoamylase transcriptional sequences are used to drive the expression of mammalian genes. In addition to this expression cassette, the plasmid molecule may contain a selectable marker, for successful transformation into the host. However, it is often unnecessary to incorporate a marker into the expression plasmid as entry can be effected using co-transformation approaches (section 4). Case situations of foreign genes expressed thus far in the aspergilli, *A. nidulans* and *A. niger* are given in Table 2.

Table 2
Foreign proteins secreted by *Aspergillus sp* (*)

Fungus	Protein	Yield	Fungal expression signals Promoter	Terminator
A.nidulans	human interferon α2	1 mg/L	alcA	glaA
A.nidulans	human lactoferrin	5 mg/L	alcA	benA
A.nidulans	human tissue plasminogen activator	100 μg/L	thiA	tpiA
A.nidulans	human tissue plasminogen activator	1 mg/L	alcA	tpiA
A.nidulans	human α-1 antitrypsin	1 mg/L	alcA	thiA
A.nidulans	human granulocyte macrophage colony stimulating factor	1 mg/L	alcA	thiA
A.nidulans	human interleukin-6	2 mg/L	complete glaA gene	
A.niger	calf chymosin	<1000 mg/L	complete glaA gene	
A.niger	hen egg lysozyme	1 mg/L	gpdA	trpC
A.niger	hen egg lysozyme	12 mg/L	glaA	glaA
A.niger	porcine pancreatic glycoprotein A[2]	10 mg/L	complete glaA gene	

(*) Highest published yields quoted. For further details of the fungal genes used in expression cassettes, see reviews (10,11,12).

From Table 2 it can be seen that a number of attempts at expressing mammalian proteins have carried out. With the exception of chymosin (for use in cheese manufacture)(13) the expression levels are relatively low. It is therefore difficult to predict if filamentous fungi will compete successfully with other hosts including bacteria and yeasts.

Some of the following possible problems with regard to low yields may be overcome.

Many factors are likely to affect the successful expression of heterologus proteins. The choice of promoter may be important. For example, the production of certain heterologous proteins may be detrimental to the host and so an inducible promoter is desirable. For most proteins expressed thus far, however, a strong constitutive promoter is sufficient. Fortunately, it appears that most filamentous fungi are capable of recognising efficiently promoter and terminator sequences of filamentous fungal origin. Therefore expression cassettes utilising promoter and terminator signals derived from a gene of one species can usually be expressed in another species. In most instances, there appears to be an, albeit loose, correlation between copy number of the expression cassette and the level of product. This is not however universally true as demonstrated in certain cases. The reason for the differences in expression levels between transformants with similar vector copy numbers is unclear but may include differences in transcriptional efficiency, in turn, dependent on the integration site, differences in mRNA stability and processing or host mutations caused by gene disruptions on integration of expression vectors. Additionally, a limitation on transcription may be placed by the availability of trans-acting regulatory proteins. Although the influence of the secretion signal sequence in filamentous fungi has not been rigoriously assessed, it appears that aspergilli are rather permissive with respect to signal sequence recognition. For example, levels of secreted chymosin were similar with either the chymosin signal sequence or that of *A. niger* glucoamylase and indeed, *A. nidulans* was capable of recognising an entirely synthetic signal for secretion of interferon.

Many of the foreign proteins of commercial interest are glycosylated. Post transcriptional modification by glycosylation is probably of importance in immunogenicity, function and stability of proteins and so accurate glycosylation by a heterologous host may be relevant particularly for pharmaceutical proteins. Filamentous fungi are capable of O- and N-glycosylation possibly by a mechanism similar to that of animal cells and extensive hyperglycosylation, which is often observed by yeast, does not apper to be as great a problem in filamentous fungi. Several foreign proteins are assumed from their size relative to authentic protein to be similarly glycosylated eg. human tissue plasminogen activator , human lactoferrin but no studies have yet empirically demonstrated this. A major problem for the expression of foreign proteins is the production by aspergilli of numerous extracellular proteases. With respect to *A. nidulans* this may be less of a problem, as proteases are repressed in cells grown under conditions of nitrogen, carbon and sulphur repression and so protease production can be manipulated by growth conditions. *A. niger* and other aspergilli, however, produce copius amounts of these extracellular enzymes in complex media. The production of the acid protease, argillopepsin, by *A. niger* probably accounted for the discrepancy between the levels of secreted chymosin expected and those observed.Other secreted or mycelium-associated proteases may also be involved in degradation of secreted foreign proteins. Currently, protease deficient strains obtained by classical mutagenesis and screening are being assessed as hosts for foreign production.

6. PROSPECTUS

The development of technology has permitted the improvement of fungal strains and their manipulation in a specific manner (Ref.1 and reviews therein). It is possible to introduce multiple copies of a desirable gene and monitor its effect on product formation. In addition to product amplification, it should be possible to produce 'designer strains' which will produce 'cocktails' of enzymes to carry out different functions. Such technology is already in progress, for example in the manipulation of *T. reesei* cellulases. Strains can likewise be programmed to produce entirely novel products by incorporation of completely new enzymes.

Already valuable heterologous proteins may be expressed in industrial filamentous fungi, but the use of these organism is, in general, hindered at present by a dearth of knowledge of the molecular processes involved in post-transcriptional modification and secretion. This is an obvious area for intensive research to obtain the same level of heterologous as that achieved with homologous proteins (ie 20-30 g/L protein).

For any given microbial industrial process, be it production of secondary metabolites, extracellular enzymes or heterologous proteins, there has to be a limit to cellular capacity. However, general opinion agrees that this limit has not yet been met and so factors such as gene expression remain of major commercial importance. Therefore, obvious target areas must in future include basic molecular research into regulation at the transcriptional and translational level, regulatory genes and secretory pathways as well post-transcriptional signals.

Acknowledgements: Research work carried out in the laboratory of J.R.K. is funded by the Science and Engineering Research Council and The Commission of Economic Communities. N.L. is funded by The Brasilian Science and Technological Research Council, CNPq.

REFERENCES

1. Kinghorn, J.R. and Turner, G. eds., (1992) Applied Molecular Genetics of Filamentous Fungi. Blackie Press, Glasgow.
2. Clutterbuck, A.J. (1974) *Aspergillus nidulans*. In Handbook of Genetics 1 (R.C.King, ed.), Plenum, New York. pp.447-510.
3. Solomons, G.L. (1980) Fermentor design and fungal growth. In: Fungal Biotechnology. (J.E.Smith, D.R.Berry and B.Kristiansen. eds.), Academic Press, London.pp. 55-90.
4. Arst, H.N. and Scazzocchio, C. (1985) Formal genetics and molecular biology of the control of gene expression in Aspergillus nidulans. In: Gene Manipulation in Fungi, (J.W.Bennett and L.L.Lasure, eds.), Academic Press, New York.
5. Ballance, D.J. (1986) Sequences important for gene expression in filamentous fungi. Yeast. **2**:229-233.
6. Gurr, S.J., Unkles, S.E. and Kinghorn, J.R. (1987) The structure and organization of nuclear genes of filamentous fungi. In: Gene Structure in Eukaryotic Microbes, (J.R.Kinghorn, ed.), IRL Press, Oxford. pp.93-138.
7. Unkles,S.E. (1992) Gene organisation in industrial filamentous fungi. In: Applied Molecular Genetics of Filamentous Fungi. (J.R.Kinghorn and G.Turner. eds.), Blackie Press, Glasgow.pp.28-53.
8. Gems, D., Johnstone, I.L. and Clutterbuck, A.J. (1991) An autonomously replicating plasmid transforms *Aspergillus nidulans* at high frequency. Gene. **98**:61-67.
9. van Brunt, J. (1986) Fungi, the perfect hosts? Bio/Technology **4**:1057-1062.
10. van den Hondel, C.A.M.J.J., Punt, P.J. and van Gorcom, R.F.M. (1991) Heterologous gene expression in filamentous fungi. In: More Gene Manipulations in Fungi, (J.W.Bennett and L.L.Lasure eds.) Academic Press, San Diego. pp.396-428.
11. Saunders,G., Picknett,T.M., Tuite,M.F. and Ward, M. (1989) Heterologous gene expression in filamentous fungi. Trends in Biotechnology, **7**:283-287.
12. Davies, R.W. (1992) Genetic engineering of filamentous fungi for heterologous gene expression and protein secretion. In: Transgenesis - Applications of Gene Transfer (J.A.H. Murray, ed.), John Wiley, New York.
13. Dunn-Coleman et al. (1991) Commercial levels of chymosin production by Aspergillus. Bio/Technology, **9**:976-982.

ECB6: Proceedings of the 6th European Congress on Biotechnology
edited by L. Alberghina, L. Frontali, P. Sensi

Strategies for the over-production of fungal proteins in *Aspergillus*

Cees A.M.J.J. van den Hondel, Peter J. Punt and Robert F.M. van Gorcom.

Medical Biological Laboratory TNO, P.O. Box 5815, 2280 HV Rijswijk, The Netherlands, Tel: +31 15 843127, Fax: +31 15 843989.

1. INTRODUCTION

In filamentous fungi, a wide variety of enzymatic activities is found, either intracellular or extracellular. An increasing number of these activities, produced by e.g. strains of the genera *Aspergillus* and *Trichoderma*, is commercially exploited after (partial) purification of the desired activity from the fermentation medium or by using the entire organism as a biocatalyst for the production of (bio)chemicals. Extracellular fungal enzymes are mainly used in wood-, paper-, textile- and leather-processing industry, as animal feed-additive or in detergents, and in the food industry. With respect to the latter industry, a number of enzymes produced by *Aspergillus* species are generally regarded as safe for human consumption (FDA GRAS approval), since already for a long time several species such as *Aspergillus oryzae* are used for the production of fermented food.

In general, extracellular fungal enzymes have attractive properties, such as low pH optimum, broad temperature optimum and high protease resistance. The production levels of these enzymes can be increased considerably by classical strain-improvement programs. In the case of glucoamylase production by *Aspergillus niger*, for example, strains have been selected which secrete over 30 grams per litre of glucoamylase under optimized conditions.

Because of the biotechnological interest in fungal enzymes and the biological attractiveness of their producers, genetic modification techniques have been developed for filamentous fungi and a number of genes has been cloned in the past 10 years [for recent reviews see 1-3]. Furthermore there is an increasing knowledge in how fungal genes are organised and expressed.

Until now a number of laboratories have investigated the possibilities of improving the current production of fungal proteins and developing systems for efficient production of new commercially interesting fungal and non-fungal proteins of which the natural source is limited. In these studies *Aspergillus niger* (and close relatives) and *Aspergillus oryzae* are most commonly used [for a review see 4].

In this report the current strategy will be illustrated which is used to obtain efficient production of intra- and extra-cellular proteins in *Aspergillus*, and some examples will be given of the construction of *Aspergillus* strains which efficiently produce (extracellular) fungal proteins.

2. EFFICIENT PRODUCTION OF FUNGAL PROTEINS

2.1 Strategies

In general the following strategy is used to achieve high level production of a extracellular (fungal) protein:

1. Isolation of the gene encoding the protein of interest,
2. Generation of strains with multiple copies of the gene,
3. Providing the gene with well characterised strong, constitutive or regulated gene expression signals and the generation of strains containing multiple copies of such expression cassette.

In the following two paragraphs the application of this strategy will be illustrated for the high level production of the extracellular enzymes glucoamylase, phytase and endoxylanase. The latter enzymes, are available as market-ready products.
Using these techniques our laboratory has also succeeded in high-level production in *Aspergillus* of, amongst others, the intracellular enzymes benzoic acid para-hydroxylase [5], and glyceraldehyde 3-phosphate dehydrogenase [6].

2.2 Production of *A. niger* glucoamylase

A good example of the approach that has been used during the last few years to obtain *Aspergillus* strains which efficiently produce interesting fungal proteins, is the construction of *A. niger* strains that overproduce the enzyme glucoamylase [7].
The *A. niger glaA* gene, which encodes glucoamylase, has been cloned and characterized [8]. Subsequent expression analysis studies showed that the *glaA* gene is proeceeded by a strong promoter [9]. To obtain an overproducing strain of *Aspergillus niger*, multiple copies of the *glaA* gene were introduced into the genome of the wild-type strain. For this purpose a cosmid vector was constructed which contains in addition to the *amdS* gene of *Aspergillus nidulans* as selection marker four copies of the *glaA* gene [10]. The latter marker facilitates selection of transformants containing multiple copies of the vector and, thus, of the *glaA* gene.

A. niger N402 [11], a commonly used laboratory strain, was transformed with this vector. Transformants were first screened for their growth behaviour under different cultivation conditions. They could be divided into three groups: (1) Growth on acetamide, (2) in addition growth on acrylamide and (3) in addition inhibition of growth on ω-amino acids. In *Aspergillus nidulans* one copy of the *amdS* gene is sufficient for growth on acetamide. Growth on acrylamide is only possible if several copies of the *amdS* gene are present or if at least one copy is very efficiently transcribed [12]. The presence of many copies of the *amdS* gene causes inhibition of growth if transformants are grown on ω-amino acids (γ-aminobutyric acid, β-alanine) as sole nitrogen and carbon source, because genes involved in the conversion of ω-amino acids are controlled by the same positive regulatory gene as *amdS*, namely the *amdR* gene. In case of multiple copies of the *amdS* gene there will be a shortage of the *amdR* protein due to titration by these copies, resulting in inhibition of growth on ω-amino acids [10]. We have shown that the phenotypic effects with multiple copies of the *A. nidulans amdS* gene in *A. niger* are identical to those described above for *A. nidulans* [7].

With the strategy described above, *A. niger* transformants were obtained carrying at least 200 copies of the *glaA* gene at a relatively high frequency. A clear relationship was found between the growth characteristics of the transformants (*amdS* copynumber) and the number of copies of the *glaA* gene.

A significant increase in extracellular glucoamylase (GLA) production was observed in most of the transformants carrying multiple copies of the *glaA* gene, based on analysis by iso-electric focusing of samples of medium and enzyme activity measurements. No simple relationship was found between the number of copies of the *glaA* gene and the production of glucoamylase. However, the level of GLA production clearly correlates with the amount of *glaA* mRNA produced in the transformants. In several transformants with different amounts of *glaA* copies, a maximal mRNA and GLA level was measured. These results indicated that GLA production is limited at the level of transcription. Other factors, such as the site of integration, apparently influence expression levels strongly as has been shown before [13]. Under the conditions used (shake flask cultures) some transformants produce at least fifteen times more glucoamylase (0.9 g/l) than the parent strain.

2.3 Production of 1,4-β-endoxylanase and phytase

Several other interesting extracelluar enzymes are secreted by *A. niger* and other filamentous fungi, often although in low amounts and/or in a mixture with other (unwanted) enzymes. An example of these kind of enzymes are xylanases. These enzymes are able to hydrolyse D-xylans consisting of 4-β-linked polymers of D-xylose as a backbone. Commercial, fungal-enzyme preparations with these activity are applied in the bakery.

The *exlA* gene encoding for this enzyme was isolated and characterized. Transformation of *A.niger* with multiple copies of the *exlA* gene, by using the *amdS* selection procedure described above, resulted in twenty- to thirty fold overproduction of 1,4-β-endoxylanase A. The same results were obtained with the *exlA* gene under control of the efficient *glaA* or *gpdA* promoters [14, Hessing *et al*, manuscript in prep.].

Another example is phytase, an enzyme that catalyses the conversion of the plant phosphate storage molecule phytic acid, to inositol and (six) phosphates. The use of phytase in the feedstuffs of monogastric animals decreases environmental problems concerning phosphorus in the manure, both by the removal of phytic acid and the absence of the need for phosphate addition to these feedstuffs.

Using similar strategies as described above *A. niger* strains have been constructed which efficiently produce phytase [15,16]. Also efficient production was achieved in this way of *A. niger* α-galactosidase A [17] as well as a number of other fungal proteins by *A. niger* and *A. oryzae* [for a review see 4].

3. CONCLUSIONS

The results obtained with the studies on the production of extracelluar fungal proteins show that, in general, these proteins are efficiently secreted by *Aspergillus* species. Indeed, during the last few years the first products made by recombinant fungal strains have been brought on the market (e.g. Lipolase, a *Humicola* lipase produced by *A. oryzae* and Natuphos, an *Aspergillus* phytase produced by *A. niger*).

Since most of the (published) results are obtained in shake flasks, additional investigations are necessary to evaluate the production of these proteins under different fermentation conditions. It is expected that alterations of the growth characteristics and/or the expression/secretion machinery of the *Aspergillus* strains, used in the studies described, are necessary to obtain the levels of production which are obtained with a number of enzymes made by classical *Aspergillus* production strains (e.g. >20 g/l).

Acknowledgements

We wish to thank our colleagues Wim van Hartingsveldt, Jan Verdoes, Hanny Hessing, Anneke Kuijvenhoven, Marian Harteveld, and Cora van Zeijl for their contribution to work presented in this paper. Part of the work described was supported by the Dutch Programme Commission Biotechnology or carried out as contract research for Gist-brocades NV, Delft, The Netherlands and Unilever NV., Vlaardingen, the Netherlands.

REFERENCES

1. J.R. Fincham, Microbiological Reviews, 53 (1989) 148.
2. W.E. Timberlake and M.A. Marshall, Science, 244 (1989) 1313.
3. C.A.M.J.J. van den Hondel and P.J. Punt, BSM Symposium Series Vol. 18, British Mycological Society, Cambridge University Press, Cambridge, UK, (1991) 1.
4. C.A.M.J.J. van den Hondel, P.J. Punt and R.F.M. van Gorcom, In: More gene manipulation in fungi, eds: J.W.Bennett & L.L.Lasure. Academic press,(1991) 396.
5. R.F.M. van Gorcom, J.G. Boschloo, A. Kuyvenhoven, J. Lange, A.J van Vark, C.J. Bos,J.A.M. van Balken, P.H. Pouwels and C.A.M.J.J. van den Hondel, Mol. Gen. Genet., 223 (1990) 192.
6. P.P.F. Hanegraaf, P.J. Punt, C.A.M.J.J. van den Hondel, J. Dekker, W. Yap, H.W. van Verseveld and A.H. Stouthamer, Appl. Microbiol. Biotech., 34 (1991) 765.
7. J.C. Verdoes, P.J. Punt, J.M. Schrickx, H.W. van Verseveld, A.H.Stouthamer and C.A.M.J.J.van den Hondel, Transgenic Research, 2 (1993) 84.
8. E. Boel, M.T. Hansen, I. Hjort, I. Hoegh and N.P. Fiil, EMBO. J., 3 (1984) 1581.
9. T. Fowler, R.M. Berka and M. Ward, Curr. Genet., 18 (1990) 537.
10. J.M. Kelly and M.J. Hynes, Curr. Genet., 12 (1987) 21.
11. C.J. Bos, PhD Thesis, Agricultural University Wageningen NL, 1986.
12. M.J. Hynes, C.M. Corrick and J.A. King, Mol. Cell. Biol., 3 (1989) 1430.
13. R.F.M. van Gorcom, P.H. Pouwels, T. Goosen, J. Visser, H.W.J. van den Broek, J.E. Hamer, W.E. Timberlake and C.A.M.J.J. vanden Hondel, Gene, 40 (1985) 99.
14. Van Gorcom, R.F.M., J.G.M. Hessing, J. Maat, M. Roza and J.M.A. Verbakel (1991) Xylanase production. PCT International Patent application, WO/91/19782.
15. R.F.M. van Gorcom, W. van Hartingsveldt, R.G.M. Luiten, P.A. van Paridon, G.C.M. Selten and A.E. Veenstra, Europ. pat. applic. EP 0420358 (1991).
16. W. van Hartingsveldt, C.M.J. van Zeijl, G.M. Harteveld, R.J. Gouka, M.E.G. Suykerbuyk, R.G.M. Luiten, P.A. van Paridon, G.C.M. Selten, A.E. Veenstra, R.F.M. van Gorcom and C.A.M.J.J. van den Hondel, Gene, (1993) in press.
17. I.F. den Herder, A.M. Mateo Rosell, C.M. van Zuilen, P.J. Punt en C.A.M.J.J. van den Hondel, Mol. Gen. Genet., 233 (1992) 404.

ECB6: Proceedings of the 6th European Congress on Biotechnology
edited by L. Alberghina, L. Frontali, P. Sensi

Chitin deacetylation by enzymatic means

D. Kafetzopoulos, A. Martinou, I. Tsigos, A. Christodoulidou, P. Kavelaki and V. Bouriotis

Enzyme Technology Division, Institute of Molecular Biology and Biotechnology, P.O. Box 1515, GR 711 10 Heraclion, Crete, Greece &
Department of Biology, Division of Applied Biology and Biotechnology, University of Crete, P.O. Box 1470, GR 711 10 Heraklion, Crete, Greece

1. SUMMARY

Chitosan, the deacetylated form of chitin, is a biopolymer with unique properties favourable for a broad variety of industrial and biomedical applications. Chitosan is presently produced by the thermochemical deacetylation of chitin. However, deacetylation of chitin can also be effected by the enzyme chitin deacetylase which occurs in several *Zygomycetes* species. Chitin deacetylase from its best defined source, the dimorphic fungus *Mucor rouxii*, has been purified to homogeneity and further characterised. The enzyme is an acidic glycoprotein of ~75000 Da. Chitin deacetylase is active on several chitinous substrates, it requires at least four N-acetylglucosamine residues for catalysis and it may lead to highly deacetylated products. It exhibits a pH optimum of 4.5 and a temperature optimum of 50°C. Furthermore, a more powerful, alternative purification procedure using immunoaffinity chromatography has also been developed for the isolation of the enzyme in one step procedure. Finally, a cDNA clone encoding *M. rouxii* chitin deacetylase has been isolated.

2. INTRODUCTION

Chitin is a homopolymer of β (1→4) N-acetylglucosamine, commonly found in the exoskeletons of many invertebrates and in the cell walls of most fungi. In the cell wall of *Zygomycetes* chitin exists in its deacetylated form, referred to as chitosan. Biosynthesis of chitosan proceeds by the enzymatic deacetylation of chitin. Chitin deacetylase, the enzyme which converts chitin to chitosan by the hydrolysis of the N-acetamido groups in the chitin chains, was first identified in the mycelial extracts of the fungus Mucor rouxii (1).

Chitin is an extremely insoluble and chemically inert material and although abundant and renewable, has yet to find an important industrial use. In contrast, chitosan is a water soluble, easily derivatisable and biologically benign material with a large number and a broad variety of potentially important applications (2-4).

Chitosan is presently produced from chitin of crab and shrimp shells by thermochemical deacetylation. This process generates environmentally significant chemical waste, is subject to seasonal variation of starting material and results in a polymer product having a broad distribution of molecular weights and a heterogeneous extent of deacetylation. However, for many important applications uniform material with specific physical and chemical properties is required.

Procedures employing either the biosynthesis of chitosan or the enzymatic deacetylation of chitinous substrates present attractive alternatives, especially when a controlled, non-degradative and well defined process is required. In an attempt to evaluate the potential use of these processes for chitosan production we have focused on the characterisation of the fungal chitin deacetylases.

3. RESULTS AND DISCUSSION

3.1. Isolation and characterisation of chitin deacetylase from M. rouxii

We employed two different assays for the determination of chitin deacetylase activity. A radiometric assay using radiolabelled glycol chitin (1) as substrate proved to be a rapid and sensitive way of screening chromatography fractions whereas a spectrophotometric method for the estimation of the released acetic acid (5) was used for monitoring the deacetylation of non radiolabelled substrates.

Mucor rouxii, the better characterised source of the enzyme, was selected for the purification of chitin deacetylase. In order to minimise protein degrading activities the fungus was grown on minimal media. The enzyme was purified to homogeneity from mycelial extracts using a combination of hydrophobic, anion exchange and cation exchange chromatography (6,7). As it is evident from the purification table (Table 1) the step which substantially increased enzyme purity is that of cation exchange chromatography exploiting the exceptionally low pI of the enzyme.

Chitin deacetylase exhibits a pI of ~3, as estimated by isoelectric focusing analysis. Its apparent molecular weight determined by SDS-PAGE analysis is

Table 1. Purification of chitin deacetylase from *Mucor rouxii*

Step	Total protein (mg)	Total enzyme activity (units)*	Specific activity (units/mg)	Yield (%)	Purification (-fold)
Crude extract	10380	317	0.0305	100.0	1
50 °C treatment	4719	221	0.048	69.7	1.6
Phenyl Sepharose	1374	138	0.100	43.5	3.3
Q Sepharose	279	73.6	0.263	23.2	8.6
S Sepharose	12.6	37.4	2.968	11.9	97.3

* One unit of enzyme activity is defined as the amount of the enzyme required to produce 1 μmole of acetic acid per min when incubated with 16.6 nmoles of hexa-N-acetyl chitohexaose under optimum pH (4.5) and temperature (50°C) conditions.

~75 kDa. The native enzyme exists as a monomer with an apparent molecular weight ~80 kDa, as determined by size exclusion chromatography.

Several pieces of evidence indicated that chitin deacetylase is a glycoprotein. The enzyme was retained on a concanavalin A-Sepharose column and then eluted by α-methyl mannoside. After electrophoresis, the enzyme band gave a positive staining with periodate-Schiff's reagent. Direct carbohydrate and amino acid analysis of the enzyme revealed that the carbohydrate part is ~30% by weight of the protein and that it consists of mannose, N-acetylglucosamine and fucose. Glycosylation may be important for the enzyme stability and for the coordination of its function in the fungal periplasmic space.

3.2. Purification of chitin deacetylase by immunoaffinity chromatography

Enzyme purification protocols employing conventional chromatography, are generally lengthy and result in low yield and specific activity for the final preparation. In order to obtain homogeneous chitin deacetylase preparations using a one step purification scheme, at a laboratory scale, we employed an immunoaffinity adsorbent (8). Chitin deacetylase-specific IgG was obtained from polyclonal sera by affinity chromatography. The immunoadsorbent was subsequently prepared by immobilising the specific antibodies on CNBr activated Sepharose. The harsh chemical conditions often required to desorb immunoadsorbents were avoided by selecting antibodies of moderate binding. Chitin deacetylase obtained by immunoaffinity chromatography was enriched ~170 times with a 30% yield and exhibited a specific activity of 13 units/mg, values much higher than the ones obtained using conventional methodology.

The same immunoadsorbent was employed for the isolation of chitin deacetylases from other *Zygomycetes* species, since an interspecies immunological similarity was observed by Western blot analysis. In this way, a comparative study of chitin deacetylases from various sources could be performed.

3.3. Characterisation of the enzyme activity

Pure enzyme preparations obtained by either the immunoaffinity procedure or the conventional chromatography method were used for the characterisation of the enzyme activity. Chitin deacetylase appears to exhibit a very stringent specificity acting only on N-acetyl glucosamine homopolymers. The enzyme is active on glycol chitin and carboxymethyl chitin (water soluble derivatives), microcrystalline chitin (colloidal chitin), chitosan and chitooligoses with more than three N-acetylglucosamine residues. The pH optimum of the enzyme is ~4.5 and the temperature optimum ~50° C, using glycol chitin as substrate. The enzyme does not require any metal ions for activity and it is not inhibited by EDTA or EGTA whereas it is inhibited by carbonylic acids, particularly acetic acid. The enzyme can be effective in converting partially deacetylated chitosans to highly deacetylated products.

3.4. Isolation of a M. rouxii chitin deacetylase cDNA clone

Based on partial aminoterminal protein sequencing data degenerate oligonucleotide probes had been designed and used for the screening of a oligo(dT) primed *M.rouxii* cDNA library. Among the positive clones the one with the longest insert was subsequently used for the determination of the DNA primary structure. The isolated cDNA clone encodes for a polypeptide of

~47 kDa with nine predicted N-glycosylation sites on its deduced sequence. Protein sequence comparisons using the deduced amino acid sequence of chitin deacetylase revealed striking sequence similarities to the rhizobial *nodB* gene product (9). These similarities highlight the part of the sequence which could be used for the isolation of chitin deacetylase genes from other species.

4. CONCLUDING REMARKS

We have described the purification, characterisation and the cDNA cloning of chitin deacetylase from the fungus *M. rouxii*. In order to effectively use chitin deacetylase in bioconversion processes further study should be carried out.

The major problem of the enzymatic deacetylation of chitin appears to be its crystalline structure. In order to improve the substrate susceptibility to the enzyme action the appropriate pretreatment procedure should be defined.

Chitin deacetylase activities from other species have been reported to exhibit different pH optimum and inhibition profiles (10). Their use in enzymatic deacetylation processes should be further evaluated.

Recombinant DNA methodology may be employed for the overproduction of the enzyme and a suitable heterologous expression system should be adopted.

Finally, chitosan production by direct fermentation of fungi could circumvent the need the complex processes outlined above. The employment of either a natural or a genetically manipulated source of chitosan need to be investigated.

REFERENCES

1. Y. Araki and E. Ito, Eur. J. Biochem. 55 (1975) 71-78.
2. C. K. Rha, D. Rodriguez-Sanchez and C. Kienzle-Sterzer, in Biotechnology of Marine Polysaccharides (R. R. Colwell, E. R. Pariser, and A. J. Sinskey, eds.) 283-311, Hemisphere Publishing Corporation, Washington/New York /London 1984.
3. D. Knorr, in Biotechnology of Marine Polysaccharides (R. R. Colwell, E. R. Pariser, and A. J. Sinskey, eds.) 283-311, Hemisphere Publishing Corporation,Washington/New York/London 1984.
4. S. Hirano, in Chitin and Chitosan (G. Skjak-Braek, T. Anthonsen, and P. Sandford, eds.) 37-43, Elsevier, Essex 1989.
5. H. U. Bergmeyer, Methods of Enzymatic Analysis 1, (1974) 112-117.
6. D. Kafetzopoulos, A. Martinou and V. Bouriotis, Proc. natn. Acad. Sci. U.S.A (in press) 1993.
7. V. Bouriotis, D. Kafetzopoulos, A. Martinou and J.Vournakis, U.S. Patent Serial No 07/773,724 (1991).
8. A.Martinou, D.Kafetzopoulos and V.Bouriotis, J.Chromatogr. (in press)1993
9. D. Kafetzopoulos, G. Thireos, J. N. Vournakis and V. Bouriotis, Proc. natn. Acad. Sci. U.S.A (in press) 1993.
10. H. Kauss, W. Jeblick. and D.H.Young, Plant Sci. Lett. 28 (1983) 231-236.

ECB6: Proceedings of the 6th European Congress on Biotechnology
edited by L. Alberghina, L. Frontali, P. Sensi

Recombinant *Trichoderma reesei* strains producing improved and "tailor made" xylanase profiles

Regina Hodits[1], Arno Butterweck[1], Sabine P. Goller[1], Anneli Törönnen[2], Robert L. Mach[1], Robert Messner[1,3], Anu Harkki [2] and Christian P. Kubicek[1]

[1] Abteilung für Mikrobielle Biochemie, Institut für Biochemische Technologie und Mikrobiologie, TU Wien, Getreidemarkt 9, A- 1060 Wien, Austria
[2] Cultor Research Ctr., Kantvik, SF-2460, Finland
[3] present adress: Institut für Angewandte Mikrobiologie, Universität für Bodenkultur, Nußdorferlände , A-1190 Wien, Austria

1. INTRODUCTION

ß-1,4-xylans are heterogeneous polysaccharides found in almost all parts of the cell walls of all land plants [1]. The ß-1,4-linked xylose monomers thereby form a characteristic backbone chain, to which several substituents are attached. Many pro- and eukaryotic microorganisms can degrade and use xylans as a carbon source. Interest in the enzymology of xylan hydrolysis has recently increased considerably because of the use of xylanases in bioconversion and biopulping [2,3]. Strains and species of the fungus *Trichoderma* have been proven to secrete high amounts of a very efficient xylanase enzyme system [4], although the relative proportions of the individual enzymes may not be ideal for all purposes [5]. Moreover, *T. reesei* is a powerful producer of cellulases, which however would harm the cellulose fiber. The genes encoding the two major xylanases [6] and the ß-xylosidase [7] from *T. reesei* have recently been cloned, thereby offering the possibility to prepare "genetically engineered xylanolytic strains. In the present paper we will describe some approaches how these genes can be used to construct improved xylanase producer strains.

2. MATERIALS AND METHODS

Organism and Conditions for Cultivation. Trichoderma reesei QM 9414 was used throughout these studies, and kept on malt agar slants. Submerged cultivation of the organism was carried out in 1-l wide mouthed Erlenmeyer flasks containing 300 ml of Mandels-Andreotti medium [8], with 1 % (w/v) carbon sources as indicated, which were incubated on a rotary shaker at 200 rpm and 28°C for the time indicated at the respective results.

Assay of enzyme activities. Xylanase activity was assayed by measuring the liberation of reducing sugars from beech wood xylan as a substrate essentially as described previously [6]. One unit (1 U) is defined as the release of 1 μmole of reducing sugars per min under these conditions.

DNA manipulation and strain transformation. All manipulations were carried out in a similar way as described previously [9,10]. A full description of the construction of the plasmids used will appear elsewhere [11].

3. RESULTS AND DISCUSSION

T. reesei secretes two low-M_r xylanases into the medium, which differ considerably in their properties (pH-optimum and mode of attack on xylan [5,12]). We have used three approaches to modify and improve the xylanolytic enzyme profile of *T. reesei*: in the first case, we amplified the activities of XYN I and XYN II by the introduction of multiple gene copies of the *xyn1* and the *xyn2* genes, respectively. In the second approach, we made advantage of the availability of *pki1* [13], encoding pyruvate kinase, which is an enzyme of glycolysis and therefore transcribed in high amounts during growth on glucose. The rationale behind this approach was that the production of xylanases on glucose would eliminate the byproduction of (glucose repressible) cellulases, which is undesireable if the xylanases are to be used for biopulping. To this end we fused the prepropeptide encoding structural genes of *xyn1* and *xyn2* between the 5' and 3' expression signals of the *pki1* and *cbh2* (cellobiohydrolase II-encoding) gene, respectively, and used these vectors (pAT4 and pAT5, respectively) for cotransformation as described above.

In a third approach, we made use of the availability of the *T. reesei bxl1* gene [7] encoding ß-xylosidase. Since this enzyme is present in culture filtrates in lower activity than the two xylanases, it was hypothesised that the introduction of multiple copies (in analogy to the first approach) might lead to xylanase preparations with higher activities.

Introduction of all genes was done by cotransformation, using plasmid pFG1 and the uridine-auxotrophic strain *T. reesei* TU-6 as selection marker and host [9], respectively. All transformants were plated on selective and non-selective medium at least three times to isolate only mitotically stable recombinants. Secretion of xylanases was tested by xylan overlay/Congo red staining, whereas a polyclonal antibody against ß-xylosidase was used to select for overproducers of this enzyme.

The results from the first two approaches are given in Table 1 and 2: Southern analysis (data not shown) proved that the transformation with the two xylanase genes resulted in the introduction of 10 - 20 copies of *xyn1* or *xyn2* into *T. reesei* DNA, which almost exclusively integrated at non-homologous loci. The same picture was obtained for the *pki:xyn* fusions. Growth of the respective transformants on xylan-containing media showed that the some transformants secereted up to 6-fold higher activities of xylanases into the medium; based on a dry weight basis, however, the effect was less dramatic, since the strains also grew much faster. This effect was more pronounced with *xyn2*. Secretion of xylanases under the glycolytic promoter was less efficient, although clearly demonstrable. Interestingly, higher values were quoted if the xylanases were assayed immunologically than when their activities was assayed (data not shown). The comparably poor yields of both xylanases under glycolytic expression signals is surprising, but not without precedent: usinga similar strategy, Vanhanen [14] reported on poor secretion of *T. reesei* EG I from a *pgk1:egl1* fusion. A repression of one of the steps in the fungus secretory pathway may be an explanation for this phenomenon, yet requires more studies to be proven.

The results from the third approach are documented in Fig. 1: it is evident that amplification of the *bxl1* gene has a dramatical effect on the total ß-xylosidase activity, and also on the xylanase activity formed. We have used a monoclonal antibody against XYN II [15] to assess that the increase in xylanase activity was not just an effect of ß-xylosidase on the activity determination, but actually enhanced XYN II protein was secreted. By the use of this simple approach, the xylanase activity in the culture fluid was more than doubled.

Table 1
Xylanase formation by *T. reesei* recombinant strains bearing multiple copies of *xyn1* or *xyn2*

Strain	Gene amplified	Xylanase activity [U/g dry weight]	
		pH 5.0	pH 3.0
TU-6	- (control)	3.1	2.8
bx4	*xyn2*	7.3	3.9
sx2	*xyn1*	3.4	4.2

Xylanase formation was quantified after 70 ours growth on beech wood xylan as a carbon source. The values for the two best transformants are quoted.

Table 2
Xylanase formation by *T. reesei* recombinant strains carrying *pki1:xyn1* or *pki1:xyn2* fusions on glucose

Strain	Fusion	Xylanase activity [U/g dry weight]	
		pH 5	pH 3
TU-6	- (control)	3.1	2.8
ATX1-2	*pki1:xyn1*	nd	3.7
ATX1-3	-"-	nd	4.4
ATX1-4	-"-	nd	4.8
ATX1-5	-"-	nd	5.1
ATX2-12	*pki1:xyn2*	5.0	nd
ATX2-13	-"-	3.0	nd
ATX2-16	-"-	3.8	nd

The activities were measured after 20 h of growth

4. ACKNOWLEDGEMENTS

AT gratefully acknowledges the receipt of a research grant from the Finnish Academy of Sciences. This study was in part supported by Biopulp Int., France.

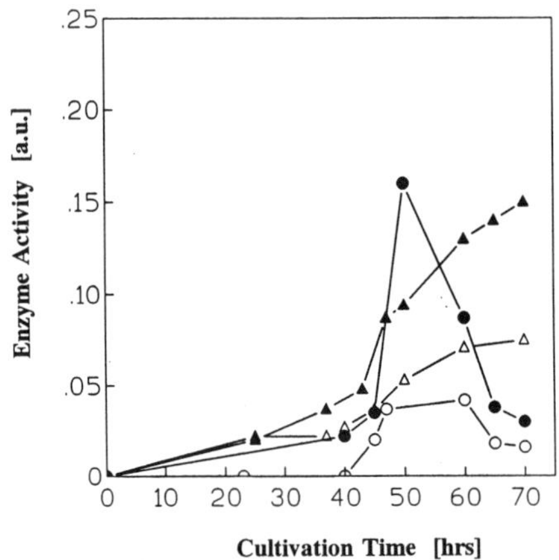

Fig. 1: Formation of xylanase (triangles) and ß-xylosidase (circles) activities during growth of *T. reesei* QM 9414 (empty symbols) and the *bxl1* multicopy strain SG-6 (full symbols) on beech wood xylan. Activities are given in arbitrary units (a.u.).

REFERENCES

1. R.F.H. Dekker. ACS Symp. Ser. 399 (1989) 619 - 631
2. L. Viikari, M. Ranua, A. Kantelinen, M. Linko, and J, Sundquist. Biotechnology in the Pulp and Paper Industry, Stockholm (1987) 67 - 69
3. A. Kantelinen, M. Rättö, J. Sundquist, M. Ranua, L. Viikari, and M. Linko. Int. Pulp Bleaching Conf., Tappi Proceedings (1988) 1 - 9
4. K.K.Y. Wong and J.N. Saddler. CRC Crit. Rev. Biotechnol. 12 (1992) 413 - 435
5. J. Buchert, M. Ranua,A. Kantelinen and L. Viikari. Appl. Microbiol. Biotechnol. 37 (1992) 825 - 829
6. A. Törrönen, R.L. Mach, R. Messner, R. Gonzalez, N. Kalkkinen, A. Harkki, and C.P. Kubicek. Bio/Technol. 10 (1992) 1461 - 1467
7. S.P. Goller, R. Hodits, and C.P. Kubicek (1993) ms in preparation
8. M. Mandels, and R.E. Andreotti. Proc. Biochem. 13 (1978) 6 - 13
9. F. Gruber, J. Visser, C.P. Kubicek, and L. De Graaff. Curr. Genet. 18 (1990) 447 - 451
10. E.M. Kubicek-Pranz, F. Gruber, and C.P. Kubicek. J. Biotechnol. 20 (1991) 83 - 94
11. A. Butterweck, A. Törrönen, R.L. Mach, R. Messner, A. Harkki, and C.P. Kubicek. (1993), ms in preparation
12. M. Tenkanen, J. Puls, and K. Poutanen. Enzyme Micr. Technol. 14 (1992) 566 - 574
13. M. Schindler, R.L. Mach, S.P. Vollenhofer, R. Hodits, F. Gruber, J. Visser, L. De Graaff, and C.P. Kubicek. (1993) Gene, in press
14. S. Vanhanen. VTT Publications 75 (1991) Valtion Teknillinen Tutkimuskeskus. Espoo, Finland.
15. A. Törrönen, K.M. Affenzeller, F. Hofer, T.A. Myohanen, D. Blaas, A. Harkki, and C.P. Kubicek. In: Xylans and Xylanases (Progress in Biotechnology Vol. 7) (1992) 501 - 505. Elsevier Press Amsterdam - New York - London

ECB6: Proceedings of the 6th European Congress on Biotechnology
edited by L. Alberghina, L. Frontali, P. Sensi

THE SIMULTANEOUS PRODUCTION OF AMYLASE AND LIPASE BY THE THERMOPHILIC FUNGUS, *THERMOMYCES LANUGINOSUS*.

A. S. Frich, B. Jensen and J. Olsen.

Department of General Microbiology, University of Copenhagen,
Sølvgade 83H, DK-1307, Copenhagen K, Denmark.

1. INTRODUCTION

The ability of the thermophilic fungus, *Thermomyces lanuginosus* to produce lipase (Liu et al., 1972) and amylase (Jensen et al., 1987) has formerly been shown by using olive oil or dextran respectively to promote the production of the two types of enzymes. The aim of the present investigation was to study the efficacy of the two carbon sources, olive oil and dextran, to induce both types of enzymes when the growth medium was formulated as well with each individual carbon source as with combinations of both.

2. MATERIALS AND METHODS

2.1 Shake-flask fermentation

Organism (*Thermomyces lanuginosus*), preparation of media and of inoculum for shake flasks have been described previously (Jensen et al., 1988).

The medium for shake flasks was modified from Haasum et al. (1991) and comprised (in g/l): Yeast Extract (Difco), 16; $K_2HPO_4 \cdot 3H_2O$, 2.0; $MgSO_4 \cdot 7H_2O$, 1.0; 0.1 ml trace element solution. The main carbon sources were either olive oil (Sigma o-1500), Dextran 4 (Serva 18688) or combinations of the two C-sources. The shake flask fermentations were performed in 300 ml erlenmeyer flasks with 50 ml medium at 50 °C and 300 rpm. Under these conditions the growth form was filamentous and the elevated concentration of yeast extract prevented the formation of conidia.

2.2 Enzyme activity

Lipase activity was assayed by titrimetry as previously described (Jacobsen et al., 1989) except for the assay temperature which was 60 °C. One unit (U) of lipase is the amount of enzyme that liberates 1 µmol equivalent of free fatty acid per min.

Amylase activity was assayed by the hydrolysis of soluble starch and measuring the production of reducing sugars (Jensen and Olsen, 1992). One unit (U) of amylase is the amount of enzyme that liberates 1µmol equivalent of glucose per min.

3. RESULTS AND DISCUSSION

3.1. Growth and enzyme production

Growth (measured as dry weight), and the concentrations of extracellular amylase and lipase activity were followed during 5 days periods in shake flasks with different compositions of the main carbon sources (Figure 1A-1D). The carbon sources were: 30 g/l dextran (Fig. 1A), olive oil 15 g/l (Fig. 1B), and in combination media 15g/l dextran + 8 g/l olive oil (Fig. 1C) and 7 g/l dextran + 12 g/l olive oil (Fig. 1D).

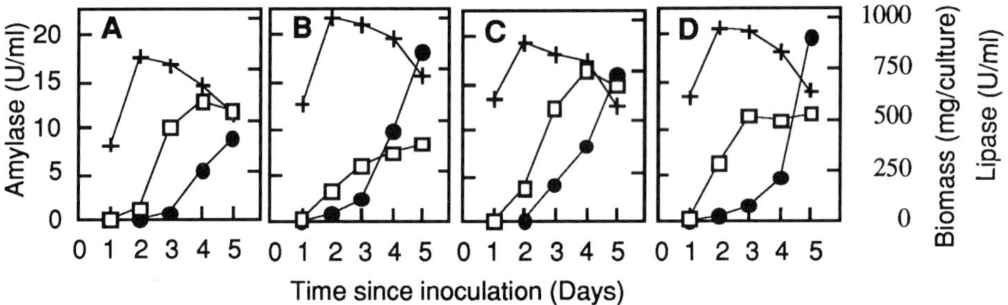

Figure 1. Time course of growth (dry weight) (+) in shake flask culture at 50 °C and 300 rpm of *Thermomyces lanuginosus* and the production of lipase (•) and amylase (□). The main carbon sources (in g/l of medium) were: (A) dextran, 30; (B) olive oil, 15; (C) dextran + olive oil, 15+8; (D) dextran + olive oil, 7+12

The biomass was maximal after 2 days of growth and as well amylase as lipase were produced on both types of carbon sources and on the combinations. The amylase activity appeared at least 24 hours before the lipase activity.

Due to the differences in biomass produced, the experiment was repeated with concentrations of olive and/or dextran in the media chosen so that the amounts of biomass obtained in the media were not significantly different. The yields of biomass, lipase activity, and amylase activity from this experiment are shown in Figure 2.

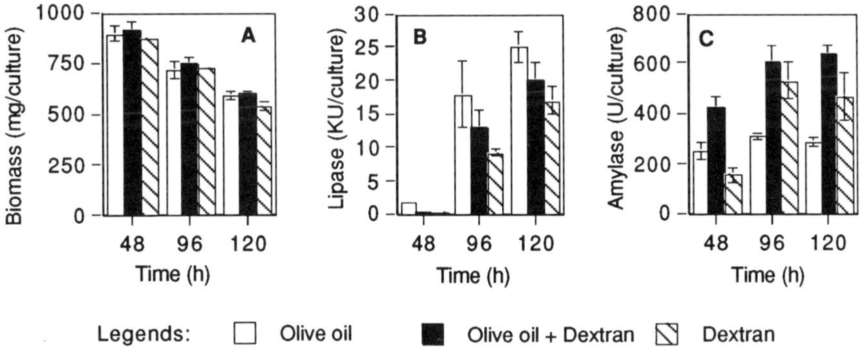

Figure 2. The yields per culture of biomass as dry weight (A), extracellular lipase (1KU = 10^3 U) (B), and extracellular amylase (C) at 48, 96, and 120 h after inoculation. The main carbon sources in the media were: 12.4 g/l olive oil ; 6.2 g/l olive oil + 15 g/l dextran ; 30 g/l dextran. Data from three replicate cultures have been averaged. Bars represent standard deviation.

The highest yields of lipase were obtained in the media containing olive oil (Figure 2B).The difference between the yields at 120 h in the media with only olive oil and dextran was significant at the 95% confidence level.

The highest yields of amylase were obtained in the media containing dextran (Figure 2C). The yield at 96 h in the medium with only olive oil was significantly different from the other media at the 95% confidence level.

That the simultaneous production of lipase and amylase in the media was not simply due to induction by some component of the yeast extract was shown in a control experiment with yeast extract being the sole carbon source for growth. In this experiment the yields of lipase and amylase on 16 g/l and 46 g/l of yeast extract were less than 10% of the lowest yields obtained with dextran or olive oil present.

3.2. Identification of enzymes

The excreted proteins in the culture filtrate from 120 h old cultures from the same growth media as used for the experiments in Figure 2 were crudely fractionated on a DEAE-Trisacryl column (Jensen et al., 1988, with modifications). The elution profiles obtained from the three different growth media were comparable, and lipase, glucoamylase, α-glucosidase, and α-amylase were identified by fused rocket immunoelectrophoresis using antibodies made against the enzymes from *T. lanuginosus* (Table 1). Furthermore, the four peak fractions of each type of enzyme from the three media were analysed against each other by fused rocket immunoelectrophoresis. The proteins of each type of enzyme from the three media fused, indicating the presence of the same epitopes.

Table 1
Identification of enzymes in fractions from DEAE-Trisacryl column

Medium (C-source)	Fractions with antigens against			
	Lipase	Glucoamylase	α-Glucosidase	α -Amylase
Olive oil	36-51	69-87	120-126	162-180
Combination	42-48	72-87	120-129	160-181
Dextran	42-54	72-90	120-144	165-183

3.3 Conclusions

The simultaneous production of lipase and amylase by *Thermomyces lanuginosus* was seen irrespective of the carbon source being olive oil, dextran or combinations of these. Although *T. lanuginosus* may produce more of a enzyme when it is cultivated in the presence of a inducing substrate for the specific enzyme, the results from growth on the combined carbon sources indicate that the production of one type of enzyme did not result in a decreased production of the other type.

REFERENCES

I.Haasum, S.H.Eriksen, B.Jensen and J.Olsen, Appl. Microbiol. Biotechnol., 34 (1991) 656-660.

T.Jacobsen, J.Olsen and K.Allermann, Enzyme Microb. Technol., 11 (1989) 90-95

B.Jensen, J.Olsen and K.Allermann, Biotechnol. Lett.,9(5) (1987) 313-316.

B.Jensen, J.Olsen and K.Allermann, (1988). Can. J. Microbiol., 34 (1988) 218-223.

B.Jensen and J.Olsen, Enzyme Microb.Technol., 14 (1992) 112-116.

W-H.Liu, T.Beppu and K.Arima, Agric. Biol. Chem. 36(11) (1972) 1919-1924.

Yeast Physiology and Genetics

ECB6: Proceedings of the 6th European Congress on Biotechnology
edited by L. Alberghina, L. Frontali, P. Sensi

A new look at carbohydrate metabolism in yeast.

Carlos Gancedo and Juana M. Gancedo.

Instituto de Investigaciones Biomédicas del C.S.I.C.
Arturo Duperier 4 . E-28029 Madrid.Spain.

The study of carbohydrate metabolism in yeast has an old story. Alcoholic fermentation has fascinated mankind both by its effects and by the mystery of its mechanism. To our knowledge the first report on alcoholic fermentation in cell free extracts was published in 1871 by a Russian lady from St Petersburg, Maria Manasseina, who observed a very active alcoholic fermentation using air-dried yeast triturated with rock crystals by a strong man in a glass mortar for 6 to 15 hours. However some experimental flaws prevented her results to be generally accepted (Fruton, 1972; Lagnado,1992). But the work that opened the field for modern biochemistry and that we may consider the first article on glycolysis *in vitro* was the revolutionary paper by Buchner in 1897.

Since that time the glycolytic pathway has been the subject of a very active study with different historical periods putting distinct emphasis on different questions. What may be called a chemical phase started soon after Buchner's report: it was a phase of identification of intermediates. It was followed by a more biochemical phase devoted to the characterization of the enzymes implicated in the pathway: from Buchner's *zymase* we have arrived to a dozen enzymes without counting the different isoenzymes and regulatory enzymes implicated. This was what may be called the classical period of glycolysis and it was practically finished in the late fifties, although the advent of molecular biology has brought new participants to the game. Then followed a phase of study of the regulation of the implicated enzymes and we are now in the phase of isolation of the genes encoding those enzymes and study of the regulation of their expression. The general impression from all this is that glycolysis, the backbone of carbohydrate metabolism in yeast is well understood and that only details remain to be filled in.

Our purpose is to show that although much is known about glycolysis recent results indicate that there are still important aspects in which our knowledge is incomplete. We will mainly consider *Saccharomyces cerevisiae* to which we will refer simply as yeast. The exposition will be centered around three main areas of glycolysis and its counterpart gluconeogenesis, namely the problem of the signal(s) that indicate to the cell the availability of glucose, the adjustement between the energy producing and energy consuming glycolytic steps and the physiological importance of the control of possible futile cycles during the transition gluconeogenesis-glycolysis.

Glucose is the sugar preferentially utilized by *S. cerevisiae* followed by fructose and mannose. It is a well established fact that glucose, and also the other easily fermentable sugars cause several effects on yeast cells. They produce an activation of the synthesis or activity of several enzymes, a repression of the expression of several genes and an inactivation of some enzymes. Gancedo and Serrano (1989) proposed the term "general catabolite control" to cover all these effects. It is obvious that the cell senses the presence of the sugars in the medium and uses this information to steer the series of phenomena described above. The question that remains unanswered is how this is achieved, i.e., what is the nature of the sensing and transducing machinery. The cell could respond to the presence of glucose itself or to changes in the amount of some metabolite which varies in the presence of glucose. One obvious candidate for the role of receptor is a protein that transports glucose. In *S. cerevisiae* glucose uptake is a facilitated diffusion process mediated by at least two systems with different affinities for glucose: a low afinity system (Km about 20 mM) and a high affinity system (Km about 2 mM). It is interesting to note that both sugar transporters of *S. cerevisiae* present a relatively high Km for their substrates (above 1 mM) , while sugar transports in bacteria show high affinities for their substrates , in the micromolar range. The low affinity system is constitutive while the high affinity system is repressed by glucose. The same picture is found in other yeast species. Genes involved in the high affinity transport like *SNF3, HXT1, HXT2, HXT3* and *HXT4* have been isolated (Celenza *et al.*, 1988 , Kruckeberger & Bisson 1990, Lewis & Bisson 1991, Ko *et al.*, 1993) while genes encoding a low affinity transport have been not yet characterized, although a *RAG1* gene homologous to that of *Kluyveromyces lactis* (see below) has been recently cloned (Prior *et al* ., 1992). Transport appears to be coupled in some way with the kinases which phosphorylate glucose, as shown by the work of Bisson and Fraenkel (1983). They found that in mutants lacking hexokinases I and II and possessing only glucokinase (unable to phosphorylate fructose) the high affinity transport for glucose was present but that for fructose was not detectable. In mutants lacking all three phosphorylating activities the high affinity systems for both fructose and glucose were not detectable. It is not clear how the kinases are mechanistically involved in the transport. Phosphorylation of the sugar itself is not required, since a sugar which cannot be phosphorylated, such as 6-deoxyglucose, is transported with different affinity depending on the presence or absence of hexokinases. Leucine zipper regions have been found in the three transporters mentioned above and it has been suggested that these regions might interact with the kinases and modulate the transport activity. This point needs to be explored further.

Two transport systems for glucose with different affinities have also been described in *Kluyveromyces lactis* and the gene *RAG1* encoding the system with low affinity has been recently cloned (Wésolowski-Louvel *et al* ., 1992).

There are at present no data on the possible participation of transport protein(s) in glucose signalling. The availability of mutants in the different genes coding for transport proteins should open new perspectives in this area.

Regarding changes in the concentration of some possible regulatory metabolite, it has been observed that immediately after glucose addition to a yeast there is a rise in the concentration of cAMP (Van der Plaat & van Solingen 1974, Eraso & Gancedo 1984, Mbonyi *et al.*, 1988). This rise in cAMP is the result of a complex series of reactions in which the products of several genes are involved. Adenylate cyclase is activated by the products of genes *RAS1* and *RAS2* that in turn are activated by the product of the gene *CDC25*. The increase in cAMP stimulates the cAMP dependent protein kinase (protein kinase A) that consist of two types of subunits: the regulatory one encoded by the gene *BCY1* and the catalytic one that may be encoded by three different genes: *TPK1*, *TPK2* and *TPK3*. Activation of protein kinase A leads directly or indirectly to modification of the activity of several proteins e.g., the activity of 6-phosphofructokinase is increased indirectly by the increase in the concentration of fructose-2,6-P2 brought about by an increase in activity of its producing enzyme due to phosphorylation (François *et al.*, 1984) while activity of neutral trehalase increases directly by phosphorylation(Panek *et al* ., 1987, Vandercammen *et al.*, 1989). But although we know the external signal, sugar, and the final result, increase in cAMP we still lack information about how this chain of events is initiated. Moreover it should be clear that in addition to cAMP there should be another way of signalling the presence of glucose since a complex process like catabolite repression may take place in the absence of a cAMP signal (Matsumoto *et al.*, 1983). A role for glucose-6-phosphate has been suggested based in the fact that extensive metabolism of the sugar is not necessary for catabolite repression, phosphorylation of the sugar being the only step of the glycolytic pathway which is required (Gancedo, 1992).

The glycolytic pathway may be divided in two parts: the initial one that consumes energy and the final one that provides it. The initial part includes transport and the steps from phosphorylation by hexose kinase to the formation of fructose-1,6-bisphosphate by phosphofructokinase; the final part comprises the reactions catalysed by glyceraldehyde-3- phosphate dehydrogenase and pyruvate kinase. Regarding transport it is important to remember that, as discussed before, it is not dependent on a unique genetic entity. Therefore transport will be used in what follows as a functional concept, referring to the process of sugar uptake as such and not to some individual protein molecule.

Sugar transport is a step with a high influence in the overall rate of glycolysis. If the initial part of glycolysis had a higher flux than the final one the cell will run out of ATP because expenditure will be higher than regeneration. In addition hexose phosphates would accumulate. In higher eukaryotes glucose-6-phosphate inhibits hexokinase thus controlling the utilization of glucose and regulating the initial flux. However in yeast, hexokinase is not inhibited by glucose 6 phosphate and therefore some other compound ought to regulate it , control the transport step or both to maintain a balance between the two segments of glycolysis. The necessity of some kind of adjustement is shown by the behaviour of certain mutants that upon addition of glucose accumulate enormous amounts of fructose 1-6 bisphosphate and are depleted of ATP (Van de Poll *et al.* 1974 , Navon *et al.*,1979, Alonso *et al.*, 1984, González *et al.*, 1992). These mutants could provide useful information on how this adjustement is carried out in wild-type cells. It should be pointed out however, that even if this regulation exists, it will work only between restricted limits and that it can be overrun if the glycolytic pathway is severely disturbed. This is deduced from the behaviour of mutants defective in phosphoglucose isomerase or in pyruvate kinase. In the first case glucose-6-phosphate accumulates to very high intracellular levels showing that transport has escaped control. Pyruvate kinase mutants are also depleted of ATP indicating problems in the proper control of glucose utilization.

A well characterized example of imbalance is provided by the behaviour of liver after an overload of fructose. In liver, fructose is phosphorylated by a fructokinase to fructose-1-phosphate and subsequently split by aldolase. If an overload of fructose is given the liver is depleted of ATP because the initial rate of phosphorylation by fructokinase cannot be matched by the lower capacity of aldolase. An accumulation of frutose-1-phosphate and a depletion of ATP without a corresponding increase of ADP or AMP ensues (Mäenpää *et al.*, 1968)) .

Among the most prominent effects of glucose are the repression of the synthesis of enzymes involved in the metabolism of alternative carbon sources (catabolite repression) and the inactivation of certain enzymes situated at key points of the metabolism of these sources (catabolite inactivation). It has been assumed that the significance of catabolite repression and inactivation is the avoidance of futile cycles that would arise by the simultaneous operation of enzymes working in opposite directions and result in a net hydrolysis of ATP. In the case of glycolysis and its counterpart gluconeogenesis there are two points where futile cycles could occur: the pairs phosphofructokinase/fructose-1,6 bisphosphatase and pyruvate kinase/phosphoenol pyruvate carboxykinase. Yeast appears to have evolved a quite elaborated mechanism to avoid futile cycles. The most elementary regulatory device has been used to keep glycolytic enzymes from functioning during gluconeogenesis, namely the built-in kinetic parameters of the enzymes combined with the differences in concentration of their substrates. During gluconeogenesis phosphofructokinase exhibits a low activity due to low concentrations of its substrate fructose-1,6-bisphosphate and of its activator fructose-2,6-bisphosphate. Pyruvate kinase in turn is kept nearly inactive due to the low concentration of its allosteric activator fructose-1,6-bisphosphate. Operation of the gluconeogenic enzymes during glycolysis is prevented by controlling the amount of enzyme by means of catabolite repression and catabolite inactivation. In addition fructose-1,6-bisphosphatase is inhibited both by AMP and by fructose-2,6-bisphosphate. We will not consider the possible mechanisms underlying catabolite repression which have been the subject of a recent review by one of us (Gancedo, 1992). The molecular mechanism of catabolite inactivation has not been yet completely clarified. The current view is that catabolite inactivation is produced by a selective proteolysis of the corresponding enzyme. The problem is then to explain how an enzyme that is normally stable is earmarked for destruction . The most plausible hypothesis is that some structural element in the protein responds to a change in the intracellular environment producing a form of the enzyme that is now either easily degraded in the cytosol by specific proteases or transported to the vacuole where it is degraded by unspecific proteases. Although recent results support vacuolar degradation (Chiang and Schekmann, 1991) the question has not been yet settled (Gamo *et al.*, 1993).

The evolutionary significance of catabolite inactivation is not straightforward. Studies with strains that have an impaired degradation of malate dehydrogenase 2 show that they need more time to adapt to grow on glucose when coming from a non-fermentable carbon source than wild type strains(Minard & McAlister-Henn,1992) Also the suggestion has been made that the aminoacids released by the proteolytic breakdown of some proteins may provide aminoacids in a situation where nitrogen sources are scarce. Finally we cannot avoid to confess our ignorance about the behaviour of yeast in its natural habitat so that our observation criteria are limited. Perhaps in some particular circumstances the lack of catabolite inactivation may really be harmful for the cell.

To try to answer the question of the relevance for the cell of catabolite repression and inactivation of the gluconeogenic enzymes we have cloned the genes encoding the gluconeogenic enzymes and we have placed the coding regions under the control of the ADH1 promoter which is not repressible by glucose. If futile cycles really matter and cause a waste of ATP the unregulated genes should be quite detrimental and the cell harboring them would grow very slowly if at all. However no significant differences with the wild type in generation time or in growth yield were found in strains expressing one or the other of these enzymes. When both were expressed an increase in generation time of about 20% and a decrease in growth yield of about 30% were found. Futile cycles were not apparent: in the case of fructose-1,6-bisphosphatase/ phosphofructokinase the cycling was less than 2%. In the case of pyruvate kinase/phosphoenolpyruvate carboxykinase the determination was not possible due to technical problems but it can be said that the recycling is less than 20%. It seems therefore that the presence of gluconeogenic enzymes in glucose-growing yeast is not associated with a high rate of cycling. This could be due to the inhibition of fructose-1,6-bisphosphatase by its effectors, particularly fructose-2,6-bisphosphate. For phosphoenolpyruvate carboxykinase no inhibitors are known but its activity could be low due to a lack of the substrate oxaloacetate. Nevertheless the presence of the gluconeogenic enzymes has a measurable effect on glucose metabolism (Navas *et al.*, 1993). We conclude that the advantages given by a tight regulation of the gluconeogenic enzymes are more subtle than expected, but provide sufficient advantage to have been selected in the course of evolution.

As Fraenkel has pointed out in a recent review (Fraenkel, 1992) metabolism is less understood than sometimes assumed and very important questions of a quantitative nature may be more difficult to answer than other questions where a yes/no answer may be expected . A century after Buchner´s report, regulation of glycolysis is still not completely solved.

References

Alonso A., Pascual C., Herrera L., Gancedo J.M and Gancedo C., (1984) Eur.J. Biochem., **138,**407-411

Bisson L. F. and Fraenkel D.G., (1983) Proc. Natl Acad. Sci. USA, 80,1730-1734

Buchner E., (1897) Ber.chem. Ges., **30**,117-124

Celenza J.L., Marshall-Carlson L., and Carlson M., (1988) Proc. Natl. Acad. Usa., **85**,2130-2134

Chiang H.L., and Schekman R., (1991) Nature, **350**,313-318

Eraso P., and Gancedo J.M., (1984) Eur.J. Biochem., **141**,195-198

Fraenkel D.G., (1992) Annu. Rev. Genet., (1992) **26**,159-177

François J., Van Schaftingen E., and Hers H.G., (1984) Eur.J. Biochem., **145,** 187-193

Fruton J.S. (1972) Molecules and life. Wiley-Interscience .New York.

Gamo F.J., Navas M.A., Blázquez M.A., Gancedo C., and Gancedo J.M. (1993) Submitted for publication.

Gancedo C., and Serrano R., (1989) The Yeasts, 2nd edition, vol 3 , pp205-260 Rose A.H., and Harrison J.S. Eds. Academic Press. New York
Gancedo, J.M. (1992) Eur. J. Biochem. **206**, 297-313.
González M.I., Stucka R., Blázquez M.A., Feldmann H., and Gancedo C. (1992) Yeast **8**, 183-192
Ko, C.H., Liang, H. and Gaber, R.F. (1993) Mol. Cell. Biol. **13**, 638-648.
Kruckeberger A.L.and Bisson L.F ., (1990). Mol.cell. biol., 10, 5093-5913
Lagnado J., (1992) The Biochemist Oct/Nov 21-22
Lewis D.A., and Bisson L.F., (1991) Mol. cel. biol., **11**, 3804-3813
Mäenpää P.H., Raivio K.O., and Kekomäki M.P., (1968) Science,**161**,1253-1254
Matsumoto, K., Uno, I., Ishikawa,T. and Oshima, Y. (1983) J. Bacteriol. **156**, 898-900.
Mbonyi K., Beullens K., Detremerie K., Geerts L., and Thevelein J.M., (1988) Mol.cell biol., **8**, 3051-3057
Minard K.I. and McAlister-Henn, (1992) J.biol.Chem., **267**,17458-17464
Navas M.A., Cerdán S. and Gancedo J.M., (1993) Proc. Natl Acad. Sci. USA. **90**,1290-1294.
Navon G., Shulman R.G., Yammame T., Eccleshall T.R., Lam K., Baronofsky J.J. and Marmur J., (1979) Biochemistry, **18**,4487-4499
Panek A.C. de Araujo P.S., Neto V.M., and Panek A.D., (1987) Curr. Genetics **11**, 459-465
Prior, C., Wésolowski-Louvel, M. and Fukuhara, H. (1992) Yeast **8** S141.
Vandercammen A., François J.,and Hers H.G.,(1989) Eur.J.Biochem., **182**,613-620
Van de Poll K.W., Kerkenaar A and Schamhart D.H.J., (1974) J. Bacteriol., **117**, 965-970
Van der Plaat J.B. and Van Solingen P. (1974) Biochem.Biophys. Res. Commun. **56**,580-587.
Wésolowski-Louvel M., Goffrini P., Ferrero I., and Fukuhara H. (1992) Mol. gen.genet., 233,89-96

ECB6: Proceedings of the 6th European Congress on Biotechnology
edited by L. Alberghina, L. Frontali, P. Sensi

Bioconversion of Xylose to Xylitol with *in situ* Generation of NAD(P)H in Recombinant *Saccharomyces cerevisiae*

Helle N. Carlsen*, Johan Hallborn, Marie-Françoise Gorwa+ and Bärbel Hahn-Hägerdal.

Department of Applied Microbiology, Chemical Center, University of Lund, P.O. Box 124, S-221 00 Lund, Sweden. Present address:*Dansk Bioprotein A/S, Stenhuggervej 7-9, 5230 Odense M, Denmark.+INSA-Département de Génie Biochimique et Alimentaire, Avenue de Rangueil, 31077 Toulouse, France.

The reduction of xylose to xylitol is an example of a reduction of industrial interest which can be enzymatically catalyzed. One important obstacle in the technical exploitation of enzymes like reductases, dehydrogenases and oxygenases is the requirement for generation of reduction equivalents consumed during the process. Different approaches including chemical, electrochemical and enzymatic reactions have been tried to establish systems where reduction equivalents are generated [1-4]. Enzymatic reactions have an advantage in beeing highly specific in the generation of reduction equivalents which is important as often either NADH or NADPH are used as cofactor. To prevent loss of cofactor, immobilization of the cofactor and use of membrane reactors, where the cofactor is held back, have been investigated [3,4]. In the present communication whole cells were used as hosts for both the bioconversion reaction and the system for generation of reduction equivalents. We discuss xylitol production in pentose fermenting yeasts and present work on recombinant *Saccharomyces cerevisiae* with emphasis on the *in situ* generation of the reduction equivalents consumed during the xylose reduction.

Xylitol

Xylitol is currently used as a sugar substitute in foods including diets for diabetics [5]. It has a sweetnes and a caloric content like glucose and in mammalian cells it is a metabolic intermediate in carbohydrate metabolism [6]. Cariogenic microorganisms have been shown to be unable to ferment xylitol and some microorganisms have been observed to be inhibited by xylitol which may be the reason for the observed dental caries reducing qualities of xylitol [7].

Xylitol is manufactured by hydrogenation of hemicellulose derived xylose at high temperature and pressure in the presence of metal catalyst [8]. The process is expensive as extensive purification and separation is required to remove undesirable byproducts [9]. An enzymatically catalyzed reaction with high specificity could lead to simpler purification of the produced xylitol and diminish the requirements for elevated temperature and pressure.

Xylitol production in pentose fermenting yeast

In pentose fermenting organisms the enzymatic reduction of xylose to xylitol is catalysed in one step by xylose reductase (alditol: $NAD(P)^+$ 1-oxidoreductase, EC 1.1.1.21) [10-12]. The xylose reductase from *Pichia stipitis* and *Candida shehatae* can use both NADH and

NADPH as reductant [13], while organisms like *Pachysolen tannophilus* contain a xylose reductase with much higher activity with NADPH than NADH [10]. No significant product inhibition by xylitol occurs indicating that high product concentrations can be achived without affecting the yield and productivity. The xylose reductase has a rather restricted specificity [10], which is an additional andvantage in an industrial process.

The production of xylitol has been studied in different xylose fermermenting organisms including *C. guillermondii* [14,15]; *C. tropicalis* [16]; *C. bondinii* [17] and *P. tannophilus* [18,19]. In these organisms xylitol accumulation occured together with ethanol under oxygen limited or anaerobic conditions. High xylitol productivity, 0.22 g/g h was obtained with *C. guillermondii* when the xylose concentration was above 110 g/l [15]. The highest conversion yield for xylitol, $Y_{xylitol/xylose}$, was obtained under carefully controlled oxygen limited conditions and it was reported to be about 0.6-0.75 g/g [15]. The value indicated that a rather large portion of the produced xylitol was further metabolized in the organism. Xylitol was converted into xylulose by a NAD^+ dependent xylitol dehydrogenase. Xylulose was further metabolized through the pentose phosphate pathway and used for maintenance, growth and production of products such as ethanol. The further metabolism was also necessary for the generation of reduction equivalents consumed during xylose reduction.

Addition of methanol together with xylose to the methylotrophic yeast *C. bondinii*, increased the xylitol yield. However, the highest xylitol yield was only 0.5 g/g [17]. Hydrogen as reductant was investigated by Nishio and colleges [20]. They immobilized and permeabilized cells of *C. pelliculosa* containing NADPH dependent xylose reductase and *Methanobacterium* sp containing hydrogenase and $F_{420}NADP^+$ oxidoreductase. A xylitol yield, $Y_{xylitol/xylose}$, close to 1.0 g/g was obtained. Some drawbacks were, however, the lability of NADPH and the requirement for continuous addition of $NADP^+$ together with the substrate.

Xylitol producing recombinant *Saccharomyces cerevisiae*

The widely studied and industrially used hexose fermenting yeast *S. cerevisiae* can not ferment xylose due to the lack of xylose reductase and xylitol dehydrogenase activity [21]. It may however be a more suitable organism for xylitol production because of its higher resistance to inhibitors present in hemicellulose hydrolysates than natural xylose fermenting organisms [23] and due to its GRAS (Generally Recogniced As Safe) status in the food industry.

Successful expression of xylose reductase in recombinant *S. cerevisiae* was achieved and thereby new posibilities for investigation of xylitol production and reduction equivalent generation [23]. The recombinant *S. cerevisiae* was constructed by cloning the XYL-1 gene coding for xylose reductase from *P. stipitis* into the multi-copy yeast expression vector pMA91 between the phosphoglycerate kinase (PGK) promoter and terminator. The plasmid was transformed into *S. cerevisiae* resulting in constitutively expressed xylose reductase.

The activity of xylose reductase in the recombinant yeast was found to be about one third of the activity observed for *P. stipitis* [23]. No change in *in vitro* xylose reductase activity was observed during fermentations for at least 200 h indicating that the plasmid was stable when the cells were kept under selection pressure [24,25]. As xylitol is not further metabolized in the recombinant yeast a co-substrate was required for generation of reduction equivalents consumed during xylose reduction. The co-substrate may also be used for maintenance and cell growth. The effect of different co-substrates including glucose, glycerol, ethanol and acetate on the xylose reduction in recombinant *S. cerevisiae* grown in batch, fed batch and continuous cultures [24-26]. As the respiratory chain under aerobic conditions is an important sink for reduction equivalents we cultivated the recombinant yeast with different aerations: aerobic, oxygen limited and anaerobic.

In aerated batch cultures glucose, ethanol and acetate but not glycerol could support growth and xylose reduction [26]. The xylitol conversion yield, $Y_{xylitol/xylose}$ was close to 1.0 g/g in all the experiments. The highest volumetric productivity of xylitol (0.14 g/l h) and highest growth rate were obtained with glucose. During the batch fermentation glucose was converted into ethanol, acetate, cell mass and carbon dioxide. The reduction of xylose to xylitol continued as long as glucose, ethanol or acetate were present. The xylitol yield in relation to consumed co-substrate, $Y_{xylitol/co\text{-}substrate}$, was for ethanol and glucose of the same magnitude, 0.8-1.5 g/g, while a significantly lower value, 0.35 g/g was obtained with acetate.

The influence of the dilution rate on the xylose conversion in a continuous culture of recombinant *S. cerevisiae* with glucose as co-substrate is show in Fig. 1. Only at low dilution rates significant amounts of xylose were converted into xylitol. The highest specific xylitol productivity, 0.03 g/g h, was observed at a dilution rate of 0.1 h^{-1}, where the specific growth rate was 0.1 g/g h and the xylitol yield in relation to consumed co-substrate, $Y_{xylitol/glucose}$, was only 0.16 g/g.

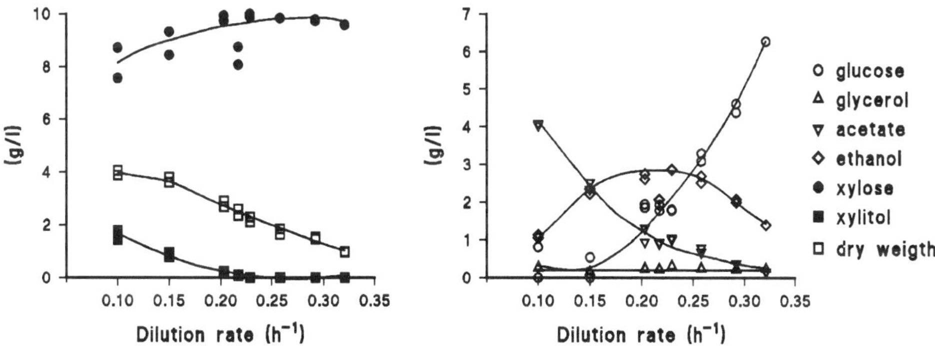

Fig 1. Continuous culture of recombinant *S. cerevisiae*. The concentation of glucose and xylose in the medium was 10 g/l.

A higher xylitol productivity and a higher xylitol yield in relation to consumed co-substrate, $Y_{xylitol/co\text{-}substrate}$, are likely to be obtained, if the co-substrate is not used for growth, and if the consumption of reduction equivalents via the respiratory chain is limited. We observed in batch fermentations a higher xylitol productivity under oxygen limited conditions than under aerated conditions [26]. The xylitol production was therfore studied in anaerobic fed batch cultures of the recombinant yeast [25]. The cells were grown in two phases. In the first phase the cells were grown aerobically on glucose to obtain a high cell density. Then xylose was added, air was interchanged with nitrogen and a slow feed of glucose (0.6 g/l h) was supplied. The xylitol conversion yield $Y_{xylitol/xylose}$ was close to 1.0 g/g, and the xylitol yield in relation to consumed co-substrate, $Y_{xylitol/co\text{-}substrate}$ was 2.2 g/g. The specific growth rate was 0.002 g/g h. Acetate and ethanol accumulated indicating that the generation of reduction equivalents was not limiting the bioconversion of xylose to xylitol under these conditions. The volumetric productivity was 1.26 g/l h, and the highest specific productivity was 0.12 g/g h, which is of the same range as reported for one of the most efficient xylitol producing yeast, *C. guillermondii* [15].

Concluding remarks

Recombinant *Saccharomyces cerevisiae* can catalyse the bioconversion of xylose to xylitol with a xylitol conversion yield, $Y_{xylitol/xylose}$ close to 1.0 g/g and a specific productivity of the order of 0.1 g/g h. The highest productivity was obtained under oxygen limited or anaerobic conditions. The cofactor, NAD(P)H, used for xylose reduction can be generated *in situ* through the oxidation of co-substrates such as ethanol, acetate or glucose.

Acknowledgments

The research on xylose fermentation at the Department of Applied Microbiology is supported financially by NUTEK (Swedish National Board for Industrial and Technical Development), STU (Swedish National Board for Technical Development), STEV (National Energy Administation), NI (The Nordic Fund for Technology and Industrial Development), NFR (Swedish Natural Science Research Council) and SSEU (Swedish Ethanol Development Foundation). H.N. Carlsen acknowledges a scholarship from NorFA (Nordisk Forskerutdannings Akademi).

References

1. A.E. Cass, M.J. Eddowes, H.A.O. Hill and K. Uosaki, Nature, 285 (1980) 673.
2. J.M. Laval, J. Moiroux and C. Bourdillon, Biotechn. Bioeng., 38 (1991) 788.
3. M.-O. Månsson and K. Mosbach, Methods in Enzymology, 136 (1987) 3.
4. M.-R. Kula and C. Wandrey, Methods in Enzymology, 136 (1987) 9.
5. T. Pepper and P.M. Odlinger, Food Technol., 10 (1988) 98.
6. A. Emodi, Food Technol., January (1978) 28.
7. S. M. Waaler, Scand. J. Dent. Res., 100 (1992) 204.
8. L. Hyvönen and P. Koivistoinen Adv. in Food Research, 28 (1983) 373.
9. A. Melaja and L. Hämäläinen, US Patent, 4, 008 (1977) 285.
10. C. Verduyn, R. van Kleff, H. Schreuder, J.P. van Dijken and W.A. Scheffers. Biochem. J., 226 (1985) 669.
11. G. Ditzelmüller, C.P. Kubicek, W. Wöhrer and M. Röhr, Can. J. Microbiol., 30 (1984) 1330.
12. M. Rizzi, P. Erlemann, N.-A. Bui-Thanh and H. Dellweg. Appl. Microbiol. Biotechnol., 29 (1988) 148.
13. P.M. Bruinenberg, P.M. de Bot, J.P. van Dijken and W.A. Scheffers, Appl. Microbiol. Biotechnol., 19 (1984) 256.
14. M.F.S. Barbosa M.B. de Medeiros, I.M de Manchila, H. Schneider and H. Lee, J. Ind. Microbiol., 3 (1988) 241.
15. V. Meyrial, J.P. Degenes, R. Moletta and J.M. Navarro. Biotechnol. Lett., 13 (1991) 281.
16. H. Horitsu, Y. Yahashi, K. Takamizawa, K. Kawai, T. Suzuki and N. Watanabe, Biotechnol. Bioeng., 40 (1992) 1085.
17. V. Vongsuvanlert and Y. Tani, J. Ferment. Bioeng., 67 (1989) 35.
18. N.E. Watson, B.A. Prior, J.C. du Preez and P.M. Lategan, Enzyme Microb. Technol., 6 (1984) 447.
19. P. Thonart, J. Gomez Guerreiro, M. Foucart and M. Paquot, Med. Fac. Landbouww. Rijksuniv. Gent, 52 (1987) 1517.
20. N. Nishio, K Sugawa, N. Hayase and S. Nagai, J. Fermen., 67 (1989) 356.
21. T. Lindén, J. Peetre and B. Hahn-Hägerdal, Appl. Envir. Microbiol., 58 (1992) 1661.
22. T. Lindén and B. Hahn-Hägerdal, Biotech. Techniques, 3 (1989) 189.
23. J. Hallborn, M. Walfridsson, U. Airaksinen, H. Ojamo, B. Hahn-Hägerdal, M. Penttilä and S. Keränen, Bio/Technol., 9 (1991) 1090.
24. N. Meinander, and H. Ojamo, Xylitol production with recombinant *Saccharomyces cerevisiae* using ethanol as co-substrate in a fed-batch system (manuscript in preparation).
25. H.N. Carlsen and B. Hahn-Hägerdal, Anaerobic xylitol production with a recombinant *Saccharomyces cerevisiae* (manuscript in preparation).
26. J. Hallborn, M.F. Gorwa, N. Meinander and B. Hahn-Hägerdal, The influence of co-substrate and oxygenation on xylitol formation with recombinant *Saccharomyces cerevisiae* (manuscript in preparation).

ECB6: Proceedings of the 6th European Congress on Biotechnology
edited by L. Alberghina, L. Frontali, P. Sensi

ETHANOL METABOLISM IN THE YEAST *Kluyveromyces lactis*

M. Saliola, S. Bellardi, I. Marta, and C. Falcone

Dept. of Cell and Developmental Biology. University of Rome "La Sapienza".
Piazzale A. Moro - 00185 ROMA

INTRODUCTION

Kluyveromyces lactis is one of the very few yeasts that can use lactose as source of carbon and energy. As a result of this property, this yeast has been used in the dairy industry for the production of cell biomass and beta-galactosidase. More recently, as a result of the possibility of introducing foreign DNA into this organism and of the availability of stable replicating vectors (1-2), increasing interest has been shown in *K.lactis* as an alternative host to *S.cerevisiae* for the expression of heterologous genes.
The ability to shift from respiration to fermentation seems to be rare among yeast species and, in this respect, *S.cerevisiae* is exceptional. In this yeast, glucose represses respiratory enzymes and cells grow mainly by fermentation even in the presence of oxygen. By contrast, *K.lactis* is essentially an aerobic yeast in which both fermentative and respiratory metabolisms can co-exist. This indicates that the latter yeast, unlike *S.cerevisiae*, can escape the repression-effect of glucose; in fact, it has been reported that glucose does not affect mitochondrial respiration (3). As a result of their predominantly aerobic metabolism, *Kluyveromyces* species are not good producers of ethanol compared to *Saccharomyces* strains.

RESULTS AND DISCUSSION

Despite its low fermentative activity, four genes encoding alcohol dehydrogenase (ADH) activities are present in *K.lactis*. The four ADH genes of *K.lactis*, named *KlADH1* to *KlADH4*, have been cloned and their nucleotide sequences have been determined (4-5-6). The four genes show strong sequence identity (near 80%) with each other and with the corresponding genes of *S.cerevisiae* both at the nucleotide and the amino acid level.
KlADH1 and *KlADH2* are essentially constitutive genes which encode two cytoplasmic activities (KlADHI and KlADHII) responsible for ethanol production, while *KlADH3* and *KlADH4* encode two activities (KlADHIII and KlADHIV) localized in mitochondria (5). The expression of the two latter genes is regulated by the carbon source and is strain-dependent. In fact, we demonstrated that *KlADH4* is induced by ethanol and no other respiratory carbon sources (5-7). Moreover, unlike the *ADH2* gene of *S.cerevisiae* (8-9-10), this gene is not sensitive to the effect of glucose repression. This suggests that *K.lactis* cells, during growth on glucose, are able to oxidize within mitochondria the ethanol produced in the cytoplasm. This

situation is completely different from the one observed in *S.cerevisiae,* where respiration and fermentation are reciprocally exclusive.

The induction of *KlADH4* can be observed only in strains that do not produce ethanol, such as those harbouring a mutation in the *PGI* gene which encodes the glycolytic enzyme phosphoglucose isomerase (11) . In such mutants, glucose is probably metabolized less efficiently through the pentose phosphate pathway, the reduced amount of pyruvate is essentially utilized aerobically and ethanol is not produced by cells. This is confirmed by the observation that the expression of *KlADH4* is constitutive in wild-type strains which can produce ethanol endocellularly (7). In fact, as is shown in figure 1, the analysis by gel electrophoresis of the ADH isozymes revealed that KlADHIV activity is present during the growth on 2% glucose in those strains such as MW121-11A (lane 1) which can produce ethanol, while it is absent in strains carrying a mutated phosphoglucose isomerase gene, such as MW98-8C (lane 2).

As can also be seen in figure 1, another band of activity, migrating slower with respect to the other ones, is present in strain MW121-11A and not in MW98-8C. This band corresponds to the product of *KlADH3* , the gene encoding the second ADH activity (KlADHIII) present in the mitochondria of *K.lactis*. Interestingly, KlADHIII was not present when MW121-11A was grown in the presence of high glucose concentrations (manuscript in preparation). This result seems to indicate that, unlike that of *KlADH4*, the expression of *KlADH3* is sensitive to the repression of glucose and that the two genes could play a different physiological role in *K.lactis* cells. KlADHIII has never been detected in MW98-8C cells grown on different fermentative and respiratory sources of carbon.

Further analysis by gel electrophoresis revealed that KlADHIII is not present in all laboratory strains and that the expression of *KlADH3* independently of their fermentative capability (not shown).

Mutants in the ADH genes obtained by gene disruption have no clear phenotype, and this could be explained by the fact that the remaining activities can compensate for the missing one (7). We crossed single adh mutants and set up a screening to detect multiple mutants. This test is based on the different intensity of color developed from parental and recombinant strains during the staining for ADH activity (to be published). In this way we obtained triple *adh* mutants and *adh*° mutants (ADH null strains), completely lacking ADH activity.

The analysis of these strains showed that all triple mutants which maintained only one residual activity were able to utilize ethanol as a source of carbon. These results indicated that all four activities can oxidize ethanol to acetaldehyde "in vivo". These strains will be very useful for the purification of the single ADH activities and for their biochemical characterization. Unlike triple mutants, *adh*° strains failed to grow on ethanol, indicating that there are no other important ADH activities in *K.lactis* cells beside the four we identified. These ADH null strains are useful tools for the study of glucose metabolism in the absence of alcohol dehydrogenase activity. We measured the concentration of some glucose metabolites in the culture supernatants of MS7-62, one of the ADH null mutants, during growth on high sugar concentration (6% glucose).

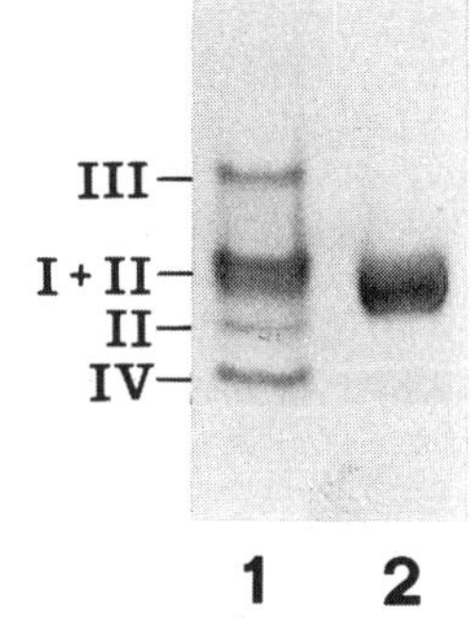

Figure 1

As can be seen in figure 2, cell biomass reached a 9.3x 10^8/ml concentration after 48 h of growth with average generation time of 100 min.

These values are very similar to those observed in the case of wild type strains grown under these conditions, indicating that the total absence of ADH activity has no influence on the rate of growth of *K.lactis* cells. Nevertheless, a residual production of ethanol can still be observed in such mutants. In fact, as is shown in table 1, ethanol was accumulated at a concentration of 1 and 2 g/l after 24 and 48 hours of growth, respectively.

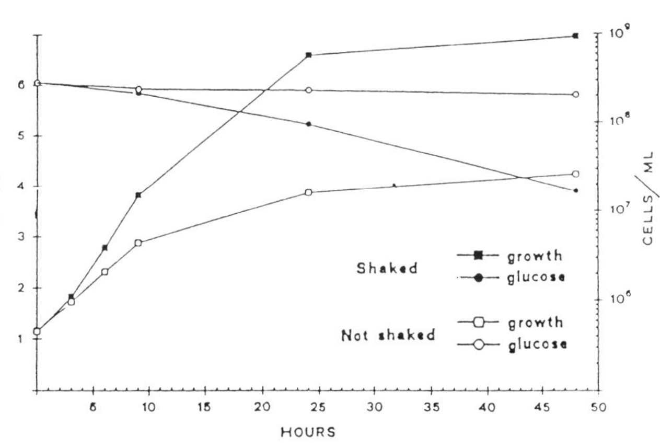

Figure 2

Acetaldehyde reached 0.33 g/l at 24 h and after decreased,having been at least in part converted into ethanol. Unexpectedly, glycerol was never accumulated during this period. After 24 h of growth under the same conditions, wild type strains accumulated about 25 g/l of ethanol and 1-2 g/l of glycerol. The concentration of these metabolites diminished after 48 h, probably as a result of their utilization.

In a parallel experiment we cultivated strain MS7-62 on 6% glucose without shaking. Under these conditions, oxygen concentration is limited and NADH is in excess. As can be seen in figure 2 and table 1, glucose consumption in the first 48 h was lower (0.5% compared to 2% in the shaked condition) and the culture reached a plateau at a concentration of $2x10^7$ cells/ml.

Table 1: Glucose metabolism in *adh°* strains of *K.lactis.*

Hours of growth		acetaldehyde (g/l)	ethanol (g/l)	glycerol (g/l)
shaked	0	0.00	0.01	0.10
	9	0.00	0.05	0.10
	24	0.33	0.87	0.10
	48	0.09	1.93	0.10
not shaked	0	0.00	0.01	0.10
	9	0.01	0.01	0.10
	24	0.05	0.02	0.20
	48	0.10	0.06	0.40

Although these growth conditions are more fermentative because of the partial anaerobiosis, less ethanol was accumulated compared to the shaked culture. This result is probably related to the lower yield in cell biomass. As far as the origin of this residual production of ethanol are concerned, we cannot exclude the existence in *K.lactis* of a new as yet unidentified alcohol dehydrogenase activity, besides the ones we isolated. In any case, such an activity seems not to be effective in the oxidation of ethanol, since the ADH null strains are not able to grow with ethanol as their sole carbon source.

A residual production of ethanol has been also observed in the *adh*° strains of *S.cerevisiae* (12). It has been hypothesized that, in this yeast, ethanol is originated from the dismutation of acetaldehyde which involves the combined activity of an aldehyde reductase (ALDH) and of an acetaldehyde oxidase and requires the presence of functional mitochondrial (13).

We are investigating whether a similar mechanism operates in *K.lactis* cells.

During the growth without shaking of MS7-62, we could also detect the presence of glycerol in culture supernatants. These results indicate that *K.lactis* cells are able to produce glycerol which is probably oxidized during the aerobic growth and can be accumulated only under non-oxidative conditions. On the contrary, glycerol is a secondary product of glucose fermentation in *S.cerevisiae* and is accumulated in ADH-deficient and null strains of this yeast as a result of the reduction of dihydroxyacetone phosphate by NADH (14-15).

In conclusion, glucose metabolism seems to differ in *K.lactis* and in *S.cerevisiae*. In the latter yeast the fermentative metabolism is preferred and ethanol is the major product. On the contrary, *K.lactis* cells can grow by respiration also in the presence of high glucose concentrations and, as a result, part of the ethanol produced is probably oxidized within the cells. KlADH IV, the activity encoded by *KlADH4*, could be involved in this process in that it is induced by ethanol also in the presence of high glucose concentrations.

A better knowledge of ethanol metabolism, the understanding of the physiological role of the single ADH activities (expecially KlADH III and KlADH IV), and the possibility of manipulating by genetic engineering the ADH system could be very useful for the construction of *K.lactis* strains with improved fermentative properties.

ACKNOWLEDGEMENTS

This work was supported by CEC (Contract Bridge n° BIOT CT91-0267-DSCN), by C.N.R. (P.F. Ingegneria Genetica), and partially by Istituto Pasteur- Fondazione Cenci Bolognetti.

REFERENCES

1. X.J. Chen, M.Saliola, C. Falcone, M.M. Bianchi and H. Fukuhara, Nucleic Acids Res., 14 (1986) 4471.
2. M.M. Bianchi, C. Falcone, X.J. Chen, M. Wesolowski, L. Frontali and H.Fukuhara, Curr. genet., 12 (1987) 185.
3. R.H. De Deken, J. Gen. Microbiol., 44 (1966) 157.
4. M. Saliola, J.R. Shuster and C. Falcone, Yeast, 6 (1990) 193.
5. M.Saliola, R. Gonnella, C. Mazzoni and C. Falcone, Yeast, 7 (1991) 391.
6. D.H. Shain, C. Salvadore and C. Denis, Mol. Gen. Genet.,232 (1992) 478.
7. C. Mazzoni, M. Saliola and C. Falcone, Molec. Microbiol.,6 (1992) 2279.
8. M. Ciriacy, Mut. Res., 29 (1975) 315.
9. M. Ciriacy, Mol. Gen. Genet.,176 (1979) 427.
10. C.L. Denis, M. Ciriacy, and E.T. Young. J.Mol.Biol., 148 (1981) 355.
11. P.Goffrini, M. Wesolowski-Louvel and I. Ferrero, Mol. Gen. Genet., 228 (1991) 401.
12. C. Drewke, J. Thielen and M. Ciriacy. J. Bacteriol., 172 (1990), 3909
13. J. Thielen and M. Ciriacy, J. Bacteriol., 173 (1991) 7012.
14. C. Neuberg and E. Rheinfurt, Biochem. Z., 100 (1918) 234.
15. C. Neuberg, Ann.Rev.Biochem., 15 (1946) 435.

ECB6: Proceedings of the 6th European Congress on Biotechnology
edited by L. Alberghina, L. Frontali, P. Sensi

SIGNIFICANCE OF RATE OF CELLULAR ADAPTATION TO ETHANOL AND RATE OF CHANGE OF ETHANOL CONCENTRATION IN FERMENTATIONS BY *ZYMOMONAS MOBILIS* AND YEAST.

T.J.Hobley, G.A.Stanley, G.Dasari, and N.B.Pamment
Department of Chemical Engineering, University of Melbourne, Parkville, Victoria 3052, Australia.

ABSTRACT

In high cell density batch fermentations by *Zymomonas mobilis* (>10^{10} cells/ml) and *Saccharomyces cerevisiae* (>10^9 cells/ml) rates of ethanol production as high as 100 $g.L^{-1}h^{-1}$ or more were achieved. Culture viability fell rapidly during the fermentations. By conducting simulated rapid fermentations in which ethanol was pumped continuously at various high rates to low cell density cultures, it was shown that yeast viability in rapid fermentations is strongly dependent upon the rate of change of ethanol concentration.[1] This effect was not observed with *Z.mobilis*. This difference appears to reflect differences in the rate at which the organisms adapt to ethanol, as revealed by the length of the lag phase of batch cultures subjected to a step change in ethanol concentration. The rate of adaptation of yeast to ethanol is strongly influenced by excreted acetaldehyde. Added acetaldehyde reduced the lag time of cultures subjected to a 4% v/v step change in ethanol concentration by as much as 60%. Acetaldehyde was not consumed during lag; its effects on yeast adaptation rates may be due to promotion of fatty acid synthesis associated with increased flux through the pentose phosphate pathway.

1. INTRODUCTION

The rates at which yeasts and ethanol-producing bacteria are capable of adapting to ethanol may be a factor influencing ethanol tolerance in fermentations. Previously, the effects of ethanol on *S.cerevisiae* and *Z.mobilis* have been studied in cultures with fixed ethanol concentrations.[6] In batch fermentations, however, the concentration of ethanol is constantly increasing: since the cells require time to respond, it is possible that they may not be fully adapted to the ambient ethanol concentration at any given point in time, especially if the fermentation is rapid.[1,2] Little has been done to measure the rates at which particular cellular adaptive responses to ethanol occur. The cell membrane is generally believed to be a primary site for the action of ethanol[3,4] and modifications to the plasma membrane which occur in *S.cerevisiae* and *E.coli* when the ethanol concentration is increased are thought to represent an ameliorative adaptive response.[3,4] Recently, excreted acetaldehyde was shown to effect markedly the rate at which cells respond to a step change in ethanol concentration.[5] In this paper we show that the rate of change of ethanol concentration affects viabilities in rapid fermentations by *S.cerevisiae* but not those by *Z.mobilis* and that this difference is attributable to the slower rate at which yeast adapts to ethanol. We also examine the effects of acetaldehyde on yeast adaptation rates.

2. MATERIALS AND METHODS

Saccharomyces cerevisiae 706800 (University of New South Wales culture collection) was grown in medium containing (g/L) glucose, 200; KH_2PO_4, 5; yeast extract (Oxoid), 10. Analytical and cultural methods were as described elsewhere.[1]

Saccharomyces cerevisiae X2180-1A (Yeast Genetic Stock Centre, University of California) was grown in medium containing (g/L) glucose, 20; tryptone (Oxoid), 10; NaCl, 9; yeast extract (Oxoid), 5. Late exponential stage parent cells were inoculated into fresh medium containing 4% v/v ethanol. Cell growth was monitored using a Coulter Counter.

Zymomonas mobilis ZM4 (ATCC 31821) was grown in medium containing (g/L) glucose, 100; KH_2PO_4, 5; yeast extract (Difco), 5; $(NH_4)_2SO_4$, 1.5; $MgSO_4$, 1. Ethanol concentration was determined as described previously.[1] Percentage cell viability was calculated from viable counts determined by triplicate spread plates and total cell counts determined with a Coulter Counter.

3. RESULTS AND DISCUSSION

In high cell density fermentations by *S.cerevisiae* or *Z.mobilis* ethanol was produced at up to 100 $gl^{-1}h^{-1}$ but culture viability fell rapidly (Hobley and Pamment, unpublished data). To determine whether this was due to the high rate of change of ethanol concentration, ethanol was pumped at different rates to low cell density cultures in which ethanol production by the cells was minimal. In simulated rapid fermentations with *S.cerevisiae*, cell viability at equivalent ethanol concentrations was always lower in cultures to which the ethanol was added more rapidly, despite the shorter time of exposure to ethanol (Fig. 1). In contrast, any effect of rate of change of ethanol concentration on *Z.mobilis* cultures was completely outweighed by the effects of ethanol exposure time (Fig.2). These data suggest that *Z.mobilis* is capable of adapting much more quickly than yeast to changing ethanol concentration.

In support of the above interpretation, the lag times of batch *Z.mobilis* cultures subjected to a step change in ethanol concentration were much shorter than those of *S.cerevisiae* (Hobley and Pamment, unpublished data). The longer lag periods of *S.cerevisiae* cultures may be due to the time taken for modifications to the lipid profile of the plasma membrane in response to shocks, such as changes in ethanol concentration, that cause disturbances in membrane integrity. *Z.mobilis* makes little change to its membranes in response to ethanol.[7,8]

The rate of adaptation to ethanol shock by *S.cerevisiae* was greatly increased by the addition of low concentrations of acetaldehyde (0.005 - 0.05 g/L) (Fig. 3). The origin of this effect is unknown. However, acetaldehyde is known to increase the metabolic flux through the pentose phosphate pathway[9] which supplies around 75% of the NADPH required for fatty-acid synthesis.[10] Its ability to increase the adaptation rate of yeast to ethanol may thus be due to promotion of fatty acid synthesis.

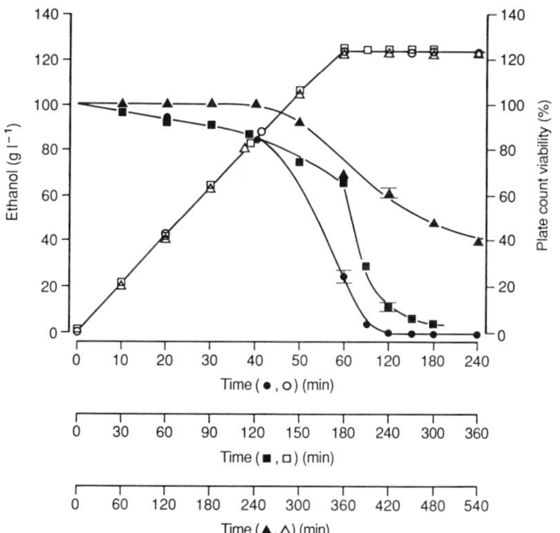

Figure 1. Viability in low cell density *S.cerevisiae* cultures to which a total of 124 g/L of ethanol was added at different rates. Open symbols, measured ethanol concentration; Closed symbols, plate count viability. Error bars indicate the average standard deviation between points from duplicate experiments. Reproduced from reference 1 with permission of John Wiley & Sons, Inc. Copyright © John Wiley, 1990.

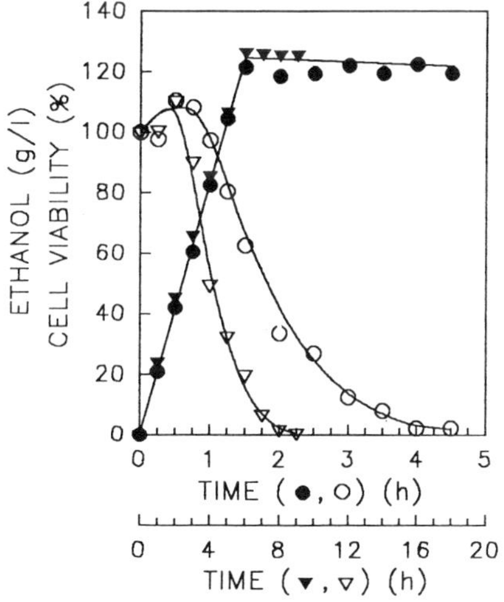

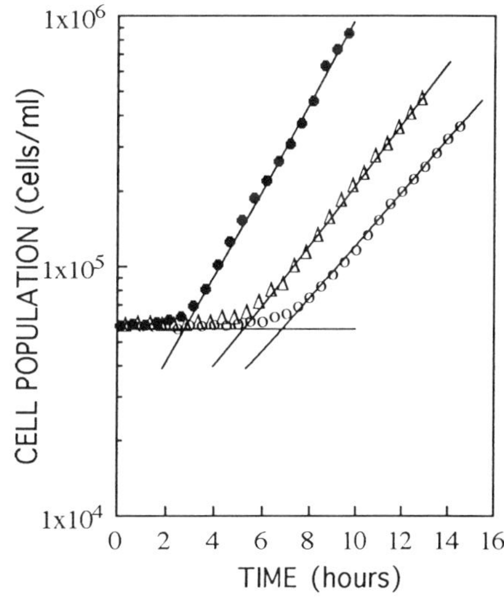

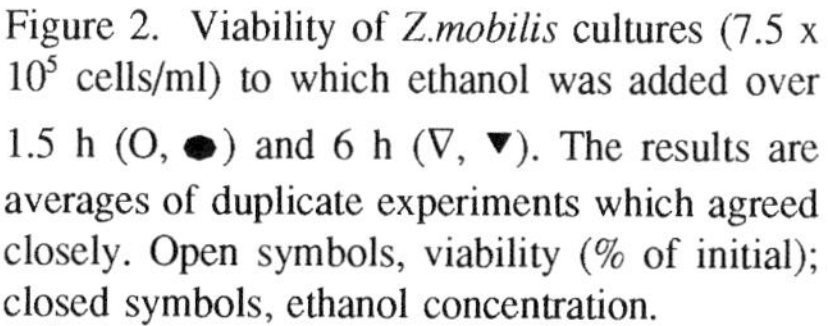

Figure 2. Viability of *Z.mobilis* cultures (7.5 x 10^5 cells/ml) to which ethanol was added over 1.5 h (O, ●) and 6 h (▽, ▼). The results are averages of duplicate experiments which agreed closely. Open symbols, viability (% of initial); closed symbols, ethanol concentration.

Figure 3. The effect of added acetaldehyde on lag periods of *S.cerevisiae* induced by a 4 % (v/v) ethanol concentration shock. Initial acetaldehyde concentrations were: 0.046 g/L (●), 0.005 g/L (**Δ**), no added acetaldehyde (O).

REFERENCES

1. Dasari, G., Worth, M.A., Connor, M.A. and Pamment, N.B. 1990. Reasons for the apparent difference in the effects of produced and added ethanol on culture viability during rapid fermentations by *Saccharomyces cerevisiae*, Biotechnol. Bioeng., 35: 109-122.

2. Jones, R.P. and Greenfield, P.F. 1985. Replicative inactivation and metabolic inhibition in yeast ethanol fermentations. Biotechnol. Lett., 7: 223-228.

3. Ingram, L.O. 1976. Adaptation of membrane lipids to alcohols. J. Bacteriol. 125: 670-678.

4. Mishra, P. and Kaur, S. 1991. Lipids as modulators of ethanol tolerance in yeast. Appl. Microbiol. Biotechnol. 34: 697-702.

5. Walker-Caprioglio, H.M. and Parks L.W. 1987. Autoconditioning factor relieves ethanol-induced growth inhibition of *Saccharomyces cerevisiae*. Appl. Environ. Microb. 53 33-35

6. Pamment, N.B. 1988. Overall kinetics and mathematical modeling of ethanol inhibition in yeasts. *In* N.F. van Uden (ed.) Alcohol Toxicity in Yeasts and Bacteria, CRC Press, Boca Raton, Florida, USA 1-75

7. Carey, V.C. and Ingram, L.O. 1983. Lipid composition of *Zymomonas mobilis*: Effects of ethanol and glucose. J. Bacteriol. 154: 1291-1300.

8. Hermans, M.A.F., Neuss, B. and Sahm, H. 1991. Content and composition of Hopanoids in *Zymomonas mobilis* under various growth conditions. J. Bacteriol. 173: 5592-5595.

9. Iscaki, M. 1975. Acetaldehyde influence of glucose degradation by bakers yeast. C.R. Acad. Sc. Paris, Series D., 280: 315-317.

10. Dickinson, F.J., and Hewlins, M.J.E. 1988. A study of the role of the hexose monophosphate pathway with respect to fatty acid biosynthesis in sporulation of *Saccharomyces cerevisiae*. J. Gen. Microbiol., 134: 333-337.

ECB6: Proceedings of the 6th European Congress on Biotechnology
edited by L. Alberghina, L. Frontali, P. Sensi

YEAST NON-REPLICATIVE PLASMIDS AND STRAIN CONSTRUCTION

Z. Zgaga, P. Koren, M. Ninkovic[1], K. Gjuracic and M. Alacevic

Faculty of Food Technology and Biotechnology, Krsnjavoga 25, Zagreb, Croatia

SUMMARY

Stable transformation with yeast non-replicative vectors relies on homologous recombination between yeast chromosome and plasmid molecule. We show that UV-irradiation of double-stranded plasmids stimulates this process. Irradiation of either double or single-stranded non-replicative plasmids does not lead to detectable mutagenesis. Transformation by conversion of point mutations is favored if single-stranded plasmids are UV-irradiated. Linearised single-stranded plasmids may even convert the insertions of the Ty element (6.1 kb).

INTRODUCTION

Yeast *Saccharomyces cerevisiae* is the first eukaryote that was successfully transformed with foreign DNA (1). Since then, transformation became one of the most powerfull tools in the yeast molecular biology. Vectors that can be used for transformation may be simply classified as replicative or non-replicative vectors. The latter are usually refered as integrative plasmids, although they can introduce a change in the yeast genome both by integration in the homologous region and by genetic conversion of the chromosomal allele. Transformation by recombination with yeast chromosome is particularly useful for replacing of the original yeast sequence with modified sequence created *in vitro* (2 and references therein). We are using transformation with yeast nonreplicative plasmids as a model system in the study of homologous and homoeologous recombination. Here we present some results that might be usefull for introducing desirable genetic changes into the yeast strains.

MATERIALS AND METHODS

Plasmid pAB218 (7.7 kb) has the *Bam*HI - *Hin*dIII fragment of the plasmid YIp5 replaced by the 2,5 kb yeast DNA containing the *CYC1* gene (3). Plasmid pULi (7.3 kb) was constructed by insertion of the 2.2 kb fragment (*Bam*HI - *Hin*dIII) containing the *LEU2* gene from the plasmid pLU (4) into the polylinker of the pEMBLYi22 (5). Plasmid pOM has the 856 bp fragment (*Hin*dIII-*Xho*I) containing the yeast *CYC1* gene inserted into the polylinker of the pEMBLYi22. pULi and pOM can be isolated from *Escherichia coli* in single-stranded form by superinfecting the plasmid-bearing cells with a f1 helper

---

[1] permanent adress: Centre for Genetical Ingeneering and Biotechnology, Filipa Kljajica 12, Sarajevo, Bosnia and Herzegovina

phage. Yeast strain used was FF18-52 (*MATa, leu2*-3,112, *ura3*-52, *ade5, can1*). DNA manipulations and yeast transformation procedure were as previously described (2,3). Linearisation of the single-stranded plasmid will be described elswhere.

RESULTS AND DISCUSSION

UV-light has been shown to induce genetic recombination in a variety of organisms, both prokaryotic and eukaryotic. Since stable transformation with nonreplicative plasmids depends on homologous recombination between yeast chromosome and plasmid molecule, we wondered whether irradiation of the plasmid could stimulate this process. Indeed, the results presented in Figure 1 show strong dose-dependent increase in the number of transformant selected either for uracil of leucine, obtained with two different plasmids. In some experiments we observed more than 50-fold stimulation of transformation (not shown). The usual way of stimulating plasmid integration relies on cutting of the plasmid molecule in the region of homology with yeast chromosome by restriction enzyme (6). Efficiency of transformation by UV-irradiated plasmids reaches values close to those obtained by plasmid linearisation and can be used as a method of choice for plasmid integration when suitable restriction enzymes are not available or/and restriction map of the fragment to be introduced is not known.

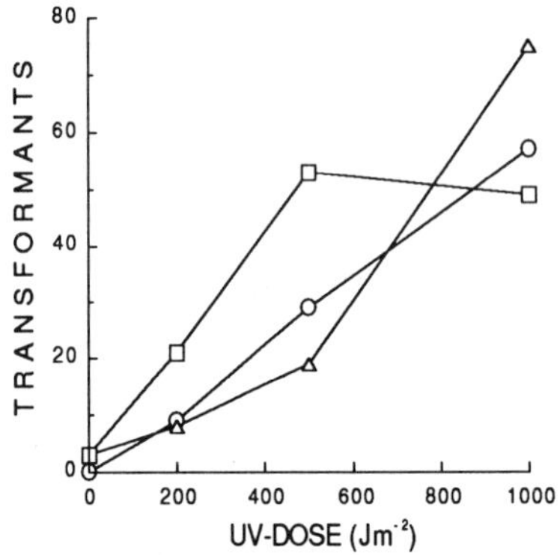

Figure 1. Transformation with UV-irradiated nonreplicative double-stranded plasmids. Triangles - pULi (Ura^+); circles - pULi (Leu^+); squares - pAB218 (Ura^+). The numbers of transformants obtained with 0,5 μg of plasmid DNA are presented.

We were also interested in the nature of recombination event (plasmid integration or gene conversion) that was stimulated by UV-irradiation of the plasmid. The transformation system using plasmid pULi, containing both the *URA3* and the *LEU2* genes and the yeast strain 18-52 as a host, makes possible to distingwish between convertants and integrants by simple replica plating. Transformants that contained both genetic makers (Leu^+ and Ura^+) have integrated the whole plasmid, while those that acquired only the marker they were selected for, arose by gene conversion. This reasoning was confirmed by Southern blot analysis of fourty Leu+ Ura- transformants (not shown). From the Table 1 it can be seen that all selected Ura+ transformants, obtained with either single or double-stranded form of the plasmid pULi, are also leucine prototrophs. This means that the transformation to uracil prototrophy requires plasmid integration, which is probably due to the nature of the *ura3-52* mutation, known to be insertion of the Ty transposon (7). More important, these data indicate that UV-irradiation does not induce mutations in the integrated plasmid at the detectable level, since no Ura+ transformants that we tested contained a mutated copy of the *LEU2* gene (Table 1). On the other hand, leucine prototrophs arose by both plasmid integration and gene conversion. The proportion of convertants obtained with double-stranded plasmid was around 10%, regardeless to the UV-dose given to the plasmid DNA, indicating that both types of transformation process are stimulated to the same extent. However, UV-irradiation of single-stranded plasmids decreases the efficiency of transformation (Ninkovic et al., submited), but increases significantly the proportion of convertants, from 10%, to almost 50% (Table1).

Table 1.
Genetic analysis of Ura+ and Leu+ transformants obtained with UV-irradiated plasmid pULi in double or single-stranded form.

	pULi, double-stranded		pULi, single-stranded	
UV-dose (Jm^{-2})	Ura+ Leu-/ Ura+ Leu+	Leu+ Ura-/ Leu+ Ura+	Ura+ Leu-/ Ura+ Leu+	Leu+ Ura-/ Leu+ Ura+
0	0/69	16/123	0/424	71/867
200	0/113	13/123	0/180	16/117
500	0/115	16/163	0/218	109/473
1000	0/127	17/148	0/25	61/129

Single-stranded nonreplicative plasmids transform yeast with 5-10 fold higher efficiency per µg of DNA than corresponding double-stranded plasmids, also giving rise to plasmid integration or gene conversion (8). We tried to further stimulate transformation by cutting the single-stranded form of the plasmid pOM with the restriction enzyme *SalI* in the region of homology with the yeast chromosome (*CYC1* gene). In order to create the substrate for the endonuclease, plasmid was first hybridised with an appropriate oligonucleotide. Linearisation was confirmed by agarose gel electrophoresis and transformation of the *E. coli*. The results obtained indicate that linearisation didn't have significant impact on the efficiency of transformation with single-stranded nonreplicative plasmid (not shown). A sample of 99 transformants was further analysed by hybridisation with different probes in order to characterise transformation events. The results are summarised as follows: only 18% gave the pattern expected from homologous integration in either *ura3* or *cyc1* gene, while 30% of transformants did not hybridise with vector sequence. The reminding transformants gave complex pattern on the Southern blots (not shown). This contrasts strongly with the results obtained with circular single-stranded plasmids, where transformation of the strain carrying the *ura3*-52 mutation is due almost exclusively to the integration by homologous recombination (8; P. Koren, unpublished observation). We explain our results by less specific integrations and/or additional recombinational events induced by linearised single-stranded plasmids. If this interpretation is correct, transformation in the presence of high amounts of denaturated carrier DNA may lead to uncontrolled changes in the yeast genome. Six transformants that did not hybridise with the vector sequence were further tested for the presence of the Ty element in the *URA3* locus and were shown to have restored the wild-type allele. To our knowledge, this is the first report of the conversion of the Ty element (6.1 kb) by plasmid transformation.

CONCLUDING REMARKS

Taken together, our data show that highly enhanced transformation with UV-irradiated nonreplicative plasmids is not coupled with significant risk of introducing unwanted changes into the yeast genome. Considering the universality of recombinational repair of DNA damages, the transformation with UV-irradiated nonreplicative plasmid may prove useful in organisms other than yeast. UV-irradiated single-stranded plasmids may be used if conversion of the chromosomal allele is prefered to integration of the whole plasmid molecule. Transformation by linearised single-stranded plasmids can lead to the conversions of large insertions in the genome, but also to the complex recombinational events.

ACKNOWLEDGEMENTS

This work was supported by the Croatian Ministry of Science and by the Commission of the European Communities Joint Research Contract CI1-0528-M (CD).

LITERATURE CITED

1. Hinnen, A., J.B. Hicks and G.R. Fink, Proc. Natl. Acad. Sci. USA, 52 (1978) 412.
2. Roitgrund, C., R. Steinlauf and M. Kupiec, Mol. Gen. Genet., 237 (1993) 306.
3. Zgaga, Z., R. Chanet, M. Radman and F. Fabre, Curr. Genet., 19 (1991) 329.
4. Zgaga, Z., Mutat. Res., 263 (1991) 211.
5. Baldari, C. and G. Cesareni, Gene, 35 (1985) 27.
6. Orr-Weaver, T., J. Szostak and R. Rothstein, Proc. Natl. Acad. Sci. USA, 78 (1981) 6354.
7. Rose, M. and F. Winston, Mol. Gen. Genet., 193 (1984) 557.
8. Simon, J.R. and P.D. Moore, Mol. Cell. Biol., 7 (1987) 2329.

Physiology and Genetics of Anaerobes

ECB6: Proceedings of the 6th European Congress on Biotechnology
edited by L. Alberghina, L. Frontali, P. Sensi

Physiology and genetics of anaerobes

R. Daniel and G. Gottschalk

Institut für Mikrobiologie der Georg-August-Universität, Grisebachstrasse 8, 3400 Göttingen, Germany

It is a characteristic of the anaerobic way of life that only one or a few molecules of ATP become available to the organisms per mol of substrate degraded. Therefore, anaerobes need to metabolize large amounts of substrates for growth and, hence, secrete corresponding amounts of fermentation products. If these products are not toxic to the organisms secreting them, they are accumulated up to relatively high concentrations. This is especially true for ethanol and certain diols such as 1,3-propanediol.

Zymomonas mobilis may serve as a first example for a microorganism secreting products up to high concentrations. This microorganism is one of the few prokaryotes which carry out a pure alcoholic fermentation:

$$\text{glucose} + 2H_2O \text{ ------> } 2 \text{ ethanol} + 2CO_2.$$

It was originally isolated from agave sap, and is somewhat adapted to high sugar concentrations in its environment. Therefore, glucose is not taken up by an active transport system but by a glucose facilitator which mediates glucose uptake by facilitated diffusion. Glucose is then degraded via the Entner-Doudoroff-pathway and the pyruvate formed is decarboxylated to yield acetaldehyde which then is reduced to ethanol (1). The ATP yield of this pathway is even lower than the one of the yeast alcohol fermentation in which glucose is degraded via the Embden-Meyerhoff-Parnas-pathway. Breakdown of glucose by *Zymomonas mobilis* is associated with the phosphorylation of one ADP to ATP whereas two ATP per glucose become available to yeasts under anaerobic conditions. Degradation of glucose by *Zymomonas mobilis* proceeds with a high rate, ethanol tolerance is comparable to the one of yeast (2). Nevertheless, *Zymomonas mobilis* has not replaced yeast as alcohol production organism. The reason being that *Zymomonas* cultures have to be operated under conditions under which lactic acid bacteria also thrive nicely so that the problem of infections is serious. In addition, the substrate spectrum of *Z. mobilis* is very narrow, only glucose and fructose are metabolized.

Applying the methods of molecular genetics efforts can be made to overcome the disadvantages of *Z. mobilis.* Its substrate spectrum could be widened by, for instance, introducing genes for the degradation of pentoses. Also, genes encoding for pyruvate decarboxylase and alcohol dehydrogenase can be cloned and transferred into another organism which may be advantageous with respect to alcohol production. The latter has been achieved, the genes have been put into *Escherichia coli* , *Klebsiella planticola* and *K. oxytoca* which then become ethanol producers (3, 4). So the techniques are available to construct an ideal ethanol-producing organism. It should be possible now to first design the optimal and cheapest fermentation broth and then construct the optimal ethanol producer for that environment.

Ethanol is tolerated by its producers in fairly high concentrations. The situation becomes different when we go to higher alcohols such as n-butanol. Because of its lipophilicity n-butanol already at fairly low concentrations deenergizes the cytoplasmic membrane and is toxic (5, 6). Therefore, a production organism such as *Clostridium acetobutylicum* can only reach final concentrations in the order of 2 % butanol and acetone (see 7, 8). *Clostridium acetobutylicum* was of industrial interest between 1915 and 1950 because of its ability to produce butanol and acetone from molasses or starch. In batch culture, *C. acetobutylicum* first produces acetic and butyric acids, the pH is lowered to values around 5, and then the culture undergoes a transition phase, as a result the above mentioned solvents are formed. It has been shown that enzyme induction is a prerequisit for the shift from acid to solvent formation (9). This refers to a coenzyme A transferase and acetoacetate decarboxylase for acetone formation and to butyraldehyde and butanol dehydrogenases for butanol formation. On the basis of changes of the intracellular ATP/ADP and $NADH/NAD^+$ ratios a two signal model for the shift has been proposed (10). It implies that accumulation of organic acids and decrease of pH result in an increase of ATP consumption. ATP is required in increasing amounts to pump protons out of the cell and to maintain a ΔpH across the cytoplasmic membrane. It is assumed that these changes finally lead to the synthesis of the enzymes involved in acetone formation. As soon as acetone formation starts cells become defficient with respect to proper hydrogen acceptors. The $NADH/NAD^+$ ratio will increase, the two mentioned dehydrogenases will be formed and butanol produced. In the solventogenic phase butanol and acetone are formed according to the following equation:

$$3 \text{ glucose} \text{ -----> } 2 \text{ n-butanol} + \text{acetone} + 7CO_2 + 5H_2.$$

Efforts to isolate mutants expressing a higher butanol tolerance as well as increased butanol production have not been very successful. Again, two possibilities exist to obtain a better production organism. One is to increase the persistance of the *C. acetobutylicum* cytoplasmic membrane by changing its composition (11). This can only be done by genetic means. The other possibility would be to screen for organisms which are a priori more resistant towards butanol and then make them a butanol producer by transferring the necessary genetic information to them (see chapter by R. Fischer et al.)

A third example to be discussed here is the fermentation of glycerol by *Citrobacter freundii*. Energy in the form of ATP can only be conserved by glycerol oxidation. Therefore, part of the glycerol is oxidized via dihydroxyacetone to dihydroxyacetonephosphate. The latter is isomerized to yield 3 phosphoglyceraldehyde; its further conversion via the glycolytic pathway to pyruvate yields ATP by substrate level phosphorylation. The reaction sequence mentioned is associated with the reduction of $2NAD^+$ to 2NADH. Some of the NADH is oxidized by the formation of ethanol, however, most of it by reduction of additional glycerol to 1,3-propanediol. The fermentation equation is rather complex. Per 100 mol of glycerol up to 66 mol 1,3-propanediol and around 10 mM each of ethanol, acetate and pyruvate are secreted (12):

$$3 \text{ glycerol} \text{ -----> } 2 \text{ 1,3-propanediol} + 1/3 \text{ pyruvate} + 2/3\ CO_2 + 1/3 \text{ acetate} + 1/3 \text{ ethanol}.$$

Four enzymes are characteristic for this fermentation: glycerol dehydrogenase, dihydroxyacetone kinase, glycerol dehydratase and 1,3-propanediol dehydrogenase. The first two mentioned enzymes are also present in *C. freundii* when growing aerobically on glycerol. The latter two enzymes are only synthesized under anaerobic conditions, and both are quite oxygen sensitive. Inactivation of propanediol dehydrogenase proceeds in a reaction requiring metal ions; the dehydratase is a protein containing coenzyme B_{12} and the dehydration of glycerol proceeds via a radical mechanism. This enzyme is extremely oxygen sensitive (13) and its activity seems to be rate-limiting when resting cells are employed for the production of 1,3-propanediol (12). The genes encoding for the four enzmyes involved in glycerol fermentation are organized in the *dha*-regulon which recently was cloned and expressed in *E. coli* (14).

Sequencing of the whole regulon is almost finished (R. Daniel, unpublished), and the construction of a strain overexpressing the enzymes for the conversion of glycerol to 1,3-propanediol is in progress. It may serve as a production organism for 1,3-propanediol using crude glycerol solutions from the fat saponification industry.

Another promissing area of application of anaerobes is the production of heat-stable exoenzymes. These thermophilic anaerobes include members of the genus Clostridium (*C. thermosulforogenes, C. thermosaccharolyticum, C. thermohydrosulfuricum,*) Thermoanaerobacter (*T. finii, T. ethanolicum, Thermoanaerobium brockii* and *Thermobacteroides acetoethylicus* (15). In addition to α- and ß-amylases the α-glucosidases and pullulanases of these organisms are of special interest. Their temperature optimum lies in the area of 70 to 90 ° C. Some of the genes encoding for these enzymes have already been cloned and sequenced, and the molecular analysis of the pullulanase-encoding gene of *Clostridium thermosulfurogenes* will be discussed in the chapter by G. Burchhard et al..

References

1. T. Conway, FEMS Microbiol Rev. 103 (1) (1992) 1.
2. S.E. Buchholz, M.M Dooley and D. E. Eveleigh, TIBTECH 5 (1987)199.
3. G. Burchhardt and L.O. Ingram, Appl. Environ. Microbiol. 58 (1992) 1128.
4. S. Feldmann, G.A. Sprenger and H. Sahm, Appl. Microbiol. Biotechnol. 31 (1989) 152.
5. L.O. Ingram, Trends Biotechnol. 4 (1986) 40.
6. A.A. Herrero, Trends Biotechnol. 1 (1983) 49.
7. H.J. Rehm and G. Rèed (eds.) Biotechnology 6b (1988) 1.
8. P.Dürre, H. Bahl and G. Gottschalk Chem.-Ing.-Tech. 64 (1992) 491.
9. W.Andersch , H. Bahl and G. Gottschalk, Eur. J. Appl. Microbiol. Biotechnol. 18 (1983) 327.
10. H. Grupe and G. Gottschalk, Appl. Environ. Microbiol. 58 (1992) 3896.
11. N.C. Johnston and H. Goldfine, J. Bacteriol. 174 (1992) 1848.
12. R. Boenigk, S. Bowien and G. Gottschalk, Appl. Microbiol. Biotechnol. 38 (1993) 453.
13, K. Ushio, S. Honda, T. Toraya and S. Fukui, J. Nutr. Sci. Vitaminol. 28 (1982) 225.
14. R. Daniel and G. Gottschalk, FEMS Microbiol. Lett. 100 (1992) 281.
15. G. Antranikian, FEMS Microbiol.Rev. 75 (1990) 201.

ECB6: Proceedings of the 6th European Congress on Biotechnology
edited by L. Alberghina, L. Frontali, P. Sensi

Regulation of solvent formation in *Clostridium acetobutylicum*

R. Fischer, U. Sauer, and P. Dürre

Institut für Mikrobiologie, Georg-August-Universität Göttingen, Grisebachstraße 8, D-3400 Göttingen, Germany

1. INTRODUCTION

Clostridium acetobutylicum is a strictly anaerobic, spore-forming bacterium showing a characteristic biphasic fermentation pattern. After producing acetate and butyrate during exponential growth, the organism switches to the formation of mainly acetone and butanol shortly before entering the stationary growth phase. Prerequisites for this shift are threshold concentrations of the aforementioned acids, a low pH, and a suitable growth-limiting factor such as phosphate or sulfate (1). The molecular mechanisms causing the onset of solventogenesis are still mostly unknown.

2. ORGANISATION OF GENES REQUIRED FOR SOLVENTOGENESIS

Acetone formation is catalyzed by the two enzymes acetoacetyl-CoA: acetate/butyrate: CoA transferase (CoA transferase or Ctf) and acetoacetate decarboxylase (Adc), whereas butanol is produced by action of butyraldehyde and butanol dehydrogenases. The *adc* gene is organized in a monocistronic operon and its transcription is induced or derepressed by conditions necessary for the onset of solvent formation (2, 3, 4). By primer extension analysis a promoter structure typical for a vegetative sigma factor could be identified, thus, the involvement of alternate sigma factors in regulation of solventogenesis appears unlikely (3). The structural genes of the two subunits of CoA transferase are part of a large operon that is contiguous to the *adc* gene, but transcribed in the different direction. Both operons are induced at the same time and are separated by a large stem-loop structure that probably functions as a rho-independent transcription terminator in both directions (2, 3, 4). The remaining part of the operon containing the *ctf* genes encodes a single large open reading frame of 2.6 kbp. The deduced amino acid sequence for this protein of app. 95 kDa showed a very high similarity (more than 50% identity) to the AdhE protein of *Escherichia coli*. This gene product represents a bifunctional enzyme exhibiting both, alcohol and acetaldehyde dehydrogenase activities (5). Thus, the operon consisting of *ctfB*, *ctfA*, and *adhE* obviously plays a crucial role in solvent formation by *C. acetobutylicum*. We therefore propose to name it *sol* operon (mnemonic: solvents). The organisation of the *adc* and *sol* operons is shown in Figure 1. It is assumed

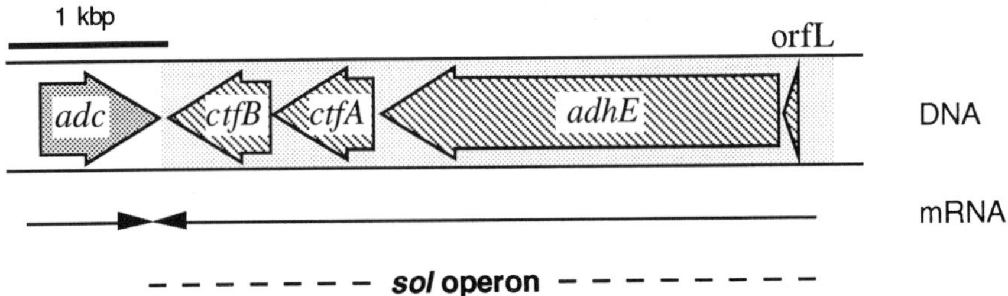

Figure 1. Organisation of genes required for solventogenesis in *C. acetobutylicum*.

that the clostridial *adhE* gene is involved in butanol formation *in vivo*. In addition, two other genes have been cloned and sequenced that encode NAD-dependent butanol dehydrogenases (5, 6). Both are also induced or derepressed immediately prior to accumulation of significant amounts of butanol in the culture (6). Their relative role is still unknown.

3. POSSIBLE FUNCTION OF A RARE tRNA IN TRANSLATIONAL CONTROL OF SOLVENT FORMATION

Recently, the Tn*916*-induced mutant AA2 of *C. acetobutylicum*, defective in the formation of acetone and butanol, has been isolated and characterized (7). Using single-specific-primer PCR the respective DNA region was cloned and subsequently sequenced (8). The insertion point of Tn*916* was shown to be 55 bp upstream of the *thrA* gene encoding a tRNA, as judged by homology studies and RNA analysis programs. The 76 nt tRNA showed the typical features of bacterial tRNAs (cloverleaf structure, CCA acceptor arm, length of stems and loops, highly conserved nucleotides in certain positions) as shown in Figure 2. Transcript length and secondary structures predicted from the computer model were confirmed by primer extension and S1 nuclease mapping assays. The tRNA is charged with threonine as deduced from the anticodon UGC. Codon usage of all sequenced genes of *C. acetobutylicum* showed

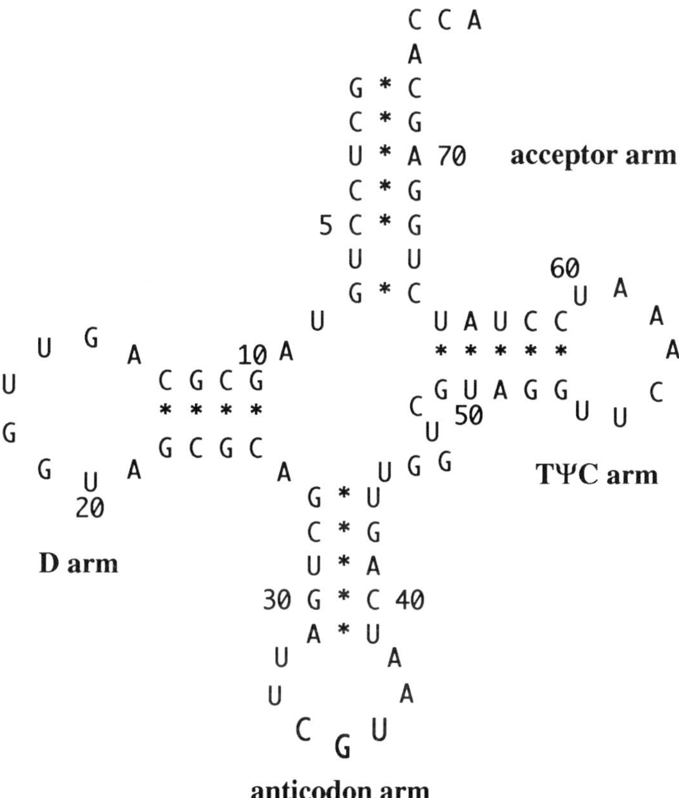

Figure 2. Deduced secondary structure from the DNA sequence of the *thrA* gene.

that the respective codon ACG is rarely used in this organism (8). With few exceptions it is exclusively found in genes related to solvent formation, spore generation, and those involved in uptake or metabolism of minor C or N substrates. However, sequenced genes of enzymes necessary for the major metabolic pathways including acid formation do not contain the ACG codon. This codon selection might reflect an evolutionary process which results in a codon usage corresponding to the most abundant isoacceptor tRNAs for highly expressed genes, thus ensuring fast translation of the messenger RNA, whereas weakly expressed genes do not show such a codon bias (9). Alternatively, the restriction of the ACG codon to the above mentioned enzymes might be part of a novel translational control system, so far only proposed for *Streptomyces* species (10), a group of mycelial Gram-positive bacteria, which use the extremely rare leucine codon TTA in a few genes necessary during differentiation, when the colonies begin to produce aerial hyphae and antibiotics. No increase of the tRNA level in solvent-producing cells of *C. acetobutylicum* could be detected using Northern blot, primer extension, and S1 mapping analysis. Thus, in addition to the transcriptional control mentioned before a potential translational regulation mechanism could exist at the level of processing the primary transcript to the mature charged tRNA, possibly involving the charging with threonine or the modification of particular bases of the tRNA. These results are consistent with those obtained by others in the better characterized regulatory system of *Streptomyces* spp. (10).

Contiguous to the *thrA* gene of *C. acetobutylicum*, but divergently transcribed, an open reading frame was found that corresponded to a 18 kDa protein with high similarity to heat shock proteins of the 20 kDa family. The respective protein is only induced at temperature increase or onset of solventogenesis (11). Its relationship to acetone and butanol formation is still obscure.

REFERENCES

1. H. Bahl and G. Gottschalk, in: H.J. Rehm and G Reed (eds.), Biotechnology, vol. 6b, p. 1-30, VCH Verlagsgesellschaft mbH, Weinheim, Germany (1982).
2. U. Gerischer and P. Dürre, J. Bacteriol., 172 (1990) 6907-6918.
3. U. Gerischer and P. Dürre, J. Bacteriol., 174 (1992) 426-433.
4. D.J. Petersen, J.W. Cary, J. Vanderleyden, and G.N. Bennett, Gene, 123 (1993) 93-97.
5. D.J. Petersen, R.W. Welch, F.B. Rudolph, and G.N. Bennett, J. Bacteriol., 173 (1991) 1831-1834.
6. K.A. Walter, G.N. Bennett, E.T. Papoutsakis, J. Bacteriol., 174 (1992) 7149-7158.
7. J. Bertram, A. Kuhn, and P. Dürre, Arch. Microbiol., 153 (1990) 373-377.
8. U. Sauer and P. Dürre, FEMS Microbiol. Lett., 100 (1992) 147-154.
9. M. Gouy and C. Gautier, Nucl. Acids Res., 10 (1982) 7055-7074.
10. B.K. Leskiw, M.J. Bibb, and K.F. Chater, Mol. Microbiol., 5 (1991) 2861-2867.
11. A. Pich, F. Narberhaus, and H. Bahl, Appl. Microbiol. Biotechnol., 33 (1990) 697-704.

ECB6: Proceedings of the 6th European Congress on Biotechnology
edited by L. Alberghina, L. Frontali, P. Sensi

Molecular analysis of the pullulanase-encoding gene of *Clostridium thermosulfurogenes* EM1

G. Burchhardt[a*], K. Sahm[a], M. Matuschek[a], H. Müller[a], K. Haeckel[a+], A. Spreinat[a§], G. Antranikian[a#], G. Steinborn[b], and H. Bahl[a]

[a]Institut für Mikrobiologie, Georg-August Universität, Grisebachstraße 8, W-3400 Göttingen, FRG

[b]Institut für Pflanzengenetik und Kulturpflanzenforschung, Corrensstraße 3, O-4325 Gatersleben, FRG

1. INTRODUCTION

Clostridium thermosulfurogenes EM1 produces and secretes high levels of thermostable α-amylase and pullulanase, when grown in continuous culture with starch as substrate [1]. Both enzymes have been purified and characterized [2]. The pullulanase of *C. thermosulfurogenes* EM1 cleaves α-1,6-linkages in pullulan, amylopectin and glycogen as well as α-1,4-linkages in amylose, amylopektin and oligosaccharides. Due to this dual specificity this enzyme was named pullulanase type II [3] and it is able to degrade branched polysaccharides completely to maltotriose and maltose.

Recently, the genes encoding the α-amylase and pullulanase of *C. thermosulfurogenes* EM1 have been cloned and expressed in *Escherichia coli* [4,5]. The nucleotide sequence of the α-amylase-encoding *amyA* gene, the processing of the enzyme, and the comparison to other α-amylases have been published elsewhere [6]. Here we report on the molecular analysis of the pullulanase-encoding *amyB* gene of *C. thermosulfurogenes* EM1.

2. CLONING AND SEQUENCING OF THE PULLULANASE GENE

It was previously shown that *E. coli* expresses an active pullulanase from a 5.3-kbp *Sau*3A DNA fragment of *C. thermosulfurogenes* EM1, cloned in the *Bam*H1 site of pUC18 to yield plasmid pCT3 [5]. The DNA sequence of this fragment was determined, and the accession number M57692 in the GenBank database has been assigned to this sequence. Surprisingly, analysis of the sequence revealed that only the 5' end (3,627 bp) of the pullulanase-encoding *amyB* gene had been cloned, coding for a protein of 143,373 Da. Since no differences in the

Present addresses
* Institut für Genetik, Ernst-Moritz-Arndt-Universität, Jahnstraße 15a , O-2200 Greifswald, FRG
\+ Staatliches Gewerbeaufsichtsamt Frankfurt, Untermainkai 27/28, W-6000 Frankfurt/Main 1, FRG
§ WISSTRANS Umwelt GmbH, Am Leinekanal 4, W-3400 Göttingen, FRG
\# Biotechnologie/Technische Mikrobiologie, Technische Universität Hamburg-Harburg, Denickestraße 15, W-2100 Hamburg 90, FRG

This research was funded by grants from the "Bundesministerium für Forschung und Technologie" to G.A. and H.B.

activity of the plasmid-encoded truncated protein and the original enzyme were observed [5], the C-terminus is obviously not necessary for enzymatic function of the protein. To obtain information on the entire gene, the 3' end was cloned by chromosome walking and by PCR technique. The isolated plasmid pCT5 contained additional 1,325 bp of *amyB,* but still not the end of the gene (Figure 1). By Southern bybridization it was shown that *amyB*, the α-amylase-encoding *amyA* gene, and the genes *amyC* and *amyD,* which encode putative maltose transport proteins [7], are located in the same region of the chromosome (Figure 1). The DNA region between *amyB* and *amyD* was amplified by PCR, and fragments of the obtained product were cloned in pUC18 to yield plasmids pCT11 and pCT12 (see Figure 1). Sequencing of this region is in progress and will define the exact size of the gene and hence the enzyme. According to our results the gene has a possible maximum size of 6.9 kbp which corresponds to a 250-kDa protein. During growth of *C. thermosulfurogenes* EM1 on starch, seven major forms of pullulanase with molecular masses between 490 kDa and 93 kDa were detected [2]. However, no structural differences could be observed between these proteins, e. g., the N-terminal sequences were identical. Association of glycocomponents to the proteins during secretion through the cell envelope, protein association, and/or proteolysis from the C-terminus are possible reasons for the multiplicity of this enzyme [2]. Interestingly, in *E. coli* a similar processing at the C-terminus of the pullulanase occurred without affecting the function of the enzyme. Analysis of the structural features of this protein domain will be the first step to elucidate its function.

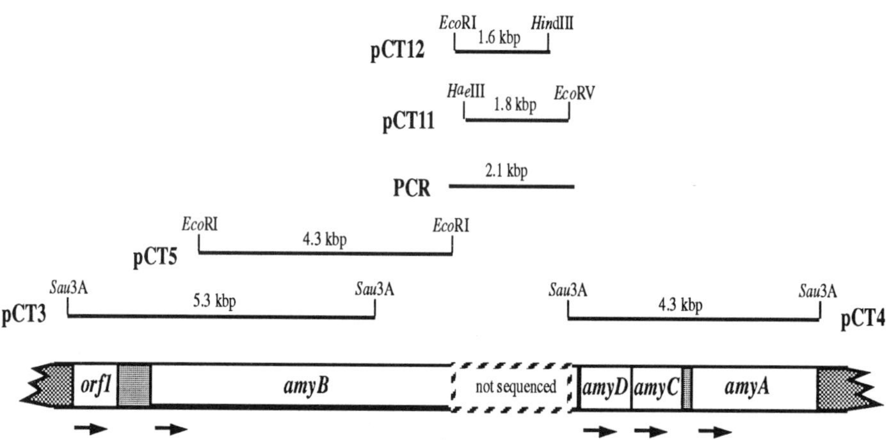

Figure 1. Schematic representation of the *amy* gene region of *C. thermosulfurogenes* EM1 and of the respective inserts of pCT3, pCT4, pCT5, pCT11, and pCT12. Restriction enzymes used for cloning are given at the end of the fragments. Arrows show the direction of transcription. A protein of unknown function is encoded by *orf1.* Further details are given in the text.

3. ANALYSIS OF MESSENGER RNA

High levels of pullulanase and α-amylase activity in *C. thermosulfurogenes* EM1 can only be detected when the cells are grown on starch in the absence of glucose. To determine whether this fact is consistent with corresponding mRNA levels, total RNA was isolated from *C. thermosulfurogenes* EM1 cells grown in a chemostat under phosphate limitation with glucose or starch as substrate and under starch or glucose limitation. Maximum *amyA*- and *amyB*-specific mRNA levels were found in cells grown with excess starch under phosphate limitation by dot blot hybridisation. Interestingly, maximum enzyme production has been observed during growth under starch limitation [1]. Northern blot analysis revealed that *amyA* and *amyB* are not organized in one operon. By primer extension analysis the transcription start points of the two genes were determined, and the promoter sequences were deduced from these sites. The promoter sequence of *amyA* was 5'-TCGATA (-35), 5'-TAATAT (-10) and of *amyB* 5'-TTCAGG (-35), 5'-CATTAT (-10). Both promoters showed reasonable similarity to the consensus promoter sequence of Gram-positive bacteria as well as *E. coli*. However, in *E. coli*, the sequence 5'-TTGACA (-35), 5'-TATAAT (-10) upstream of *amyB* was used as a promoter. This indicates that in *Clostridium* special sequences in addition to the conserved -10 and -35 regions have to be present for a functional promoter. This is important for organisms with a low G+C content, in which sequences like 5'-TATAAT frequently occur, especially in non-coding regions of the chromosome.

4. SEQUENCE COMPARISON

The enzymatically active N-terminus of the pullulanase (130 kDa) had most similarity to the α-amylase-pullulanase of *C. thermohydrosulfuricum*, an enzyme also cleaving α-1,4- and α-1,6-linkages in branched polysaccharides [8]. It contained 72 % identical and 83 % similar amino acids. Furthermore, the N-terminus of the pullulanase of *C. thermosulfurogenes* EM1 showed several regions similar to the four conserved regions of amylases [9]. Comparison of the consensus regions of several pullulanases suggested that enzymes, which split α-1,4- and α-1,6-linkages in polysaccharides have specific amino acid sequences in these consensus regions, which are different from those of pullulanases, which only cleave α-1,6-linkages like the enzyme of *Klebsiella aerogenes* (Table 1). The most striking differences between the two types of pullulanases were found in region 2 and region 4. The consensus region 2 of type II pullulanases is divided. The last part of this region is located between region 3 and 4. Whether this observation is significant for the specificity of the enzyme has to be demonstrated by appropriate mutation experiments.

5. SYNTHESIS OF PULLULANASE IN *BACILLUS SUBTILIS*

As a basis for an overproduction of the commercially valuable pullulanase from *C. thermosulfurogenes* EM1, the *amyB* gene was cloned and expressed in *B. subtilis*. Therefore, a 4-kbp fragment of pCT3 containing only the necessary 5' end of *amyB* was ligated into plasmid pSB319, a derivative of pIB501 [10], resulting in plasmid pKH1. During growth in minimal medium with starch as substrate *B. subtilis* GSB26 transformed with pKH1 produced 700 U/l of pullulanase. One unit of pullulanase is defined as that amount of enzyme which is able to liberate 1 µmol of reducing sugar as a maltose standard per minute. Approximately 70 % of the enzyme was secreted into the culture medium by a signal peptide-dependent pathway. Analysis of the N-terminal amino acid sequence of the mature proteins revealed that the signal peptide was cleaved at exactly the same site between two alanine residues in *C. thermosulfurogenes* EM1, *B. subtilis*, and also in *E. coli*, where export of the enzyme to the periplasmic space could be observed (Figure 2).

```
α-amylase                  MKKTFKLILVLMLLSLTLVFGLTAPIQA ↓ ASD
pullulanase     MNKKLFTNRFISFNMSLLLVLTAVFSSIPLHSVHA ↓ ADN
```

Figure 2. Signal peptides of α-Amylase [6] and pullulanase of *C. thermosulfurogenes* EM1. Arrows show the cleavage sites in *C. thermosulfurogens* EM1, *B. subtilis*, and *E. coli*.

Table 1
Consensus regions of amylases [9] in pullulanases

Enzyme/Organism	Region 1	Region 2	Region 3	Region 4
	D A V I N H	G F R L D A A K H	E V I D	F V D N H D
PUL C. thermosulf.	486 D G V F N H	589 G W R L D V E N E 664 H N P I D A A K L	622 E N W G D	694 L L G S H D
APU C. thermohydr.	488 D G V F H H	624 G W R L D V A N E 669 H N P I D A A K L	627 E N W N D	699 L L G S H D
NPL B. stearo.	242 D A V F N H	324 G W R L D V A N E 382 G V L R F F A K E	357 E I W H D	419 L L G S H D
PUL K. aero.	600 D V V Y N H	671 G F R F D L M G Y	704 E G W D	827 Y V S K H D
PUL Thermus sp.	338 D A V Y N H	406 G F R F D L M G V	438 E G W D	522 Y V E C H D

Active-site residues and those involved in substrate binding are surrounded by a circle and a rectangle, respectively. The position of the amino acids in the enzymes is indicated by the numbers. C. thermosulf., *C. thermosulfurogenes* EM1 (this work); C. thermohydr., *C. thermohydrosulfuricum* [8]; B. stearo, *B. stearothermophilus* [11], K. aero. *Klebsiella aerogenes* [11], *Thermus* spec. [Sashihara, personal communication]. PUL, pullulanase; APU, α-amylase-pullulanase; NPL, neopullulanase.

6. CONCLUSION

The data presented on the pullulanase of *C. thermosulfurogenes* EM1 are valuable for understanding the synthesis, the structure-function relationship, and the secretion of this enzyme. The elucidation of the function of the large C-terminus of the enzyme not necessary for activity will be an important aspect of future experiments.

REFERENCES

1. E. Madi, G. Antranikian, K. Ohmiya, and G. Gottschalk, Appl. Environ. Microbiol., 53 (1987) 1661.
2. A. Spreinat and G. Antranikian, Starch/Stärke, 44 (1992) 305.
3. G. Antranikian, FEMS Microbiol. Rev., 75 (1990) 201.
4. K. Haeckel and H. Bahl, FEMS Microbiol. Lett., 60 (1989) 333.
5. G. Burchhardt, A. Wienecke, and H. Bahl, Curr. Microbiol., 22 (1991) 91.
6. H. Bahl, G. Burchhardt, A. Spreinat, K. Haeckel, A. Wienecke, B. Schmidt, and G. Antranikian, Appl. Environ. Microbiol., 57 (1991) 1554.
7. H. Bahl, G. Burchhardt, and A. Wienecke, FEMS Microbiol. Lett., 81 (1991) 83.
8. H. Melasniemi, M. Paloheimo, and L. Hemiö, J. Gen. Microbiol. 136 (1990) 447.
9. Y. Matsuura, M. Kusunoki, W. Harada, and M. Kakudo, J. Biochem., 95 (1984) 697.
10. F. Horodniceanu, P.H. Bouanchaud, G. Bieth, and Y.A. Chabbert, Antimicrob. Agents Chemother., 10 (1976) 795.
11. T. Kuriki and T. Imanaka, J. Gen. Microbiol., 135 (1989) 1521.

Taxonomical Aspects in Biotechnology

ECB6: Proceedings of the 6th European Congress on Biotechnology
edited by L. Alberghina, L. Frontali, P. Sensi

IMPLICATION OF BACTERIAL TAXONOMY ON BIOTECHNOLOGY

Karl Heinz Schleifer

Lehrstuhl für Mikrobiologie, Technische Universität,
Arcisstr. 21, D-8000 München 2, Germany

Biotechnologists are mainly interested in special biochemical properties, physiological activities and growth behaviour of bacteria. The need to classify and identify organisms has rarely been seen as an important task in biotechnology. Therefore, biotechnologists have neglected bacterial taxonomy for a long time. Conventional techniques used for the identification of microorganisms were considered dull and uninteresting.However, the new safety regulations require a detailed characterization of the organisms, their taxonomic position has to be known and therefore the identification of bacteria has become more important in recent years. Progress in molecular biology, in particular in the sequencing of genetic material, has revolutionized the taxonomy of bacteria. For the first time it is now possible to infer the phylogenetic relationships of bacteria. Moreover, nucleic acid probes can now be designed and used for rapid and reliable identification of organisms.

Phylogenetic relationship

A breakthrough in the study of the phylogenetic relationships of prokaryotes was achieved by Carl Woese and coworkers who compared first partial and later total sequences of 16S ribosomal ribonucleic acids (rRNA). Most of the current information on the phylogenetic relatedness of bacteria from comparative 16S rRNA analysis (Woese, 1987). Comparative sequence studies on genes coding for other conserved macromolecules such as 23S rRNA, elongation factor Tu and ß-subunit of ATP-synthase are in very good agreement with the 16S rRNA analysis (Schleifer and Ludwig, 1989; Ludwig and Schleifer, 1993). This indicates that the trees reflect actually the relationship of the organisms and not that of the genes per se. On the basis of these analyses it is now widely accepted that all living organisms can be divided into three domains (former kingdoms): *Archaea, Bacteria* and *Eukarya* (Woese et al., 1990). *Bacteria* can be subdivided into at least 13 lines of descent, so-called phyla. The most important ones are the *Proteobacteria* which contains most of the well-known Gram-negative bacteria and the Gram-positive bacteria which can be clearly distinguished into two phyla. One phylum comprises all Gram-positives with a high guanine plus cytosine (GC) content in their DNA (higher than 50 mol%, e.g. all actinomycetes), the other phylum consists of Gram-positives with a low GC-content (below 50 mol%, e.g. clostridia, bacilli). Other phyla are cyanobacteria, chlamydiae, bacteroides-flavobacteria, green sulfur bacteria, spirochaetes, radioresistant deinococci, green non-sulfur bacteria and the extreme thermophiles Thermotoga and Aquifex. The *Archaea* can be subdivided into the two kingdoms: *Euryarchaeota*(comprising methanogens and extreme halophiles) and the *Crenarchaeota* (encompassing extreme thermophiles).

Improved identification by using nucleic acid probes

The information on the 16S and 23S rRNA sequences cannot only be used for inferring the phylogenetic relationships but also for designing specific nucleic acid probes (Schleifer et al., 1992; 1993). A nucleic acid probe is a fragment (preferably single stranded) of a nucleic acid that will bind to complementary regions of DNA or RNA (target nucleic acid). Probes can either be isolated from the DNA of the target organisms or synthesized as oligonucleotides in the laboratory. The reactions are highly specific so that probes will bind to their complementary sequences even when these sequences account for only a small fraction of the available nucleic acids.

The nucleic acid probe technique consists of four steps (Schleifer, 1991). First, the probe has to be labelled with a radioactive or non-radioactive detector group. Second, the target nucleic acid has to be extracted or the cells have to be made permeable for the labelled probe. Third, the probe has to react (hybridisation) with the target nucleic acid and the unbound or non-specifically bound probe has to be removed. Fourth, the specific hybrids are detected by measuring the amount of labelled probe bound to the target nucleic acid.

There are different approaches to design nucleic acid probes. The three most important ones are summarized in Table 1.

Table : Design of nucleic acid probes

Origin	Design	Specificity
Randomly cloned DNA fragment	Isolation and cloning of non cross-hybridizing fragments	Strain to species
Gene or gene fragment	Isolation and cloning of genes or fragments of genes	Organisms containing these genes
16S/23S rRNA	Comparative sequence analysis, synthesis of oligonucleotides	wide specificity, species to universal

I shall concentrate on rRNA-targeted oligonucleotide probes. The 16S and 23S rRNA molecules contain regions of highly conserved and less conserved sequences. Therefore, it is possible to design probes of a wide specificity ranging from species specificity to a universal probe that will react with any cellular life form. Moreover, using rRNA as target nucleic acid increases considerably the sensitivity of the test since rRNAs are present in high copy numbers (up to 10^5 copies per cell) in growing bacterial cells.

Different formats can be used for probe assays. For the **dot-blot** hybridization assay small amounts of target nucleic acids have to be extracted and immobilized onto nylon, nitrocellulose or teflon membranes. **Colony hybridization** is a convenient, rapid and simple method for analysing mixed cultures. The colonies are transferred to membranes and are lysed. Gram-negative bacteria can be easily lysed on filters using simple alkali treatment, whereas Gram-positive bacteria require an additional heating step. For the *in situ* identification of individual **whole cells** it is necessary to make the cells permeable for the oligonucleotide probes reacting with rRNA. The whole cell hybridization technique even allows the identification and *in situ* detection of individual cells without prior cultivation using fluorescently labelled rRNA-targeted oligonucleotide probes (Amann et al., 1992). The labelled cells can be analyzed by epifluorescence microscopy or flow cytometry (Amann et al., 1990). This approach is also helpful for the identification and detection of non-culturable bacteria (Amann et al., 1991).

Identification of non-culturable bacteria

The so far cultured bacteria represent only a minor portion of the species living on Earth. It is estimated that about 95% are currently not known. From most habitats less than 10% and sometimes even less than 1% of the actual occurring bacteria can be cultivated. The non-culturable microorganisms represent a huge genetic and also biotechnological potential that currently cannot be used since these organisms are not available as pure cultures. The application of molecular technologies allows the detection of hitherto completely unknown groups of microorganisms (DeLong, 1992; Fuhrman et al., 1992; Schmidt et al., 1991; Spring et al., 1992). Various non-culturable endosymbionts, pathogenic bacteria and aquatic bacteria could be identified by 16S rRNA sequence analyses and some of them could be *in situ* detected by using specific fluorescently labelled rRNA-targeted oligonucleotide (Amann et al, 1991; Britschgi and Giovannoni, 1991; Embley et al., 1992; Relman et al., 1992; Springer et al., 1992).

Implication on biotechnology

Modern molecular taxonomic methods, in particular rRNA sequencing, are powerful tools to provide insight into microbial diversity. Moreover, knowledge on the phylogenetic relationship of non-culturable bacteria to culturable bacteria may facilitate their cultivation in the laboratory and make them available for biotechnological exploitation.

Nucleic acid probes are very helpful for the rapid detection and identification of novel and target bacteria on primary isolation plates by applying colony hybridization techniques or even without prior cultivation by using whole cell hybridization. These probes can be also used for monitoring growth of target organisms or detecting contaminants. Moreover, it should be possible to develop nucleic acid probes for *in situ* or laboratory-based screening.Using probes which are reacting with genes coding for certain enzymes, toxins or physiological properties one could search for organisms carrying these properties.

As mentioned in the Introduction, new biosafety regulations and directives require a comprehensive characterization of organisms used in biotechnological processes. Therefore, a thorough taxonomic study of these organisms is a prerequisite. Usually systematists identify microorgansims at the species level. However, in biotechnology it is often necessary to distinguish individual strains. This could be achieved by designing nucleic acid probes that will react with specific genes, plasmids or characteristic DNA sequences of the corresponding strain. If the necessary sequence information is not available it might be also possible to distinguish individual strains or at least groups of closely related strains by restriction fragment length polymorphisms of rRNA genes (Grimont and Grimont, 1986; Köhler et al., 1991) or by randomly amplified polymorphic DNA analysis (Welsh and McClelland, 1990; Aznar et al., 1993).

References

R.I. Amann, B.J. Binder, R.J. Olson, S.W. Chisholm, R. Devereux, and D.A. Stahl. Appl. Environ. Microbiol., 56 (1990) 1919.

R.I. Amann, N. Springer, W. Ludwig, H.-D. Görtz and K.H. Schleifer. Nature, 351 (1991) 161.

R.I. Amann, W. Ludwig and K.H. Schleifer. FEMS Microbiol. Letters, 100 (1992) 45.

R. Aznar, W. Ludwig and K.H. Schleifer. Syst. Appl. Microbiol., 16 (1993) in press.

T.B. Britschgi, and S. J. Giovannoni. Appl. Environ. Microbiol., 57 (1991) 1707.

E.F. Delong. Proc. Natl. Acad. Sci. USA, 89 (1992) 5685.

T.M. Embley, B.J. Finlay, R.H. Thomas and P.L. Dyal. J. Gen. Microbiol., 138 (1992) 1479.

J.A. Fuhrmann, K. McCallum and A.A. Davis. Nature, 356 (1992) 148.

F. Grimont and P.A.D. Grimont. Ann. Inst. Pasteur, 137B (1986) 165.

G. Köhler, W. Ludwig and K.H. Schleifer. FEMS Microbiol. Letters, 84 (1991) 307.

W. Ludwig and K.H. Schleifer. Antonie van Leeuwenhoek J. Microbiol., in press.

D.A. Relman, P.W. Lepp, K.N. Sadler and T.M. Schmidt. Mol. Microbiol., 6 (1992) 1801.

K.H. Schleifer. Mitteil. Gebiet Lebensmittel-Untersuchung und -Hygiene, 82 (1991) 513.

K.H. Schleifer and W. Ludwig. K.H., W. Ludwig and R. Amann. **In:** B. Fernholm, K. Bremer and H. Jörnvall (eds.), Hierarchy of Life, (1989) 103, Excerpta Medica, Amsterdam.

K.H. Schleifer, W. Ludwig and R.I. Amann. Naturwissenschaften, 79 (1992) 213.

K.H. Schleifer, W. Ludwig and R.I. Amann. **In:** M. Goodfellow and A.G. O'Donnell (eds.), Academic Press, in press.

T.M. Schmidt, E.F. DeLong and N.R. Pace. J. Bacteriol., 173 (1991) 4371.

S. Spring, R.I. Amann, W. Ludwig, K.H. Schleifer and N. Petersen. Syst. Appl. Microbiol., 15 (1992) 97.

N. Springer, W. Ludwig, W. Drozanski, R.I. Amann and K.H. Schleifer. FEMS Microbiol. Letters, 96 (1992) 199.

J. Welsh and M. McClelland. Nucleic Acids Res., 18 (1990) 7213.

C.R. Woese. Microbiol. Rev., 51 (1987) 221.

C.R. Woese, O. Kandler and M.L. Wheelis. Proc. Natl. Acad. Scie. USA, 87 (1990) 4576.

ECB6: Proceedings of the 6th European Congress on Biotechnology
edited by L. Alberghina, L. Frontali, P. Sensi
1994 Elsevier Science B.V.

The importance of correct nomenclature in biotechnology: a brief history of *Saccharomyces sensu stricto*

A. Vaughan-Martini and A. Martini
Dipartimento di Biologia Vegetale, Università di Perugia, I-06121 Perugia, Italy

1. INTRODUCTION

Those yeasts commonly associated with the production of bread, wine, beer and other less familiar alcoholic beverages, collectively indicated as the *sensu stricto* group of the genus *Saccharomyces* and comprising as their main representative the champion fermenter of all yeast species (*S. cerevisiae*), are the most extensively studied and widely exploited organisms known to man. After briefly illustrating the uncertainty in the taxonomy of these species and the nomenclatural confusion over the last 150 years, we will discuss the possible implications and consequences of a correct classification in relation to their use in contemporary biotechnology.

2. THE BEGINNING OF YEAST TAXONOMY

In 1838, Meyen (1) described yeast cells in fermenting malt and introduced the name *Saccharomyces cerevisiae,* literally "fungus of sugar". Reess (2) later provided a thorough morphological description, while Emyl Christian Hansen (3) applied the pure culture technique to the study of yeast, describing two species: *S. cerevisiae* and *S. ellipsoideus,* isolated respectively from fermenting malt and grape must. This was the first valid description of *S. cerevisiae* since, though limited to morphological features, it was performed on a pure culture. In the years to follow, many new epithets such as *S. pastorianus, S. carlsbergensis, S. chevalieri* , *S. uvarum* and *S. oviformis* were introduced for fermenting yeasts.

3. CLASSIFICATION AND NOMENCLATURE OF FERMENTATION YEASTS (1912-1984)

In 1912 Guilliermond (4) separated 20 different species among the oval, elliptical-shaped, highly fermenting cultures isolated from grape must and breweries. The designation of these species underwent many modifications in the various editions of the monograph of the Dutch School of Yeast Taxonomy (The Yeast-a Taxonomic Study): 16 species were recognized in the first edition of 1952 (5) while, in the second edition of 1970, van der Walt (6) placed the 21 fermentation-associated species into the expressly created *sensu stricto* group I. Later on, it became clear that conventional taxonomy based solely upon phenotypic characteristics did not sufficiently clarify taxa (7, 8) and in the third 1984 edition (9) all 21 species of the *sensu stricto* group (6) were reduced into one omnicomprehensive taxon, the species *Saccharomyces cerevisiae.*

4. MOLECULAR TAXONOMY OF *SACCHAROMYCES SENSU STRICTO*

The shortcomings of traditional classification in *Saccharomyces sensu stricto* were largely confirmed by the advent of molecular taxonomy. In 1982, Rosini *et al.* (8) found by nDNA/nDNA reassociation that the type strains of the five wine species: *S. bayanus, S. cerevisiae, S. chevalieri, S. italicus* and *S. uvarum* belong to only two separate, unrelated taxa. These findings were later confirmed for several *S. cerevisiae* synonyms (10, 11) while at the same time another species, *S. pastorianus,* intermediate between the two unrelated groups, was identified. A later study (12) revealed the presence of a fourth species: *S. paradoxus,* as another intermediately related member of the group.

As a result of the above studies, in the next edition of the The Yeast-a Taxonomic Study (to be published in early 1994), *Saccharomyces sensu stricto* will include four species: *S. bayanus, S. cerevisiae, S. paradoxus* and *S. pastorianus,* characterized by different genotypes even though often indistinguishable by conventional taxonomic criteria (see Figure 1 for details).

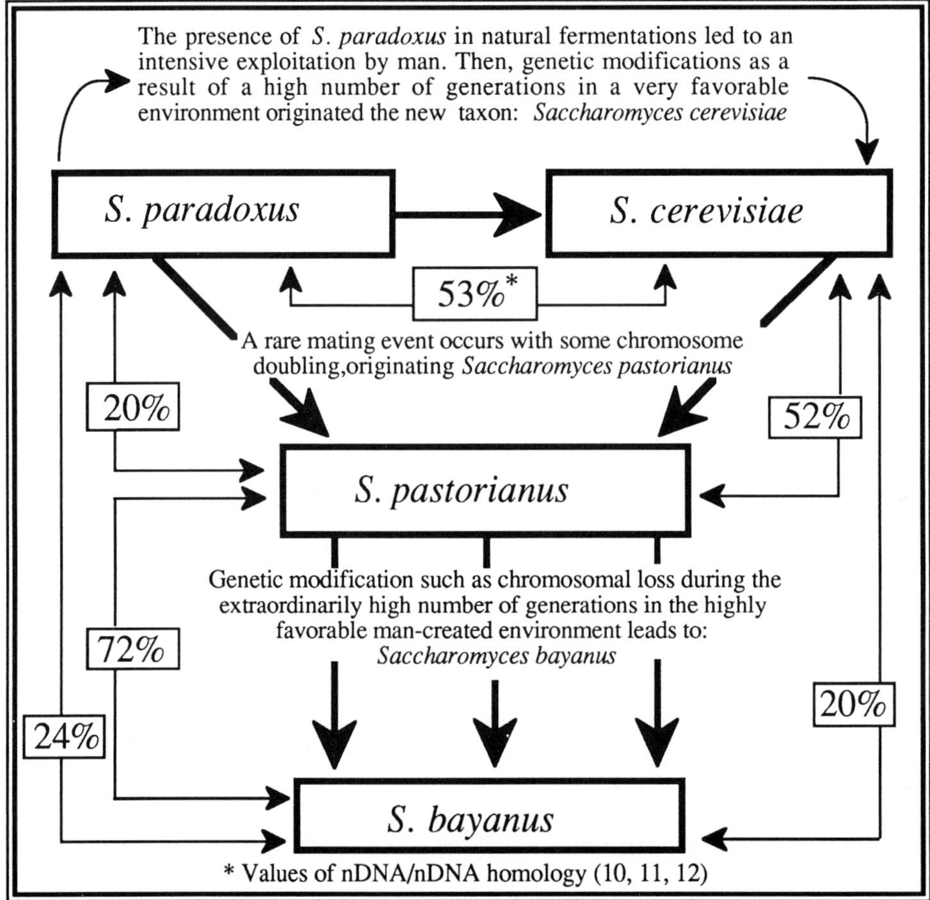

Figure 1. Possible evolution of the species within *Saccharomyces sensu stricto*

5. NOMENCLATURE PROBLEMS WITHIN *SACCHAROMYCES SENSU STRICTO*

In the molecular taxonomy studies cited above more than 45 type strains of the most famous epithets of alcoholic fermentation were examined and reassigned to one of the four resulting taxa according to their nucleotide sequence relationships. The actual names assigned to the four species were dictated by the rules of the International Code of Botanical Nomenclature decreeing that the "oldest validly published epithet" has precedence. This mandatory renaming has given rise to many apparent nomenclature contradictions, very hard to understand or accept by non yeast taxonomists (the majority of those who study and utilize yeast). A few examples are summarized in Table 1 where some traditional epithets of the wine industry are reviewed along with their popularly believed technological attributes.

Table 1
Past and present names of well-known epithets in the wine-making industry

---

The wine yeast "par excellence"
S. ellipsoideus (1883) ---->*S. cerevisiae* var. *ellipsoideus* (1931) ----> *S. cerevisiae* (1952)

The agent of fermentation in cold climates
S. pastorianus (1904) ---> *S. uvarum* (1970) ---> *S. cerevisiae* (1984) --->*S. pastorianus* (1987)

The species endowed for refermentation processes
S. oviformis (1924) --------> *S. bayanus* (1970) --------> *S. cerevisiae* (1984)

The most problematic example is represented by *S. bayanus* that was formally reinstated as a species in 1985 (10) since its type strain is the oldest described epithet of the group within which it fell. However, the type strain of *S. oviformis*, formerly reclassified as *S. bayanus* in 1970 and commonly considered the yeast most indicated for refermentation, was not included in this group, because homologous to the type strain of *S. cerevisiae* (11). As a consequence, this new designation will undoubtedly create problems and confusion since many wine-makers will automatically deduce from the name *S. bayanus* an ability to tolerate high concentrations of alcohol. On the contrary, it was found that strains classified by nDNA/nDNA reassociation as *S. bayanus* are the worst fermenters of the four *sensu stricto* species (Vaughan-Martini & Martini, unpublished data).

6. THE NEW NOMENCLATURE OF FERMENTATION YEASTS

This new information supported by methods of molecular taxonomy, although apparently confusing at first glance, should be regarded with optimism. On one hand, the once mighty complex of fermentation-related species of the genus *Saccharomyces* which went from 20 (4) to 16 (5) to 21 (6) to 1 species (9), has now been restored to a more realistic dimension: 4 variably related species

(12) . This in many ways corresponds to the technological differences which actually exist between these species; something yeast technologists have known all along, in spite of the names that the strains happened to have. These differences reflect, in fact, genetic variation within *Saccharomyces sensu stricto,* a potentially useful trait for strain improvement and development.

Another positive aspect of these findings is that by using only those strains that have been clearly defined on the macromolecular level as a model, it has been possible to return to inexpensive and less complicated phenotypic tests for the definition of these species (13). This is especially important for those yeast users without access to expensive apparatuses, elaborate methods or particular expertise, who should now be able to verify which of the four species they are actually using.

REFERENCES

1. J. Meyen, Wiegmann Arch. Naturgesch. 4, Bd. 2, (1838) 100.
2. M. Reess, Botanische Untersuchungen über die Alkoholgärungspilze. A. Felix, Leipzig. (1870)
3. E.C. Hansen, Medd. Carlsgerg Lab., 2 (1883) 29.
4. A. Guilliermond, Les Levures. Encyclopédie Scientifique. O. Doin et Fils, Paris (1912)
5. J. Lodder and N.J.W. Kreger-van Rij, The Yeasts-a taxonomic study. North-Holland, Amsterdam (1952)
6. J.P. van der Walt, The Yeasts-a taxonomic study J. Lodder (ed.). Elsevier Science Publishers, Amsterdam, (1970) 555.
7. R. Scheda and D. Yarrow , Arch. Mickrobiol., 61 (1968) 310.
8. G. Rosini, F. Federici, A. Vaughan, and A. Martini, Europ. J. Appl. Microbiol. Biotechnol., 15 (1982) 188.
9. D. Yarrow, The Yeasts, a taxonomic study. N.J.W Kreger-van Rij (ed.)Amsterdam, Elsevier Science Publishers. B.V. (1984) 379.
10. A. Vaughan-Martini, A. and C. Kurtzman, Int. J. Syst. Bacteriol., 35 (1985) 508.
11. A. Vaughan-Martini and A. Martini, Antonie van Leeuwenhoeck, 53 (1987) 77.
12. A. Vaughan-Martini, A., System. Appl. Microbiol., 12 (1989) 119.
13. A. Vaughan-Martini, A., System. Appl. Microbiol., 16 (1993) in press.

ECB6: Proceedings of the 6th European Congress on Biotechnology
edited by L. Alberghina, L. Frontali, P. Sensi

Direct Automated Solid Phase Sequencing of an In Vitro Amplified Segment of 16S rDNA from Bacteria

B. Pettersson,[a] K.-E. Johansson,[b] and M. Uhlén[a]

[a] Department of Biochemistry and Biotechnology, The Royal Institute of Technology, S-100 44 Stockholm, Sweden,

[b] The National Veterinary Institute, S-750 07 Uppsala, Sweden

1. SUMMARY

A fast and reliable method based on PCR for partial determination of 16S rRNA sequences is described. The PCR primers were designed to generate a 605 base pair long biotinylated amplicon for immobilization to streptavidin-coated para-magnetic beads. The strands were separated after treatment with alkali resulting in two single stranded sequencing templates. One strand was immobilized onto the beads and the other was neutralized and recovered in the supernatant, which enabled sequencing of both strands. This strategy has the following advantages (i) reverse transcription is not needed (ii) cloning procedures are not required (iii) analysis of non-cultivable species is possible and (iv) the solid-phase procedure is suitable for automation. The method was applied to the 16S rRNA gene of five mycoplasma species which have not been sequenced earlier. Phylogenetic relationships were established from the partial sequences and it was found that a properly chosen stretch of the 16S rRNA gene, in many cases, is adequate for species identification. The results suggest that the direct automated solid-phase procedure is a useful tool for diagnostic and phylogenetic studies.

2. INTRODUCTION

Comparison of 16S rRNA sequences has proven to be extremely valuable for studies of bacterial evolution and phylogenetic relations (1,2). Sequence information of 16S rRNA genes and 16S rRNA can be generated by conventional dideoxynucleotide sequencing of cloned rRNA genes and by direct rRNA sequencing with reverse transcriptase (3), respectively. However, comparatively large amounts of rRNA is required in the latter technique which is time-consuming and not suitable for uncultivable organisms.

In vitro amplification by PCR (4) has revolutionized life sciences and can be used for amplification of rDNA if the primer annealing regions are properly chosen (5). The products can then be cloned and subsequently sequenced. This procedure is, however, both cumbersome and susceptible to nucleotide incorporation errors during the PCR step. Direct sequencing of rDNA amplified by PCR has been reported (4,6), requiring removal of mineral oil, nucleotides and PCR primers. The concept for determination of 16S rDNA sequences described in the present work is based on generation of an immobilized PCR product from microliter amounts of a cell suspension. Para-magnetic beads with covalently coupled streptavidin were used to capture the biotin labeled PCR product which can easily be purified and made single stranded for solid phase sequencing (7). This approach enables automation of both the template preparation (8) and the sequencing reactions (9) resulting in few manipulations, high reproducibility and results obtained within one working day.

Mycoplasmas were selected as model for direct automated solid-phase sequencing of in vitro amplified 16S rDNA in this study. Mycoplasma is the common name for a group of microorganisms which have been arranged in a separate class, the Mollicutes, because of some unusual phenotypic characteristics, the most important being the lack of a rigid cell wall (10). However, phylogenetically, the mycoplasmas are closely related to gram-positive bacteria (11) (with a genome size of 1/6 to 1/3 of that of *E. coli*). Many mycoplasmas are pathogenic and, therefore, of great interest in human and veterinary medicine. The 16S rRNA sequences have been completely or partially determined for more than 40 of about 120 mycoplasmal species and arranged into 5 phylogenetic groups (12). In the present work, partial sequences of the 16S rRNA gene were determined by automated solid phase sequencing of several other mycoplasma species. The sequence data were used for phylogenetic classification.

3. MATERIAL AND METHODS

3.1. Mycoplasma strains

All mycoplasma species were type strains obtained from Dr G. Bölske. Appropriate media for cultivation of mycoplasmas were used.

3.2. Amplification by PCR

PCR-primers used for in vitro amplification of the 16S rDNA fragment, corresponding to the nucleotides between base 334 to base 939 in the 16S rRNA gene of *E. coli* according to the IUB numbering (13), were RIT 388: 5′-CGT TGT AAA ACG ACG GCC AGT CCA RAC TCC TAC GGR AGG CAG C-3′ and RIT 390B: 5′-Biotin-CTT GTG CGG GYY CCC GTC AAT TC-3′. The reaction mixtures were cycled 30 times (15 s denaturation at 96°C and 2 min primer annealing/elongation at 70°C) with DNA from 1µl of a prelyzed mycoplasma culture as template.

3.3. Immobilization and sequencing reactions

A robotic work station (Biomek-1000 from Beckman Instruments Inc., Fullerton, Calif.) was equipped with: 1) a magnetic separation unit (8) for immobilization of the biotinylated amplicons onto streptavidin coated para-magnetic beads from DYNAL AS, Norway and to generate single stranded sequencing templates and 2) a heating/cooling device (HCB-100, Beckman Instr.) for the automated solid phase sequencing reactions (7,9). The sequencing primers used in concert with the A.L.F. system (see below) were synthesized with a fluorescent label, the universal sequencing primer (USP): 5′-FITC-CGT TGT AAA ACG ACG GCC AGT-3′and the RIT390F: 5′-FITC-CTT GTG CGG GYY CCC GTC AAT TC-3′). All reactions were carried out in 96-well microtiter plates. The separation of the sequencing products were monitored on-line, by using an automated laser fluorescent DNA sequencer, A.L.F., from Pharmacia LKB Biotechnology AB, Sweden.

3.4. Sequence analysis

The sequences of 16S rRNA from several mycoplasmas were retrieved from the EMBL nucleotide sequence data base. Dendrograms were drawn by using the UPGMA (Unweighted Pair Group Method with Arithmetic Mean Average) Tree Window of the IntelliGenetics software (IntelliGenetics, Mountain View, Calif.).

4. RESULTS AND DISCUSSION

The procedure is schematically outlined in Fig. 1. A bacterial colony was picked and resuspended in PCR buffer and heated at 99°C for 10 minutes. Alternatively, bacterial suspensions can be used, which might be prelyzed or treated with various extraction procedures. After the PCR, the samples were taken to a magnetic separator (8) where the biotinylated DNA fragments were immobilized onto para-magnetic beads with covalently coupled streptavidin. The strands were separated by alkali treatment. This technique yields clean, reproducible single stranded material suitable for sequencing, without the need for precipitation, filtration or centrifugation. The material in a microtiter plate was subsequently moved to another robotic workstation (9) with a heating/cooling device. The sequencing reactions, including primer annealing and elution of the extended strand by treatment with formamide were performed by the workstation. The samples were then loaded on the sequencing gel and electrophoresis was then carried out by an automated laser fluorescent sequencing apparatus to allow on-line detection and computerized sequence evaluation.

The target gene was amplified with 16S rRNA gene specific primers. The primers were designed to give a Tm-value compatible with a 2-step PCR to decrease steric hindrance due to secondary structure and to generate a sufficiently long fragment of an informative region to be useful for identification. PCR primers complementary to U2 and U5 were chosen because the evolutionarily variable regions V6, V7 and V8 (14) will then be amplified. The primers were constructed to amplify primarily mycoplasmas. However, the generality of the system was increased by introducing a few wobbling bases and amplicons were obtained from all species tested. PCR products and furthermore, reliable sequence data, were obtained from gram-positive bacteria like *Bacillus*, *Lactobacillus*, *Staphylococcus* and *Streptococcus* species and gram-negative bacteria belonging to the genera *Chlamydiae* and *Vibrionaceae* (data not shown).

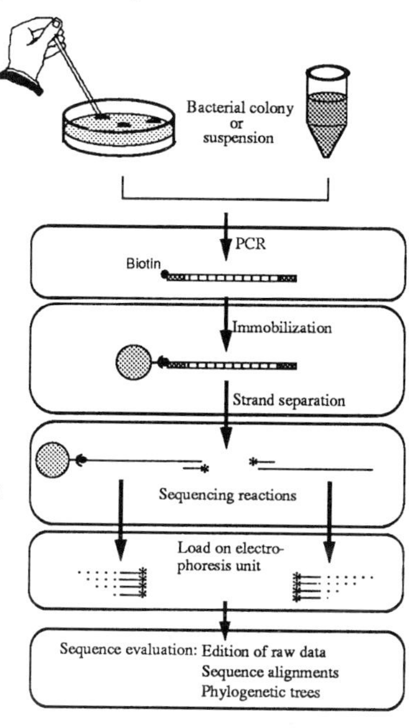

Fig.1. The concept of the solid phase sequencing procedure.

The complete procedure requires little manual operation and the whole procedure, from colony to edited sequence, can be performed within a day. As an example of this direct genomic sequencing method we decided to study five mycoplasma species (*M. meleagridis*, *M. ovipneumoniae*, *M. mycoides* subsp. *capri*, *M. canadense* and *Acholeplasma axanthum*). These have not been sequenced before, on the 16S rRNA level. Specific PCR bands corresponding to the expected length of the partial 16S rDNA were obtained in all cases (data not shown). A typical sequencing output from the direct sequencing of one of these bacteria is shown in Fig. 3. Note, that clear sequence can be observed up to at least 500 nucleotides, although a stretch of six T residues at a position around 455 causes some problems. The low signal to noise ratio shows, which was also often observed, that sequencing of just one strand is sufficient. However, confirmation by sequencing in both directions is recommended to avoid mis-interpretation of problematic regions due to secondary structures.

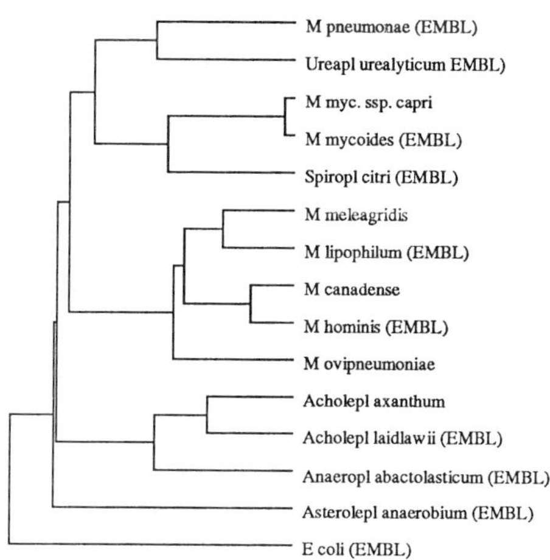

Fig. 2. Dendrogram showing the phylogenetic relations between the mycoplasmas sequenced in this work.

The new partial 16S rRNA sequences were compared by aligning them to a few of their phylogenetic representatives, in each of the five phylogenetic groups as defined by Weisburg et al. (12). The phylogenetic relation ship is presented in Fig. 2 with the corresponding stretch of the 16S rRNA gene from *E. coli* as outgroup. The results show that phylogenetic inferences can be drawn from partial sequence data of the U2 to U5 region of 16S rRNA.

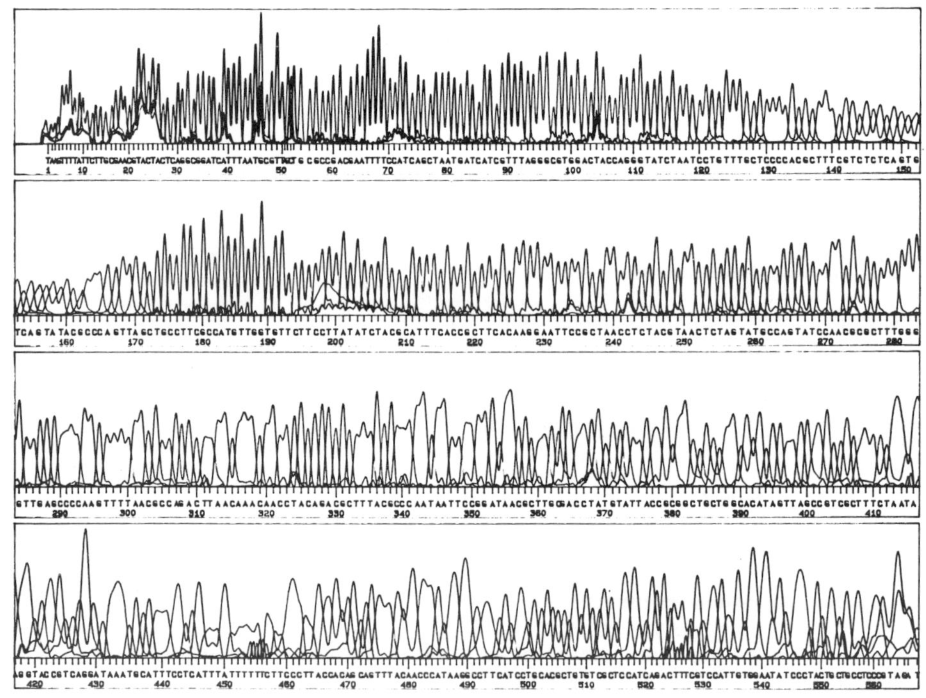

Fig.3. The partial 16S rDNA sequence of *M. meleagridis*.

REFERENCES

1. G.J. Olsen and C.R. Woese, FASEB J., 7 (1993) 113-123.
2. C.R. Woese, Microbiol. Rev., 51 (1987) 221-271.
3. D. Lane, B. Pace, G. J. Olsen, D. A. Stahl, M. L. Sogin and N. R. Pace, Proc. Natl. Acad. Sci. USA, 82 (1985) 6955-6959.
4. K.B. Mullis and F.A. Faloona, Methods Enzymol., 155 (1987) 335-350.
5. E. Böttger, FEMS Microbiol. Lett., 65 (1989) 171-176.
6. M. Dorsch and E. Stackebrandt, J. Microbiol. Methods, 16 (1992) 271-279
7. T. Hultman, S. Ståhl, E. Hornes and M. Uhlén, Nucleic Acids Res., 17 (1989) 4937-4946.
8. J. Wahlberg, A. Holmberg, S. Berg, T. Hultman and M. Uhlén, Electrophoresis, 13 (1992) 547-551.
9. T. Hultman, S. Bergh, T. Moks, and M. Uhlén, BioTechniques, 10 (1991) 84-93.
10. S. Razin, Microbiol. Rev., 49 (1985) 419-455.
11. J. Maniloff, R. N. McElhaney, L. R. Finch and J. B. Baseman (ed.), Mycoplasmas: Molecular Biology and Pathogenesis. American Society for Microbiology, Washington, D.C., 1992.
12. W.G. Weisburg, J.G. Tully, D.L. Rose, J.P. Petzel, H. Oyaizu, D. Yang, L. Mandelco, J. Sechrest, T.G. Lawrence, J. van Etten, J. Maniloff and C.R. Woese, J. Bacteriol., 171 (1989) 6455-6467.
13. J. Brosius, J.L. Palmer, J.P. Kennedy and H.F. Noller, Proc. Natl. Acad. Sci. USA, 75 (1978) 4801-4805.
14. M.W. Gray, D. Sankoff and R.J. Cedergren, Nucleic Acids Res., 12 (1984) 5837-5852.

ECB6: Proceedings of the 6th European Congress on Biotechnology
edited by L. Alberghina, L. Frontali, P. Sensi

A GENERAL METHOD FOR GENERATING SPECIFIC DNA PROBES BY ARBITRARILY-PRIMED PCR (AP-PCR)

Renato Fani, Giuseppe Damiani [a], Claudia Di Serio, Enzo Gallori, Annamaria Grifoni and Marco Bazzicalupo

Dipartimento di Biologia Animale e Genetica, Università degli Studi, Via Romana 17, 50125, Firenze, Italy and [a]I.D.V.G.A.- C.N.R., Via Celoria 10, 20133 Milano, Italy

1. SUMMARY

We report the rapid generation of DNA probes for several *Azospirillum* strains.This method is based on the generation of random amplified polymorphic DNA (RAPD) fingerprints using primers with an embedded restriction site.

2. INTRODUCTION

In many fields of microbiology there is increasing demand for methods for the identification and detection of microorganisms. One of the DNA techniques most frequently used is the Arbitrarily Primed PCR (AP-PCR) and its variant RAPD (Random Amplified Polymorphic DNA) where very small amounts of total DNA are subjected to PCR using a single synthetic oligonucleotide of random sequence as primer. The amplification products are resolved on agarose gels giving rise to a pattern specific for a given genome [1-3]. In a search for a strategy to generate specific probes for bacteria identification, as an alternative to the rRNA sequences which are the most frequently used [4, 5], we have exploited the ability of RAPD to generate markers specific for a given genome [6]. Depending on the type of primer used and on the conditions of the reaction, random amplification of bacterial genomes can generate DNA fragments which are diagnostic for a genus, for a species, or for a particular strain. We have tested the strategy described above by applying it to the genus *Azospirillum*. *Azospirillum* strains are microaerobic bacteria, fixing nitrogen both as free-living and in association with the roots of grasses, including important cereals like maize and wheat [7].

3. MATERIALS AND METHODS

3.1 Bacterial strains used

The bacterial strains used in this work are *A.brasilense* DSM1859, DSM2287, SPF94, *A.lipoferum* SP242, ATCC29707, ATCC29708, ATCC29731, SpBr17, F, PA1, 133/1 and 137/5, *A.halopraeferens* Au5, *A.amazonense* Y1 and Y2.

3.2 Primer

The oligonucleotide used as primer was NP2, a 23-mer (5'CGGGGGACTGTTGGGCGCC ATCT) containing a NarI site near the 3' of the molecule.

3.3 DNA amplification conditions

Amplification reactions and the analysis of the products of amplification reactions were performed as previously reported (8).

3.4 DNA manipulation

All DNA manipulations were carried out as described [9]. Sequencing of DNA, extraction of total DNA from bacterial cells and transformation of *Escherichia coli* cells with plasmid DNAs were carried out as described [10, 11, 12].

4. RESULTS AND DISCUSSION

4.1 Overall strategy

The first step of the overall strategy (figure 1) adopted by us is the extraction and purification of total DNA from cultures of *Azospirillum* strains either from culture collections or isolated from soil. DNA is then amplified using a random primer and the products of the reaction analyzed by agarose gel electrophoresis. The amplification band(s) peculiar to a given strain or common to two or more strains are identified, purified from agarose gel, cloned in a plasmid vector, sequenced and finally the sequences (or part of them) utilized as specific molecular probe.

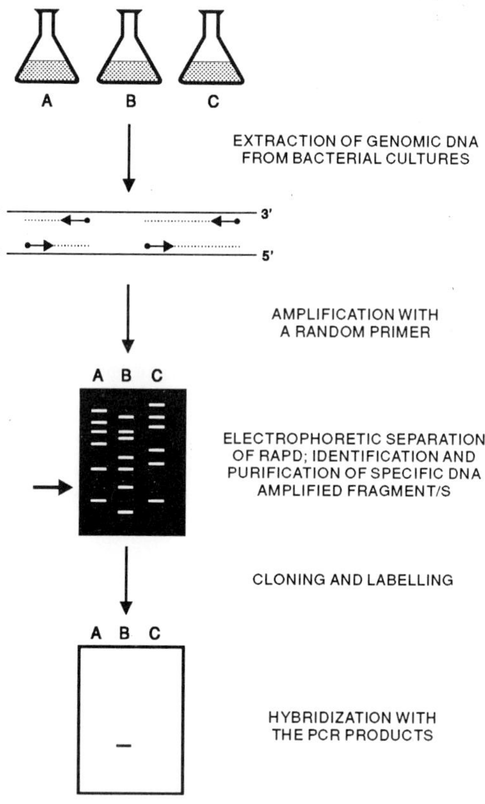

Figure 1. Overall strategy for the identification of specific DNA probes for microorganisms

4.2 Generation of RAPDs, identification, cloning and sequencing of specific amplification products

Total DNA from fifteen Azospirillum strains belonging to the four species *brasilense*, *lipoferum*, *amazonense* and *halopraeferens* and from other eubacteria (*Bacillus subtilis* , *Streptomyces rochei*, *Pseudomonas spp*, *Spirulina platensis*, *Ectothiorodospira spp*, *Rhizobium meliloti*, *Staphylococcus aureus*, *S. epidermidis* and *Enterobacter spp*) was extracted and amplified by using the primer NP2. Results obtained showed that each *Azospirillum* strain possesses a unique and distinctive amplification pattern with bands peculiar to itself plus others common to other *Azospirillum* strains, but with no bands common to the patterns showed by the other eubacterial strains tested (data not reported). These results confirmed the ability of this method to differentiate among genus, species and strains of the same species and also suggested the possibility of identifying and isolating specific molecular markers. Among them we identified in three *A. lipoferum* strains, ATCC 29708, ATCC 29731 and ATCC29709 a common band of about 400 bp; a 560 bp DNA fragment was present only in the amplification pattern of the *A.amazonense* Y2 strain; finally three *A. brasilense* strains, DSM 2287, DSM 1859 and Sp242 showed a common band of 920 bp. These three fragments were digested with the restriction endonuclease NarI, which has a recognition site near the 3' end of the primer molecule and cloned into the ClaI site of the plasmid vector pGEM7-Zf(+).

Three of the recombinant plasmids obtained, called pAFD11, pAFD32 and pAFD42 carrying DNA sequences from *A. lipoferum*, *A. amazonense* and *A. brasilense* respectively, were chosen as molecular probes and their nucleotide sequence determined.

4.3 Hybridization experiments

The plasmids pAFD11, pAFD32 and pAFD42 containing the DNA fragment from *A. lipoferum*, *A. amazonense*, and *A. brasilense*, respectively, were labelled with digoxigenin-dUTP and used as probes in hybridization experiments using the RAPD products amplified with the NP2 primer of all the eubacterial strains tested. Results obtained (not reported) showed that plasmid pAFD11 containing the DNA fragment from *A. lipoferum* ATCC29708 hybridized with a fragment of about 400 bp present in the RAPD products of the *A. lipoferum* strains ATCC29708, ATCC29731 and ATCC29709. Plasmid pAFD32 containing the DNA insert from *A. amazonense* Y2 hybridized with a DNA fragment from strain Y2. Finally plasmid pAFD42 containing the DNA fragment from *A. brasilense* DSM2287 hybridized with a fragment of about 920 bp of strains DSM2287, DSM1859 and Sp242. The same probes were used in slot-blot hybridizations with total genomic DNA from all the *Azospirillum* strains. The results obtained (not reported), showed that each probe gave hybridization signals of varying intensity with every *Azospirillum* genome tested. These data suggest that sequences homologous to the probes are present in all the *Azospirillum* genomes tested

4.4 RAPD under different reaction conditions

We tested the ability of the RAPD markers to identify a given genome also when the RAPD conditions were changed. For this purpose the Y2 DNA was amplified changing the parameters of PCR; the products of all the amplification reactions were spotted on membranes and hybridized to the plasmid pAFD32, carrying the DNA probe of *A.amazonense* Y2. Results obtained (not reported) showed that: a) the reducing of reaction volume down to 25 μl, the lowering of template DNA concentration down to 1pg and the lowering of average molecular weight of DNA due to the digestion of genomic Y2 DNA with EcoRI did not affect the intensity of the signal. The latter results can be particularly relevant if this methodology is applied to microorganisms from which it is not always possible to obtain high molecular weight DNA molecules; moreover it can be also relevant for DNA extracted from environmental samples; b) the lowering of primer concentration and the raising of $MgCl_2$ concentration resulted in a decrease of the intentsity of the hybridization signal; c) the presence of a competitor DNA (of similar complexity) did not affect the intensity of the hybridization signal even when the ratio between the Y2 DNA and the competitor DNA was 1:100.

In conclusion we adopted RAPD fingerprinting to the nitrogen fixing soil bacterium *Azospirillum* to generate probes with different degree of specificity, useful in microbial ecology studies. Data reported in the present paper demonstrate that some of the DNA fragments amplified with random primer PCR could effectively be used as strain or species specific probe provided that the DNA to be probed had been previously amplified with the same primer.

We propose that a strategy for producing probes from RAPD as described in the present paper could be appropriately applied to the analysis of the bacteria living in particular environments. We suggest an application, as other PCR based probes, to detect the occurrence of particular strains in environmental samples without isolation (figure 2). Total DNA extracted from the sample could in fact be amplified, labelled and hybridized to a membrane filter where all the probes have been previously spotted. Positive signals indicate the occurrence of the corresponding strain. For the detection of very low number of the target genome in mixed biosamples a precise knowledge of the competition process in amplification is required.

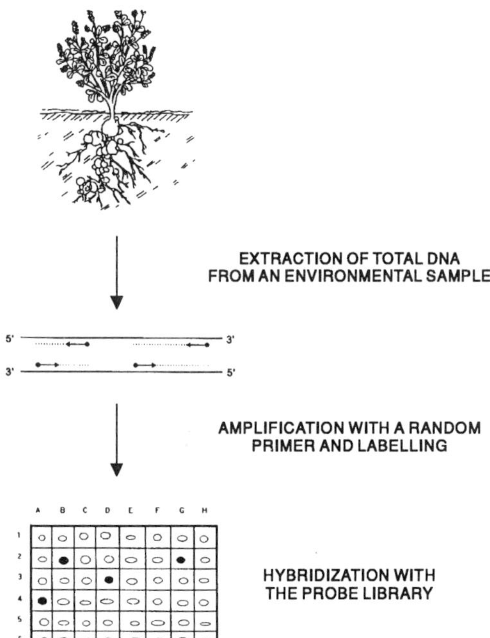

Figure 2. Strategy to detect microorganisms in the environment

REFERENCES

1. J.Welsh and M. McClelland, Nuc Acids Res 18 (1990) 7213.
2. J.G.K.Williams, A.R.Kubelik, R.J.Livak, J.A.Rafalski and S.V.Tingey, Nuc Acids Res 18 (1990) 6351.
3. G. Caetano-Anolles, G.J.Bassam and P.M.Gresshof, Biotechnology 9 (1991) 553.
4. B.L.Tsai and B.H.Olson, Appl Environ Microb, 58 (1992) 754.
5. T.Barry , R.Powell and F.Gannon, Biotechnology 8 (1990) 233.
6. H.Hadrys, M.Balick and B.Schierwater, Mol Ecology 1 (1992) 55.
7. J.Dobereiner and J.M.Day, In Newton WE, Nyman JC (eds)Proceedings of the 1st International Symposium on Nitrogen Fixation, Pullman: Washington State University Press (1975) 518.
8. R.Fani , C.Bandi, M.G.Bardin, S.Comincini,, G.Damiani, A.Grifoni and M.Bazzicalupo, . Microb Rel (1993) (in press).
9. J.Sambrook, E.F.Fritsch and T.Maniatis, Molecular cloning. A laboratory manual. Second edition.(1989) Cold Spring Harbor Laboratory Press. Cold Spring Harbor, NY.
10. F.Sanger, S.Nicklen and A.R.Coulson, Proc Nat Acad Sci 74 (1977) 5463.
11. L.Giovannetti, S.Ventura, M.Bazzicalupo, R.Fani and R.Materassi, J.Gen Microb 136 (1990) 1161.
12. D.Hanahan, J Mol Biol 166 (1983) 557.

ECB6: Proceedings of the 6th European Congress on Biotechnology
edited by L. Alberghina, L. Frontali, P. Sensi
1994 Elsevier Science B.V.

Protein electrophoretic patterns for high-resolution identification of micro-organisms

K. Kersters, B. Pot, D. Dewettinck, U. Torck and L. Vauterin

Laboratorium voor Microbiologie, Universiteit Gent, 35 Ledeganckstraat, B-9000 Gent, Belgium

1. INTRODUCTION

Polyacrylamide gel electrophoresis (PAGE) of cellular proteins and DNA restriction fragments are powerful tools for rapid, accurate and sensitive fingerprinting of various micro-organisms. The complex electrophoretic banding patterns are useful for high-resolution characterization and identification of e.g. bacterial strains of biotechnological interest.

A microbial cell expresses some 2000 different proteins constituting a rich source of information for the characterization, classification and identification of micro-organisms. The utility of PAGE of cellular proteins in microbial systematics has been established for over 20 years (for reviews see [1-4]). Protein electrophoretic fingerprints are highly reproducible, provided that strains are cultivated in reproducible conditions and standardized electrophoretic techniques are used. Bacterial strains with at least 70% DNA relatedness tend to display similarities in their protein electrophoregrams and bacteria sharing 90 - 100% DNA-hybridization display only minor differences in their protein patterns. Hence, electrophoresis of bacterial proteins is a sensitive technique yielding information on similarities of strains within the same (sub)species, i.e. at the infraspecific level. Depending on the protein electrophoretic variation within a given taxon individual strains can often be recognized by small, but specific and highly reproducible differences in part of their protein fingerprints.

In biotechnology, electrophoresis of microbial proteins is a useful and inexpensive technique in the following areas:

- verification of strain authenticity
- differentiation between colony variants and contaminants
- monitoring changes during storage of bacterial cultures
- screening and selection of strains
- classification and identification of bacteria and yeasts

Some of these applications are discussed below.

2. EXPERIMENTAL PROCEDURES

Several types of electrophoretic separations in polyacrylamide gels are applied in bacterial systematics: continuous PAGE, discontinuous PAGE, iso-electric focusing, two-dimensional PAGE and gradient PAGE. The following types of microbial proteins can be compared by PAGE: soluble, native proteins; proteins solubilized by e.g. sodium dodecyl sulphate (SDS); cell-envelope and ribosomal proteins; radiolabeled proteins. The present contribution deals only with the most commonly applied technique of SDS-PAGE of whole-cell bacterial proteins. The advantages of one-dimensional SDS-PAGE are: (i) the technique is fairly simple and inexpensive to perform; only ± 50 mg (wet weight) of cells is required; (ii) in a single experiment a complex pattern and hence large numbers of characteristics per strain are obtained; (iii) the same or at least very similar experimental procedures can be applied for almost every micro-organism; (iv) the patterns can be digitized and stored in data banks for objective comparisons and subsequent identifications.

The SDS-PAGE procedure is performed in a discontinuous polyacrylamide gel system. This consists of a large pore stacking gel (concentrating gel or spacer gel) and a small pore separating gel (resolving gel) with different buffer concentration. The SDS-PAGE procedure considered here is based on the techniques described by Laemmli [5], Jackman [1], Costas et al. [6] and our own experience [7,8]. It is of course important to keep cultivation conditions as standardized as possible (growth medium, growth time and temperature). Usually washed bacterial cells are lysed by heat treatment in the presence of 2% SDS. However, various Gram-positive bacteria (and yeasts) require an ultrasonic treatment prior to the addition of SDS.

Estimation of resemblance between protein electrophoretic patterns was originally done visually by comparing electrophoregrams or their (normalized) photographs. Appropriate grouping of protein fingerprints upon visual inspection is suitable when only a few strains need to be compared. However, when large numbers of strains are investigated, it becomes impossible to compare all patterns with one another by eye and to group them in a reliable manner. Patterns can be digitized by densitometers or video-cameras. Computer-aided techniques are required for correction of discrepancies within and between gels. The reliability of databases and computer-assisted identification depends largely on the accuracy of the standardization method. The software package GelCompar [9] uses an advanced pattern recognition algorithm that compares whole contours rather than peaks of the densitometric curves. Libraries of normalized reference patterns can be generated and used for subsequent identification of unknown patterns.

3. APPLICATIONS

An interesting example of successful verification of strain authenticity by SDS-PAGE concerns the type strain NCIB 4112 of the cellulose-producing species *Acetobacter xylinum* (the original sorbose bacterium). Protein electrophoresis combined with DNA:rRNA hybridizations revealed that the type strain kept at the ATCC, CIP and NCIB bacterial culture collections belonged to the genus *Gluconobacter* and did not correspond anymore to its original description. Comparative protein electrophoresis showed that most likely an inadvertent interchange of strains took place at the NCIB between 1933 and 1957, because the original *Acetobacter xylinum* type strain could be revived at NCIB from an old stock-culture that was Sordelli-dried in 1949. Consequently a new type strain of *Acetobacter xylinum* (NCIB 11664) was proposed [9].

Other examples of applications of the SDS-PAGE technique concern the differentiation and identification of lactobacilli occurring in fermented milk products and the human intestinal flora. Identification of lactic acid bacteria is traditionally based on physiological and biochemical criteria. These procedures are not only time-consuming, but are quite often ambiguous due to the increasing number of lactic acid bacterial species. Recently we have demonstrated the use of a large-scale SDS-PAGE fingerprinting study for the development of rRNA-based ologonucleotide probes. Both techniques proved to be useful to study e.g. the relationships within the heterogeneous *Lactobacillus acidophilus* complex [10]. Numerical analysis of SDS-PAGE protein fingerprints was also very helpful to verify the specificity of rRNA-targeted oligonucleotide probes for the identification of *L. delbrueckii*, *L. helveticus* and *L. paracasei* [11].

REFERENCES

1. P.J.H. Jackman. In: R.R. Colwell and R. Grigorova (eds.), Methods in Microbiology, Vol. 19, pp. 209-225, Academic Press, London, 1987.

2. K. Kersters. In: M. Goodfellow, D. Jones and F.G. Priest (eds.), Computer-assisted bacterial systematics, pp. 337-368. Academic Press, London, 1985.

3. K. Kersters. In: Z. Klement, K. Rudolph and D.C. Sands (eds.), Methods in Phytobacteriology, pp. 191-198, Akadémiai Kiado, Budapest, Hungary, 1990.

4. L. Vauterin, J. Swings and K. Kersters. In: M. Goodfellow and A.G. O'Donnell (eds.), The new Bacterial Systematics, Academic Press, London, 1993 (in press).

5. U.K. Laemmli, Nature (London) 227 (1970) 680.

6. M. Costas, B.D. Cookson, H.G. Talsania and R.J. Owen, J. Clin. Microbiol. 27 (1989) 2574.

7. B. Pot, P. Vandamme and K. Kersters. In: M. Goodfellow and T. O'Donnell (eds.), Chemical Methods in Bacterial Systematics, J. Whiley & Sons, Chichester, UK, 1993 (in press).

8. L. Vauterin and P. Vauterin, Europ. Microbiol. 1 (1992) 37.

9. M. Gillis, K. Kersters, F. Gosselé, J. Swings, J. De Ley, A.R. MacKenzie and I.J. Bousfield, Int. J. Syst. Bacteriol. 33 (1983) 122.

10. B. Pot, C. Hertel, W. Ludwig, P. Descheemaeker, K. Kersters and K.-H. Schleifer, J. gen. Microbiol., 139 (1993) 513.

11. C. Hertel, W. Ludwig, B. Pot, K. Kersters, P. Krusch and K.-H. Schleifer, Syst. Appl. Microbiol. (1993) (submitted for publication).

Expression of Proteins in transgenic Animals

ECB6: Proceedings of the 6th European Congress on Biotechnology
edited by L. Alberghina, L. Frontali, P. Sensi
1994 Elsevier Science B.V.

The production of recombinant proteins in transgenic animals

L. M. Houdebine

Unité de Différenciation Cellulaire, Institut National de la Recherche Agronomique, 78352 Jouy-en-Josas Cédex - France -

The production of recombinant proteins for various purposes and particularly for pharmaceutical use was one the earlier success of biotechnology. Although several recombinant proteins such as human growth hormone and insulin are routinely prepared and used for human therapy it is clear that many problems remain to be solved before all proteins can be obtained in large amounts and in functional state.

The genetic code being the same for all living organisms, a gene from animals or plants can be expressed in bacteria as soon as regulatory regions from bacterial genes are fused to the foreign cDNA to be expressed. Bacteria harbouring such gene constructs are commonly being used to prepare recombinant proteins for multiple uses, including pharmaceutical use. The limits of this production system were rapidly identified. Most of the bacteria strains used to prepare recombinant proteins are unable to secrete these proteins. Thus, when strong promoters are used, bacteria can synthetize huge amounts of the foreign proteins which of ten form aggregates of insoluble material. The collection of the recombinant proteins is then difficult and it often requires denaturing conditions [1].

It is also well-known that bacteria cannot proceed to post-translational modifications in a correct manner : folding, specific cleavage, glycosylation,... Bacteria can therefore be used only to prepare a certain number of relatively simple proteins. Lower eucaryote cells such as yeast share many advantages with bacteria. They can be maintained in large number in conventional fermentors. These cells can secrete proteins and they are expected to be able to modify proteins post-transcriptionnally. Indeed in most cases, they can do a correct folding and particularly the formation of disulfide bridges in the recombinant proteins. Yeast can also add carbohydrates to the polypeptide chains, but the sugar composition of the carbohydrate side-chain is generally not similar to that of the higher vertebrates natural proteins [2]. Plants in which foreign genes have been transferred can synthetize significant amounts of functional recombinant proteins. In this respect, it has been recently shown than transgenic tobacco leaves contain hepatitis B antigen which has good antigenic properties and might be used for vaccination [3]. It remains to be proved however that large quantities of recombinant proteins for pharmaceutical use can be easily synthetized, extracted and purified from transgenic plants. On the other hand, plant cells are unlikely to be able to do post-translational modifications of all recombinant proteins in a proper way.

Animal cells are therefore expected to be more appropriate to direct the synthesis of human recombinant proteins. Several *in vitro* systems are being used. Cultured insect cells infected by recombinant baculovirus proved to be a very efficient system. It is relatively easy to prepare the vectors and to maintain the infected cells. However, some of the recombinant proteins form aggregates which heavily complicate their recovery. Moreover, the insect cells are not able to do all the biochemical modifications of the recombinant proteins. This is particularly true for glycosylation which takes place in the insect cells but in a manner which is generally different of that observed in higher vertebrate cells. A striking exemple is the envelope HIV gp160 glycoprotein. This protein is being studied to be tentatively used for the vaccination of human beings again AIDS disease. The gp160 glycoprotein prepared from insect cells infected by a recombinant baculovirus showed antigenic properties markedly different of the same protein prepared from cultured animal cells [4]. Recombinant baculovirus can be used to infect living insects rather than cultured cells. This procedure is not expected to obtain recombinant proteins having been more properly matured.

Cells from higher vertebrates, namely from mammals, are therefore required to synthetize the proteins which are highly glycosylated such as erythropoietin or which has to be γ-carboxylated such as the coagulation factor IX. Mammalian cells in culture can thus be the source of recombinant human erythropoietin [5]. The capacity of the mammalian cells in culture remains however rather low and costly despite recent improvements of vectors and culture methods.

These observations logically led to consider whole transgenic animals as a possible source of recombinant proteins [6]. Indeed, cells are very numerous in an animal and they are in an ideal metabolic situation, as opposed to cultured cells. Moreover, transgenesis in mammals proved to be possible more than ten years ago and many transgenic animals can be bred. Attempts to generate transgenic animals synthesizing recombinant proteins have thus been undertaken several years ago.

THE CHOICE OF THE BIOLOGICAL FLUID

Ideally, the transgenic animals should be obtained and bred easily and collection of the recombinant proteins should not be a complicated process. Several methods having quite different advantages and drawbacks are being studied. All of them rely on the collection of a biological fluid containing the recombinant proteins.

Secretion from silk gland and haemolymph from transgenic insects have been shown to contain significant amounts of recombinant proteins [7, 8]. This procedure is certainly limited by the relatively small amount of available biological fluid per animal.

Egg white is expected to be an excellent source of recombinant proteins. Strong promoters and particularly ovalbulmin gene promoter should be able to direct very efficiently the synthesis of foreign proteins in the oviduct of transgenic chicken. This will not be the case as long as transgenesis will remain so inefficient in birds.

Blood from transgenic animals may be an excellent source of recombinant proteins. It is abundant, as being a by-product from slaughterhouse. In two cases at least, blood was shown to contain relatively large amounts of foreign proteins. Biologically active human α-1-antitrypsin was obtained at the concentration of 1 mg/ml in blood of transgenic rabbits [9]. Human hemoglobin is being synthesized by red blood cells from transgenic pigs [10]. Althrough attractive in some cases, blood cannot be used as the source of recombinant proteins for several reasons. Most of the proteins are unstable in blood and cannot accumulate in large amount. Moreover many of the recombinant proteins which are being prepared to be used in humans or animals and their biological activites are expected to alter health of the transgenic animals in most cases. This was not the case for rabbit expressing human α-1-antitrypsin which is naturally present at a high level in blood with marked physiological changes in concentration and for pigs expressing the human hemoglobin genes.

Some proteins have to be subjected to specific post-translational modifications which can take place in only a very limited number of cell types. The whole organs from transgenic animals might then be the source of a protein instead of the human organs. This possibility does not seem to have been envisaged so far.

Milk appears to everybody as the best biological fluid for the preparation of recombinant proteins from transgenic animals. It is abundant, easily collected and milk protein genes are being studied by several groups in the world.

THE SYNTHESIS OF RECOMBINANT PROTEINS FROM MILK

Milk contains a large amount of proteins (up to 130 mg/ml in rodents) and only about six of them, according to species, represent up to 80 % of the proteins : caseins are the most abundant. Four or less caseins encoded by distinct genes are found at different concentrations according to species. Several other proteins synthesized by the mammary gland are found in the whey from different species : α-lactabulmin (1 mg/ml) in most species, β-lactoglobulin (4 - 5 mg/ml) in ruminants, whey acidic protein (WAP) (up to 15 mg/ml in rabbit milk) in rodents transferrin and lactoferrin (0 - 20 mg/ml). Most of these genes have been isolated and sequenced in several species.

Targeting foreign gene expression specifically in the mammary gland was expected to be possible using regulatory regions from a milk protein gene fused to the coding region of the gene to be expressed. The first demonstration that this goal could be reacted was given in 1987 by Simons et al [11]. Who showed that ovine β-lactoglobulin gene was expressed at a very high level in the milk of transgenic mice. Similar experiments using various whole milk proteins were carried out successfully by several groups [12 - 16]. Hybrid genes containing the regulatory regions from milk protein genes and whole foreign genes [17 - 22] or cDNAs [23 - 34] were also expressed in the milk of different transgenic animals. Several of the recombinant proteins found in milk proved to be active. The idea that milk can be an important source of recombinant proteins is therefore firmly established.

THE PROBLEMS OF GENE CONSTRUCTS

Although many groups succeeded in obtaining foreign proteins in milk it is essential to note that the foreign genes were expressed at high or low level and in an unpredictable manner. A general rule emerges somewhat however. Whole milk protein genes were generally expressed at a high level. Whole foreign genes fused to milk protein gene promoters were also efficiently expressed in several cases whereas cDNAs were in most cases only poorly active. This fact is not really surprising since introns have been shown to play an essential role in the expression of transgenes with various promoters and cDNAs [35]. Generally speaking, it seems that some introns are more potent that others to favour transgenes expression although the efficiency of a given promoter-intron-cDNA association remains largly unpredictable. This seems to be the case for a foreign gene associated with β-lactoglobulin promoter [36]. Transcription terminators also seem to play an important role. Short rat WAP gene terminator was more efficient in transgenic mice than a longer DNA fragment or the late gene SV 40 terminator [15]. The promoter used may be of major importance. This seems to be the case for the rabbit WAP gene promoter which showed high efficiency when fused to α_1-antitrypsin gene [22], human and bovine growth hormone genes [21] whereas the mouse WAP gene promoter was much less efficient [19, 20]. This difference may be due to the fact that WAP is a particularly abundant protein in rabbit milk [37] but more likely to the fact that a long DNA fragment containing important regulatory regions was used [38]. Unexpected observations have been done with several gene constructs. It is for instance striking that the mouse WAP promoter fused to protein cDNA was much more active in transgenic pigs than in mice [25]. This clearly means that a vector or a family of vectors capable of expressing at a reasonnably high level in most cases foreign cDNAs and even genes remains to be found.

The expression of a transgene in any tissue seems in most cases to be dependent upon its site of integration in the host genome rather than upon the number of integrated copies. It is generally admitted that a transgene is highly influenced by its chromatin environment which controls to a large extent its expression. Experiments carried out by several groups have led to identify AT-richDNA sequences which insulate transgenes and greatly favour their expression [39]. These sequences act when they are fused to the transgene before transfert or interestingly when coinjected with the transgenes [40]. The exact mechanism of action of these AT-richDNA sequences is not known. It is admitted that they specifically bind to nuclear matrix generating a open region in chromatin which favours the expression of a gene it their vicinity [41]. The use of such matrix attached regions (MAR) should thus very significantly enhance foreign protein secretion in milk.

Another possible approach may be to use embryonic stem cells (ES cells) and homologous recombination [42] to replace a milk protein gene by a foreign gene which would be located in this way in a ideal chromatin environment. This will remain impossible as long as ES cells use will be restricted to mouse.

The bacterial Cre/lox system which induces efficient targeted recombinations might also be used [43]. The lox sequence might be introduced in the vicinity of a given milk protein gene promoter to target the insertion of foreign genes.

THE CHOICE OF THE TRANSGENIC MAMMALS

Transgenic mice are routinely used to evaluate the efficiency of gene constructs to be transferred into larger animals. It is uneasy to milk a mouse which is generally considered as inappropriate to provide milk in sufficient quantity for biochemical or clinical studies. A recent work has shown that isolated mammary glands kept on ice for a few hours spontaneously released the milk they contain. A lactating mouse separated from its offspring for one day can thus provide 1.5 ml milk containing for example active human growth hormone [44]. Using one or a few hundred transgenic mice it is thus possible to obtain 100 mg of a recombinant protein.

For amounts not higher than 1 kg per year, transgenic rabbits seem appropriate. Transgenesis is easy in this species and rabbit can be milked. For higher quantities pigs [25], sheeps [45], goats [23] or cows [46] are required. Transgenesis is much more costly in these species the *in vitro* production of mature oocytes, fertilization and embryo development is a considerable progress. This procedure was used to generate transgenic cows [46].

THE PURIFICATION OF RECOMBINANT PROTEINS FROM MILK

It is essential that the recombinant proteins can be easily isolated from milk in an undegrated and active state. Experiments carried out by several groups suggest that this should be the case in most cases [47]. The large majority of the foreign proteins are found in a soluble state in milk and they can be recovered after elimination of caseins. Interestingly, in one case, with the CFTR (Cystic Fibrosis Transmembrane Conductance Factor), the protein was found associated with the lipid membranes [32]. This was of course due to its hydrophobic nature. Milk can thus be used to prepare this kind of proteins as well. A large proportion of the recombinant proteins obtained in milk are prepared to be injected into human beings. The biochemical purity of the final product must therefore be high. The proteins must also be devoid contaminants such as virus or prions. The transgenic animals will have therefore to be kept is good breeding conditions and treatments to inactivate virus will have to be applied to the recombinant proteins.

THE FUTURE OF LIVING FERMENTORS

As long as the whole procedure allowing the preparation of recombinant proteins will not have been optimized, it will remain difficult to evaluate the cost of a protein prepared in this manner in comparison to other methods. Whatever happens, optimism seems to be the rule and it has been suggested that 10 % of the recombinant proteins might be prepared from milk by the end of this century. This would correspond to a market of about 100 millions dollars per year [48]. The cost of the procedure is in fact not the only point to be taken into consideration. Indeed, proteins which are presently extracted from human blood with good efficiency will most likely have to be replaced by products from other sources for safety reasons. In this regard, milk appears an attractive alternative.

A particular point which may be crucial in some cases must also be taken into consideration. Several authors noticed that the foreign proteins were present not only in the milk but also to a very significant extent in the blood of their transgenic animals. This was due in most cases to the fact that the expression of the transgene was not strictly mammary specific but also that the mammary epithelium is not completely closed. A small proportion of milk proteins are thus transferred from milk to blood circulation. This seems to be a physiological process, at least for small proteins, since rabbit WAP was found in blood of normal lactating rabbits [37]. Health of transgenic animals expressing a foreign protein specifically in their milk may be heavily altered by the presence of this protein in blood during lactation. It may be predicted that in some cases, although successful from a biochemical point of view, this method may disturb too much the animals to be utilizable.

The mammary cell can do some of the post-translational modifications. It would not be surprising however if some of these modifciations are not correctly performed in some cases. This is expected, due to the great complexity of these processes [49]. It is concievable, althrough most likely not easy to do, to endow mammary cells of living animals with new cellular enzymes through transgenesis making them capable of proceeding to additional post-translation modifications.

The present report is no more than an overview of the state of art on living fermentors. The four speakers of this symposium will give additional details on the methods showing their success and the problems they face.

REFERENCES

1 - P. Blum, Velligan M., Lin N., A. Matin. Bio/Technol., 10 (1992) 301-304.
2 - R. G. Buckholz, M. A. G. Gleeson. Bio/Technol., 9 (1991) 1067.
3 - Mason H. S., D. Man-Kit Lam, C. Arntzen. Proc. Natl. Acad. Sci. USA, 89 (1992) 117-145.
4 - J. Moore, G. K. Lewis, J. Robinson. Nature, 361 (1993) 503.
5 - D. T.-Y. Liu. TIBTECH, 10 (1992) 114-120.

6 - J. Hodgson. Bio/Technology, 10 (1992) 863-866.
7 - S. Maeda, T. Kawai, M. Obinata, H. Fujiwara, T. Horiuchi, Y. Saeki, Y. Sato, M. Furusawa. Nature, 315 (1985) 592-594.
8 - D. E. Rancourt, I. D. Peters, V. K. Walker, P. L. Davies. Bio/Technology, 8 (1990) 453-457.
9 - M. Massoud, R. Bischoff, W. Dalemans, H. Pointu, J. Attal, H. Schultz, D. Clesse, M. G. Stinnakre, A. Pavirani, L. M. Houdebine. J. Biotech., 18 (1991) 193-204.
10 - M. E. Swanson, M. J. Martin, J. K. O'Donnell, K. Hoover, W. Lago, V. Huntress, C. T. Parsons, C. A. Pinkert, S. Pilder, J. S. Logan. Bio/Technology, 10 (1992) 557-559.
11 - J. P. Simons, M. McClenaghan, A. J. Clark. Nature, 328 (1987) 530-532.
12 - K. Lee, F. J. De Mayo, S. H. Atiee, J. M. Rosen. Nucleic Acids Res., 16 (1988) 31027-31041.
13 - M. A. Persuy, M. G. Stinnakre, C. Printz, M. F. Mahé, J. C. Mercier. Eur. J. Biochem., 205 (1992) 887-893.
14 - S. Soulier, J. L. Vilotte, M. G. Stinnakre, J. C. Mercier. FEBS, 297 (1992) 213-218.
15 - T. C. Dale, M. J. Krnacik, C. Schmidhauser, C..L. Yang, M. J. Bissel, J. M. Rosen. Mol. Cell. Biol., 12 (1992) 3905-3914.
16 - A. Shamay, V. G. Pursel, E. Wilkinson, R. J. Wall, L. Hennighausen. Transgenic Res., (1992)
17 - G. Wright, A. Carver, D. Cottom, D. Reeves, A. Scott, P. Simons, I. Wilmut, I. Garner, A. Colman. Bio/Technology, 9 (1991) 830-834.
18 - C. Tomasetto, C. Wolf, M. C. Rio, M. Mehtali, M. LeMeur, P. Gerlinger, P. Chambon, R. Lathe. Mol. Endocrinol., 3 (1989) 101579-101584.
19 - V. B. Reddy, J. A. Vitale, C. Wei, M. Montaya-Zavala, S. L. Stice, J. Balise, J. M. Robl. Anim. Biotechnol., 2 (1991) 115-129.
20 - W. H. Günzburg, B. Salmons, B. Zimmermann, M. Müller, V. Erfle, G. Brem. Mol. Endocrionol., 5 (1991) 1123-1133.
21 - E. Devinoy, submitted
22 - R. Bischoff, E. Degryse, F. Perraud, W. Dalemans, D. Ali-Hadji, D. Thépot, E. Devinoy, L. M. Houdebine, A. Pavirani. FEBS, 305 (1992) 3265-3268.
23 - K. M. Ebert, J. P. Selgrath, P. DiTullio, J. Denman, T. E. Smith, M. A. Memon, J. E. Schindler, G. M. Monastersky, J. A. Vitale, K. Gordon. Bio/Technology, 9 (1991) 835-838.
24 - S. Yu, K. C. Deen, E. Lee, L. Hennighausen, R. W. Sweet, M. Rosenberg, H. Westphal. Mol. Biol. Med., 6 (1989) 255.
25 - W. H. Velander, J. L. Johnson, R. L. Page, C. G. Russell, T. Morcol,A. Subramanian, T. D. Wilkins, R. Canseco, B. L. Williams, F. Dwazdauskas, J. W. Knight, C. Pittius, J. M. Young, W. N. Drohan. Proc. Natl. Acad. Sci. USA, 89 (1992) 12003.
26 - A. J. Clark, H. Bessos, J. O. Bishop, P. Brown, S. Harris, R. Lathe, M. McClenaghan, C. Prowse, J. P. Simons, C. B. A. Whitelaw, I. Wilmut. Bio/Technology, 7 (1989) 487-492.
27 - M. McClenaghan, A. L. Archibald, S. Harris, J. P. Simons, C. B. A. Whitelaw, I. Wilmut, A. J. Clark. Anim. Biotechnol., 2 (1991) 2161-2176.
28 - H. Meade, L. Gates, E. Lacy, N. Lonberg. Bio/Technology, 8 (1990) 443-446.
29 - T. A. Bühler, T. Bruyère, D. F. Went, G. Stranzinger, K. Bürki. Bio/Technology, 8 (1990) 140-143.
30 - N. M. Greenberg, J. W. Anderson, A. J. W. Hsueh, K. Nishimori, J. J. Reeves, D. M. DeAvila, D. N. Ward, J. M. Rosen. Proc. Natl. Acad. Sci. USA, 88 (1991) 8327-8331.

31 - M. G. Stinnakre, J. L. Vilotte, S. Soulier, R. L'Haridon, M. Charlier, P. Gaye, J. C. Mercier. FEBS, 1 (1991) 19-22.
32 - P. DiTullio, S. H. Cheng, J. Marshall, R. J. Gregory, K. M. Ebert, H. M. Meade, A. E. Smith. Bio/Technology, 10 (1992) 74.
33 - M. Shani, I. Barash, M. Nathan, G. Ricca, G. H. Searfoss, I. Dekel, A. Faerman, D. Givol, D. R. Hurwitz. Transgenic Res., 1 (1992) 195-208.
34 - R. Halter, J. Carnwath, G. Espanion, D. Herrmann, E. Lemme, H. Niemann, D. Paul. Theriogenology, 39 (1993) 137-149.
35 - R. D. Palmiter, E. P. Sandgren, M. R. Avarbock, D. D. Allen, R. L. Brinster. Proc. Natl. Acad. Sci. USA, 88 (1991) 478-482.
36 - C. B. A. Whitelaw, A. L.Archibald, S. Harris, M. McClenaghan, J. P. Simons, A. J. Clark. Transgenic Res., 1 (1991) 3.
37 - H. Grabowski, D. Le Bars, N. Chêne, J. Attal, R. Maliénou-N'Ghassa, C. Puissant, L. M. Houdebine. J. Dairy Sci., 74 (1991) 4143-4150.
38 - E. Devinoy, R. Maliénou-N'Ghassa, D. Thépot, C. Puissant, L. M. Houdebine. Mol. Cell. Endocrinol., 81 (1991) 185-193.
39 - C. Bonifer, A. Hecht, H. Saueressig, D. M. Winter, A. E. Sippel. J. Cell. Biochem., 47 (1991) 99-108.
40 - R. A. McKnight, A. Shamay, L. Sankaran, R. J. Wall, L. Hennighausen. Proc. Natl. Acad. Sci. USA, 89 (1992) 6943-6947.
41 - E. Käs, L. Poljak, Y. Adachi, U. K. Laemmli. EMBO J., 12 (1993) 115-126.
42 - J. Travis. Science, 256 (1992) 1392-1394.
43 - M. Lasko, B. Sauer, B. Jr. Mosinger, E. J. Lee, R. W. Manning, S. H. Yu, K. L. Mulder, H. Westphal. Proc. Natl. Acad. Sci. USA, 89 (1992) 6232-6236.
44 - M. G. Stinnakre, E. Devinoy, D. Thépot, N. Chêne, M. Bayat-Sarmadi, H. Grabowski, L. M. Houdebine. Anim. Biotechnol., 3 (1992) 245-255.
45 - G. Wright, A. Carver, D. Cottom, D. Reeves, A. Scott, P. Simons, I. Wilmut, I. Garner, A. Colman. Bio/Technology, 9 (1991) 830-834.
46 - P. Krimpenfort, A. Rademakers, W. Eyestone, A. Van der Schans, S. Van der Broek, P. Kooiman, E. Kootwijk, G. Platenburg, F. Pieper, R. Strijker, H. de Boer. Bio/Technology, 9 (1991) 844-847.
47 - R. J. Wilkins. Proceedings of the New Zealand Society of Animal Production, 51 (1991) 211-217.
48 - Markets for transgenic animals. Technology Management Group. Genetic Technology news (1990) 8-11.
49 - S. M. Edgington. Bio/technology, 10 (1992) 1413-1420.

ECB6: Proceedings of the 6th European Congress on Biotechnology
edited by L. Alberghina, L. Frontali, P. Sensi

THE APPLICATION OF AQUEOUS TWO PHASE PARTITIONING TO THE PURIFICATION OF PROTEINS FROM TRANSGENIC SHEEP MILK

P.Harris, G.Wright*, A.T.Andrews, D.L.Pyle, J.A.Asenjo
Biotechnology and Biochemical Engineering Group, University of Reading, Reading RG6 2AP, England.

*Pharmaceutical Proteins Limited, Edinburgh EH9 3JQ, Scotland.

Summary

Three of the main milk proteins in sheep's milk, ovine serum albumin (OSA), ovine casein and ovine ß-Lactoglobulin (ß-Lg), were individually partitioned in aqueous two phase systems composed of polyethylene glycol (PEG) and phosphate. The effects of PEG molecular weight, pH and addition of NaCl on the partitioning of these proteins were studied. Results showed that increasing the NaCl concentration increased the partition coefficient (K). This trend was more pronounced at pH 5 than at pH 7 or 9. Higher partition coefficients were obtained for OSA and casein than for ß-Lg. Transgenic milk containing human α1-Antitrypsin (AAT), was also studied in these systems. Conditions were found where AAT would partition to one phase and ß-Lg to the other; most casein precipitated in the interface.

Introduction

α1-Antitrypsin (AAT) is one of several protease inhibitors found in mammalian blood. Low levels of AAT can cause pulmonary emphysema (degenerative lung disease) and juvenile cirrhosis of the liver.

Quite large quantities ($\approx$200g/patient/year) of AAT are required for replacement therapy. Such large amounts of protein can only be obtained via recombinant DNA technology. However, large scale production of mammalian cell cultures is expensive and technically demanding. Other methods, such as using *E.coli*, cannot provide the post-translational modifications necessary for the production of a stable protein. For these reasons, an expression system ideally suited to producing bioactive AAT is transgenic animals.

Sheep have been generated that are transgenic for the protein human AAT and are producing levels of up to 35g/L in their milk. It has been shown that subsequent generations also possess the gene for AAT thus establishing a successful line of transgenic animals[1].

Aqueous two phase systems (ATPS) were chosen to evaluate as a method for protein purification because of the mild separation conditions, low operating costs, simplicity of handling and suitability for large scale and continuous applications. ATPS are formed when two incompatible polymers, or a polymer and a high concentration of salt are mixed together. Commonly used systems are combinations of PEG with Dextran, or PEG with salts such as Phosphate or Sulphate.

Separation of proteins is obtained by altering parameters such as PEG molecular weight, pH, type of ions present and ionic strength. These parameters affect the hydrophobic and electrostatic interactions between the proteins and the system[2]. In an ideal purification scheme conditions are chosen such that the contaminant proteins will partition to one phase and the protein of interest will partition to the other phase.

The aims of our work are to study the partitioning of main ovine milk proteins in selected ATPS to develop generic procedures for separation of transgenic proteins from sheep's milk. Then to investigate the partitioning of AAT in the selected ATPS's in order to find conditions where the contaminant milk proteins partition to one phase and AAT to the other phase.

Materials and Methods

Transgenic sheep milk containing 35g/l AAT was supplied by Pharmaceutical Proteins Ltd. Ovine serum albumin (OSA) was purchased from Sigma, casein and β-Lactoglobulin (β-Lg) were fractionated from sheep milk.

ATPS were chosen at a relatively fixed distance from the binodial. From binodials, concentrations of PEG and salt were chosen so that equal volumes of top (PEG) and bottom (salt) phases were obtained.

Samples from the top and bottom phases were analysed for total protein concentration using a dye-binding method (Bradford). Samples were blanked against a control sample which consisted of the same ATPS with buffer in place of protein.

The partition coefficient K was calculated as the concentration of protein in the top phase divided by the concentration of protein in the bottom phase.

Results and Discussion

An initial screening of a number of PEG-Phosphate systems revealed that PEG of MW's 4000 and 1500 were suitable for partitioning the proteins of transgenic sheep milk. The effects of pH and addition of NaCl on these systems were studied[3,5].

Four milk proteins were identified as being the major contaminants of transgenic milk. These were casein 50%, β-Lg 6%, α-La 1.5% and OSA 0.5%. AAT was 42% of the protein content. The partition coefficients of casein, β-Lg and OSA were determined. It was found that with PEG 1500 the partitioning of the proteins were more extreme (i.e. higher $\pm$Log K values) and it is these results that are discussed further. Figures 1, 2 and 3 show the partitioning of these proteins in PEG 1500-Phosphate.

At pH 5 and with 0% NaCl all three proteins partition to the bottom phosphate phase. As the % NaCl is increased, the proteins partition to the top PEG phase. This trend is more notable for OSA which has a Log K value of -1.58 at 5% NaCl, and 3.42 at 18% NaCl. This means at 18% NaCl the partition coefficient K is about 2600. A similar degree of partitioning is observed for casein.

For ß-Lg and casein, at pH 7 and 9, the Log K remains relatively unchanged as the % NaCl is increased. However, OSA at pH 7 and 9, partitions from the bottom phase at low NaCl to the top phase at high NaCl. This change in partitioning is greater at pH 7 than at pH 9.

The partitioning of proteins in ATPS have been shown to be influenced by surface hydrophobicity and charge[3]. Addition of a salt, such as NaCl, increase the effect of the proteins hydrophobicity differences in the partitioning between the PEG and the phosphate

Figure 1. Partitioning of B-Lg in PEG 1500-Phosphate

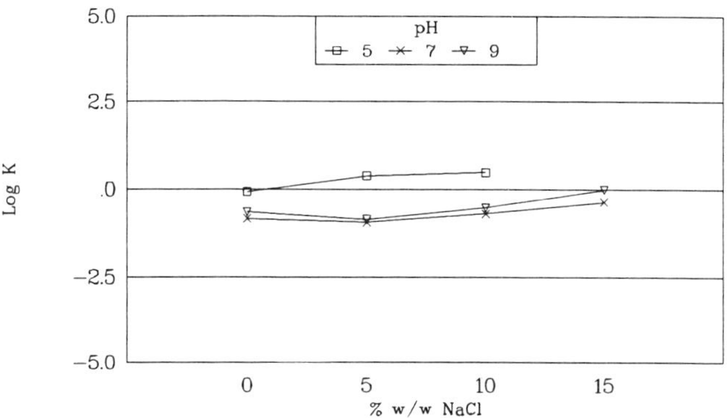

Figure 2. Partitioning of Casein in PEG 1500-Phosphate

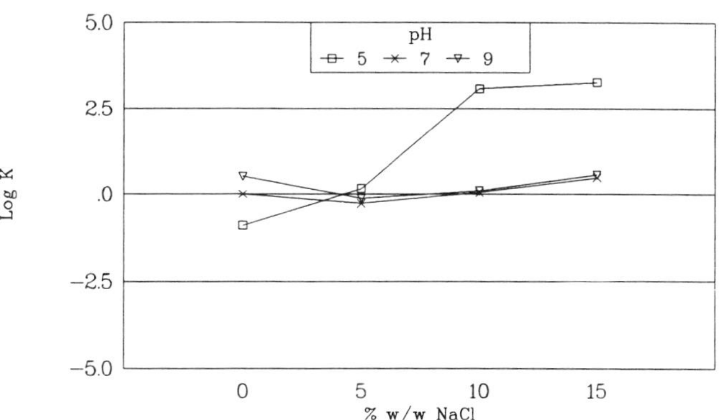

Figure 3. Partitioning of OSA in PEG 1500-Phosphate

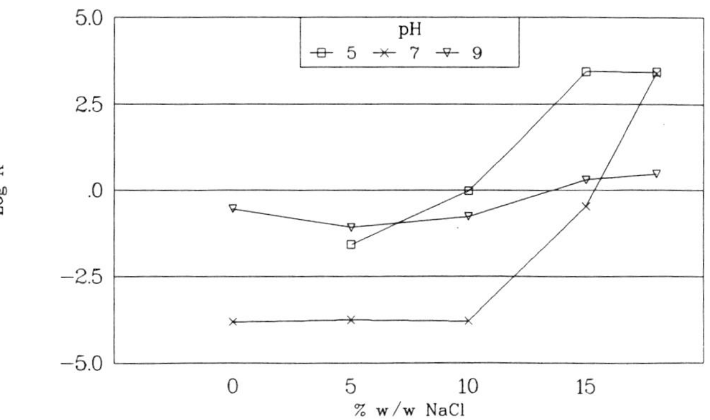

phase[4]. This has been explained as a result of hydrophobicity and salting out effects. Analysis of OSA and ß-Lg on hydrophobic interaction chromatography using an alkyl superose column showed that OSA was eluted at a much lower salt concentration than was ß-Lg, making OSA the more hydrophobic protein. Casein also has a high hydrophobicity. This would explain why the partitioning of OSA and Casein to the top PEG phase is much stronger than that for ß-Lg.

The isoelectric points of OSA, casein and ß-Lg are 4.7, 4.7 and 5.3 respectively. Therefore, in systems at pH 5 these proteins will have very little surface charge. In this case, electrostatic interactions between the proteins and the system will be minimal and the hydrophobic effects will be predominant.

The partitioning of whole transgenic milk was studied and the phases were analysed on gel electrophoresis. Conditions were found where AAT partitioned to the phosphate phase and ß-Lg to the PEG phase. With both PEG 4000 and 1500, addition of small amounts of NaCl forced ß-Lg to partition to the PEG phase whereas greater amounts of NaCl were needed to partition AAT to the PEG phase. In these systems, the casein had precipitated from the systems and formed around the interface of the two phases. Therefore using systems with low NaCl concentrations could provide a suitable mechanism for separating ß-Lg from AAT.

Conclusions

Greater $\pm$ Log K values were obtained with the lower molecular weight PEG 1500 than with PEG 4000-phosphate systems.

For all three milk proteins examined, increasing the NaCl concentration resulted in an increase in partition coefficient. This trend was more pronounced at pH 5 than at pH 7 or 9. These results can be explained by hydrophobic effects. The more hydrophobic proteins OSA and casein showed higher Log K values than the less hydrophobic ß-Lg protein.

Systems using PEG 4000 or 1500 with low concentrations of NaCl could provide a suitable mechanism for separating ß-Lg from AAT.

References

1. Wright,G. et al. Bio/Technology **9** (1991) "High Level of Expression of Active Human α1-Antitrypsin in the Milk of Transgenic Sheep"
2. Asenjo,J.A, Franco,T., Andrews,A.T. & Andrews,B.A.: "Affinity Separation of Proteins in Aqueous Two Phase Systems" in: Biologicals from Recombinant Micro-organisms and Animal Cells: Production and Recovery. Ed. M.White et al.
3. Cascone,O., Andrews,B.A. and Asenjo,J.A. (1991) Partition and Purification of Thaumatin in Aqueous Two-Phase Systems. Enz. Microb. Technol. **13**, 629-635.
4. Hachem,F.M. JCH publishers p439-454: "Hydrophobic Behaviour of Five Model Proteins in Chromatography, Precipitation, and Aqueous Two Phase Systems". 1992 Ph.D thesis. University of Reading, U.K.
5. Schmidt,A.S., Ventom,A.M. and Asenjo, J.A. (1993). Partition and Purification of α-amylase in Aqueous Two Phase Systems. (Submitted for publication).

Gene Expression

in

animal cell Cultures

ECB6: Proceedings of the 6th European Congress on Biotechnology
edited by L. Alberghina, L. Frontali, P. Sensi

Relationships between heterologous protein synthesis and the culture environment of Chinese hamster ovary cells

Alan T. Bull

Biological Laboratory, University of Kent, Canterbury, Kent, CT2 7NJ, United Kingdom.

1. INTRODUCTION

In 1991 sixteen recombinant proteins were being marketed as pharmaceutical products; of these, nine were products of animal cell biotechnology and had an estimated sales value of $1600 million, comprising about 43% of the world sales of such proteins. The major proteins in this category in sales terms are erythropoietin, tissue plasminogen activator and Factor VIII. Mammalian cells have been chosen frequently as hosts for recombinant protein synthesis because it is generally considered that they have the most appropriate metabolic machinery for producing and secreting *authentic* recombinant human proteins. Of the nine proteins referred to above, five are produced in Chinese hamster ovary cell cultures, whilst the remainder are produced by hybridomas, Namalwa and other human cells.

As Gray *et al.* [1] have recently declared, the earlier mystique associated with the large scale growth of animal cells is fast disappearing and being replaced by a realization that it is possible to design, operate and optimize animal cell-based processes using much of the established fermentation technology developed for microorganisms. However, it is crucial to appreciate that recombinant proteins produced in animal cell cultures are likely to be subject to variable post-translational processing thereby resulting in varying degrees of product heterogeneity. Accordingly, optimization of the production of therapeutic proteins using animal cell systems needs to be cognizant not only of:

(1) improving volumetric productivity, by means such as increasing biomass concentration, increasing specific productivity, stabilizing and increasing heterologous gene expression; but also of:

(2) protein authenticity, that is the ability to produce the recombinant protein in its actual and exact native form; and

(3) protein fidelity; that is the ability to reproduce accuracy of recombinant protein synthesis in each and every production operation.

The achievement of protein authenticity is affected very largely by the choice of the host cell (or organism) and stability of the transformant, while fidelity is critically dependent on the definition of and adherence to optimized fermentation conditions.

Process intensification, therefore, must consider both the *quantity* and the *quality* of recombinant proteins. We have addressed these issues during a study of human interferon-gamma (IFN-γ) production by Chinese hamster ovary (CHO) cells.

1.1 Experimental System

Human IFN-γ has a core peptide of 146 amino acids and is N-glycosylated at asparagines 28 and 100. This lymphokine, which is secreted by T lymphocytes and has antiviral potency [2], has been cloned and expressed in CHO cells [3]. Our experiments have been made with a mutant CHO-K1 cell line (DHFR$^-$; CHO-320) in which co-amplification for IFN-γ expression was achieved by methotrexate selection [4]. Recombinant IFN-γ synthesis was completely stable for periods greater than 2000h (*ca.* 50 generations) when the cell line was grown in chemostat cultures [5,6].

The fully glycosylated human IFN-γ (2N) has a molecular weight of 25kDa but heterogeneity of both the native and the recombinant protein is known. The major heterogeneity results from incomplete glycosylation (1N, M_r 20kDa; 0N, M_r 17.1kDa) and from proteolysis of the core polypeptide at the C-terminus. We have assessed IFN-γ authenticity and fidelity by a number of analytical techniques: SDS-polyacrylamide electrophoresis of immunoprecipitated product for glycoform macroheterogeneity; mass spectrometry (fast atom bombardment, electrospray) of proteolytic digests for analysis of the core peptide; and laser desorption mass spectrometry of exoglycosidase digests for analysis of oligosaccharides.

1.2 Investigative Strategy

Two complementary approaches have been made to investigating the causes of IFN-γ heterogeneity and hence devising process conditions that will lead to high authenticity and high fidelity product:

(1) The definition of factors that influence recombinant protein heterogeneity is complicated by simultaneous changes in several physico-chemical environmental parameters when growth is made in batch cultures. Chemostat culture, on the other hand, allows one to examine systematically and unequivocally the effect(s) of any factor; we have deployed chemostats to determine the relationship between physiological state and IFN-γ heterogeneity.

(2) The composition of production media for animal cell processes is highly complex and manipulation of the nutritional environment is likely to effect significant changes in the quality of recombinant proteins. The use of statistical experimental designs becomes obligatory for screening large numbers of nutritional variables and we have applied the Plackett-Burman [7] methodology for such multifactorial analysis.

2. RESULTS

Good suspended growth is produced when CHO-320 is cultured in shaken flasks, spinner vessels or stirred tank reactors. Under these conditions the maximum specific growth rate (μ_{max}) is about 0.025 to 0.03h^{-1} and the maximum IFN-γ titre may be up to 9000 IU ml^{-1}. The specific rates of growth and IFN-γ production were positively correlated when the cell line was grown in an RPMI 1640-based medium (see below) but when glutamine was replaced with glutamate growth and IFN-γ synthesis become uncoupled. Following the omission of glutamine from the medium growth is halted but CHO cell viability remains high and IFN-γ synthesis proceeds at a rate comparable to that in the presence of glutamine [8]. The most dramatic event accompanying batch growth is the increased heterogeneity of the IFN-γ. The percentage of 2N variants fell from 60-70% during exponential growth to 30-40% by the end of growth which coincided with glucose exhaustion; the corresponding change in 0N variants was from <5% to 20-30% of the total IFN-γ [4]. The shift towards 1N and especially 0N variants of IFN-γ during batch cultivation was not prevented or diminished by increasing the medium glucose concentration at the start of the culture [8]. Up to 12 IFN-γ variants are detectable by SDS-gel electrophoresis: three major peaks detected by scanning densitometry represent 0N, 1N and 2N glycoforms and each contains minor variants resulting from proteolytic cleavage of the polypeptide. The changes in glycosylation did not occur following addition of IFN-γ to CHO cell cultures suggesting that the heterogeneity was a consequence of intracellular post-translational processing and not of extracellular glycosidase digestion. Similarly significant proteolysis occurred before IFN-γ was secreted. *In vitro* translation studies [9] showed that proteolysis was not associated with processing of IFN-γ within the endoplasmic reticulum but probably occurred following transport of the nascent glycosylated IFN-γ into the Golgi apparatus.

At the outset of our studies growth was made in an RPMI 1640 medium containing 7% adult bovine serum. Serum-free culture was achieved by sequential passage in the presence of reduced concentrations of serum [8]. We have used statistical optimization techniques to identify those medium constituents which are the most important for CHO cell growth and IFN-γ production, and to move towards the use of more well defined culture media. One goal has been the development of protein-free medium; bovine serum albumin (BSA) in particular is a constituent that introduces significant variability and is usually present in high concentration. An initial Plackett-Burman analysis evaluated the effects of 20 medium variables (full experimental details and data analysis are reported by Castro *et al.*, 1992 [10]). The analysis enabled us to identify a group of positive medium constituents which, in combination, enhanced the specific growth rate, maximum cell and IFN-γ concentrations by 43, 41 and 45% respectively above values typical of the original medium formulation [10]. We have shown subsequently that the glycosylation pattern also is improved when CHO-320 is grown in this modified medium: the proportion of the 0N glycoform never rises above 10% of the total even by the end of batch culture [11]. BSA was one of the positive effectors. Given the advantage which could result from reducing the BSA concentration (5g L^{-1}) a search was made for BSA substitutes. When the BSA concentration was reduced to 1g L^{-1} and the medium supplemented with Pluronic F68 and a commercial lipid mix growth patterns almost identical to control cultures containing 5g L^{-1} BSA were achieved. IFN-γ production was adversely effected, however (35% reduction). Growth on 1g L^{-1} BSA alone resulted in approximately 40% reductions in cell growth and IFN-γ production. A Plackett-Burman experiment was set up to further elucidate the effects of the lipid supplement. The lipid mix was replaced by various combinations of oleic acid, linoleic acid, cholesterol, choline and ethanolamine in a 7-variable matrix. The analysis revealed that only linoleic acid had a significantly positive effect on cell growth rate and number but cholesterol had a significantly negative effect on both parameters. Oleic acid, choline and ethanolamine produced no significant effects although it has been shown recently [12] that choline and etholamine can substitute for lipoproteins in hybridoma growth. The effects of Pluronic F68 and linoleic acid supplementation of 1g L^{-1} BSA medium is shown in Table 1.

Table 1
CHO-320 cell growth and IFN-γ production for cultures with 5g L^{-1} BSA (I), 1g L^{-1} BSA and 0.1% Pluronic (II) and 1g L^{-1} BSA and 0.1% Pluronic + 8mg L^{-1} linoleic acid (III).

	I	II	III
Specific growth rate (h^{-1})	0.024 ± 0.002	0.024 ± 0.001	0.026 ± 0.001
Viable cell number ($x10^{-5}$)	9.7 ± 0.2	8.1 ± 0.01	9.7 ± 0.3
IFN-γ titre (IU ml^{-1})	8540 ± 900	6890 ± 675	4870 ± 400

Although linoleic alone could substitute for the high BSA concentration in terms of growth it was unable to restore IFN-γ production to that of the high BSA medium. Moreover, linoleic acid supplementation had a profoundly adverse effect on the glycosylation profile such that the 0N glycoform accounted for 30% of the IFN-γ [11]. Work is continuing in order to understand the interaction of BSA and lipids in this system but it is clear, however, that extrapolations to other systems at this time are unwarranted. Thus Gray *et al.* [1] have shown that lipid precursors such as mevalonate, and cholesterol, significantly enhance recombinant human follicle stimulating hormone production by CHO cells.

In order further define relationships between CHO cell physiology and IFN-γ production and processing experiments were made in chemostats so that specific growth rate could be controlled. We have shown previously [5] that glucose-limited growth of CHO-320 cells can be maintained for very long periods, and that IFN-γ productivity and glycoform pattern were stable when the dilution rate was held constant. The population of 2N IFN-γ under these conditions was less (57%) than that produced in the early stage of batch growth (*ca.* 70%). Step increases and pulse additions of glucose to steady state, glucose-limited cultures only caused transitory increases in the percentage of 2N IFN-γ suggesting that factors other than the availability of glucose could be affecting glycosylation. Recently [6] we have analysed CHO physiology and IFN-γ production over a wide range of dilution rates (D) in glucose-limited chemostats. Cell viability fell at low values of D and a minimum specific growth rate of 0.011h^{-1} was extrapolated from the observed data.

Extrapolation of the Monod model suggests that at the minimum specific growth rate the steady state glucose concentration is 47μM which is very close to the calculated saturation constant for glucose [6] and thereby supports the hypothesis [13] of a threshold concentration for the limiting substrate below which animal cell growth will not occur. The IFN-γ production kinetics appeared to be growth-associated [6] and at all but the lowest values of D 55% of the product was doubly glycosylated. Thus, in comparison to glucose-excess batch culture [4] and the *in vivo* condition [14] IFN-γ glycosylation is impaired under glucose-limited chemostat culture. Work is currently in progress to examine the effect of pulse additions of oligosaccharide and lipid precursors on the glycoform profile in an attempt to identify limiting factors in the glycosylation process.

Purified IFN-γ was digested with Staphylococcal V8 protease, the peptides were separated by reverse phase HPLC. Following removal of sugars with endoglycosidase F the peptides were analyzed by FAB-MS. The removal of sugar residues from Asn sites results in an increase in mass of 1Da (asparagine converted to aspartic acid), a change that can be detected by FAB-MS. The molecular mass of all peptides containing Asn-28 increased by 1Da; only a proportion of Asn-100 containing peptides revealed such a mass difference [15]. Thus, we conclude that Asn-28 is occupied *exclusively* in 1N IFN-γ and is the preferred site for N-glycosylation. Bause [16] has shown that the tripeptide Asn-X-Thr (equivalent to the IFN-γ Asn-28 site) is glycosylated 40 times more rapidly than the Asn-X-Ser tripeptide (equivalent to the IFN-γ Asn-100 site). Collectively these findings are consistent with the 3-dimensional structure of recombinant human IFN-γ [17] which predicts that Asn-28 is the more accessible glycosylation site. These analyses now enable a fuller interpretation of the glycoform profiles of batch cultures (see above) in terms of Asn-28 and Asn-100 occupancy. Glycosylation at Asn-100 falls preferentially towards the end of batch growth whereas the Asn-28 glycoform remains intact throughout almost the entire product run.

Finally we have begun to reveal the magnitude of the microheterogeneity of recombinant IFN-γ produced in CHO cells. ES-MS of IFN-γ revealed a high number (>60) of molecular mass species some of which suggested microheterogeneity of oligosaccharides. Dr. David James in our laboratory has recently started to apply laser desorption-mass spectrometry to the analysis of the microglycoforms. Preliminary studies show a large degree of heterogeneity (>20 glycoforms) and variable proportions of the different glycoforms (individual types ranged from <1 to 67% of the total glycosylated IFN-γ). Discrimination of Asn-28 and Asn-100 glycosylation was evident in terms of fucosylation (87% of Asn-28 glycoforms, 4% of Asn-100) and sialicylation (30% of Asn-28 glycoforms, 63% of Asn-100). These data simply indicate the extensive diversity of glycoforms that can arise during recombinant synthesis. The development of rapid affinity chromatography and CZE for 0N, 1N and 2N purification and preparation (D. James, unpublished) will now permit us to examine systematically the effects of culture environment on such microheterogeneity.

3. CONCLUSIONS

Although we remain somewhat distant from a full understanding of the basis of IFN-γ heterogeneity when produced in CHO cells we are now in a position to recognize and quantify such heterogeneity and soon to be able to devise culture conditions that will ensure product fidelity. Post-translational modifications of the type discussed in this paper can affect many properties of recombinant therapeutic proteins (solubility, antigenicity, *in vivo* half-life, folding). There are major implications for producers of these proteins in defining and delivering product fidelity because this can lead to the more desirable commercial position of licensing products rather than licensing processes. To this end the fullest possible detail regarding glycoforms and other types of heterogeneity is required, a feature which re-emphasizes the necessity of what objectives and what criteria to establish for optimization programmes [18].

REFERENCES

1. P.P. Gray, K. Jivasripongpun and C. Gebert, in Harnessing Biotechnology for the 21st Century, 197, American Chemical Society, Washington D.C., 1992.
2. H. Johnson, Lymphokines, 11 (1985) 33.
3. J.H.G.M. Mutsaers, J.P. Kamerling, R. Devos, Y. Guisez, W. Fiers and J.F.G. Vliegenhart, Eur. J. Biochem., 156 (1986) 651.
4. E.M.A. Curling, P.M. Hayter, A.J. Baines, A.T. Bull, P.G. Strange and N. Jenkins, Biochem. J., 272 (1990) 333.
5. P.M. Hayter, E.M.A. Curling, A.J. Baines, N. Jenkins, I. Salmon, P.G. Strange, J.M. Tong and A.T. Bull, Biotechnol. Bioeng., 39 (1992) 327.
6. P.M. Hayter, E.M.A. Curling, M.L. Gould, A.J. Baines, N. Jenkins, I. Salmon, P.G. Strange and A.T. Bull, Biotechnol. Bioeng., (1993) submitted.
7. R.L. Plackett and J.P. Burman, Biometrica, 33 (1946) 305.
8. P.M. Hayter, E.M.A. Curling, A.J. Baines, N. Jenkins, I. Salmon, P.G. Strange and A.T. Bull, Appl. Microbiol. Biotechnol., 34 (1991) 559.
9. N.J. Bulleid, E.M.A. Curling, R.B. Freedman and N. Jenkins, Biochem. J., 268 (1990) 777.
10. P.M.L. Castro, P.M. Hayter, A.P. Ison and A.T. Bull, Appl. Microbiol. Biotechnol., 38 (1992) 84.
11. P.M.L. Castro and A.T. Bull (1993) unpublished data.
12. B. Maiorella, in Harnessing Biotechnology for the 21st Century, 26, American Chemical Society, Washington D.C., 1992.
13. K.K. Frame and W.-S. Hu, Biotechnol. Bioeng., 37 (1991) 55.
14. E. Rinderknecht, B.H. O'Connor and H. Rodriguez, J. Biol. Chem., 259 (1984) 6790.
15. C. Wingrove, N. Jenkins and R.B. Freedman (1992) unpublished data.
16. E. Bause, Biochem. Soc. Trans., 12 (1984) 514.
17. S.E. Ealick, W.J. Cook, S. Vijaykumar, M. Carson, T.L. Nagabhushan, P.P. Trotta and C.E. Bugg, Science, 252 (1991) 698.
18. A.T. Bull, T.A. Huck and M.E. Bushell, in Microbial Growth Dynamics, 145, IRL Press, Oxford, 1990.

ACKNOWLEDGEMENTS

The findings discussed in this paper is the work of a multidisciplinary team and it is a pleasure to acknowledge the major contributions of Paula Castro, Lis Curling, Paul Hayter, David James and Callum Wingrove; the excellent technical support of Lucy Gettins and Malcolm Gould; and the stimulating co-direction of the programme provided by Nigel Jenkins, Anthony Baines, Ian Salmon and Philip Strange. I thank the SERC Biotechnology Directorate, British Biotechnology, Celltech, Glaxo, ICI, Porton

International, Smith-Kline Beecham and Wellcome for support of the research programme of which the work described here was a part.

The Biosynthesis of Ribosomes: Translational regulation of the ribosomal protein genes

P. Pierandrei-Amaldi and B. Cardinali

Institute of Cell Biology CNR, Via Marx 43, 00137 Rome, Italy

Some features of the regulation of expression of the protein component of the ribosomes (r-proteins) are reported. The mRNAs for these class of proteins are controlled at the translational level that represents an important regulatory step in their coordinate synthesis. Some studies of regulatory mechanisms involved in this process are described.

1. INTRODUCTION

Ribosome production is a housekeeping function present in all cells and involves the coordinate expression of genes coding for the ribosomal components, some molecules of ribosomal RNA and numerous ribosomal proteins (r-proteins); this function is directly related to cellular growth.

The expression of the r-protein genes has been studied in the frog *Xenopus laevis,* a system particularly suitable for studies on the regulation of ribosome biosynthesis, as it presents two particularly convenient developmental situations for such analysis: the oogenesis, when a huge amount of ribosomes is produced and stored for use in early development, and the embryogenesis, when the maternal store is used up and the embryo begins to synthesize its own ribosomes. We have studied the regulation of expression of the genes coding for the numerous ribosomal proteins which are produced in a balanced amount to form ribosomes.

The approaches used in our studies to find out about this regulation were the analysis of the expression of the r-protein genes in oocytes and in normal developing embryos and in the anucleolate embryos. These are homozygous mutants for a deletion of the r-RNA gene cluster, which are unable to make ribosomal RNA, but can undergo embryogenesis using maternal ribosomes, thus giving information on the influence of the rRNA on r-protein gene expression. Once known how these processes were normally going on, the pattern of expression of the r-protein genes during development was modified by microinjection experiments in oocytes and fertilized eggs in order to be able to correlate the changes observed in comparison with the normal developmental pattern and possible regulatory mechanisms (1).

This work was supported by "Target Project Biotechnology and Biostumentation" and by "Target Project Genetic Engeneering", CNR.

The results of these studies had shown that at least two types of regulation are involved in the production of r-proteins in the *Xenopus* system. A post-transcriptional regulation, operated by feedback of the proteins themselves, controls processing and stability of r-protein transcripts, and a translational regulation that controls the efficiency of utilization of the class of r-protein mRNAs (rp-mRNAs) (2-4).

In this presentation we will consider only the translational control of the r-protein genes, which was found to be active also in other developmental systems as *Drosophila*, *Dictyostelium* and also in cultured mammal cells and operates as a rapid response to changes in the requirement for ribosomes in the cell due to physiological variations in cell growth conditions.

2. FEATURES OF THE TRANSLATIONAL REGULATION OF rp-mRNA

The translational regulation is clearly observed during *Xenopus* development, when the accumulation of rp-mRNA is uncoupled from its utilization. In fact newly synthesized rp-mRNA is mostly kept unused as free mRNPs for several hours and it is mobilized onto polysomes when the maternal store of ribosomes is exhausted (5). The partition of the mRNA between polysomal and subpolysomal compartments, analyzed by sucrose gradient sedimentation, is typical of the developmental stage and it is specific for this class of mRNAs (5). As previously shown for mouse cells a rp-mRNA translational regulation has been observed also in a *Xenopus* cultured cell line which promptly responds to nutritional downshift and upshift by rapid changes of the percent of rp-mRNA loaded onto polysomes (6).

The distribution pattern of the rp-mRNA along the gradient is clearly bimodal, being the rp-mRNA engaged in translation fully loaded with ribosomes. This means that each molecule of the rp-mRNA is either translationally inactive or fully active with about one ribosome for every one hundred nucleotides of the translated region of the mRNA (6).The differently efficient utilization of the rp-mRNA cannot be justified by changes in the protein synthesis initiation rate which would result in a shift of rp-mRNA in more or less heavy polysomes.

Looking for effectors involved in the translational control, we have at first investigated if, by analogy with *E.coli*, a feedback inhibition by the r-proteins on their own mRNA was working also in this eukaryotic system (7). Evidence has been obtained that rule out this mechanism: *in vivo* and *in vitro* experiments have shown that the presence of an excess of free r-proteins has no effect on the translation of rp-mRNA (8). However, there are indications that the efficiency of utilization of rp-mRNA might respond somehow to the demand of the cell for new ribosomes, being the ribosome the effector of such process. In fact either in normal and more impressively in the anucleolate embryos (3), and also in cultured cells (6), there is a relationship between the availability of ribosomes and the utilization of rp-mRNA. To test if ribosomes were involved in this process the amount of ribosomes in the embryo was experimentally modified: it was found that this change affects specifically the utilization of rp-mRNA. Whether this effect is direct or mediated by other agents is not known (9).

3. *CIS* AND *TRANS* ACTING ELEMENTS IN THE rp-mRNA TRANSLATIONAL CONTROL

The observation that the translational behavior is specific and common to the mRNA for the numerous r-proteins, resulting in their coordinate synthesis, suggested a search for particular sequences and structures that could differentiate this class of mRNA from others conferring to them their typical translational properties. Sequences comparison revealed the presence of some common features in the 5' untranslated region (5'UTR) of the mRNAs for the *Xenopus* r-proteins analyzed. As shown in Figure 1, most typically they are short and start with an oligopyrimidine tract at the 5' terminus. It is possible that these characteristics, shared also with mammalian rp-mRNAs, might have a role in the translational regulation. Support for this hypothesis comes from experiments of microinjection into Xenopus fertilized eggs of a fused gene which produces a mRNA containing the 5' UTR of Xenopus r-protein S19 joined to the coding portion of a reporter gene. The translation pattern of mRNA in the developing embryo mimicked that of r-protein mRNA (10). Similar results were obtained in mammal cells, where it was also shown that the pyrimidine tract has a critical role in the control (11).

```
L1    CCTTTTCTCTTCGTGGCCGCTGTGGAGAAGCAGCGAGGAGATGGCGGTCAGCATG...
L14   CCTTTCCTCCCCGAGAAGCAGCTGCTGCTACAGCCGCCATCATG...
L22   CCTCTTTTCTCTTCCACCGCCGACATG...
L32   CCTTTTCCTCCATCTTGGATACCAGTGCGGTCCTGCCTGCATCTGGACATTAGGAGAGCAAGATG...
S1    CCTTTCCTGCCAGCGGCGCTTAGCAAAGATG...
S8    CCTCTTCTAGGCCGTCACTGAGAAAGGACGAAAGGCCATG...
S19   CCTTTCCTTCGTCACCGTGAGAGATAGCCGGCAAGATG...
```

Figure 1. 5'UTR sequences of different *Xenopus* r-protein mRNAs.

Translational regulation, in analogy to transcriptional control, can depend on specific interactions between *cis* regulatory elements in the mRNA and *trans*-acting factors. In order to find out if RNA/protein interactions are involved in the rp-mRNA control mechanism, we have characterized the particles containing the translationally repressed rp-mRNA *in vivo* and we have searched for possible factors that specifically interact *in vitro* with the 5' UTR typical of these mRNAs. The studies were mostly carried out for rp-L1 mRNA. By sedimentation analysis and isopycnic centrifugation we have found that the repressed rp-mRNAs in the embryo are assembled in slow sedimenting complexes, where the RNA is prevalent over the protein with a ratio of 2.3 to 1. The same composition is maintained also after *in vitro* reconstitution of the particle. A detailed analysis of *in vitro* RNA/protein complex formation was carried out by focusing our attention on the 5' UTR. *In vitro* binding experiments followed by gel shift assay and UV crosslinking have shown a specific interaction of the L1 rp-mRNA with four proteins (Figure 2). The binding site for two

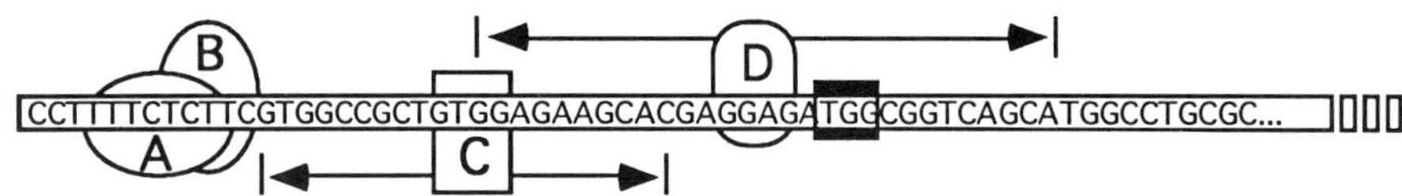

Figure 2. Position of binding proteins on the 5' UTR of L1 mRNA

of them, 57kD and 48 kD, is in the typical pyrimidine sequence at the 5' end and it is position dependent. Proteins of the same size interact also with the analogous region of r-protein S1 and L14 mRNA, but not with unrelated RNAs. Binding of two other proteins, 31kD and 24kD, in the downstreem region was also observed and seems to be affected by alternative conformation of the RNA. The fact that the described proteins are mostly bound to the 5' UTR could suggest a role in translational regulation of this class of mRNA, as occurs for some other mRNAs, the best studied example being represented by the ferritin mRNA (12). However, at the moment, there is no proof that the described proteins themselves act as translational repressors; in fact it is also possible that they are always associated with the 5' UTR of rp-mRNA to form a peculiar 5' end structure sensitive to translational control by some other factor(s). In fact a complex situation is expected as the binding activity was found to be equally present in developmental stages when the repression of rp-mRNA translational declines.

REFERENCES

1. F. Amaldi, I. Bozzoni, E. Beccari and P. Pierandrei-Amaldi. TIBS, 14 (1989) 175.
2. I. Bozzoni, P. Fragapane, F. Annesi, P. Pierandrei-Amaldi, F. Amaldi and E. Beccari. J. Mol. Biol., 180 (1984) 987.
3. P. Pierandrei-Amaldi, E. Beccari, I. Bozzoni and F. Amaldi. Cell, 42 (1985) 317.
4. P. Pierandrei-Amaldi, I. Bozzoni and B. Cardinali. Genes Dev., 2 (1988) 23.
5. P. Pierandrei-Amaldi, N. Campioni, E. Beccari, I. Bozzoni and F. Amaldi. Cell, 30 (1982) 163.
6. F. Loreni and F. Amaldi. Eur. J. Biochem., 205 (1992) 1027.
7. M. Nomura, R. Gourse, G. Baugham. Annu. Rev. Biochem., 53 (1984) 75.
8. P. Pierandrei-Amaldi, N. Campioni, P. Gallinari, E. Beccari, I. Bozzoni and F. Amaldi. Dev. Biol., 167 (1985) 281.
9. P.Pierandrei-Amaldi, N. Campioni and B. Cardinali. Cell.Mol. Biol., 37 (1991) 227.
10. P. Mariottini, and F. Amaldi. Mol. Cell. Biol., 10 (1990) 816.
11. S. Levy, D. Avni, N. Hariharan, R.P. Perry and O. Meyuhas, Proc. Natl. Acad. Sci. USA, 88 (1991) 3319.
12. M.W. Hentze, T.A. Rouault, J.B. Harford and R.D. Klausner. Science, 244 (1989) 357.

ECB6: Proceedings of the 6th European Congress on Biotechnology
edited by L. Alberghina, L. Frontali, P. Sensi

In situ hybridization and image analysis for mRNAs evaluation on single cells.

S. Barlati*, N. Zoppi, A. Ghinelli, and M. Colombi.

Division of Biology and Genetics, Department of Biomedical Sciences and Biotechnology, University of Brescia, Via Valsabbina 19, 25123 Brescia, Italy.

1. INTRODUCTION.

The analysis of purified mRNAs by filter hybridization (i.e. Northern blotting and dot-blot) with specific probes allows the evaluation of levels and types of a given mRNA, but does not give information on the possible heterogeneity of expression in the cell population under study. In situ hybridization (ISH) represents an invaluable tool for gene expression studies either in cultured cells or in histological tissue sections since it allows the identification and the localization of specific mRNAs present in single cells allowing therefore the characterization of cell populations expressing variable levels and/or different mRNAs isotypes. ISH has mainly been used as a qualitative approach for the detection of infectious agents' genomes or for the study of gene expression.

We have combined ISH with Image analysis (IA) and developed a method which gives, in an interactive mode, the measure of the autoradiographic grains' areas, obtained by ISH with radiolabelled probes, in relationship with the cellular, nuclear and cytoplasmic surface in cultured cells or, in tissue sections, with specific cells of different tissues.

This approach allowed the detection of cell populations heterogeneously expressing fibronectin (FN) mRNA (1), the evaluation of FN mRNA isoforms generated by alternative splicing in single cells (2), the semiquantitative evaluation and comparison of FN mRNA levels expressed by different cell types in human normal and tumor tissue sections (3) and the study of given mRNAs in the different phases of the cell cycle (manuscript in preparation).

* Work supported by C.N.R. Target Projects "Biotechnolgy and Bioinstrumentations", "Genetic Engineering" and "Clinical application of oncological research" and by A.I.R.C.

2. MATERIALS AND METHODS.

The quantitative evaluation of ISH, performed using tritiated FN cDNAs (1), was carried out using the Magiscan Image analysis System (Joyce Loebl, Gateshed, England). A task-list file was developed from the IN SITU software available on the Magiscan, which could be executed in a semi-automatic mode on different images (1, 3). The sequence of options available in the task-list was repeated until a sufficient number of cells or tissue areas were measured, then a data file containing the measurements was created to be analysed using the RESULTS program. In the list of the available measurements we considered the field area (FA), corresponding to the cell area or tissue field, the hybridization grains' area in given cells compartments or tissue portions (GA) and the fraction region (FR) defining the ratio GA/FA corresponding to the "specific activity" of the cell compartment or tissue area under study.

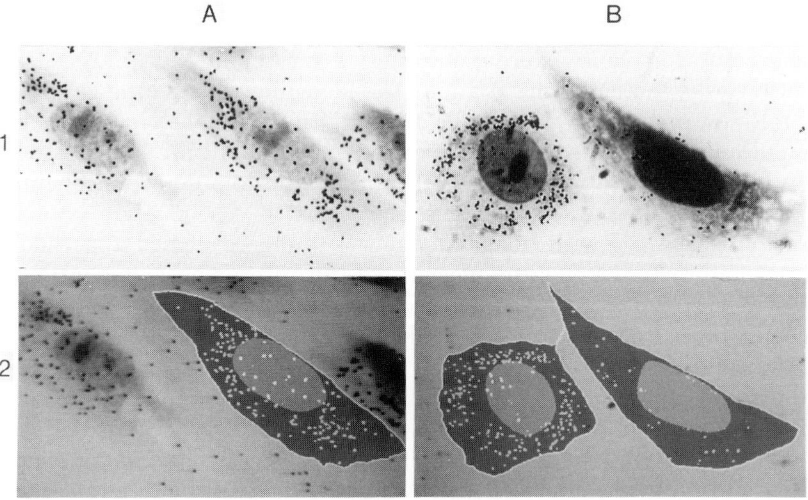

Figure 1. Cells expressing homogeneously (A) and heterogeneously (B) FN mRNA as detected after ISH (1) and IA(2).

3. RESULTS AND DISCUSSION.

ISH–IA performed on human skin fibroblasts (Flow7000) using a radiolabelled FN cDNA probe, showed that the level of FN mRNA expressed per units of area is homogeneous in different cells and is directly related (coefficient of correlation equal to 1.0) with the cellular and nuclear area. These results indicate that in fibroblasts the nuclear size is related to the cell surface and that the expression of FN mRNA is related to the different phases of the cell cycle: G2 cells produce a double amount of FN mRNA if compared with G1 cells. This observation also suggests that the level of FN mRNA expressed by skin fibroblasts is related to the gene dosage and that, after DNA duplication, all genes (four) present in G2 cells may be actively transcribed. This mechanism could provide the increased amount of FN required by G2 fibroblasts having a cell surface which is the double of G1 cells (manuscript in preparation).

ISH–IA was applied for the study of FN mRNA expression in several tumor-derived cell lines. In one of these cell types (RD, rhabdomyosarcoma) we identified two subpopulations: RD− (93% of the cells) expressing low levels of FN mRNA and RD+ expressing a seven-fold amount of FN mRNA if compared with RD− cells (Table 1). The ISH–IA therefore represents an invaluable tool for the study of gene expression in heterogeneous populations of cells.

Table 1

Levels of FN mRNA in Flow7000 and RD cells.

		Flow7000	RD+	RD−
GA	$\bar{x}$	870.1	737.5	98.7
	σ	239.5	343.2	36.0
	CV	27.5	46.5	36.5
FA	$\bar{x}$	21467	12652	11194
	σ	4502	4600	1925
	CV	21.0	36.3	17.2
FR	$\bar{x}$	40.5	58.0	8.8
$\bar{x}10^3$	σ	8.7	22.0	2.9
	CV	21.5	37.7	32.9

$\bar{x}$: average out of 20 cells; σ: standard deviation; CV: coefficient of variation.

ISH–IA was also applied for the study of FN mRNA isoforms arising from alternative splicing of the EDA region. For this purpose, two FN cDNA probes of equal length (180 bp) radiolabelled to the same specific activity, one containing a portion of the EDA region and the other hybridizing with all FN mRNA isoforms, were used. With this approach it was possible to conclude that EDA+ and EDA−

FN mRNA isotypes are produced in the same cell and that skin fibroblasts derived from Ehlers–Danlos syndrome (EDS) patients show a decrease of EDA+ FN mRNA if compared with control fibroblasts. In EDS cells in fact, 45% of FN mRNA contain the EDA region, while in control fibroblasts 80% of FN mRNA is EDA+ (Table 2).

Table 2

Levels of total and EDA+ FN mRNA in control and EDS skin fibroblasts.

			GA	%EDA+	FA	FR($\bar{x}10^3$)
Flow7000	Total	$\bar{x}$	744.6	100	2001	37.0
		σ	236.7		4184.3	7.9
		CV	31.8		20.9	21.3
	EDA+	$\bar{x}$	573.5	77	20174	28.5
		σ	148.4		4290.7	5.0
		CV	25.9		21.3	17.5
EDS	Total	$\bar{x}$	397.1	100	21698	18.6
		σ	61.5		3019.9	3.7
		CV	15.5		13.9	19.9
	EDA+	$\bar{x}$	179.2	45	19799	9.1
		σ	47.6		2726.9	2.2
		CV	26.6		13.8	24.1

$\bar{x}$: average out of 20 cells; σ: standard deviation; CV: coefficient of variation.

These levels of EDA+ FN mRNA correlate with the defective FN matrix assembled by in vitro grown EDS skin fibroblasts (4).

Finally, ISH–IA was used for the evaluation of FN mRNA present in histological sections of human ectocervical and laryngeal carcinomas (3). This method allowed the identification of the cell types producing FN mRNA (stromal fibroblasts surrounding tumors and capillary endothelial cells) and the semiquantitative evaluation of FN mRNA induced in the peritumoral stroma by the developing tumors.

REFERENCES.

1. L. Moro, M. Colombi, S. Crespi, R. Di Lernia and S. Barlati, Cell Biol. Int. Rep. 14 (1990) 701.

2. M. Colombi, L. Moro, N. Zoppi, A. Ghinelli and S. Barlati, Cell Biol. Int. Rep. 15 (1991) 1195.

3. L. Moro, M. Colombi, M.P. Molinari–Tosatti and S. Barlati, Int. J. Cancer 51 (1992) 692.

4. S. Barlati, L. Moro, N. Zoppi, R. Gardella and M. Colombi, Cell Biol. Int. Rep. 14 (1990) 1183.

ECB6: Proceedings of the 6th European Congress on Biotechnology
edited by L. Alberghina, L. Frontali, P. Sensi

A Cancer Sensitivity Test Based on Expression of Mutant p53 in Cultured Human Cells.

C. Mothersill[a], J. Harney[b] and C.B. Seymour[a].

[a]Physics Department, Dublin Institute of Technology, Kevin Street, Dublin 8, IRELAND*

[b]Department of Urology, Beaumont Hospital, Dublin 9, IRELAND.

1. SUMMARY

The development of a test which could identify persons at risk for developing cancer due to environment carcinogen exposure is a major goal in medical biotechnology. Such a test would allow sensitive subgroups to be protected, screened or counselled more intensively than is cost effective for the general population. One of the major candidates to form the basis for such a test is the oncosuppressor gene p53. This exists in both mutant and wild type forms and the mutant form is associated with many tumours. Li-Fraumeni syndrome is a cancer syndrome associated with congenital p53 mutation, but little is known of the sensitivity to mutation of the gene in normal humans.

In this paper we report on the results of a study where tissue from 25 normal humans was cultured and exposed to radiation and/or to a chemical carcinogen (n-ethyl nitrosamine). The results revealed considerable differences in the susceptibility of individual human tissue specimens to induction of mutant p53 following carcinogen exposure. The population segregated into four major subgroups. Group 1 (20%) expressed mutant p53 following culture alone of their tissue; Group 2 (15%) required exposure to radiation (<5 Gy); Group 3 (43%) required exposure to nitrosamine in addition to radiation, and Group 4 (22%) did not express mutant p53, in spite of high dose radiation and carcinogen exposure. All the intact tissues from which cultures were established were negative for mutant p53. It is suggested that sensitivity of tissue to induction of mutant p53 expression in vitro might form the basis of an eventual "cancer sensitivity" test.

* Work supported by Grants from the EC Radiation Protection Programme and the Irish Cancer Research Advancement Board.

2. INTRODUCTION

New ideas concerning the multistage nature of the carcinogenic process make it unlikely that traditional approaches to the development of a "transformation assay" will have much relevance either to the detection of carcinogenic chemicals or to the identification of cancer sensitive subgroups in the population.

Tests which focus on identification of particular genes are more promising, and in the case of familial breast and ovarian cancer, appear to be almost available for screening programmes. A general test for cancer proneness is not yet available. This would have to focus on identification of a "cancer enabling" gene/gene defect.

Obvious candidates for this would be global proliferation genes (such as cmyc) or repair monitoring genes (such as p53) which, when defective, would allow abnormal cells to persist unrepaired or to grow out of control.

This study focuses on the status of the p53 oncoprotein, p53-240, in urothelial tissue of 25 normal humans.

3. METHODS

Culture methods have been published in detail elsewhere (1).

Briefly multiple explant cultures were established from normal urothelium obtained from patients undergoing reconstructive surgery. Explants 2 mm^2 were initially plated in 2 ml RPMI 1640 containing 20% foetal calf serum, insulin and hydro-cortisone.

After 48 hours, cultures were transferred to keratinocyte serum free media and challenged with n-ethyl nitrosamine (0.05 μg/ml) and/or 5 Gy Cobalt 60 gamma rays.

After 2 weeks, cultures were fixed in formalin and assayed for expression of p53-240 protein in situ. Histological sections retained from each sample permitted the p53 status and morphology of the intact tissue to be determined.

Positive controls for the test were the SV80 cell line and a bladder cancer explant culture previously determined as positive.

4. RESULTS

Fig. 1 shows the % p53 positive cells in cultures derived from normal ureters, from patients with no cancer, following exposure to 5 Gy Cobalt 60 gamma rays with or without chronic exposure to 0.05 μg/ml n-ethyl nitrosamine. The means ± SE and significance are shown on Table 1. Clearly, carcinogen challenge with either agent significantly increases the % of cells expressing p53-240. The combination of both carcinogens significantly increases expression over either agent separately. The most interesting result, however, is seen when patient variation is examined.

Table 1

p53 Positive Cells in Cultures of Normal Urothelium, 14 days following Carcinogen Challenge

Treatment	**% Positive Cells**
Control	13.8 ± 4.1
Radiation (5 Gy)	44.7 ± 5.3
Nitrosamine	45.3 ± 7.0
Radiation + Nitrosamine	65.3 ± 7.5

Each test group significantly different at $p < 0.001$ level

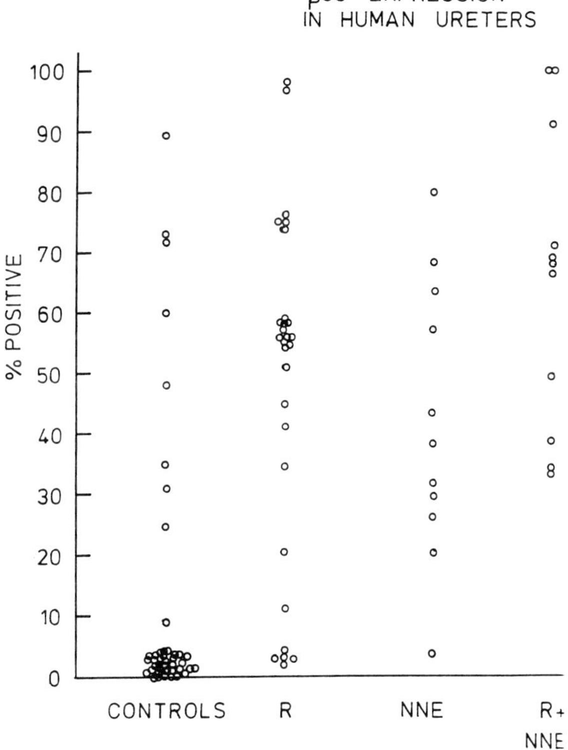

Fig. 1 p53 Expression in 25 Normal Human Patients Following Carcinogen Challenge.

Table 2 shows the % of patients whose cultures express p53 following culture alone, either carcinogen, or both carcinogens. These results suggest a heterogeneous response among patient tissues. It is notable that 8% of patients did not become positive even after radiation and chronic nitrosamine exposure.

Table 2

Segregation of 37 normal urothelial cultures based on the level of carcinogen challenge required to induce p53 protein expression using the p53-240 antibody.

Treatment	Number of patients with p53 positive cells	
	Total Number	Percent
None (sections of tissue)	None	0
Culture only	9	24%
Culture + radiation <u>or</u> Nitrosamine	20	54%
Culture + radiation <u>and</u> Nitrosamine	29	78%

22% (8 patients) never showed positive p53 expression.

5. DISCUSSION

The results clearly show segregation of 25 normal humans into 4 groups based on the level of carcinogen challenge needed to induce the cancer associated form of p53. There is a group requiring nothing more than the stimulus to proliferate provided by in vitro culture. The majority of samples can be induced by carcinogens, but there is a significant residue which remained negative in spite of chemical and radiation challenge.

It is clearly important to establish the reasons for the patient variation in susceptibility to p53-240 protein induction by carcinogens. Presumably, this results from heterozygosity in the genome coupled with the increase in proliferation induced in surviving cells. The particular qualities of the group of patients which were resistant should also be investigated.

If the results can be shown to be similar for a more accessible tissue, such as skin keratinocytes, it could form the basis of a cancer susceptibility test in the future.

REFERENCE

1. C. Mothersill, A. Cusack, M. MacDonnell, T.P. Hennessy and C.B. Seymour, Acta Oncologica 27 (1988) Fasc 3 275-280.

ECB6: Proceedings of the 6th European Congress on Biotechnology
edited by L. Alberghina, L. Frontali, P. Sensi
1994 Elsevier Science B.V.

Expression of exogenous genes in porcine somatic cells using recombinant retroviral and episomal vectors

P. K. Nandi and O. Grépinet

INRA, Unité de Pathologie Infectieuse et Immunologie, 37380 Nouzilly, France

1. INTRODUCTION

The concept of germ line gene transfer to produce transgenic animals (gene farming) has been mostly applied to pigs [1]. This technique is also attractive to produce proteins of phamaceutical interest. Recently, somatic cell transgenesis in pigs has been considered as a model for human gene therapy [2]. This procedure involves modification of lymphocytes, keratinocytes, fibroblasts, endothelial cells and myoblasts by genetic engineering to produce therapeutic proteins of interest followed by their implantation at the site of the disease or release of the therapeutic gene products by the modified cells in general circulation after implantation [3]. The major vector for modification of these above cells to produce therapeutic proteins has been the recombinant retrovirus (RRV).

Here we describe the use of recombinant retroviral vectors and an episomal plasmid vector to transfer genes into porcine somatic cells with a histocompatible pattern (d/d haplotype).

2. INFECTION WITH RECOMBINANT RETROVIRUS

Most recombinant retroviral vectors described to date are replication defective and are propagated through the use of (i) wild type viruses or (ii) helper-free packaging cell lines which provide all of the functions in trans necessary for packaging a recombinant genome. We have produced a Moloney murine leukaemia virus (MMuLV)-derived xenotropic RRV by infecting an established swine testis (ST) cell line by the process (i) [4]. This RRV produced by STBF11 cell line contains an indicator bacterial lacZ gene coding for β-galactosidase fused to a nuclear location signal (nls) which is under the transcriptional control of both SV40 early promoter and retroviral LTR. By definition, the amphotropic RRV derived from MMuLV should infect all species from mouse to man. The xenotropic

RRV, on the other hand, should infect all species except the species of origin, i.e. mouse and closely related rodents. The infection of porcine primary and immortalized endothelial cells from vessels as well as aorta showed ≈1.5 % of the cells expressing the marker gene. The lymphoblastoid cell line L52 (B-cell lineage) [5] expressed a higher level (≈4 %) of the same gene. In Table 1, we present the titre values of the same RRV on fibroblastic cells from different animals.

Table 1. Titres of SVnlslacZ RRV produced by the porcine STBF11 packaging cell line [4].

Species	Cell lines	Origin	STBF11
Mouse	3T3	Foetus fibroblasts (NIH)	< 1
Rabbit	RL307	Kidney	1.3×10^5
Cat	G355.5	Lung	2×10^4-1×10^6
Sheep	PFM	Foetal lung (PC)	2×10^4
Beef	PV	Foetal lung (PC)	1×10^2
Pig	ST	Swine testis	4×10^4
Goat	PFC	Foetal lung (PC)	1×10^4
Horse	NBL	Skin	3×10^2

Titres are expressed as blue colony-forming unit (bcfu)/ml; cells except rabbit RL307 are of fibroblastic origin. PC, primary cells.

In subsequent studies, we obtained the evidence that STBF11 cell line also produced wild virus along with the RRV. The helper-free packaging cell lines producing pure amphotropic (NB5) and pure xenotropic (ΨX8) RRV, constructed by procedure (ii), were a gift from Dr. O. Danos (Institut Pasteur, Paris). These RRV contained the above indicator nlslacZ gene under the transcriptional control of the mouse phosphoglycerate kinase promoter.

Table 2. Titres of pure SVnlslacZ amphotropic and xenotropic RRV produced respectively by NB5 and ΨX8 packaging cell lines.

Species	Cell lines	NB5	ΨX8
Mouse	3T3	1×10^5	0
Pig	ST	0	2×10^5
Pig	iPEC	0	5×10^4

Titres are expressed as bcfu/ml. iPEC, immortalized porcine endothelial cells.

The xenotropic RRV could infect the porcine cells effectively whereas the amphotropic RRV was incapable of infecting the porcine cells studied here (Table 2). Further, the xenotropic RRV could infect both the primary and immortalized porcine cells to a similar extent. In separate experiments, the pure xenotropic RRV was found to infect immortalized human umbilical chord endothelial cell line (Eahy, ATCC). The extent of infection was similar to those observed for porcine primary and immortalized endothelial cells. The pure amphotropic RRV was unable to infect the human endothelial cells to any significant extent.

3. EPISOMAL PLASMID VECTOR

The bovine papilloma virus-derived vector BCMGSNeo [6] carries a neomycin resistance gene as a dominant selectable marker and a subgenomic fragment of bovine papilloma virus which enables a stable multicopy extrachromosomal replication of the plasmid in mammalian cells. The transcription of the cDNA's inserted into this vector is under the control of the human cytomegalovirus promoter and enhancer.

Table 3. Secretion of hIL6 by the porcine lymphoblastoid cell line L35 (T-cell lineage) stably transformed with pIL63.

Clones	ng hIL6/10^6 cells/72 h (a)	
	with G418 (b)	without G418 (c)
L35(BCMGSNeo)	0	0
L35(pIL63)		
L35-D3	77	163
L35-D5	153	124
L35-A2	81	135

(a) The amount of hIL6 produced is expressed as ng secreted by 10^6 cells for 72 h; cells were cultured in presence or absence of antibiotic for 72 h and numbered before recovery of the supernatant.
(b) Cells were grown in presence of antibiotic G418 (1000 μg/ml).
(c) Cells were maintained in culture without G418 for 2 months.

Plasmid pIL63 was constructed by subcloning a 850 bp EcoRI-DraI fragment of human interleukin-6 (hIL6) cDNA into BCMGSNeo vector.

This fragment contains the whole protein-coding region, but lacks most of the 3' untranslated AT-rich sequences involved in selective degradation of transiently expressed messengers. Plasmid pIL63 and BCMGSNeo vector were introduced in the porcine lymphoblastoid cell line L35 [5] by electroporation. Antibiotic G418 (1000 μg/ml) was added 48 h after to select stable transformants. L35 cells transformed with pIL63 were cloned by limiting dilution. Three high-producer clones, L35-D3, L35-D5 and L35-A2 were isolated. Amount of hIL6 secreted was measured by ELISA test using recombinant hIL6 as standard. The three clones (10^6 cells) produced 80 to 150 ng of hIL6 for 72 h (Table 3), which made these supernatants usable as a source of hIL6. Moreover, this production was stable for 2 months when the cells were cultured without antibiotic, which is similar to previously obtained results [7]. According to the presently accepted level of human gene (immuno) therapy protocol of cancer, this amount of cytokine secretion is sufficient for treatment [8].

The results we obtained in our laboratory show that xenotropic RRV could be used to transfer exogenous genes into porcine somatic cells. Work is in progress to engineer endothelial and myoblastic cells from histocompatible pig (d/d haplotype) to produce various cytokines with the help of pure xenotropic RRV. The implantation of these genetically modified cells into pig of the same haplotype and study of the kinetics of cytokine biosynthesis could be used to develop model for both production of pharmaceutical proteins and human gene therapy.

ACKNOWLEDGEMENTS

We thank Drs. H. Karasuyama and F. Melchers for providing BCMGSNeo vector and Dr. P. Galanaud for the kind gift of hIL6 cDNA. We also thank Dr. J-C Lecron for performing hIL6 ELISA test.

REFERENCES

1. V.G. Pursel, C.A. Pinkert, K.F. Miller, D.J. Bolt, R.G. Campbell, R.D. Palmiter, R.L. Brinster and R.E. Hammer, Science, 244 (1989) 1281.
2. B.J. Culliton, Science, 246 (1989) 749.
3. M. Hoffman, Science, 254 (1991) 1455.
4. B. Charreau, O. Grépinet, C. Delouis and P.K. Nandi, Res. Virol., 142 (1991) 343.
5. B. Kaeffer, E. Bottreau, L. Phan Thanh, M. Olivier and H. Salmon, Int. J. Cancer, 46 (1990) 481.
6. H. Karasuyama, A. Kudo and F. Melchers, J. Exp. Med., 172 (1990) 969.
7. H. Karasuyama and F. Melchers, Eur. J. Immunol., 18 (1988) 97.
8. P.T. Golumbek, A.J. Lazenby, H.I. Levitsky, L.M. Jaffee, H. Karasuyama, M. Baker and D.M. Pardoll, Science, 254 (1991) 713.

Expression of heterologous Genes in yeast Systems

ECB6: Proceedings of the 6th European Congress on Biotechnology
edited by L. Alberghina, L. Frontali, P. Sensi

Biotechnology in Yeasts: Alternative Protein Production Systems

Jeffrey R. Shuster*

Chiron Corporation, 4560 Horton St., Emeryville, CA 94608

The field of yeast biotechnology may be considered to be a few thousand years old, originating in the optimization of the cellular products, ethanol and carbon dioxide. Thus yeast systems have had a long history of technology advancement and safe use. Current yeast biotechnology as defined by the expression of heterologous proteins using recombinant DNA techniques began less than twenty years ago. In this period, *Saccharomyces cerevisiae* has been utilized extensively to produce a number of heterologous proteins with commercially successful results. Some examples include, a vaccine for hepatitis-B, human insulin for treatment of diabetes, a diagnostic reagent for hepatitis-C viral infection, and a human colony stimulating factor for cell proliferation. Although there are a plethora of heterologous proteins that have been expressed in *S. cerevisiae* at detectable levels, many of these proteins have not become biotechnology successes. This is in part due to the fact many new and investigatory proteins have not demonstrated efficacy for their proposed uses when tested in clinical trials. However, it is also true for other significant proteins that there has been an inability to produce sufficient quantities and appropriate quality materials in yeast. It is the latter issue that is the subject of the session at this meeting, "Expression of heterologous genes in yeast systems". Results from studies of a number of different expression systems will be presented with an emphasis on a comparative analysis of the systems and their potentials to produce heterologous proteins.

The successful expression of heterologous proteins from any organism can be defined by two interrelated parameters. These are the production of protein in sufficient quantity and the obtainment of appropriate quality. The eukaryotic microorganism most often utilized for the expression of heterologous proteins is the yeast *Saccharomyces cerevisiae*. In this organism, different protein production systems have been created by the combination of molecular biology, genetics, and fermentation methodologies. Interestingly, any particular successful (as defined by quantity and quality) system for one protein product may not

---

*current address: Novo Nordisk Biotech, 1445 Drew Ave., Davis, CA 95616

function satisfactorily for others. In fact, significant differences in the expression levels of different proteins are often observed when the expression systems are comprised of identical RNA transcriptional control elements, DNA vectors, protease deficient host strains, fermentation and growth conditions, and other relevant parameters. This puzzling protein specific attribute of expression systems, and the inability to produce some proteins successfully in *S. cerevisiae*, has led to question whether other yeast expression systems might produce successful results. This should not be construed to imply that *S. cerevisiae* should always be the yeast of first choice. Rather, *S. cerevisiae*, is currently the best characterized, safe, and most easily manipulated system. Clearly, if one can produce a product in a well characterized and generally regarded as safe host organism (*S. cerevisiae*), one will choose to do so. The strength of using alternative, and less well characterized, yeast hosts lies in their abilities to satisfy the two critical parameters, quantity and quality, for commercial processes when *S. cerevisiae* is unable to do so. It is important to bear in mind that the ability of an alternative host system to produce a higher amount of protein when an *S. cerevisiae* system is considered successful for a particular need, does not in itself, justify use of the new system for production.

The first experiment performed when comparing the production of proteins in different host systems, is usually a measurement of product quantity in impure extracts or fermentation broth cultures. A comparative analysis is relatively easy to perform if standard analytical techniques are employed. However, when different laboratories report expression levels in weight amounts of, for example, *mg/liter*, one should pay close attention to the methodology of the assays. Different assays can yield very different results. For example, assays of a specific protein that is based upon high-pressure liquid chromatography analysis could be expected to result in different numbers if the same samples were assayed by a polyclonal antibody ELISA technique. In fact, different ELISA results may be obtained when different sources of antibody are used in the assays. Thus, when comparing expression levels of proteins from different hosts, one should (1) always use the same analytical procedures and (2) if there is a choice of analytical methods, use the one most closely reflecting the final product use. For example, a receptor binding assay is more appropriate assay for a production of a mammalian peptide growth substance than an SDS polyacrylamide gel analysis. In addition, attention should be made to the scale of the experiments in different systems. If it is general knowledge that fermenter growth results in five-fold higher yields than shake-flasks, then a comparison of protein yields in different organisms should be compared using this knowledge and appropriate growth technologies.

In the event that an initial encouraging result is obtained, a comparison of the quality of the product usually follows. A comparison of the quality of the protein produced in different host systems should be instigated as soon as possible following a positive result in expression quantity. Particular attention should be paid to possible secondary modifications of the protein that may render it suitable or unsuitable for its intended purpose. For example, a change in protein glycosylation pattern in an enzyme produced in a heterologous host could alter its stability so as function inappropriately in an industrial process. Similarly, a post-translational modification of a protein intended for final use as an injectable substance in humans, could render the protein antigenic and result in an inappropriate and perhaps life-threatening response.

One of the advantages of a microbial expression system is the ability to grow very large quantities of recombinant organisms. Therefore, even a discussion concentrating on yeast should include a brief comparison of the well characterized prokaryotic *E. coli* expression systems. Many heterologous proteins have been expressed in the *"grams/liter"* range in this organism rather easily. However, these very high expression levels more often than not, result in the precipitation of the proteins into inactive intracellular inclusion bodies. Efficient technologies for the recovery, solubilization, and refolding of proteins make the high-level production of proteins from *E. coli* a very attractive alternative to eukaryotic systems for the production of some proteins that can be refolded to active form *in vitro*.

Yeast systems for the expression of heterologous proteins have many advantages over prokaryotic microbial systems. In general, yeasts can provide for correct folding of many proteins by use of a secretion expression system. Unlike prokaryotic hosts, yeasts are also capable of many post-translational modifications including glycosylation, myristylation, ubiquitin cleavage, *etc.* In addition to *S. cerevisiae*, the yeast hosts used in biotechnology include *Kluyveromyces lactis*, *Pichia pastoris*, *Hansenula polymorpha*, and *Yarrowia lipolytica*. As described earlier, alternative yeast host systems are being developed to provide for advantages where *S. cerevisiae* systems fail. *Kluyveromyces* has proven to be a yeast more capable of high efficiency secretion, *Pichia* and *Hansenula* are methylotrophs with a demonstrated ability to be grown to very high cell densities and also may have advantages for the secretion of proteins. *Yarrowia* yeast are capable of secreting homologous protein in high amounts and are therefore being investigated for secretion of heterologous proteins.

Given that research resources are rarely limitless, how does one decide whether to use the *S. cerevisiae* expression systems or to investigate some of the less well characterized alternatives? The papers in the following session will present results that can aid in this decision making process. In the work presented, the listeners will hear of recent results in the field of heterologous protein expression in yeasts. One paper compares the expression of a protein in *S. cerevisiae* versus that in *E. coli*, another paper contrasts protein expression in *S. cerevisiae* and *Hansenula polymorpha*. A third paper describes new technology for the alternative yeast, *Kluyveromyces lactis*, and its use for the expression of human serum albumin (a protein that can be made at high levels in *S. cerevisiae*). A final paper describes a recent work concerning the glycosylation pattern of a protein produced in *a pmr1* mutant of *S. cerevisiae*. The *pmr1* mutation has been reported previously to alter glycosylation of proteins secreted from yeast. The secondary modification of proteins by glycosylation, and the significant effect that glycosylation may have on the quality of any specific protein makes this a key issue in yeast biotechnology.

The listener and reader are advised to bear in mind the following question for each system described as an alternative system to *S. cerevisiae*. Does the system include the necessary elements for a successful heterologous protein expression system as described by the following:

(1) methods to introduce exogenous DNA and stably maintain it through fermentation in large scale

(2) strong (and regulatable) transcription machinery

(3) molecular biological, genetic, and/or physiological methods of controlling undesirable proteolytic degradation of the protein of interest

(4) appropriate secondary protein modification

(5) the absence of toxins or other undesirable biological by-products

(6) appropriate fermentation, scale-up, recovery, and purification technologies

(7) satisfactory resolution of pending or granted patent issues

(8) regulatory approvals by government agencies.

ECB6: Proceedings of the 6th European Congress on Biotechnology
edited by L. Alberghina, L. Frontali, P. Sensi
1994 Elsevier Science B.V.

Expression of a synthetic acylphosphatase gene in *E.coli* and *S.cerevisiae*

A. Modesti, G. Raugei, M. Vecchi, R. Marzocchini, G. Manao, G. Camici, G. Cappugi and G. Ramponi

Dipartimento Scienze Biochimiche, Universita' di Firenze,
Viale Morgagni 50, 50134 Firenze, Italy.

1. SUMMARY

The initial purpose of this study was to establish a strategy for investigating the function of acylphosphatase, with the use of protein engineering techniques. For this reason a chemically synthesized DNA sequence coding for human muscle acylphosphatase has been cloned in two different expression vectors and introduced in *E.coli* and in *S.cerevisiae*. In both cases high level expression of acylphosphatase has been observed. After purification to homogeneity, the recombinant proteins have been assayed in comparison with the natural acylphosphatase, using benzoylphosphate as a substrate and phosphate as a competitive inhibitor: the recombinant enzymes maintain the kinetic properties of the natural protein.

In addition, preliminary results indicate that the expression of high level of acylphosphatase in *S. cerevisiae* may enhance the production of ethanol in fermentative conditions.

2. INTRODUCTION

Acylphosphatase (EC 3.6.1.7) is a hydrolase that specifically catalyzes the hydrolysis of the carboxyl-phosphate bond of acylphosphates. It is a small (Mr. about 11,000) and stable enzyme which is distributed among a wide variety of species and tissues. The enzyme has been purified from skeletal muscle of various mammals and birds and the primary structures have been determined [1]. The primary structure of the enzyme is well conserved among different species. The physiological function of acylphosphatase is still debated. The enzyme is active on several important physiological substrates such as 3-phosphoglyceroyl phosphate, carbamoyl phosphate, and succinoyl phosphate. Moreover it can hydrolyze the phosphorylated intermediates (aspartylphosphate) of (Na^+-K^+)-ATPase and of Ca^{2+}-ATPase of sarcoplasmic reticulum [2].

It has been demonstrated that the addition of acylphosphatase to cell-free extracts of *S.cerevisiae* increased the rate of alcoholic fermentation and glucose degradation [3].

This effect could be due to the hydrolysis of 3-phosphoglyceroyl phosphate: acylphosphatase may increase the rate of glycolysis at the expense of ATP formation. in condition where glycolysis may be limited by low concentrations of inorganic phosphate and ADP.

3. RESULTS

3.1 Expression of recombinant proteins.

In order to express human muscular acylphosphatase in *E.coli* and *S.cerevisiae* we have

constructed a synthetic 300 bp long DNA sequence coding for the enzyme and we have inserted it in two different expression vectors [4]: 1) in pT7.7 (obtaining the pTAP1 construct) for the expression in *E.coli* and 2) in pYEpSec1 (obtaining the pYAP1 construct) for expression in *S.cerevisiae*. In both cases cloning were done in such a way that the protein was accumulated in the cytoplasm rather than secreted in the medium.
To control the expression of acylphosphatase in transformed cells, Western blot analysis of *E.coli* and *S.cerevisiae* lysates was performed: a protein of the expected molecular weight was recognized by the antibodies against the human enzyme (data not shown).
Immunoaffinity purification of recombinant acylphosphatases from crude lysates was performed and purity checked by SDS-PAGE (Fig.1). In addition to the expected 11 kd species, a minor slow moving band is also recognized by the antibodies: this form should represent the dimer of the protein, due to the presence of a single cysteine in the acylphosphatase molecule. Treating the samples with the reducing agent mercaptoethanol (lanes 4 to 6) only the fast moving band is present, due to the reduction of the cysteine presents in the molecule.
With this method of purification it was possible to obtain 3 mg of purified protein from 1 l culture of *E.coli* and 1 mg from a 1 l culture of *S.cerevisiae.*
Analysis of the amino acid content of the two proteins showed no significant discrepancy with the natural human muscle acylphosphatase (data not shown). The possibility to perform NH_2 terminal sequence of the protein expressed in *E.coli* demonstrated the absence of acetylation on the first aminoacid, a modification that , in contrast, is present on the natural protein and on the recombinant one obtained in *S.cerevisiae.*

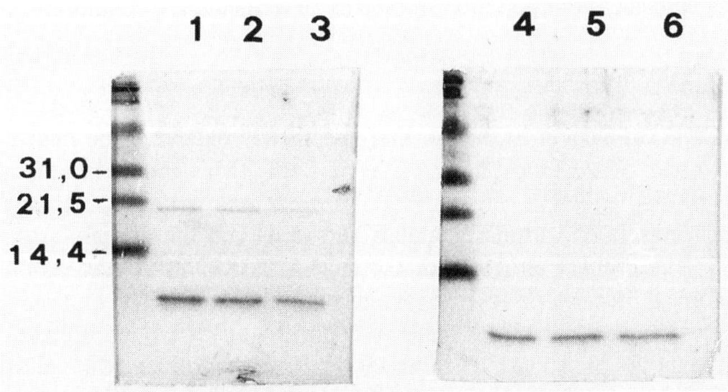

Figure 1. SDS-PAGE of acylphosphatase. Lanes 1 and 4: native. Lanes 2 and 5: from *E.coli* trasformed with pTAP1. Lane 3 and 6: From *S. cerevisiae* transformed with pYAP1. Lanes 4 to 6 : samples were treated with mercaptoethanol.

3.2 Catalytic activity.

The kinetic properties of the two recombinant acylphosphatases were determined in comparison with the native protein purified from human muscle (Tab. 1). The Km values of the recombinant proteins from *E.coli* and *S.cerevisiae* were determined using benzoylphosphate as a substrate and Ki values were calculated using phosphate as a competitive inhibitor. The results show that Km and Ki values were in the same order of magnitude, suggesting that the affinity of the two recombinant acylphosphatases for these molecules was substantially unchanged in respect to that of the natural protein.

The specific activity of the recombinant enzyme purified from *E.coli* using benzoylphosphate as a substrate resulted to be two times higher than that of the enzyme obtained from yeast and from horse muscle. This difference seems to be correlated with the acetylation on the NH_2 terminal residue, present in the native enzyme and in the recombinant acylphosphatase obtained in *S.cerevisiae*, but absent in the protein obtained from *E.coli*. Structural studies of the enzyme indicate that the first six aminoacid of the protein form a mobile tail with unknown function: acetylation in that position could play a role in modulating the activity of the enzyme.

The possibility of obtaining large amounts of recombinant acylphosphatase will facilitate further structural and enzymatic studies of this protein, mainly with the creation of mutated versions of the enzyme.

Table 1. Kinetic properties of recombinant and native acylphosphatase. A: from *S. cerevisiae*. B: from *E.coli*. C: human

	A	B	C
Specific activity (U/mg)	3600	7000	3600
pH optimum	5.3	5.3	5.3
Km (BP) mM	0.3	0.5	0.8
Km (Pi) mM	1.0	1.0	2.7

3.3 Ethanol fermentation

The observation that acylphosphatase increased the rate of alcoholic fermentation in cell-free extracts of *S.cerevisiae* [3], suggested us to evaluate ethanol production in *S.cerevisiae* transformed with the pYAP1 construct and expressing high level of human muscle acylphosphatase, in comparison with a strain transformed with the vector alone. For this purpose, ethanol production was measured in a late log culture, maintained in nitrogen atmosphere and supplemented with 0.8% of glucose. In these conditions we can reproducibly observe an enhancement of about 30% in the rate of ethanol production in the strain expressing human acylphosphatase. We can also observe that consumption of glucose in the medium is significantly augmented

In Fig. 2 (next page) the results of a typical experiment are presented: a time course of ethanol and glucose concentration in the culture is shown, together with other control parameters (acetic acid concentration, number of viable cells, total protein concentration).

More accurate measurements of this phenomena are needed, together with the determination of other parameters, like ATP and ADP concentration, in order to clarify the mechanism of action of acylphosphatase. We think that this finding could be potentially very important, giving the possibility to develop more efficient fermentation processes of natural substrates.

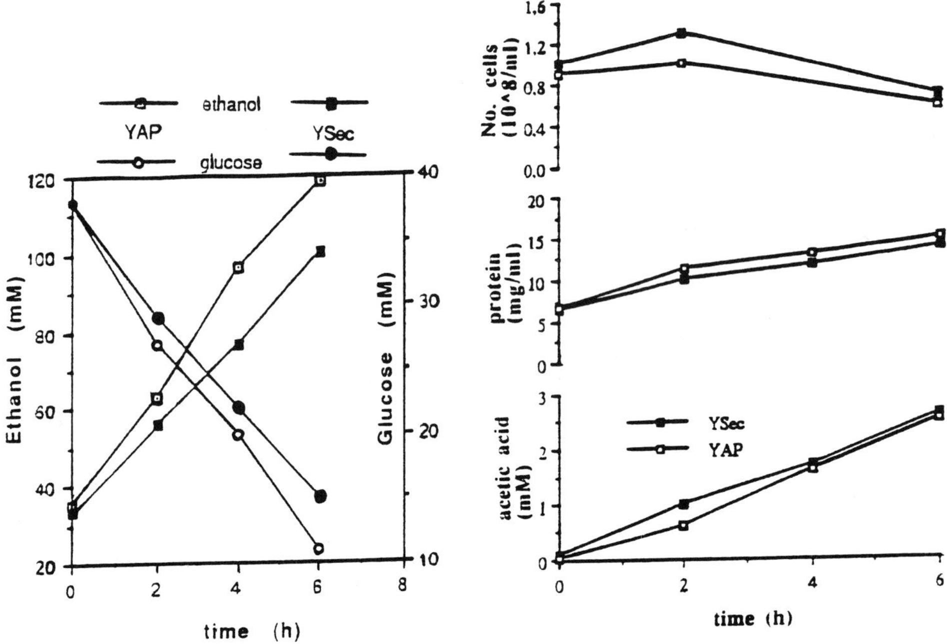

Figure 2. Time course of: left, ethanol and glucose concentration in the medium of culture; right, number of viable cells, total protein and acetic acid concentration. YSec: control strain. YAP: S. cerevisiae expressing human acylphosphatase.

REFERENCES

1. G. Manao, G. Camici, G. Cappugi, M. Stefani, G. Liguri, A. Berti and G: Ramponi, Arch. Biochem. Biophys, 241 (1985) 418.
2. A. Berti, M. Stefani, G. Liguri, G. Camici, G. Manao and G. Ramponi, J. Biochem., 26 (1977) 377.
3. C. Nediani, E. Marchetti, P. Nassi, G. Liguri and G. Ramponi, Biochem. Int., 24 (1991) 959.
4. A. Modesti, G. Raugei, N, Taddei, R. Marzocchini, M. Vecchi, G. Camici, G. Manao and G. Ramponi, (1993) submitted for publication.

ACKNOWLEDGMENTS

This work was supported in part by C.N.R. Target project on Biotechnology and Bioinstrumentation of the Italian Research Council.

Patent pending in Italy, N. RM 92 A 000232

ECB6: Proceedings of the 6th European Congress on Biotechnology
edited by L. Alberghina, L. Frontali, P. Sensi

A comparison of guar α-galactosidase production by *Saccharomyces cerevisiae* SU50B and *Hansenula polymorpha* 8/2 in continuous cultures

M.L.F. Giuseppin[a], J.W. Almkerk[a], J.C. Heistek[b], C.T. Verrips[a]

[a]Unilever Research Laboratorium Vlaardingen, Olivier van Noortlaan 120, 3133 AT Vlaardingen, The Netherlands.
[b] TNO Centre for Phytotechnology, Wassenaarseweg 64, 2333 AC Leiden, The Netherlands.

Two yeast strains have been compared in their ability to produce guar α-galactosidase. A model study was carried out with *S.cerevisiae* SU50B and *H.polymorpha* 8/2 both carrying a multi copy integrated α-galactosidase ex guar. The comparison of productivities in terms of specific production rate allows of a better comparison with other heterologous expression systems in literature. The secreted α-galactosidas in the yeasts differ and show remarkable differences in glycosylation patterns.

1. INTRODUCTION

The enzyme α-galactosidase has many potential applications in the modification of galactomannans e.g. guar gum as gelling agents (10) and oligosaccharides containing galactose e.g. raffinose and stachyose. It can remove 1-6 linked α-D-galactopyranosyl groups. Compared to various other α-galactosidases, the α-galactosidase from the plant *Cyamopsis tetragonoloba* (guar) has the special property of being able to work effectively at low water activities and on polygalactans (2).

In the past few years the α-galactosidase gene from the guar plant has been cloned (20) into *Bacillus subtilis* (11), *H.polymorpha* (3) and *S.cerevisiae* (19,21). In *H.polymorpha* the expression of heterologous genes is under the control of the very strong inducible methanol oxidase (MOX) promoter (7). MOX enzyme levels can account up to 40 % of the total soluble protein. The MOX promoter can be induced by methanol, but also very efficiently by formaldehyde or formic acid (5). In *S.cerevisiae* the production of α-galactosidase is under control of the strong inducible GAL7 promoter.

For effective product recovery a secreted α-galactosidase is needed using relatively clean media. Recent publications show the potential expression of homo- and heterologous genes located on plasmids in *S.cerevisiae* using galactose as an inducing substrate in fed batch fermentations (1,8). Continuous culture studies provide more defined information than batch or fed-batch processes on e.g. biomass yields, maximum or critical growth rates, specific substrate/inducer consumption rates, optimal induction levels and productivity of enzymes. Continuous culturing of *H.polymorpha* has been studied extensively (4,5).

In earlier publications we have introduced the simple formula $P/V = D{\cdot}X{\cdot}E{\cdot}Se$ to describe the productivity of homo- and heterologous proteins by microorganisms. With the terms P/V (free α-galactosidase $\cdot l^{-1} \cdot h^{-1}$), the growth rate μ (D), the biomass concentration (X), the expression per biomass (E), and the fraction secreted (Se). The results have been used to quantify the above productivity formula. Furthermore, continuous cultivation is the most suitable tool to study the genetic stability under well-defined selective pressure.

2. MATERIALS AND METHODS

Strains and genetic constructs. Strain *S.cerevisiae* SU50, YT6-2-1L, CBS 235.90 has been used in this study. The phenotype : MEL^-, $HIS4^-$ $LEU2^-$ CIR^0. (9,19). The strain SU50 has been transformed with the integration plasmid pUR2774. The integration took place on the ribosomal DNA locus (9). The transformant used for this study had 100-110 copies of pUR2774 integrated with α-galactosidase gene under control of the inducible GAL7 promoter. *H.polymorpha* 8/2 was a transformant derived from strain A16 ($LEU2^-$) and selected according to Veale (3,17). It contains three copies of α-galactosidase under control of the MOX promoter integrated in the MOX locus (14).

Media. All media were designed to obtain a carbon-limited chemostat. Deproteinized hydrolysed whey (DHW) was obtained from DMV Campina, The Netherlands. All media are described in reference 6.

Fermentation equipment. Continuous cultivations were carried out in 1 L fermenters with a magnetic bottom drive. Most studies were carried out in two parallel fermenters. The working volume was 0.85-1.0 L. See reference 6 for other conditions.

Assays. Fermenter inlet and exhaust gases were analyzed automatically using a mass spectrometer MM8-80 (Fisons VG Instruments, UK) for O_2, CO_2, ethanol and N_2 (for volume corrections). All metabolites, proteins and enzymes are determined according to reference 6.

3. RESULTS

3.1 Continuous cultivation of S.cerevisiae SU50B strain in chemically defined media

Several transformants were tested in continuous cultures using a range of galactose/glucose ratios that allows of economic production of α-galactosidase. The data obtained from one selected transformant will be given. The best results were obtained using a fixed galactose/glucose ratio of 1:2. The biomass formation drops dramatically above D = 0.1 h^{-1}. This is correlated with the formation of ethanol and a reduced consumption of galactose. This reflects a strong Crabtree like-effect in this yeast strain, which may be related to the low leucine and oxygen uptake capacities.

Metabolic fluxes. The oxygen uptake is limited to 4.1 mmol $O_2 \cdot gX^{-1} \cdot h^{-1}$. This value is lower than that the values of 6-12 mmol $O_2 \cdot gX^{-1} \cdot h^{-1}$, usually found for bakers's yeast.

Enzyme secretion. The production of α-galactosidase varies as a function of the growth rate The specific enzyme production rate, which facilitates the comparison at various dilution rates. The production rate decreases above 0.225 h^{-1}. The ratio between intra- and extracellular α-galactosidase is strongly dependent on the growth rate. Whereas at low growth rate (D= 0.05 h^{-1}) about 50 % of the α-galactosidase was secreted, the secretion increased to more than 75 % at a dilution rate of 0.1 h^{-1}. Three discrete glycosylated forms were found.

Effect of galactose/glucose ratio. Three ratios galactose/glucose were tested at a D of 0.1 h^{-1} using medium 4 with 20 $g \cdot litre^{-1}$ glucose. The highest production of α-galactosidase of 403 $mg \cdot l^{-1}$ was found at a galactose/glucose ratio of 0.5. The yield of the strain, 0.52 $gX \cdot g\ glucose^{-1}$, is relatively high.

3.2 Production of α-galactosidase *H. polymorpha* by with formaldehyde/glucose

The production of α-galactosidase was tested using formaldehyde, a good inducer of the MOX promoter (5). The effect of formaldehyde/glucose ratio has been reported previously (5) for wild-type *H.polymorpha*. On the basis of former experiments, formaldehyde/glucose feeds with molar ratios of 0, 0.5 and 1.0 were tested.

The best productivities were obtained using a formaldehyde/glucose ratio of 0.5. Although this ratio was sub-optimal with the wild type strain (5), growth in this medium resulted in a high expression of both MOX, α-galactosidase and induction of the dissimilatory pathway.

Enzyme synthesis. MOX expression shows two optima. The highest optimum is about 80% of that in the untransformed strain. The first optimum is at D=0.08 h^{-1}. The second and lower one, is at 0.17 h^{-1}. A high extracellular level of 426 mg·$litre^{-1}$ at D=0.08h^{-1} is found. Judging from the electrophoresis gels and the total extracellular protein, more than 87% of the extracellular protein is α-galactosidase. This represents 16.5% of the total amount of protein. The enzyme is produced as in 7-8 discrete glycosylated forms.

4 GENERAL DISCUSSION AND CONCLUSIONS

Fermentation behaviour. Despite the production of vast amounts of protein, biomass yield on glucose remains to a constant high value. This is in contrast to other expression systems using intracellular accumulation of proteins. The formation of ethanol at high growth rates is a known limitation for baker's yeast biomass production. In contrast to *S.cerevisiae*, *H.polymorpha* showed constant high biomass yields without significant by-product formation over the dilution range tested.

Stability and variability. The selected transformants were extremely stable and gave reproducible results in various parallel experiments. Experiment runs took regularly 500 to 1000 hours without decrease in α-galactosidase production or gene copy number.

Induction and inducer consumption. The induction of α-galactosidase synthesis with galactose in *S.cerevisiae* SU50B is effective despite the low galactose uptake rate. However, at high dilution rates galactose uptake is reduced by glucose repression (12). Other systems have been described with improved galactose induction and less glucose repression (1, 8). In *H.polymorpha*, the induction of α-galactosidase synthesis besides MOX synthesis using formaldehyde is very effective. Under induced conditions 8 U·mg $protein^{-1}$ is found. The parent strain A16, produces more MOX under similar conditions, 14.3 U·mg $protein^{-1}$, which indicates a negative effect of α-galactosidase expression on MOX expression.

Repression. The GAL-genes were induced under the conditions tested. However, the capacity of galactose permease (encoded by GAL2) seems to be too limited for a complete uptake of galactose at higher dilution rates. The galactose uptake fluxes are low 0.4 mmol galactose·gX^{-1}·h^{-1}, which is seven fold lower than that for industrial strains (13). The low uptake rate is in effect an advantage, as it reduces the costs of galactose for the process.

Secretion. The secretion in both systems is efficient (41-89%) but depends strongly on the dilution rate. In *S.cerevisiae* the secretion decreased at high dilution rates with an optimum at 0.1 h^{-1}. The production depends strongly on the transformant used but in all cases a two peaked optimum is found in the production. The total synthesis of more than 5.8 mg α-gal·g X^{-1}·h^{-1} is considerable. This is half the maximal rate of MOX synthesis in the wild type strain. The rate of MOX synthesis, also shows a two peaked profile, which suggests that secretion does not limit the production rate.

Specific enzyme production rates. Comparison of productivities in terms of specific production rate allows of a good comparison with other heterologous expression systems in the literature. The summary described by (18) shows that the productivities of both *S.cerevisiae* and *H.polymorpha* are better than those of the other systems based on e.g. *Pichia*, which yields 2.9 mg product·gX^{-1}·h^{-1}. The absolute yield cited may be higher but the process times are longer and biomass concentrations are higher. Higher biomass levels showed similar productivity rates.

Productivity of continuous cultures. α-galactosidase is the main protein product in cultures of both *H.polymorpha* and *S.cerevisiae*. The productivity per amount of biomass can be used to compare the cultures cited in literature. e.g. Sreekrishna et al. found 1-2.9 mg product·gX^{-1} l^{-1}·h^{-1} (15). The growth rate dependence of the specific enzyme production rate shows the need to study the effect of the dilution rate for optimization.

REFERENCES

1. **Alberghina, L., D. Porro, E. Martegani, B.M. Ranzi.** 1991. Biotechnol. Appl. Biochem. **14**:82-92.

2. **Bulpin, P.V., M.J. Gidley, R. Jeffcoat, D.J. Underwood.** 1990. Carbohydrate Polymers **12**:155-168.

3. **Fellinger, A.J., J.M.A. Verbakel, R.A. Veale, P.E. Sudbery, I.J. Bom, N. Overbeeke , C. T. Verrips.** 1991. Expression of the a-galactosidase from *Cyamopsis tetragonoloba* (Guar) by Hansenula polymorpha. Yeast **7**:463-473.

4. **Giuseppin, M.L.F., H.M.J. van Eijk, B.C.M. Bes.** 1988. Biotech. Bioeng. **32**:577-583.

5. **Giuseppin, M.L.F., H.M.J. van Eijk, C. Verduyn, I. Bante, J.P. van Dijken.** 1988. Appl. Micriobiol. Biotechnol. **28**:14-19.

6. **Giuseppin, M.L.F., J.W. Almkerk, J.C. Heistek, C.T. Verrips.** 1993. Appl. Envron. Microbiol. **59**:52-59.

7. **Ledeboer, A.M., L. Edens, J. Maat, C. Visser, J.W. Bos, C.T. Verrips, Z. Janowicz, M. Eckart, C.P. Hollenberg.** 1985. Nucl. Acids. Res. **13**:3063-3082.

8. **Lin, K.H., S. Iijima, S.Y. Huang, F. Hishinuma , Y. Kobayashi.** 1991. J. Fermt. Bioeng. **72**:187-192.

9. **Lopes, T.** 1990. Ph.D. thesis. Vrije Universiteit Amsterdam, The Netherlands.

10. **Meier, H., J.S.G. Reid.** 1983. Reserve polysaccharides other than starch in higher plants. *In* F.A. Loewes, W. Tanner (ed.), Encyclopedia of plant physiology. Springer Verlag., New York.

11. **Overbeeke, N., G.H.M. Termorshuizen, M.L.F. Giuseppin, D.R. Underwood, C.T. Verrips.** 1990. Environ. Microbiol. **56**:1429-1434.

12. **Sierkstra, L.N., N.A. Nouwen, J.M.A. Verbakel, C.T. Verrips.** 1992. Yeast. **8**:1077-1087.

13. **Sierkstra, L.N., J.M.A. Verbakel, C.T. Verrips.** 1992. by J. Gen. Microb. **138**:2559:2566.

14. **Sierkstra, L.N., J.M.A. Verbakel, C.T. Verrips.** 1991. Curr. Genet. **19**:81-87.

15. **Sreekrishna, K., R.H.B. Potenz, J.A. Cruze, W.R. McCombie, K.A. Parker, L.Nelles, P. K. Mazzaferro, K.A. Holden, R.G. Harrison, P.J. Wood, C.E. Hubbard, M. Fuke.** 1988. J. Basic. Microbiol. **28**:265-278.

16. **Strasser, A.W.M., K. Melber, A. Merckelbach, U. Weydemann, U.M. Dahlems, C.P. Hollenberg, Z.A. Janowicz.** 1990 Bioengineering **5**:20-26.

17. **Veale, R.A.** 1989. Ph.D. thesis. University of Sheffield, United Kingdom.

18. **Veale, R.A., M.L.F. Giuseppin, H.M.J. van Eijk, P.E. Sudbery , C.T. Verrips.** 1992. Yeast **8**:361-372.

19. **Verbakel, J.M.A.** 1991. Ph.D. thesis. Rijks Universiteit Utrecht, The Netherlands.

20. **Uniliver patent**, EP-B-0121690.

21. **Unilever patent**, EP-A-0225153, US-A-5082778.

ECB6: Proceedings of the 6th European Congress on Biotechnology
edited by L. Alberghina, L. Frontali, P. Sensi

Processing and secretion of *Cyamopsis tetragonoloba* α-galactosidase by *Saccharomyces cerevisiae* wild-type and *pmr1* strains[1]

M.M. Harmsen[a], R.H. Geerse[a], H.A. Raué[a] and J. Maat[a,b]

[a] Laboratorium voor Biochemie en Moleculaire Biologie, Vrije Universiteit, de Boelelaan 1083, 1081 HV Amsterdam, The Netherlands,

[b] Unilever Research Laboratorium, P.O. Box 114, 3130 AC Vlaardingen, The Netherlands.

1. INTRODUCTION

The yeast *Saccharomyces cerevisiae* has frequently been used for the production and secretion of heterologous proteins (for a review see ref. 1). However, the secretion of foreign polypeptides is often inefficient, even when an authentic yeast signal peptide is applied for the targeting of the heterologous protein to the secretory pathway. One of the reasons may be saturation of the secretory pathway, due to the high-level expression of the heterologous gene. Indeed, in several cases (*e.g.* prochymosin; ref. 2) increased synthesis of the foreign protein resulted in intracellular accumulation without an increase in the level of the secreted product. Secretion of prochymosin and some other heterologous proteins was stimulated by inactivating the *PMR1* gene, which encodes a protein with strong sequence homology to membrane-localized, ATP-dependent Ca^{2+}-pumps (2,3). Recently, it has been shown that PMR1 localizes to a Golgi-like organelle (4). A *pmr1* null mutant shows pleiotropic effects, most of which are modulated by external Ca^{2+}-concentrations, including a strongly reduced outer-chain glycosylation of secretory proteins (3,4). However, the mechanism underlying the enhanced secretion of heterologous proteins in a *pmr1* mutant is not yet clear.

We have studied the secretion of heterologous proteins by *S. cerevisiae,* using the *Cyamopsis tetragonoloba* (guar plant) α-galactosidase (αGal) as a model protein. We addressed the question whether the pro-part of the prepro-α-factor (MFα1) signal peptide influences the secretion efficiency of αGal, as compared to different signal peptides lacking a pro-part. Another point of interest concerned the effect of a *pmr1* disruption on the intracellular processing and level of secretion of αGal.

2. RESULTS AND DISCUSSION

2.1 Effect of signal peptide on levels of secreted and intracellular αGal

In order to determine the effect of the nature of the signal peptide on the efficiency of secretion of a heterologous protein [*i.e.* the ratio between the amount of secreted enzyme and the total amount of enzyme (intracellular plus extracellular)], we constructed various expression plasmids, containing a 2μm-based origin of replication, a *LEU2* gene as the selection marker, and the sequence encoding guar αGal, flanked by the *GAL7* promoter and the terminator from the *PGK1* gene. The promoter and the αGal-encoding sequence were separated by a synthetically prepared fragment encoding the 5'-untranslated region of yeast *GAPDH* mRNA and one of a number of different yeast-derived signal sequences (Table I). In plasmid pSY16 the 5'-terminal portion of the coding region specifies the prepro-sequence of MFα1, including

[1] Part of this material is published in Gene (1993). This work was financially supported by Unilever Research.

Table I
Effect of the nature of the signal peptide on the secretion of αGal*

Strain	Plasmid	Signal peptide	αGal (mg/g dry weight)		% Secreted
			Medium	Cells	
SU10	pSY1	*SUC2*	11.3 ± 1.4	1.0 ± 0.1	91.9
SU10	pSY12	*PHO5*	15.6 ± 0.4	1.9 ± 0.5	89.1
SU10	pSY16	*MFα1* (prepro)	18.6 ± 2.3	0.19 ± 0.02	99.0
SU10	pSY49	*MFα1* (pre)	19.7 ± 4.0	3.0 ± 0.6	86.8
MR1	pSY1	*SUC2*	14.0 ± 1.2	2.0 ± 0.25	87.5
MR1	pSY16	*MFα1* (prepro)	17.6 ± 2.1	0.04 ± 0.01	99.8

* The amount of αGal present in culture supernatants and cell extracts prepared after growth for 48 h at 30° C in YEPG (1% Yeast extract, 1% Bacto-peptone, 2% galactose) was determined by a guar αGal-specific ELISA. MR1 is an isogenic *pmr1::URA3* derivative of SU10 (α, *leu2*, *ura3*, *his3*, *cir*$^+$).

the (Glu-Ala)$_2$ spacer peptide to ensure proper KEX2 processing (1).

Each of the plasmids was used to transform yeast strain SU10 (5) as well as its isogenic *pmr1::URA3* derivative MR1. The resulting transformants were grown to stationary phase in medium containing 2% galactose and the αGal content of the culture supernatants and cell extracts was determined by a guar αGal-specific ELISA. As shown in Table I, the secretion efficiency of the enzyme was essentially the same for all fusions containing one of the three 'simple' secretion signals, *i.e.* the MFα1 pre-peptide, the invertase or the acid phosphatase signal peptide. However, the secretion efficiency of the fusion carrying the complete MFα1 leader was consistently found to be about 10% above that of its 'simple' counterparts (99% vs. 90%). Apparently, the pro-part has a stimulatory effect on the secretion of αGal.

We did not observe any substantial increase in the level of secreted αGal in strain MR1. However, we did observe a several-fold higher secretion efficiency for prochymosin when we replaced the αGal-encoding gene in plasmid pSY1 by the gene encoding bovine prochymosin. Therefore, MR1 does display the *pmr1* phenotype as described by Smith *et al.* (2).

2.2 Intracellular localization of αGal

While passing through the secretory pathway proteins are sequentially modified by signal peptide removal (ER), the addition of core oligosaccharide (ER) and outer-chain oligosaccharide (early Golgi) and, in the case of secretion directed by the MFα1 prepro-peptide, removal of the pro-peptide by the *KEX2* gene product (late Golgi). Because the presence of each of these modifications is an indication for the subcellular localization of intracellular proteins, we performed a Western blot analysis of cell extracts before and after treatment with endo-β-*N*-acetylglucosaminidase H (EndoH), which removes *N*-linked oligosaccharides from secretory proteins. As shown in Fig. 1 (lanes 1 and 2), treatment of the αGal secreted by SU10 cells with EndoH reduces its M_r by approximately 2.5 kDa. Since one core *N*-linked oligosaccharide has a M_r of about 2 kDa, this suggests that αGal is *N*-glycosylated at only a single site.

When secretion was directed by the invertase signal peptide, the intracellular αGal was glycosylated to a lesser extent than secreted αGal (*cf.* lanes 3 and 4 to lanes 1 and 2). This indicates that the intracellular enzyme is predominantly core-glycosylated and probably resides in the ER. This localization was confirmed by immunogold electron microscopy of SU10 [pSY1] cells (Fig. 2), where the nuclear envelope clearly shows the highest labeling. Thus transport from ER to Golgi appears to be the rate-limiting step in the secretion of αGal directed by the invertase signal peptide.

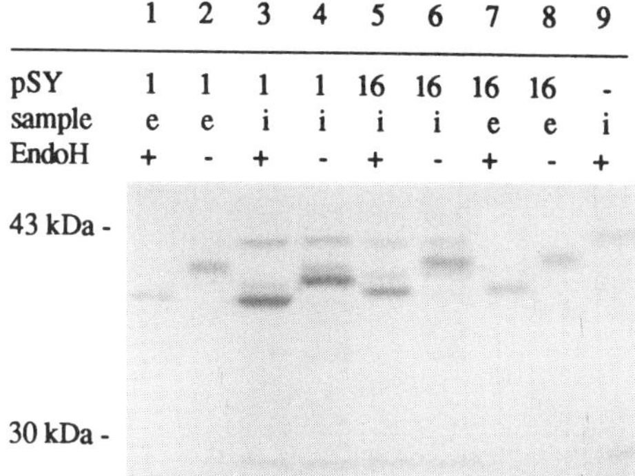

Fig. 1. Western blot analysis of αGal produced by SU10 cells carrying plasmids pSY1 (lanes **1-4**), pSY16 (lanes **5-8**) or YEplac181 (lane **9**). Transformants were grown for 48 h, after which samples of cell extracts (**i**) and culture supernatants (**e**) were prepared. Samples were treated with (+) or without (-) EndoH. Positions of protein standards are indicated.

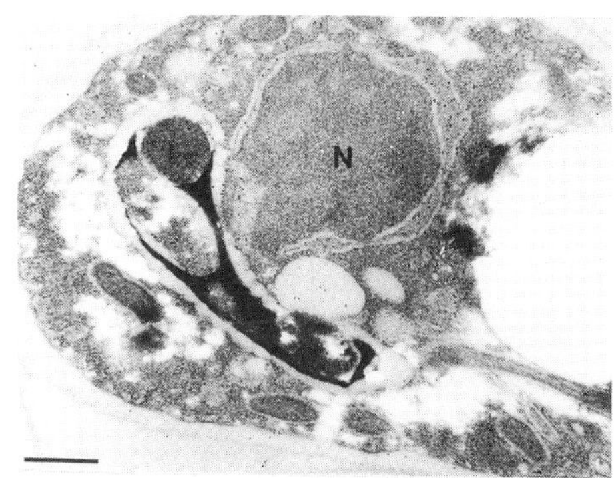

Fig. 2. Subcellular localization of cell–associated αGal of SU10 [pSY1] cells. Electron microscopy and immunological labeling were performed by E. van Tuinen (University of Amsterdam) according to Van Tuinen and Riezman (6). N; nucleus. Bar = 0.2 μm.

When the MFα1 prepro-peptide is used, the cell-associated αGal migrates at the same position in SDS-PAGE as its secreted counterpart (Fig. 1; *cf.* lanes 6 and 8), which as shown in section **2.3**, is outer-chain-glycosylated and KEX2-processed. This indicates that all cell-associated αGal is located in or after the late Golgi compartment containing KEX2. Possibly, this αGal is retained in the cell-wall. It is striking that the intracellular αGal is predominantly present in the ER when the invertase signal peptide was used, whereas secretion directed by the MFα1 signal peptide does not result in typical ER-forms representing αGal fused to the MFα1 (pre)pro-peptide. Together with the lower secretion efficiency of αGal when the invertase signal peptide was used (Table I), these data suggest that the MFα1 pro-peptide stimulates the efflux of αGal from the ER.

2.3 *N*-glycosylation of secreted αGal

In order to study the effect of the *pmr1* disruption as well as the nature of the signal peptide on the glycosylation of secreted αGal, we analysed culture supernatants of SU10 and MR1 cells transformed with either pSY1 or pSY16 by immuno-blotting. A comparison of extracellular with intracellular αGal of pSY1-transformed SU10 cells (Fig. 1; *cf.* lanes 1 and 2 to lanes 3 and 4) shows a slightly increased M_r of the secreted protein. Since the intracellular αGal is localized in the ER (Fig. 2) and, therefore, must represent core-glycosylated enzyme, the secreted αGal must be outer-chain glycosylated, albeit to only a limited extent.

When secretion of αGal was directed by the MFα1 prepro-peptide, EndoH-treatment of the secreted enzyme caused the same reduction in M_r as observed in the case where secretion was governed by the invertase signal peptide (Fig. 1, lanes 7 & 8). Thus, glycosylation is not detectably affected when the 'simple' invertase signal peptide is replaced by the more complex MFα1 prepro-peptide. However, there is a slight decrease in electrophoretic mobility of both the glycosylated and deglycosylated forms of the enzyme in the latter case (Fig. 1, *cf.* lanes 7 & 8 to lanes 1 & 2; also Fig. 3, *cf.* lanes 2 & 8 to lanes 4 & 10). This difference in mobility indicates that KEX2 processing of the MFα1 fusion protein has occurred, but that no processing by STE13 did take place, leaving the $(Glu\text{-}Ala)_2$ spacer peptide still attached to the secreted αGal. The absence of processing by STE13 was confirmed by N-terminal sequence analysis of the secreted enzyme (data not shown).

The αGal secreted by MR1 [pSY1] cells has a slightly higher electrophoretic mobility than the fully glycosylated enzyme secreted by wild-type cells (Fig. 3, lanes 10-13). On the other hand, it migrates somewhat more slowly than the core-glycosylated intracellular enzyme from SU10 cells. Therefore, the αGal secreted by the *pmr1* mutant is outer-chain-glycosylated but to a lesser extent than its counterpart from wild-type cells. The nature of the signal peptide is unimportant in this respect since the enzyme from MR1 [pSY16] cells shows an identical behavior (Fig. 3, lanes 6-9).

	1	2	3	4	5	6	7	8	9	10	11	12	13	14	15	16
strain	S	S	M	S	M	S	M	S	M	S	M	S	M	S	M	S
pSY	1	16	16	1	1	16	16	16	16	1	1	1	1	1	1	1
sample	e	e	e	e	e	e	e	e	e	e	e	e	e	i	e	i
EndoH	±	+	+	+	+	-	-	-	-	-	-	-	-	-	-	-

Fig. 3. Western blot analysis of αGal produced in yeast cells. Strains SU10 (S) or MR1 (M) transformed with either pSY1 or pSY16 were cultured to the stationary phase, after which cell extracts (i) and culture supernatants (e) were prepared. Samples were treated with (+) or without (-) EndoH (±; partial EndoH digestion).

Rudolph *et al.* (3) have suggested that in *pmr1* mutants vesicles derived from the ER might circumvent the Golgi apparatus and instead fuse directly with the plasma membrane. Since the enzymes involved in outer-chain glycosylation and the KEX2 enzyme are known to be located in the early Golgi compartments and a late Golgi compartment, respectively, this hypothesis would predict absence of both outer-chain glycosylation and KEX2 processing of secreted proteins in the *pmr1* disruption mutant. However, our results show that some outer-chain glycosylation of a secretory protein does still occur in a *pmr1* mutant. Moreover, since the electrophoretic mobility on an SDS-polyacrylamide gel of αGal secreted by mutant and wild-type cells is identical after deglycosylation with EndoH (Fig. 3, *cf.* lane 2 to lane 3), the *pmr1* mutation does not affect KEX2 processing of the αGal fusion protein. Therefore, our results indicate that a bypass of the Golgi is unlikely to occur. Further research is currently being undertaken to elucidate the basic mechanism underlying the *pmr1*-enhanced secretion efficiency of heterologous proteins.

REFERENCES

(1) M.A. Romanos, C.A. Scorer and J.J. Clare, Yeast 8 (1992) 423-488.

(2) R.A. Smith, M.J. Duncan and D.T. Moir, Science 229 (1985) 1219-1224.

(3) H.K. Rudolph, A. Antebi, G.R. Fink, C.M. Buckley, T.E. Dorman, J. LeVitre, S. Davidow, J. Mao and D.T. Moir, Cell 58 (1989) 133-145.

(4) A. Antebi and G.R. Fink, Mol. Biol. Cell 3 (1992) 633-654.

(5) J.M.A. Verbakel, Heterologous gene expression in the yeast *Saccharomyces cerevisiae*, PhD thesis, Rijksuniversiteit Utrecht, The Netherlands, 1991.

(6) E. van Tuinen and H. Riezman, Cytochem. 35 (1987) 327-333.

Enabling Technologies

RATIONAL DESIGN OF NOVEL BIOCATALYSTS

ECB6: Proceedings of the 6th European Congress on Biotechnology
edited by L. Alberghina, L. Frontali, P. Sensi

THE BLIND WATCHMAKER AND PROTEIN ENGINEERING

Steffen B. Petersen and Finn Drabløs
MR Center, SINTEF-UNIMED, N-7034 Trondheim, Norway
Phone 47-7-997700
FAX 47-7-997708
E-MAIL : sbp@marvin.mr.sintef.no

1. INTRODUCTION

The present paper will address the exciting possiblities that the groving number of available protein and nucleic acid sequences present. Today more than 40.000 protein sequences have been determined or deducted from nucleic acid sequence information. Less than 400 truly different 3-D protein structures have been determined. Fortunately the same basic protein fold seem to occur in many functionally different proteins. Since the function of a protein is intimately linked with its structure any methodology that facilitates that we can build a reasonably accurate model of the proteins 3-D structure is very important. In the present paper we will analyze such a tool : MULTIM.

2. METHOD

The identification of parts of the amino acid sequence which are partially or fully conserved is of great importance during the study or identification of a particular class of enzymes. Such conservation may indicate regions of structural or functional importance. Several computer based methodologies are available for the alignment of multiple biological sequences (Carillo 88, Altschul 89). In general terms these methods make use of a scoring scheme that is based on a pairwise comparison between the amino acids in the sequences which are being aligned. The optimal alignment is that which yields the highest score. Insertions and deletions are introduced at the cost of a reduced total score. Application of the correct gap-penalty is especially difficult when the homology between sequences is weak . A central feature in all such multiple alignments is the choice of scoring scheme. Three major scoring schemes (George, 1990) are commonly in use : 1) Identity : non-zero scores are only given if a pair of amino acids are identical, 2) Similarity : scoring is based upon the degree of similarity in physical-chemical properties such as size, polarity and hydrophobicity (described by several different scales (George,1990)) between the residues being compared and 3) Genetic : scoring is based on the number of genetic mutations nescessary to transform one amino acid to the amino acid it is being compared to. The amino acid serine, which plays an important role in the present study, is represented by two very different codons TCN and AGN and thus provides additional scoring information.

In many cases similarities in single letter sequence representations can be readily detected by eye, but this is not possible for sequences where the relationship is more subtle. The program MULTIM (Petersen 1992) has been developed to enhance these more subtle features and thus utilize the remarkable ability of the human brain to detect similarities in visually presented information.

MULTIM (Drabløs, 1992) is a program suite which detects concurrent motifs in a set of sequences. These are then evaluated for statistical significance in the context of the protein sequence before visualization of the motifs, color-coded for significance. The motif may be continous or disperse, e.g. containing 3 out of 6 residues conserved. The extent of the conservation is defined by the user. It is important to note that MULTIM does not make use of gap-penalties, so its ability to detect conserved motifs is in no way impaired by the presence of insertions and deletions. The full implementation is available for Silicon Graphics Workstations and a PC-version is available for applications on IBM compatible PC's having a VGA graphical adapter.

3. DISCUSSION

Throughout the last 3.5 billion years nature has experimented with proteins, whereas mankinds conscious effort started only a few decades ago. It is a central theme in the present paper to highlight the vast resource of information that 'the blind watchmaker' nature can provide us with with respect to protein and genetic engineering. It is crucial in any protein engineering project to screen for sequences related to the one in question in order to identify conserved regions or residues in the protein sequence. These regions may have structural or functional importance for the protein.

We will present a novel way of identifying and highlighting such residues using the program MULTIM. An example is shown in figure 1, where the alignment of various amino acid decarboxylases have been displayed graphically. Each horizontal bar represents a protein sequence, and each vertical tick marks 50 AA. Lines connecting the sequences denotes homology regions that are shared among all the sequences. As is seen from this figure two sequence fragments are shared among the decarboxylase. When searching against the whole protein sequence database one of the motifs : W_H_DAA uniquely identifies the amino acid decarboxylases. We will report similar observations for other protein families, such as the trypsin family where the *combined* presence of both the small and seemingly inconspicously motifs CGG and the GWG apparently identifies a very substantial subgroup of trypsin like proteins. If several sequences from a certain family of proteins are available, a plot of the number of observed mutations at a given sequence location versus sequence position is extremely informative. In figure 2 is shown how the number of observed number of mutations correlate with sequence position and water accessibility for the lysozyme family of enzymes.

By studying the substitution patterns that are found in nature for a given protein we can gain knowledge about possible strategies for modifying proteins towards specific purposes. We may for example study how proteins from bacteria living under extreme temperature conditions , differ from similar proteins from organisms living under more normal conditions. Throughout evolution we generally assume that a large number of mutants have been generated for all proteins in living cells. Today we only find traces of the successfull experiments on the genetic information in living beings, but undoubtedly many unsuccessfull ones have been attempted by nature. Through natural selection the organism with the most efficient protein content will be selected for (assuming everything else is equal). Eventually, the less successfull design will then become extinct. The present paper is not intended to be a review of "protein archaeology", rather it attempts to provide the reader with a few hopefully helpfull suggestions of how to optimize the starting point for a protein engineering project, simply by using all the available relevant sequence information that we have available in the sequence data banks.

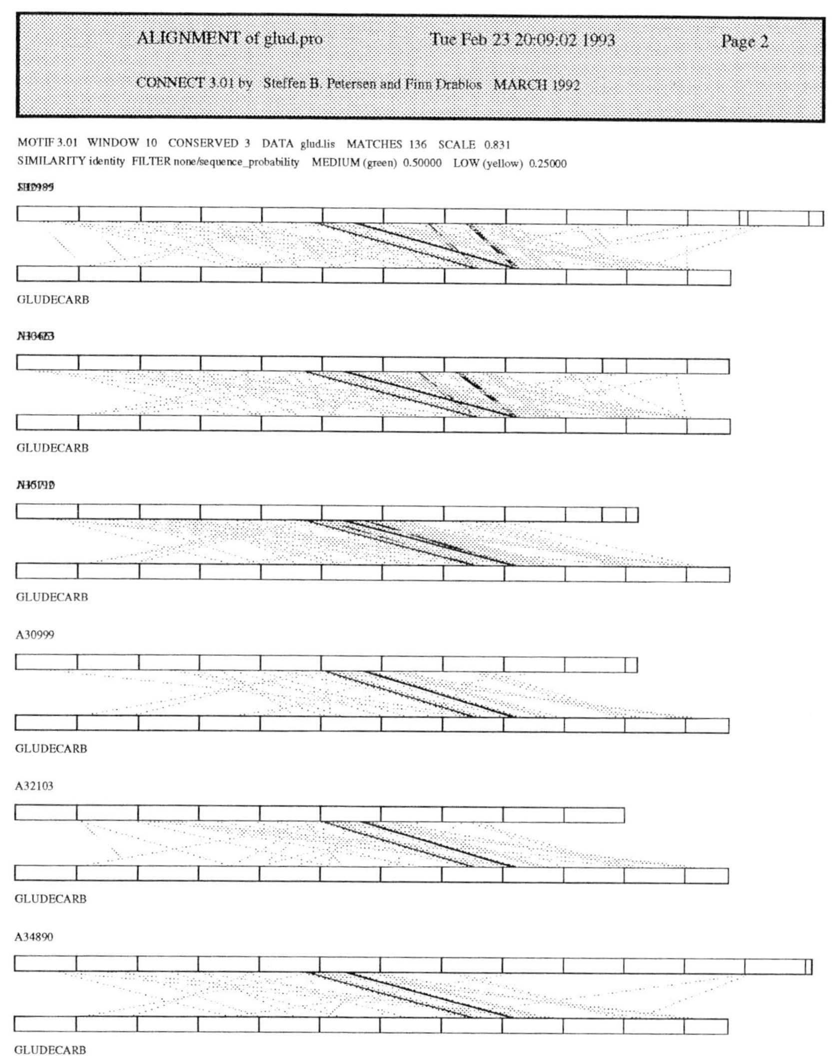

Figure 1. MULTIM alignment of amino acid carboxylases. Note that two positions in the sequence are highly conserved throughout this group of sequences.

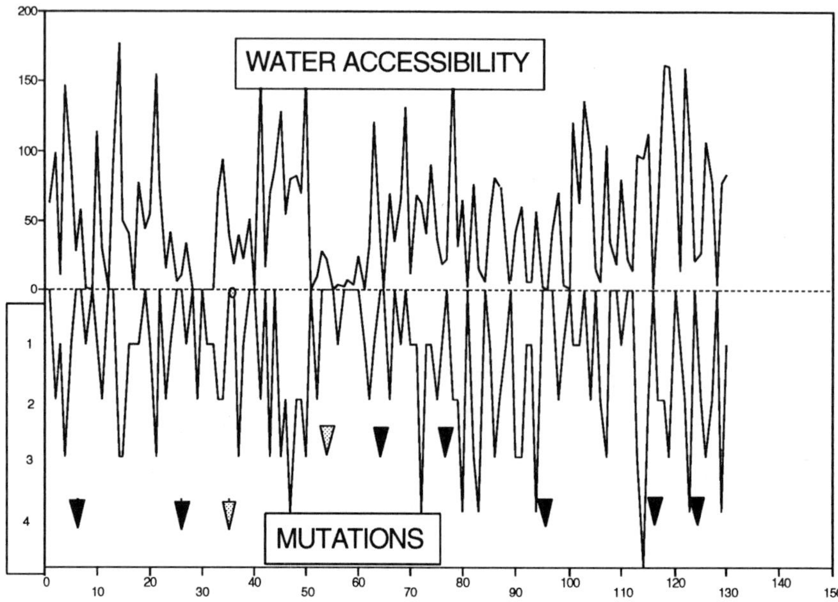

Figure 2 :

Water accessibilty and number of mutations versus sequence position in lysozyme. The water accessibility corresponds to human lysozyme. The dark triangles pointing downward show CYS positions involved in disulphide bridge formation. The light triangles pointing downward indicate active site residues positions (GLU 35 and ASP 52). Triangles pointing upwards show residues involved in substrate binding.

4. REFERENCES

Altschul, S.F. "Gap costs for Multiple Sequence Alignment", J. Theor. Biol. 138, 297-309, 1989.

Benson, D.G. "Fourier methods for biosequence analysis", Nucleic Acids Research, 18, 6305-6310, 1990.

Carillo,H. and Lipman,D. "The multiple sequence alignment problem in biology", SIAM J. Appl. Math., 48, 1073-1082,1988.

Cheever, E.A., Overton, G.C., Searls, D.B. "Fast Fourier transform-based correlation of DNA sequences using complex plane encoding" CABIOS, 7, 143-154, 1991.

Drabløs, F. and Petersen, S.B., "Multim : Theory and applications (???)" , manuscript under preparation.

George, D.G., Barker, W.C. and Hunt, L.T. "Mutation Data Matrix and its Uses" p 333-352 in Doolittle R.F (Ed) : Molecular Evolution: Computer Analysis of Protein and Nucleic Acid Sequences" vol 183 of Methods in Enzymology, Academic Press, San Diego, 1990.

ECB6: Proceedings of the 6th European Congress on Biotechnology
edited by L. Alberghina, L. Frontali, P. Sensi
1994 Elsevier Science B.V.

Evidence of catalytic antibodies endowed with spermidine synthase-like activity

G. Pontoni, A. Borriello, M. Carteni-Farina and V. Zappia

Institute of Biochemistry of Macromolecules, The Second University of Naples, via Costantinopoli 16, 80138 Naples (Italy)

L. Lopalco and A.G. Siccardi

Department of Biology and Genetics for Medical Sciences, University of Milan, via G.B. Viotti, 3, 20100 Milan (Italy)

1. ABSTRACT

The transition state analog (TSA) S-adenosyl-(5')-1,8-diamino-3-thiooctane (AdoDATO), an inhibitor of spermidine synthase, duly conjugated with keyhole limpet hemocyanin (KLH), was used in the immunization of mice. Spermidine synthase-like activity has been detected in sera as well as in purified IgG fractions of immunized mice. The generation of new catalytic activities in the blood by means of suitable immunizations, as here reported, can be envisioned as a future powerful tool in biotechnology as well as in metabolic engineering.

2. INTRODUCTION

Antibodies endowed with catalytic activities (abzymes) represent a very promising achievement of biotechnology (1-3). The main advantage of catalytic antibodies is that such proteins can be designed as catalysts on the basis of the mechanism known for the target reaction (4-7). Although so far abzyme-dependent rate enhancements were never found comparable with those of related enzymes, all the characterized abzymes are just as regio- and stereoselective as enzymes are. For this reason, many research groups are now trying to obtain, in this way, "tailor-made" abzymes as biocatalysts designed to speed up many chemical reactions of biotechnological interest, either when enzymes and other catalysts are unavailable (8-10), or within biological systems where abzymes may be useful when enzymatic activities are lacking or impaired (11).

The preparation of catalytic antibodies involves the synthesis of special TSA haptens, that can be designed based on the mechanism of the target reaction, followed by the immunization of mice with the conjugated TSA and by the characterization of the desired catalytic activity (4). The target enzymes here presented, spermidine synthases, catalyze the transfer of an aminopropyl group from decarboxylated adenosylmethionine (decAdoMet) to putrescine, yielding spermidine and methylthioadenosine (MTA) (fig. 1).

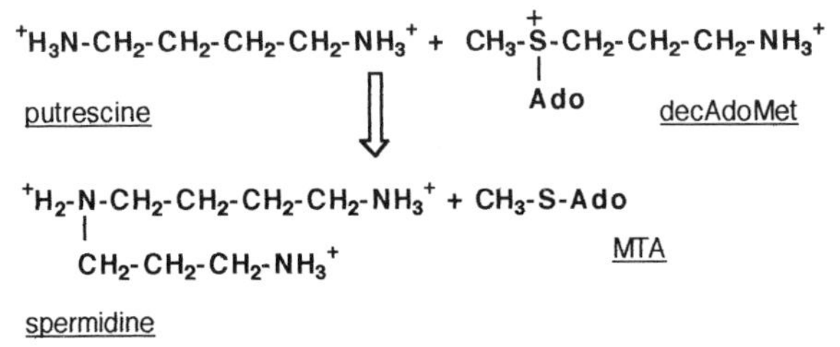

$$^{+}H_3N\text{-}CH_2\text{-}CH_2\text{-}CH_2\text{-}CH_2\text{-}NH_3^{+} + CH_3\text{-}\overset{+}{S}(\text{Ado})\text{-}CH_2\text{-}CH_2\text{-}CH_2\text{-}NH_3^{+}$$

putrescine decAdoMet

$$\Downarrow$$

$$^{+}H_2\text{-}N(CH_2\text{-}CH_2\text{-}CH_2\text{-}NH_3^{+})\text{-}CH_2\text{-}CH_2\text{-}CH_2\text{-}CH_2\text{-}NH_3^{+} + CH_3\text{-}S\text{-}Ado$$

spermidine MTA

Fig.1 : Reaction of spermidine synthases

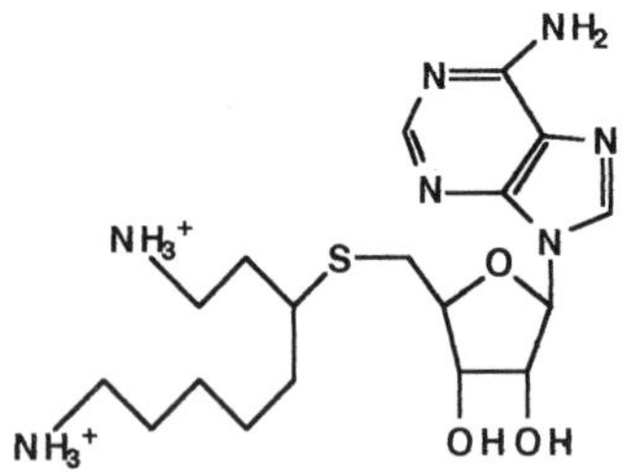

Fig.2: Structure of AdoDATO

These enzymes have been purified and characterized from different sources (12); the mechanism has also been identified, by means of kinetic (13) and nuclear magnetic resonance (14,15) approaches, as a single displacement bimolecular nucleophilic substitution (SN2) involving a direct attack of a putrescine nitrogen to the sulphur-adjacent methylene carbon. Based on this mechanism, AdoDATO (see fig.2) was synthesized by J.K.Coward and coworkers as a TSA (16), and characterized as a potent inhibitor of these enzymes (17,18). In this paper we describe the use of such TSA as a hapten with the aim of raising polyclonal antibodies whose catalytic activity is mimic of that of spermidine synthase.

3. MATERIALS AND METHODS

3.1 Materials

AdoDATO was a kind gift of prof. J.K. Coward of the University of Michigan at Ann Arbor, MI, USA. All other chemicals were of the highest purity grade available. Radiochemicals were from Amersham, Bucks, U.K.

3.2 Immunization and purification of IgG's.

AdoDATO conjugate with KLH was obtained by direct condensation with 1-ethyl-3-(3-dimethylaminopropyl)carbodiimide according to standard procedures (19). Ten Balb/c mice, 6-8 weeks old, were immunized i.p. with either 50 (mice A and B) or 100 μg (mice C to L) of such conjugate in complete Freund's adjuvant; the second dose, in incomplete Freund's adjuvant, was given on day 15. Each dose was given in PBS every 30 days, thereafter. The anti-AdoDATO titer was evaluated by standard ELISA assays. Purification of IgG from mouse sera was performed by protein A/G affinity chromatography (Pierce) to SDS-PAGE homogeneity.

3.3 Spermidine synthase assay

The aminopropyltransferase activity was detected by cation exchange HPLC, as pmol of (1,4 ^{14}C)spermidine formed from (1,4 ^{14}C)putrescine as previously described (18) Controls were performed by adding sera from mice immunized with non AdoDATO bound KLH. The possible interference due to formation of putrescine from spermidine via the polyamine oxydase pathway was ruled out by the lack of any detectable putrescine formation when adding radiolabelled spermidine instead of putrescine under the assay conditions.

4. RESULTS

When tested for aminopropyltransferase activity, the sera from control mice never presented any evidence of spermidine synthesis under the assay conditions. Blood sample were drawn from treated mice after four and seven immunizations.

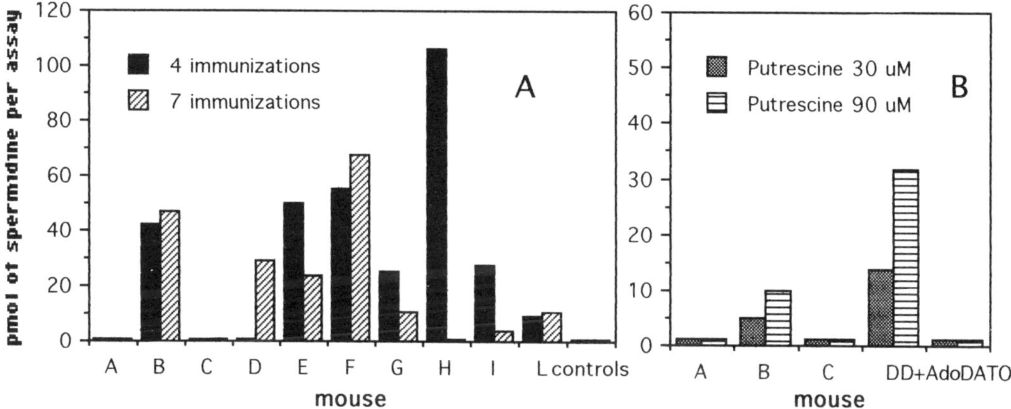

Fig. 3. Spermidine synthase-like activities: A: in sera from AdoDATO immunized mice, after 4 and 7 immunizations, and B: in purified IgG's from the same mice after 10 immunizations; AdoDATO acts as an inhibitor of IgG from mouse D.

As shown in fig 3, sera from several mice showed detectable catalytic activity, while a few of them were found to be inactive. In order to assign directly the catalytic activity to the antibodies, the IgG fractions from sera of mice A,B,C and D, drawn after 10 immunizations, were purified by A/G protein affinity chromatography to SDS PAGE homogeneity. IgG's from both catalytic antisera were confirmed active (see fig.3B). The addition of 30 μM AdoDATO completely inhibited the activity of IgG from mouse D, thus acting as expected by a TSA enzyme inhibitor. Spermidine forming activity was also measured with respect to putrescine concentration ranging from 30 to 120 μM in IgG fractions from sera of mouse D. Saturation kinetics was observed, with an apparent K_M value of 127 μM and a V_{MAX} of 32 nM/h.

5. DISCUSSION

The reported results are consistent with the hypothesis that spermidine synthase-like catalytic antibodies were elicited in mice by TSA immunization. Taking into account the specific activities of purified IgG's, and very conservatively assuming that 5% of the IgG's are active, turnover numbers from 2 to 6 can be calculated in the assays, thus demonstrating that IgG's play a catalytic rather than a stoichiometric role in spermidine formation.

The appearance of active abzymes in the blood can be envisioned as a very promising perspective for the near future, in that it might be exploited in stimulating *in vivo* catalytic activities that might affect metabolism and find biotechnological applications. Moreover, the detection of the polyclonal antibody dependent catalysis here reported might even suggest the therapy by TSA-immunization as a future topic deserving feasibility studies and further research efforts. On the other hand, while most well characterized abzymes are indeed monoclonals, only few catalytic polyclonal antibodies have already been described in literature (20,21), with no clear evidence of their *in vivo* action.

Future perspectives of the present work include the characterization of several anti-AdoDATO monoclonal antibodies that have already been prepared and are now being screened, and the approach to similar SN2 reactions by means of cognate TSA designs.

ACKNOWLEDGEMENTS

The authors are thankful to prof J.K. Coward of The University of Michigan at Ann Arbor, MI, for the kind gift of AdoDATO, to the National Council of Research , Target Project "Biotechnology and Bioinstrumentation" for financial support and to Dr F. Della Ragione for very helpful discussions.

REFERENCES

1) A. Tramontano, K.D. Janda and R.A. Lerner, Proc. Natl Acad. Sci. USA, 83 (1986) 6736
2) J. Jacobs and P.G. Schultz, J. Am. Chem. Soc., 109 (1987) 2174
3) R.J. Massey, Nature, 326 (1987) 457
4) G.M. Blackburn, A.S. Kang, G.A. Kingsbury and D.R. Burton, Biochem. J., 262 (1989) 381
5) P.G. Schultz, R.A. Lerner and S.J. Benkovic, Chem. and Eng. News, May 28, 1990, 26
6) S.J. Benkovic, Annu. Rev. Biochem., 61 (1992) 29
7) M.J. Powell and D.E. Hansen, Protein Eng., 3(2) (1989) 69
8) D. Hilvert, K.W. Hill, K.D. Nared and M.T.M. Auditor, J. Am. Chem. Soc., 111 (1989) 9261
9) A.C. Braisted and P.G. Schultz, J. Am. Chem. Soc., 112 (1990) 7430
10) K.D. Janda, C.G. Shevlin and R.A. Lerner, Science, 259 (1993) 490
11) Y. Tang, J.B. Hicks and D. Hilvert, Proc. Natl. Acad. Sci. USA, 88 (1991) 8784
12) G. Cacciapuoti, M. Porcelli, G. Pontoni and V. Zappia, in *The Physiology of Polyamines*, U. Bachrach and Y.H. Heimer (eds), CRC Press, New York, 47
13) V. Zappia, G. Cacciapuoti, G. Pontoni and A. Oliva, J. Biol. Chem. 255 (1980) 7276
14) G. Pontoni, J.K. Coward, G.R. Orr, and S.J. Gould, Tet. Lett., 24 (1983) 151
15) G.R. Orr, D.W. Kullberg, G. Pontoni, P.C. Prabhakaran, S.J. Gould and J.K. Coward, J. Am. Chem. Soc., 110 (1988) 5791
16) K.C. Tang, R. Mariuzza and J.K. Coward, J. Med. Chem., 24 (1981) 1277
17) A.E. Pegg, K.C. Tang and J.K. Coward, Biochemistry, 21 (1982) 5082
18) G. Pontoni, G.Cacciapuoti, M. Porcelli and V. Zappia, in Perspectives in Polyamines Research, A. Perin, G. Scalabrino, A. Sessa and M.E. Ferioli (eds), Wichtig, Milano, 29
19) G. Pontoni, A. Borriello, C. Martuccio, L. Parravicini and V. Zappia, Life Chem. Rep., 10 (1992) 163
20) G. Gallacher, M. Searcey, C.S. Jackson and K. Brocklehurst, Biochem. J., 284 (1992) 675
21) S. Paul, S. Mei, B. Mody, S.H. Eklund, C.M. Beach, R.J. Massey and F. Hamel, J. Biol. Chem.,266 (1991) 16128

ECB6: Proceedings of the 6th European Congress on Biotechnology
edited by L. Alberghina, L. Frontali, P. Sensi
1994 Elsevier Science B.V.

Modification of substrate specificity of haloalkane dehalogenase by site-directed mutagenesis

J. P. Schanstra, F. Pries and D. B. Janssen

Department of Biochemistry, Groningen Biomolecular Sciences and Biotechnology, University of Groningen, Nijenborgh 4, 9747 AG Groningen, The Netherlands

INTRODUCTION

Haloalkane dehalogenase from *Xanthobacter autotrophicus* GJ10 is involved in the bacterial detoxification of 1,2-dichloroethane. The enzyme catalyses the release of a chloride ion from the substrate with water as the sole cosubstrate, resulting in the replacement of the chlorine by a hydroxyl group. Haloalkane dehalogenase is capable of hydrolyzing terminally chlorinated alkanes with chain lengths up to 4 carbon atoms and also brominated alkanes with chain lengths up to 10 carbon atoms (1).

Haloalkane dehalogenase has a molecular weight of 35,143 D and is composed of 310 amino acids. The gene, *dhlA*, was cloned and sequenced (2) and the X-ray structure of the enzyme is known (3). The protein consists of 2 domains: a cap domain on top of an α/β hydrolase fold globular main domain (4). Between these, there is a hydrophobic cavity which is the putative active site (3). Haloalkane dehalogenase belongs to a group of hydrolytic enzymes that have a similar α/β hydrolase fold and a conserved arrangement of the catalytic triad (4). The structure suggests that the enzyme catalyses the hydrolysis of halogenated compounds via nucleophilic displacement of the halogen by Asp_{124}, followed by hydrolysis of the alkyl-enzyme intermediate by activated water attack on the carbonyl carbon of Asp_{124}. Activation of water probably proceeds via donation of a proton to His_{289}, which is enhanced by Asp_{260} (Figure 1).

The affinity of the enzyme towards 1,2-dichloroethane is not high (K_M=0.68), and also the turnover number is low (k_{cat}=6 s^{-1}). The recent construction of an expression and site-directed mutagenesis system for haloalkane dehalogenase in *Escherichia coli* (5) together with the availability of a refined 3D-structure at 1.9 Å (6) serves as a firm basis for rational design and construction of mutant proteins with improved activity and broadened substrate specificity.

In order to broaden the substrate specificity *in vivo*, mutants were selected on 1-chlorohexane, which is a poor substrate for the wild type enzyme. These mutants pointed towards the importance of the Lys_{261} - Asp_{170} salt bridge which seems to modify the structure of a part of the cap domain and thereby alters the substrate range (7). To obtain more information about the importance of this salt bridge we have investigated the roles of Lys_{261} and Lys_{259} by site-directed mutagenesis and enzymological studies.

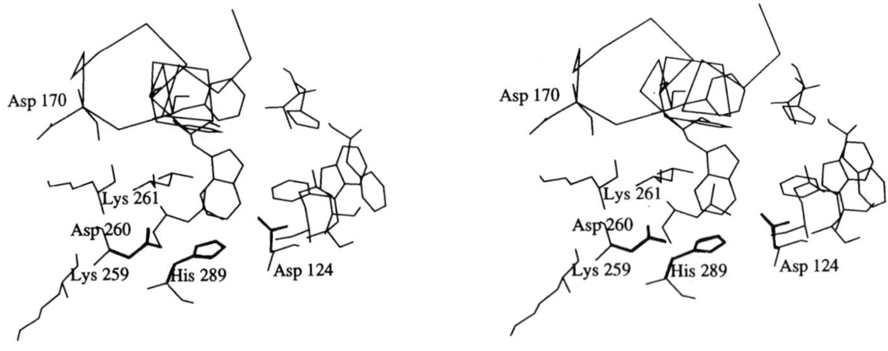

Figure 1. Stereo representation of the active site cavity of haloalkane dehalogenase.

RESULTS AND DISCUSSION

Lys_{259} and Lys_{261} were randomly mutagenized to investigate their involvement in the salt bridge to Asp_{170}.

Activity analysis of 9 different mutants showed that 1,2-dichloroethane dehalogenase activity was retained in all mutants. Analysis of the kinetics of 4 mutants (Table 1) showed that all mutants had a higher affinity for 1-chlorohexane than the wild type enzyme. Substitution of Lys_{261}→Ile (K4) resulted in an affinity for 1-chlorohexane which was intermediate to that of the wild type and the Asp_{170}→His mutant U5, which was selected *in vivo* (7). To check the role of Lys_{259} in the Lys_{261} - Asp_{170} salt bridge, we mutated this residue too. Substitution of Lys_{259} to Leu resulted in a mutant with an affinity similar to that of U5. The double mutants of Lys_{259}/Lys_{261} showed improved binding of 1-chlorohexane compared to mutant K4 and LY15 (Table 1). The highest affinity for 1-chlorohexane was obtained with double mutant Lys_{259}→Glu/Lys_{261}→Ala (KK16), which had a 3-fold lower K_M than the wild type enzyme.

Although the binding of 1-chlorohexane increased in all mutants, the V_{max} of most of the mutants decreased. Only substitution of Lys_{261}→Ile resulted in a 2-fold increase of the V_{max} for 1-chlorohexane compared to the wild type. The double mutants KK10 and KK16 had a 3 fold decrease in V_{max} for this substrate. In mutant LY15, only a small decrease of V_{max} was found (Table 1).

Table 1
Activities of mutants of haloalkane dehalogenase with a disturbed Lys_{261} - Asp_{170} or Lys_{259} - Asp_{170} salt bridge.

Mutant	Mutation	1-Chlorohexane	
		K_m (mM)	V_{max} (mU/mg protein)
LY15	Lys_{259}→Leu	0.87	132
K4	Lys_{261}→Ile	0.97	324
KK10	Lys_{259}→Gly/ Lys_{261}→Arg	0.63	45
KK16	Lys_{259}→Glu/ Lys_{261}→Ala	0.45	41
U5	Asp_{170}→His	0.88	590
Wild type		1.4	150

The above results show that the salt bridge influences the substrate specificity. This may be caused by its effect on the N-terminal part of the cap domain of the dehalogenase. The N-terminal part has only a few stabilizing elements and the Lys_{261} - Asp_{170} salt bridge is one of them (7). Apparently, modification of this salt bridge changes the N-terminal part of the cap domain in a way that favours binding of 1-chlorohexane. The different affinities of the *in vitro* mutants thus could be caused by different interactions with the Asp_{170}, thereby changing the strength and/or position of the salt bridge.

Another possible cause of the observed differences between the mutants and the wild type could be an effect on Asp_{260}, which is involved in catalysis. This residue lies between Lys_{259} and Lys_{261} (Figure 1) and its position can probably be influenced by mutation of the surrounding residues. Asp_{260} is connected *via* a salt bridge to His_{289}, which is involved in catalysis and located in the wall of the active site cavity. Therefore, changing residues Lys_{259} and Lys_{261} could also affect the formation of the alkyl-enzyme intermediate. Since it is the rate-limiting step for 1-chlorohexane hydrolysis, the rate of formation of the alkyl-enzyme has a strong influence on the overall V_{max}.

We conclude that modification of the Lys_{261} - Asp_{170} salt bridge in haloalkane dehalogenase can have two effects: 1) Modification of the N terminal part of the cap domain. Upon removal, accommodation of a larger substrate (i.e. 1-chlorohexane) is favoured; 2) Modification of the salt bridge influences the position and/or mobility of Asp_{260}, which in its turn might affect the formation of the alkyl enzyme-intermediate.

REFERENCES

1. S. Keuning, D. B. Janssen and B. Witholt, J. Bact. Vol. 163 (1985) 635.
2. D. B. Janssen, F. Pries, J. R. van der Ploeg, B. Kazemier, P. Terpstra and B. Witholt, J. Bacteriol. Vol. 171 (1989) 6791.
3. S. M. Franken, H. J. Rozeboom, K. H. Kalk and B. W. Dijkstra, EMBO J. Vol. 10 (1991) 1297.
4. D. L. Ollis, E. Cheah, M. Cygler, B. W. Dijkstra, F. Frolow, S. M. Franken, M. Haral, S. J. Remington, I. Silman, J. Schrag, J. L. Sussman, K. H. G. Verschueren and A. Goldman, Prot. Engin. Vol. 5 (1992) 197.
5. J. P. Schanstra, R. Rink, F. Pries and D. B. Janssen, submitted.
6. K. H. G. Verschueren, S. M. Franken, H. J. Rozeboom, K. H. Kalk and B. W. Dijkstra, submitted.
7. F. Pries, A. J. van den Wijngaard, R. Bos and D. B. Janssen, submitted.

ECB6: Proceedings of the 6th European Congress on Biotechnology
edited by L. Alberghina, L. Frontali, P. Sensi

Replacement and introduction of COOH-groups involved in catalysis, acarbose binding, and stability in *A. niger* glucoamylase by site-directed mutagenesis

T. P. Frandsen[a], J. Lehmbeck[b], C. Dupont[a], B. Stoffer[a], M. R. Sierks[a,c], M. T. Hansen[b] and B. Svensson[a]

[a]Department of Chemistry, Carlsberg Laboratory, Gamle Carlsberg Vej 10, DK-2500 Copenhagen Valby, Denmark

[b]Fungal Gene Technology, Bioindustrial Group, Novo Nordisk, Novo Allé, DK-2880 Bagsvaerd, Denmark

[c]Department of Chemical and Biochemical Engineering, University of Maryland, Baltimore County, Baltimore, MD 21228, U.S.A.

Site-directed mutagenesis of glucoamylase from *Aspergillus niger* of the proposed catalytic base Glu 400 resulted in approx. 50-fold reduction in k_{cat} and 15-fold increase in K_M compared to the wild-type enzyme. Replacement of the adjacent Gln 401 by Glu shifted the pH-activity profile slightly and had only minor effects on activity. Replacement of Asp 112 and Ala 392, tentatively aligned with active site regions from α-amylases, had essentially no effect on activity, but caused some decrease in the thermostability.

1. INTRODUCTION

Fungal glucoamylases are widely used industrially in the production of glucose syrups from starch. Investigations of the structure/function relationships in the glucoamylase from *Aspergillus niger* has been carried out by using chemical modification and sequence comparisons in conjunction with site-directed mutagenesis (Clarke & Svensson, 1984; Svensson, 1988; Svensson et al., 1990; Sierks et al., 1989, 1990, 1993; Sierks and Svensson, 1992, 1993). Recently a three-dimensional model has become available of the catalytic domain of the homologous glucoamylase from *Aspergillus awamori* var. *X100* (Aleshin et al., 1992). Prior to this information mutants were constructed of the *Aspergillus niger* enzyme in regions with sequence similarity to α-amylases or at residues protected against chemical modification by the transition-state analog, acarbose. Mutants were later made at the Glu 400, proposed to be the catalytic base by the crystal structure (Harris et al., 1993) and at the adjacent Glu 401.

2. MATERIALS AND METHODS

Mutant genes were constructed by a modified PCR reaction (Nelson and Lange, 1989) and subcloned into an *Aspergillus* expression vector containing the AMG promoter and terminator. The plasmids obtained were used to transform an *Aspergillus niger* strain lacking its natural AMG gene by cotransformation with the *Aspergillus nidulans* amdS gene as marker (Christensen et al., 1988). Wild-type and mutant enzymes were purified from the culture liquid by affinity chromatography on acarbose-Sepharose and the G1 and G2 forms of glucoamylase were separated by ion exchange chromatography (Stoffer et al., 1993). In all cases the G1 form was chosen for further characterization.

The enzyme kinetic studies were performed in 0.05 M sodium acetate pH 4.5 at 45°C. The substrate concentrations range from 0.125 times K_M to 8 times K_M. Aliquots were removed and initial rates of glucose production determined by using a glucose oxidase method (Fleming and Pegler, 1963) adapted to a microtiter plate reader (Fox and Robyt, 1991). The kinetic parameters were calculated using the computer program Enzfitter (Leatherbarrow, 1987). The pH activity profile was determined essentially as described (Sierks et al., 1990). The thermostability was determined by incubating the enzyme (0.9μM) in 0.05 M sodium acetate pH 4.5 at temperatures in the range 25-80°C as described (Svensson et al., 1986).

3. RESULTS

3.1. Kinetic properties

The kinetic parameters, k_{cat} and K_M, were determined for the different mutants and wild-type glucoamylase on the α-1,4-linked substrates maltose and maltoheptaose and the α-1,6-linked substrate isomaltose (Table 1).

Very little effect was obtained for the mutants of the acarbose-protected groups (Svensson et al., 1990), Asp112→Tyr, Asp153→Asn, Glu259→Asp/Gln, and Glu389→Asp/Gln. The replacement of Asp 112 by Tyr was guided by tentatively aligned active-site sequences from α-amylases and related enzymes (Svensson, 1988). The higher activity of Glu389→Asp on isomaltose and the lower activity of Glu389→Gln on maltoheptaose are noticable. Another mutation Ala392→Asp was guided by regional sequence similarity to the active site from α-amylases. Also in this case there was little effect on activity, except for the 1,4-fold increase in k_{cat} on maltoheptaose (Table 1). Both this mutant and the Asp112→Tyr were made to increase the relative specificity for maltose over isomaltose. The $[(k_{cat}/K_M)_{maltose}/(k_{cat}/K_M)_{isomaltose}]$ values, however, had not changed significantly from wild-type.

The Glu 400 was identified as a good candidate for the catalytic base from the crystal structure of the complex with the inhibitor 1-deoxynojirimycin (Harris et al., 1993). To test this role we replaced the group by the corresponding amide. This caused a dramatic decrease in activity on both of the disaccharides, which was not compensated for in maltoheptaose (Table 1) believed to fill the entire binding site. Since the neighbouring Gln 401 was shown by alignment to be conserved in all but one of eleven glucoamylase sequences (Stoffer, unpublished), we speculated that it might be important for activity. The replacement of Gln

Table 1
Kinetic parameters[a] for the hydrolysis of maltose, isomaltose and maltoheptaose by mutant and wild-type *A. niger* glucoamylases

	Maltose			Isomaltose			Maltoheptaose		
Mutant	k_{cat} (s^{-1})	K_M (mM)	k_{cat}/K_M ($mM^{-1}s^{-1}$)	k_{cat} (s^{-1})	K_M (mM)	k_{cat}/K_M ($mM^{-1}s^{-1}$)	k_{cat} (s^{-1})	K_M (mM)	k_{cat}/K_M ($mM^{-1}s^{-1}$)
D112Y	7.62	1.12	6.80	0.43	26.5	0.016	67.1	0.14	479
D153N	8.49	1.20	7.08	0.37	22.7	0.016	60.1	0.18	334
E259D	9.34	0.88	10.6	0.46	23.6	0.020	65.3	0.094	695
E259Q	8.58	0.91	9.43	0.37	23.5	0.016	69.3	0.093	745
E389D	9.33	1.16	8.04	0.78	22.6	0.035	61.9	0.14	442
E389Q	10.0	1.09	9.17	0.50	22.0	0.023	46.1	0.14	329
A392D	9.35	1.52	6.15	0.44	27.8	0.016	74.4	0.17	437
E400Q	0.30	14.8	0.020	n.d.			1.06	0.38	2.79
Q401E	3.49	1.09	3.20	0.14	19.4	0.0072	20.9	0.084	249
wt G1	11.4	1.13	10.1	0.42	22.2	0.019	56.7	0.11	515

a) standard deviations are within 6% for k_{cat} and 12% for K_M
n.d.) not determined due to lack of detectable activity

401 by Glu attempted at increasing the activity by an electrostatic effect. Introduction of a negative charge in the microenvironment of Glu 400, might influence its activation of the catalytic water molecule. The Gln401→Glu, however, was found for all substrates tested to have k_{cat} reduced approx. 3-fold, while K_M remained as for the wild-type enzyme (Table 1).

3.2. pH-activity dependence

Asp112→Tyr, Asp153→Asn, Ala392→Asp, and Gln401→Glu had slightly reduced pH-optimum of pH 3.8, 3.8, 3.8 and 4.0, respectively, relative to the wild-type value of pH 4.4. The pK_a values of the catalytic groups were shifted a little in Asp153→Asn and drastically for the mutant Glu400→Gln, especially in the low pH region.

3.3. Thermostability

All purified mutants were tested for thermostability, expressed as the T_m value. A decrease of 5°C was found for Asp112→Tyr, probably reflecting small conformational changes leading to destabilization. An interesting difference in thermostability was found between the two mutants of Glu389. While the replacement to Asp has retained the thermostability, the mutation to the corresponding amide, caused a decrease in T_m of 4°C.

4. DISCUSSION

Analysis of the three-dimensional structure (Aleshin et al., 1992; Harris et al., 1993) reveals that none of the groups Asp 112, Asp 153, Glu 259 and Glu 389 are in immediate contact with the bound inhibitor or the catalytic groups. This explains why only minor changes were obtained in the kinetic parameters of the corresponding mutants. The observed protection by acarbose against chemical modification (Svensson et al., 1990) probably stems from ligand binding outside of the active site, perhaps involving *O*- or *N*-glycosylated residues.

The replacement of Glu 400 in combination with the three-dimensional structure (Aleshin et al., 1992; Harris et al., 1993) provides further support for its proposed role as catalytic base. The kinetic parameters obtained with Glu400→Gln, thus demonstrate that Glu 400 is very important for the mechanism. The replacement of the catalytic base is not, however, as critical for activity as replacement of the general acid, Glu 179 (Sierks et al., 1990). This has been found in general for active site mutants at the catalytic nucleophile and general acid groups, respectively, of different carbohydrases (for a review, see Svensson and Søgaard, 1993).

REFERENCES

A. Aleshin, A. Golubev, L.M. Firsov and R.B. Honzatko, J. Biol. Chem. 267 (1992) 19291.
A.J. Clarke and B. Svensson, Carlsberg Res. Commun. 49 (1984) 559.
T. Christensen, H. Woeldike, E. Boel, S.B. Mortensen, K. Hjortshoej, L. Thim, and M.T. Hansen, Bio/Technology 6 (1988) 1419.
I.D. Fleming and H.F. Pegler, Analyst 88 (1963) 967.
J.D. Fox and J.F. Robyt, Anal. Biochem. 195 (1991) 93.
E.M.S. Harris, A.E. Aleshin, L.M. Firsov, and R.B. Honzatko, Biochemistry 32 (1993) 1618.
R.J. Leatherbarrow, Enzfitter, a non-linear regression data analysis program for IBM PC. Elsevier Science Publishers BV, Amsterdam, The Netherlands (1987).
R.M. Nelson and G.L. Lange, Anal. Biochem. 180 (1989) 147.
M.R. Sierks, C. Ford, P.J. Reilly and B. Svensson, Protein Eng. 2 (1989) 621.
M.R. Sierks, C. Ford, P.J. Reilly and B. Svensson, Protein Eng. 3 (1990) 193.
M.R. Sierks, C. Ford, P.J. Reilly and B. Svensson, Protein Eng. 6 (1993) 75.
M.R. Sierks and B. Svensson, Protein Eng. 5 (1992) 185.
M.R. Sierks and B. Svensson, Biochemistry 32 (1993) 1113.
B. Stoffer, T.P. Frandsen, P.K. Busk, I. Svendsen, P. Schneider and B. Svensson, Biochem. J. 292 (1993) 197.
B. Svensson, FEBS Lett. 230 (1988) 72.
B. Svensson, A.J. Clarke and I. Svendsen, Carlsberg Res. Commun. 51 (1986) 61.
B. Svensson, A.J. Clarke, I. Svendsen and H. Møller, Eur. J. Biochem. 188 (1990) 29.
B. Svensson and M. Søgaard, J. Biotechnol. (1993) in press.

BIOCATALYSIS IN ORGANIC SYNTHESIS: THE POTENTIALITY OF LYASES

ECB6: Proceedings of the 6th European Congress on Biotechnology
edited by L. Alberghina, L. Frontali, P. Sensi

THE POTENTIAL OF LYASES IN ORGANIC SYNTHESIS

W.J.J. van den Tweel[1], M.J. van der Werf[2], S. Hartmans[2],
H.E. Schoemaker[1], J. Kamphuis[1] and J.A.M. de Bont[2]

[1]Bio-organic Chemistry Section, DSM Research,
P.O. Box 18, 6160 MD Geleen, The Netherlands
[2]Division of Industrial Microbiology, Department of Food Science, Agricultural University,
P.O. Box 8129, 6700 EV Wageningen, The Netherlands

1. INTRODUCTION

Lyases are enzymes cleaving C-C, C-O, C-N, and other bonds by elimination, leaving double bonds or rings, or conversely adding groups to double bonds. The lyase enzyme class comprises many interesting enzymes which can be very useful for synthesizing various chiral chemical compounds.
As a result of the biocatalyst availability, stability, stereospecificity (almost 100% yield of the desired enantiomer can be achieved), and the fact that no cofactor recycling is required, some biocatalytic processes based on lyases are competitive to chemical processes. Consequently, several lyase-catalyzed processes have currently been commercialized.
To exemplify the potential of lyases an overview is given of the commercialized processes as well as some R&D developments.

2. COMMERCIALIZED PROCESSES

R-phenylacetyl carbinol
Pyruvate decarboxylase, a thiamine pyrophosphate dependent lyase, catalyzes the conversion of pyruvate to acetaldehyde and CO_2, a key step in the alcoholic fermentation of sugar. This enzyme, however, also catalyzes the condensation of thiamine pyrophosphate bound acetaldehyde with benzaldehyde yielding optically pure R-phenylacetyl carbinol (Scheme 1). Already in 1921, Neuberg and Hirsch[1] reported on this condensation reaction catalyzed by fermenting brewer's yeast. Due to the fact that R-phenylacetyl carbinol is an important intermediate for both L-ephedrine and L-pseudoephedrine, and no economically attractive alternative chemical production routes exist, this conversion achieved great practical importance. Nowadays R-phenylacetyl carbinol is produced on a multi-tonne scale by Knoll (Germany) and Malladi (India) by means of yeast cells containing the pyruvate decarboxylase.
More recent work[2] has also shown that in the presence of other aldehydes many other acyloin type of compounds can be produced by organisms containing the pertinent pyruvate decarboxylase. The resulting chiral synthons can be used for numerous applications.

Scheme 1

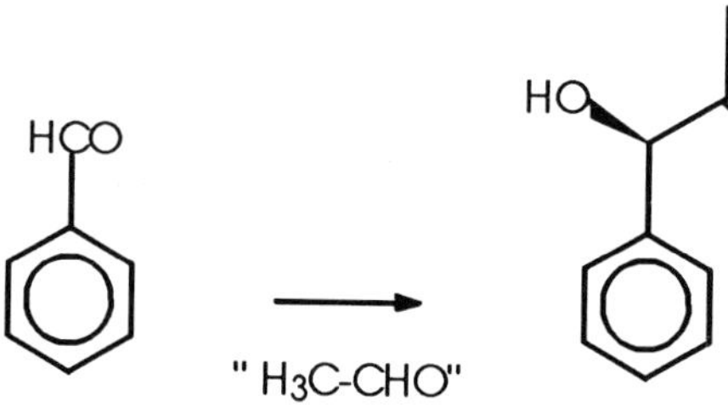

" H3C-CHO"

L-Aspartic acid

L-Aspartic acid is currently widely used as an intermediate for pharmaceuticals and food additives (e.g. for the production of the intense sweetener aspartame). Already in 1960 Tanabe (Japan) started producing L-aspartic acid from fumaric acid using intact *Escherichia coli* cells having high aspartase (aspartate ammonia-lyase) activity[3] (Scheme 2A). In 1973 this batchwise production using free cells was replaced by a continuous process using immobilized cells packed in a column. This industrial process was the first application of immobilized cells as a solid catalyst. Initially cells were immobilized in polyacrylamide, however, later on cells were entrapped in K-carrageenan (less loss of activity during immobilization and a higher operational stability). By incubating the culture broth (pH 4.9) containing L-aspartic acid at 45 C for 1 h the interfering fumarase activity - resulting in the undesired L-malic acid - was almost completely eliminated without inactivating aspartase. The current process, which employs immobilized cells having a very high aspartase activity and a half-life time above 100 days, yields L-aspartic acid in high yields and possesses a high volumetric productivity[3]. Nowadays also other companies are running this process on a commercial scale.

Scheme 2

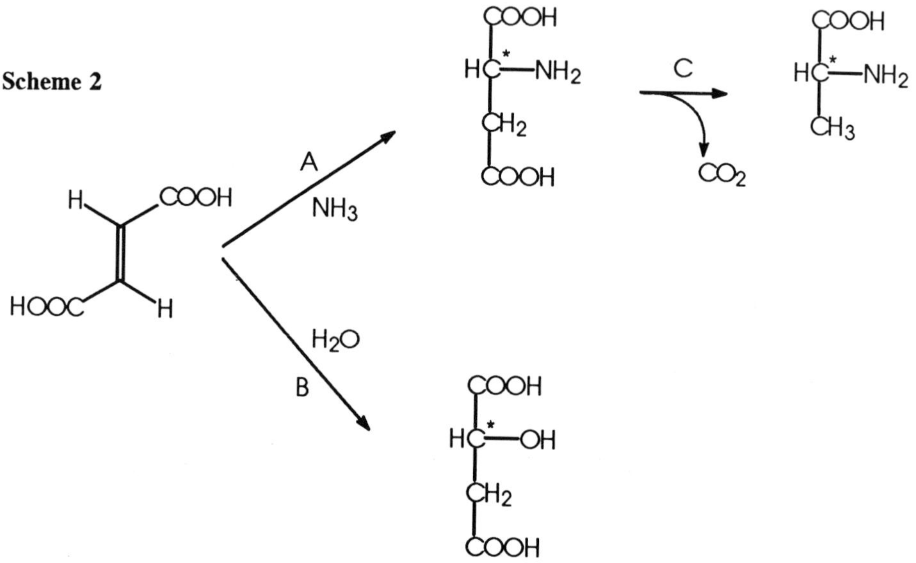

A NH_3

B H_2O

C CO_2

L-Alanine and D-aspartic acid

Another well-known lyase is the L-aspartate 4-decarboxylase. This enantioselective enzyme decarboxylates L-aspartic acid yielding L-alanine (Scheme 2C). Currently Tanabe is producing L-alanine in a continuous way using immobilized *Pseudomonas dacunhae* cells in a packed column reactor[3]. To maintain plug-flow of the substrate solution and to keep a constant pH in the reactor the formation of CO_2 gas is prevented by performing the enzyme reaction at an elevated pressure. Currently L-alanine is produced most efficiently by the sequential use of an *E. coli* column, for converting fumaric acid to L-aspartic acid, followed by a *Pseudomonas dacunhae* column converting L-aspartic acid to L-alanine.
As this decarboxylase enzyme is highly selective for the L-enantiomer (D-aspartic acid is no substrate for this enzyme), this enzyme is also used for the simultaneous production of both L-alanine and D-aspartic acid from racemic aspartic acid[3]. D-Aspartic acid is a useful intermediate for a synthetic penicillin.

L-Malic acid

L-Malic acid is an important acidifier in the food industry and L-malate salts of basic amino acids are used in infusion fluids. The lyase-enzyme fumarase very efficiently catalyzes the asymmetric hydratation of fumaric acid yielding optically pure L-malic acid[3] (Scheme 2B). Initially Tanabe employed *Brevibacterium ammoniagenes* cells immobilized in polyacrylamide for this conversion. Currently, however, *Brevibacterium flavum* cells immobilized in K-carrageenan in the presence of Chinese gallotanin are used mainly because of an enhanced activity and stability in a continuous process[3]. Due to the thermodynamic constants of this reaction the maximal conversion yield of this reaction is about 80%. However, substrate and product are easily separated using HCl precipitation.

L-Phenylalanine

L-Phenylalanine is the other building block for the low-calorie sweetener aspartame. Some micro-organisms possess an ammonia-lyase which is able to convert trans-cinnamic acid to L-phenylalanine. In the beginning of the eighties the U.S. firm Genex has commercialized this biotransformation[4] using phenylalanine ammonia-lyase from *Rhodutorula rubra.* However, mainly due to some intrinsic disadvantages of this process - low yields as a result of the unfavourable equilibrium, low operational stability of the biocatalyst and severe inhibition by trans-cinnamic acid - this process was stopped shortly after completion of the plant. Nowadays L-phenylalanine is preferably made by fermentation.

Scheme 3

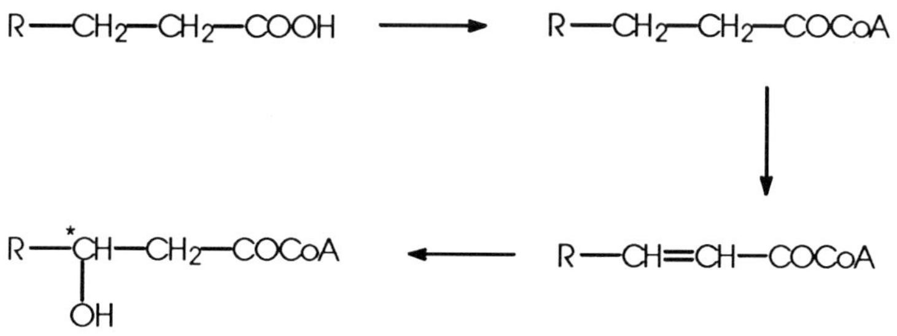

$$R{-}CH_2{-}CH_2{-}COOH \longrightarrow R{-}CH_2{-}CH_2{-}COCoA$$

$$\downarrow$$

$$R{-}\overset{*}{C}H(OH){-}CH_2{-}COCoA \longleftarrow R{-}CH{=}CH{-}COCoA$$

ß-Hydroxylation of carboxylic acids

For the production of some chiral ß-hydroxycarboxylic acids the enzymes of the ß-oxidation pathway can be very useful (Scheme 3). This ß-oxidation route, the pathway for fatty acid oxidation, proceeds in two stages: an initial dehydrogenation to the α,ß-unsaturated carboxylic acid and a subsequent hydration by means of an enantioselective lyase. The saturated carboxylic acids as well as their α,ß-unsaturated counterparts can be used as raw materials. Before the initial dehydrogenation can occur the carboxylic acid is activated ATP-dependently via an esterification with the thiol group of coenzyme A. All subsequent oxidative steps take place with the carboxylic acid in the form of its CoA thioester. Due to cofactor recycling and the energy dependent uptake process the production of ß-hydroxycarboxylic acids can only be accomplished using growing/resting cells. This in striking contrast to the forementioned biotransformation of fumaric acid to L-malic acid which is catalysed by only one lyase-enzyme. Consequently for the production of ß-hydroxy carboxylic acids the biocatalyst is only used once, and no processes using immobilized cells have been developed.

Nowadays two ß-hydroxycarboxylic acids are produced in such way: L-carnitine out of γ-butyrobetaine by Lonza (Switzerland)[5] and the conversion of isobutyric acid to D-ß-hydroxy-isobutyric acid by Kaneka (Japan)[6] (Scheme 4). For the first process various enterobacteria can be used, however, for the latter conversion only a few yeasts were able to produce the desired unnatural D-enantiomer of ß-hydroxy-isobutyric acid. To optimize these processes mutants were selected which are unable to metabolize the desired product. Although active cells are needed for these bioconversions, the volumetric productivity for such processes is comparable to the productivity of chemical processes (about 100 kg m^{-3} day^{-1})[7]. Consequently these processes yielding chiral products with 100% yield are economically very attractive.

Scheme 4

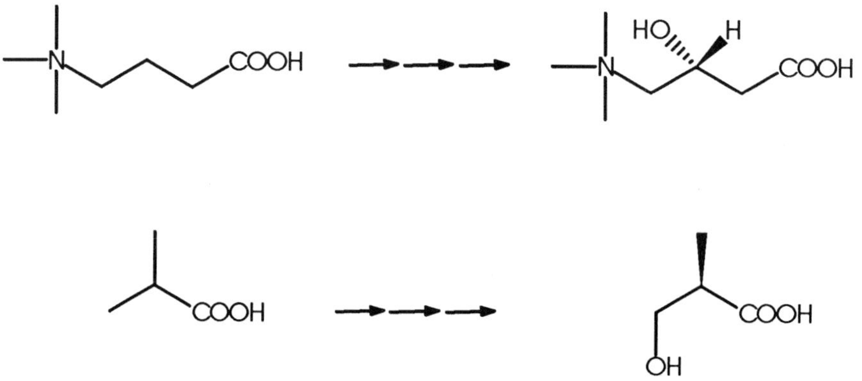

COOH → → → HO H COOH

COOH → → → COOH OH

3. R&D DEVELOPMENTS

R- and S-Cyanohydrins

The known lyases R- and S-oxynitrilase catalyze the enantioselective addition of HCN to aldehydes to give R- and S-cyanohydrins respectively[8,9] (Scheme 5). The R-specific enzyme is readily available from bitter almonds, whereas the S-oxynitrilase can be isolated from sorghum. Although both enzymes are highly enantioselective, the formed products often are not optically pure. The poor enantiomeric ratio is attributed to a competing non-enzymatic addition of HCN to the aldehyde. By lowering the pH and temperature and/or using an organic solvent this competing chemical reaction is almost completely circumvented. Based on some recent developments, it is likely that efficient, cost-effective production methods for a wide variety of optically pure cyanohydrins on a multikilogram scale will soon be within reach.

Scheme 5

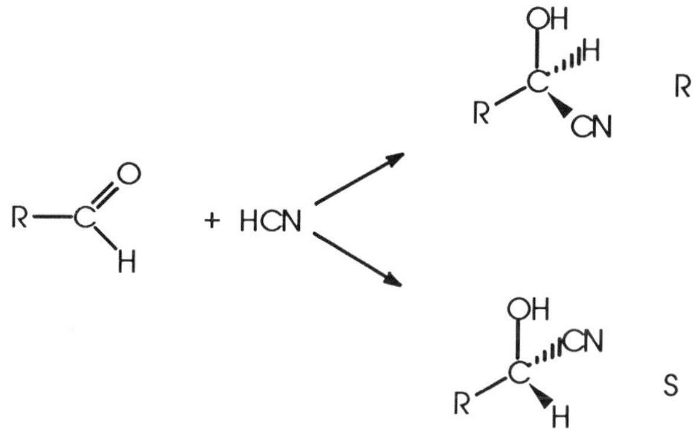

R—CHO + HCN → R (R-cyanohydrin) / S (S-cyanohydrin)

Acyloin formation mediated by benzoylformate decarboxylase

Similar to the pyruvate decarboxylase the thiamine pyrophosphate-linked benzoylformate decarboxylase can be applied to catalyze the formation of acyloins[10]. Using this enzyme or cells containing this enzyme benzoylformate in the presence of acetaldehyde was converted in almost 60% yield to S-2-hydroxypropiophenone (≥98% e.e.) (Scheme 6). As the product concentration (about 10 g/l) and specific productivity compare rather well with the industrial R-phenylacetyl carbinol production this biotransformation might become economically attractive.

Scheme 6

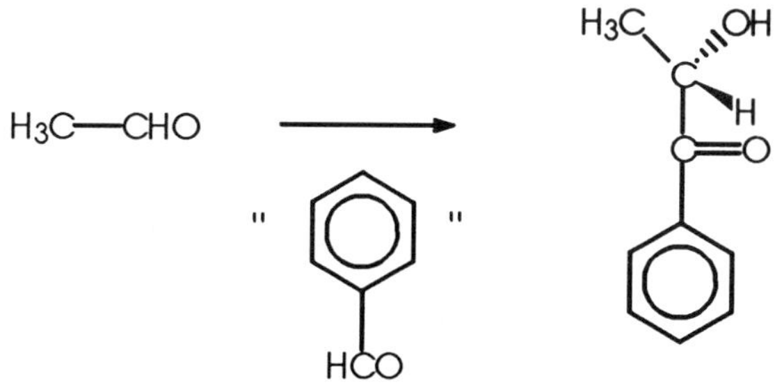

H_3C—CHO → (" benzaldehyde ") → S-2-hydroxypropiophenone

D-malic acid and D-citramalic acid

Recently it has been shown that also D-malic acid and D-citramalic acid (e.e.'s $\geq$ 99%) can be made from respectively maleic acid and citraconic acid, using a malease from *Pseudomonas pseudoalcaligenes* NCIMB 9867[11]. In contrast to the fumarase-catalyzed formation of L-malic acid from fumaric acid, a conversion yield of nearly 100% can easily be obtained. Moreover, in view of the high specific malease activity of the *Pseudomonas pseudoalcaligenes* cells this biotransformation process may be very attractive for producing the pertinent chiral D-hydroxy acids.

Formation of carbon-carbon bonds using aldolases

Aldolases form a sub-class of lyases that catalyze the stereoselective construction and degradation of C-C bonds in monosaccharides and amino acids. Resulting from the fact that enzymatic aldol reactions, which are highly enantioselective, can be performed in aqueous solution at neutral pH without extensive protection, these cofactor independent bioconversions attract increasing interest. In view of the amount of classified aldolases (more than 40) and their generally relaxed substrate specificity (especially for the acceptor) many interesting molecules[12,13,14] can be synthesized by these enzymes. Although no conversions have been commercialized yet, in the near future economically feasible processes will be developed.

Synthesis of ß-hydroxynitriles

Recently Japanese researchers have isolated bacteria which transform prochiral 1,3-dichloro-2-propanol into optically active R-3-chloro-1,2-propanediol. This transformation proceeds via an intramolecular dehalogenation yielding epichlorohydrin[15]. The conversion of 1,3-dichloro-2-propanol to epichlorhydrin is catalyzed by a halohydrin-halide lyase. Detailed studies on the catalytic properties of this enzyme have recently revealed that this enzyme also catalyzes the synthesis of ß-hydroxynitriles from epoxides and cyanide[15] (Scheme 7). So far nothing is known about the enantioselectivity and stability of this conversion. However, this remarkable conversion might be very useful for synthesizing specific ß-hydroxy carboxylic acids.

Scheme 7

$$R{-}\underset{\backslash\,O\,/}{CH{-}CH_2} + HCN \longrightarrow R{-}\underset{OH}{\underset{|}{CH}}{-}CH_2{-}CN$$

Carbon-carbon formation by Transketolases

In addition to the lyase enzymes also some other enzymes can catalyze the formation of C-C bonds. One nice example in this respect is the transketolase enzyme which requires the cofactor thiamine pyrophosphate for activation of the ketone[16] (Scheme 8). Recently it was shown that L-2-hydroxyaldehydes with a great variety of substituents in 3-position are obtained in good chemical and excellent optical yields by kinetic resolution in the transketolase-catalyzed reaction of racemic 2-hydroxyaldehydes with lithiumhydroxypyruvate[17]. Only the R-enantiomers reacted to the corresponding 5-substituted 5-deoxy-D-xyluloses.

Scheme 8

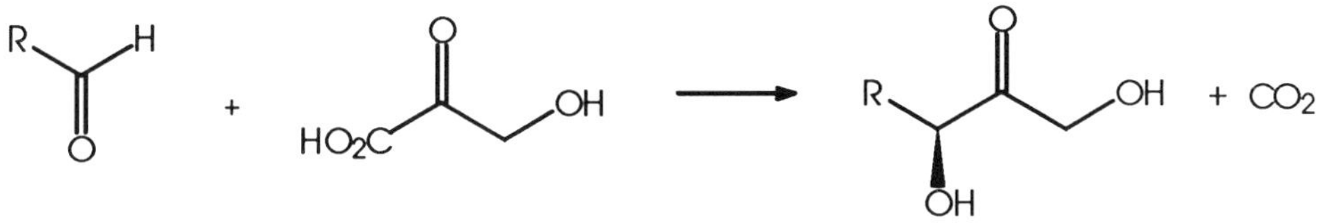

$$\text{RCHO} + \text{HO}_2\text{C-CO-CH}_2\text{OH} \longrightarrow \text{R-CH(OH)-CO-CH}_2\text{OH} + CO_2$$

Transketolase, Mg^{2+}, thiamine pyrophosphate

The addition of H_2S to double bonds

ß-Mercapto-2-methylpropionic acid is a potential precursor for the synthesis of the antihypertensive agent Captopril. Recently we have shown that some micro-organisms possess sulphur lyase type enzymes which might be applicable for the direct addition of bisulphite to methacrylic acid[18]. Further research is currently being performed to investigate the feasibility of these enzymes for producing ß-mercapto acids.

4. CONCLUSIONS AND FUTURE PERSPECTIVES

As can be concluded from this paper lyases are very useful enzymes which catalyze numerous interesting addition reactions.

Simple conversions like the formation of either L-malic acid and L-aspartic acid out of fumaric acid are industrially very attractive. However, in contrast to the enzymes which are used for producing L-carnitine and D-ß-hydroxy-isobutyric acid, fumarase and aspartase are very substrate specific. Consequently for hydratation and amination of similar compounds other enzymes have to be found. To accomplish this, extensive screening has to be performed.

The enantioselective formation of carbon-carbon bonds is chemically rather difficult. In this respect enzymes can be very useful. During the coming years more enzymes will be found and their intrinsic characteristics identified giving rise to an even bigger collection of industrially useful biocatalysts.

In addition to the mentioned hydratase reactions yielding chiral compounds also the production of non-chiral compounds can be commercially feasible. A nice example in this respect is the enzymic hydratation of acrylonitrile yielding acrylamide[19]. Nowadays Nitto Chemical Industry (Japan) produces over 15,000 tonnes per year using a nitrile hydratase of a *Rhodococcus rhodochrous* strain.

In view of these developments lyases also in the future will be very important enzymes for the production of chemicals.

ACKNOWLEDGEMENTS

The assistance of Mr. W. Kortenoeven with the preparation of this manuscript is greatly appreciated.

REFERENCES

1. C. Neuberg and J. Hirsch, Biochem. Z., 115 (1921) 282-286
2. V. Kren, D.H.G. Crout, H. Dalton, D.W. Hutchinson, W. Konig, M.M. Turner, G. Dean and N. Turner, J. Chem. Soc., Chem. Commun., (1993) 341-343
3. I. Chibata, T. Tosa and T. Shibatani, in: A.N. Collins, G.N. Sheldrake and J. Crosby (eds.), Chirality in Industry, John Wiley and Sons, Chichester, 1992, pp. 351-370
4. European Chemical News, August 29 (1983) 15
5. C. Chassin, Chirality 1990 Symposium Manchester, Proceedings 107-108
6. T. Ohashi and J. Hasegawa, in: A.N. Collins, G.N. Sheldrake and J. Crosby (eds.), Chirality in Industry, John Wiley and Sons, Chichester, 1992, pp. 249-268
7. H-P. Meyer, Biotechnol. Forum Europe, 10 (1991) 602-606
8. C.G. Kruse, in: A.N. Collins, G.N. Sheldrake and J. Crosby (eds.), Chirality in Industry, John Wiley and Sons, Chichester, 1992, pp. 279-302
9. F.X. Effenberger, in: S. Servi (ed.), Microbial Reagents in Organic Synthesis, Kluwer Academic Publishers, Dordrecht, 1992, pp. 25-33
10. O.P. Ward, R. Wilcocks, E. Prosen, S. Collins, N.J. Dewdney and Y. Hong, in: S. Servi (ed.), Microbial Reagents in Organic Synthesis, Kluwer Academic Publishers, Dordrecht, 1992, pp. 67-75
11. M.J. van der Werf, W.J.J. van den Tweel and S. Hartmans, Appl. Environ. Microbiol., 58 (1992) 2854-2860
12. C-H. Wong, Y. Ichikawa, T. Kajimoto, K.K-C. Liu, D.P. Dumas, Y-C. Lin and G.C. Look, in S. Servi (ed.), Microbial Reagents in Organic synthesis, Kluwer Academic Publishers, Dordrecht, 1992, pp. 35-43
13. M.D. Bednarski, in: H.W. Blanch and D.S. Clark (eds.) Applied Biocatalysis Vol. 1, pp. 87-116
14. K. Faber, Biotransformations in Organic Chemistry, Springer-Verlag, 1992, 204-220
15. T. Nakamura, T. Nagasawa, F. Yu, I. Watanabe and H. Yamada, Biochem. Biophys. Res. Commun., 180 (1991) 124-130
16. G. Hobs, ECB6 Conference Proceedings, 1993
17. F. Effenberger, V. Null and T. Ziegler, Tetrahedron Lett., 33 (1992) 5157-5160
18. M.R. Smith, W.J.J. van den Tweel and J.A.M. de Bont, submitted for publication in J. Gen. Microbiol.
19. Y. Ashina and M. Suto, in: A. Tanaka, T. Tosa and T. Kobayashi (eds.), Industrial Application of Immobilized Biocatalysts, Marcel Dekker, New York, 1992, pp. 91-108

ECB6: Proceedings of the 6th European Congress on Biotechnology
edited by L. Alberghina, L. Frontali, P. Sensi

Development of transketolase as a catalyst for carbon-carbon bond formation

G.R. Hobbs*[,a], M.D. Lilly[a], N.J. Turner[b], J.M. Ward[c], and J.M. Woodley[a]

[a] Advanced Centre for Biochemical Engineering, Department of Chemical and Biochemical Engineering, University College London, Torrington Place, London WC1E 7JE, UK.

[b] Department of Chemistry, University of Exeter, Stocker Road, Exeter EX4 4QD, UK.

[c] Department of Biochemistry and Molecular Biology, University College London, Gower Street, London WC1E 6BT, UK.

1. INTRODUCTION

The use of biotransformations to carry out a range of synthetically useful organic reactions is becoming increasingly popular [1]. Enzymes from microorganisms can perform in one step a transformation that may require a number of chemical steps, and often to high enantiomeric purity. Stereospecific carbon-carbon bond formation is an important reaction in synthetic organic chemistry and may be catalysed by several enzymes, including intracellular transketolase (TK) (E.C.2.2.1.1).

In vivo transketolase catalyses a vital step in the pentose phosphate pathway, namely the transfer of a two carbon ketol group from D-xylulose-5-phosphate to D-ribose-5-phosphate thereby generating D-sedoheptulose-7-phosphate [2], in the presence of the cofactors thiamine pyrophosphate (TPP) and magnesium (II) ions (Mg^{++}). Using β-hydroxypyruvate (**2**) as the ketol donor (scheme 1) carbon dioxide is released on reaction with the acceptor aldehyde (**1**). This ensures that the reaction is irreversible [3].

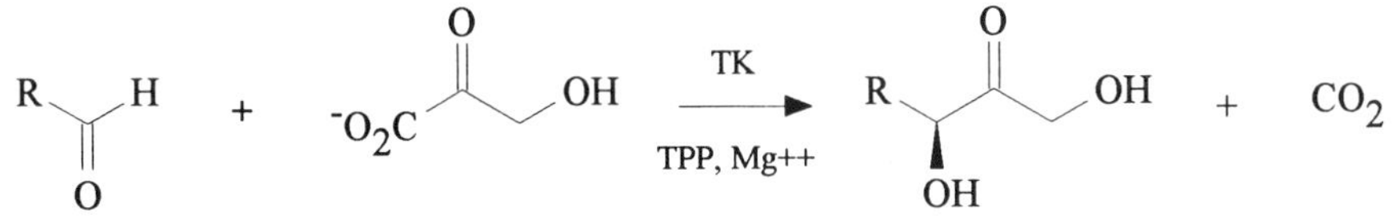

1 + **2** $\xrightarrow[\text{TPP, Mg++}]{\text{TK}}$ **3** + CO_2

Scheme 1

Until recently it was believed that only α-hydroxyaldehydes of the D-configuration (**1**) were possible aldehyde acceptors, leading to compounds (**3**) containing the D-*threo*-configuration. However, a recent report [4] as described the use of some simple α-unsubstituted aldehydes. Thus the low specificity of TK for the aldehyde component (**1**) together with the high stereoselectivity for the carbon-carbon bond forming step combine to make this enzyme a potentially useful catalyst for asymmetric synthesis.

This paper describes work on a programme in which the aim is to study the fundamental principles required to develop an integrated process for carbon-carbon bond formation on a large scale using TK as the catalyst.

2. MATERIALS AND METHODS

A recombinant form of *E. coli* (JM107/pQR183) [5] was used to prepare TK used in all experiments. Culture medium A contained: $Na_2HPO_4 \cdot 12H_2O$ (15 g/L), KH_2PO_4 (3 g/L), NH_4Cl (1 g/L),NaCl (0.5 g/L), $MgSO_4 \cdot 7H_2O$ (0.25 g/L), thiamine (0.001 g/L), glycerol (2.5 g/L), and ampicillin (0.1 g/L).

2.1 Preparation of the biocatalyst

Cultures of *E. coli* JM107/pQR183 were prepared by inoculating flasks containing medium A (100 mL) with a starter culture of the organism (10 mL, LB broth) and incubated for 16 hours on an orbital shaker (150 rpm) at 37 °C. Cells were harvested by centrifugation and resuspended in glycyl-glycine buffer (0.1 M, pH 7.6). A crude extract was subsequently prepared by sonication followed by centrifugation to remove cell debris.

2.2 Relative reaction rate of acceptor aldehydes

The reaction mixture contained the following expressed as a final concentration: TPP (0.2 mM), $MgCl_2 \cdot 6H_2O$ (0.9 mM), aldehyde (100 mM), hydroxypyruvate (7.5 mM), glycyl-glycine buffer (70 mM, pH 7.6), TK (~3 Units crude extract) in a final volume of 1 mL. TK was added last to initiate the reaction. Aliquots (30 μL) were taken at 4 minute intervals, deproteinized immediately and assayed for residual hydroxypyruvate [6].

2.3 Preparative TK catalysed reactions

To a solution of $MgCl_2 \cdot 6H_2O$ (1 mg, 6 mmol), TPP (12 mg, 0.025 mmol), and TK (50 U) in glycyl-glycine buffer (0.5 mL, 0.1 M, pH 7.6) was added a solution of lithium hydroxypyruvate (105 mg, 1 mmol) and the aldehyde (3 mmol, 3 equivalents) in glycyl-glycine buffer (20 mL, 0.1 M, pH 7.6). The pH of the solution was then adjusted to pH 7.6 with NaOH (0.1 M). The reaction was allowed to proceed at 37 °C and monitored by t.l.c. (ethyl acetate-methanol = 95:5). After the reaction was complete, silica (~4 g) was added to the reaction mixture and the water removed under reduced pressure. Flash chromatography (ethyl acetate-petrol = 1:1) afforded the product.

a) Using propionaldehyde the reaction yielded (3S)-1,3-*dihydroxypentan*-2-*one* (27 mg, 23%); characterization data as peviously reported [5].

b) Using pyruvaldehyde the reaction yielded (3S)-1,3-*dihydroxypentan*-2,4-*one* (21 mg, 16%); characterization data as peviously reported [5].

Table 1
Relative rates (V_{rel}) of aldehydes in their reaction with hydroxypyruvate using crude extracts of *E. coli* JM107/pQR183.

Aldehyde (**1**)	V_{rel}	Aldehyde (**1**)	V_{rel}
	100		**41**
	82		**70**
	13		**4**
	24		**21**
	3		**~0**
	~0		**~0**

3. RESULTS AND DISCUSSION

The results of this initial investigation into the specificity of *E. coli* TK for the aldehyde acceptor are illustrated in Table 1. Rates are given relative to glycolaldehyde, the fastest reacting substrate we have found [5]. The low rate of DL-glyceraldehyde highlights the enzyme's specificity for α-hydroxyaldehydes of the D-configuration. This is further exhibited by the difference between the rates of glucose and mannose (which differ only in the stereochemistry at the α-hydroxycentre); the rate of both also being affected by the equilibrium between their chain and ring forms.

Cyclic aldehydes were not substrates for this form of TK presumably due to unfavourable interactions at the active site. It is interesting to note the similarity between this substrate specificity profile and those reported for yeast TK [4,7] which may indicate a similarity between the enzyme from the two different sources. The relative rate of reaction of the aldehyde does not have to be particularly high, however, for it to be used in preparative organic synthesis. This is demonstrated by the preparations described in the previous section, illustrating the high synthetic potential of TK.

This project ultimately aims to carry out the biotransformation on a pilot scale. The large-scale production of catalyst will be reported shortly. Current research is examining the nature of those TK reaction characteristics which pose particular challenges for the selection, design, and operation of an appropriate bioreactor for this biotransformation [8]. In particular, the production of large quantities of carbon dioxide will significantly influence the bioreactor design. Futhermore, in batch operation the pH of the medium will change through consumption of hydroxypyruvate and production of carbon dioxide (the solubility of which will depend on pH). Hence, the control of pH in the reactor presents an interesting challenge which we are currently addressing.

REFERENCES

1. H.G. Davies, R.H. Green, D.R. Kelly, & S.M. Roberts, Biotransformations in preparative organic chemistry, Academic Press, London, 1989.
2. E. Racker, In The Enzymes, P.D. Boyer, H. Lardy, & K. Myrzback (eds.), vol. 5, p. 397, Academic Press, New York, 1961.
3. P. Srere, J.R. Cooper, M. Tabacknick, &.E. Racker, Arch. Biochim. Biophy., 236, (1958), 624.
4. C. Demuynck, J. Bolte, L. Hecquet, & V. Dalmas, Tetrahedron Lett., 32, (1991), 4835.
5. G.R. Hobbs, M.D. Lilly, N.J. Turner, J.M. Ward, A.J. Willets, & J.M. Woodley, J. Chem. Soc. Perkin Trans. 1, 2, (1993), 165.
6. A.W. Holldorf, In Methods of Enzymatic Analysis, third edition, vol. 5, Bergmeyer, 1966, p578
7. Y. Kobori, D.C. Myles, & G.M. Whitesides, J. Org. Chem., 57, (1992), 5899.
8. J.M. Woodley & M.D. Lilly, In Applied Biocatalysis, J. Cabral, D. Best, L. Boross, & J. Tramper (eds.), Harwood Academic, in press.

ECB6: Proceedings of the 6th European Congress on Biotechnology
edited by L. Alberghina, L. Frontali, P. Sensi

Application of microbial transformation of prochiral 1,3-dichloro-2-propanol into (R)-3-chloro-1,2-propanediol

T. Nagasawa[a], T. Nakamura[b] and H. Yamada[c]

[a]Department of Food Science & Technology, Nagoya University, Furo-cho, Chikusa-ku, Nagoya 464, Japan

[b]Central Research Laboratory, Nitto Chemical Industry Co., Ltd. 10-1 Daikoku-cho, Tsurumi-ku, Yokohama 230, Japan

[c]Department of Biotechnology, Kansai University, Suita, Osaka 564, Japan

Chiral C3 compounds, e.g., optically active glycerol derivatives such as (R)-3-chloro-1,2-propanediol ((R)-MCP), have their fundamental structure as chiral compounds and are very important chiral building reagents. So far, it has been reported that (R)-MCP could be prepared from racemic MCP by optical resolution using bacteria or from D-mannitol through chemical synthesis. However, in the former case the conversion yield of (R)-MCP is less than 50% unless the residual substrate is racemized, and the procedure in the latter case is very complicated and not suitable for the large scale production.

1. Enzymes involved in enatioselective transformation of 1,3-dichloro-2-propanol to (R)-3-chloro-1,2-propanediol by Corynebacterium sp. strain N-1074

We have attempted microbial production of (R)-MCP from prochiral 1,3-dichloro-propanol (DCP) expecting its quantitative yield (Equation 1). Microorganisms which utilize MCP or epichlorohydrin (ECH) as a sole carbon source were isolated from soil samples by means of an enrichment culture technique, and then some strains with MCP-formation activity were obtained (1).

$$\underset{\substack{\text{1,3-Dichloro-2-propanol}\\ \text{DCP}}}{ClCH_2\text{-}\underset{|\atop OH}{CH}\text{-}CH_2Cl} \longrightarrow \longrightarrow \underset{\substack{\text{(R)-3-Chloro-1,2-propanediol}\\ \text{(R)-MCP}}}{ClCH_2\text{-}\underset{|\atop OH}{\overset{*}{C}H}\text{-}CH_2OH} \quad (1)$$

Using one of these strains, Corynebacterium sp. N-1074, we examined the production of

(R)-MCP from DCP by the resting cell reaction, however, the optical purity (about 70% e.e.) was unsatisfactory at the beginning of this study. To accomplish further high optical purity of (R)-MCP, at first, we carried out the enzymatic studies on the transformation of DCP into (R)-MCP, at first, we carried out the enzymatic studies on the transformation of DCP into (R)-MCP. When enzyme activities involved in the transformation were fractionated by HPLC on a DEAE column, two kinds of DCP-dechlorinating and two kinds of ECH-hydrolyzing activities were observed. This result indicated that MCP was formed from DCP via ECH.

We purified two enzymes (enzyme Ia and Ib) with DCP-dechlorinating activity. Enzyme Ia had a molecular mass of about 108 kDa and consisted of four subunits identical in molecular mass (about 28 kDa). Enzyme Ib was a protein of 115 kDa and composed of two different subunits in molecular mass (about 32 and 35 kDa). The specific activity of enzyme Ib for DCP was about 30-fold higher than that of enzyme Ia. Both enzymes catalyzed the transformation of several halohydrins into the corresponding epoxides with liberation of hydrogen halide into its reserve reaction (Equation 2), so we designated these enzymes "halohydrin hydrogen-halide-lya se".

$$R\text{-}CH(OH)\text{-}CH_2Cl \underset{+HCl}{\overset{-HCl}{\rightleftharpoons}} R\text{-}CH\text{—}CH_2 \ (\text{epoxide, }O) \quad (2)$$

Halohydrin hydrogen-halide-lyase

Two halohydrin hydrogen-halide-lyase genes from Corynebacterium sp. N-1074 were cloned and expressed in E. coli. For further studies on the enzymatic properties, both lyases were purified from the two recombinants (1), because an enormous amount of the enzymes was formed in each recombinant. The identification of their immunological properties and N-terminal amino acid sequences indicated that hydrogen-halide-lyase A and B of the recombinants were enzyme Ia and enzyme Ib of Corynebacterium sp. N-1074, respectively. Their enzymatic properties, e.g., subunit structure, substrate specificity, thermal stability, optimum pHs and kinetics of the reaction, were different each other. Enantioselectivity for the conversion of DCP into (R)-ECH was observed in the reaction with halohydrin hydrogen-halide-lyase B but not in that with lyase A (3). To our knowledge this is the first case of carbon-halogen cleavage enzyme which can recognize the prochiral stereospecificity.

Two kinds of ECH-hydrolyzing enzyme also seemed to be different with respect to the enantioselectivity for the conversion of ECH to MCP. Based on these results, it can be concluded that the optical purity of (R)-MCP formed from DCP may result from a conjugated system of at least these four enzymes involved in the transformation of DCP into MCP (Equation 3).

$$ClCH_2\text{-}CH(OH)\text{-}CH_2Cl \underset{+HCl}{\overset{-HCl}{\rightleftharpoons}} ClCH_2\text{-}\overset{*}{C}H\text{—}CH_2 \ (\text{epoxide, }O) \xrightarrow{H_2O} ClCH_2\text{-}\overset{*}{C}H(OH)\text{-}CH_2OH \quad (3)$$

Lyase I_A, I_B — ECH-hydrolyzing enzyme II_A, II_B

DCP — Epichlorohydrin ECH — (R)-MCP

2. A new catalytic function of halohydrin hydrogen-halide-lyase, synthesis of β–hydroxynitriles from epoxides and cyanide

Halohydrin hydrogen-halide-lyase catalyzes the transformation of various halohydrins into the correspodning epoxides with liberation of halide and its reverse reaction. Further studies on the catalytic properties of the enzyme revealed that the enzyme also catalyzes the synthesis of β-hydroxynitriles from epoxides and cyanide (Equation 4). Once formed β-hydroxynitriles were not attacted by the enzyme (4).

$$\underset{\text{(epoxide: R-CH—CH}_2\text{ bridged by O)}}{\text{R-CH}\overset{O}{—}\text{CH}_2} + H^+ + CN^- \longrightarrow \text{R-CH(OH)-}\overset{*}{C}H_2CN \quad (4)$$

Halohydrin hydrogen-halide-lyase

The effect of chloride on the formation of β-hydroxyvaleronitrile by the enzyme was examined. The presence of chloride in the reaction mixture caused by an inhibition for the formation of β-hydroxyvaleronitrile, and the inhibition was found to be competitive with cyanide. Cyanide also inhibited competitively the enzymatic formation of 1-chloro-2-butanol from 1,2-epoxybutane and chloride. These facts suggest that both reactions are catalyzed at the same active site of the enzyme although the affinity for each ion for the enzyme is fairly different. This suggests that reactions with epoxides and some other nucleophilic reagents also might be catalyzed by the enzyme. These reactions by the enzyme might follow S_N2 type nucleophilic substitution like a chemical reaction, but the difference in the affinity of the enzyme for chloride and cyanide strongly suggests that the interaction between the enzyme and these ions affects the reactions. The enantioselectivity for the formation of β-hydroxynitriles was observed with hydrogen-halide-lyase B but not with lyase A.

REFERENCES

1. T. Nakamura, F. Yu, W. Mizunashi and I. Watanabe, Agric.Biol. Chem., 55 (1991) 1931.
2. T. Nagasawa, T. Nakamura, F. Yu, I. Watanabe and H. Yamada, Appl. Microbiol. Microbiol. Biotechnol., 36(1992) 478.
3. T, Nakamura, T. Nagasawa, F. Yu and I. Watanabe, J. Bacteriol., 174 (1992) 7613.
4. T. Nakamura, T. Nagasawa, F. Yu and I. Watanabe, Biochem. Biophys. Res. Commun., 180 (1991) 124.

ECB6: Proceedings of the 6th European Congress on Biotechnology
edited by L. Alberghina, L. Frontali, P. Sensi

D-malate and D-citramalate production with maleate hydro-lyase from *Pseudomonas pseudoalcaligenes*

M.J. van der Werf[a], W.J.J. van den Tweel[b], J. Kamphuis[b], S. Hartmans[a] and J.A.M. de Bont[a]

[a] Division of Industrial Microbiology, Department of Food Science, Wageningen Agricultural University, P.O. Box 8129, 6700 EV Wageningen, The Netherlands

[b] Bio-organic Chemistry Section, DSM Research, P.O. Box 18, 6160 MD Geleen, The Netherlands.

1. INTRODUCTION

Biological formation of optically active synthons for the fine chemicals industry is one of the major application areas in biotechnology. Lyases (e.g. hydratases, ammonia-lyases) are very interesting enzymes in this respect. They do not require cofactor recycling (since addition does not involve a net reduction or oxidation), are quite stable, show an almost absolute stereospecificity/stereoselectivity and have a theoretical yield of 100% compared to only 50% for enantiomeric resolutions.

Lyases have already been applied on a commercial scale for some decades. L-Malate and L-aspartate are both produced from fumarate in a continuous process by immobilized cells of *Brevibacterium flavum* containing fumarase activity and immobilized cells of *Escherichia coli* containing aspartase activity respectively, by the Japanese firm Tanabe Seiyaku [1].

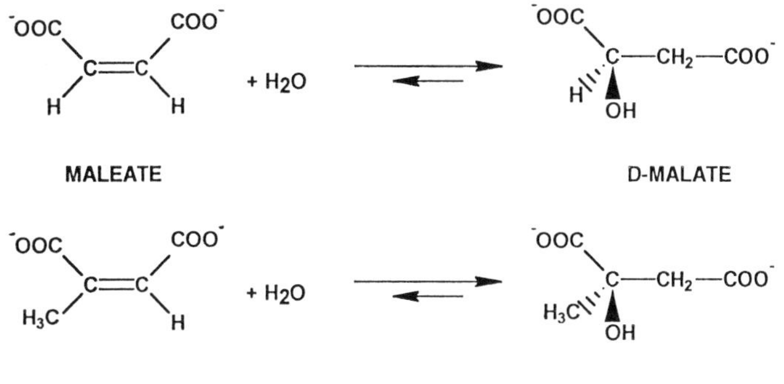

MALEATE + H_2O ⇌ D-MALATE

CITRACONATE + H_2O ⇌ D-CITRAMALATE

Figure 1. Reactions catalysed by malease from *Pseudomonas pseudoalcaligenes*

Other chiral α-hydroxy acids with a great commercial potential are D(*R*)-malate and D(*R*)-citramalate (Fig. 1). These hydroxy acids can be used as chiral synthons or as resolving agents in resolutions of racemic compounds. Hydration of maleate to D-malate [2] and of citraconate to D-citramalate [3] has been described. Here we present the results from an extensive study on the production of D-malate and D-citramalate from maleate and citraconate respectively, catalysed by malease from *Pseudomonas pseudoalcaligenes.*

2. RESULTS AND DISCUSSION

2.1 Screening

We have screened over 300 microorganisms for their ability to convert maleate into D-malate due to the action of maleate hydratase [4]. Fumarate accumulation was used as an indication for the enantiomeric composition of the malate formed, since fumarate is converted into L-malate by the citric acid cycle enzyme fumarase. Many strains (128) were found to be capable of converting maleate into D-malate with an enantiomeric purity of more than 97%. Malease containing *Pseudomonas pseudoalcaligenes* NCIMB 9867 was selected for more detailed studies.

2.2 Enzyme purification

Before malease was applied in the production of D-malate, it was purified from *Ps. pseudoalcaligenes* [5] to gain more biochemical knowledge about the enzyme. The enzyme consists of two subunits (24 and 57 kDa) and no cofactor is required for full enzymatic activity. This is in contrast to malease from an *Arthrobacter* sp. which also consists of two subunits but which needs a divalent metal ion and a sulphur compound for full enzymatic activity [6]. Malease from *Ps. pseudoalcaligenes* also has a broader substrate specificity: it also catalyses the hydration of citraconate at a considerable rate (56% of the maleate hydration activity). The K_m for maleate and citraconate are 0.35 and 0.2 mM respectively. The product of the maleate hydration reaction, D-malate, is not only a competitive inhibitor of malease (K_i = 0.63 mM) but it also stabilizes the enzyme.

2.3 Thermodynamics

With the thermodynamic constants ($\Delta G^{o\prime}$, $\Delta H^{o\prime}$ and $\Delta S^{o\prime}$) of a reaction the maximal obtainable yield and the influence of temperature on the maximal obtainable yield of the reaction can be calculated. Therefore these constants were determined for the maleate and citraconate hydration reactions catalysed by malease from *Ps. pseudoalcaligenes* (Table 1)[7]. The maleate and citraconate hydration reactions have quite a large negative $\Delta G^{o\prime}$'s which correspond to yields >99.9% (25°C). The fumarate hydration reaction catalysed by the commercialized enzyme fumarase has a much lower negative $\Delta G^{o\prime}$(Table 1) and consequently a much lower yield can be reached.

Table 1. $\Delta G^{o\prime}$, $\Delta H^{o\prime}$ and $\Delta S^{o\prime}$ for the maleate and citraconate hydration reactions and related hydration reactions (25°C, I=0.1).

reaction	$\Delta G^{o\prime}$ (kJ/mol)	$\Delta H^{o\prime}$ (kJ/mol)	$\Delta S^{o\prime}$ (J/mol*K)	Reference
$maleate^{2-} + H_2O \rightleftharpoons D\text{-}malate^{2-}$	-19.28	-18.07	2.94	
$citraconate^{2-} + H_2O \rightleftharpoons D\text{-}citramalate^{2-}$	-11.65	-22.55	-35.92	
$fumarate^{2-} + H_2O \rightleftharpoons L\text{-}malate^{2-}$	-3.69	-16.55	-43.15	[8]
$mesaconate^{2-} + H_2O \rightleftharpoons L\text{-}citramalate^{2-}$	-4.31	-	-	[9]

2.4 Optical purity of the produced hydroxy acids

The optical rotation of the hydroxy acids produced from maleate and citraconate were determined and compared with commercially available D-malate and D-citramalate (Table 2). Both malate and citramalate produced with *Ps. pseudoalcaligenes* have a negative rotation indicating that the D-enantiomer of the hydroxy acids was formed. Based on the optical rotation data (Table 2) the hydroxy acids produced with *Ps. pseudoalcaligenes* appeared to have a higher optical purity than the commercially available isomers. The optical purity of D-malate was determined enzymatically and was found to be 99.99% [4].

Table 2. Optical rotation of the produced hydroxy acids determined by the method of Krebs and Eggleston [10](20°C).

	$[\alpha]^T_D$
Product maleate	-111.8
Product citraconate	-133.5
D-malate (Janssen)	-111.6
D-citramalate (Aldrich)	-119.8

2.5 Optimization malease induction

Although *Ps. pseudoalcaligenes* was not able to grow on maleate, probably because it lacks a transport mechanism for maleate uptake [4], this microorganism contains high malease activity. The highest malease activity was observed in permeabilized cells grown on 3-hydroxybenzoate (2.0 $\mu mol*min^{-1}*(mg\ protein)^{-1}$). This might be explained by the fact that maleate is an intermediate in the degradation pathway of 3-hydroxybenzoate in this microorganism [2]. Malease activity of cells grown on 3-hydroxybenzoate was constant during the logarithmic phase, but dropped rapidly as soon as all the substrate was consumed. Consequently, cells need to be harvested before all the substrate is consumed, to obtain cells with maximal malease activity.

2.6 D-malate production

Cells of *Ps. pseudoalcaligenes* used to transform maleate into D-malate were permeabilized with 0.5% Triton X-100 to overcome transport problems of substrate and product over the celmembrane which could limit the maleate hydration rate. Permeabilized cells showed activity towards maleate at concentrations up to 2 M. In Fig. 2 the formation of D-malate from 800 mM maleate is followed in time. 105 g/l D-malate was formed in 15 hours with a yield in excess of 99% (see Fig. 2).

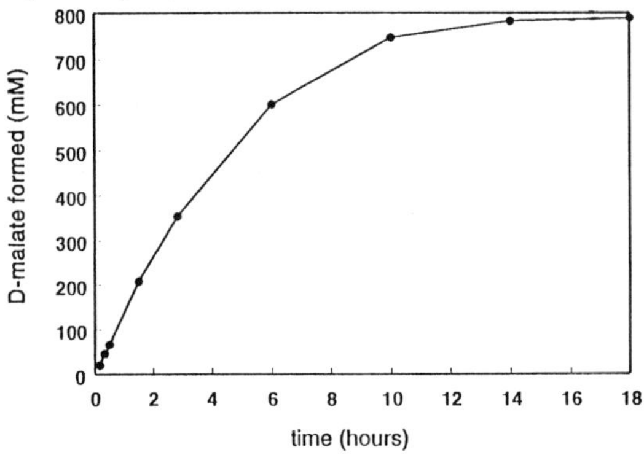

Figure 2. Production of D-malate from 800 mM maleate (2,3 mg/ml protein)

3. CONCLUSIONS

Based on the optical purity of the D-malate and D-citramalate formed, the favourable equilibrium of the reactions and the relatively high specific activity of malease from *Ps. pseudoalcaligenes* a process using this strain appears to be very promising. Further research will focus on the optimization of the reaction conditions for the production of D-malate and D-citramalate.

REFERENCES

1. Chibata, I. et al. (1992) *In*: Chirality in Industry (Collins, A.N. et al. eds) John Wiley & Sons Ltd. pp. 351-370.
2. Hopper, D.J. et al. (1968) Biochem. J. 110: 798-800.
3. Subramanian, S.S. and Raghavendra Rao, M.R. (1968) J. Biol. Chem. 243, 2367-2372
4. Werf, M.J. van der et al. (1992) Appl. Environ. Microbiol. 58, 2854-2860.
5. Werf, M.J. van der et al. (1993) Submitted for publication.
6. Yamagata, H., Personal communication.
7. Werf, M.J. van der et al. (1993) To be submitted for publication.
8. Bock, R.M. and Alberty, R.A. (1953) J. Am. Chem. Soc. 75, 1921-1925.
9. Buckel, W. and Miller, S.L. (1987) Eur. J. Biochem. 164, 565-569.
10. Krebs, H.A. and Eggleston, L.V. (1943) Biochem. J. 37, 334-338.

PHYSIOLOGICAL AND ENGINEERING PROBLEMS LINKED TO IMMOBILIZED CELL FERMENTATIONS

ECB6: Proceedings of the 6th European Congress on Biotechnology
edited by L. Alberghina, L. Frontali, P. Sensi

IMMOBILIZED VIABLE CELL FERMENTATIONS AS APPLIED TO MICROBIAL ENZYME PRODUCTION - PROBLEMS AND PROMISES

S. Linko, R. Haapala and P. Linko

Helsinki University of Technology, Department of Chemical Engineering, Laboratory of Biotechnology and Food Engineering, SF-02150 Espoo, Finland

1. INTRODUCTION

After immobilized enzymes had already been well established with a number of industrial applications,[1] the potential of immobilized viable cells in fermentations was discussed in detail at the Third European Congress on Biotechnology in Münich in 1984, nearly 10 years ago.[2] That microorganisms are often attached to surfaces in nature was well known, and the principle had for long been applied in the 'Orleans' process for vinegar production, in microbial leaching, and in waste water treatment. Consequently, it was not surprising that microorganisms could be actively immobilized on or within various carrier materials, and maintained either in resting or growing state in bioreactors of varying design. A large number of papers on immobilized viable cells have since been published, with most work on ethanol[3] and lactic acid[4] fermentations. Immobilized brewer's yeast has recently been introduced to industrial scale beer production.[5]

Nevertheless, discussion on the viability of immobilized cells has continued until recently. Harder[6] pointed out the lack and importance of knowledge of physiology of immobilized cells, which led to a special symposium on the subject organized in 1989 in Wageningen by the EFB Working Parties of Applied Biocatalysis and Microbial Physiology.[7] For the first time cell physiology was comprehensively discussed in the context of immobilized animal, microbial and plant cells by experts of various disciplines. Inasmuch as the given constraints do not allow a complete review of the various aspects of immobilized viable cells the paper is limited to the production of microbial enzymes as a case study.

2. IMMOBILIZED VIABLE CELLS

2.1. The rationale

Cleraly immobilized viable cell systems should present techno-economic advantages over conventional fermentations in order to be of commercial interest. Cells may be passively attached on carrier surfaces with relative ease, and a number of either inexpensive or regenerable carrier materials are available. The techno-economic viability of active immobilization within a gel matrix has also been demonstrated in a large scale.[3] And inasmuch as the immobilization allows very high biomass densities, the result is considerable savings in capital costs and space. So far so good, but there is more. Immobilization protects fragile cells against shear forces and other environmental hazards. This is most evident with animal and plant cells, but fungal mycelia have also been shown to be sensitive against high agitation rates. With viable cells the total multienzyme system complte with coenzyme regeneration mechanisms is available. At a controlled resting state immobilized viable cells can thus provide a very efficient, stable biocatalyst system. And just as with immobilized enzymes, continuous or repeated batch processing becomes practical. When a continuous viable yeast bioreactor was applied to commercial scale beer brewing the time required for the secondary fermentation could be reduced from weeks to hours! And this without loss in quality. When one still adds the ease of controlling an immobilized viable cell bioreactor, are there any doubts left?

2.2. The problems

The techno-economic situation has to be always evaluated case by case. After all, the immobilization is an additional, often not an inexpensive step. The bioreactor configuration may or may not require special considerations when immobilized viable cells are concerned. The methods of immobilization of viable cells are usually limited to relatively gentle surface attachment and entrapment in order to preserve viability. However, in some cases a partial loss of viability may even be an advantage as only the most active and resistant cells are retained. An increased diffusion resistance may be a problem but usually less so than with immobilized enzymes, because the cells can be grown on or near the biocatalyst surface. Unwanted byproducts such as indesirable enzymes may be a real problem and difficult to deal with, but similar problems may be encountered also with conventional fermentations. In the case of food and beverage applications in particular, the methods are further limited to food grade GRAS materials. Possible effects of novel technologies such as the use of immobilized viable cells on the quality of the final product is also always a real concern.

3. ENZYME PRODUCTION

With the increasing importance of enzymes as industrial biocatalysts considerable efforts have been directed towards improved and novel enzyme production methods.[8] Screening, mutation, and genetic engineering have become invaluable in microbial strain development for enzyme production. However, most microbial enzymes are still produced by conventional submerged batch and, sometimes, solid state surface cultures. Immobilized biocatalyst technology has only recently been applied in production of extracellular enzymes, taking advantage of the ability of microorganisms to secrete extracellular enzymes while growing attached on surfaces.[8,9]

3.1. Amylolytic enzymes

About 15 years ago Suzuki and coworkers[10] of Tokyo Institute of Technology were the first to demonstrate that a bacterium, *Bacillus subtilis* excretes α-amylase while entrapped within polyacrylamide gel matrix. Shinmyo et al.[11] investigated the physiology of κ-carrageenan entrapped *B. amyloliquefaciens* when used in α-amylase production. Oxygen diffusion was found to be the growth limiting factor, and α-amylase was excreted after the completion of the exponential growth phase. More recently Chevalier and Noüe[12] were able to improve oxygen transport in α-amylase production with κ-carrageenan entrapped *B. subtilis*. Li et al.[13] showed in our laboratory that calcium alginate gel entrapped mutant strain of the fungus *Aspergillus niger* produces efficiently both extracellular α-amylase and glucoamylase. When fungal spores used in the immobilization are germinated within the alginate gel, mycelium is grown as a thin layer near the biocatlyst bead surface, and diffusional limitations were thus minimized. Figure 1 shows a relatively stable enzyme production in at least 11 about 4-day batches using the same biocatalyst. The number of batches could be extended to at least 20 without difficulty.[14] The presence of calcium and magnesium ions in the medium further increased the α-amylase activity.

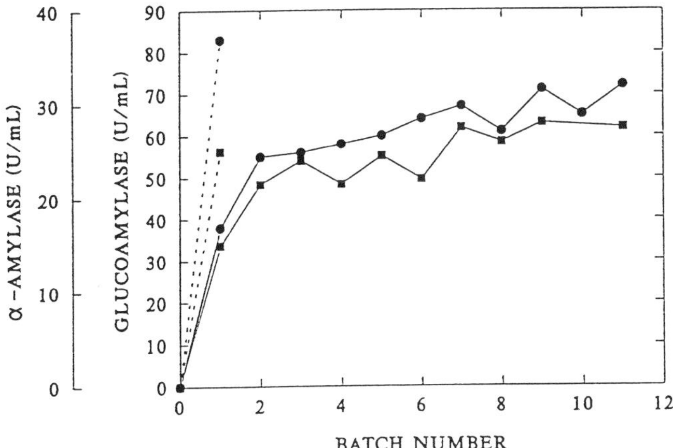

Figure 1. α-Amylase (■) and glucoamylase (●) production by *Aspergillus niger* in repeated 4-day batch fermentations (immobilized fungus —; free mycelium ...). Enzyme production was carried out in 250 ml Erlenmeyer flasks on a starch based-medium[13] of pH 6.0 at 28°C and 175 rev/min agitation. Values are averages of three independently performed fermentations.

3.2. Cellulolytic and hemicellulolytic enzymes

Frein et al.[15] demonstrated that extracellular cellulolytic enzymes can be produced by κ-carrageenan entrapped *Trichoderma reesei*. Türker and Mavituna,[16] and Kumakura et al.[17] have further investigated the production of cellulolytic enzymes employing *T. reesei* immobilized on various carriers. Türker and Mavituna[16] used reticulated polyurethane foam matrix with similar results than was obtained by using free mycelium. Polyurethane foam was used as carrier also in the present work. Table 1 shows typical endoglucanase, xylanase and filter paper hydrolyzing activities obtained when polyurethane foams of varying densities were used. The filter paper hydrolyzing activities obtained were of the same order of magnitude or higher than those reported by Türker and Mavituna[16], and Kumakura et al.[17]. In the present work marked xylanase activities were also obtained with the immobilized fungus. This is to our knowledge the first time xylanase has been reported to be produced by an immobilized biocatalyst system. The polyurethane foam of the lowest density as carrier resulted in the highest enzyme activities.

Trichoderma reesei is known to excrete only small quantities of ß-glucosidase, an enzyme necessary for the complete hydrolysis of cellulose. Linko et al.[18] showed that immobilized *Penicillium funiculosum* excreted high quantities of ß-D-glucosidase while the culture medium also exhibited satisfactory filter paper hydrolyzing activity. Figure 2 illustrates that the hydrophobic polyurethane gel entrapped fungus was a better enzyme producer than the free mycelium.

Table 1
The effect of carrier porosity on endo-1,4-ß-glucanase, xylanase and filter paper activity by polyurethane foam immobilized *Trichoderma reesei*[a]

Time (d)	Endoglucanase[b] (ECU/ml)			Xylanase[c] (BXU/ml)			FPA[d] (IU/ml)		
	PU-16	PU-23	PU-30	PU-16	PU-23	PU-30	PU-16	PU-23	PU-30[e]
3	151	160	46	740	708	581	0.9	0.9	0
5	231	211	87	993	960	563	1.6	1.1	0.6
7	360	287	130	1710	925	790	2.4	1.2	2.2
8	350	341	134	1962	1752	966	nd	2.5	1.8

[a] Fermentations were carried out in 500 ml Erlenmeyer flasks containing the carrier as a form of cubes ($7 \times 6 \times 4$ mm^3) and 100 ml of a synthetic medium (initial pH 5.3) with 27 g/l of lactose and 3 g/l of Solka Floc (BW-40) as the carbon source. During the enzyme production the temperature was kept at 27°C and the shaking speed was 150 rev/min. Values given are averages of two parallel fermentations.

[b] Endo-1,4-ß-glucanase activity was assayed at 50°C as described by Bailey and Nevalainen[19] by measuring the release of reducing sugars from hydroxyethylcellulose in 50 mM sodium citrate buffer, pH 4.8.

[c] Xylanase activity was assayed at 50°C using 1% birch xylan (Roth) as substrate in 50 mM sodium citrate buffer, pH 5.3.[20]

[d] Overall cellulolytic activity was determined as filter-paper units in 50 mM sodium citrate buffer, pH 4.8, as described by Ghose.[21]

[e] Porosity of the polyurethane foams (Espe Oy, Kouvola, Finland) used as a carrier are given as densities: PU-16 16 g/l; PU-23 23 g/l; PU-30 30 g/l.

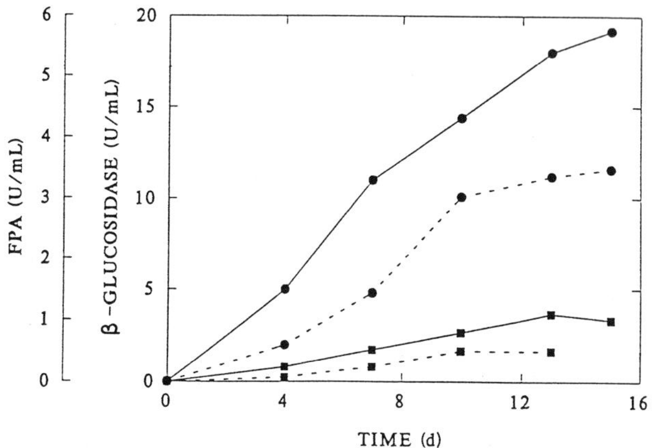

Figure 2. ß-Glucosidase (●) and filter paper activity (■) as the function of time by hydrophobic polyurethane gel cube immobilized *Penicillium funiculosum* (—) and by free *P. funiculosum* cells (- - -). PU-3 entrapped spores were germinated and grown in a synthetic 2% glycerol medium. The enzyme production medium contained 2.5% Avicel SF and 0.75% Bacto Casitone in place of glycerol.

3.3. Ligninolytic enzymes

The production of lignin peroxidase (ligninase) by surface attached white rot fungus *Phanerochaete chrysosporium* is an interesting case. Lignin peroxidase was first reported only 10 years ago, with great difficulties in large scale production by conventional fermentation techniques.[22] First the discovery that high lignin peroxidase activities can be obtained by immobilized *P. chrysosporium* opened up the way to industrial scale production. Table 2 lists some milestones in lignin peroxidase production by the immobilized fungus, from the first laboratory scale trials reported in 1986[23,24] through the first

Table 2
Examples of immobilized biocatalyst systems for lignin peroxidase production by white rot fungus *Phanerochaete chrysosporium*

Year	Carrier	Activity (U/l)	Remarks
1986[23]	Agarose gel beads	245 (23°C)	Erlenmeyer flask cultivation, repeated batch
1986[24]	Poly(methylmethacrylate)	130 (37°C)	Rotating disc reactor, 2.5 l working volume
1987[25]	Nylon-web cubes	370 (23°C)	Erlenmeyer flask culture
1987[26]	Polyurethane foam cubes	100	Erlenmeyer flask culture, repeated batch
1987[27]	Silicon tubing	232 (37°C)	Stirred tank reactor, 1 l working volume
1988[28]	Nylon-web cubes	730 (23°C)	Conventional tank reactor, mixing with air and/or oxygen flow, 8 l working volume, semicontinuous
1990[29]	Alumina spheres	183 (37°C)	Erlenmeyer flask culture, repeated batch
1990[30]	Polyurethane foam cubes	950 (30°C)	Erlenmeyer flask culture
1990[31]	Polyurethane foam cubes	780 (23°C)	Airlift reactor, 8 l working volume
1990[32]	Nylon-web sheet	1035 (23°C)	Stirred tank reactor, 8 l working volume, temperature decreased from 37°C to 30°C for enzyme production phase
1990[33]	Sintered glass Raschig® rings	2200 (37°C)	Fixed bed loop reactor, 40 l working volume, 39°C→30°C
1991[34]	Nylon-web sheet	432 (37°C)	Stirred tank reactor, 800 l working volume

successful 8-liter bench scale trials[28] to the first industrial scale production in 1991[34] by nylon-web attached *P. chrysosporium*. The industrial production was carried out in a fed batch semicontinuous mode by employing the nylon-web immobilized viable biocatalyst system developed in our laboratory, separating the produced lignin peroxidase by an ultrafiltration loop and returning the permeate to the bioreactor. It should be noted that Table 2 gives the maximum ligninase activities as reported in the literature, and that the temperature of the determination markedly affects the activity obtained.[35]

An immobilized biocatalyst system is particularly suitable for repeated batch enzyme production. Table 3 demonstrates that high lignin peroxidase activities can be repeatedly obtained with the same immobilized biocatalyst in succesive batch fermentations. Similar

Table 3
Lignin peroxidase production by nylon-web cube immobilized *Phanerochaete chrysosporium* in repeated batch shake cultures[a]

Batch number	1	2	3	4	5	6	7
Maximum ligninase activity[b] (U/l)	499	683	576	576	750	650	480
Time[c] (d)	1.8	1.6	1.6	1.0	3.4	1.5	1.8

[a] Fermentations were carried out under carbon limitation in 2-liter Erlenmeyer flasks with 400 ml of the medium and about 120 g (wet weight) of the biocatalyst. The temperature was 37°C and shaking speed 60 rev/min (rotational).

[b] Lignin peroxidase activity was assayed essentially according to the method of Tien and Kirk[36] using 0.4 mM veratryl alcohol in 100 mM sodium tartrate buffer of pH 3.0 as substrate at 23°C.

[c] Time for the maximum activity from the beginning of a given batch.

results have also been obtained with stirred 8 liter working volume immobilized fungus cultures.[32] The biocatalyst retains its lignin peroxidase production ability for extended periods when stored at +4°C which further illustrates the stability.

4. EPILOGUE

In a little more than a time span of 10 years immobilized biocatalyst systems have become an interesting alternative to the conventional production of a number of enzymes. Immobilized biocatalyst technology made the first reported industrial scale production of lignin peroxidase possible, in quantities sufficient for field trials in the forest industry. An other enzyme of recent interest to the forest industry, apparently benefiting from fungal immobiliztion is xylanase. The production of amylolytic and cellulolytic enzymes may also benefit from the immobilization of the production organism. The overview given is far from complete. Among other enzymes reported to have been produced by immobilized viable cells are protease[37] and lipase.[38-40] It is likely that a number of important enzymes will be commercially produced in the near future employing immobilized viable cell technology.

REFERENCES

1. P. Linko and Y.-Y. Linko, CRC Crit. Rev. Biotechnol., 1 (1984) 289.
2. P. Linko, Proc. Third European Congress on Biotechnology, 4 (1985) 357.
3. M. Nagashima, M. Azuma and S. Noguchi, Ann. New York Acad. Sci., 413 (1983) 457.
4. P. Linko, In Enzymes and Immobilized Cells in Biotechnology, A.I. Laskin (ed.), The Benjamin/Cummings Publishing Company, Menlo Park, California (1985) 25.
5. E. Pajunen, A. Grönqvist and B. Ranta, Proc Eur. Brew. Conv. 23rd Congress Lisbon, Oxford University Press, Oxford (1991) 361.
6. W. Harder, Proc. Fourth European Congress on Biotechnology, 4 (1987) 109.

7. J.A.M. De Bont, J. Vissert, B. Mattiasson and J. Tramper (eds.), Physiology of Immobilized Cells, Elsevier Science Publishers, Amsterdam, 1990.
8. S. Linko, Food Biotechnol., 3 (1989) 31.
9. Y.-Y. Linko, G.-X. Li, L.-C. Zhong, S. Linko and P. Linko, Methods Enzymol., 137 (1988) 686.
10. T. Kokubu, I. Karube and S. Suzuki, Eur. J. Appl. Microbiol. Biotechnol., 5 (1978) 233.
11. A. Shinmyo, H. Kimura and H. Okada, Eur. J. Appl. Microbiol. Biotechnol. 14 (1982) 7.
12. P. Chevalier and J. de la Noüe, Enzyme Microb. Technol., 9 (1987) 53.
13. G.-X. Li, Y.-Y. Linko and P. Linko, Biotechnol. Lett., 6 (1984) 645.
14. P. Linko, Y.-Y. Linko and G.-X. Li, Kemia-Kemi, 12 (1985) 203 (abstract in English).
15. E.M. Frein, B.S. Montenecourt and D.E. Eveleigh, Biotechnol. Lett., 4 (1982) 287.
16. M. Türker and F. Mavituna, Enzyme Microb. Technol., 9 (1987) 739.
17. M. Kumakura, M. Tamada, N. Kasai, I. Kaetsu and S. Kanno, Biotechnol. Bioeng., 33 (1989) 1358
18. Y.-Y. Linko, G.-X. Li and P. Linko, Ann. New York Acad. Sci., 501 (1987) 381.
19. M.J. Bailey and K.M.H. Nevalainen, Enzyme Microb. Technol., 3 (1981) 153.
20. M.J. Bailey and K. Poutanen, Appl. Microbiol. Biotechnol., 30 (1989) 5.
21. T.K. Ghose, Pure Appl. Chem., 59 (1987) 257.
22. S. Linko, Biotechnol. Adv., 10 (1992) 191.
23. Y.-Y. Linko, M. Leisola, N. Lindholm, J. Troller, P. Linko and A. Fiechter, J. Biotechnol., 4 (1986) 283.
24. T.K. Kirk, S. Croan, M. Tien, K.E. Murtagh and R.L. Farrell, Enzyme Microb. Technol., 8 (1986) 27.
25. S. Linko and L.-C. Zhong, Biotechnol. Techn., 1 (1987) 251.
26. N. Kirkpatrick and J.M. Palmer, Appl. Microbiol. Biotechnol., 27 (1987) 129.
27. H. Willershausen, A. Jäger and H. Graf, J. Biotechnol., 6 (1987) 239.
28. S. Linko, Enzyme Microb. Technol., 10 (1988) 410.
29. K.L. Cornwell, M.-F. Tinland-Butez, P.J. Tardone, I. Cabasso and K.E. Hammel, Enzyme Microb. Technol., 12 (1990) 916.
30. D. Ma, T. Hattori, M. Shimada and T. Higuchi, Wood Research, No 77 (1990) 35.
31. P. Bonnarme and T.W. Jeffries, J. Ferm. Bioeng., 70 (1990) 158.
32. S. Linko and M. Hujanen, Ann. New York Acad. Sci., 613 (1990) 764.
33. A.G. Jäger and C. Wandrey, In Physiology of Immobilized Cells, J.A.M. debont, J. Vissert, B. Mattiasson and J. Tramper (eds.), Elsevier Science Publishers, Amsterdam (1990) 433.
34. K. Polvinen, P. Lehtonen, M. Leisola and K. Visuri, ACS Symp. Ser. 460 (1991) 225.
35. S. Linko and R. Haapala, Biotechnol. Techn., 7 (1993) 75.
36. M. Tien and T.K Kirk, Proc. Natl. Acand. Sci. U.S.A., 81 (1984) 2280.
37. T. Kokubu and S. Suzuki, Biotechnol. Bioeng., 28 (1981) 29.
38. P. Ferrer and C. Solà, Appl. Microbiol. Biotechnol., 39 (1992) 737.
39. T. Nakashima, H. Fukuda, S. Kyotani and H. Morikawa, J. Ferment. Technol., 66 (1988) 441.
40. T. Nakashima and H. Fukuda, J. Ferment. Bioeng., 70 (1990) 355.

ECB6: Proceedings of the 6th European Congress on Biotechnology
edited by L. Alberghina, L. Frontali, P. Sensi

Semi-continuous Production of the Enzymes by A Novel Solid-State Fermentation System using Urethan Foam as Carriers

S. Ozawa*, K. Amada*, K. Sato*, T. Nagamune and I. Endo

Chemical Engineering Laboratory, RIKEN(Institute of Physical and Chemical research), Hirosawa, 2-1, Wako-shi, Saitama, 351-01, Japan

*Department of Industrial Chemistry, Faculty of Engineering, Science University of Tokyo, Kagurazaka, 1-3, Shinjuku-ku, Tokyo, 160, Japan

1. INTRODUCTION

Conventional solid-state fermentations of fungi are usually carried out to produce some enzymes using agricultural products like wheat bran as solid medium. However, the conventional fermentations have several problems concerning quality control of solid medium and scale up of processes[1]. Recently we have been developing a novel solid-state fermentation method which can produce enzymes more efficient than the conventional one. In this method, uses a solid-state medium made of urethan foam carrier(UFC) impregnated with the liquid medium. The produced enzymes could be simply recovered by pressing the UFC. Furthermore, the growing fungi on UFC could be reused to produce enzymes again by adding fresh liquid medium. Repeating these two operations, semi-continuous production of enzymes was possible. As previously reported, we have found that the productivity of glucoamylase or alkaline protease in batch cultivation of *Aspergillus oryzae* by this method was remarkably higher than that by the conventional solid-state fermentation[2][3]. In this paper, we have investigated the effects of both volume and composition of exchange medium on the semi-continuous productions of glucoamylase and alkaline protease.

2. MATERIALS AND METHODS

Aspergillus oryzae B-3 and *Aspergillus oryzae* 460 (ATCC 20386) were used as test organisms for the productions of glucoamylase and alkaline protease, respectively.

Hydrophilic poly urethan foam was used as carrier of liquid medium, which is made from prepolymer, isocyanate and polypeptide (Nitto Denko Co., Ltd., 5 mm in side length). The pore size was 0.2 to 0.5 mm. The apparent density of the carrier was about 0.04 g/mL and the porosity was about 0.95.

Liquid medium for production of glucoamylase was composed of: soluble starch 120 g/L; yeast extract 120 g/L; K_2HPO_4; 30 g/L; $MgSO_4 \cdot 7H_2O$, 3 g/L and initial pH was adjusted at 6.5. The UFC was impregnated with 0.32 L/L-bulk volume of the medium. For alkaline protease production, liquid medium was composed of: soluble starch 75 g/L; polypepton 75 g/L; K_2HPO_4; 30 g/L; $MgSO_4 \cdot 7H_2O$, 3 g/L and initial pH was adjusted at 3.5. The UFC was impregnated with 0.30 L/L-bulk volume of this medium. After the impregnation of liquid medium to UFC, these semi-solid media were sterilized at 121 °C for 30 min.

A medium for seed culture was prepared as follows: after adding both 10 g of powdered wheat bran and 100 mL of distilled water into a 500-mL conical flask, initial pH was adjusted at 4.5 with 2 N HCl, then it was sterilized at 121 °C for 20 min. Seed culture was started by inoculating three platinum loops spores of each strains and was carried out for 2 days at 30 °C on a rotary shaker at 250 rpm. The seed culture broth of 1.7 V/V% was inoculated to a 300-mL of conical flask containing 2 g of UFC impregnated with liquid medium, whose bulk volume was about 120 mL. Temperature was maintained at 30 °C. After 36 or 48 hr cultivation, a semi-continuous production of enzyme was initiated, and the culture broth was partially replaced by soluble starch solution every 12 hr. In the case of the semi-continuous production of glucoamylase, the exchange volume of culture medium(V) and the concentration of soluble starch in the exchange medium(Cs) were varied from 0.025 to 0.067 L/L-bulk volume · dose and from 80 to 140 g/L, respectively. In the case of the alkaline protease, V and Cs were changed from 0.025 to 0.058 L/L-bulk volume · dose and from 80 to 120 g/L, respectively. When an effect of polypepton was studied, this component was added to the exchange medium with the concentration(Cp) of 10, 20 or 30 g/L, while both V and Cs were fixed at 0.025 L/L-bulk volume · dose and 120 g/L, respectively.

Culture broth containing enzymes was directly obtained by squeezing the UFC. Glucoamylase activity in the harvested broth was directly assayed, adding 10 mL of a 5 g/L soluble starch solution to 1 mL of the culture broth. This enzymatic reaction was carried out in 1 N sodium buffer at pH 5 for 30 min at 40 °C. The residual starch was measured by iodometry. One unit of glucoamylase activity was defined as the amount of enzyme which produces 1 μmol glucose from soluble starch per minute at 40 °C. In the case of the measurement of alkaline protease activity, 1 mL of the culture broth was added to 5 mL of 6 g/L casein solution. The reaction was carried out at pH 10.5(42 mM sodium carbonate buffer), for 10 min at 37 °C. Produced amino acids were measured using the Lowry method. One unit of alkaline protease activity was defined as the amount of enzyme which produces 1 μg of amino acids from casein per minute at 37 °C.

Sugar concentration was determined by the Bertrand method. Concentration of glucosamine, contained in cell wall of fungi, was measured by the Elson-Morgan reaction to estimate cell mass concentration.

3.RESULTS AND DISCUSSION

3.1. Semi-continuous production of glucoamylase

Figure 1 shows time course of semi-continuous production of glucoamylase. In this

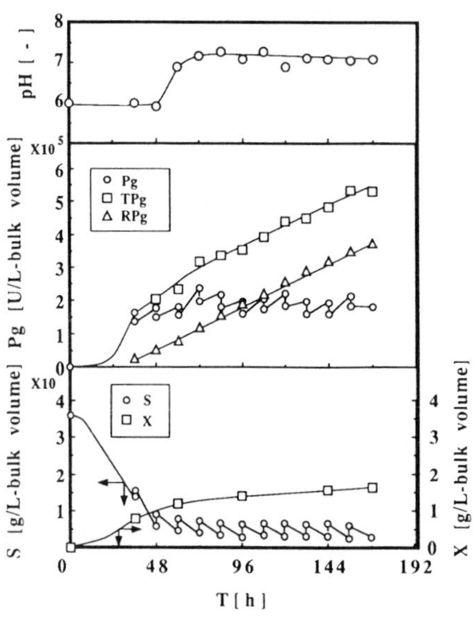

Figure 1. Time course of the semi-continuous production of glucoamylase.

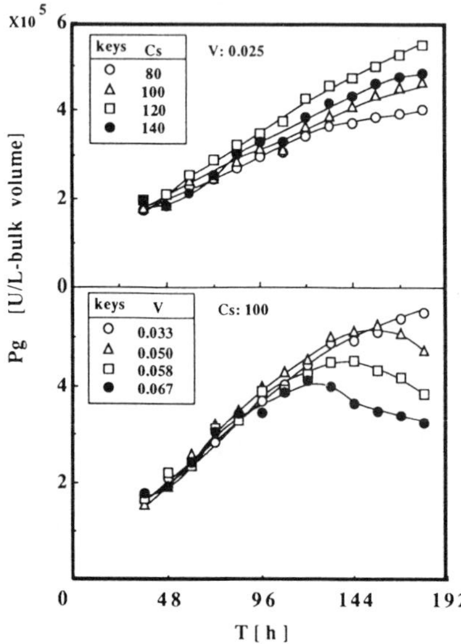

Figure 2. Effects of V and Cs on the semi-continuous production of glucoamylase.

cultivation, V and Cs were set at 0.041 L/L-bulk volume · dose and 100 g/L, respectively. The concentration of sugar(S) was maintained at approximately 5 g/L-bulk volume by adding soluble starch solution every 12 hr. pH of the culture broth showed almost constant value of 7.0 as long as glucoamylase was produced. Each time when medium was exchanged, Pg firstly decreased as a consequence of the enzyme harvest, then gradually rose up to the initial level. It followed that total activity of glucoamylase produced(TPg) and activity of glucoamylase recovered(RPg) increased linearly during the fermentation.

We investigated the effects of V and Cs on the semi-continuous production of glucoamylase by varying V and Cs from 0.025 to 0.067 L/L-bulk volume · dose and from 80 to 140 g/L, respectively(Fig. 2). From this figure, it appears that Cs controls a production rate of enzyme, while V strongly effects on a production period. It was suggested that the early ceasing of glucoamylase production occurring under high value conditions of V is due to the dilution effect on medium components except soluble starch. The maximum productivity of glucoamylase by the semi-continuous process was obtained when V is 0.025 L/L-bulk volume dose and Cs is 120 g/L. Under these culture conditions, 520,000 U/L-bulk volume activity of glucoamylase was obtained after 168 hr cultivation.

3.2. Semi-continuous production of alkaline protease

Figure 3 shows time course of semi-continuous production of alkaline protease. In this cultivation, V and Cs were adjusted at 0.041 L/L-bulk volume · dose and 75 g/L, respectively. Until cultivation time 96 hr, all variables show the same tendency as those of the semi-continuous production of glucoamylase. After 96 hr, pH of the culture broth , Pg and X began to decrease. It was considered that these decrease reflected a lowering of a physiological activity of fungi and it started when polypepton concentration decreased considerably due to a dilution effect of semi-continuous medium exchange.

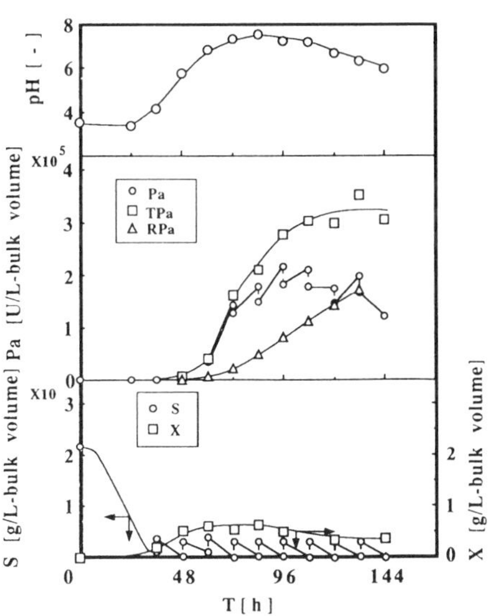

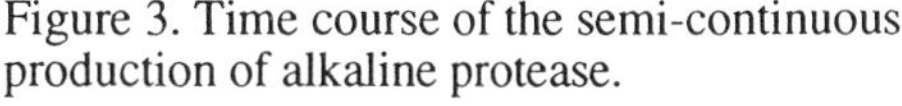

Figure 3. Time course of the semi-continuous production of alkaline protease.

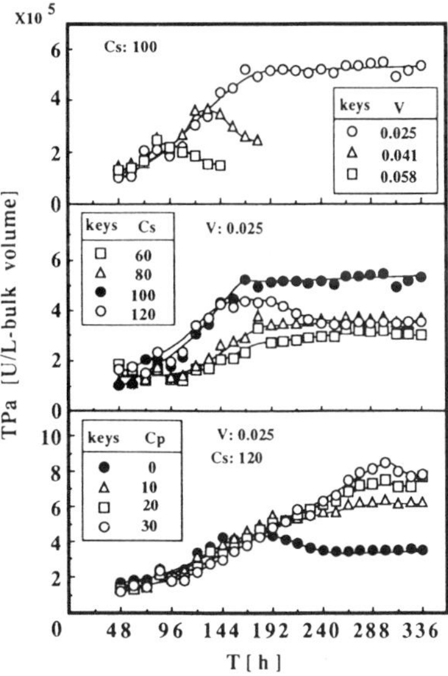

Figure 4. Effects of V and Cs on the semi-continuous production of alkaline protease.

The effects of V and Cs on the semi-continuous production of alkaline protease were investigated by changing V and Cs from 0.025 to 0.058 L/L-bulk volume · dose and from 80 to 120 g/L, respectively. Moreover, in order to ascertain the effect of polypepton dilution on the physiological activity of fungi. 10,20 or 30 g/L of polypepton was added to the exchange medium. In these experiments, V and Cs were fixed at 0.025 L/L-bulk volume · dose and 120 g/L, respectively. Figure 4 shows the time courses of TPa. Tendency of the effects of V and Cs on the semi-continuous production of alkaline protease was almost similar to those obtained in the semi-continuous production of glucoamylase. It was observed that the production activity of alkaline protease were prolonged to 300 hr by keeping concentration of polypepton in the culture broth at a certain level. It was found that the maximum productivity of alkaline protease by the semi-continuous process was obtained when V and Cs were set at 0.025 L/L-bulk volume dose and 100 g/L, respectively. Under these culture conditions, 520,000 U/L-bulk volume activity of alkaline protease was obtained after 168 hr cultivation.

4. CONCLUSION

In the novel solid-state fermentation system using UFC with liquid medium as solid-state medium, enzymes can be produced semi-continuously by exchanging the culture broth with fresh liquid medium periodically. The results are summarized as follows:

i)the maximum productivity of glucoamylase was obtained under the culture conditions where V was 0.025 L/L-bulk volume · dose and Cs was 120 g/L. It was observed that the production of glucoamylase stopped earlier with increment of V.

ii)the maximum productivity of alkaline protease was obtained when the cultivation was carried out under the conditions; V was 0.025 L/L-bulk volume · dose, Cs was 100 g/L. It was elucidated that the production period of alkaline protease was prolonged by addition of polypepton to the exchange medium.

NOMENCLATURE

Cp: concentration of polypepton in exchange medium	[g/L]
Cs: concentration of soluble starch in exchange medium	[g/L]
P: total activity of enzyme in unit bulk volume	[U/L-bulk volume]
RP: activity of enzyme recovered per unit bulk volume	[U/L-bulk volume]
S: amount of sugar in unit bulk volume	[g/L-bulk volume]
TP: total activity of enzyme produced per unit bulk volume	[U/L-bulk volume]
V: exchange volume of medium during repeated batch operation	[L/L-bulk volume · dose]
X: amount of glucosamine in unit bulk volume	[g/L-bulk volume]

<Subscripts>

a: semi-continuous production of alkaline protease

g: semi-continuous production of glucoamylase

REFERENCES

1. R. Madgett (1986) Solid-state fermentation. In: A. Demain, N. Solomon (eds.) Manual of industrial microbiology and biotechnology. American society for microbiology. Washington, D.C. pp66-83
2. T. Kobayashi, S. Ozawa, K. Sato, T. Nagamune, I. Endo: Production of glucoamylase by solid-state fermentation using urethane foam as a semi-solid medium, Kagaku Kogaku Ronbunshu, 17 (1991) 491-496
3. S. Ozawa, K. Amada, K. Sato, T. Kobayashi, T. Nagamune, I. Endo: Production of alkaline protease by semi-solid-state fermentation using urethan foam as carrier, Proceeding of Asia-Pacific Biochemical Engineering Conference 1992, (1992) pp405-408

ECB6: Proceedings of the 6th European Congress on Biotechnology
edited by L. Alberghina, L. Frontali, P. Sensi

Specific ATP Concentrations in Freely Suspended and Immobilized Cells Degrading 3,4 Dichloroaniline

P. Gikas and A.G. Livingston

Department of Chemical Engineering, Imperial College, London SW7 2BY, UK

ABSTRACT

Dry biomass is commonly used to measure the quantity of active biomass in microbial systems. However in retained cell systems a significant fraction of the cells could be dead or inactive. In these experiments the specific concentration of ATP [mg(ATP) g^{-1}(biomass)] of a mixed culture growing on 3,4 dichloroaniline was determined, at varying dilution rates, in (a) a reactor where the cells were freely suspended, and (b) in a three phase air lift reactor where the cells were immobilized on silica particles. Standard techniques were used to measure variables of interest. The specific concentration of ATP is compared between the two reactors, and it is found that the immobilized cells exhibit a significantly lower ATP concentration per unit of dry biomass.

1. INTRODUCTION

The use of bioreactors in which cells are retained (immobilized cells and membrane bioreactors) is becoming increasingly popular across the biotechnology industry. This is attributable to a number of operational advantages and to the enhanced productivity offered by these retained cell systems. Typically when designing and modelling such systems the kinetic parameters used to describe freely suspended cells are employed in describing the reaction kinetics of the immobilized biomass. The classical techniques for calculating the quantity of biomass (turbidity, dry weight) do not allow a ready distinction to be drawn between viable and non-viable cells. Thus a given quantity of immobilized biomass measured as dry weight is often assumed to have the equivalent biocatalytic activity as the same quantity of freely suspended biomass. However, immobilized cells often grow in biofilms wherein dead and inactive cells are bound into the matrix, extracellular materials may be present binding the cells together, and where diffusional limitations may exist. In view of these factors, it is possible that equivalent amounts of dry biomass from freely suspended and immobilized systems will have significantly differing biocatalytic activities.

Adenosine 5' triphosphate (ATP) exists primarily in viable cells (Holm-Hansen and Booth (1966)) and so the specific concentration of ATP in biomass (mg (ATP) mg^{-1}(dry biomass) gives a measure of the activity of a culture sample. The amount of ATP existing in a culture sample can be measured through bioluminescence assays, in which the ATP reacts with luciferin (LH_2) in a reaction catalysed by luciferase and which produces photons ($h\nu$).

The reaction which takes place is:

$$LH_2 + ATP \rightarrow LH_2{\cdot}AMP + P\text{-}P$$
$$LH_2{\cdot}AMP + 1/2\ O_2 \xrightarrow{\text{luciferase}} L{\cdot}AMP^* + H_2O$$
$$L{\cdot}AMP^* \rightarrow L{\cdot}AMP + h\nu$$

In this paper the specific ATP concentration in a mixed culture growing on 3,4 dichloroaniline (34DCA) has been determined for (a) a chemostat culture of freely suspended cells growing at varying dilution rates (b) immobilized cells grown in a Three Phase Air Lift (TPAL) bioreactor at various dilution rates above μ_{max}.

2. EXPERIMENTAL

A 2L working volume continuous stirred tank (CST) bioreactor was used to culture the freely suspended cells. An especially designed impeller was employed in conjunction with glass beads and vigorous agitation in order to minimize wall growth in the reactor. The pH was controlled at 7.0 and the temperature at 30°C. A 3L working volume TPAL bioreactor was employed to culture the immobilized cells. The culture from the CST bioreactor was used to inoculate this reactor and the cells were immobilized by adsorption on porous silica biomass support carriers (Celite R-633, Manville, USA). The pH was controlled at 7.0, the temperature was held at 30°C, and the dissolved oxygen concentration at 80% of it's air saturation level through addition of pure oxygen to the reactor sparge gas. The gas flow (air) through the reactor was maintained at 0.6 L min^{-1}, this being sufficient to create circulation of the liquid and solid particles in the reactor. The original inoculum for the CST bioreactor was taken from a continuous enrichment reactor which has been fed a synthetic 34DCA containing wastewater for a period of several months. This culture degrades 34DCA but is undefined. The sole carbon source in both reactors was 34DCA at a concentration of 250 mg L^{-1} dissolved in a minimal salts medium.

Standard techniques (GC, TOC, colourimetric assays) were used to measure variables of interest during reactor operation. The ATP was extracted from the cells using boiling Tris-EDTA buffer. This extracted ATP was reacted via the luciferin-luciferase reaction and the $h\nu$ produced were detected using a bioluminometer. The CST bioreactor was run continuously for approximately 20 days, with the dilution rate being changed every several days from an initial value of 0.027 h^{-1} to a final value of 0.115 h^{-1}. The fermentation in the TPAL bioreactor was run continuously for about 30 days at dilution rates between 0.111 and 0.500 h^{-1}.

3. RESULTS AND DISCUSSION

Figure 1 shows the dependence of biomass concentration and 34DCA concentration in the fermenter liquid on the dilution rate. At dilution rates between approximately 0.12 and 0.13h^{-1}, washout occurred. This data enables the estimation of μ_{max} as approximately 0.115h^{-1}.

Figure 2 shows the variation of the average specific ATP concentration observed as a function of dilution rate in the CST bioreactor. The specific ATP concentration appears to increase with dilution rate varying between 0.44 and 1.86 mgATP g^{-1}dry biomass. For dilution rates over 0.066h^{-1} in the CST bioreactor, the specific ATP concentration did not significantly increase further, reaching a limiting value of 1.73 mgATP g^{-1}dry biomass just prior to wash out. This value is assumed to be the specific ATP concentration in 100% viable cells, since at dilution rates close to wash out the cells are growing at the maximum rate (Weddle and Jenkins

(1971), Roe et al. (1982), Nelson and Lawrence (1980)). Weddle and Jenkins (1971), working with activated sludge culture in a cell recycling system, reported specific ATP concentrations in the range 1.0-1.5 mgATP g^{-1}VSS (Volatile Suspended Solids), for high dilution rates, where they assume that the cell viability is 100%. According to Van de Werf and Verstraete (1984), the specific ATP concentration fluctuates between 0.1 and 10 mgATP g^{-1} dry biomass for different strains with an average value of 2 mg ATP g^{-1}dry biomass. The same parameter has been calculated to have an average value of 2.8 (Jørgensen et al. (1992)) for cultures growing on wastewater during the exponential growth phase. The results obtained for this freely suspended culture are thus in good agreement with those of other researchers.

Figures 3 and 4 show the results of measurements made on the TPAL bioreactor. Figure 3 presents the variation of the biomass loading on the support particles as a function of experimental time. The biomass loading on the silica particles appears stable for approximately the first 20 days, after which it increases in an approximately linear fashion over time. Livingston and Chase (1991) have shown that the biomass loading on porous support particles in a TPAL bioreactor is a function of the rate of biomass growth, and the rate of biomass attrition due to turbulence in the reactor. The gas velocity was maintained constant during the present experiment, but the 34DCA concentration in the bioreactor liquid, and therefore the average specific growth rate of the immobilized biomass, was increasing with dilution rate, thus leading to the net accumulation of immobilized biomass observed. Figure 4 shows the average specific ATP concentration of the immobilized and floc biomass in the TPAL bioreactor as a function of dilution rate. The specific ATP concentration in the immobilized biomass varies between 0.03 and 0.14 mgATP g^{-1}dry biomass for dilution rates between 0.111 and 0.500 h^{-1}, and appears to be increasing with dilution rate. The specific ATP concentration in the floc biomass in the TPAL bioreactor was found to lie in the same range as for the immobilized cells. This was evidence that the floc biomass in the TPAL bioreactor was comprised of biomass fragments which part from the support particles due to attrition. The dilution rate in the TPAL bioreactor was above the wash out point in the CST bioreactor, and so the floc biomass is not capable of maintaining itself in the TPAL system, which is consistent with the view above.

4. CONCLUSIONS AND IMPLICATIONS

(a) The specific ATP concentration in a microbial culture growing freely suspended with 34DCA as a growth substrate, in a CST bioreactor, increases with dilution rate. For dilution rates higher than 0.066h^{-1} the specific ATP concentration reached a limiting value of 1.75 mgATP g^{-1}dry biomass [the culture washed out at dilution rate 0.115h^{-1}]. Assuming that immediately prior to wash out the biomass is 100% viable, the specific concentration of 100% viable cells is 1.75 mg ATP g^{-1}.(b) The specific ATP concentration in immobilized biomass increases with dilution rate in an approximately linear fashion, fluctuating between 0.03 and 0.14 mg ATP g^{-1}dry biomass.(c) The ATP level in the suspended cells in the TPAL bioreactor lay in the same range as for the immobilized cells.(d) The fraction of viable cells in the immobilized cell bioreactor was approximately 3% of the dry biomass for dilution rates between 0.111-0.200 h^{-1}. For dilution rates between 0.300-0.500 h^{-1}, approximately 7% of the dry biomass was viable.

Expressions for reaction kinetics developed in freely suspended cell systems may not be appropriate for use in conjunction with immobilized cell systems. It may be that immobilized cell systems can be modelled by simply assuming a proportion of the biomass present functions identically to their freely suspended counterparts, or it may be that immobilized cell reaction kinetics require a fundamentally different analysis due to differences in microbial physiology between freely suspended and immobilized cell systems.

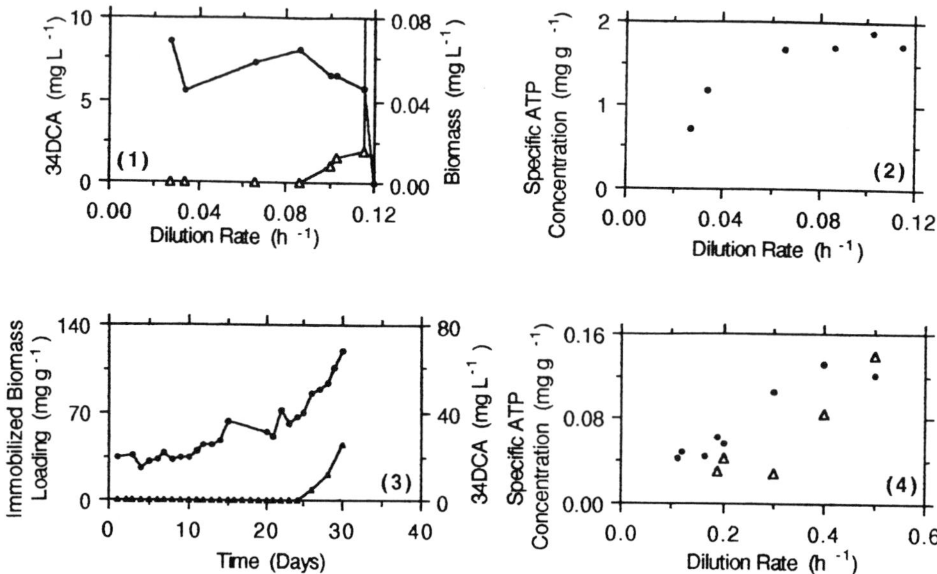

Figure 1: Dry biomass concentration (-•-) and 34DCA concentration (-Δ-) in the CST bioreactor versus dilution rate. The 34DCA concentration approaches the inlet concentration as the dilution rate approaches μ_{max}.

Figure 2: Average specific ATP concentration (mgATP g^{-1}dry biomass) (•) in the CST bioreactor at various dilution rates.

Figure 3: Biomass loading on the support particles (mg biomass g^{-1} silica) (-•-) and 34DCA concentration (-Δ-) in the TPAL bioreactor versus time.

Figure 4: Specific ATP concentration (mgATP g^{-1}dry biomass) in the TPAL bioreactor, in the immobilized biomass (•) and in the suspended biomass (Δ) as a function of dilution rate.

REFERENCES

Atkinson B, Mavituna F (1991) Biochemical engineering and biotechnology handbook, 2nd Edn. Stoklon press, New York

Jørgensen PE, Eriksen T, Jensen BK (1992) Estimation of viable biomass in wastewater and activated sludge by determination of ATP, oxygen utilization rate and FDA hydrolysis. Water Res. **26**: 1495-1501

Holm-Hansen O, Booth CR (1966) The measurement of adenosine triphosphate in the ocean and its ecological significance. Limnol. Oceanogr. **11**: 510-519

Livingston AG, Chase HA (1991) Development of a phenol degrading fluidized bed bioreactor for constant biomass holdup. Chem.Eng. J. **45**: B35-B47

Nelson PO, Lawrence AW (1980). Microbial viability measurements and activated sludge kinetics. Water Res. **14**: 217-225

Roe PC, Surinder J, Bhagat K (1982) Adenosine triphosphate as a control parameter for activated sludge processes. J Water Pollut. Control Fed. **54**: 244-254

Weddle CL, Jenkins D (1971) The viability and activity of activated sludge. Water Res **5**: 621-640

Van de Werf H, Verstraete W (1984) ATP measurement by bioluminescence: environmental applications. p. 33-48. In Kricka LJ, Stanley PE, Thorpe GHG, Whitehead TP (ed), Analytical applications of bioluminescence and chemiluminescence, Academic Press INC, London.

ECB6: Proceedings of the 6th European Congress on Biotechnology
edited by L. Alberghina, L. Frontali, P. Sensi

Immobilization of yeast cells in spherical porous glass particles: consequences of intraparticle convection

L. De Backer and G. Baron

Department of Chemical Engineering, Vrije Universiteit Brussel,
Pleinlaan 2, B-1050 Brussel, Belgium.

Spherical porous glass particles with a porosity of 60 % were used as an immobilization matrix for Saccharomyces cerevisiae cells. Residence time distribution experiments were performed in a packed bed and revealed the existence of intraparticle convection at high liquid flow rates. When cells were immobilized on the porous glass particles, a dependence of the biomass loading on the flow rate was observed: the biomass loading increased with flow rate until a maximum was reached for a Reynolds number (Re) equal to 5. Another striking consequence of the intraparticle convection effect emerged from fermentation experiments: at low flow rates of the fermentation medium, the cells grow mainly in the outer matrix region, whereas at the higher flow rates a perceptible cell growth can also be observed in the centre of the porous particles.

1. INTRODUCTION

Inert carriers, such as porous glass beads, are of great technological interest, and are easy to use in continuous systems because of their stability to microbial attack and changes in pH and solvent conditions. They can be used as a carrier for the immobilization of enzymes [1], yeast cells [2-4], bacterial cells [5,6] and animal cells [7]. Cells are easily immobilized and only have to be brought into contact with the support, and when used in a bioreactor in continuous operation it is possible to retain a high cell density, even beyond wash-out conditions.

Another advantage of porous supports is the possibility of an enhanced mass transfer between the extraparticle and intraparticle fluid phase through internal convection, in contrast to the popular gel immobilized cell systems, where nutrient transport to the immobilized cells is only possible by diffusion. The existence of intraparticle convection in the porous glass beads first emerged in an experiment to determine the effective diffusivity of glucose [8].

Here we present some important consequences of this intraparticle convection effect: influence of flow rate on the biomass loading during the immobilization of yeast cells on porous glass beads, and influence of the medium flow rate on the intraparticle cell growth during fermentation experiments in a differential bioreactor.

2. RESULTS AND DISCUSSION

2.1. Residence time distribution experiments

To evaluate the mass transfer parameters, RTD (residence time distribution) experiments were performed in a column (length and diameter equal to 0.5 m and 2.55 cm, respectively) containing porous glass beads (diameter = 2.4 mm), with calcium chloride as tracer. The tracer concentration at the column outlet (C) was monitored on-line, after a step change in tracer concentration at the column inlet ($0 \rightarrow C_o$). For Re less than 1, mass transfer by intraparticle convection is negligible and a model including axially dispersed flow for the extraparticle fluid phase and mass transfer by intraparticle diffusion predicts experimental

RTD data satisfactorily, as can be seen from Figure 1. The parameters used in the model to predict the RTD profile for Re = 0.8 were:
D_{ax} (the axial dispersion coefficient in the extraparticle zone) = 3.75 10^{-6} m^2/s
D_{eff} (effective intraparticle diffusion coefficient of the tracer) = 5 10^{-10} m^2/s.

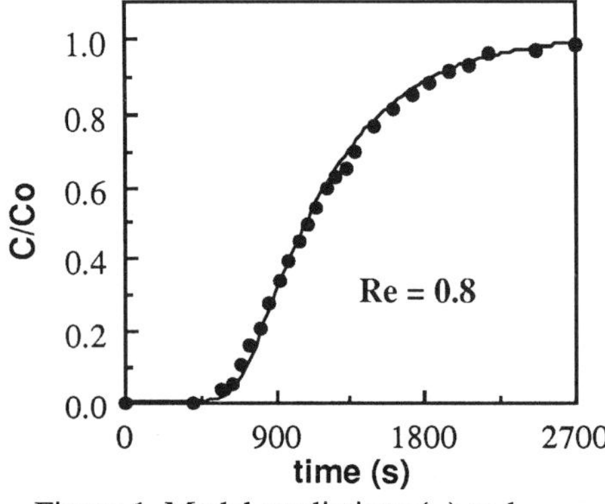

Figure 1. Model predictions (–) and experimental data (•) of the RTD profile for Re = 0.8.

Figure 2. Model predictions (–) and experimental data (•) of the RTD profile for Re = 8.

At high fluid flow rates, intraparticle forced convection can be important, especially for large pore porous particles [9]. Since no good model for intraparticle fluid flow is available, we lumped all mass transfer resistances, namely external film, internal diffusion, and internal convection in an overall effective mass transfer coefficient. We represented the mass transfer rate by a linear rate expression, and coupled this expression with the mass balance for the external fluid phase. The model equations in dimensionless units can be found from a differential tracer balance in the extraparticle (C_1) and in the intraparticle zone (C_2):

$$\lambda \frac{\partial C_1}{\partial \theta} = \frac{1}{Bo} \frac{\partial^2 C_1}{\partial \xi^2} - \frac{\partial C_1}{\partial \xi} - \gamma (C_1 - C_2) \qquad (1-\lambda)\frac{\partial C_2}{\partial \theta} = -\gamma (C_2 - C_1)$$

The initial and boundary conditions are:

$$\theta \le 0,\ C_1 = C_2 = 0 \qquad \xi = 0,\ \frac{\partial C_1}{\partial \xi} = Bo\,(C_1 - 1) \qquad \xi = 1,\ \frac{\partial C_1}{\partial \xi} = 0$$

In these equations λ is the part of the liquid volume belonging to the extraparticle zone and was calculated from the packed bed and particle properties (λ = 0.51), C_1 and C_2 are the dimensionless tracer concentrations in the extraparticle and the intraparticle zone respectively, θ (= t / t_s) the dimensionless time with t_s the space time, ξ the dimensionless axial co-ordinate, Bo (= v L / D_{ax}) the Bodenstein number for the extraparticle zone with v the interstitial fluid velocity and L the column length, and γ the number of mass transfer units for the mass exchange between the extraparticle and the intraparticle zone. The Bodenstein number for the extraparticle fluid phase was determined in an independent experiment using non-porous glass particles of the same size as the porous particles (Bo = 120).
The value of the parameter γ was found by least squares curve fitting the experimental RTD data to the results predicted by the model equations.

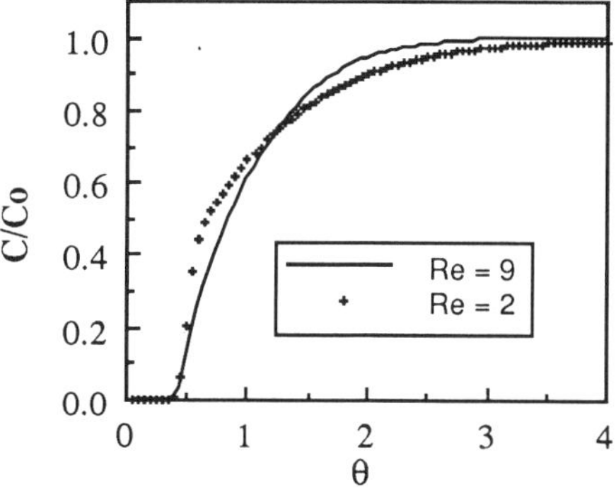

Figure 3. RTD profiles for different values Reynolds number.

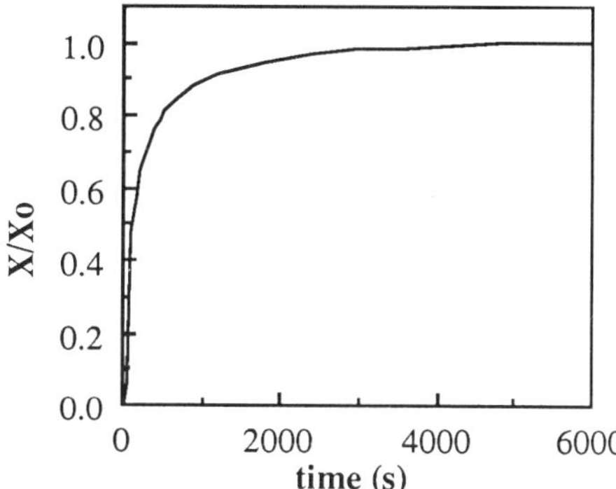

Figure 4. Dimensionless yeast cell of the concentration as function of time

The model equations were solved numerically yielding the tracer concentration at the column outlet and over the experimental flow rates range, the model was adequate to predict the experimental data, as can be seen in Figure 2. Figure 3 shows the effect of the liquid flow rate on the mass exchange rate between the extraparticle and intraparticle fluid phase. Clearly the mass exchange is faster if the Reynolds number (Re) is increased, as a consequence of intraparticle convection.

2.2. Immobilization experiments

Since cells have to be transported by convection, a dependence of the biomass loading on the flow rate was expected. To investigate this influence, yeast cell suspensions were pumped through a packed bed of porous glass particles at different flow rates. The immobilization procedure was considered as completed if the yeast cell concentration at the column outlet (X) was equal to the one at the column inlet (X_0). The cell concentration at the column outlet was monitored on line using a fibre optics spectrophotometer. A typical immobilization profile is shown in Figure 4, the experiment was performed at a flow rate of 87 ml/min.

Experimental verification of the amount of cells (gram dry weight) immobilized per gram porous glass revealed that the biomass loading increased with flow rate, until a maximum was reached for Re = 5, as is shown in table 1.

Table 1.
Influence of the flow rate (ml/min) on the biomass loading (g D.W./g carrier)

Flowrate	Re	Biomass loading
4 7	0 4	0.014
6.6	0.5	0.015
47	3.6	0.022
65	5.0	0.025
87	6.7	0.018

At low flow rates, cells do not easily enter the interior of the porous glass beads and as a consequence mainly the outer part of the matrix is colonised by cells. Increasing the flow rate leads to higher intraparticle fluid velocities, so that the cells can penetrate further into the immobilization matrix. At a certain flow velocity however, hydrodynamic forces are sufficient to detach the cells from the support, especially in the outer matrix region, and this results in a decrease in biomass loading.

2.3 Fermentation experiments

To study the influence of the medium flow rate on mass transfer under fermentation conditions, a packed bed differential reactor containing porous glass particles with immobilized cells was continuously fed with fermentation medium. Thin cross sections of the porous glass particles were examined by microscopy before and after the fermentation phase. A schematic representation of what can be seen under the microscope is shown in figure 5.

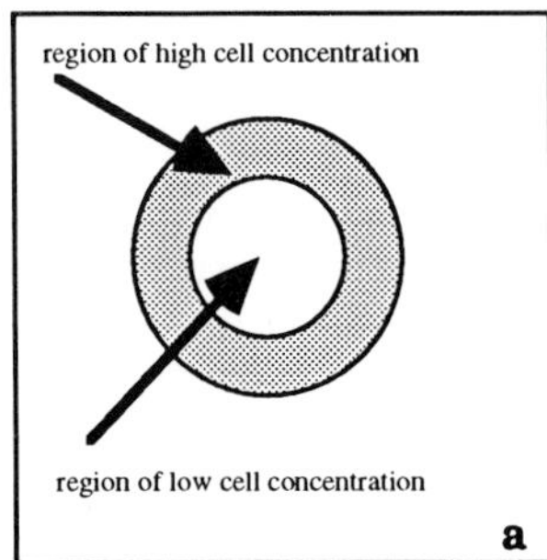

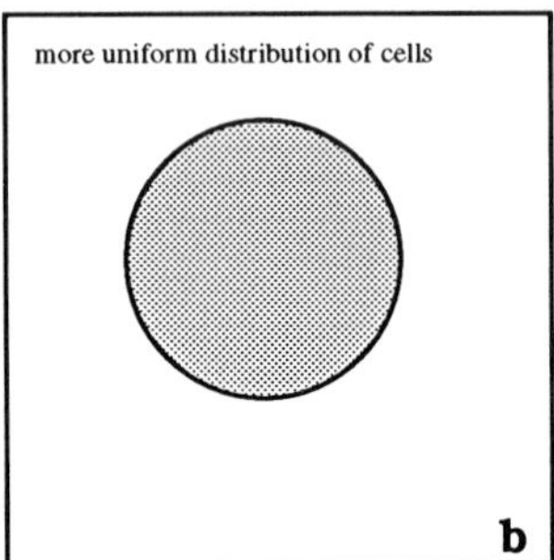

Figure 5. Schematic representation of two cross sections of the porous glass beads, after continuous fermentation at a low medium flow rate (a) and at a high medium flow rate (b)

At low flow rates of the fermentation medium (Re < 1), mainly the cells in the outer matrix region have grown, which leads to a situation identical to what can be observed with cells immobilized in spherical gel particles. Even with some seeding in the centre of the particles, dense cell growth in the peripheral area blocks substrate transport to the bead centre.

At high flow rates (Re > 1), however, cell growth could be observed even in the centre of the porous particle, as a consequence of intraparticle convection. The more homogeneous distribution of cells at the higher flow rates can be attributed to a number of complementary effects: if the medium flow rate is increased, mass transport of the nutrients towards the bead centre is enhanced. Moreover, the cells immobilized in the outer bead region become more susceptible to detachment so that excess cells are removed, and finally these detached cells can eventually migrate to the bead centre.

REFERENCES

1. A. Manjón, J.L. Iborra, J. Bastida and A. Bódalo, Biotechnol. Bioeng., 30 (1987) 491
2. J.M. Navarro and G. Durand, Eur. J. Appl. Microbiol. Biotechnol. 4 (1977) 243.
3. B. Bisping, U. Baumannn and H.-J. Rehm, Appl. Microbiol. Biotechnol. 32 (1990) 380.
4. A. Mörsen and H.-J. Rehm, Appl. Microbiol. Biotechnol. 33 (1990) 206.
5. A. Aivasidis, Bioprocess Eng. 4 (1989) 19.
6. C. Krekeler, H. Ziehr and J. Klein, Appl. Microbiol. Biotechnol. 35 (1991) 484.
7. D. Looby and B. Griffiths, Tibtech 8 (1990) 204.
8. L. De Backer and G. Baron, Appl. Microbiol. Biotechnol. (1993), in press.
9. A.E. Rodrigues, B.J. Ahn and A. Zoulalian, AIChE Journal 28 (1982) 541.

ECB6: Proceedings of the 6th European Congress on Biotechnology
edited by L. Alberghina, L. Frontali, P. Sensi

Production of actinorhodin by immobilised and freely suspended *Streptomyces coelicolor*

K. Ozergin* and F. Mavituna

Department of Chemical Engineering
University of Manchester Institute of Science and Technology
Manchester, M60 1QD, UK.

1. INTRODUCTION

Streptomyces spp. are well known for their capacity to synthesize an enormous variety of antibiotics. *Streptomyces coelicolor* A3(2) produces four known antibiotics; methylenomycin, calcium dependent antibiotic, undecylprodigiosin and actinorhodin. The latter two of these antibiotics are pigmented.
Self-immobilisation of microorganisms as films, flocs and pellets is a widespread phenomenon in nature. Our aim was to compare the activity of "naturally" immobilised and freely suspended *Streptomyces coelicolor* in defined and complex media. In the experiments presented here, *Streptomyces coelicolor* cells were immobilised by physical entrapment in the open pore network of reticulated polyurethane foam slabs. There are several advantages of this "natural" immobilisation technique over other methods of immobilisation:
- It is biologically harmless to cells,
- There is no need for the pretreatment of cells,
- It involves mechanically strong and reusable biomass support materials,
- It can be performed under sterile conditions *in situ* in the bioreactor as an integrated stage of bioreactor operation.

2. MATERIALS AND METHODS

Organism: Streptomyces coelicolor A3(2) was provided by G Hobbs, UMIST, UK and was maintained as frozen spore suspension in 20% glycerol at -20 ^{o}C. Inocula were prepared according to the procedure described by Hobbs et al. (1989).
Media: The defined medium was based on that described by Hobbs et al. (1989). The modifications were the exclusion of the polymer Junlon and the increase in

* Present address: Chemical Engineering Department, Bogazici University, 80815 Bebek, Istanbul, Turkey.

glucose concentration to 10 - 20 g/l. The complex medium was the YEME medium of Hopwood et al. (1985) excluding $MgCl_2.6H_2O$.

Growth: A series of 500 ml shake flasks were prepared each containing 100 ml medium for freely suspended cell experiments and 200 ml medium for immobilised cell experiments. Each flask was inoculated with 750 μl spore suspension and placed on a New Brunswick orbital shaker at 30 °C and 200 rpm. Two-day old freely suspended culture from shake flasks was used to inoculate the 20-l bioreactor with 12 l working volume at 8 - 15% (v/v) level.

Immobilisation: S. coelicolor was immobilised within reticulated polyurethane foam slabs (Declon, Corby, UK) of 11-13 cm^3 total void volume (97% porous). Each slab was held stationary at the bottom of a 500 ml shake flask. After sterilisation in an autoclave at 15 psig for 15 min with 200 ml of medium, each flask was inoculated with the spore suspension. *Streptomyces coelicolor* spores and mycelia were then immobilised "naturally" in the open pore network of the foam matrix initially through filtration brought about by mixing in the bulk liquid. Further growth took place essentially within the supports consolidating the immobilisation by a combination of phenomena such as the increase in mycelial aggregate size, adhesion of cells to the foam fibres and entanglement of mycelia around the foam fibres.

Bioreactor: The 20-l computer controlled bioreactor had an aspect ratio of 2. The dissolved oxygen concentration in the medium was kept at 90-100 % of saturation by automatic control of agitation at 200-300 rpm and aeration between 2-6 l/min. The temperature was kept at 30 °C and pH was initially adjusted to 7.2 and then allowed to follow its natural course.

Analyses: The growth of free and immobilised biomass was measured as the dry weight of cell material. For comparison, both free and immobilised biomass concentrations were based on unit volume of liquid medium and expressed as gram dry weight per litre. The residual glucose concentration in the medium was measured by Beckman Glucose Analyzer. Actinorhodin concentrations were determined as described by Doull and Vining (1990).

3. RESULTS AND DISCUSSION

The kinetic parameters were determined from batch experiments performed in the 20-l computer controlled bioreactor with freely suspended cultures using defined medium. Trends in all the batches were the same (Figure 1). The onset of actinorhodin accumulation in the defined medium coincided with the beginning of the exponential growth phase. In the complex medium however, phases of cell growth and actinorhodin production were separate. For the determination of the values of growth kinetic parameters the methodology of Knights was employed (Atkinson and Mavituna, 1991). In the defined medium, depending on the initial conditions, some of the kinetic parameters were: maximum specific growth rate, $\mu_m = 0.03\ h^{-1}$; death rate constant, $k_d =$ $1.4 - 6.3 \times 10^{-3}\ h^{-1}$; observed biomass yield, $Y_{x/s} = 0.21$ g cells/ g glucose and the

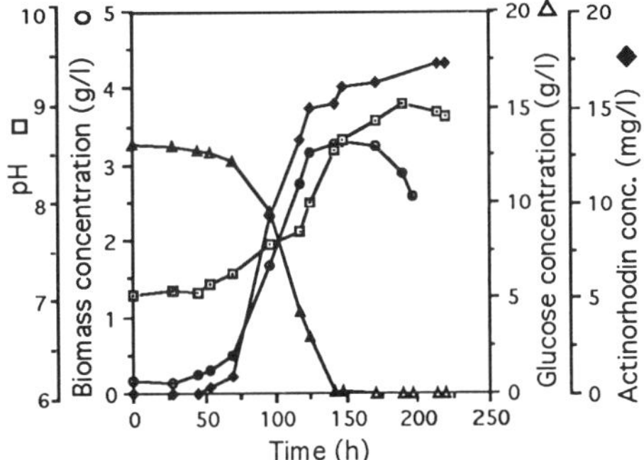

Figure 1. Time course of a batch run in the 20-l bioreactor using defined medium.

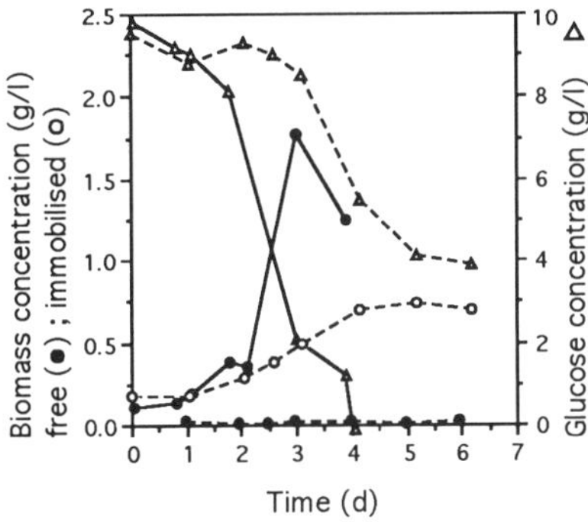

Figure 2. Batch growth and glucose uptake in defined medium.
- - - - immobilised culture
—— free culture

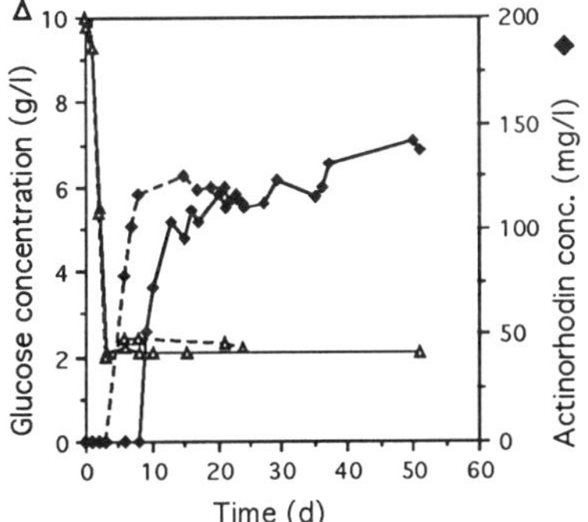

Figure 3. Glucose uptake and actinorhodin production in complex medium.
- - - immobilised culture
—— free culture

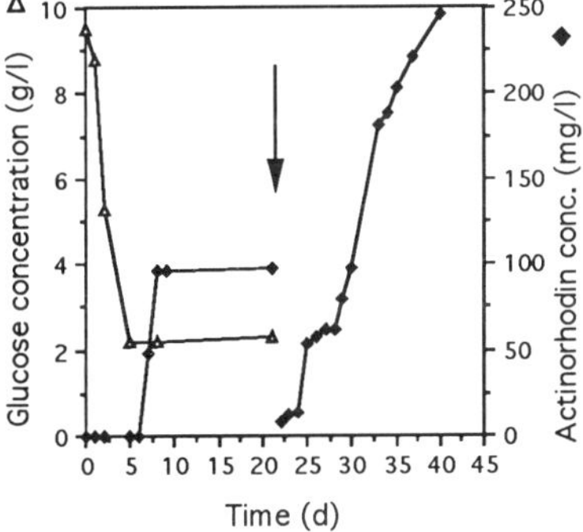

Figure 4. Repeated batch with immobilised culture in complex medium. Arrow indicates replacement medium without glucose.

maintenance coefficient for the cells, m_S= 0.0448 g glucose/ g cells-h.

A series of flasks was prepared to compare batch growth, glucose consumption and actinorhodin production in freely suspended and immobilised *Streptomyces coelicolor* cultures using defined medium (Figure 2). Extracellular actinorhodin concentration in immobilised culture was very low compared with that in free culture. The intracellular concentration of actinorhodin on the other hand, was equivalent to 2.32 mg/l. The specific yield of actinorhodin was 1.2 mg/g biomass in freely suspended culture compared to 3.8 mg/g biomass in immobilised culture which indicated that the immobilised cells were more efficient producers of actinorhodin.

When complex medium was used, actinorhodin was excreted into the medium by both free and immobilised cultures. There was no significant difference in glucose uptake and product formation pattern in both cultures except for the earlier appearance of actinorhodin in the medium in immobilised culture (Figure 3). In both cultures actinorhodin production coincided with the exhaustion of glucose indicating non-growth associated nature of the product. The average product concentration in the medium was 125 mg/l at the end of both batches.

One of the advantages of immobilised cultures is their possible repeated use with media replacements especially in the case of secondary metabolite production. Since in the complex medium actinorhodin was excreted, its production followed a non-growth associated pattern, and it attained much higher concentrations than those in the defined medium, the repeated batch experiments with immobilised cells were therefore performed using the complex medium. The glucose-free replacement medium did not support growth but apparently kept the immobilised culture viable as evidenced by continued production of actinorhodin (Figure 4). The highest actinorhodin concentration of 246 mg/l was obtained with this regime of repeated batch with glucose-free complex medium. Immobilised cells were stable for at least three months.

4. CONCLUSION

In the defined medium, the most pronounced difference between the two cultures was the accumulation of actinorhodin extracellularly in freely suspended and intracellularly in immobilised cultures. In the complex medium however, actinorhodin was excreted by both cultures. Furthermore, actinorhodin production was growth associated in the defined and non-growth associated in the complex medium. In addition, the complex medium yielded much higher actinorhodin concentrations. Further increase in product concentration was obtained by repeated batches of immobilised culture.

5. REFERENCES

B. Atkinson and F. Mavituna, Biochemical Engineering and Biotechnology Handbook, 2nd ed., Macmillan Publ. Ltd, Basingstoke, UK, 238 (1991).

J.F. Doull and L.C.Vining, Appl. Microbiol. Biotechnol. 32 (1990) 449.

G. Hobbs, C.M. Frazer, D.J.C. Gardner, J.A. Cullum and S.G. Oliver, Appl. Microbiol. Biotechnol. 31 (1989) 272.

D.A. Hopwood, M.J. Bibb, K.F. Chater, T. Kieser, C.J. Bruton, H.M. Kieser, D.J. Lydiate, C.P. Smith, J.M. Ward, H. Schrempe, Genetic Manupulation of Streptomyces : a Laboratory Manual, Norwich, UK, 1985.

NEW SEPARATION METHODS IN BIOTECHNOLOGY

ECB6: Proceedings of the 6th European Congress on Biotechnology
edited by L. Alberghina, L. Frontali, P. Sensi

NEW SEPARATION METHODS IN BIOTECHNOLOGY

J.A. Asenjo

Biochemical Engineering Laboratory
University of Reading
Reading RG6 2AP, England

1. Introduction

Discoveries and achievements in modern biology and recombinant DNA technology in the last few years have resulted in the development of a number of new therapeutics for human use such as insulin, human growth hormone (hGH), tissue plasminogen activator (tPA) for cardiac disease, erythropoietin (EPO) and hepatitis B vaccine and thus the possibility of their industrial large scale production. This poses a tremendous challenge for the chemical and biochemical engineer in terms of developing efficient separation processes for these new proteins. As they are intended for human use the levels of purity required are of the order of 99.9% or 99.98% or even higher (depending on dosage) and they have to be separated from a very large number of contaminants, other proteins, nucleic acids, polysaccharides and many other components present in the cell culture or cell lysate used to manufacture these proteins. Competitive advantage in production will depend not only on innovations in molecular biology and other areas of basic biological sciences but also on innovation and optimization of separation and downstream processes.

The main issues important for the development of novel separation techniques to give **improved resolution**, **simplicity**, **speed**, **ease of scale-up** and possibly **continuous operation** are presented and discussed. The assesment of the state of the art as well as promising future developments concentrate on the separation and purification of proteins from complex mixtures. The present trend to develop techniques that exploit fundamental physico-chemical principles more efficiently is emphasized. This includes the analysis of the physico-chemical properties of proteins such as pI, charge as a function of pH, biological affinity (including metal ion and dye affinity), hydrophobicity and size and its relation to efficiency in a bioseparation. Some properties (e.g. charge and affinity) can show extremely high resolution in purification operations, whereas others (e.g. molecular weight), show much less resolution.

2. System Characteristics

2.1. Physico-Chemical Basis for Separation Operations

Development of new and efficient separation processes will be based on more efficiently exploiting differences in the actual physicochemical properties of the product (surface charge/titration curve, surface hydrophobicity, M.W., biospecificity towards certain ligands (e.g. metal ions, dyes), pI, stability) compared to those of the contaminant components in the crude broth. The main physicochemical factors involved in the development of separation processes are shown in Table 1.

Table 1. Physicochemical basis for the development of separation processes

Physicochemical Basis	Separation Process
Charge	Ion Exchange Cromatography Electrodialysis Aqueous Two-Phase Partitioning Reverse Micelle Extraction
Hydrophobicity	Hydrophobic interaction Chromatography Reversed Phase Chromatography Aqueous Two-Phase Partitioning
Specific Binding	Affinity Chromatography
Size	Gel filtration Ultrafiltration Dialysis
Electric Mobility	Electrophoresis
Isoelectric Point	Chromatofocusing Isoelectric Focusing
Sedimentation Rate	Centrifugation
Surface Activity	Adsorption Foam Fractionation
Solubility	Precipitation Solid-Liquid Extraction Supercritical Fluid Extraction

2.2. Kinetics and Mass Transfer

The physical behaviour of the system has an effect on the development of novel separation processes. Processes can be divided into **equilibrium** and **rate** processes. In equilibrium processes selective separation depends on the attainment of a favourable equilibrium state. This for example includes liquid-liquid extraction and ion exchange chromatography. Rate processes, on the other hand, will separate different proteins on the basis of their response to an imposed field. Mobility and other similar properties determine the selectivity of this type of operation; a successful process will be one in which the proteins have markedly different mobilities (e.g. electrophoresis).

In a number of protein separation processes the residence time in the reactor is insufficient for equilibrium to be achieved and the **kinetics** of adsorption play an important role such as in affinity chromatography and in the CARE (continuous adsorption recycle extraction) process. New developments of materials have recently shown dramatic advances in overcoming **mass transfer limitations** (e.g. perfusion and membrane chromatography and adsorption) resulting in extremely fast separations. Some recent examples of novel techniques, that exploit the principles discussed above, and useful analyses for optimal design of operations include expanded bed (fluidised bed) adsorption of proteins and inverted matrix chromatography which allow direct broth extraction, cross flow electrofiltration of disrupted microbial cells and for improved ultrafiltration of proteins, mathematical modelling of partitioning and phase behaviour in liquid-liquid extraction, centrifugal partitioning chromatography, mathematical modelling of chromatographic columns, perfusion and membrane chromatrography, and advanced reversed phase chromatography using HPLC. The potential for scale-up of many of these systems is analyzed and discussed.

3. Liquid-Liquid Extraction

Liquid-liquid extraction as a technology has been used in the antibiotics industry for several decades and it is now beginning to be recognized as a potentially useful separation step in protein recovery and separation, particularly because it is readily scalable, can, if necessary be operated on a continuous basis, and we are starting to understand the physico-chemical factors of the protein that determine partitioning. It is a reasonable high capacity process and can offer good selectivity for the desired protein product. However, the lack of solubility of the large protein molecules in typical organic solvents restricts the range of solvents available for use in such a separation process.

Two classes of solvents that appear to offer advantages for protein recovery for protein separations are aqueous polymer/salt (in some cases also polymer/polymer) systems and reverse micellar solutions. In both cases two phases are formed and the separation exploits the difference in partitioning of the proteins in the feed and extraction phases. In the aqueous polymer/salt separation systems the partitioning of the protein occurs between two immiscible aqueous phases; one rich in a polymer (usually PEG) and the other in a salt (e.g. phosphate or sulphate). These systems show a non-denaturing solvent environment, small interfacial resistance to mass transfer, relatively high protein capacity and high selectivity. On the other hand reversed micelles, exploit the solubilizing properties of surfactants that can aggregate in organic solvents to form so called inverted or reversed micelles. These aggregates consist of a polar core of water and the solubilized protein stabilized by a surfactant shell layer. For protein extraction, one phase is the aqueous feed solution, the other the reversed micellar phase that acts as the extractant. They have several of the advantages quoted for aqueous two-phase systems.

3.1. Aqueous Two-Phase Separation

Partitioning in two aqueous phases can be used for the separation of proteins from cell debris as well as for purification from other proteins. Partitioning can be done in a single step or as a multistage process. Difference in partition coefficients, however, between the different proteins can be high, hence one step tends to be sufficient (usually one for extraction and one for elution or back-extraction). The use of affinity partitioning can greatly enhance the specificity of the extraction.

Most soluble and particulate material will partition to the lower, more polar (e.g. salt) phase and the protein of interest will partition to the top less polar phase, usually PEG. Separation of actual proteins in such systems is based on manipulating the partition coefficient by altering parameters such as average molecular weight of the polymer, type of phase forming salt used for the heavy phase, ions included in the system and ionic strength of added salts (e.g. NaCl).

The partition coefficient (K) is defined as the concentration of a particular protein in the lighter phase divided by the concentration in the heavier phase. The main factors which determine partition will depend on the type of system used:

1. Hydrophobicity: Differences in the surface hydrophobicity beetween proteins are exploited when partitioning them in PEG/salt two-phase systems. Typical systems that exploit a protein's hydrophobicity are PEG/phosphate and PEG/sulphate with addition of a high concentration of NaCl (eg. 10%).

2. Size dependent partition: molecular size of the proteins or surface area of the particles to be partitioned is the dominating factor. It has been shown that for PEG/Dextran systems a protein's m.w. is inversely proportinal to its partition coefficient.

3. Electrochemical: electrical potential between the phases is used to separate molecules or particles according to their charge. This is demonstrated in PEG/Dextran systems with addition of small concentration of salts (following the Hofmeister series) whose charged ions will partition between the phases (e.g. 0.1 M NaCl or 0.05 M Na_2SO_4) and also by manipulating the pH of the system (in PEG/salt systems an increase in the pH will usually increase the value of K).

4. Biospecific affinity: the affinity between sites on the proteins and ligands attached to one of the phase polymers is used for separation. Dyes, inhibitors, fatty acids, glutathione, Protein A and several other ligands have been used. Particularly impressive results for selective partitioning have been obtained with metal ions (eg. Cu^{++}) both in PEG/Dextran but also in PEG/salt systems.

5. Conformation dependent: conformation of the proteins is the determining factor.

According to this, it is possible to split the logarithm of the partition coefficient into different terms:

$$\ln K = \ln K_{hfob} + \ln K_{mw} + \ln K_{el} + \ln K_{aff} + \ln K_{conf}$$

where hfob, mw, el, aff and conf stand for hydrophobicity, size (mw), electrostatic, affinity and conformational contributions to the partition coefficient.

In practical terms partitioning in aqueous two-phase systems is influenced by many system variables. Generally, the higher the molecular weight of the polymers the lower the concentration needed for the formation of two phases. Also, the larger the m.w. of the PEG, the lower the value of K. Work is presently being carried out on elucidating how the different physico-chemical properties of individual proteins determine their partition behaviour in two phase systems.

In addition to the partition coefficient (K) it is important to know the actual concentration of the protein in the extractant phase as this will be essential for the sizing of the phase system. In typical PEG/salt systems protein solubility tends to be higher in the PEG and lower in the salt phase. It is clear that in the region near saturation of the protein in one of the phases a constant partition coefficient is no longer observed.

3.2. Reverse Micelle Extraction

The extraction and separation of proteins using reversed micelle systems has been the subject of extensive study in recent years. Reversed micelle systems have interesting potential for the liquid-liquid extraction of biomolecules. Protein partitioning is dependent upon many factors which are directly related to the interaction of those inherent to the system itself and those pertaining to the protein under investigation (e.g. charge, m.w., hydrophobicity). Fundamental studies of the factors determining selective separation of biomolecules are necessary to establish correlations between the physico-chemical properties of the proteins and the reversed micelle system. These factors include pH, ionic strength and type of ions present in the system, type of surfactant and organic solvent used and physico-chemical properties of the proteins such as isoelectric point, hydrophobicity, size, charge density and charge distribution.

Reversed micelle systems formed from AOT, isooctane and water are typical. AOT is an anionic surfactant which will form micelles in apolar solvents without the need for a cosurfactant. In reversed micelle systems the polar head groups of the surfactant molecules are directed towards the interior of the micelle and form a polar core which can solubilize water, the lipophilic chains are exposed to the solvent. Water content of a micellar system and phase ratio (V_{org}/V_{aq}) both have a strong effect on protein solubilization and function.

Increasing ionic strength can affect micelle size, charged surfactant headgroups repel each other and larger micelles are formed at lower ionic strengths. Also at higher ionic strengths there is a shielding effect between the charged surfactant head groups and the protein making solubilization in the micelles more electrostatically unfavourable.

The typical behavior of a protein is that at relatively low ionic strength (eg. 0.1M KCl) and at a pH below its isoelectric point (positively charged) the protein solubilizes in the reversed micelle. If the pH is increased above the pI of the protein or the ionic strength is increased the protein is backextracted into the aqueous phase. This 'electrostatic' behaviour however is not always observed due to hydrophobic and size effects amongst others.

3.3. Other Liquid-Liquid Systems

An interesting novel approach for the separation of proteins in liquid-liquid systems is the use of a liquid phase in which the solutes are insoluble and where the selective adsorption occurrs at the surface of the droplets and not in the bulk of the liquid. The principle is similar to that using non-porous solids. **Perfluorocarbons** have been used for such an application. The surfaces of perfluorocarbon liquid droplets have been coated with ligands so that selective interactions are created. The main advantage would be the high density of the extracting liquid phase and the fact that both liquid phases are insoluble in one another.

Liquid membranes represent another type of liquid-liquid system, however they are mainly used for smaller molecules produced by fermentation such as organic acids. They can be used as emulsion or as supported liquid membranes. A liquid membrane is an insoluble liquid, usually an organic solvent which is selective for a solute, separating the two other liquid phases. Separation occurs by selective transport of solutes through the membrane. These are called **emulsion liquid membranes** and are of potential interest in biotechnology as separation and concentration can be achieved in a single step and scale-up is straightforward. Such systems can be divided into unfacilitated and facilitated transport ones. In the first, separation occurs exclusively due to differential solubility in the different phases. Facilitated or carrier mediated extraction involves the formation of a carrier-solute complex, which is specific for the solute, reacts reversibly with it, and is present in the liquid membrane but totally insoluble in both aqueous phases.

Supported liquid membranes containing a carrier to facilitate extraction has been successfully used for the selective separation of citric acid in dilute aqueous solutions.

Whereas in an emulsion liquid membrane the extracting phase is encapsulated by emulsification in the organic solvent, supported liquid membranes consist of a thin porous support retaining the solvent (membrane), which may also contain a carrier.

The suitability of using **foam separation** as a novel separation technique for proteins is presently under investigation and shows promise.

4. Solid-Based Separations

A variety of new and unconventional technologies for biological molecules are presently being developed. These include systems that facilitate the handling of materials by the use of membrane, fludized bed and novel chromatographic matrix technologies, improving separation specificity and efficiency by using metal, dye and other ligands and also by developing techniques used very efficiently at the analytical scale such as isoelectric focusing now at preparative scale.

4.1. Adsorption Systems

Expanded bed adsorption is a new technique that in one step accomplishes removal of whole cells and cell debris, concentration and initial purification of the target protein. It is based on the design of an adsorbent whose density enables the formation of a stable expanded bed (2-3 fold expansion) that allows the establishment of a plug-flow concentration profile across the column. Binding capacities for proteins are similar to those obtained in fixed bed adsorption and linear velocities can be higher than 300 cm/h.

Continuous adsorption recycle extraction was recently adapted for biotechnological applications as a downstream process for efficient separation of proteins from crude feedstocks to be carried out continuously. Modern biotechnology involves relatively small scale processes that favours batch procedures, usually following the concept that a batch is well defined (from beginning to end) and that this will pose fewer problems with regulatory approval for therapeutic use. However, it is well known that in the chemical and biochemical process industries batch to batch variations can be substantial. It is also well known that a continuous process can operate under steady state conditions minimising batch to batch variations. In addition, a continuous process is much more appropriate for implementing a control strategy in order to maintain constant conditions more readily.

Continuous adsorption recycle extraction, also called **CARE**, is based on the continuous protein adsorption to solid phase supports and features two well mixed reactors with solids recycle, as shown in Figure 1. The adsorbing stage takes place in the first reactor (adsorber) where the liquid feed is contacted with the adsorbent; desorption of the protein to be purified is obtained in the second reactor (desorber) by maintaining appropriate conditions and by the action of a suitable eluting solvent. The adsorbent beads are recycled to the adsorber reactor while the product is continuously removed. The system's performance is determined by the nature of the adsorbent and the feed material, flow rates and reactor volume and the conditions of the adsorption and desorption stages. An accurate mathematical model of the system has recently been developed which can be used for investigating optimisation criteria, but also is important for system design and process control.

4.2. Chromatographic and Adsorption Materials

Perfusion chromatography represents an interesting approach to overcoming speed and throughput limitations in chromatography caused by diffusive mass transport. The packing used for perfusion chromatography contains two classes of pores: *throughpores* (see Fig.2) to allow convective flow through the particles and smaller, *diffusive pores* lining the throughpores to provide high adsorption surface area. Conventional materials contain only diffusive pores and thus solute transport to the binding sites on the medium is limited by the slow diffusion process.

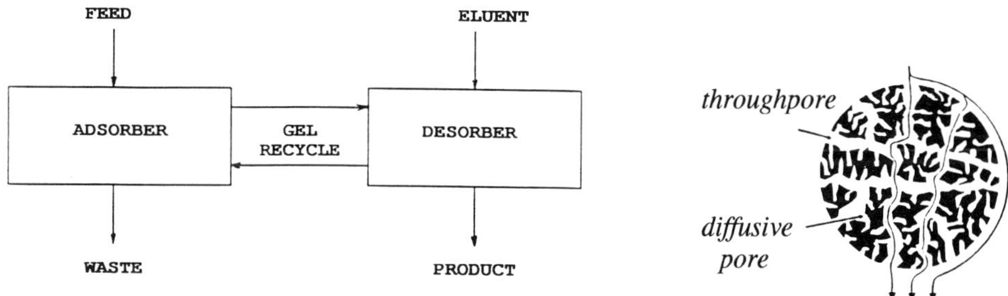

Fig. 1. Continuous Adsorption Recycle Extraction.

Fig. 2. Perfusion Cromatography.

In perfusive media, when the mobile phase reaches a sufficiently high velocity, convection in the throughpores takes place. In this regime, the combined rapid convective transport and ultra short path length in the diffusive pores make resolution and dynamic loading capacity virtually independant of flow rate with linear velocities ca. ten times or even higher than with conventional materials.

A different approach is the attachment of the functional group to the inner surface of synthetic **microporous membranes**. This has been applied for any type of **adsorption** type **chromatography** (dyes, ion exchange, IMAC, etc) and the main advantage, as in the previous case, is the convective transport of the medium through the porous structure (by filtration) minimising diffusion and thus mass transfer limitation. A membrane adsorber can be regarded as a column with large cross sectional area and extremely small height which also results in a very low pressure drop at high flow rates and hence can be used to process large volume of dilute solutions.

Inverted matrix chromatography uses a resin matrix where the resin is the continuous phase in the matrix and the void spaces are distributed. This is in contrast to the normal resins which consist of discrete particles. Very low pressure drops are possible and their use in direct broth extraction is presently being investigated.

Affinity chromatography has been used for a number of years both at the laboratory as well as at the large scale. Most early industrial processes for therapeutic proteins included one or more affinity steps. This corresponds to a very high resolution purification even in the absence of physicochemical knowledge about the main contaminants (see Table 1). These however relied in many cases on very large ligand molecules (eg. antibodies) and were usually exhorbitantly expensive at either the large scale or even the laboratory scale. At present there is a tendency to use less expensive more 'generic' ligands. Important advances in this area have been obtained in the last few years which include the synthesis and use of specific **dyes** and also of **metal ions**. This technique is also known as **IMAC** (Immobilized

Metal Affinity Chromatography).

Recombinant DNA techniques have also allowed the development of highly specific separations. This has been carried out by adding a specific tail to the target protein by genetic engineering techniques (e.g. the zz domain of protein A) which binds a specific ligand (eg. IgG or a more inexpensive such as metal ion). After purification of the protein the tail is removed usually by enzymatic cleavage.

5. Electrically Enhanced Separations

Electrophoresis is capable of resolving biological molecules on the basis of differences in their molecular weights, isoelectric points and mobilities in an electric field with a very high resolution. At the laboratory scale this includes some of the most powerful techniques available for the purification of biologically active molecules. Isoelectric focusing (**IEF**) and Recycle Isoelectric Focusing (**RIEF**) are used in the laboratory as efficient preparative techniques to fractionate complex protein mixtures. The main advantage is the very high resolution, however removal of ampholites and the low protein solubility at its isoelectric point are disadvantages for large scale use. Other techniques that have been at least partially scaled up are Recycle Continuous Flow Electrophoresis (**RCFE**) and Counteracting Chromatographic Electrophoresis (**CACE**).

As electrophoresis is seldomly employed at the commercial scale primarily due to the convective mixing problems caused by ohmic heating it has recently been shown that an aqueous two-phase system can be used as a medium for electrophoretic transport and that the liquid-liquid interphase provides stability against convection and facilitates product recovery. The method combines the scalability of aqueous two-phase systems with the selectivity of electrophoresis.

Electrically enhanced membrane processes are presently being investigated as potentially useful techniques both for resolution of protein mixtures and for the separation of disrupted microbial cells.

6. Purification of Intracellular Proteins

A large number of proteins synthesized in *E. coli* and yeast are intracellular. These include intracellular protein particles such as recombinant vaccines. Hence the first step in their separation consists in their extraction from the contaminant cell material. Techniques for **selectively** carrying out the **solubilization** or **extraction** of the protein are presently being developed. These include the use of solvents (e.g. toluene), chelating agents (EDTA) detergents (eg. Triton X-100) and chaotropic agents (guanidine and urea). Also, the use of a pure lytic glucanase to selectively release protein particles from yeast has recently been reported.

References

1) Asenjo, J.A. (Ed.) Separation Processes in Biotechnology, Marcel Dekker, N.Y (1990).
2) Abstracts submitted to ECB6 (1993).
3) Recovery of Biological Products VI, Abstracts, Interlaken, September (1992).
4) Rodrigues, M.I., et al., *Chem. Eng. Sci.*, **47**, 263-269 (1992).

ECB6: Proceedings of the 6th European Congress on Biotechnology
edited by L. Alberghina, L. Frontali, P. Sensi

Cross-Flow Electrofiltration of disrupted microbial cells

A.Brors, K.H.Kroner and W.D.Deckwer

GBF, Gesellschaft für Biotechnologische Forschung mbH, Mascheroder Weg 1, 3300 Braunschweig, Germany

The influence of an superimposed electric field during cross-flow microfiltration of disrupted microbial cells has been analyzed. Results from a case study of the separation of malate dehydrogenase from cellular debris of Baker's Yeast show, that the overall performance can be significantly enhanced by means of a pulsed AC-field. A two to five fold increase of the specific enzyme transport rate was obtained at field strenghts of about 20 - 40 V/cm, showing no or only small effects on the stability of the enzyme.

1. INTRODUCTION

Cross-flow microfiltration is tend to be used increasingly for the separation of intracellular enzymes from suspensions of disrupted microbial cells. However, the method is limited in its performance due to sublayer formation and membrane fouling, resulting in low flux and a significant rejection of proteins in solution. Beside other means, such as hydrodynamic improvements or surface modifications, the application of an electric field perpendicular to the membrane may help to overcome the limitations [1]. Studies on the filtration of suspensions of whole microbial cells have shown substantial flux increases against conventional tangential filtration; the best results were obtained with alternating current fields [2]. Based on this findings we have tested further application for the separation of intracellular enzymes from cell homogenates. As a case study the separation of malate dehydrogenase (MDH) from Baker's Yeast has been choosen. Beside filtration performance the effect of the electric field on enzyme recovery (transmission, stability) was evaluated.

2. EXPERIMENTAL

Experiments were carried out with a self-made flat-channel membrane module from perspex as earlier described [1,2]. The dimensions of the flow channel were: L 275 cm; H 1 mm; W 4 cm; Area 96 cm^2. Titanium-electrodes were placed paralell to the membrane in the retentate and filtrate channel, respectively, directly exposed to the liquid. The electrodes were connected to a DC power supply; AC-fields were generated by means of a computer controlled switch. Through all experiments a pulsed AC-field was used, with polarity switches of the electrodes each after 5 seconds and an off-time between two pulses of 10 seconds (typed as 5/5/10). As membrane a 0.2 μm microporous nylon sheet (Ultipore, Pall Corp.) was used. The experimental runs were generally performed in a 'batch-recycle' mode. Enzyme and total protein concentrations were measured by standard laboratory assays taken samples from the process streams. Bakers's Yeast cells, suspended in 0.05 M potassium phosphate buffer (pH 7.5), were mechanically disrupted by high pressure homogenization at 700 bar (three times). The concentration for the cross-flow experiments was set as 10 % (wet mass/ volume), final pH 7.0. During parameter studies data were generally taken after 30 minutes run-time.

3. RESULTS AND DISCUSSION

In order to describe the performance of enzyme separation the specific enzyme transport rate Q_s (U/sm^2), given as the product of flux and enzyme concentration in the filtrate, was determined. Data shown in Fig.1 indicate, that in the presence of the electric field the overall performance was significantly increased (about two-fold). It can also be seen that the enzyme transport rate was stabilized over time in contrast to a run without an electric field under otherwise identical conditions.

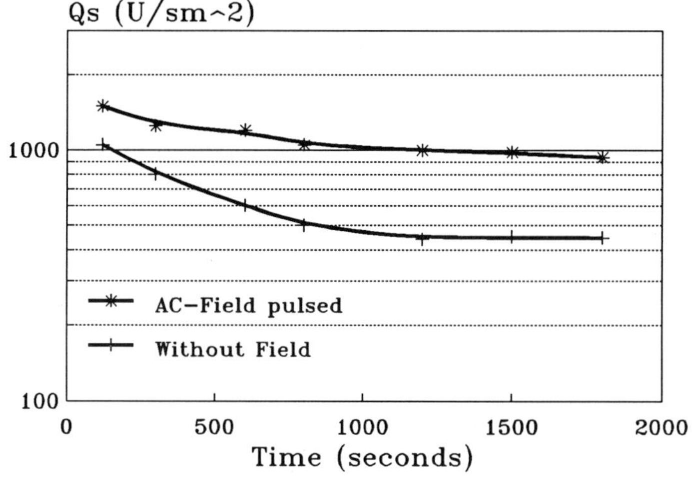

Figure 1: Specific Enzyme Transport Rate Q_s vs. Time. Conditions: Transmembrane pressure 0.5 bar; Tangential velocity 2.5 m/s; Temperature 17°C; pH 7.0; AC-field 5/5/10.

Analyzing the influence of the field strenght an optimum for Q_s was found between 20 - 40 V/cm as can be seen from Fig.2. At higher field strenght the transport rate tend to decrease, as also the flux values and the transmission (transmission of total protein T_p = ratio of concentrations between filtrate and retentate). The increase in Q_s seems to be mainly associated with the increase in flux and not of changing the enzyme rejection at the membrane. This can be explained by the assumption, that the electric field enhances the mass-transfer in the boundary layer, thus reducing the total filtration resistance (e.g. layer height, polarization), but obviously not the separation characteristics of the still remaining (stationary) layer attached to the surface of the membrane. The properties of the layer are mainly determined by surface interactions (adsorption), particle and pore size distribution (porosity, specific resistance). In this respect the transmembrane pressure is still a critical parameter, influencing the permeability and the sieving properties of such layers.

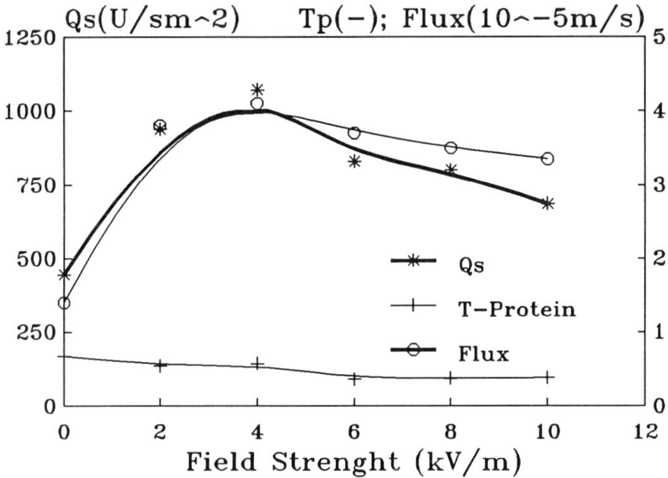

Figure 2: Influence of Field Strenght on Flux, Q_s and Protein Transmission. Conditions: Transmembrane pressure 0.5 bar; Tangential velocity 2.5 m/s; Temperature 16°C; pH 7.0; AC-field 5/5/10. Values taken from separate runs with fixed field strenghts after 30 min.

The decrease of flux, Q_s and protein transmission at field strengts > 40 V/cm seems to be correlated with the onset of electrolytic water dissoziation characterized by a significant increase in current flow at voltages > 4 V (the critical voltage depends on the electro-chemical system parameters, which determine the degree of electrode polarization). This could alter the properties of the biological materials leading to some denaturation or precipitation effects; already small amounts of precipitated proteins may lead to a pore blockadge. But also the increase in total protein mass-flux towards the membrane (polarization) and through the pores of the membrane (pore fouling) must be taken into account. Thus, an optimum of Q_s in dependence of the field strenght is reasonable.

The influence of pH (charge variations) on performance has been checked separately in the range of pH 5.5-8.0 (stability range of the enzyme). The results are shown in Fig.3. A slight increase of Q_s with pH has been found for the run at zero field strenght. This may indicate a reduced particle/solute interaction due to an increase in charge densities. The curves at 20 and 100 V/cm show a much steeper increase, reflecting the strong influence of the electrophoretic properties of the particles and the proteins. There seems to be an optimum at pH 7, which is close to the pI of the enzyme (6.8-7.1). But it is assumed that the observed enhancement of mass-transfer is mainly related to charge properties of the particles, which changes only slightly with pH (zeta-potential about -10mV at pH 7), and of the bulk proteins (pI < 6). It should further be considered that at the stability limits of pH some denaturation may occour, even at higher voltages (pH-changes

changes due to electrolysis). In Table 1 data of overall performance and recovery yields are summarized for comparison. It can be seen that at moderate field strenght there was nearly no loss of enzyme detected. The increase in flux and consequently of Q_s was about 4-5 fold at a transmembrane pressure of 0.2 bar, which was more favourable than higher pressures (compare Fig.1 and 2). Concluding our findings Cross-Flow Electrofiltration seems to be an interesting mean to improve the primary separation of intracellular enzymes. The method seems to be technical feasiable by using AC-fields.

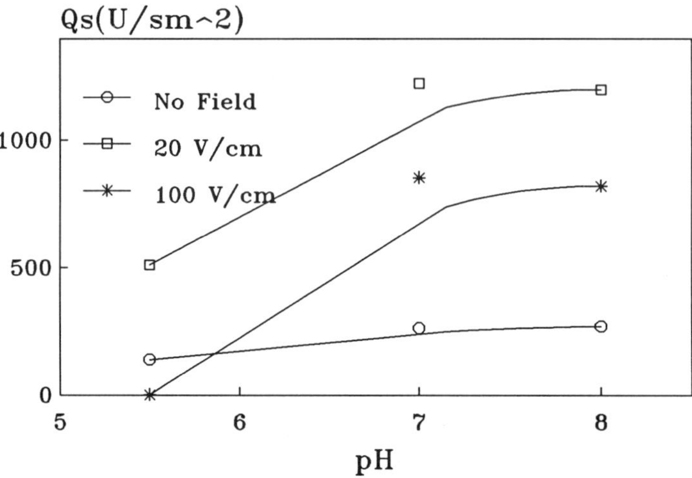

Figure 3: Infuence of pH Variation on Performance. Conditions as before except transmembrane pressure (0.2 bar).

Table 1
Comparison of performance data for the separation of MDH from disrupted Baker's Yeast

Field Strength* V/cm	Q^s U/m^2*s	Transmission %	Recovery %	Flux L/hm^2
0	263	75	98	32
2	1224	76	97	146
10	853	60	94	128

* AC-Field-pulsed (0.15 Hz); v= 2.5m/s; 0.2bar; C_{cells}=10% wet; pH=7; T=16°C

REFERENCES:

1. A.Brors, K.H.Kroner, H.Hustedt, W.D.Deckwer. In: Biochemical Engineering Stuttgart, M.Reuss, H.Chmiel, E.D.Gilles (eds.), Gustav Fischer, Stuttgart 1991, pp.287-292.
2. A.Brors, K.H.Kroner. In: Harnessing Biotechnology for the 21st Century, M.R.Ladisch, A.Bose (eds.), ACS Conf.Proceedings Series, Washington DC, 1992, pp.254-257.

ECB6: Proceedings of the 6th European Congress on Biotechnology
edited by L. Alberghina, L. Frontali, P. Sensi

Modelling of Centrifugal Partition Chromatography

M.J. van Buel[a], L.A.M. van der Wielen[a], R. Meester[a], R. Verpoorte[b] and K.Ch.A.M. Luyben[a]

[a]Department of Biochemical Engineering, Delft University of Technology, Julianalaan 67, 2628 BC Delft, The Netherlands

[b]Division of Pharmacognosy, Gorlaeus Laboratory, Leiden University, PO Box 9502, 2300 RA Leiden, The Netherlands

1. INTRODUCTION

Centrifugal Partition Chromatography (CPC) is a separation technique which is based on the difference in partitioning of components over two immiscible liquid phases. One of these phases is kept stationary in a column while the mobile phase flows through. To keep the stationary phase in the column, it has a specially designed internal geometry. As shown in Figure 1, a CPC-column consists of channels engraved in plates of polymer. The channels are connected by narrow ducts. Several plates are put together to form a cartridge. Several of these cartridges are placed in the rotor of a centrifuge and are connected to form the chromatographic column. If two immiscible liquids are present in the channel, the more dense liquid moves away from the axis because of the difference in density of the two liquids. Therefore the less dense liquid is pushed towards the axis. The mobile phase can either be the lighter or the denser phase. In the latter case, the mobile phase flows through the channels from the inside to the outside. The ducts are solely filled with mobile phase. This situation is depicted in Figure 1.

Two solutes with different partition coefficients introduced into the column via the mobile phase, distribute differently over the two phases, and consequently develop different migration rates and leave the column separated. The exact geometry of the channels might deviate from the one shown in Figure 1.

There are several reasons why CPC might gain in importance. A variety of liquid two-phase systems is available (any immiscible two-phase solvent combination may in principle be used), ranging from organic solvents to aqueous two-phase systems. This makes it possible to adapt the phase system to a specific mixture of components to be separated. In addition, the solvents are relatively cheap compared to solid chromatographic column fillings and can easily be replaced or reused. The amount of stationary phase in the channel can be as high as 80 % depending in operation conditions. The CPC cartridges are commercially available in volumes ranging from 125 ml up to 100 l. Several pilot scale processes have been developed and scale-up from laboratory to full production scale seems fairly simple (Cazes, 1988).

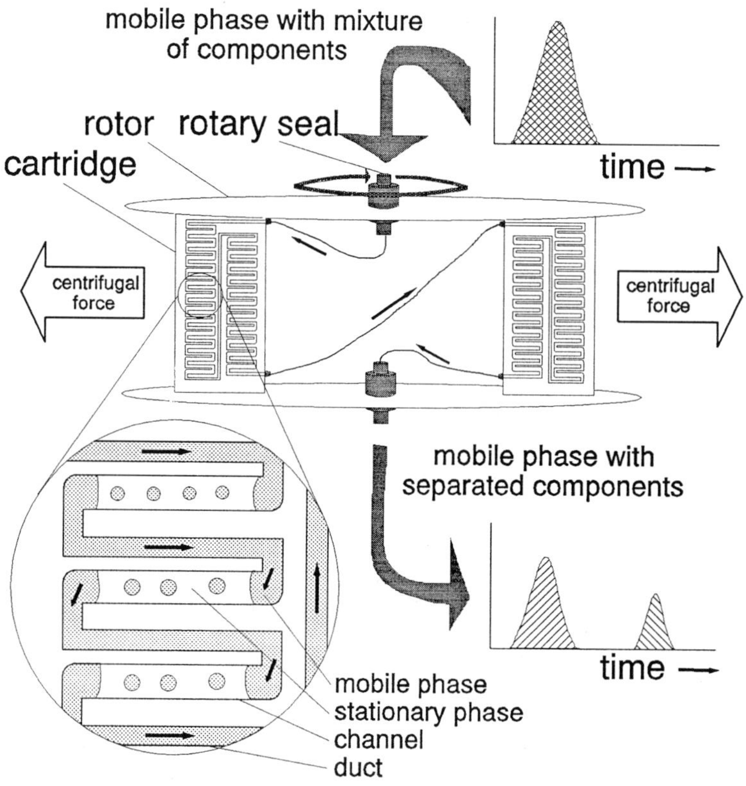

Figure 1. Schematic representation of a CPC-apparatus.

2. APPLICATIONS

CPC can be used for the separation of biomolecules with similar molecular structures and hydrophobicities (Foucault, 1991). For compounds such as proteins, which are unstable in systems with organic solvents, aqueous two-phase systems can be used, providing a mild environment (Foucault and Nakanishi, 1990a). The selection of a biphasic system in which the hyhobicity of the two phases differ too much, will not be optimal for the separation of related components. In this case, the components will mainly solve in one of the phases which results in a poor resolution. By adding a cosolvent, which is partially miscible with both phases, their hyhobicity differences are moderated, thus improving the resolution of the separation. The cosolvent can be applied both in an isocratic or gradient mode. Because of the changing composition of the phases in the latter case, a sound thermodynamic model is required to describe the phase behaviour of these biphasic ternary solvent systems (Foucault and Nakanishi, 1990b). Another way to achieve a separation is by adding affinity or ion pairing modifiers.

3. MODELLING OF CPC

For modelling of CPC, knowledge is required of the hydrodynamics, thermodynamics and mass transfer in a CPC cartridge. Figure 2 shows roughly the interactions between the relevant variables and phenomena in CPC.

The input variables are the flowrate, rotation frequency and temperature. The apparatus variables are the (internal) geometry and number of cartridges. By choosing a temperature and a two-phase system, the system variables (density, viscosity, interfacial tension, phase composition and partition coefficients of the solutes) are set. A sound thermodynamic basis is needed for the calculation of phase compositions and partition coefficients. The exact influence of mass transfer is not known. This article will emphasize on the flow regimes and mobile phase holdup in a CPC channel.

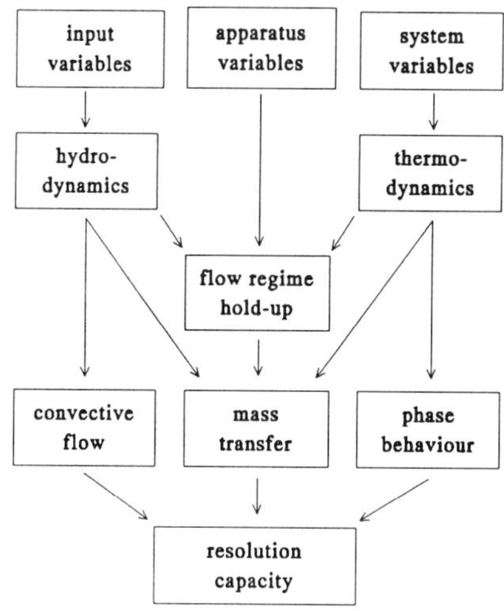

Figure 2. Engineering approach towards CPC showing the relations between variables and phenomena.

3.1 Flow regime and mobile phase holdup in a CPC channel

The flow rate, rotation frequency and the physico-chemical system variables (viscosity, density and interfacial tension) control the type of flow regime and thereby the maximum amount of mobile phase present in the column (mobile phase holdup). It is generally assumed (Armstrong *et al.*, 1988) that at low flow rates the mobile phase flows through the stationary phase in the form of droplets. The diameter of the droplets (d) in the channel is estimated by equating the centrifugal force and the interfacial tension:

$$V \omega^2 R \Delta\rho = \pi \sigma D \qquad (1)$$

in which V is the volume of the droplet, ω is the rotational frequency, R is the distance from the axis to the middle of the cartridge, $\Delta\rho$ is the density difference, σ is the interfacial tension and D is the hydraulic diameter of the duct. It should be stressed that the diameter of the droplet does not depend on the flowrate. If the droplet is a rigid sphere, its diameter can be calculated from its volume. At normal operating conditions the droplets have a diameter of 0.3 to 1.1 mm.

By using Stokes law corrected for wall effects (Strom and Kintner, 1958), the velocity of the droplets (v_d) is estimated and depends on the rotational frequency and the densities and the interfacial tension of the phases but does not depend the flowrate.

By increasing the flowrate at a fixed rotational frequency, more droplets are formed with this flowrate independent velocity. It is assumed that a jet is formed when two successively formed droplets touch. The critical flowrate ($\phi_{v,crit}$) of the transition from the droplet to the jet regime is reached when the time to form a droplet (dependent on the flowrate) is equal to the time in which a droplet travels his own diameter:

$$\frac{\frac{1}{6}\pi d^3}{\Phi_{v,crit}} = \frac{d}{v_d} \tag{2}$$

4 PRELIMINARY RESULTS

In Figures 3 and 4 some experimental results of the mobile phase holdup as a function of the flowrate are shown for a water-octanol and a hexane-ethylacetate-methanol-water (HEMH) system, respectively (Meester *et al.*, 1993). These results have been obtained with an analytical scale CPC type LLN (Sanki Engineering) with six cartridges. The mobile phase holdup depends on the number of droplets in the channel, which is a linear function of the flowrate. This results in a linear relation of holdup and flowrate at low flow rates. The critical flowrates at the conditions given in Figures 3 and 4 are $1.1 \cdot 10^{-7}$ and $0.6 \cdot 10^{-7}$ m/s, respectively. The predicted critical flowrates agrees with the point where the holdup deviates from this initially linear relationship, and supports the assumption of a flow regime transition.

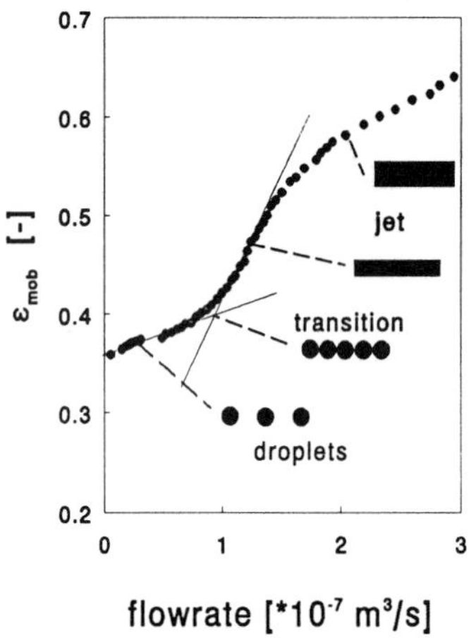

Figure 3. Mobile phase holdup as a function of the flowrate for octanol(stat)-water(mob) (experiment at 917 rpm, $\Delta\rho$=169 kg/m^3,σ= $2.7 \cdot 10^{-2}$).

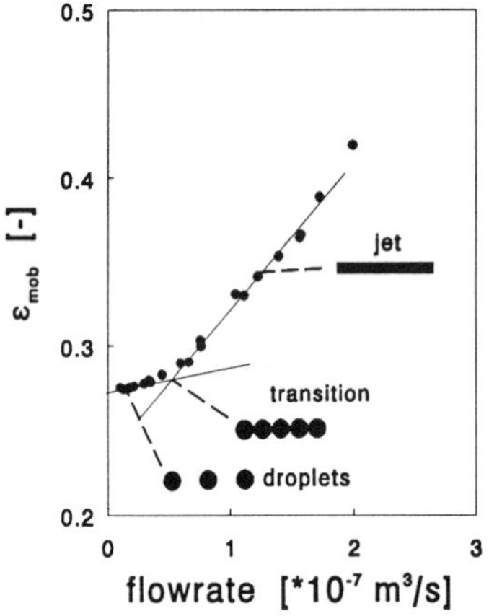

Figure 4. Mobile phase holdup as a function of the flowrate for HE(mob)-MH(stat) (experiment at 713 rpm, $\Delta\rho$=251 kg/m^3, σ=$6 \cdot 10^{-3}$).

At the transition point, the flow rate of the droplets should equal that of an equivalent jet. The linear velocity of the droplets, however, when compared according to Stokes law exceeds the linear velocity of an equivalent jet. According to equation 3, which is obtained by equating the flow rate expressions in each regime, the holdup in the jet-regime exceeds the holdup of the droplet regime.

$$\varepsilon_{m,d} v_d = \varepsilon_{m,jet} v_{jet} \quad (3)$$

in which $\varepsilon_{m,d}$ and $\varepsilon_{m,jet}$ are the mobile phase holdups in the droplet regime and the jet regime respectively and v_{jet} is the linear velocity of the jet. This increase in mobile phase is observed experimentally. A prediction of the holdup in the jet regime lacks due to the absence of an adequate model.

Experiments at gravitational conditions show the jet regime and a bypassing regime in which the mobile phase flows along the wall instead of flowing through the stationary phase (Meester *et al.*, 1993). With equation 1 it is calculated that the diameter of the droplet at these conditions is larger than the size of a channel, which might be the explanation for the bypassing regime. As no visual evidence of the type of flow regime in a channel occurring at centrifugal conditions has been reported as yet, a transparent cartridge has been constructed, and experimental data are being collected.

5 CONCLUSIONS

CPC is an attractive technology for the chromatographic separation of biomolecules. An engineering approach leading to an adequate description of CPC-contactors, concerning hydrodynamics, thermodynamics and mass transfer is presented. A first simple model concerning flow regimes and flow regime transition describes the preliminary results qualitatively. This encourages the engineering approach.

NOTATION

d	diameter of droplet	[m]
D	hydraulic diameter of duct	[m]
R	distance from axis to middle of cartridge	[m]
v_d, v_{jet}	velocity of droplet and jet	[m]
V	volume of the droplet	[m^3]
$\Delta\rho$	density difference	[kg/m^3]
σ	interfacial tension	[N/m]
ε_{mob}	mobile phase holdup	[-]
$\varepsilon_{m,d}, \varepsilon_{m,jet}$	mobile phase holdups in droplet and jet regime	[-]
$\phi, \phi_{v,crit}$	flowrate and critical flowrate	[m^3/s]
ω	rotational frequency	[rad/s]

REFERENCES

Armstrong, D.W., Bertrand, G.L., Berthod, A., (1988), Anal. Chem., **60**, 2513-2518.
Cazes, J., (1988), *Bio/technol.*, **6**, 1398-1400.
Foucault, A., Nakanishi, K., (1990a), *J. Liq. Chromatogr.*, **13**, 2421-2440.
Foucault, A., Nakanishi, K., (1990b), *J. Liq. Chromatogr.*, **13**, 3583-3602.
Foucault, A., (1991), *Anal. Chem.*, **63**, 569A-579A.
Meester, R., van Buel, M.J., van der Wielen, L.A.M., Verpoorte, R., Luyben, K.Ch.A.M., (1993), Proceedings of the symposium Precision Process Technology, eds. M.P.C. Weijnen en A.A.H. Drinkenburg, Delft.
Strom, J.R., Kintner, R.C., (1958), *A.I.Ch.E. J.*, **4**, 153-156.

ECB6: Proceedings of the 6th European Congress on Biotechnology
edited by L. Alberghina, L. Frontali, P. Sensi

High Performance Membrane Chromatography (HPMC), Superior to Conventional Protein-HPLC ?

Oscar-Werner Reif, Volker Nier, Ruth Freitag

Institut für Technische Chemie (Prof. Dr. Karl Schügerl), University of Hannover, Callinstrasse 3, 3000 Hannover 1, Germany

1. INTRODUCTION

Todays analytical and preparative protein separations depend heavily on various column based chromatographic methods. From a biotechnologists point of view this entails some major drawbacks. Columns are too expensive for one time use only. The high back pressure of efficient HPLC-columns limits the speed of analysis, a problem if process control is to be based on the analytical data. The scale up of an analytical method is difficult and requires the acquisition of a new column with the suitable dimensions. Furthermore, the problem of how to provide a sufficient adsorptive surface without loosing efficiency due to intraparticular mass transport resistance, has only been alleviated but not solved by the arrival of perfusion chromatography, i.e. the use columns packed with gigaporous particles where convective mass transport occurs within the large throughpores.

Many of the requirements of analytical biotechnology are better met if a cheap but efficient membrane is substituted for the column. Membranes have no interparticular void, mass transport through the pores is always convective. The plate number seems to be nearly independent of the flow rate. Concomitantly, membranes can be used at extremely high flow rates, due to their negligible flow resistance. Fast analytical separation become thus possible. The membranes cheapness allows one-time use. The scale up is easy, as a simple arrangement of several membranes in a row will suffice.

2. MEMBRANES

Sartobind Q anion-exchange membranes (Sartorius, Göttingen, Germany) and IDA-membranes (Sartorius, Göttingen, Germany) suitable for immobilized metal affinity chromatography were studied. The membranes surface area was 5.4 cm^2, their thickness 170-190 µm. Average pore sizes were given as 0.45 µm by the manufacturer. The membranes consist of a synthetic copolymer. They are stable at pH-values between 2 and 13. The commercially available syringe filter-membrane holders (SM 16517, Sartorius, Göttingen, Germany) used in the experiments, can withstand operating pressures of up to 600 kPa. The cartridges are equipped with female and male luer connectors at the inlet and the outlet respectively, so connection to standard HPLC and FPLC hardware becomes possible.

3. MEMBRANE PROPERTIES

Plate numbers and capacities as a function of flow rates between 1 ml/min and 50 ml/min were established for the SartobindQ membranes using toluene and BSA (bovine serum albumin, Sigma) as respective probes. The membrane capacity decreased by approximately 10 % if the flow rate was increased from 1 ml/min to 10 ml/min. Then a plateau was observed for flow rates up to 35 ml/min. Within the flow rate range defined by this plateau the most efficient separations in terms of speed and resolution are possible. The total protein concentration of a sample applied to a single membrane should not exceed 1mg/ml. If this is found too low for a given application, the systems capacity can be increased linearly by adding up to 9 further membranes, before the flow resistance becomes a problem.

The plate number, on the other hand, did not depend significantly on the flow rate of the mobile phase. High flow rates should therefore be possible in membrane chromatography without loss in efficiency. It was also verified, that the signal tailing was not increased by a change of the flow rate from 1 ml/min to 20 ml/min. The main contribution to the signal broadening stemmed from dispersion occuring at the cartridge outlet, which, other than the inlet where the problem of distributing the incoming flow has been preempted by the manufacturer, cannot be considered well suited to membrane chromatography. By halving the length of the outlet tube or, better, by connecting the tubing, leading to the detector, directly to the cartridge after total removal of the outlet, the width of the signals could be reduced considerably.

4. MEMBRANES COMPARED TO COLUMNS IN IMAC

In IMAC (Immobilized Metal Affinity Chromatography) protein separations are mainly based on the affinity of Histidin residues at the surface of the protein to metals such as Cu^{2+}, Zn^{2+}, Ni^{2+}, or Co^{2+}. The metal ions are immobilized at the surface of the chromatographic support by chelating agents, often IDA (imino diacetic acid) groups. IMAC has been used for the efficient removal of even trace amounts of recombinant proteins carrying a $(HIS)_6$-tag, i.e. six Histidin residues in a row, from cell lysates in a single step operation.

IDA-membranes and conventional columns (7.1ml) made by slurry packing of Chelating Sepharose Fast Flow (Pharmacia), an agarose based material also carrying IDA-groups, were used for the isolation of $(HIS)_6$-tagged EcoR V, an *E.coli* restriction endonuclease, from cell lysates of *E. coli*. Table 1 lists the capacities of the naked supports for Ni^{2+}, and the capacities of the Ni^{2+}-saturated supports for EcoR V.

	Membrane	Column
Ni^{2+}	13 - 14 μmol/cm^2	66 - 70 μmol/ml
EcoR V	93 μg/cm^2	181 μg/ml

Table 1 Capacities of IDA-membranes and IDA-columns for Ni^{2+}-ions and for the EcoR V on supports saturated with Ni^{2+}-ions.

The general retention properties of both systems are similar. Proteins could be desorbed in gradients of decreasing pH, increasing salt, or increasing imidazole concentration in the eluting buffer. EcoR V, however, was separated from the other *E. coli* proteins only in the last case. Figure 1a shows the separation on the IDA-column under optimized conditions. EcoR V is found only in the second peak, while the first contains a number of proteins. When the same linear gradient was used in case of the IDA-membranes, Figure 1b ensued. Analysis of the fractions by gel-electrophoresis showed, however, that a certain amount of separation did take place, i.e. fractions taken after approximately 40 minutes contained high amounts of EcoR V. Better results in terms of resolution were obtained, when the final buffer concentration was reached in a step gradient, Figure 1c.

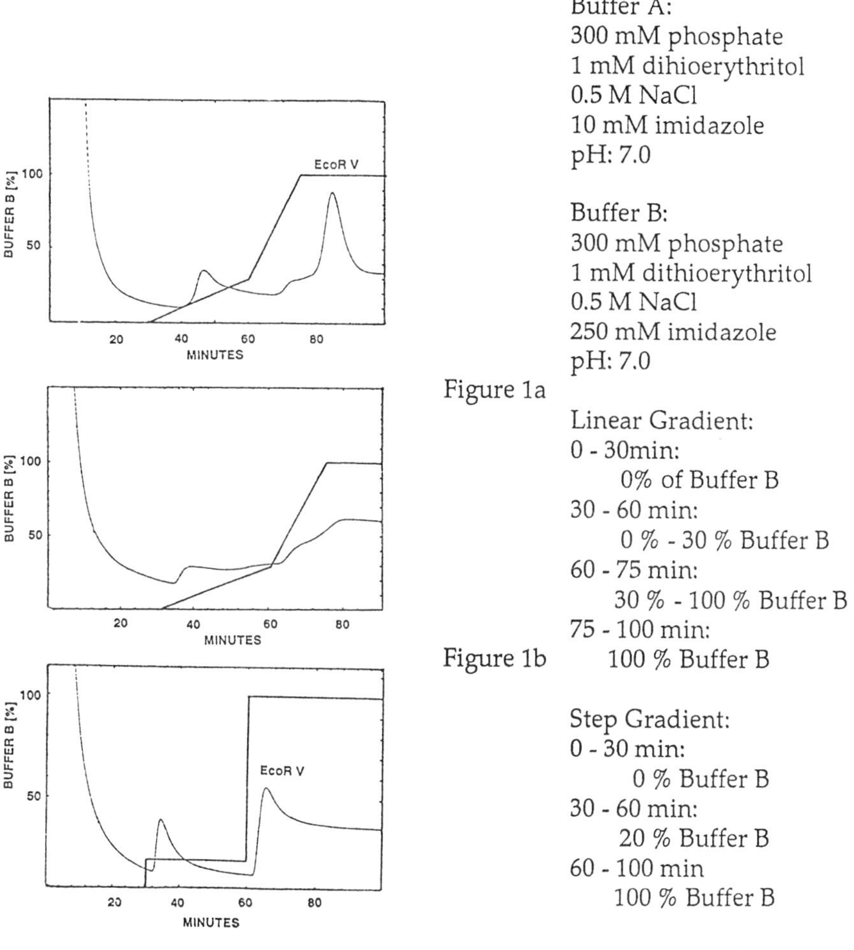

Figure 1a

Figure 1b

Figure 1c

While a linear gradient has a sharpening effect on the analyte zones in column gradient elution, in membrane gradient elution a step gradient is more suitable, else radial differences in the volumetric flow rate across the membrane will lead to zone broadening. The IDA-membranes are similar to the columns in terms of resolution. They can be used instantaneously, whereas the columns have to be prepared in a time consuming, not always reproducible, suspending/settling process. Presumably a flow rate of more than 1 ml/min could be used in case of the membranes.

5. SEPARATION OF PROTEIN MIXTURES ON SARTOBIND Q MEMBRANES

SartobindQ membranes were used to separate increasingly complex protein mixtures. Conditions for protein binding and elution were found to be similar to those described in conventional ion-exchange chromatography of proteins. Step gradients instead of linear gradients should be used. Protein recovery was usually better than 90 %. Trypsin inhibitor, α-Chymotrypsinogen, and Lysozyme (0.25 mg/ml each) were separated in less than 1 min in a three step pH/salt-gradient using flow rates of 30 ml/min. 1. step: 100 % Buffer A (10 mM borate, pH 10.2), 2. step: 20 % Buffer B (10 mM acetate, 1 M KCl, pH 4.0), 3. step 100 % Buffer B. Resolution was as good as when an idetical gradient was run at a flow rate of 1ml/min (~ 15 minutes).

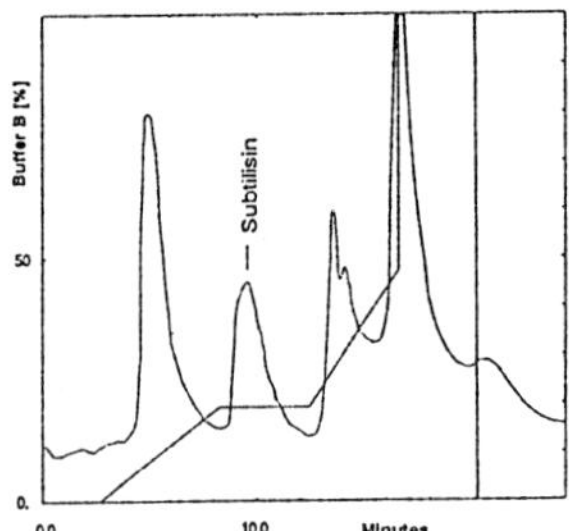

Figure 2

A combination of step and linear salt gradients was used in the separation of the protease *Subtilisin Carlsberg*, Figure 2, from culture supernatants of *B. licheniformis*. The Subtilisin was nearly pure, other fractions did not contain the enzyme. No loss of activity was seen.

Buffer A: 15 mM Tris/HCl, pH9.0

Buffer B: 15 mM Tris/HCl, 0.8 M KCl, pH 9.0

Flow Rate: 2 ml7min

Figure 3 shows the fractionization of human serum proteins in a salt/ pH-step gradient.

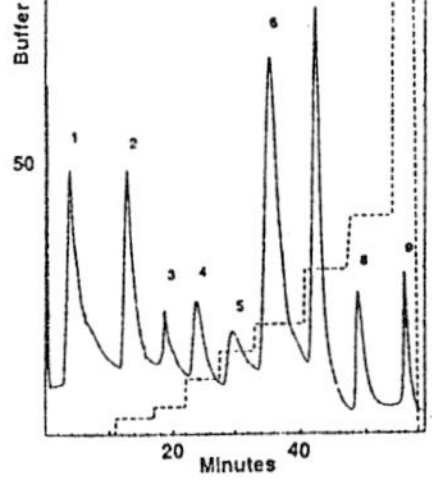

Buffer A:
10 mM Tris/borate, pH 8.7

Buffer B:
10 mM Tris/borate, 1 M KCl, pH 6.0
Flow Rate:
1 ml/min

Figure 3

Affinity based Processes

in

Bioseparations

ECB6: Proceedings of the 6th European Congress on Biotechnology
edited by L. Alberghina, L. Frontali, P. Sensi

Affinity methods in bioseparation

Y. D. Clonis

Enzyme Technology Laboratory, Department of Agricultural Biology & Biotechnology, Agricultural University of Athens, Iera Odos 75, Athens 118 55, Greece

Chromatography methods are used extensively for manufacturing several proteins and enzymes, and where substantial purity is demanded affinity chromatography is often employed [1,2]. The affinity concept is realised by forming reversible specific complexes between the (macro)molecule to be purified and the "ligand" immobilized on a suitable polymer support. Since its introduction, the development of this technology paralleled that of support materials, immobilization chemistry and ligand design. There have been used various types of porous beaded supports, for example, natural polysaccharides, synthetic organic polymers, inorganics, and composite materials. Following the advancement of monosized/monodispersed particles, a breakthrough is now offered by perfusion chromatography. In this technique the pores are wide enough to allow convective flow *through* the particles. When affinity ligands are coupled to such supports advanced bioselective adsorbents should be produced. Other materials for ligand coupling and use in non-column configurations are hollow fibres, flat membranes, phase-forming water soluble polymers and perfluorocarbon emulsions. The coupling chemistry of ligands is diverse and established, the simplest and safest being the immobilization of reactive dyes, whereas physical immobilization approaches have been reported as well. Numerous ligands are used in affinity separations; those of high-selectivity (e.g. antibodies, peptides, biomimetic dyes, and multisequence-specific polynucleotides) and those of lower selectivity (e.g. triazine dyes, nucleotides, immobilized metal ions, lectins, phenylboronate, DNA, heparin, protein A and G).

1. DYE-LIGAND CHROMATOGRAPHY

Perhaps it is the most promising affinity method of large-scale potential. The ligands employed are commercial textile reactive chlorotriazine polysulfonated dyes [3,4] which are readily immobilized on supports with hydroxyl groups. Unlike the various adsorbents based on biological affinity ligands, the stability of dye-adsorbents is limited only by the support itself. Dyes offer clear advantages, they are inexpensive and available in bulk quantities, chemically and biologically stable, can easily and safely be immobilized to a large variety of supports, and the corresponding adsorbents exhibit high protein binding capacity. The overall size, shape and distribution of charges and hydrophobic groups enable these molecules to interact with the binding site(s) of target proteins. Although the binding is often dominated by electrostatic forces, the dye-protein interaction should not be compared to simple ion-exchange type since binding is frequently possible at pHs greater than the pI of the protein. Furthermore, elution is often effected by soluble competing ligands which suggests interaction with protein at discrete sites. This view is supported by spectra difference, kinetic and inactivation studies all showing that many triazine dyes interact with several dehydrogenases and kinases in a specific manner at the nucleotide binding sites. Although the most widely used dye, Cibacron blue 3GA (Reactive blue 2) as well as other dye structures, generally, bind proteins with moderate specificity, this has not prevent them from acting as effective ligands in numerous purification cases. Immobilized Cibacron blue 3GA is used for the large-scale purification of human albumin where trace contaminants of leached dye can be quantified by an ELISA method. Several recent purification protocols may be found where immobilized dyes are employed, nevertheless, to address the problem of moderate dye's selectivity, two procedures are available. In the first, screening of numerous dye-adsorbents takes place to establish these exhibiting the highest purification factor for the target macromolecule. The second approach has opened a new era to enzyme downstream processing and is based on a new generation of dyes, the so-called biomimetic dyes. Purpose-made structural changes on the parent dye lead to new redesigned dyes mimicking naturally occuring biological ligands, thus, exhibiting increased selectivity towards target enzymes. Immobilized Cibacron blue 3GA of which the terminal 2-aminobenzylsulfonate ring is substituted with 4-aminobenzylphosphonic acid is able to purify calf intestinal alkaline phosphatase by 330-fold in a single step. Horse liver alcohol dehydrogenase displays a 7-fold decrease of its dissociation constant with Cibacron blue 3GA which terminal aryl-*ortho*- sulfonic acid group has been substituted by a *meta*- carboxyl group. The two main proteolytic constituents of crude bovine pancreatic extract, typsin and chymotrypsin, may be totally separated on agarose-immobilized cationic benzaminido-triazine dye where only trypsin binds. This new field of biomimetic dye technology is expected to follow rapid development and solve problems associated with the selectivity of commercial textile dyes.

2. SEQUENCE-SPECIFIC DNA AFFINITY CHROMATOGRAPHY

Earlier purification methods of DNA-binding proteins were based on immobilized DNA adsorbents. The idea of applying sequence-specific DNA synthetic fragments as ligands for isolating DNA-binding proteins and enzymes attracts a lot of attention. The principle can be applied, for example, to the purification of two DNA-binding proteins from nuclear extracts of HeLa cells implicated in transcription control. The adsorbent employed is Sepharose CL-4B-immobilized 26-mer duplex form synthetic oligonucleotide sequence-specific ligand. A group of enzymes where this technology holds great potential is that of restriction endonucleases [5,6]. For example, Hpa I, Sph I, EcoR V and EcoR I, can be purified to near homogeneity in two steps, DNA-cellulose affinity chromatography and sequence-specific DNA affinity chromatography. Three self complementary oligonucleotides were synthesised, each one was annealed with itself, 5' phosphorylated and ligated. Each of the resulting sequence-specific dsDNA oligomers, after purification, was immobilized to CNBr-activated beaded agarose. Likewise, an agarose-immobilized sequence-specific dsDNA synthetic ligand containing recognition sequences for 34 different restriction enzymes was proved effective in the purification of 6 restriction endonucleases.

3. IMMUNOAFFINITY CHROMATOGRAPHY

Immunoaffinity chromatography (IAC) [7] exploits the highly specific interactions between antigen-antibody and the technique may be used with either polyclonal or monoclonal antibodies (MAb). Antibody fragments alone can interact with antigens, thus, by solid-phase chemical synthesis, it is possible to create ligands consisting of amino acid sequences similar to a suitable part of MAb directed against the molecule to be isolated, e.g. lysozyme. Based on the same logic, F_V antibody fragments have been cloned and expressed in *E. coli* and the resulted binding domain was used in IAC purification of hen-egg lysozyme. Another approach to IAC purifications is the use of cross-reactive antibodies. Anti-barley a-amylase antibody was used to purify rice a-amylase cloned, expressed and secreted by yeast cells. Also, purification of the same enzyme was effected by anti-peptide antibody ligand that was prepared against the eight amino acid sequence of the C-terminal region of rice a-amylase.

4. AFFINITY TAILS

The large-scale production and downstream processing of heterologous proteins can be facilitated by a genetic approach which is based on the attachment of an affinity tail at the target protein by covalent bond. In principle, the target-protein gene is fused after a synthetic fragment encoding the "affinity tail", e.g. binding domains derived from staphylococcal Protein A which are known to exhibit affinity for IgG. The fusion product is secreted to the growth medium, e.g. of *E. coli*, and purified on an IgG-affinity column which binds specifically the affinity tail of the fused product [8]. Afterwards, the affinity tail is disconnected by chemical cleavage from the fused protein. Taking this concept further, a dual affinity fusion approach is designed [9]. Accordingly, the gene encoding the target product is fused between two flanking heterologous genes encoding IgG- and albumin-binding domains, respectively, of protein A and protein G. The tripartite fusion product is purified by affinity chromatography on HSA-agarose and IgG-agarose columns using both the N- and the C-terminal domains. Immunoaffinity chromatography is employed in separations of tail-fused proteins where a small peptide, e.g. of eight amino acids, is engineered onto the N-terminal of the target-product. An antibody specific for the engineered sequence is used as affinity immobilized ligand to purify the target-protein by binding at its tail. Another type of affinity tail is the poly-histidine peptide which is also fused to the protein of interest. In this case, the affinity adsorbent used to purify the fused protein is a metal-chelator.

5. ANTISENSE PEPTIDE LIGANDS

Amino acid sequences encoded in the antisense strand of DNA are called "antisense peptides" and may be used as ligands apparently with appreciable selectivity against the corresponding "sensepeptides" or "senseregions" of proteins [10]. Following computer-predicted exposure of protein fragments on the surface of β-interferon, were selected the corresponding antisense peptides encoded in the antisense strand of DNA corresponding to three fragments of the target-protein. A somewhat different approach is offered by the so-called "paralog" peptide ligands which provide molecular scale multi-mode adsorbents of appropriately diversified characteristics (charge, hydrophobicity).

6. IMMOBILIZED-METAL AFFINITY CHROMATOGRAPHY (IMAC)

The IMA adsorbents [11] are designed to chelate metal cations that exhibit selectivity for specific groups, e.g. the imidazole of histidine residues, in peptides and on protein surfaces. Immobilized iminodiacetic acid (IDA) remains the most popular adsorbent, however, tris(carboxymethyl)ethylenediamine, nitrilotriacetate and 8-hydroxyquinoline are some alternatives. By varying the metal-ion, one can adjust the strenght of interaction between protein and IMAC adsorbent, whereas elution is effected by pH change, inclusion of EDTA or competing species .

7. GLYCOCONJUGATE-BINDING LIGANDS

Lectines are glycoproteins whose affinity for the sugar moiety of target protein varies with their origin [12]. Immobilized concanavalin A has been commercially available as affinity adsorbent for years and still finds several applications. Boronic acid affinity adsorbents [13] offer an alternative to lectins. Aminobenzeneboronic acid immobilized via its arylamino group gives affinity adsorbents which form stable complexes with sugar diols. The pK_a of the acid is about 9, therefore, the working pH cannot be much lower than 8,0-8,5 in order for the boronic acid to be in its functional boronate anion form. In certain occasions use is made of individual sugars as affinity ligands, e.g. immobilized lactose adsorbent purifyies toxins and agglutinins.

8. HIGH-PERFORMANCE AFFINITY CHROMATOGRAPHY (HPAC)

Modern support materials consist of macroporous hydrophilic beaded particles, usually bearing free hydroxyl groups available for ligand attachment. When the support is made of noncompressible spherical particles of small diameter and narrow size distribution the technique is termed high-performance affinity chromatography [14]. Silica-based supports have been used widely in HPAC though they present serious drawbacks such as poor chemical stability in alkaline conditions, need for derivatisation with organofunctional silanes prior to ligand immobilization, and relatively small pore size. It appears that such problems are now solved with the introduction of suitably coated silica and purpose-designed synthetic high-performance particles with free hydroxyls or activating groups. In spite of its problems silica is still employed in HPAC, mainly at analytical-scale, with just only one large-scale example, the purification of rabbit muscle lactate dehydrogenase on a 3.3-lit column of silica-immobilized triazine dye Procion MX-R. Several other HPAC separations with silica-based adsorbents have been reported. The affinity ligand 8-[(6-aminohexyl)amino]-2'-phosphoadenosine-5'-diphosphoribose immobilized to epoxy-activated silica has been used to purify two $NADP^+$-dependent enzymes from various sources. Resolution of adenylic acid oligomers, $(dA)_{19-24}$, and poly(A) mRNA from yeast can be accomplished on a 50-mer of thymidylic acid, $(dT)_{50}$, immobilized to silica. Simultaneous extraction of albumin and transferrin from immunoglobulin G is performed on a mixed-function HPAC adsorbent consisting of immobilized anti-albumin and anti-transferrin antibodies. Synthetic supports offer a potential alternative to silica: evaluation of high-performance supports for purifying lactate dehydrogenase, revealed that two synthetic HPAC materials exhibited superior performance. High-performance IMAC with synthetic particles is an effective technique in evaluating the adsorption properties of 67 synthetic, biologically active, peptides of different molecular size. Two biomimetic dyes, a phosphonic acid-blue analogue and a benzamidino-cationic which bind, respectively, alkaline phosphatase and urokinase were immobilized on a monosized synthetic support. The resulting dye-affinity adsorbents purified the two target enzymes from crude extracts in good yield and specific activity. A different approach to column packing has been introduced by the concept of deformed non-porous agarose high-performance beads. Non-porous quartz fibres are new material which requires silylation prior to ligand coupling.

9. OTHER AFFINITY CHROMATOGRAPHY-BASED TECHNIQUES

Centrifugal affinity chromatography (CAC) [15] combines the high flow-rate, created by centrifugation force, with the specificity of affinity chromatography and it is a less expensive technique than HPAC. Rather than packing conventional adsorbent particles in a fixed bed, solid/liquid contact can be achieved in well-mixed reactors. Continuous operation is effected by recirculation of the adsorbent between two or more contactors. This approach is termed continuous affinity recycle extraction (CARE) [16]. In affinity-repulsion chromatography [17] electrostatic charges, of the same sign as that of the target protein, are present on the affinity adsorbent. The strength of such electrostatic interactions between target protein and adsorbent are minimized in the presence of salt. Therefore, the protein mixture is loaded on the column in a buffered salt solution and adsorbed molecules are eluted at low ionic strenght or with deionized water. Combined use of a heterobifunctional affinity ligand [18], whose one of the two different affinity moieties is reacting in-solution with the target protein to be purified, while the second one is left to bind to a suitable affinity adsorbent, presents yet another affinity chromatography technique.

10. AFFINITY MEMBRANES

Covalent attachment of an affinity ligand to a filtration membrane provides the basis of affinity-membrane fractionation. Central to the technique's speed, as it was first introduced, is a hollow-fibre microporous membrane to the interior of which the ligand is immobilized [19]. The membrane pores are large enough to permit convective flow, so the binding of the target protein to ligand is not limited by diffusion. Affinity membranes allow rapid processing of large volumes even right after the bioreactor harvesting, eliminating concentration and partial-purification steps. Protein A-membrane is employed to purify mouse IgG monoclonal antibodies from serum. A module of 10-ml hollow fibre affinity membrane processes in 15 min 1.2-lit crude monoclonal leading to purification by 95% in nucleic acids and 99.9% in albumin. Membrane discs and modules of flat membrane sheets bearing immobilized ligands have been used in downstream processing. The purification of formate dehydrogenase from *Candida boidinii* is realised in a three-step membrane filtration process [20]; cation-exchange, Procion red HE-3B and anion-exchange. Hydrophilic modified polyvinylidene difluoride, activated nylon, cellulosic polymers, polysulphone and polyacrolein mixtures are some of the membrane material range.

11. AQUEOUS TWO-PHASE AFFINITY PARTITIONING

Affinity partitioning [21] is regarded as the non-expensive approach to affinity-based purification techniques. When two aqueous solutions each containing a high molecular weight polymer, e.g. dextran and polyethylene glycol (PEG), are mixed and then left to settle, depending on the molecular weight of the polymers, their concentration and temperature, two phases may be formed. Because the upper phase is PEG-rich and the bottom phase dextran-rich, when some of the PEG is replaced by PEG-immobilized affinity ligand, the latter is restricted at the upper phase along with the bound target protein. Unbound substances remain at both phases and at different relative concentrations, therefore, some purification of the target protein is achieved but, of course, not as high as that obtained with other affinity methods. The advantage of affinity partitioning, especially important at industrial scale, is that it can be used at the first stage of a purification process, immediately after cell disintegration, without need for centrifugation or debris removal. Nevertheless, the presence of PEG carried along at subsequent fractionation stages may interfere with column performance. Immobilized triazine dyes on PEG have been widely exploited in two-phase affinity systems for partial purification of enzymes. When working with dye-ligands the usual choice of target enzymes are those binding nucleotides, e.g. glucose-6-phosphate dehydrogenase, phosphofructokinase, restriction endonucleases etc. Finally, affinity partitioning is also useful in ligand-receptor recognition separations.

It is believed that the areas of biomimetic dyes, sequence-specifid DNA, anti-peptides, affinity tails, perfusion chromatography and affinity membranes will play an important role in downstream processing.

REFERENCES

1. Y. Clonis, Bio/Technology, 5 (1987) 1290.
2. Y. Clonis, in: Separation Processes in Biotechnology (J. Asenjo, ed.), Marcel-Dekker, New York, p. 401, 1990.
3. Y. Clonis, A. Atkinson, C. Bruton and C. Lowe (eds), Reactive Dyes in Protein and Enzyme Technology, Macmillan Press, Basingstoke, 1987.
4. Y. Clonis, in: HPLC of Proteins, Peptides and Polynucleotides (M. Hearn, ed.), VCH, New York, p. 453, 1991.
5. G. Vlatakis and V. Bouriotis, Anal. Biochem., 195 (1991) 352.
6. C. Pozidis, G. Vlatakis and V. Bouriotis, J. Chromatogr., (1993) in press.
7. H. Ehle and A. Horn, Bioseparation, 1 (1990) 97.
8. T. Monks, L. Abrahmsen, B. Osterlof, S. Josephson, M. Ostling, S.-O. Enfors, I. Persson, B. Nilsson and M. Uhlen, Bio/Technology, 5 (1987) 379.
9. B. Hammarberg, P.-A. Nygren, E. Holmgren, A. Elmbland, M. Tally, U. Hellman, T. Momks and M. Uhlen, Proc. Natl. Acad. Sci. USA, 86 (1989) 4367.
10. I. Chaiken, J. Chromatogr., 597 (1992) 29.
11. F. H. Arnold, Bio/Technology, 9 (1991)151.
12. D. O'Shannessy and M. Wilchek, Anal. Biochem., 191 (1990) 1.
13. J. Hageman and G. Kuehn, in: Practical Protein Chromatography (A. Kenney and S. Fowell, eds.), Humana Press, Totowa, p.45, 1992.
14. Y. Clonis, in: Practical Protein Chromatography (A. Kenney and S. Fowell eds), Humana Press, Totowa, p.105, 1992.
15. R. Slingerland and W. Scouten, J. Chromatogr., 510 (1990) 205.
16. N. Gordon, H. Tsujimura and C. Cooney, Bioseparation, 1 (1990) 9.
17. V.I. Teichberg, J. Chromatogr., 510 (1990) 49.
18. B. Mattiasson and U. Olsson, J. Chromatogr., 370 (1986) 21.
19. B.J. Spalding, Bio/Technology, 9 (1991) 229.
20. B. Champluvier and M.-R. Kula, Bioseparation, 2 (1992) 343.
21. H. Walters, G. Johansson and D. Brooks, Anal. Biochem., 197 (1991) 1.

ECB6: Proceedings of the 6th European Congress on Biotechnology
edited by L. Alberghina, L. Frontali, P. Sensi

Kinetics and equilibria of interactions between lysozyme and dye/dextran conjugates

A.G. Mayes[+], R. Eisenthal[+] and J.Hubble[*]

[+] Department of Biochemistry and [*]School of Chemical Engineering, University of Bath, Bath, BA2 7AY, U.K.

Interactions between the triazine dye Cibacron Blue F3GA and the protein lysozyme have been investigated by monitoring the characteristic spectral shift which accompanies the binding phenomenon. This approach has been used to compare experimental results obtained with both free and dextran coupled dye allowing the effects of dye conjugation to be quantified in the absence of diffusional limitations seen with insoluble supports. Changes in absorbance have also be used to follow binding kinetics using a stopped flow system.

1. Introduction

To assess the consequences of ligand immobilisation on binding phenomena in the absence of diffusional limitations[1] it is possible to use soluble supports so that binding occurs in a continuous liquid phase[2]. In such a system there will be negligible mass transfer limitations, and kinetic and equilibrium binding results can be used to provide a quantitative assessment of the effects of ligand immobilisation on the intrinsic parameters of the interaction.

Measurement of ligand-protein binding by spectrophotometry is a well established technique[3], and is attractive as it can be applied to equilibrium and kinetic measurements with equal facility. In a previous report we showed that spectral titration can be used to compare interactions of both free and supported dye with the protein lysozyme[4]. The results obtained demonstrated the potential of the technique but highlighted problems caused by the stacking interactions known to occur between planar dye molecules in close proximity[5,6]. Dye stacking can have two significant effects on the results obtained from spectral studies, arising from 1) the difference in extinction coefficient between stacked and unstacked dye, and 2) from the effect of stacking on dye availability for binding.

2. Methods

Dextran-dye conjugates of different dye-dextran ratios were prepared and characterised using methods described previously[2,4]. For equilibrium titrations, sample and reference cuvettes each containing a solution of approximately 70μM (with respect to dye) conjugate in 10mM tris 50mM NaCl pH8.0 were equilibrated at 25C in a Cecil 272 spectrophotometer with a thermostatted cuvette holder. Small volumes of 5mM lysozyme solution were added to the sample cuvette and equal volumes of buffer added to the reference cuvette. Reaction kinetics were followed using a stopped flow apparatus with associated computerised data capture.

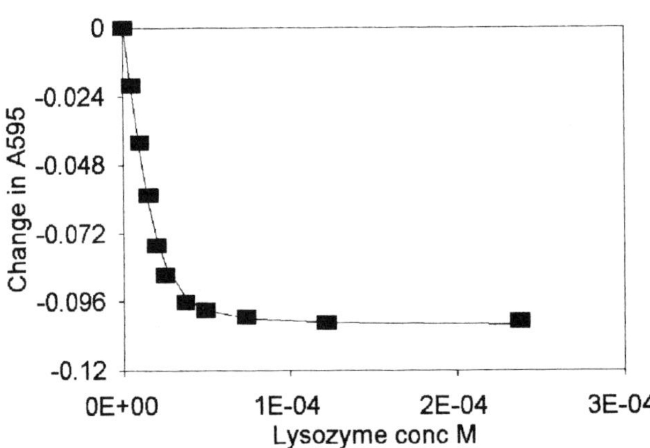

Fig. 1. Absorbance change during equilibrium titration of lysozyme with 83 uM 20D10 dye-dextran conjugate

3. Results

Equilibrium and kinetic experiments were conducted with conjugates based on 5 x 10^5 Dalton dextran (5D series) and 2 x 10^6 Dalton dextran (20D series). The experimental data were fitted to the appropriate equations[7] to give estimates of the dissociation constant, K_d, the stacking constant, K_s, and the dye available for binding, D_a. Figures 1 and 2 show typical fits obtained.

4. Discussion

In deriving the theoretical relationships used in this study the aim was to take account of the high affinity interaction of protein with dye, and also dye/dye stacking interactions. Comparison of the values obtained from the different experiments shows that all give a similar value for the extinction coefficient for unstacked dye of 12000 M^{-1}

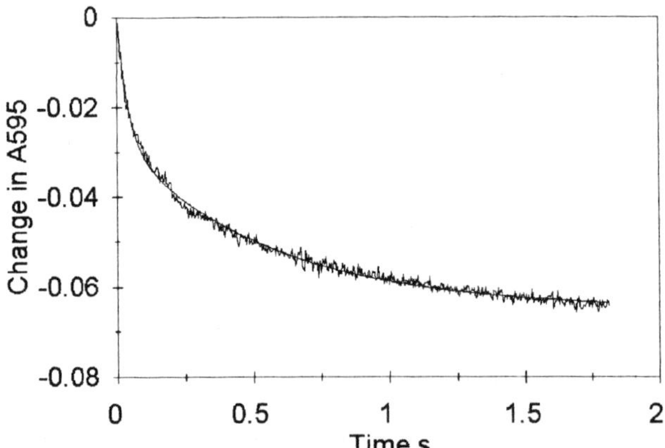

Fig. 2. Kinetics of interaction of 20D10 dye-dextran conjugate (43uM) with lysozyme (250uM)

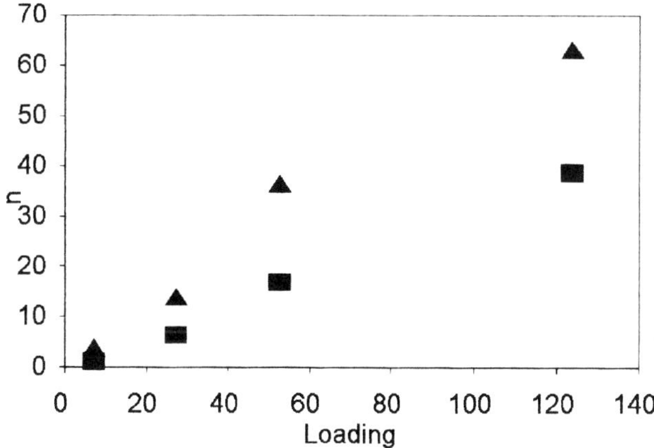

Fig. 3. Variation of n with dye loading (moles dye/mole dextran) for 5D dye-dextran conjugates (▲kinetic estimates ■ equilibrium estimates)

(mean 11960, s.d. 87) which is in good agreement with literature values reported for the molar extinction coefficient of Cibacron blue (13000 M^{-1})[8]

As expected the values of both stacked and protein bound dye are markedly lower than this value. This finding is consistent with the assumptions of the model. The fact that the estimates are in the main similar from both kinetic and equilibrium data is also reassuring.

Estimates of available dye concentrations also show little variation from preparation to preparation. The value of n (the maximum number of dye moieties per conjugate molecule capable of participating in binding) as a function of loading is shown in figure 3. However, kinetic studies suggest a higher availability, which requires explanation.

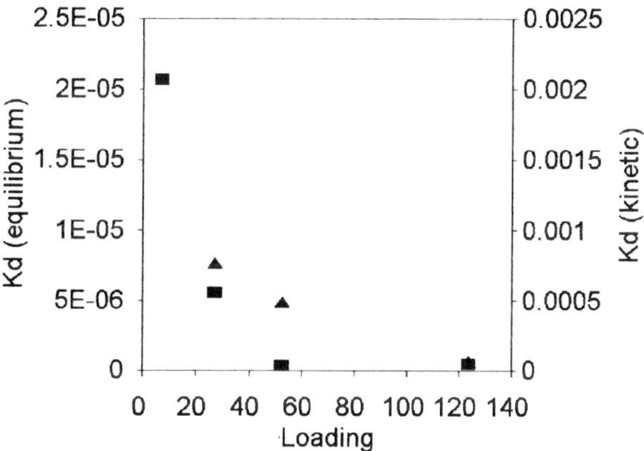

Fig. 4. Variation of Kd with dye loading, conditions and symbols as for figure 3.

Our previous work suggested that K_d should decrease with increasing dye dextran ratio. A similar effect might be expected for stacking interactions where stacking might be enhanced by the closer proximity of adjacent dye molecules. In fact the results do show that K_d values decrease with loading (figure 4). However, there is not good agreement between the estimates of K_d obtained from kinetic studies with equilibrium determinations. Two explanations may be advanced.

i) The results obtained from our earlier gel permeation studies[2] suggested that there are two classes of interaction between lysozyme and Cibacron Blue. Both show saturation phenomena, but may be distinguished on the basis of their half saturation constants. It is probable that the spectral changes described in this report are a consequence of the higher affinity interactions, and that the lower affinity interactions are not directly quantified by this technique. Although not directly apparent in spectral titration studies, low affinity binding effects could lead to changes in ligand availability on a timescale longer than those employed for the kinetic studies and hence might explain the lower D_a estimates obtained from equilibrium results. Similarly the K_d values obtained from equilibrium spectral changes are 4-10 fold lower than those from gel permeation studies providing further evidence for this effect.

ii) It is possible that the interaction of protein with ligand is a two-stage process with the complex undergoing a transitional change between an initial and a stabilised state. Given that the transitional change is an equilibrium process and that the extinction coefficients of the two forms are identical it can be shown algebraically that this would lead to a higher value for the apparent K_d in the case of kinetic experiments.

The value of the stacking constant K_s again appears to show little systematic variation with ligand loading. However, other results[7] suggest that the degree of ligand loading significantly affects the size of the resultant complex and it may be that this complicates the effect of ligand/ligand proximity on the apparent value of K_s.

Acknowledgement

The authors thank the SERC for support through Grant No. GR/F/90967, and a studentship for A.G.M.

5. References

1 S.L Fowell and H.A. Chase, J. Biotechnol., 4 (1986) 355.
2 A.G. Mayes, J.D. Moore, R. Eisenthal and J. Hubble, Biotech. Bioeng., 36 (1990) 1090.
3 S.T. Thompson and E. Stellwagen, Proc. Nat. Acad. Sci., 73 (1976) 361
4 A.G. Mayes, J. Hubble and R.Eisenthal, Biotech. Bioeng., 40 (1992) 1263.
5 M.M Federici, P.B. Chock and E.R. Stadtman, Biochemistry 24 (1985) 647.
6 E. Stellwagen and Y-C. Liu, in I.M. Chaiken (Ed.), I.M. Analytical Affinity Chromatography, CRC Press, Boca Raton, 1987, p170.
7 J. Hubble, A.G. Mayes, and R. Eisenthal, Anal. Chim. Acta (1993) in press.
8 G.K. Chambers (1977) Anal. Biochem., 83 (1976) 551.

ECB6: Proceedings of the 6th European Congress on Biotechnology
edited by L. Alberghina, L. Frontali, P. Sensi

Purification of secreted proteins by immunoaffinity chromatography with cross-reactive and anti-peptide antibodies

Shigeo Katoh and Masaaki Terashima

Dapartment of Synthetic Chemistry and Biological Chemistry
Kyoto University, Yoshida Hon-machi, Sakyo-ku, Kyoto 606, Japan

1. INTRODUCTION

In recent years many high-value bioproducts have been produced by use of microorganisms genetically designed. However, the concentrations of these products, especially secreted ones, in fermented broth are usually very low. For the purification in these cases, immunoaffinity chromatography would be very effective because of the high affinity and specificity between antigens and antibodies. Although this method was often used in the final stage of purification steps, it may be much effective to use it at an early stage for concentrating and purifying secreted proteins of very low concentrations in fermentation broth.

In immunoaffinity chromatography it is very important to obtain antibody ligands with suitable characteristics for purification, *i.e.* high affinity and selectivity at the adsorption step and mild desorption conditions at the elution step. In most cases, enough amount and purity of antigens (target proteins) for immunization are not available in production of recombinant proteins at the first stage. Selection of monoclonal antibodies showing these characteristics is one approach, but it might be a hard task. As alternatives we propose two methods, *i.e.* utilization of cross-reactive antibody and anti-peptide antibody for the immunoaffinity purification of rice α-amylase produced by secretion from recombinant yeast cells. α-Amylases from various species show high homology in their amino acid sequences, and so antibodies against these α-amylases are expected to show cross-reactivity. Recently it was shown that antibodies against short peptides which consist of parts of sequences of proteins can bind to native proteins [1]. Anti-peptide antibodies with these characteristics may be used as ligands for immunoaffinity purification of native proteins. In this work by use of cross-reactive or anti-peptide antibodies, which can be obtained without target proteins, rice α-amylases secreted from yeast cells were purified.

2. MATERIALS AND METHODS

The cloned genes of two rice α-amylase isozymes (RAmy 1A and RAmy 1A/3D) were expressed by yeast cells and secreted in fermentation broth [2]. These α-amylases were purified from the concentrated broth by immunoaffinity chromatography with two types of antibody ligands. One was anti-barley α-amylase antibody, which is expected to show cross-reactivity with rice α-amylase because of high homology between barley and rice α-amylases. The other was anti-peptide antibody against the eight amino acids of the C-terminal region (RVPAGRHL, PC-Am3D) of RAmy 1A/3D.

The antigens used for immunization were crude barley α-amylase partially purified by salt-out and gel chromatography and the peptide synthesized by the solid-phase method. Instead of production and purification of recombinant proteins, these antigens were easily obtained . These antigens were immunized to rabbits, and specific antibodies were purified from pooled sera by affinity chromatography using the antigens as ligands [3]. The immunoadsorbents were prepared by coupling these antibodies to CNBr-activated Sepharose 4B.

The yeast was grown in 2L of YEPD media (1 % yeast extract, 2 % bacto peptone, 8 % glucose, 5 mM $CaCl_2$) with a jar fermentor for 24 h at 30 °C. After cells were separated by centrifugation, the supernatant containing α-amylase was ultrafiltrated with a hollow fiber module.

The immunoadsorbent prepared was packed in an adsorption column of 1.26 cm in diameter, and the column was equilibrated with a buffer solution (50 mM Tris-HCl + 5 mM $CaCl_2$, pH 7.6). The concentrated fermentation broth was applied to the column at a flow rate of 2 ml/min. After washing, adsorbed α-amylase was eluted by 2.5 M NaSCN and dialysed against a sodium acetate buffer (50 mM, pH 5.0). The activity of α-amylase was determined from the rate of increase in reducing ends from potato soluble starch. One enzyme unit was defined as the activity to liberate 1 μmol of maltose per minute. Its purity was examined by SDS-PAGE stained with silver.

3. RESULTS AND DISCUSSION

In fermentation broth, α-amylase was secreted with increase in the cell concentration. After centrifugation of cells and ultrafiltration, α-amylase was purified by immunoaffinity chromatography, as shown in Figure 1, in which a conventional method was also shown for comparison. Immunoaffinity chromatography adopted at an early stage of the purification process reduced the number of stages to attain required specifications.

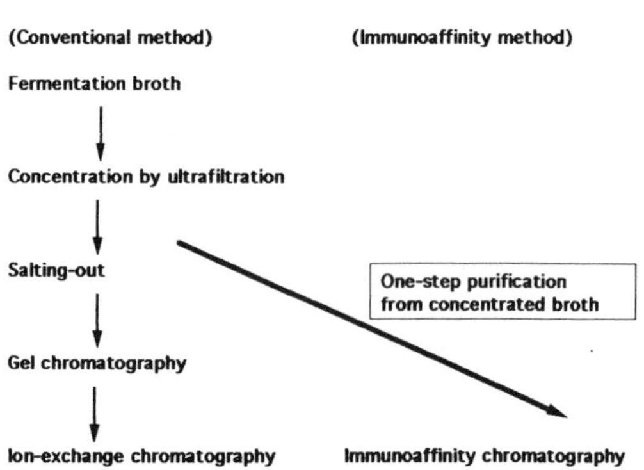

Figure 1. Purification processes for α-amylase

As shown in Figure 2, anti-barley α-amylase adsorbed rice α-amylase by cross-reaction and showed an adsorption isotherm of the Freundlich type .

The activity of α-amylase (RAmy 1A) in the effluent solution from the anti-barley α-amylase column was shown in Figure 3. At first the activity of α-amylase was not detected in the effluent solution and then gradually increased (breakthrough). After washing, α-amylase was eluted as a sharp peak by 2.5 M NaSCN. In 8 ml of the eluent, 70 % of adsorbed α-amylase was recovered. The product showed a single band by SDS-PAGE, which indicates the high purity of obtained α-amylase.

Since impurities contained in the antigen (crude barley α-amylase) were different from those in fermentation broth of yeast, antibodies raised against the impurities in the antigen little affected the purity of rice α-amylase purified from the broth. RAmy 1A/3D was similarly purified by this anti-barley α-amylase column. Thus, utilization of cross-reactive antibodies are very useful for the purification of recombinant proteins foreign to host cells.

In the case of the anti-peptide antibody (anti-PC-Am3D), RAmy 1A/3D was adsorbed and showed a similar adsorption isotherm to anti-barley α-amylase-RAmy 1A. Naturally RAmy 1A, which has the different C-terminal sequence with that of RAmy 1A/3D, was not adsorbed by the column. These results show that with the antibody against 8 amino acids at the C-terminal re-

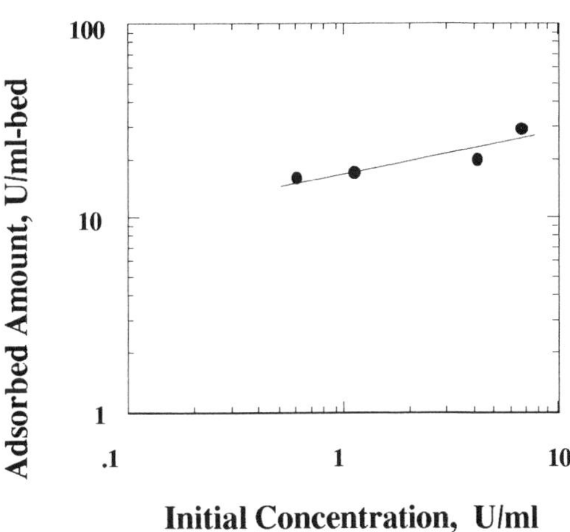

Figure 2. Adsorption equilibrium anti-barley α-amylase - RAmy 1A

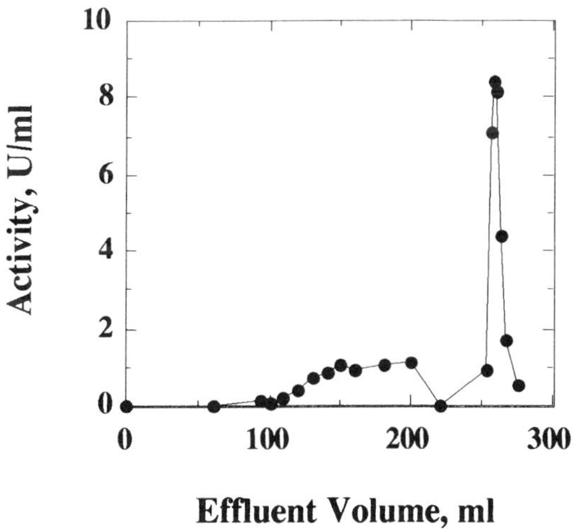

Figure 3. Breakthrough and elution curves from anti-barley α-amylase RAmy 1A, 1.09 U/ml

gion, the α-amylase molecule of RAmy 1A/3D consisting of more than 400 amino acids was selectively adsorbed. Figure 4 shows the activity of RAmy 1A/3D in the effluent from the column of anti-PC-Am3D. RAmy 1A/3D was adsorbed to the adsorbent but slight leakage of the activity from the column suggested possibility of some processing at the C-terminal region of RAmy 1A/3D. RAmy 1A/3D was also eluted by 2.5 M NaSCN, and the purity was again very high.

By use of these antibodies (cross-reactive and peptide antibodies) α-amylases were successfully purified to a high purity from the fermentation broth in one step. These result show that cross-reactive and anti-peptide antibodies have wide applicability for the purification of recombinant proteins secreted from microorganisms.

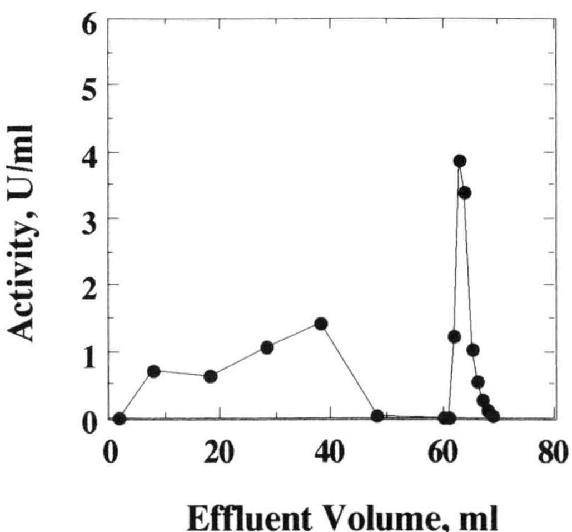

Figure 4. Breakthrough and elution curves from anti-PC-Am3D column RAmy 1A/3D, 2.25 U/ml

References

1. R.A. Lerner, Adv. Immunol., **36**, (1984) 1.
2. M.H. Kumagai, M. Shah, M. Terashima, Z. Vrkljan, J.R. Whitaker and R.L. Rodriguez, Gene, **94** (1990) 209.
3. A. Kondo, H. Takamatsu, S. Katoh and E. Sada, Biotech. & Bioeng., **35**, (1990) 146.

ECB6: Proceedings of the 6th European Congress on Biotechnology
edited by L. Alberghina, L. Frontali, P. Sensi

Sequence-specific DNA affinity chromatography: Application of a group specific adsorbent for the isolation of restriction endonucleases

C. Pozidis, G. Vlatakis and V. Bouriotis

Enzyme Technology Division, Institute of Molecular Biology and Biotechnology, P.O. Box 1515, Heraklion 711 10 Crete, Greece and Department of Biology, Division of Applied Biology and Biotechnology, University of Crete, P.O. Box 1470, Heraklion 71110, Crete, Greece

1. ABSTRACT

Several rapid and effective methods have been described to obtain restriction endonucleases suitable for commercial exploitation. However lengthy and laborious protocols have been necessary to obtain homogeneous enzymes.

The use of sequence-specific DNA affinity chromatography to purify restriction endonucleases *Eco*R I and *Sph* I to near homogeneity in a two step procedure has been recently reported. However, the high cost of these adsorbents is a limiting factor for their wider application. The application of a sequence-specific DNA affinity ligand containing recognition sequences for 34 restriction endonucleases as group-specific ligand for the isolation of restriction endonucleases is now reported. Crude samples of six restriction endonucleases namely *Bsh*F I, *Bam*H I, *Sma* I, *Sac* II, *Pvu* II and *Sal* I were shown to bind to this adsorbent and could be eluted at different KCl concentrations with purification factors obtained varying from 8 to greater than 300 fold and recoveries from 75-94%. Furthermore, restriction endonuclease *Bsh*F I, an isoschizomer of *Hae* III from the microorganism *Bacillus sphaericus* was purified to near homogeneity employing a two step procedure which involves DNA cellulose chromatography and oligonucleotide ligand affinity chromatography.

2. INTRODUCTION

Development and improvement of technologies related to restriction endonuclease downstream processing are becoming key factors for the successful exploitation of genetic engineering. Not surprisingly therefore great attention is paid to engineering and screening new microorganisms as well as to improving purification methodology for restriction endonucleases.

Simplified purification methods based on dye-ligand chromatography, HPLC and affinity partitioning have been described for obtaining partially purified restriction enzymes (1-3).

It is known that proteins which recognize specific sequences within the DNA molecule exhibit high affinity for interaction with these sequences. Therefore, it is conceivable that these proteins could be isolated on affinity adsorbents bearing as ligands suitable base sequences. Sequence-specific DNA affinity chromatography has been successfully applied to the isolation of proteins that interact with specific DNA sequences (4- 6).

Certain restriction endonucleases in the absence of Mg^{2+} ions have the ability to interact specifically with the sequence they recognise without hydrolyzing the DNA molecule. Therefore, DNA segments could be used as specific ligands for isolating restriction enzymes which recognise the corresponding specific sequences. The synthesis of sequence-specific DNA affinity adsorbents and their application in the purification of two restriction endonucleases, namely EcoR I and Sph I has been recently reported (7).

Although sequence-specific DNA affinity chromatography compares favourably with other adsorbents used for the isolation of restriction endonucleases to homogeneity, the high cost of this technique, resulting from the different synthesis of the ligand for each restriction endonuclease, is a limiting factor for its wider application. The development of oligonucleotide affinity adsorbents containing sites for a number of restriction enzymes which can therefore be used as group specific adsorbents could potentially solve this problem.

The synthesis of a group-specific DNA affinity adsorbent and its potential application in the isolation of a number of restriction endonucleases recognizing sequences within the oligonucleotide ligand is now reported. Furthermore, the purification of restriction endonuclease *BshF* I, an isochizomer of *Hae* III (8) to homogeneity employing a two step procedure is also presented.

3. RESULTS AND DISCUSSION

An oligonucleotide affinity ligand containing recognition sequences for 34 restriction endonucleases was constructed after annealing of two oligonucleotides (Fig. 1) and was subsequently coupled to cyanogen bromide

5'-GATCGCATGCCGCGGATCCCGGGCCCAGGTGGCCAGCTGTCGAC-3'

+

3'-CGTACGGCGCCTAGGGCCCGGGTCCACCGGTCGACAGCTGCTAG-5'

5'-GATCGCATGCCGCGGATCCCGGGCCCAGGTGGCCAGCTGTCGAC-3'
3'-CGTACGGCGCCTAGGGCCCGGGTCCACCGGTCGACAGCTGCTAG-5'

Figure 1. Preparation of a sequence-specific DNA affinity ligand containing recognition sequences for 34 restriction endonucleases: *Acc* I, *Alu* I, *Alw* I, *Apa* I, *Ava* I, *Bal* I, *Bam*H I, *Ban* II, *Bgl* I, *Bsp*1286 I, *Bst*N I, *Bst*U I, *Bst*Y I, *Eae* I, *Fnu*4H I, *Hae* III, *Hinc* II, *Hpa* II, *Mbo* I, *Nci* I, *Nla* III, *Nla* IV, *Nsp* I, *Nsp*B II, *Pvu* II, *Sac* II, *Sal* I, *Scr*F I, *Sec* I, *Sfi* I, *Sma* I, *Sph* I, *Taq* I, *Xma* I. The two oligonucleotides were annealed and subsequently immobilised on CNBr activated Sepharose.

Table 1. Elution of restriction endonucleases from the oligonucleotide affinity adsorbent at various KCl concentrations

Enzyme	Elution [KCl (M)]	Purification (-fold)	Recovery (%)
*Bsh*F I	0.3	48	86
*Bam*H I	0.3	>300	75
Sma I	0.2	43	88
Sac II	0.2	122	94
Pvu II	0.1	108	86
Sal I	0.1	8	80

activated Sepharose. The oligonucleotides were designed in such a way such as that the resulting DNA affinity ligand contains 5′-single strand overlapping termini in order that primary amines of the free bases could be coupled to activated agarose.

In order to hydrolyse DNA, restriction enzymes require the presence of divalent metal ions usually Mg^{2+}. For this reason the addition of EDTA in all buffers for chelating the metal ion is necessary. The same method is used for DNA-cellulose or DNA-agarose in order to avoid hydrolysis from non specific nucleases [9].

In order to examine the potential application of the oligonucleotide affinity adsorbent as a group specific adsorbent for the isolation of restriction endonucleases we have used partially purified enzyme preparations of *Bsh*F I, *Bam*H I, *Sma* I, *Sac* II, *Pvu* II, and *Sal* I which have recognition sequences within the oligonucleotide ligand. All enzyme preparations tested did bind to this adsorbent and were subsequently eluted at different salt concentrations by step gradients of KCI. The ability of the affinity adsorbent which was designed in order to specifically interact with restriction enzymes was evaluated on the basis of KCI concentration which was necessary for the desorption of the enzymes. *Bsh*F I and *Bam*H I were eluted at 0.3 M, *Sma* I and *Sac* II at 0.2 M while *Pvu* II and *Sal* I at 0.1 M KCI, which is an indication of different affinities of the enzymes for the ligand in the absence of Mg^{2+} (Table 1). The purification factors obtained varied from 8 to greater than 300 fold while recoveries varied from 75 to 94% (Table 1). Despite the liability of the DNA ligand, the column showed very good stability and could be used more than 10 times for the purification of the above enzymes.

A possible explanation for the weak interaction of *Pvu* II and *Sal* I with the affinity adsorbent is that the specific interaction of the enzymes with their recognition sequence might require the presence of Mg^{2+} ions. The only enzymes which have been studied and for which is supported that the specific interaction with DNA is independent from the presence of Mg^{2+} ions are the enzymes *Eco*R I and *Bgl* II [9,10]. Furthermore, it has been reported that in the absence of Mg^{2+} the binding affinities of *Eco*R I for specific and non specific DNA differ by a factor of $1x10^5$ [10-12]. Similar results were obtained with *Rsr* I an isoschizomer of *Eco*R I [13] while with *Eco*R V the ratio of binding affinities is 1 [14].

Table 2. Summary of the purification protocol for *Bsh*FI

Step	Protein (*mg*)	Enzyme activity (*units*)	Specific activity (*units/mg*)	Purification (*-fold*)	Yield (*%*)
Crude extract	160	-	-	-	-
DNA-Cellulose	1	$1.92x10^6$	$1.92x10^6$	1	100
Affinity adsorbent	$1.35x10^{-3}$	$1.46x10^6$	$1.08x10^9$	562	76

Finally, employing a two step procedure including DNA-cellulose and the oligonucleotide affinity adsorbent we have purified restriction endonuclease *Bsh*FI to homogeneity. The successful isolation of the enzyme is mainly due to the effectiveness of the affinity chromatography column. The enzyme preparation was enriched 562 times while a 76% yield was obtained (Table 2). The enzyme exists in its native form as a monomer with an apparent molecular weight of 34 kDa as determined by both sodium dodecyl sulfate polyacrylamide gel electrophoresis and size exclusion chromatography.

This study has demonstrated the effectiveness of a group specific oligonucleotide affinity adsorbent for the isolation of restriction endonucleases. It is expected that restriction endonucleases which exhibit higher affinity for specific rather than non specific DNA sequences will be successful candidates for purification by sequence specific DNA affinity chromatography.

REFERENCES

1. G. Vlatakis, G. Skarpelis, I. Stratidaki, Y.D. Clonis and V. Bouriotis, Appl. Biochem. Biotechnol. 15 (1987) 201
2. V. Bouriotis, A. Zafeiropoulos and Y.D. Clonis Anal. Biochem. 160 (1987) 127
3. G. Vlatakis and V. Bouriotis, J. Chromatogr. 538 (1991) 311
4. J.T. Kadonaga and R. Tjian, Proc. Natl. Acad. Sci U.S.A., 83 (1986) 5889
5. K.A. Jones, J.T. Kadonaga, P.J. Rosenfeld, T.J. Kelly and R. Tjian, Cell, 48 (1987) 79
6. W. Lee, P. Mitchell and R. Tjian, Cell 49 (1987) 741
7. G. Vlatakis and V. Bouriotis, Anal. Biochem 195 (1991) 352
8. G. Vlatakis, D. Clark and V. Bouriotis, Nucleic Acids Res, 17 (1989) 8882
9. R.M. Litman, J. Biol. Chem. 243 (1968) 6222
10. B.J. Terry, E.W. Jack, R.A. Rubin and P. Modrich, J. Biol. Chem., 258 (1983) 9820
11. G.M. Clore, A.M. Gronenborn and R.W. Davies,J. Mol. Biol., (1982) 447
12. D.R. Lesser, M.R. Kurpiewski and L. Jen Jacobson, Science, 250 (1990) 776
13. C.R. Aiken, E.W. Fisher and R.I. Gumport, J. Biol. Chem., 266 (1991) 19063
14. D.J. Taylor, J. I. Badcoe, R.A. Clarke and E.S. Halford, Biochemistry 30 (1991) 8743

ECB6: Proceedings of the 6th European Congress on Biotechnology
edited by L. Alberghina, L. Frontali, P. Sensi

New diffusive matrix to improve throughput in affinity chromatography applications.

E. Boschetti

Sepracor S.A., 35 avenue Jean Jaurès, 92395 Villeneuve la Garenne Cédex, France.

I. INTRODUCTION

Increased purity and competitive costs in large scale proteins purification stimulate the development of more effective chromatographic sorbents (for review, see ref. 1).

In the last decade, progresses have been made in different directions such as pressure resistance, separation efficiency, molecular selectivity. On the capacity level, however, the progression was not really spectacular. The productivity of a chromatographic sorbent has been consequently improved more by the diminution of the separation cycle (higher linear speed) than by the intrinsic binding capacity.

Homogeneous cross-linked polysaccharides (e.g. agarose) have been succesfully developed for large scale users.

Macroporous materials based on synthetic polymers with a large amount of hydroxyls and designed to have large pores, were then proposed to overcome some limitations of homogeneous networks (2). In some cases, the macroporosity concept was pushed beyond certain limits where the separation of molecules is defined as "perfusive chromatography" (3). Here the main advantage is the high speed permitted by a rapid mass transfer.

Adsorption-desorption kinetics has been enhanced by the introduction of so-called "tentacular sorbents" (4). Active groups are attached to linear flexible polymeric chains that increase the mobility of interacting groups and the chances of rapid interactions with the macromolecules to be separated.

For preparative applications where productivity is the most important notion to consider, increasing the flow rate results generally in a reduction of the sorption capacity.

It can be observed in fact that some sorbents have quite low dynamic sorption capacity decrease when increasing the flow rate while other show a rapid decrease of dynamic binding capacity.

In this paper, we describe a new concept of sorbent constituted of two main parts with independent role to permit a good dynamic sorption capacity at high linear velocities.

2. DEVELOPMENT OF THE HyperD™ CONCEPT

Information on all parameters having a direct impact on the productivity of a chromatogaphic sorbent are not easily available.

It can be admitted that the sorption capacity is induced by the number of active sites and their availability. Here three groups of parameters can be analyzed as described on the following table.

Table 1
Selective factors linked to the chromatographic sorbent that influences the binding capacity.

Parameters governing the binding capacity	Definition of the packing material
Chemical Parameters	* Numerous active sites * Dense distribution of active site * High polymer flexibility
Physical Parameters	* Large pore and low pore size distribution * Small particles * Narrow particle size distribution * Very high mechanical resistance
Thermodynamic Parameters	* High diffusivity properties * High association constants * High level of active site availability

Binding capacity for very low molecules is classically constant at any flow rate but for proteins, the accessibility to active sites is much lower, due to steric hindrances and pore size distribution within the polymeric network.

Sorption capacity decreases rapidly when the pore size is close to the protein hydrodynamic diameter .

This phenomenon is schematically illustrated on figure 1. The slope of the dynamic sorption capacity curve depends on few factors identified as the particle diameter, the effective diffusivity and the permeability factor. When modifying the physiochemical properties of the polymer, it is then possible to optimize sorbents in order to limit the degradation of the sorption capacity at high linear velocities.

High linear velocities for a better throughput induce a certain level of backpressure. To maintain a linear relationship between the flow rate and the back pressure, the particles constituting the sorbent must obviously be uncompressible.

HyperD™ concept is based on the above-mentioned assumptions. It is constitued of a very large open pore uncompressible material and a soft gel with active adsorbing sites distributed homogeneously inside this rigid skeleton (5).

The active cross-linked polymer is characterized by a low density, homogeneous pore size, and a high level of flexibility to enhance the interaction between the protein to be separated and the active sites.

From this concept, ion exchangers have been developed (1) with very high productivity properties ; affinity media are currently under evaluation.

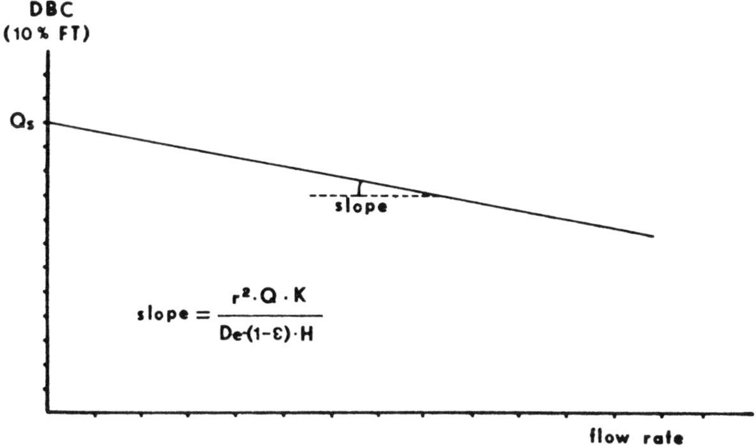

Figure 1. Schematic representation of dynamic binding capacity (DBC) variation as a function of flow rate. The curve is characterized by a given static capacity Q_S and a typical slope. This latter depends on the radius of the sorbent particles (r^2) , the value of the DBC in equilibrium with the solute (Q_o) , the effective diffusivity (D_e), the intraparticle porosity (ε) and the length of the column (H). K is a constant from the breakthrough curve.

3. AFFINITY DERIVATIVES AND RESULTS

Non ionic composite material HyperD was taken as basic material to develop affinity derivatives.

Among a large number of ligands attached chemically on this material, we can mention small size ligands such as dyes, medium macromolecules such as heparin and proteins (e.g. Protein A).

Coupling reaction has been realized in all cases by condensation reactions between primary amino groups from the modified packing and carboxyl groups of the ligands with the formation of a stable peptidic bond.

The most important properties of these affinity sorbents is to present a high level of sorption capacity with a relatively small amount of ligands. Due to the very flexible property of the polymer, proteins to purify are rapidly interacting with active site, maximizing the sorption efficiency (ratio between protein capacity and number of active sites) .

Moreover the sorption capacity was maintained at quite high levels at high flow rates, demonstrating a strong potential of productivity with short contact time.

Figure 2 illustrates this phenomenon of dynamic sorption capacity versus flow rate for Protein-A-HyperD compared to more classical sorbents.

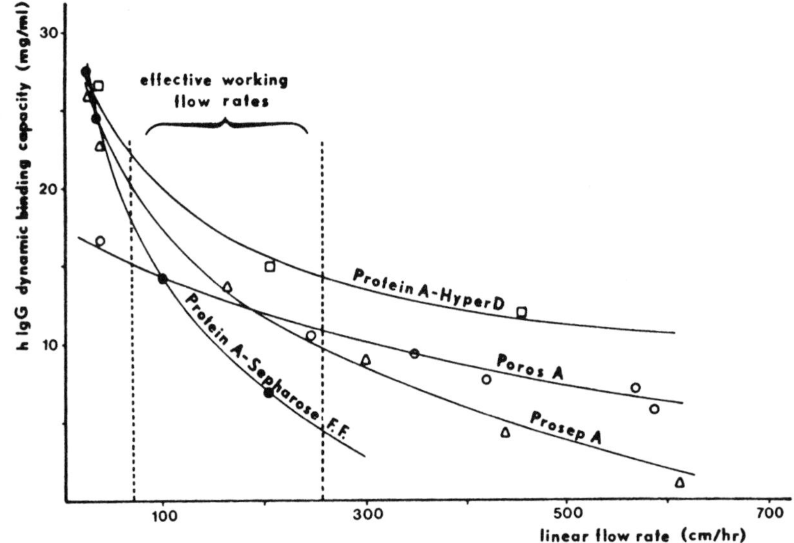

Figure 2. Variation of the effective dynamic binding capacity of Protein A-HyperD for human IgG compared to other classical Protein A-packing materials. Measurements have been done from breakthrough curves using a solution of 10 mg/ml human IgG in a physiological buffer (PBS) on 2 ml columns. Linear flow rate was between 20 and 600 cm/hour.

HyperD affinity material is particularly well designed for preparative applications where capacity and productivity are important objectives. Their high density with rapid adsorption-desorption kinetics make them usable in fluidized or expanded bed capture operations at the early stages of downstream bioprocessing.

REFERENCES

1. E. Boschetti, J. Chromatogr., (1993), submitted.
2. P. Girot and E. Boschetti, J. Chromatogr., 213 (1981) 389.
3. N. B. Afeyan, N. F. Gordon et al., J. Chromatogr., 519 (1990) 1.
4. W. Muller, J. Chromatogr., 510 (1990) 133.
5. E. Boschetti, Proc. Symp. Bio-East, Washington, USA, January 1993.

BIOSENSOR DESIGN

ECB6: Proceedings of the 6th European Congress on Biotechnology
edited by L. Alberghina, L. Frontali, P. Sensi

Report on Biosensor Technology in Europe

M. Mascini

Dip. di Sanita' Pubblica, Epidemiologia e Chimica Analitca Ambientale, Sezione di Chimica Analitica
Via Gino Capponi 9, 50121 Florence, Italy

The importance of Biosensor Research is growing enormously and is now appearing at National as well at Community levels.

At present Biosensors are included in several Research Programs in different technological areas from Medicine to Environmental, from Food to Biotechnology, because this technology can solve important problems related to Safety, Testing, Monitoring and Process Control.

However the fundamental problems of Biosensor research are common regardless of scientific field or applications and therefore these problems could be better solved by the integration of different actions.

This Report is a survey of the Biosensor Research in Europe and of its strategic importance for National and Community Research Programs.

The Report has been prepared by 17 scientists working experimentally in the field of Biosensors, they can be considered the most expert in their country in Europe, USA and Japan.

It reports the activity of the European Research in each country in important fields of application like Clinical Chemistry, Agriculture, Food Chemistry, Environmental, Process Control Monitoring; it is an update inventory of this relatively new technology and its application in different fields in Europe.

Important features like "*in vivo* monitoring", mass production, miniaturization techniques, self monitoring and food testing are highlighted.

The position of Research in Europe is compared to those of USA and Japan from the point of view of scientific productivity as well as applied technology and achieved results. Europe has a high level of activity in Biosensor research compared to USA and Japan. However Japan has the highest number of patents and USA is the most important single market. Europe with different rules in each state suffers from lack of interest from industrial partners.

Moreover Biosensor development requires a range of expertise from Analytical Chemistry to Biochemistry, from Electronics to Microengineering to Genetic Engineering and only in a few organized centers can experts in all disciplines

TOTAL PUBLICATIONS

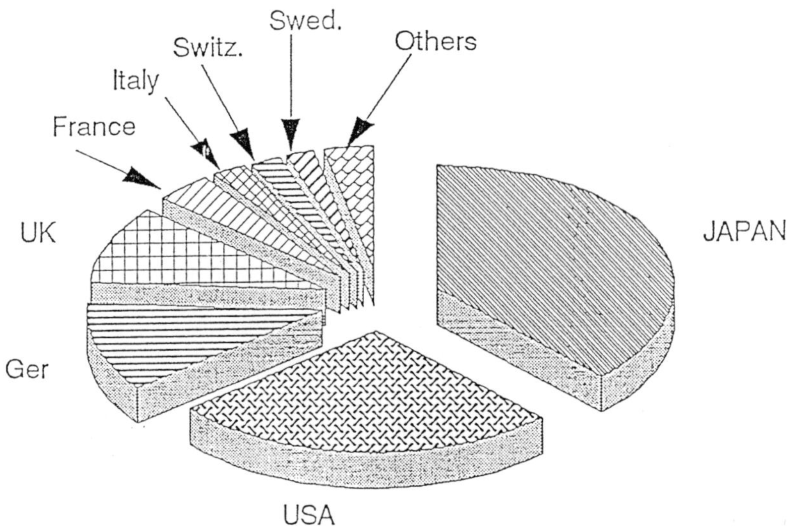

PATENTS

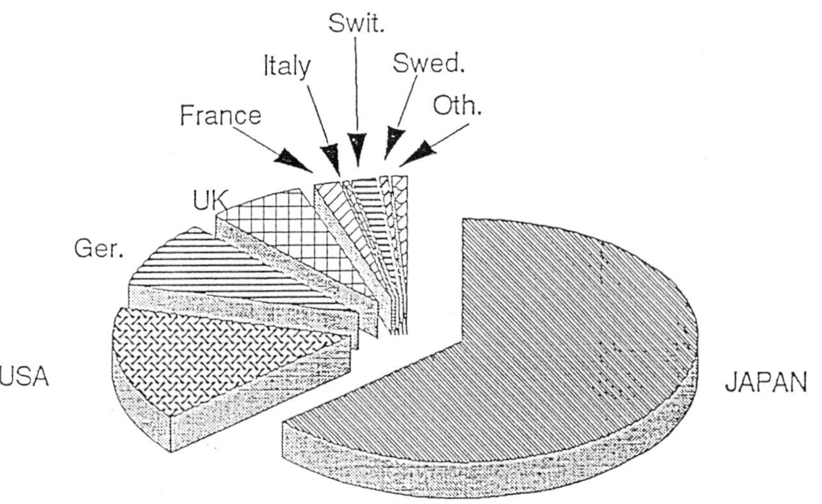

succesfully interact together. In Europe now we assist the formation of small companies where the number of experts and funds involved are limited (there are of course some exceptions). EC, in concertation with national governments should play an important role in this strategic sector by integrating industrial interest and academic knowledge to obtain better exploitation in the development of Biosensor technology in Europe (See Figure).

In the Report the national programmes on Biosensors are reviewed and large differences are apparent in the attitude of the governments in Europe; some nations have large research programmes, others have only limited funds in this area.

In the Report specific recommendations for public policy in Europe at a National and Community level are formulated to develop a Community strategy for Biosensor Research which will help national industries.

They include:

-to fund a Biosensor program independent from Biotechnology and Microelectronics

-to enable the establishment of a Network between European scientists in this field

-to consider three large areas: 1) a basic research where fundamental issues are considered (Enabling Technologies) 2) targeted research with specific projects programmed in conjuction with industrial partners 3) training young scientist in biosensor research and to establish scientist mobility in Europe to exchange information and to work on a targeted project.

The Report is free available at:

Dr. P. de Taxis du Poet
Commission of the European Communities
Directorate General for Science, Research and Development
Joint Research Centre
Cube
Rue de la Loi 200
B-1049 Brussels
Belgium

ECB6: Proceedings of the 6th European Congress on Biotechnology
edited by L. Alberghina, L. Frontali, P. Sensi
1994 Elsevier Science B.V.

Environmental monitoring of pesticides using biosensors

J-L Marty

Université de Perpignan, Groupe d'Etudes et de Recherches Appliquées Pluridisciplinaires, URA CNRS 461, 52 avenue de Villeneuve, 66860 Perpignan France.

1. INTRODUCTION

The protection of environment draws great attention and has become a world wide problem. Monitoring of physical and chemical parameters of the environment is one of the most important challenge. Although classical analytical parameters are continuously improved, these laboratory based methods are generally expensive and time consuming. The use of electrochemical or optic systems in combination with selected biological receptors offers considerable promise in developping specific, sensitive and inexpensive biosensors. A variety of biological macromolecules (antibodies, enzymes) and cells (microorganisms, cell tissues) are candidates for environmental biosensors able to detect : heavy metals, organic pollution, xenobiotic compounds such as organic compounds or pesticides.

The intensive agriculture has highlighted the need of procedures to study the impact of pesticides in water and soil. This paper outlines the development of biosensors for the detection of pesticides with reference to the choice of biological receptors. We present our results on the detection of selected herbicides and fungicides.

2. WHOLE CELL AND CELL TISSUE SENSORS

Rawson *et al.* (1) exploit photosynthetic cyanobacteria such as the receptor. The biosensor is able to detect a wide range of herbicides with sites of action on the photosynthetic electron transport chain. Redox couple $Fe(CN)_6^{3-/4-}$ is employed as an electrochemical mediator and the magnitude of the current due to the reoxidation at the working electrode is related to the photosynthetic activity. Less than 200 ppb of linuron and ioxynil can be detected. Purcell and Carpentier (2) developped a biosensor using immobilized thylakoids which measures the perturbation of photosynthetic activity by anodic oxidation of hydrogene peroxide. Some toxicity tests have been based on the luminescent activity of *photobacterium phosphoreum*; there are tentatives to develop biosensors with modified bacteria obtained by cloning the gene encoding luciferase for the detection of pollutants and pesticides (3).

3. IMMUNOSENSORS

Within the past 15 years, there has been a heightened interest in generating antibodies to pesticides for trace residue analysis. A lot of immunoassays (4) based upon polyclonal or monoclonal antibodies have been developped for the detection of herbicides (chlorsulfuron, 2,4,5T, 2,4D, diclofopmeyhyl, atrazine...), insecticides (parathion, permethryn....) and fungicides (iprodione, metalaxyl...). The immunoreactions are well known but the development of immunosensors implies to solve the problems to achieve robust, cheap and reusable devices with a wide measuring range and amplification of the signal.

4. ENZYME-SENSORS

There are a lot of enzyme sensors for the detection of various substrates : glucose, sucrose, lactate.... In all these cases, the amperometric signal is related to the concentration of substrate. For the pesticides, the enzyme sensor can be only based on enzyme inhibition due to their very low concentration. It is well known that cholinesterase is an excellent target for the determination of organophosphorus and carbamate insecticides such as parathion, malathion........ Several enzyme sensors for the detection of such products have been developped (5-8) because the enzyme is commercially available with very high stability and sensitivity and its activity can be measured using potentiometric or amperometric transducers. For the other pesticides the enzyme target is not commercially available as acetolactate synthase for sulfonylureas or ESPS for glyphosate or unknown for dithiocarbalmate fungicides. Elsewhere it is impossible to find a transducer for the determination of enzyme reaction. We are going to present preliminary results concerning the detection of sulfonylureas herbicides and dithiocarbamte fungicides.

4.1 Dithiocarbamate fungicides

The target of dithiocarbamate fungicides is not well known, although Wiegand-Rosinus (9) demonstrated that these compounds inhibit aldehyde dehydrogenase. We are developping a bienzyme sensor based on the following reactions :

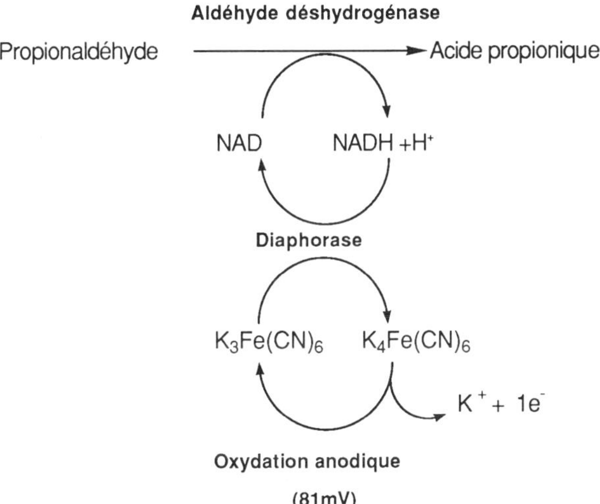

The fungicides inibit aldehyde dehydrogenase and the decrease in current allows the determination of their concentration. The lowest detection limit is 0.05 ppm and 0.25 ppm for maneb and zineb although results obtained with the official chemical method is 0.4 ppm

4.2 Sulfonylureas herbicides

The target of sulfonylureas herbicides in plants is acetolactate synthase (ALS-EC 4.1.3.18) (10). They are applied at very low rates; less than 10 g a.i. ha^{-1} (11) because they have high herbicidal activity and the classic analytical methods are not sensitive enough for their detection (500 ppb using H.P.L.C.). Preliminary experiments based on the inhibition of ALS show that we can detect less than 20 ppb. The aim of our work is to solve the problems due to the stabilization of the enzyme and connection with potentiometric transducers.

5. CONCLUSION

There is an urgent need to develop cheap, sensitive and specific methods for monitoring pollutants in the environment. Biosensors have attracted high interest. Whole cell and tissue cell sensors can be used for the determination of general toxicity of an effluent (pollutants inhibiting photosynthesis...), enzyme sensors can be used for the detection of family of pollutants (organophosphorus, dithiocarbamates, sulfonylureas....) and immunosensors can exploit the sensitivity and the specificity afforded by antibodies (atrazine, chlorsulfuron....)

REFERENCES

1. D. M. Rawson and A.J. Willmer, Biosensors, 4 (1989) 299.
2. M. Purcell and R. Carpentier, Water Poll. Res. J. Canada, 25, 2 (1990) 175.
3. S.M. Lee, M. Suzuki, E. Tamiya and I. Karube, Anal. Chim. Acta, 244 (1991) 201.
4. J.M Van Emon, J.N. Seiber and B.D. Hammock, In Analytical method for Pesticides and plant growth regulators.J. Sherman (ed), Academic Press,New York, 1989.
5. V.J. Razumas, J.J. Kulys and A.A. Malinauskas, Env. Scien. Technol., 15, 3 (1981)360.
6. P. Durand and D. Thomas, JEPTO., 5-4/5 (1984) 51.
7. M. Bernabei, C. Cremisini, M. Masumi and G. Palleschi, Anal. Letters, 24, 8 (1991) 1317.
8. J-L. Marty, K. Sode and I. Karube, Electroanalysis 4 (1992) 249.
9. M. Wiegand-Rosinus, K. Haberer, U. Obst and A Wild, Z. fur Wasser und Abwasser Forsch., 23 (1990) 98.
10. R.S. Chaleff and C.J. Mauvais, Science, 224 (1984) 1443.
11. E.M. Beyer, M. Duffy, J.V. Hay and D.D. Schlueter, In sulfonylureas herbicides. P.C. Kearney and D.D. Kaufman (Eds), Marcel Dekker Inc., New York, 1987.

Quantitative Flow Microfluorimetry in Biotechnology

ECB6: Proceedings of the 6th European Congress on Biotechnology
edited by L. Alberghina, L. Frontali, P. Sensi

Flow Cytometry in Biotechnology: Potential and Limitations

F. Srienc

Department of Chemical Engineering and Materials Science, University of Minnesota, 151 Amundson Hall, Minneapolis, MN 55455, USA

Institute for Advanced Studies in Biological Process Technology, University of Minnesota, 240 Gortner Laboratory, St. Paul, MN 55108, USA

A brief overview is given on the capabilities of flow cytometric measurements in biotechnological applications. After a short introduction of the instrumentation and some cell preparation techniques, selected examples are presented with particular emphasis on the use of flow cytometry in applications relevant to bioprossesses involving eucaryotic cells.

1. INTRODUCTION

Flow cytometry is an established measuring technique in basic biological and biomedical research. An important application of flow cytometry is in the area of cancer research, and an extensive body of literature exists, therefore, that describes many applications of this measuring approach in studies of basic cell biology, such as cellular immunology, and in investigations with a focus on cancer research.

Biotechnological applications have been rather rare. This is partially due to the significant cost associated with the purchase and maintenance of such an instrument. In addition, it requires considerable training in the operation of the instrument and some experience in the handling, analysis and interpretation of the large amount of data that can be obtained in a short period of time. Also, direct applications of flow cytometry that demonstrate an immediate benefit to bioprocesses are perhaps much less obvious than in the medical field where, for instance, the early detection of some aberrant cell forms might lead to increased chances of diagnosing and curing certain diseases. However, if one has the motivation and opportunity to elucidate the basic nature of problems associated with the dynamics of growth and product formation of cell cultures, then flow cytometry becomes an indispensible tool. The reward of its proper use is an insight into the mechanisms that govern the dynamics of cell culture processes.

Flow cytometry, along with the vast repertoire of cell preparation techniques, measuring procedures and data analysis techniques developed for mammalian cell systems by biomedical researchers, currently represents an under-utilized tool that could be exploited in many biotechnology applications involving cell cultures. This is particularly true if it is realized that advances can be made only if one is able to understand phenomena on a mechanistic basis.

Cell cultures consist of cell populations, and the complication one encounters is that cells within the culture are different from each other. This variability can be traced to different origins. First, cells might differ from each other because they represent different species as defined by an inheritable, genetic material. This includes mutants or cell types altered with recombinant DNA. Second, a cell line with homogeneous and identical genetic information typically consists of cells that are different from each other. This variability is due to the inherent nature of cell growth, that is, the operation of the cell cycle which drives cells from one cell division to the other. Cells at different positions in the cell cycle typically have significantly different properties or physiological states. Finally, influences from the environment such as nonideal mixing conditions in a bioreactor might lead to different development and growth of certain cells, and this also contributes to heterogeneous cell populations in the culture.

Heterogeneities in cell populations can be identified if cell cultures are studied at the single-cell level and if distributions of cellular properties are measured. Other sensors employed in bioreactor applications to monitor cell activities are usually population average sensors, i.e. the sum of responses obtained from each individual in the cell population is detected. Thus, the interpretation of the response on a mechanistic level is difficult, if not impossible, without additional information. From population average responses one can sometimes infer certain heterogeneities in a cell population if the effect of the heterogeneity is well established. In general, however, population average sensors preclude insight into the structure of a growing cell population. This makes it impossible to explain the mechanistic basis of cellular phenomena from population average data.

2. PRINCIPLE OF FLOW CYTOMETRY

Flow cytometry can only be applied to cell samples consisting of single cells. Cells grown in tissue must be dispersed. This is due to the principle of measurement. A sample stream containing the individual cells is injected into a sheath stream, typically water, saline or a dilute buffer. The pressure difference between the two streams can be adjusted such that the laminar sample stream is reduced to a very small diameter and cells are spacially separated along the axis of this stream. A light source, generally a laser beam, is focussed on this sample stream. Individual cells that flow through the light beam generate absorbance, scattering and fluorescence signals which are individually detected by photomultiplier tubes and registered in a pulse height analyzer. The cells can be analyzed at very high rates of up to several thousand cells per second. The result obtained from such analysis is a histogram that shows how the measured single-cell parameters are distributed within the cell population. If more than one parameter has been measured simultaneously for each cell, then multi-dimensional histograms, or frequency distribution functions, are obtained giving a detailed picture of the property space occupied by the cell population [1]. A more detailed description of the instrumentation can be found in several excellent books [2-4] and review articles [5] and it will not be discussed further in this article.

3. CELL PREPARATION

Flow cytometry is typically an off-line instrument. The analysis requires cell preparation procedures that can involve numerous staining steps and can last several hours. In principle, however, some of the simpler staining procedures could be automated and flow cytometry could be done on-line. Although such procedures are being discussed in several research laboratories, currently the main power of the instrument is as an off-line analyzer simultaneously serving multiple purposes.

The objective in cell preparation is to obtain a single-cell suspension with quantitative fluorescent staining of the cellular component of interest. This is straight forward if the cells are already growing as a single-cell suspension such as in the case of hybridoma cells or many other microbial cells. Cell tissues and cells growing in aggregates require disaggregation by mechanical or enzymatic means. In addition, a permeabilization step is usually required because most fluorescent dyes and large molecules such as antibodies do not cross the functional external membranes of viable cells. Cell permeabilization is typically achieved in fixation steps using either methanol, ethanol, acetone, or formaldehyde alone or in combination with detergents. During the cell staining step a particular component of the cells associates with fluorochromes in a specific mannerand ideally in stoichiometric amounts. Antibodies or other proteins such as lectins covalently linked to fluorochromes are used for this purpose. Certain fluorescent dyes have by themselves a high affinity to certain cell components. Other fluorochromes might be generated in situ by the action of certain enzymes and ezyme activities can be measured in single cells in this way. Generally, the staining conditions are dependent on the cell material and on the cellular target to be quantified. Specific protocols can be found in several books [2,4,6] or must be developed or adapted from original articles.

4. PROCESS ANALYSIS WITH FLOW CYTOMETRY

The outcome of a bioprocess is determined by the interactions of the cell population with its surrounding. The cellular environment is subject to external manipulations. Therefore, cellular activity can be directed if the interactions between the cells and their environment are well understood. The reactions of the cell population which govern interactions with the cellular environment are determined by the distribution of physiological states in a cell population. The physiological state is defined by the composition of a cell and by the spatial arrangement of cellular components within the cells [7,8]. In order to determine the physiological states of cells, it is necessary to obtain data on the cellular composition and how this composition is distributed within the growing cell population. In principle, determination of the amounts of all cellular components is of interest because they all contribute to the physiological state of the cell. In reality, however, the number of parameters must be reduced to the major, measurable components of a cell such as the total protein content, the DNA content, the RNA content, the amounts of individual proteins. These parameters, either individually or in combination, more or less describe the physiological state of a cell.

With flow cytometry one is able to measure for single cells (i) the intensity of scattered light at different angles from the incoming light beam, (ii) axial light loss (extinction) at different wavelengths, and (iii) the intensity of fluorescent light emitted by cells. While the intensity of scattered light at low angles can be related to cell size and indicates in many cases the cell volume, light scattered at a right angle serves as an indicator for the internal structure or granularity of a cell [9]. Light scattering measurements are especially attractive because no cell pretreatment is required and cells remain viable. Viable subpopulations can be physically separated with a cell sorter which provides a means for rapidly isolating certain desired cell types. However, in the light scattering measurement there is usually no clear correlation of the signal and a biological function of the cell. Fluorescence measurements, in contrast, enable a quantitative determination of specific cellular components related to well defined biological functions. Using appropriate staining procedures, subcellular components such as DNA, RNA, protein content, cytoplasmic enzyme content as well as the amount of surface components can be determined [1]. Some selected examples of flow cytometry applications are given in the next sections.

4.1 Cell count and cell viability

The increase in cell number over time is often the first parameter that is determined in a cell cultivation process. The light scattering properties of a cell is one of the most accessible parameters that can be measured with flow cytometry. The cells do not require any pretreatment and can be analyzed in growth medium if they grow in a single-cell suspension as do hybridomas. Two light scattering parameters, defined by the angle of the scattered light relative to the laser beam are measured with a flow cytometer. The disadvantage of the parameters is that they cannot be associated with a clear biological function of the cell and their interpretation is therefore restricted. Also, the forward angel light scattering intensity (FALS) is affected by the shape and the refractivity of the cells. Therefore, a more easily interpreted volume distribution can be obtained with the Coulter counter. The parameter is still useful, however, as it allows identification of subpopulations which have distinctly different light scattering properties. Such differences are very often seen among viable and dead cells during the time course of a cultivation.

Normally the measured sample volume is not known. Therefore, the absolute cell concentration in the sample cannot be obtained from this measurement. The quantity can be rapidly determined, however, by adding a defined amount of fluorescent tracer particles to the sample and by measuring the fluorescence together with the light scattering intensities [10]. The detected number of fluorescent particles relative to the number of cells measured yields the concentration of cells in the sample.

4.2 Genetically unstable cells and cell cycle analysis

It is essential that a productive cell culture maintain its characteristics throughout the cultivation period. Unfortunately this is not the case for instance for many mammalian cell lines. Some cell's genetic makeup may change such that the cell line loses its high level of productivity [11,12]. Such changes can be detected with flow cytometry at a very early stage when the undesired, altered cells are still at a very low concentration in the population. Moreover with multiple parameter flow cytometry typically several subpopulations of cells can be identified. If the DNA content is measured at the same time cell cycle analysis can be carried out on the individual subpopulations. Evaluation of the cell cycle fractions may show that for instance in comparison to the producing cell line, the non producing cell line has a significantly lower fraction of cells in the G1 cell cycle phase and a higher cell fraction in the S cell cycle phase. This would indicate a stronger proliferative activity of the nonproducing cell population under the given cultivation conditions. This information is of course very valuable for the planning and operation of a production process with such cell lines.

4.3. Cell Surface fluorescence and protein secretion

Many proteins produced by mammalian cell culture are secreted into the medium. The basis for efficient production is a stable cell line with high specific productivity. Flow cytometry can be used in several ways to maintain high-productivity cell lines. It has been demonstrated that the antibody molecules secreted by and associated with the surface of hybridoma cells can be labelled immunofluorescently using antibodies directed against the hybridoma antibodies [13]. This can be used in combination wth cell sorting to isolate and clone specific cell lines producing specific isotypes of IgG [14]. Moreover, it has been found that the measured surface-associated fluorescence is correlated with the specific antibody production rate [15,16]. This observation might apply only to certain cell lines as other researchers were not able to find such a correlation [17]. However, it points to a potentially useful applications of flow cytometry. The surface-associated fluorescence does not require cell fixation and thus can be obtained with live cells after short cell preparation procedure. The cell preparation takes less than 60 minutes. Using surface-associated cell fluorescence, higher-productivity variants can be rapidly isolated using cell sorting. Conventional cloning by limiting dilution has shown that a large clonal variation can exist in a hybridoma cell line [18]. Furthermore, if surface-associated cell fluorescence is correlated with production rate then this provides a means to obtain information on the dynamic properties of the cell population from a single measurement. This measurement can be used to study effects of cultivation conditions on the kinetics of product synthesis.

4.4. Detection of intracellular pH, membrane potential, and calcium

Maintenance and regulation of intracellular pH is important for central biochemical pathways such as glycolysis, gluconeogenesis, and protein synthesis and secretion. The intracellular pH can be measured with flow cytometry using fluorescent dyes that are sensitive to hydrogen ion concentrations [19,20]. It could represent a useful parameter for studying metabolic responses of cell populations. In particular as it has been shown that the intracellular pH changes depending on nutritional conditions of the cell population [21]. Similarly, membrane potential is an important physiological property of cells and some of the organelles. The electrical potential difference across the plasma membrane is directly involved in the transport of ions and nutrients across this membrane. The membrane potential dye Rhodamine123 can be used for staining mitochondria [22] and has been used to characterize hybridoma cells in different growth stages [23]. Also, intracellular calcium ion concentration can be measured with fluorescent probes and flow cytometry [24,25]. This would be important in studies of the effect of environmental conditions on cell growth as calcium has an important role in the transduction of signals across the cell membrane. The use of these dyes for monitoring the state of a cell population is somewhat restricted, however, as cells need to be loaded with the appropriate dye before the measurement can be made. The measured metabolic state of the cells may not represent the actual conditions in the bioreactor because the cells are most likely perturbed during the loading process.

4.5. Single-cell growth rates of cellular components

The goal of mathematical models that describe growing cell populations is to connect the kinetics of cell populations with the biological processes in the cell and with the environmental conditions that surround the cells. Ultimately reliable mathematical models can be used to provide a quantitative basis for the design, optimization and control of bioprocesses. Many kinetic models of cell growth that have been developed for biotechnology have a limited utility as they are not based on a rigorous theory. In particular, many of these

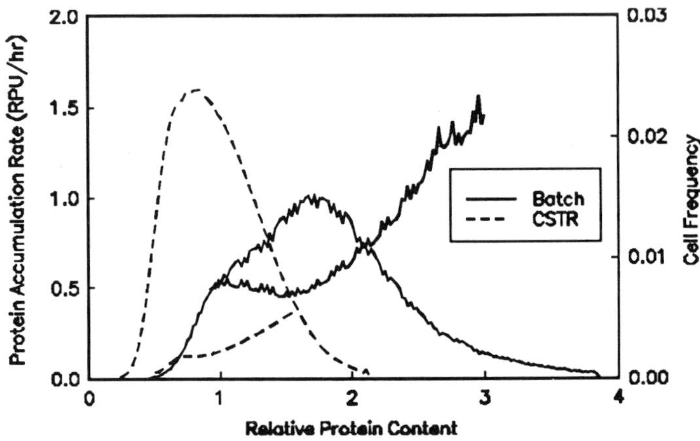

Figure 1 Single-cell growth rate and cell fractions as a function of protein content for two *Saccharomyces cerevisiae* cell populations growing at different growth rates.

models neglect that a cell culture represents a segregated system and that cellular properties are distributed. An exact description of the dynamics of a cell population has to take into account the distribution of the physiological state of the cell population. Although the framework for a rigorous theory of cell growth has been developed in the 1960's [7,26], little use has been made of this approach, presumably because of the lack of instrumentation that would provide the necessary experimental data.

Knowledge of three physiological functions are needed to exactly describe the evolution in time of a cell population. These are (i) the rate pattern during the cell cycle, (ii) the transition rates between temporally separated cell cycle stages, and (iii) the partitioning function which describes how cellular components are divided among the daughter cells that are generated at cell division. If the way in which these functions depend on the physiological state of a cell and the surrounding medium is known, the model is able to predict how a cell culture will develop in time. These three functions are unknown for most cell populations.

Recently it has been shown for growing hybridoma [27] and yeast [28] cell populations that flow cytometry can provide the experimental data to determine these intrinsic physiological functions. In this approach it is essential that one realizes that cells move in time through the cell cycle. The compartmentalization of cellular functions in time establishes a temporal order for specific morphological and biochemical events that every cell must follow between birth and division. The relative temporal order of some of these events during the cell cycle can be determined for an asynchronous culture with flow cytometry. For instance cellular DNA content before and after the S phase can be quantified with flow cytometry and this provides a basis for classifying cells into three categories: cells that have not yet duplicated their DNA (G_1 phase), cells which are in the process of duplicating their DNA (S phase) and cells which have completed DNA duplication (G_2 and M phase). The beginning and ending of each of these three cell cycle compartments is defined by two transitions which represent the state of the cells at entry into and exit from that cell cycle phase. For instance, the G_1 phase is bounded by cell birth and the initiation of DNA replication, and when cells are growing and dividing, they cannot enter or leave the G_1 phase by any other biological means. Similarly the S phase is bounded by initiation of DNA synthesis and completion of DNA synthesis, and the G_2+M cell cycle phase is bounded by the completion of DNA synthesis and cell division. It has been shown in this work that distributions of cellular state properties in specific cell cycle phases and at transition points between cell cycle phases can be experimentally determined. These data in combination with population balance equations can be used to compute single-cell growth rates of the measured state properties.

5. *SYNOPSIS*

Flow cytometry is being increasingly used to answer numerous questions that occur in connection with the cultivation of cells. This includes monitoring of the cells during the cultivation process to obtain a better idea of the structure of the cell population [29-32]. The interaction of cells with their enviroment can significantly affect the productivity of a cell culture. For instance, the serum level in a growing culture of hybridoma cells can change the ratio between producing and non-producing cells [33]. Flow cytometry is being used to isolate and characterize cell lines with a high copy number of a desired gene [34,35], or to determine the harvesting point of a baculovirus-infected insect cell culture [36]. It has been applied to study in great detail the effect of shear stress on DNA synthesis rates in populations of CHO cells exposed to different agitation intensities [37]. The questions that can be addressed using this instrumentation are numerous. The discussed examples demonstrate that cell populations are inherently variable, and that an understanding of this variability is needed if we want to understand the behavior of cell cultures in a cultivation process.

Acknowledgment

The support for this work by the National Science Foundation is gratefully acknowledged (Grant BCS 9100385).

REFERENCES

1. H.A. Crissman, H.A., Darzynkiewicz, Z.., Tobey, R.A. and Steinkamp, J.A. (1985) Science 228,1321-1324.

2. Melamed, M. R., Lindmo, T., and Mendelsohn, M. L. (1990) Flow cytometry and sorting.

Wiley-Liss, New York. Ed. 2nd
3. Shapiro, H. M. (1988) Practical flow cytometry. Alan R. Liss, Inc., New York. Ed. 2nd
4. Watson, J. V. (1991) Introduction to flow cytometry. Cambridge University Press, Cambridge.
5. Steinkamp, J. A. (1984) Rev. Sci. Instrum. 55,1375-1400.
6. Darzynkiewicz, Z. and Crissman, H. A. (1990) Flow cytometry. Methods in cell biology Vol.33, Academic Press, San Diego.
7. Fredrickson, A. G., Ramkrishna, D., and Tsuchiya, H. N. (1967) Math. Biosci.1,327-374.
8. Ramkrishna, D. (1979) In:Advances in biochemical engineering.Vol.11. Ghose, T. K., Fiechter, A., and Blakebrough, N. , eds. Springer-Verlag, New York., p 1-47.
9. Salzman, G. C., Singham, S. B., Johnston, R. G., and Bohren, C. F. (1990) In:Flow cytometry and sorting. Melamed, M. R., Lindmo, T., and Mendelsohn, M. L., eds. Wiley-Liss, New York. Ed. 2nd, p 81-107.
10. Sen, S., Srienc, F., and Hu, W. -S. (1989) Cytotechnol 2,85-94.
11. Frame, K. K. and Hu, W. S. (1991) Enzyme Microb. Technol. 13,690-696.
12. Ozturk, S. S. and Palsson, B. O. (1990) Hybridoma 9,167-175.
13. Parks, D. R., Bryan, V. M., Oi, V. T., and Herzenberg, L. A. (1979) Proc. Natl. Acad. Sci. US 76,1962-1966.
14. Dangl, J. L. and Herzenberg, L. A. (1982) J. Immunol. Meth. 52,1-14.
15. Sen, S., Hu, W. -S., and Srienc, F. (1990) Enzyme Microb. Technol. 12,571-576.
16. McKinney, K. L., Dilwith, R., and Belfort, G. (1991) Biotechnol. Prog. 7,445-454.
17. Meilhoc, E., Wittrup, K. D., and Bailey, J. E. (1989) J. Immunol. Meth. 121,167-174.
18. Kitano, K., Iwamoto, K., Shintani, Y., and Akiyama, S. (1988) J. Immunol. Meth. 109,9-16.
19. Visser, J. W. M., Jongeling, A. A. M., and Tanke, H. J. (1979) J. Histochem. Cytochem. 27,32-35.
20. Musgrove, E., Rugg, C., and Hedley, D. (1986) Cytometry 7,347-355.
21. Hernlem, B. J. and Srienc, F. (1989) Biotechnol. Techn. 3,79-84.
22. Johnson, L. V., Walsh, M. L., Bockus, B. J., and Chen, L. B. (1982) J. Cell Biol. 88,526-535.
23. Al-Rubeai, M., Chalder, S., Bird, R., and Emery, A. N. (1991) Cytotechnol. 7,179-186.
24. Grynkiewicz, G., Poenie, M., and Tsien, R. Y. (1985) J. Biol. Chem. 260,3440-3450.
25. Rabinovitch, P. S. and June, C. H. (1990) In:Flow cytometry and sorting. Melamed, M. R., Lindmo, T., and Mendelsohn, M. L. , eds. Wiley-Liss, New York. Ed. 2nd, p 651-668.
26. Bell, G. I. and Anderson, E. C. (1967) Biophys. J. 7,329-351.
27. Kromenaker, S. J. and Srienc, F. (1991) Biotechnol. Bioeng. 38,665-677.
28. Srienc, F. and Dien, B. S. (1992) Ann. N. Y. Acad. Sci. 665,59-71.
29. Al-Rubeai, M., Emery, A. N., and Chalder, S. (1991) J. Biotechnol. 19,67-82.
30. Dalili, M. and Ollis, D. F. (1990) Biotechnol. Bioeng. 36,64-73.
31. Heath, C., Dilwith, R., and Belfort, G. (1990) J. Biotechnol. 15,71-90.
32. Leno, M., Merten, O. -W., Vuillier, F., and Hache, J. (1991 J. Biotechnol. 20,301-312.
33. Chuck, A. S. and Palsson, B. O. (1992) Biotechnol. Bioeng. 39,354-360.
34. Kaufman, R. J., Bertino, J. R., and Schimke, R. T. (1978) J. Biol. Chem.,253,5852-5860.
35. Pendse, G. J., Karkare, S., and Bailey, J. E. (1992) Biotechnol. Bioeng. 40,119-129.
36. Lazarte, J. E., Tosi, P. -F., and Nicolau, C. (1992) Biotechnol. Bioeng. 40,214-217.
37. Lakhotia, S., Bauer, K. D., and Papoutsakis, E. T. (1992) Biotechnol. Bioeng. 40,978-990.

ECB6: Proceedings of the 6th European Congress on Biotechnology
edited by L. Alberghina, L. Frontali, P. Sensi

Flow Cytometry of Micro-colonies as an Approach for Mutant Selection and Industrial Strain Improvement

E. Sahar, R. Nir*, T. Molcho and R. Lamed

Department of Molecular Microbiology and Biotechnology, Tel Aviv University, Tel Aviv, Israel.

1. INTRODUCTION

Screening and selection of biological cells for use in the biotechnological industry is of great economical importance. The methods currently used to achieve this are tedious and time consuming when selective growth conditions cannot be devised. They involve growing the cell population to be screened, usually after mutagenesis, to form separate clonal subpopulations which are then each tested separately for the desired trait. This procedure not only increases the cost and duration of the screening process but often prevents achieving the goal if the desired cell is rare. Therefore, an approach for performing this screening process with increased efficiency and speed is highly desirable.

Flow cytometry (FCM), with its capacity to analyze and sort cells one-by-one at rates greater then 10^7/hour, appears to be ideally suited for this task. In spite of this, the application of FCM in this field has been rather limited, for a number of reasons. First, most of the organisms used in the biotechnological industry are microbial. Cells of these organisms are typically 100-1000 fold lower in volume and in biochemical content then the mammalian cells for which FCM is routinely used. Thus, the optical signals generated by these cells are relatively weak, making their measurement more difficult. Additionally, many of the commercially important species of microorganisms are filamentous, prohibiting their analysis by FCM. Perhaps the most important limitation of screening on the single cell level is the biotechnological interest in secreted products which cannot be measured this way.

We discuss here an approach which overcomes these obstacles by analyzing encapsulated micro-colonies and illustrate the potential with examples of systems where it has been applied.

2. METHODS AND RESULTS

The method described here is based on entrapping single cells in small, homogeneous microbeads (MB's) in which each cell can proliferate to form a micro-colony (1-3). The MB's can be produced with highly uniform, precisely controlled diameters in the range of 10-50 micrometers rendering them amenable to FCM analysis. The properties of the bead-entrapped micro-colonies can thus be analyzed in situ by FCM and colonies with desired properties can be isolated.

----

* present address: Schering-Plough Research, 1011 Morris Ave., Union, New Jersey 07083.

The details of the methods used to entrap individual cells in suitable microbeads have been published elsewhere (1,2). Briefly, the cells are suspended in a solution of a gellable polymer such as agarose or alginate and the suspension is ejected through a vibrating orifice which breaks the emanating liquid into identically-sized droplets which are then caused to gel. The average number of cells entrapped per bead follows (2) the Poisson distribution (for non-aggregating cells). Table 1 summarizes the experimental steps involved in utilizing the MB strategy.

TABLE 1

GENERAL STRATEGY FOR MICROBEAD UTILIZATION IN FLOW CYTOMETRIC SCREENING

1. **Immobilize single cells in uniform microbeads.**
2. **Incubate MB's under desired conditions.**
3. **Label product or cell property (if necessary).**
4. **Analyze and sort by flow cytometry.**
5. **Examine isolates (e.g. productivity).**

Entrapment and cultivation of E. coli in beads prepared from 0.6% agarose (type VII, low melting point, Sigma Chemicals) is illustrated in Fig. 1. The bacterial cells were suspended in the molten agarose at 39^{0}C to a final concentration of 10^7/ml. The beads were collected in growth medium and incubated for 9 hrs. Results of FCM analysis (A,C) and phase contrast photomicrography (B,D) are depicted in the illustration, showing the uniformity of the beads and the appearance of colonies (subpopulation 2 in C) which are characterized by increased light scatter and are shown in the micrograph (D) after their sorting.

Perhaps the most promising uses of the MB technology are in the measurement of secreted products. Many important biotechnological processes rely on strains that secrete the product into the fermentation medium. However, secretion rate and productivity per cell cannot be measured flow cytometrically on the single cell level. The MB approach allows performance of these measurements. In these cases special provisions are necessary to insure that the secreted product will be retained within the bead. One general approach for achieving this (shown in Fig. 2), is based on introducing antibodies directed against the specific product into the MB, e.g., by direct coupling to the polymeric matrix (1) or other means of immobilization (4,1). The secreted product, diffusing through the MB, is captured by these antibodies and can then be quantitated by various means depending on the nature of the product. For example, (Fig. 2), one can use a second, labelled antibody which diffuses into the MB and binds to the trapped product. Fig. 2b demonstrates the application of this approach to measurement of a polymer, termed biodispersant (5), secreted by a strain of *Acinetobacter calcoaceticus*. The fluorescence extends beyond the limits of the micro-colony, demonstrating that the secreted product was trapped by the antibodies distributed throughout the bead volume.

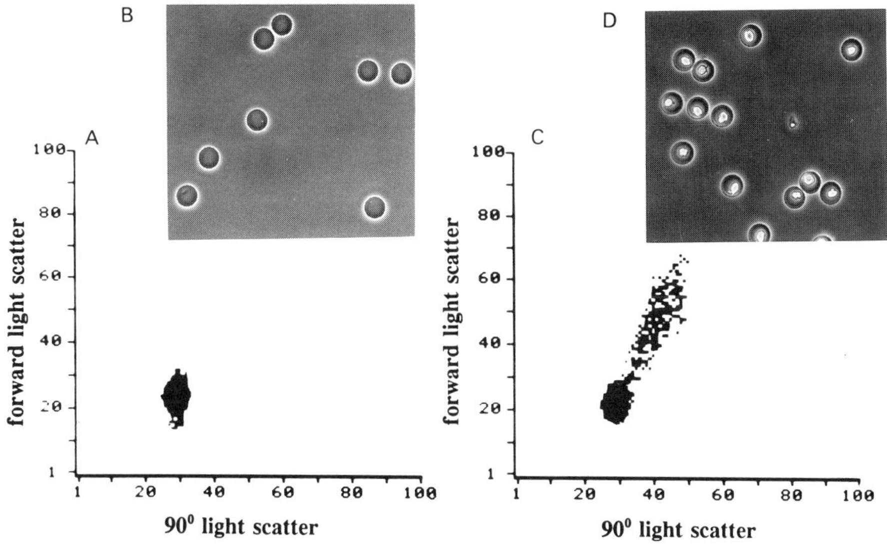

Fig. 1: Analysis of MB-entrapped E. coli before (A,B) or after (C,D) micro-colony development.

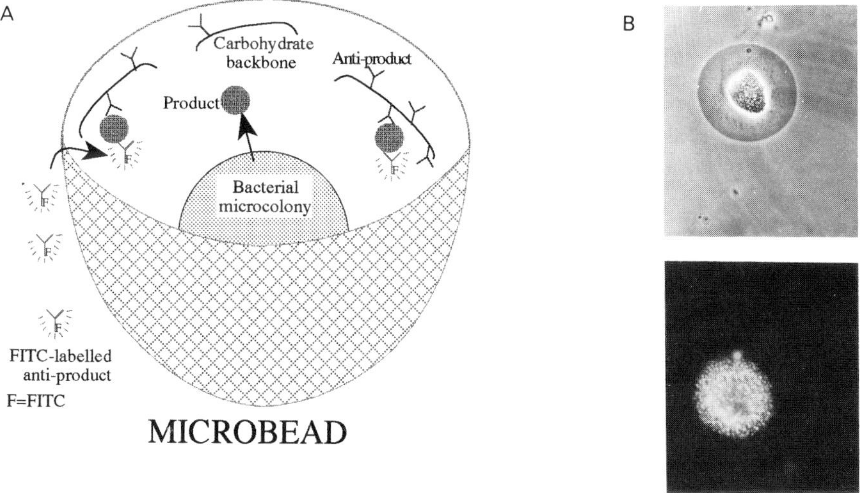

Fig. 2: Immunochemical product capture and labelling. A. Schematic presentation of the principle. B. A colony of *Acinetobacter calcoaceticus* (top) secreting a biodispersing polymer, captured and labelled with a polyclonal antiserum (bottom).

The MB approach has been applied to a variety of problems involving bacterial, yeast and mammalian cells. Table 2 lists some of these applications.

STUDIES PERFORMED USING THE MICROBEAD APPROACH

APPLICATION	ORGANISM	DETECTION PRINCIPLE	REF.
selection of over-producing mutants			
1. Extra-cellular product:			
-lipase	P. aeruginosa	antibodies	1
-monoclonal antibody	Hybridoma	antibodies	4
-glycosidase	Bacillus spp.	antibodies	-
-collagenase	P. aeruginosa	loss of fluorescence	1,3
2. Intracellular product (viably):			
-β galactosidase	E. coli, Candida pseudotropicalis, M. xanthus	fluorogenic substrate	1,4
selection of conditional mutants			
1. Auxotrophs	E. coli	light scatter 2 stage sort	7
growth rate mutants	Candida pseud.	light scatter	1,8
cell-cell interaction	Myxococcus xanthus	cell morphology	1,9

3. DISCUSSION.

The MB system described here has opened new possibilities for applying FCM to the selection and improvement of both eukaryotic and prokaryotic strains important to the biotechnological industry. Each microbead, harboring a separate clone, is essentially equivalent to a culture vessel employed in the classical screening methods. Importantly, it allows selection under conditions which more closely mimic the conditions of fermentation then colonies on a petri-dish. The method not only allows FCM analysis and sorting on the basis of product secretion but also permits intracellular product measurement combined with survival of at least some representative cells in each colony. In addition, the signal measured relative to that of single cells is highly amplified, since a bacterial microcolony can contain several hundred cells. Problems which occur during the phenotypic expression phase after mutagenesis, namely loss of the desired mutant due to slower growth rate and multiplicity of identical mutants, can be prevented if mutagenesis is afflicted after entrapment. The MB method may also allow FCM analysis of filamentous organisms, but this aspect has not yet been tested.

REFERENCES

1. R. Nir, Ph.D. Thesis, Tel Aviv University, 1990
2. R. Nir, R. Lamed, L. Gueta and E. Sahar, Appl. Environ. Microbiol., 56 (1990) 2870.
3. J. C. Weaver, J.G. Bliss, G.I. Harrison, K.T.Powel and G.B. Williams, Methods, 2 (1991) 234.
4. K.T. Powell and J.C. Weaver, Biotechnology, 8 (1990) 333.
5. E. Rosenberg, C. Rabinovitz, A. Gottlieb, S. Rosenhak and E. Z. Ron, Appl. Environ. Microbiol., 54 (1988) 317.
6. R. Nir, Y. Yisraeli, R. Lamed and E. Sahar, Appl. Environ. Microbiology, 56 (1990) 3861.
7. E. Zudicker, M.Sc. Thesis, Tel Aviv University, 1991.
8. R. Nir, R. Lamed, Y. Shabtai and E. Sahar, J. Microbiol. Methods, 14 (1992) 247.
9. A. Rosenbluh, R. Nir, E. Sahar and E. Rosenberg, J. Bact., 171 (1989) 4923.

ECB6: Proceedings of the 6th European Congress on Biotechnology
edited by L. Alberghina, L. Frontali, P. Sensi

Analysis of the respiratory activity in growing budding yeast by flow cytometry.

D, Porro, C. Smeraldi, B.M. Ranzi, E. Martegani and L. Alberghina.

Dipartimento di Fisiologia e Biochimica Generali, Sez. Biochimica Comparata, Università di Milano, Via Celoria 26, 20133, Milan, Italy.

Introduction. Advanced biotechnological productions require control, optimization and validation of bioreactor operations [1-3]. Since the performance of the cell population in a biotechnological process depends upon the metabolic activities of individual cells, the availability of fast and sensitive procedures to monitor such activities could be useful. Flow cytometry allows to measure physico/chemical parameters on the single cell, yielding their distribution in the population. Among the various cytometric applications applied to higher eukaryotic cells, increasing attention is now being devoted to the study of mitochondria and mitochondrial activities using a lipophylic cationic dye, Rhodamine 123 (Rh123)[4].
With the aim to develop a new procedure for the monitoring and validation of yeast fermentations, we extended flow cytometric analysis to growing cultures of *Saccharomyces cerevisiae* cells stained with Rh123. Budding yeast is an eukaryotic microorganism widely used in an increasing number of industrial applications, ranging from traditional fermentations (wine, beer, vitamins, biomass, ethanol, etc...) to advanced productions (chemicals, heterologous proteins such as vaccines, pharmaceutical agents, etc....). All these processes are or could be strongly influenced by the respiratory activity of the yeast populations. The results presented in this paper demonstrate that flow cytometry is a fast and very sensitive method to evaluate the respiratory activity in yeast cells and that it is suitable for the determination of rapid changes in yeast metabolism.

Material and Methods. The haploid *S.cerevisiae* strain X4004 (MAT a, lys5, ura3, met2, trp1) was used for this work. Yeast cells were grown in conical flasks by shaking in YEP-complete medium. The carbon sources used were glucose (GLU), galactose (GAL), lactate (LACT) and ethanol (ETOH); 2% w/v carbon source was normally used. Cell number was determined by counting the number of cell per milliliter with a Coulter Counter ZBI [1]. Prior to staining, two million of cells from the culture were shifted into a conical flask containing 30 ml of medium; cells were stained by adding Rh123 (Sigma) for 15 min (final concentration 10 μg/ml). Then, 2 ml aliquots were harvested by centrifugation and washed once with sorbitol buffer (140 mM KCl, 1.2 M Sorbitol, 20 mM Tris/HCl-pH 7.2). For flow cytometric analysis 1 ml of PBS and 0.5 ml of sorbitol buffer were added and samples immediately analysed. The incubation temperature for the growth and the staining procedure (30°C), as well as the interval between the dye addition and the introduction of the sample into the flow cytometer, were kept constant for every sample.

Flow cytometric analysis was performed with a FACStar^{plus} (Becton & Dickinson) equipped with 2W argon-ion laser (excitation wavelength 488 nm, laser power 200 mW). Rh123 fluorescence intensity was measured at 520 nm. The flow rate was approx. 300 cells per sec and the number of cells analyzed was set to 15000.

Results and Discussion. The energetic metabolism of *S. cerevisiae* cells during aerobic growth is differentially modulated by the carbon source employed. In particular, high glucose concentrations can be metabolized by yeast cells through aerobic fermentation (Respiratory Quotient: >1), while there is only one possible pathway in which yeast cells can grow on ethanol, and it is purely oxidative (R.Q.=0.5)[5]. In order to gain insight the analysis of the respiratory activity, initially we evaluated whether the presence of Rh123 in the culture medium affects the growth rate of the cells, obtaining indications of no appreciable effects (data not shown).
Typical Rh123-distributions obtained by flow cytometric analysis of yeast cells growing on GLU, GAL, LACT, or ETOH based media (from a fermentable to a fully oxidable carbon source), are shown in **Fig.1**. The distributions are clearly unimodal with a large coefficient of variation (CV= 0.60 +/- 0.05), reflecting a sizable heterogeneity in the uptake of the dye by the cells of the population. Furthermore, both shape and CV were found to be quite reproducible. A quantitative analysis of these distributions is reported in **Tab.1**, showing that there is a substantial agreement between flow cytometric data and classical biochemical determinations.

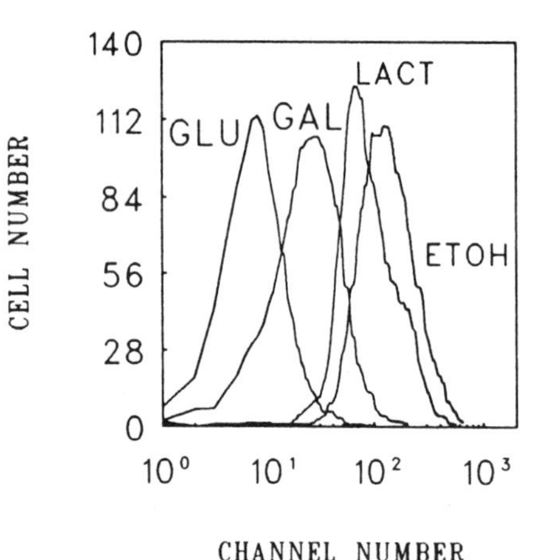

Fig.1: Rh123 fluorescence of yeast cells growing on different substrates.

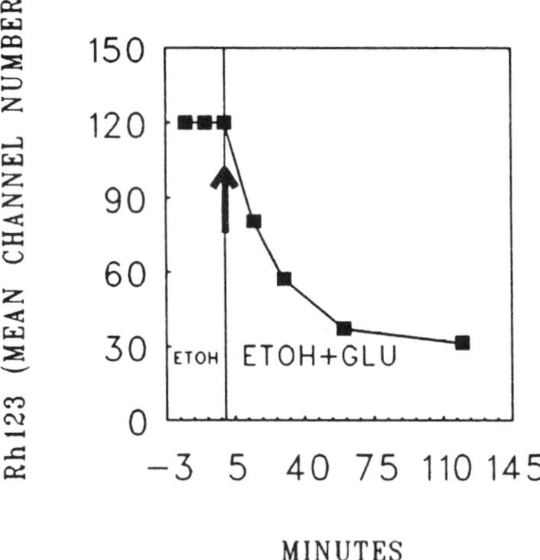

Fig.2. Effect of the addition of glucose to yeast cells growing on YEP-ETOH.

However, to reduce misinterpretations of the flow cytometric analyses, we have designed an experiment of control strongly related to the environmental conditions of the growing cells. In agreement with the hypothesis that accumulation of the dye is related to the mitochondrial activity, we observed a significant and very fast (in a few minutes) extrusion of the fluorochrome when a repressing carbon source (glucose, 2% w/v the final concentration) was added to yeast cells growing on YEP-ETOH containing Rh123, **Fig.2.**

Table 1. Rh123 fluorescence and respiratory activity of budded cells growing on different carbon sources.

Carbon Source	A	B*	C*
GLU	1.00	1.00	1.00
GAL	3.23	3.26	2.80
LACT	9.55	N.D.	N.D.
ETOH	13.71	12.21	10.11

The data represent the intensity of Rh123 fluorescence (A) or the respiratory activity (B and C) expressed as ratio between the values found using different carbon sources and that observed during growth on glucose.
A: Results found after Rh123 staining; the mean channel number of each distribution was considered; B and C: Values calculated from data of Gancedo et al. 1989 [6] and from Fraenkel, 1982 [7], expressing moles of oxygen used per hour per gram of dry weight (B) or μmole used per min per gram of dry weight (C); (*) data were normalized to our conditions by dividing for the specific growth rate of the single cultures.

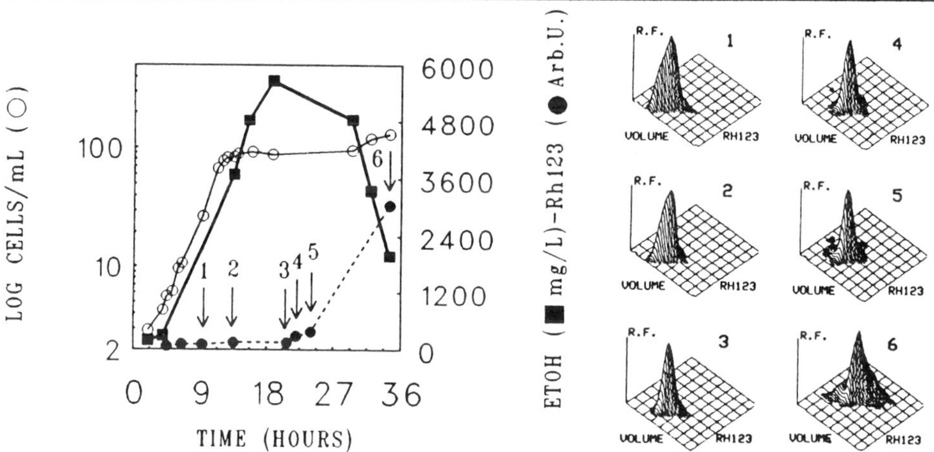

Fig.3. Diauxic growth pattern of X4004 yeast cells during growth on YEP-GLU-Rh123 medium. Right: data displayed as tridimensional histograms: Volume (FSC) - Rh123 flourescence - Relative Frequency (R.F.)

One has to recall that the addition of a high glucose concentration induces the so called Crabtree or glucose effect [5] that is the block, in spite of the presence of oxygen, of the mitochondrial activities. Extensive studies [5-7] have proved it is accompanied by a variety of biochemical responses such as the repression of several tricarboxylic acid cycle enzymes, the decreasing in the mitochondrial cytochrome content, the repression and inactivation of several mitochondrial enzymes, etc... From a qualitative point of view, similar results were obtained after a glucose addition to stained cells growing on YEP-LACT and YEP-GAL (data not shown). The time-course measurement technique described is also applicable to kinetic studies of the Crabtree effect. In fact, we employed the

experimental procedure described in Fig.2 to determine the minimal concentration of glucose that can repress the respiratory activity. This kind of investigation indicates for X4004 cells a strong repression can be obtained using glucose concentrations higher than 500 mg/L (Approx. 3mM)(data not shown). This value is quite close to the value of 6 mM reported earlier for the Crabtree effect [6-7].
In conclusion, since Rh123 is a supervital dye, we have evaluated the intensity of Rh123-fluorescence during a typical diauxic batch growth of X4004 cells on YEP-GLU-Rh123 medium, **Fig.3**. As expected, cells growing on glucose or on the produced ethanol have a very different tridimensional plot (FSC> <FL1-Rh123> <Relative Frequency, R.F.). It is interesting to note the tridimensional plots begin to change earlier than a diauxic pattern of growth can be detected by counting the cell number concentration.
Flow cytometric analyses of Rh123-stained cells might be also employed to detect differences in the genetic background of different yeast strains (data not shown).

Conclusions. In this paper we have shown that the analysis of the respiratory activity of *S. cerevisiae* cells can be performed by flow cytometry in Rh123-stained cells. This method appears to be rather fast (maximum 20-25 minutes) and quite sensitive. In this light, given the relevance of *S.cerevisiae* for basic research and biotechnological applications, the results reported in this paper could assume in perspective a great interest. In fact, the technique described can be used also to analyse the behaviour of mutants, to study the effect caused from the (over)expression of heterologous genes as well as for the isolation of single cells with different mitochondrial number or with altered mitochondrial membrane potential by means of a fluorescence activated cell sorter.
Furthermore, it could be possible to develop methodologies for the monitoring and for a direct control of the metabolic state of a growing population. New improvements of the technique (i.e., allowing the analysis of other fluorescent probes in addition to Rh123) will give access to more dynamic information for the development of more realistic models able to account for links between the respiratory activity and other metabolic functions.

Acknowledgements. This project was supported from CNR, Progetto Finalizzato Biotecnologie e Biostrumentazioni, subproject #3, to B.M.R.

References.

1. L. Alberghina D. Porro, *YEAST*, in press. (1993).
2. L. Alberghina, B.M. Ranzi, D. Porro and E. Martegani, *Biotechnol. Progress.* 7, (1991), 299.
3. L. Alberghina, D. Porro, E. Martegani and B.M. Ranzi, *Biotechnol. Appl. Biochem.* 14, (1991), 82.
4. L.V. Johnson, M.L. Walsh and L.B. Chen, *Proc. Natl. Acad. Sci. USA*, 77, (1980) 990.
5. A. Fiecther, G.F. Fuhrmann and O. Kappeli, *Adv. Microbial. Phys.* 22, (1981) 123.
6. C. Gancedo and R. Serrano, *in* THE YEAST, Vol.3, (1989) Academic Press Limited.
7. D.G. Fraenkel, *in* "The molecular biology of the yeast *Saccharomyces cerevisiae.*" Metabolism and gene expression (J.N. Strathern, E.W. Jones and J.R. Broach, Eds) Cold Spering Harbor Laboratory, N.Y., (1982), 1.

ECB6: Proceedings of the 6th European Congress on Biotechnology
edited by L. Alberghina, L. Frontali, P. Sensi

A New Method for Measurement of poly-ß-hydroxybutyrate inside *Alcaligenes eutrophus* During Cultivation by flow cytometry and spectrofluorometry

T. Scheper[1], A. Degelau[1], M. Ballmaier[2]
[1]**Institut für Biochemie, Wilhelm-Klemm-Str. 2, 4400 Münster, Germany**
[2]**Blood Transfusion Service of Lower Saxony, Department of Research and Developement, Eldagsener Str. 38, 3257 Springe 1, Germany**

1. Introduction

Poly-hydroxy-butyrate (PHB) is a thermoplastic and biologically degradable polymer that can be produced e.g., by the bacterium *Alcaligenes eutrophus* when fed with fructose. Other organisms can produce copolymers of similar type. Methods to monitor the PHB-content during cultivation processes are mostly very complicated and require a large amount of sample. (Bitar, 1990; Guske, 1990; Schlegel, 1985, 1990; Sonnleitner *et al.*, 1979; Srienc *et al.*, 1984; Steinbüchel, 1989).

An easier method for quantitative PHB detection is the fluorescence method with Nile Red. This dye easily penetrates the cells in suspension and the measurement can be performed after a short time. Nile Red is a fluorescent lipid dye (Fowler, 1985; Greenspan, 1985). The fluorescence intensity of Nile Red stained cells results from the PHB concentration; non-PHB staining is low because the bacteria have almost no lipids (Schlegel, 1985, 1990). The aim of our research was to investigate the relationship between the fluorescence data obtained from cytometric or spectrofluorometric measurements and the values obtained via conventional gas chromatographic assays.

2. Material and Methods

PHB staining with Nile Red

For all staining assays, a stock solution of Nile Red was prepared containing 100 mg Nile Red in 100 ml acetone. Nile Red can be dissolved easily in lipophilic phases. Its solubility in water is poor and its fluorescence is quenched by water (Fowler and Greenspan, 1985; Linz, 1989). Nile Red, which is nontoxic, can stain both fixed and unfixed cells; however the fluorescence intensity of fixed cells is higher. For staining, cells were resuspended in saline solution. Previous investigations (Scheper, 1991; Linz, 1989) showed that 10 μl of Nile Red stock solution stain 1 ml cell suspension up to OD 5 with a PHB content up to 70 % of cell dry weight. After 30 or 40 minutes, the cells are totally stained and the fluorescence maximum is reached. This assay can be used to detect PHB quantitatively.

Analysis with a laser flow cytometer

In our investigations, a FACS Scan cytometer (Becton Dickinson) was used. In previous studies on the determination of PHB in *Alcaligenes eutrophus*, light scatter signals were used because PHB forms intracellular refractile bodies that alter the light scattering properties (Srienc *et al.*, 1984). Before being analyzed in the cytometer, the cell samples were diluted so that the counting rate was up to 1000 cells per second. An argon ion laser was used for excitation (488 nm) and a photomultiplier with a 600 nm filter was used for detection of the fluorescence. Cells were stained as described above. The fluorescence signals obtained from the photomultiplier are proportional to the component concentration.

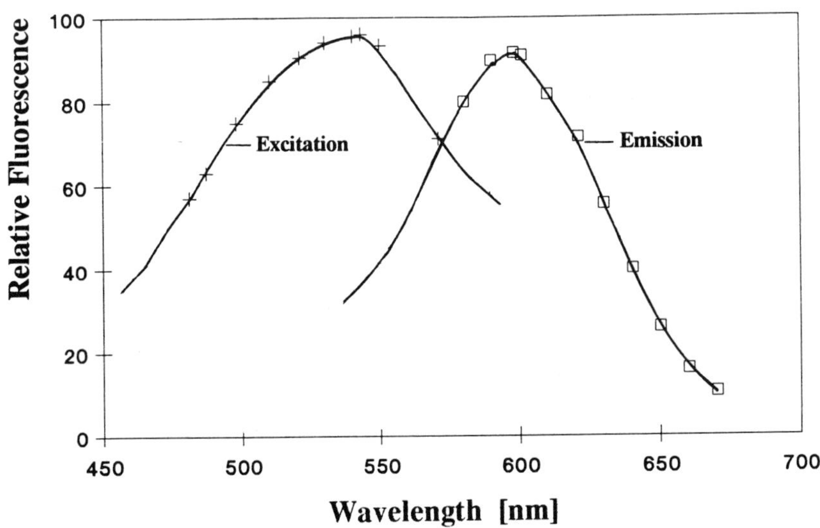

Figure 1. Fluorescence spectrum of *A. eutrophus* stained with Nile Red (maximum excitation wavelength for emission scan: 543 nm; maximum emission wavelength for excitation scan: 598 nm)

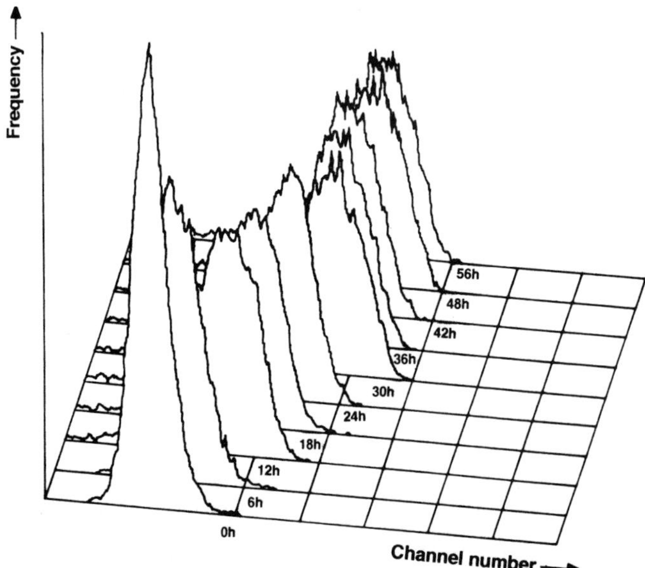

Figure 2. Cytograms of Nile Red stained *A. eutrophus*. Frequency gives the number of cells belonging to a certain channel; channel number represents the fluorescence intensity, related to the PHB content (logarithmic scale)

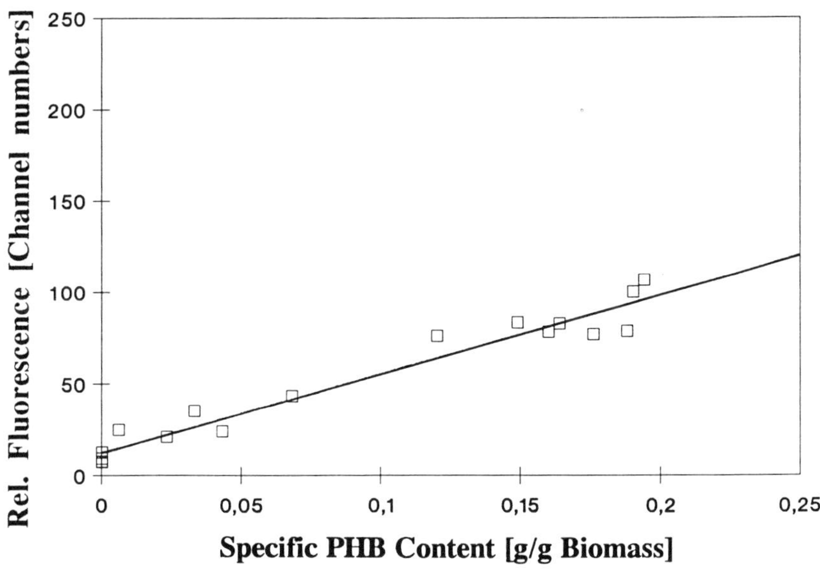

Figure 3. Linear relationship between gas chromatographic (specific PHB content) and cytometric data

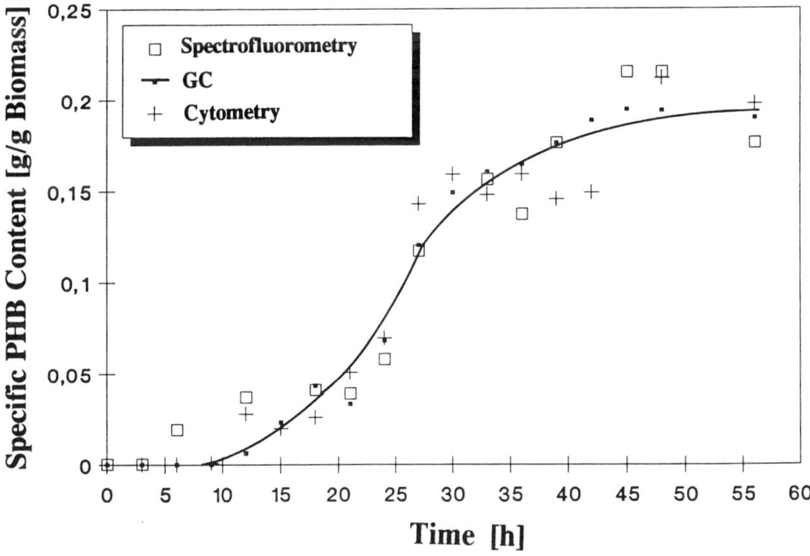

Figure 4 Recalculated PHB content from the spectrofluorometric data and the cytometric data plotted against the GC data as a function of cultivation time.

Analysis with a spectrofluorometer

The fluorescence spectrum of *A. eutrophus* stained with Nile Red is given in Figure 1. The maximum of excitation wavelength for emission scan is 543 nm and the emission maximum for excitation scan is 598 nm for stained PHB-containing *Alcaligenes eutrophus*. Figure 2 shows the relationship between the optical density of the cell suspension and the Nile Red fluorescence, after addition of a constant volume of the stock staining solution. In the range of 0 to 5 OD units, the fluorescence intensity is linear. The data which were measured by the spectrofluorometer were kept at a constant optical density of 0.3 (excitation wavelength 543 nm /emission wavelength 598 nm).

3. Results and discussion

A batch cultivation of *A. eutrophus* was performed over a period of 56 hours. Every three hours, samples were withdrawn for the measurement of the optical density, cell dry weight and also for cytometric, spectrofluorometric, (sample volume 2 ml) and GC analysis (sample volume 10 ml).

The cell dry weight reached 1.5 g/l, which was equivalent to an optical density of approximately 9. The PHB content of the bacteria was up to 20% of the cell dry weight in this cultivation. In the spectrofluorometric assay, the fluorescence of all cells is measured, while in the flow cytometric assay the PHB content in single cells is monitored (discrete analysis). The cytometer determines the PHB-content of single cells and not the PHB content per volume.

Figure 5 shows the results of cytometric measurements with Nile Red stained cells. In this cytogram, the channel number is plotted against the frequency for samples taken at various times. The whole fluorescence intensity range (logarithmic scale) is divided into 254 channels, each representing a certain fluorescence intensity value. The frequency indicates the number of cells belonging to a certain channel. Since the peaks shift to higher channel numbers with time it is clear that the PHB content increased during the first 30 hours of the cultivation.

Figure 3 shows the linear relationship between GC and cytometric data (correlation coefficient of 0.978). The spectrofluorometric data plotted against the GC data result in a similar plot (data not shown). In Figure 4, the recalculated specific PHB content values from the spectrofluorometric and the cytometric data are plotted in comparison with the GC data. Obviously there is good agreement of the GC data as a reference measurement with the fluorescence data of cytometric and fluorometric measurements.

References

Bitar, A., Underhill, S., (1990) Effect of Ammonium Supplementation on Production of Poly-β-Hydroxybutyric acid by*Alcaligenes eutrophus* in Batch Culture, Biotechnol. Letters 12 (8), 563-68.

Fowler, S.D., Greenspan, P., Application of Nile Red, a Fluorescent HydrophobicProbe, for the Deteaion of Neutral Lipids Deposits in Tissue Sections, J. Histochem. 33, 833-36.

Greenspan, P., Fowler, S D., (1985) Biological Applications of the Fluorescent Lipid Probe, Nile Red, Kodak Bulletin 56, 3.

Guske, CJ., (1990) Metabolic modeling of Growth and Poly-β-Hydroxybutyrate (PHB) production *inAlcaligenes eutrophus* H16, PhD thesis, California Institute of Technology Pasadena, California, USA.

Linz, F., (1989) Durchflußcytometrie zur Prozessbeobachtung in der Biotechnologie, PhD thesis, Universität Hannover Schlegel, H.G., (1990)*Akaligenes eutrophus* and its scientific and industrial career, Novel Biodegrad. Microb. Polym., 133-41.

Schlegel, H.G., (1985) Allgemeine Mikrobiologie, 6. überarbeitete Auflage, Stuttgart, New York, Thieme.

Sonnleitner, B., Heinzle, E., Braunegg, G., Lafferty, R.M., (1979) Formal Kinetics of Poly-β-Hydroxybutyric Acid (PHB) production in *Alcaligenes eutrophus* H16 and *Mycoplana rubra* R14 with Respect to the Dissolved Oxygen Tension in Ammonium-Limited Batch Cultures, Europ. J. Applied Microbiol. Biotechnol., 7,1-10.

Srienc, F., Arnold, B., Bailey, J.E., (1984) Characterization of intracellular accumulation of Poly-β-Hydroxybutyrate (PHB) in Individual Cells of *Alcaligenes eutrophus* H16 by Flow Cytometry, Biotechnol. and Bioengineering, 26, 982-87.

Steinbüchel, A., (1989) Poly(hydroxyalkanoates)- Storage Materials of Bacteria: Biosynthesis and Genetics, Forum Mikrobiol. 4, 190-8.

ECB6: Proceedings of the 6th European Congress on Biotechnology
edited by L. Alberghina, L. Frontali, P. Sensi

Process application of flow cytometry to the analysis of growth and physiological response in animal cells

M. Al-Rubeai and A.N. Emery

SERC Centre for Biochemical Engineering, School of Chemical Engineering, University of Birmingham, Birmingham, B15 2TT, United Kingdom

Optimisation of animal cell culture processes demands a better understanding of the proliferative and physiological states of cultured cells. Detailed information collected with efficiency, accuracy and speed can increase the precision of process control through the provision of improved data quality and can enhance product safety. Flow cytometry (FC) is a relatively new technique which can offer such information process monitoring and therefore can be valuable for the study of animal cell population dynamics. The technique produces measurements with high accuracy and reproducibility. Different constituents can be measured simultaneously, the results being available not simply as mass averages but also in terms of distributions across the cell population. Furthermore the power of FC can be further enhanced with the capability for physical selection of a cell population having a desired structural and functional properties. Sorting of antigen markers on the membrane has already provided the basis for the isolation of high producer hybridoma cell lines.

1. Process identification of animal cell culture by Flow Cytometry

The work that has been done so far has demonstrated very clearly the importance of FC as a means of animal cell culture process identification and as a sophisticated early process sensor. FC allows one to follow the temporal changes in DNA, RNA, mitochondrial activity, membrane integrity, intracellular pH, mitotic index, protein content cell size during the growth cycle of batch culture [1,2]. This information can provide sufficient details to describe batch culture processes of animal cells in terms both of the characteristics of individual cells and of the biomass of the whole population. One of the most interesting features of FC analysis is its ability to provide advanced and reliable information on proliferation dynamics and product synthetic stability of cell culture [2,3]. The differences in growth rate, death rate and maximum cell number which in turn depend upon the rate of cell division are predictable from measurements of the fraction of S-cells in mammalian culture and of G1-cells in insect culture [2].

One of the earlier interesting findings was that the IgG synthesis and secretion in murine hybridomas peaked in the G1 phase [5]. To achieve higher specific productivity the cell were in a later study arrested in the G1 phase using excess dosage of thymidine. Maximum enhancement of MAb productivity (100%) was obtained at the slowest growth rate (0.04/day) using 2mM thymidine [4]. It has also been found that cell size and viability were affected by the thymidine addition.

Most recently we have used FC to study the mechanisms of damage to animal cells by hydrodynamic forces [6]. Damage to surface associated receptors together with increasing leakage and passive transport were observed. The mitochondrial trans-membrane potential, which is linked to the cells ATP requirements, and also the mean cell size remained unchanged in intensive agitation. It was also

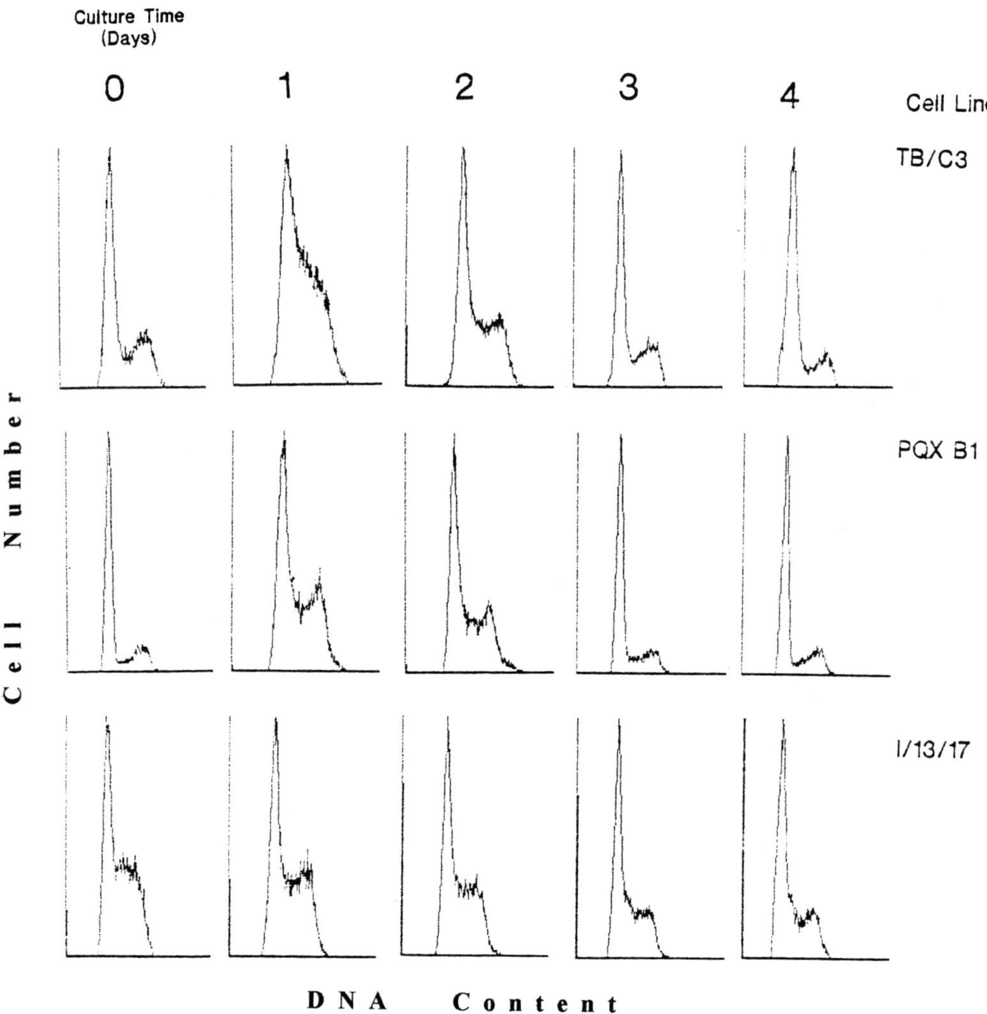

Fig. 1. Cell cycle distribution of 3 hybridoma cell lines during batch cultivation.

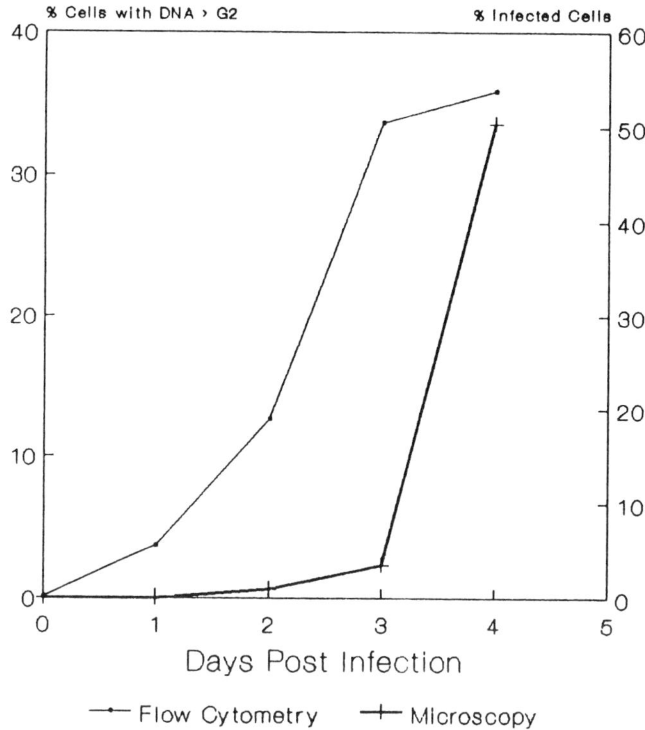

Fig. 2. Monitoring of infection of *Spodoptera frugiperda* (SF-9) cells with baculovirus.

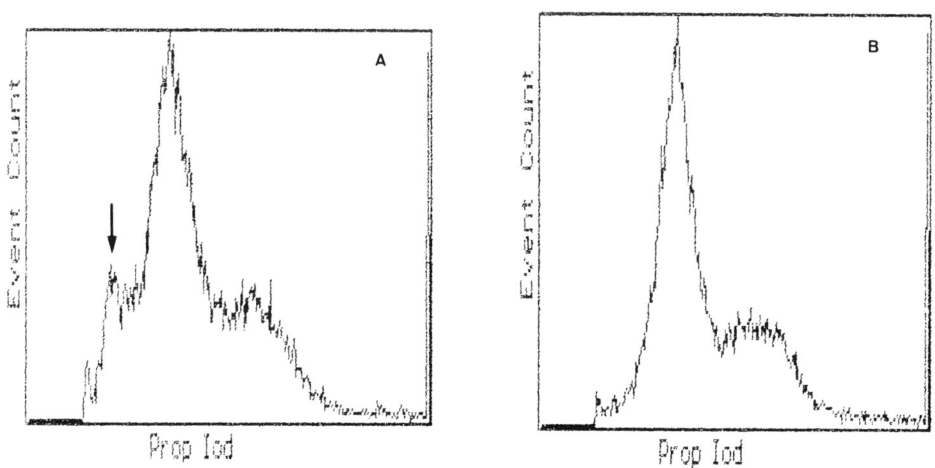

Fig. 3. DNA histograms of hybridoma cells obtained from perfusion culture. Cells were stained with Propidium Iodide without (A) or with (B) DNase treatment. Arrow indicates dead cells and nuclear fragments.

demonstrated [6] that cells were equally affected in all phases of the cell cycle. F- actin filaments were found to play an important role in determining the shear sensitivity of cells.

2. **Differences in cell cycle kinetics between cell line**

In analyses of batch cultures of 3 hybridoma cell lines, representing high and low levels in the range of productivity, growth rate and maximum population densities, the patterns of changes in cell cycle phases were essentially similar (Fig.1). However a difference was manifested in the cell cycle kinetics of the 3 cultures. Cells with a higher growth rate during the exponential phase showed a relatively higher fraction of S-cells during and before this phase. In contrast, during the decline phase the culture with a higher death rate exhibited a higher fraction of G1 - cells. In all 3 cell lines the level of activity of cells in the inoculum culture (as determined by the fraction of S-cells) had a significant effect on the proliferative rate during the growth phase of the inoculated culture. It was also demonstrated that, as the intracellular monoclonal antibody content varied between cell lines, so it also varied during culture time [4].

3. **Monitoring of Baculovirus infection**

Monitoring of Baculovirus infection of insect cells is routainly done by microscopic observation of the morphogenesis of the polyhedral inclusion bodies. However FC can be used to monitor such process more efficiently and accurately with the advantage of obtaining reliable advanced information on the consequence of infection. Fig.2 shows that flow cytometric monitoring of cellular DNA can give an earlier identification of infectivity and the subsequent development of viral particles. In this particular infected culture the polyhedra were only microscopically observed at day 4 post infection when these structures became fully developed and abundant. FC could detect the increasing DNA content as the cells moved from the G1 phase into the G2 and > G2 phases of the cell cycle. By day 3 all infected cells showed maximum DNA content. However as the polyhedrin protein only assembles late in the nucleus of the infected cells (at which stage the virus particle become embedded in the proteinaceous occlusions) then it only at this stage that it becomes possible to assess infectivity by microscopic observation.

4. **Monitoring of cellular activity of viable cells**

Monitoring of high cell density perfusion culture where cells (and cellular debris) are retained in the bioreactor can pose a problem for assessing cell cycle, chemical composition and the physiological activity of viable cells. Non viable cells and cellular debris can bind antibodies or other markers non-specifically.Therefore,it is important that nonviable cells are removed from the analysis. We have used a method based on the treatment of unfixed cells with DNase to remove DNA from nonviable cells and staining of cells with Propidium Iodide after washing out DNase and fixing cells with 70% ethanol. As can be seen in Fig.3 cells in G1, S and G2/M phases of the cell cycle can be recognised. An extra peak with lower DNA than G1 which is associated with DNA degradation characterised by dead cells is seen. This peak probably arises from cells which have undergone apoptotic death (Singh et al., 1993). The use of DNase has eliminated this peak as well as reduced the coefficient of variation of the G1 peak. This technique is currently in use for analysis of the relationship between cell cycle phases and antibody content of hybridoma cells in a high density retention system.

References

1. M. Al-Rubeai, A.N. Emery, J. Biotechnol., 19 (1991) 67.
2. M. Al-Rubeai, M. Kloppinger, G. Fertig, A.N. Emery and H.G. Miltenburger, in Animal cell Technology-Developments, Processes and Products, R.E. Spier et al. (eds.), Butterworths, Heinemann, Oxford, 1992, p. 301.
3. M. Al-Rubeai, S. Chalder, R. Bird and A.N. Emery, Cytotechnol., 7 (1991) 179.
4. M. Al-Rubeai, A.N. Emery, S. Chalder and D.C. Jan, Cytotechnol., 9 (1992) 85.
5. M. Al-Rubeai, and A.N. Emery, J. Biotechnol., 16 (1990) 67.
6. M. Al-Rubeai, A.N. Emery, S Chalder and M.H. Goldman, J. Biotechnol. (1993) in press.
7. R.P. Singh, M. Al-Rubeai, C.D. Gregory and A.N. Emery, 12th meeting ESACT, Wurzburg, May 1993.

Genetic and physiological Approaches for the Production of Amino Acids

ECB6: Proceedings of the 6th European Congress on Biotechnology
edited by L. Alberghina, L. Frontali, P. Sensi

Biosynthesis and Secretion of Amino Acids in Coryneform Bacteria

R. Krämer, A. de Graaf, L. Eggeling, B. Eikmanns and H. Sahm

Institut für Biotechnologie I, Forschungszentrum Jülich, Jülich, FRG

In 1957, a glutamic acid producing bacterium, *Micrococcus glutamicus*, later renamed *Corynebacterium glutamicum* was discovered [1]. Since that time numerous organisms which can excrete amino acids have been identified. Consequently, the production of amino acids by fermentative processes has become an important aspect of industrial microbiology. The major amino acids produced by these processes are glutamate, lysine, phenylalanine, glutamine, arginine, tryptophane, threonine, isoleucine and histidine [2,3]. Besides coryneform bacteria (*Corynebacterium* and *Brevibacterium* species), a variety of different microorganisms are used, e.g. *Escherichia coli*, *Bacillus subtilis*, *Serratia marcescens*, as well as various yeast species. The economic impact is impressive, currently the annual production of glutamate is about 400 000 t and that of lysine around 140 000 t. The different amino acids produced have a wide variety of applications. Glutamic acid is mainly used as flavoring agent, lysine as animal feed additive, whereas other amino acids are also used for pharmaceutical and cosmetic purposes [2,3].

1. STRATEGIES FOR CHARACTERIZATION AND OPTIMIZATION OF AMINO ACID PRODUCING BACTERIA

Since the discovery of glutamic acid producing bacteria significant progress has been made in breeding microorganisms capable of "hyperproduction" of various kinds of amino acids. Although the results of classical methods (random mutagenesis and selection), as well as the application of recombinant DNA technology led to impressive improvement of the productivity, there is still a significant lack of knowledge in major fields of principle interest with respect to the physiology and biochemistry of amino acid producing microorganisms. As an example, in *Corynebacterium glutamicum*, which is a well known and widely used micro-organism for amino acid production, the knowledge of even its central metabolic pathways as well as the corresponding regulatory patterns are not sufficiently understood. The various aspects of major importance for understanding the metabolic basis of amino acid production and for development of appropriate optimization strategies can be listed as follows:

(a) basic metabolic pathways and properties of the corresponding enzymes,
(b) regulation patterns of enzymes/pathways,

(c) quantitative flux analysis, and
(d) knowledge about transport reactions (uptake and secretion).

Fig. 1 shows schematically an amino acid producing bacterial cell and the various important aspects involved in amino acid production, as listed above.

(a) It seems trivial to state that the knowledge of the correct biosynthetic pathways is an essential prerequisite for basic understanding of amino acid production. We will, however, show in the following both for lysine and for glutamate, that even in these two cases representing the most important amino acids produced, several fundamental aspects have been completely neglected or incorrectly interpreted. Knowledge of the relevant metabolic pathways should of course include not only those routes which directly lead to the synthesis of a certain amino acid, but also the properties of pathways providing the basic building blocks, as well as redox equivalents or energy. Furthermore all pathways branching off from that biosynthetic route which leads to the respective amino acid may be important (Fig. 1), since they are supposed to cause increased formation of byproducts under conditions of increased metabolic flux to the particular amino acid. Finally, also pathways possibly responsible for degradation of the respective amino acid under study, as well as those for degradation of important precursor molecules should be known, since an increased concentration of those metabolites and end products will inevitably lead to activation of degradation pathways, if present.

(b) It is furthermore essential to know the regulatory properties of all the enzymes involved in the biosynthetic pathway of a particular amino acid, as well as those involved in all other connected metabolic pathways, as discussed above. This holds true for regulation both on the activity level and with respect to expression. Especially the possible feedback inhibition loops of intermediates and end products effecting key enzymes at the beginning of a pathway or at branching points must be known in detail (Fig. 1).

(c) In addition to the metabolic data, as discussed above, which mainly represent static information, also dynamic information about the metabolic pathways is essential. This means quantitative data concerning metabolic fluxes through the relevant pathways, both for the central metabolism which provides building blocks, redox equivalents and energy, as well as for those anabolic pathways leading to particular amino acids. Sophisticated methods have been developed to obtain these data on the dynamics of metabolism, some of them will be again discussed with respect to lysine production. There are basically two different approaches to obtain a quantitative and sufficiently complete picture of the metabolic network in the cell: (I) Methods based on kinetic data of individual steps and (II) methods based on the direct measurement of the concentrations of substrates, products and possibly also of intermediates. The ultimate description of the metabolic network would of course need the knowledge of the particular properties (including regulation on activity and expression level) of every single enzyme (complete model of type I) or the flux distribution in every single pathway (complete model of type II), which is of course not possible. However, practical approaches using simplifying strategies have already shown to be relevant and applicable for quantiative description of metabolic fluxes. In the case of the kinetic approach (I), the metabolic control theory [4,5], has been successfully used in several cases [6]. The basic principle is the investi-

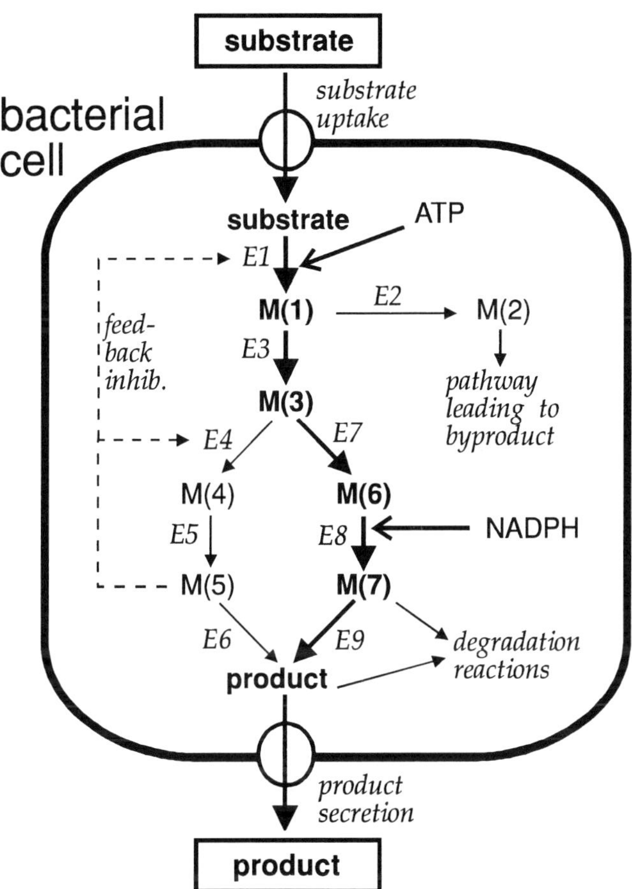

Fig. 1: Scheme of a bacterial cell with some steps important for amino acid secretion. After uptake, the substrate is converted by various pathways, involving different metabolites (M) and enzymes (E), finally leading to the desired product. These pathways may be split, other routes may branch off, leading to byproducts and to degradation of the product or of intermediates. In addition to the supply with the appropriate carbon sceleton, in general energy (ATP) and reducing equivalents (NADPH) have to be provided by the cell. All pathways in the metabolic network are controlled by complicated patterns of feedback inhibition and induction/repression mechanisms. The presence of branch points in metabolic pathways leads to various possibilities of flux distribution, which are regulated by the cell. Finally, the desired end product of the metabolic pathway has to be secreted into the culture medium.

gation of the effects of (generally minute) changes in enzyme activity (or concentration) on the flux and on metabolite concentrations. These analyses then lead to determination of specific "flux control coefficients" which indicate the significance of a given reaction in controlling a complex pathway. A related but significantly more general approach is the biochemical systems theory [7]. On the other hand, derivation of the models of type (II) is based on direct determination of substrates, products and eventually intermediates of importance. A valuable approach is the metabolite balance technique, which has recently been successfully applied to lysine fermentation in *Corynebacterium glutamicum* [8](see also below). Provided the correct pathway pattern of the metabolic network under study is known, and under the assumption of a steady state metabolic situation, the correct distribution of fluxes through particular pathways can be derived on the basis of the exact analysis of all fluxes into and out of the cell. These type of approaches may also include other methods based on the measurement of metabolite concentrations and their respective time dependent changes, e.g. tracer flux experiments and application of NMR-techniques. These measurements are especially useful to provide additional information with respect to flux partitioning within the metabolic network.

(d) It is obvious that the vectorial reactions of substrate uptake and product secretion are of significant relevance in amino acid production (Fig. 1). Whereas the different mechanisms of solute uptake have been studied to great detail in some bacteria, especially in *Escherichia coli* [9], the knowledge of sugar or carboxylic acid uptake in *Corynebacterium glutamicum* is very limited [10-12]. Moreover, the question how the product (amino acid) is secreted through the plasma membrane is a more or less completely neglected field of investigation [13]. As will be shown in the following, it is now clear that the generally accepted assumption of passive permeation of these amino acids through the bacterial plasma membrane under conditions of amino acid production is incorrect. Thus the knowledge of mechanism, energetics and regulation of amino acid secretion is of utmost importance in understanding the process of amino acid production.

When discussing the strategies which can be applied in order to force microorganisms to produce certain amino acids or to enhance their production, two principally different cases have to be discriminated. Most production strains have some alterations in the central metabolism and/or in an anabolic pathway. This alteration may consist in the introduction of auxotrophy for a given metabolite, connected to the biosynthesis of desired amino acid, the induction of resistance to toxic analogues which may lead to the deregulation of a key enzyme in the biosynthetic pathway (e.g. in the case of lysine, see below), or the modification of branching pathways which lead to desired (up-modulation) or undesired (down-modulation) alternative pathway and byproducts. These changes in the activity and/or regulation of crucial enzymes in general lead to a significant increase in the cytosolic concentration of the end product (amino acid) which is then finally excreted. On the other hand, production of the most important amino acid in economic terms, i.e. glutamic acid, is based on a completely different metabolic situation. Excretion of glutamic acid is induced by several very different procedures, including biotin starvation (*Corynebacterium glutamicum* is biotin auxotrophic), addition of amine surfactants, addition of penicillin, and growth of glycerol or fatty acid auxotrophic strains under limiting conditions. Obviously no general alteration in metabolism is involved but the procedures used have some impact on the plasma membrane and/or cell wall. In the following chapters, excretion of lysine and glutamate, as representative examples of the two types of strategies to induce secretion will be discussed in detail, based mainly on recent data from our group and also from other laboratories.

2. PRODUCTION OF LYSINE BY CORYNEBACTERIUM GLUTAMICUM

Lysine is an essential amino acid for animals and is found in limiting amounts in animal feed. Addition of lysine therefore significantly improves the nutritional value of various feed grains. Lysine biosynthesis starts from aspartate which is phosphorylated by the key enzyme of this pathway, the aspartate kinase. The anabolic pathway to lysine (diaminopimelate pathway) is split at the level of aminoketopimelate, either via succinylated intermediates or directly to m-diaminopimelate [14]. Biochemical and genetic information is now available for the com-plete set of the ten enzymes involved in the biosynthetic pathway from aspartate to lysine [14-16]. Several surprising findings were obtained by these studies. (i) The variation of enzyme activities between fully induced and uninduced state was less than threefold, which is much lower than that observed for example in *Bacillus subtilis* or *Escherichia coli* [17,18]. (ii) Besides the well known regulation at the level of aspartate kinase via concerted feedback inhibition by lysine and threonine, a second enzyme (dihydrodipicolinate synthase) was found to be involved in the control of lysine biosynthesis. As a consequence, overexpression of this enzyme led to enhanced secretion of lysine [19]. (iii) It was not possible to clearly define the significance of either of the two alternative routes leading to lysine in *Corynebacterium glutamicum* by measuring enzyme activites or using recombinant DNA techniques. If one pathway is blocked by gene disruption, the alternative route can compensate the block to a significant extent [14]. By application of NMR-techniques and use of 13-C-labelled glucose, we could show that both pathways are used in *Corynebacterium glutamicum*. The flux distribution between the two alternative routes depends on the availability of ammonium in the culture medium [20].

An elaborate approach in the direction of detailed flux analysis was recently published [9,21] (cf. also an article later in this book). These authors successfully used the metabolite balance technique to calculate metabolic fluxes through central pathways in *Corynebacterium glutamicum*, in particular with respect to lysine production from glucose. The various contributions of different pathways (e.g. glycolysis and pentose phosphate cycle) and the importance of metabolic branching points ("nodes") were elucidated in detail in these studies. However, in order to solve the complex network of equations used, several significant simplifications had to be introduced. Hopefully, a further improvement may be possible by a combination of different approaches, most desirable a combination of the metabolite balance method with powerful NMR-techniques. These results clearly demonstrate the importance of combining data from different experimental approaches, i.e. from studies of the basic properties of the enzymes involved in a particular biosynthetic pathway, and results concerning their regulatory behaviour as well as results with respect to flux distribution.

Lysine production is a highly appropriate example to discuss the various approaches important for complete elucidation of the physiological, biochemical and biotechnological background in amino acid synthesis not only with respect to properties of biosynthetic enzymes and flux parameters, but also with respect to the significance of the secretion step in the overall sequence of amino acid production (cf. Fig 1). Besides the generally accepted hypothesis of passive efflux of amino acids in amino acid producing microorganisms, in the case of lysine, specific efflux pores were suggested which were supposed to be osmotically controlled [22]. In contrast to these hypotheses, we could demonstrate that lysine is secreted in *Corynebacterium glutamicum* by a carrier mediated process. The lysine secretion carrier is energy dependent. It is driven by the electrical potential across the bacterial plasma membrane and is modulated by the pH-gradient and the availability of internal lysine [23,24]. The lysine secretion carrier can be regulated by the cell and it is an important controlling step in lysine production, at least under certain metabolic conditions. This was clearly demonstrated by the construction of an recombinant strain of *Corynebacterium glutamicum* which combined the metabolic state of a lysine hyperproducer and the membrane state of the wild type [25]. Surprisingly this strain turned out to be very limited in its ability to secrete lysine. It was thus concluded that an effective producer strain differs from the wild type not only by its biosynthetic capabilities but also by the properties of its lysine secretion carrier system. This emphasizes the relevance of investigating the secretion step in constructing an optimum producer organism.

An example for the strategy to breed an amino acid producer by modification of the flux in an important branching point is the conversion of a lysine producing *Corynebacterium glutamicum* strain to a threonine and isoleucine producer. The initial two reactions of L-lysine biosynthesis are common for biosynthesis of L-threonine and L-isoleucine too and, by introduction of plasmid-borne threonine biosynthesis genes, the carbon flux from aspartate to lysine could be directed to threonine (and isoleucine) [26].

3. PRODUCTION OF GLUTAMATE IN CORYNEBACTERIUM GLUTAMICUM

In contrast to the production of other amino acids (see preceding chapters) effective glutamic acid production from glucose is a property of *Corynebacterium glutamicum* wild type cells under certain metabolic conditions. Each of the widely different methods applied to force *Corynebacterium glutamicum* to secrete glutamate (see above) are supposed to have a direct influence on the plasma membrane and/or the cell wall. This may be a direct effect (e.g. by the added fatty acids, fatty acid esters, or penicillin) or may be caused indirectly (e.g. biotin limitation leading to inhibition of lipid synthesis). Consequently these findings were interpreted by the following hypothesis which is still widely acknowledged: Glutamate is synthesized via glycolysis, PEP-carboxylase, citric acid cycle and ultimately glutamate dehydrogenase. Membrane perturbation by the above mentioned methods leads to an increased permeability for glutamate and thus to excretion of this amino acid. Consequently the internal level of glutamate decreases and the feedback inhibition of glutamate biosynthesis is released. This leads to continuous synthesis and secretion of glutamate [27,28].

This hypothesis is not valid in several respects. We showed that the plasma membrane of *Corynebacterium glutamicum* is fully intact under conditions of glutamate secretion triggered by biotin limitation [29], although the membrane is in fact altered in its composition [30]. For a completely different method of inducing glutamate secretion, i.e. addition of amine surfactants, it was shown that glutamate secretion, also in this case, is not mediated by simple passive diffusion [31]. In biotin limited cells, glutamate is actively secreted into the surrounding medium, the carrier system catalyzing this process is energy dependent and is tightly controlled by the cell [32]. Furthermore, also the implication that glutamate secreted by *Corynebacterium glutamicum* is synthesized via the glutamate dehydrogenase became questionable by recent results. We could show that a recombinant strain in which this enzyme was completely inactivated by gene disruption, efficiently secreted glutamate under biotin limitation with unchanged rate as compared to the parent strain [33].

In summary, our recent findings on efflux of glutamate in *Corynebacterium glutamicum* can be interpreted in the following way. The basis for effective glutamate secretion is a particular metabolic situation which is in general called "overflow metabolism" [34]. This condition is characterized by the simultaneous presence of an excess of carbon and energy source and a limitation in some essential compound. This leads to a metabolic imbalance which is circumvented by the cell in secreting a particular metabolite, in general a carboxylic acid, e.g. oxoglutarate, pyruvate or lactate. In the case of glutamate production, according to our results the metabolic imbalance (presumably in the redox status) triggers the glutamate secretion carrier which becomes activated by a change in the physical state of the membrane. The combined factors (overflow metabolism and change in membrane state) ultimately lead to effective glutamate secretion.

It is thus obvious that after 35 years of research in glutamate production by *Corynebacterium glutamicum*, even fundamental steps in the metabolic sequence from the substrate glucose which is taken up to the product glutamate which is secreted need further elucidation.

REFERENCES

1. Kinoshita, S., Tanaka, K., Udaka, S., and Akita, S., Proc. Intern. Symp. Enzyme Chem., 2, (1957) 464
2. Aida, K., Chibata, I., Nakayama, K., Takinami, K., and Yamada, H. (eds.), Biotechnology of amino acid production, Kondansha, Tokyo & Elsevier, Amsterdam, 1986.
3. Nakayama, K., in Industrial Microbiology (Reed, C. ed.), Prescott & Dunn 1982, 748.
4. Kacser, H. and Burns, J.A., Symp. Soc. Exp. Biol., 27 (1973) 65.
5. Heinrich, R. and Rapoport, T.A., Eur. J. Biochem., 42 (1974) 89.
6. Niederberger, P., Prasad, R., Miozzari, G., and Kacser, H., Biochem. J., 287 (1992) 473.
7. Savageau, M.A., Voit, E.O., and Irvine, D.H., Math. Biosci,. 86 (1987) 127.

8. Vallino, J.J. and Stephanopoulos, G., Biotechnol. Bioengin., 41 (1993) 633.
9. Maloney, P.C., in E.coli and S.typhimurium. Cellular and molecular biology (F.C. Neidhardt et al., eds.) ASM Washington 1987, 222.
10. Marauska, D.F., Ruklish, M.P., and Galynina, N.I., Microbiology, 50 (1982) 763.
11. Malin, G.M. and Bourd, G.I., J. Appl. Bacteriol., 71 (1991) 517.
12. Ebbighausen, H., Weil, B. and Krämer, R., Arch. Microbiol., 155 (1991) 505.
13. Milner, J.L., Vink, B., and Wood, J.M., CRC Crit. Rev. Biotechn., 5 (1987) 1.
14. Schrumpf, B., Schwarzer, A., Kalinowski, J. Pühler, A., Eggeling, L., and Sahm, H., J. Bacteriol., 173 (1991) 4510.
15. Cremer, J., Treptow, C., Eggeling, L., and Sahm, H., J. Gen. Microbiol., 134 (1988) 3221.
16. Yeh, P., Sicard, A.M., and Sinskey, A.J., Mol. Gen. Genet., 212 (1988) 112.
17. Cohen, G.N. and Saint-Girons, I., in E.coli and S.typhimurium. Cellular and molecular biology (Neidhardt et al., eds.), ASM Washington 1987, 429.
18. Graves, L.M. and Switzer, R.L., J. Bacteriol., 172 (1990) 218.
19. Cremer, J., Eggeling, L., and Sahm, H., Appl. Environ. Microbiol., 57 (1991) 1746.
20. Sonntag, K. Eggeling, L. De Graaf, A. and Sahm, H., Eur. J. Biochem. (1993) in the press
21. Stephanopoulos, G. and Vallino, J.J., Science, 252 (1991) 1675.
22. Luntz, M.G., Zhdanova, N.I., and Bourd, G.I., J. Gen. Microbiol., 132 (1986) 2137.
23. Bröer, S. and Krämer, R., Eur. J. Biochem., 202 (1991) 131.
24. Bröer, S. and Krämer, R., Eur. J. Biochem., 202 (1991) 137.
25. Bröer, S., Eggeling, L., and Krämer, R., Appl. Environm. Microbiol., 59 (1993) 316.
26. Eikmanns, B.J., Metzger, M., Reinscheid, D., Kircher, M., and Sahm, H., Appl. Microbiol. Biotechnol., 34 (1991) 617.
27. Demain, A.L. and Birnbaum, J., Curr. Top. Microbiol. Immunol., 46 (1968) 1.
28. Bunch, A.W. and Harris, R.E., Biotech. Gen. Eng. Rev., 4 (1986) 117.
29. Hoischen, C. and Krämer, R., Arch. Microbiol., 151 (1989) 342.
30. Hoischen, C. and Krämer, R., J. Bacteriol., 172 (1990) 3409.
31. Duperray, F., Jezequel, D., Ghazi, A., Letellier, L., and Shechter, E., Biochim. Biophys. Acta, 1103 (1992) 250.
32. Gutmann, M., Hoischen, C., and Krämer, R., Biochim. Biophys. Acta, 1112 (1992) 115.
33. Börmann-El Kholy, E.R., Eikmanns, B.J., Gutmann, M., and Sahm, H. (1993) submitted for publication
34. Tempest, D.W. and Neijssel, O.M., FEMS Microbiol. Lett., 100 (1992) 169.

ECB6: Proceedings of the 6th European Congress on Biotechnology
edited by L. Alberghina, L. Frontali, P. Sensi

ADVANCED GENETIC TECHNIQUES FOR AMINO ACID PRODUCING CORYNEBACTERIA

A. Pühler, W. Jäger, F. Kassing, A. Schäfer, A. Schwarzer, A. Tauch, and J. Kalinowski

Department of Genetics, Faculty of Biology, University of Bielefeld, P.O.Box 100131, D-33501 Bielefeld

The Gram-positive *Corynebacterium glutamicum* is well-known for its use in industrial production of amino acids. Up to now, amino acid producing strains were routinely obtained by the classical approach of mutation and subsequent selection for more efficient production. An alternative way to construct amino acid producing strains is now offered by genetic engineering (Wohlleben et al., 1993). This article deals with the development of genetic methods to analyze and to manipulate *C.glutamicum* genes involved in amino acid production. In particular, the introduction of plasmid vectors into *C.glutamicum* and methods for gene disruption and gene replacement are presented. In addition, the development of a transposon mutagenesis system is described. To demonstrate the usefulness of the different genetic techniques, examples dealing with the lysine biosynthetic pathway are included in the article.

Introduction of plasmid vectors into *C.glutamicum* can be achieved by the polyethylene glycol induced transformation of osmotically sensitive spheroplasts or cell wall less protoplasts (Thierbach et al., 1988). A more advanced method makes use of electroporation which can be carried out with intact *C.glutamicum* cells (Wolf et al., 1989). A completely different method for the introduction of plasmid DNA into *C.glutamicum* is based on the conjugal transfer from the Gram-negative *Escherichia coli* to the Gram-positive *C.glutamicum* (Schäfer et al., 1990). The conjugal transfer could be enhanced tremendously by a short heat shock treatment of the recipient cells. In the meantime, it became clear that the heat shock treatment inactivates the *C.glutamicum* restriction barrier.

The conjugation system is not only suitable for the introduction of replicating plasmids into *C.glutamicum* but also for the effective transfer of non-replicating *E.coli* plasmids (Fig. 1). Non-replicating plasmids are extremely useful for gene disruption and gene replacement experiments in *C.glutamicum* (Schwarzer and Pühler, 1991). The gene disruption and replacement techniques in *C.glutamicum* were first used to analyze several genes of the lysine biosynthetic pathway. We were successful in disrupting the *lys*A gene by inserting a non-replicating plasmid via a single crossover or by inserting an antibiotic resistance cassette via a double crossover. Gene replacement with the *lys*A gene resulted in the introduction of a defined deletion in the *lys*A coding region (Schwarzer and Pühler, 1991). Another application of gene disruption solved the question whether *meso*-diaminopimelate, the immediate precursor of lysine, can be synthesized in *C.glutamicum* by two alternative pathways. Disruption of the *ddh* gene was successful and delivered a prototrophic strain which was still able to synthesize lysine indicating the existence of two alternative pathways (Schrumpf et al., 1991)

In the meantime, the integrative vector system was modified in order to use it for promoter studies. The promoterless *lac*Z gene was combined with a multiple cloning site and inserted into a mobilizable *E.coli* plasmid. Cloning of a DNA fragment carrying part of *dap*A in front of *lac*Z and integration of the vector into the *C.glutamicum* chromosome resulted in the tandem arrangement of the wildtype *dap*A gene and a *dap*A promoter fused to the *lac*Z coding region. This arrangement allowed transcriptional studies of the *dap*A promoter while minimizing titration and position effects as encountered with systems based on replicating plasmids. It was found that the *dap*A promoter is most probably repressed by diaminopimelate.

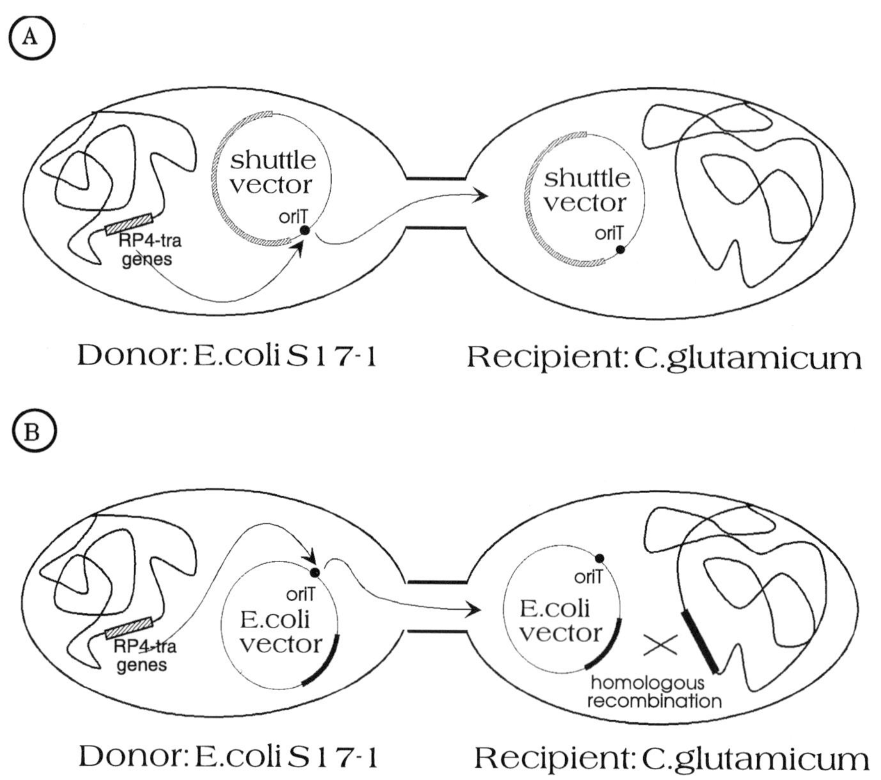

Figure 1: Conjugal transfer of replicating (A) and non-replicating (B) vector plasmids. *ori*T: origin of transfer replication; *tra* genes: transfer genes

The first detection of transposable elements in *C.glutamicum* was just by chance. We concentrated on the molecular analysis of the lysine import gene *lys*I (Seep-Feldhaus et al., 1991) and found that one *lys*I mutation was due to the insertion of the IS-element ISCg1 which is present several times in *C.glutamicum* (Fig. 2). We then constructed a positive selection system making use of the *Bacillus subtilis sac*B gene encoding levansucrase, the action of which is lethal for *C.glutamicum* when grown on sucrose (Jäger et al., 1992). A shuttle plasmid carrying the *sac*B gene was introduced into several corynebacterial strains and the transconjugants were tested for resistance to sucrose. Analysis of the residing plasmids revealed at a high percentage the presence of different insertion elements (Table 1).

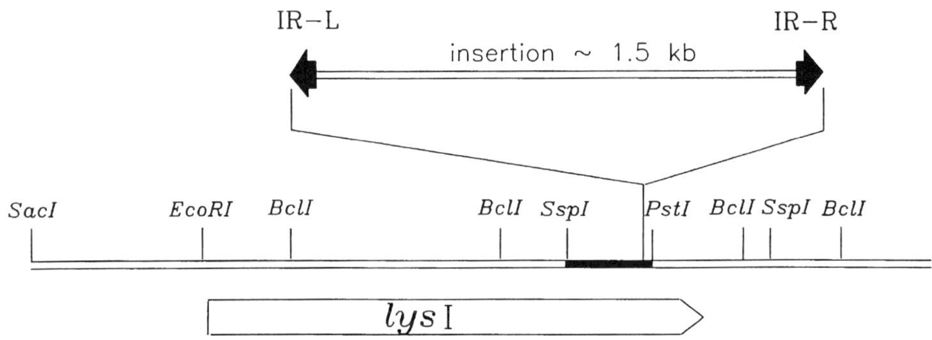

Figure 2: Localization of the insertion of ISCg1 within the *C.glutamicum lys*I gene. Abbreviations: *lys*I: lysine import gene; IR-L and IR-R: left and right inverted repeat; *Sac*I, *Eco*RI, *Bcl*I, *Ssp*I, *Pst*I: endonuclease restriction sites

Table 1: Different insertion elements from corynebacteria.

Strain	element	length	inverted repeat	target duplication	copies
Corynebacterium glutamicum	ISCg1	1.45 kb	24 bp	8 bp	4 - 7
Rhodococcus fascians	ISRf1	1.3 kb	13 bp	3 bp	3
Brevibacterium lactofermentum	ISB11	1.45 kb	26 bp	8 bp	4

Of special interest was the development of a transposon mutagenesis system for *C.glutamicum*. We analyzed large antibiotic resistance plasmids for the presence of transposable resistance genes. A resistance plasmid found in *Corynebacterium xerosis* turned out to carry resistance genes against kanamycin, chloramphenicol, tetracycline and erythromycin. Testing all these genes for transposition, the erythromycin resistance gene was detected to transpose in *C.glutamicum*. Nucleotide sequence analysis of the DNA fragment carrying the erythromycin resistance gene revealed the presence of a 4.5 kb class II transposon composed of two insertion elements flanking the antibiotic resistance gene (Fig. 3). At the moment, this transposon, designated TnCx1, is employed to analyze mutations leading to non-production in *C.glutamicum* strains originally constructed to excrete lysine.

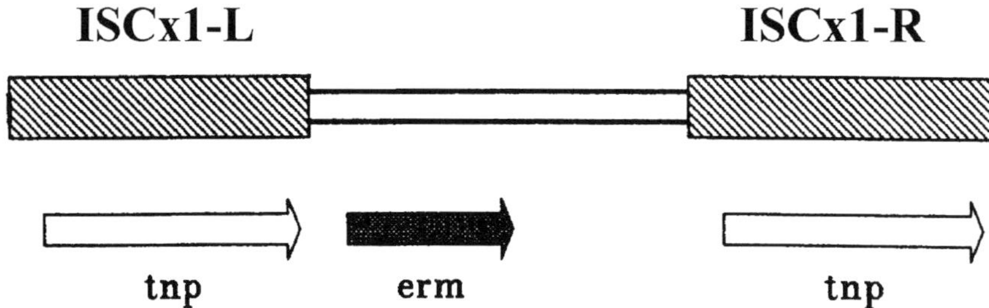

Figure 3: Structure of the transposon TnCx1 from *C.xerosis*. The putative genes encoded by the transposon are shown.
ISCx1-L and ISCx1-R: left and right insertion element forming the composed transposon TnCx1; *tnp*: transposase gene; *erm*: erythromycin resistance gene

To summarize: In this article advanced molecular genetic techniques were presented which can be used for the analysis and modification of *C.glutamicum* genes involved in amino acid production. In particular, the conjugal transfer of plasmids from *E.coli* to *C.glutamicum* turned out to be extremely useful for gene disruption and gene replacement but also for the application of integrative vectors designed for promoter studies. Finally, the conjugation system was also employed to isolate insertion elements and to develop a transposon mutagenesis system.

REFERENCES

1. G. Thierbach, A. Schwarzer and A. Pühler, Appl. Microbiol. Biotechnol. 29 (1988) 356
2. H. Wolf, A. Pühler and E. Neumann, Appl. Microbiol. Biotechnol. 30 (1989) 283
3. A. Schäfer, J. Kalinowski, R. Simon, A.-H. Seep-Feldhaus and A. Pühler, J. Bacteriol. 172 (1990) 1663
4. A. Schwarzer and A. Pühler, BIO/Technology 90 (1991) 84
5. B. Schrumpf, A. Schwarzer, J. Kalinowski, A. Pühler, L. Eggeling and H. Sahm, J. Bacteriol. 173 (1991) 4510
6. A.-H. Seep-Feldhaus, J. Kalinowski and A. Pühler, Mol. Microbiol. 5 (1991) 2995
7. W. Jäger, A. Schäfer, A. Pühler, G. Labes and W. Wohlleben, J. Bacteriol. 174 (1992) 5462
8. W. Wohlleben, G. Muth and J. Kalinowski, Biotechnology, Vol. 2, A. Pühler, ed. (1993) 455

ECB6: Proceedings of the 6th European Congress on Biotechnology
edited by L. Alberghina, L. Frontali, P. Sensi

Expression of *Vitreoscilla* Hemoglobin in *Corynebacterium glutamicum* Increases Final Concentration and Yield of L-lysine

Fred C. Sander,* Roger A. Fachini,* Dallas E. Hughes,* Jorge L. Galazzo,* and James E. Bailey.§

* Exogene Corporation, 222 E. Huntington Dr., Monrovia, California 91016, USA

§ Institut für Biotechnologie, ETH-Hönggerberg, CH-8093, Zürich, Switzerland

The glutamic acid-producing coryneform bacteria comprising the genera *Corynebacterium* and *Brevibacterium* are the most important organisms used for the production of amino acids. The two major amino acids produced in these organisms, glutamic acid and L-lysine, are used as feed supplements and flavor enhancers. Traditional methods for improving amino acid yields have relied on the use of highly mutagenized strains in which feedback repression of amino-acid synthesis is relieved, allowing the accumulation of high amounts of product. More recently, the development of host-vector systems have allowed the use of recombinant DNA techniques by which many of the genes involved in the regulation and synthesis of amino acid have been cloned.

Genetic manipulation to increase cofactor supply to amino acid biosynthesis pathways complements other approaches aimed at overexpressing pathway enzymes and altering their regulation. In studies pursuing this strategy, the gene vhb encoding the 15,775 kD subunit of the homodimeric hemoglobin (VHb) from *Vitreoscilla* has been cloned behind the tac promoter in a corynebacteria vector subsequently designated pFS1. *Corynebacterium glutamicum* ATCC 13287 (hse$^-$) was employed in this study since it is a well-characterized strain that produces elevated levels of L-lysine as a result of a mutation in homoserine dehydrogenase. This mutation prevents conversion of L-aspartate semialdehyde to L-threonine and instead shunts it into the diaminopimelate-lysine (dap) biosynthetic pathway. *C. glutamicum* 13287 transformed with pFS1 by electroporation expressed VHb as detected by Western immunoblot analysis. To test for biologically active hemoglobin, cell extracts were tested with a carbon monoxide binding assay. Samples from VHb-expressing and non-expressing strains were reduced with dithionite in the presence and absence of carbon monoxide, then scanned by spectrophotometer. The difference spectrum of the transformant, 13287:pFS1, shows the characteristic hemoglobin profile of a peak at 419 nm followed by a minimum at 436 nm, a feature which is not present in similar assays in the non VHb-expressing control (Figure 1). These data show that cells transformed with the VHb gene are able to express biochemically active hemoglobin.

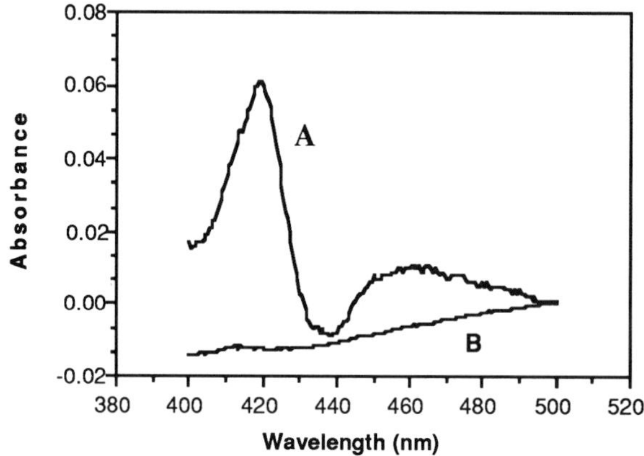

Figure 1: The carbon monoxide (CO) difference spectra of cell extracts from *C. glutamicum*. (A) 13287:pFS1; (B) 13287:no plasmid. The difference in absorbance between CO-dithionite-reduced and dithionite-reduced samples is plotted

VHb-expressing transformants were compared to control plasmid-containing and plasmid-free strains in small-scale fermentors. The control plasmid, designated pDLT1, has a deletion in the VHb-gene region and thus does not express VHb. Batch fermentations were conducted in 3-L bench-top fermentors. Dissolved oxygen concentration was controlled at 5% air saturation by adjusting the fermentor's agitation rate. Temperature was kept at 30°C, and pH was maintained at 7 by addition of 4N NaOH. L-lysine concentration was determined in the sample supernatants by using HPLC, and glucose concentration by using an enzymatic kit. Off-gas composition was monitored by mass spectrometry. Colony isolation and replica plating were used throughout the fermentation to test for reversion frequency of the hse^- phenotype and for plasmid stability.

Figures 2, 3, and 4 show results of a typical 2-L batch fermentation. There was no significant difference in cell growth which was limited by threonine and methionine concentration in the medium. L-lysine production and oxygen uptake rate were higher in VHb-expressing cells than in control cells.

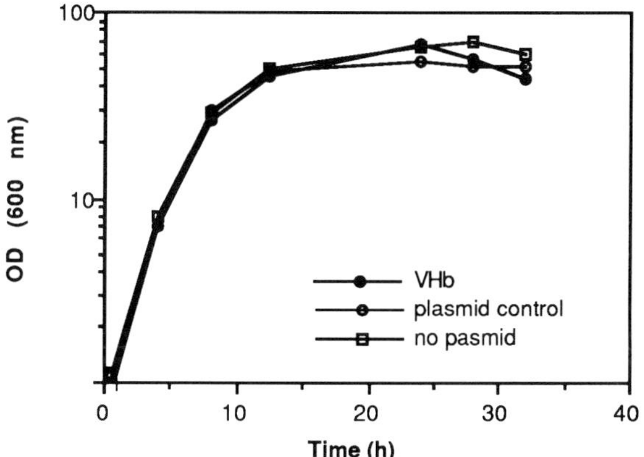

Figure 2: Cell growth as measured by optical density (OD) in *C. glutamicum* batch fermentations. VHb: 13287:pFS1; plasmid control: 13287:pDLT; no plasmid: 13287 host.

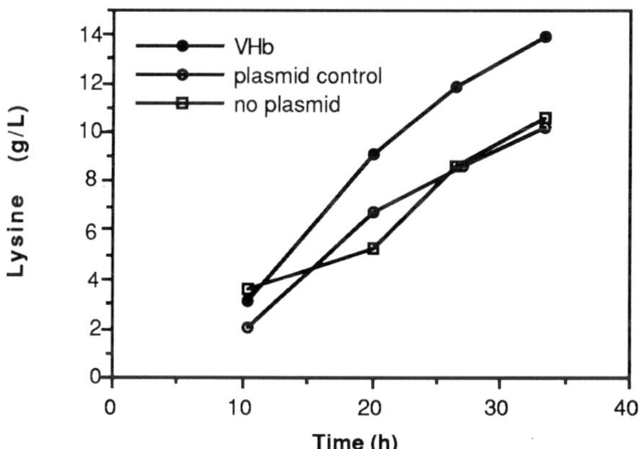

Figure 3: L-Lysine production in *C. glutamicum* batch fermentations. VHb: 13287:pFS1; plasmid control: 13287:pDLT; no plasmid: 13287 host.

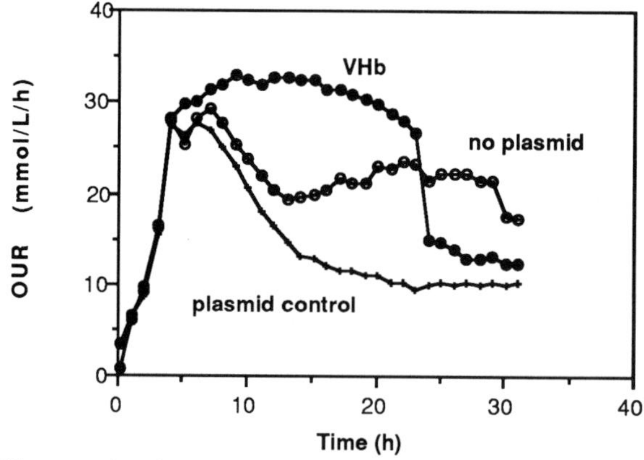

Figure 4: Oxygen uptake rate as a function of time in *C. glutamicum* batch fermentations. VHb: 13287:pFS1; plasmid control: 13287:pDLT; no plasmid: 13287 host.

The following table summarizes the results of three independent fermentations.

	pFS1(+VHb)	pDLT(-VHb)	no-plasmid
L-lysine·HCl (g/L)	11.5	7.9	8.8
Lysine yield (g/g)	0.23	0.16	0.18
Plasmid stability (%)	100	100	---
Reversion frequency (%)	0	0	0

Final L-lysine concentration and yield from glucose were about 30% higher in VHb-expressing cells than in control cells.

Physiological studies of *E. coli* expressing VHb indicate an increase in proton pumping efficiency, ATP production rate, and ATP concentration relative to VHb-free controls. Analogous effects in *C. glutamicum* may contribute to increased flux through the ATP-demanding lysine synthesis pathway.

Biotechnological Innovations in Food Preservation Processing and Analysis

ECB6: Proceedings of the 6th European Congress on Biotechnology
edited by L. Alberghina, L. Frontali, P. Sensi

Recent development in yeast strains for baking

Tahía Benítez, A.C.Codón and J.M.Gasent-Ramírez

Departamento de Genética, Facultad de Biología, Aptdo. 1095,
41080-SEVILLE, SPAIN

1.INTRODUCTION

The use of biotechnology in the manufacture of food and beverages has been practised for over 8000 years with bread, vinegar, alcoholic beverages (beer, wine, cider), sourdough and cheese production. Biotechnological processes are now used to produce other fermented products, food and feed additives (amino acids, vitamins, bakers' yeast) and processing aids (citric acid, enzymes). An important issue today is the impact that modern biology (molecular biology, fermentation science, bioengineering) has on the food industry, more precisely on baking. Major advances in the understanding of the biochemistry and genetics of bakers' yeast allow these microorganisms which are used extensively in various aspects of food processing to reduce costs (by increasing yield and productivity) and to improve the quality of new products.

Most breads are fermented with the bakers' yeast *Saccharomyces cerevisiae*. Wheat and rice flour are used, together with selected bakers' yeast, proteases to degrade gluten and amylases to reduce dextrin concentrations. The liberated peptides and amino acids together with the dextrins contribute to the flavour (1). For economic reasons, there is a tendency to reduce the time of fermentation, therefore reducing flavour and aroma. Nowadays there is also a tendency to enrich bread and baking products either by restoration (which consists in giving back to the bread those nutritional properties which are lost during the process) and fortification techniques (to add nutrients lacking in the bread in its natural form). There exist breads enriched in B vitamins, calcium, phosphorus and iron, fibre, egg and milk, and breads for special diets (hipoallergens, hipocalorics, for diabetics, for children, etc). Lipooxygenases to bleach and phospholipases and pentosanases to retard the hardness of the bread are recently used too (1, 2).

Bakers' and fodder yeasts are undoubtedly the most important microorganisms exploited by mankind. Bakers' yeast production is actually 5 million tons per year. Research on bakers' yeast is mainly concerned with nutrition qualities, possibilities of genetic manipulation, and biomass preservation. In addition, bakers' yeast is also employed in the preparation of yeast extract, as a supplement in animal feed and as a source of vitamins in human diets (1).

2.CHARACTERIZATION OF YEAST

Traditionally, industrial strains have been characterized according to their morphological and fermentation attributes; these parameters are readily accessible and easily evaluated. By those methods several yeast genera and groups of species can be recognized but they are not suitable to differentiate individual strains unequivocally. The application of more reliable criteria such as the interfertility of strains for delimitation of species has led to situations where well-known names of species has passed into synonymy with names published earlier (i.e. *Saccharomyces cerevisiae*). However, industrial microbiologists who deal with relatively few yeasts, the names of which have come to be associated with a particular fermentation process, are interested in names with industrial connotations. It has therefore been inevitable (even desirable) that industries were biased in favour of the use of industrially relevant descriptive characters (3). At the same time industries are obliged to differentiate between industrial strains and distinguish between strains used in various industrial applications. For delimitation of such industrially characterized individual strains, industries have to rely on characters and criteria which allow delimitation of typological species unequivocally.

At present, strains can be identified more precisely because newer methods and more advanced techniques of identification which differ from traditional techniques have been devised. These techniques are based on fingerprinting of nucleic acids, electrophoresis of chromosomal DNA, proteins and enzymes, electrophoretic karyotype, restriction endonucleases and the identification of unique restriction fragment lengths polymorphism, and others, (Fig.1). For instance, karyotype analysis by pulsed field technique has allowed separation of individual chromosomal DNA molecules of bakers' strains of *Saccharomyces* and constitutes a reliable tool for characterization of the numerous commercial strains (4). The electrophoretic profiles have provided qualitative evidence for aneuploidy and, together with further analysis, allow determination of chromosome copy number, translocations, deletions or variability in chromosome length for homologous chromosomes.

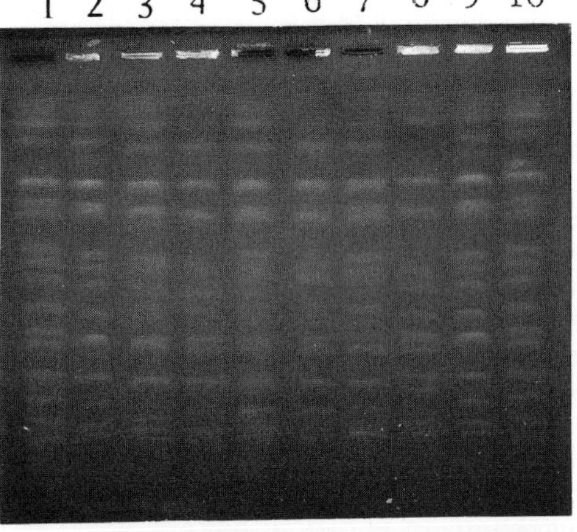

Fig. 1: Electrophoretic profiles of laboratory and bakers' yeasts. Lanes 1 and 10, YNN 295; lanes 2 and 9, DS81; lanes 3-7 bakers' yeast: DADI [3], VS [4], CT [5], SB2 [6], SB11 [7]; and lane 8 K55A.

An impressive congery of varied polymorphisms has been detected by employing probes specific for some known yeast genes, indicating that industrial strains contain variously sized loci indicating the strain to be a mixture composed of two or more cell types. These evolutionary changes are apparently in contradiction with the fact that -because of the need for constant properties in each strain - there seems to be a widespread use of polyploid yeasts, which are genetically more stable and less susceptible to mutational forces, and a selection against efficient sexual reproduction. This makes breeding and genetic analysis difficult.

3. IMPROVEMENT OF GENETIC MANIPULATION

Although bakers' yeasts seem to have been selected for those characteristics which render them so unamenable to genetic manipulation - polyploidy, lack of a mating type ability, poor sporulation frequency, low viability of the spores - there have recently been trials to increase the possibilities of genetic manipulation of the strains. The reason is that bakers' yeasts need to be genetically modified for obtaining either new products or cheaper ones. At the same time, most features to be improved are polygenic and therefore not amenable by simply mutation and selection. Improvement of genetic manipulation is absolutely necessary if hybridization is to be carried out.

3.1.Changes in the culture conditions

Nutritional and other environmental factors have a marked influence on ascosporogenesis. Improvement in the sporogenic ability of bakers' yeasts has been made evident when cells have been cultivated in liquid rather than in agar culture media (3, 4). Also, analysis of the effects on ascosporogenesis of temperature and presporulation and/or sporulation carbon source (fermentative vs. respiratory carbon source) has led to the identification of culture conditions of improved tetrad production in several bakers' yeasts (1). Sporulation at 22ºC instead of 30ºC gives significant increases in ascus formation. Substitution of acetate for glucose as the sporulation carbon source, and exponential rather than stationary phase cultures give a further increase in total sporulation and a very high induction of tetrads is noted (2, 3).

3.2.Mitochondria transfer

The temperature optimum for sporulation, which is apparently lower in industrial yeasts than in laboratory strains seems to reflect differences in temperature requirements for optimal mitochondrial protein synthesis, since the optimal temperatures for sporulation and for mitochondrial protein synthesis are coincident. In addition, data from comparison of effects on sporulation of fermentable vs. non-fermentable carbon sources suggest tetrad formation in industrial strains to depend closely on derepression for respiratory gene functions. This, together with the fact that petite yeasts do not sporulate offer the possibility of inducing a very high yield of asci and tetrads by transferring mitochondria from strains selected by their ability to produce the highest percentage of both asci and tetrads. Cytoductans can be obtained in which mitochondria from laboratory strains are tranferred to bakers' yeast recipients. In order to do so, the specialyzed form of rare mating involving *kar* donor strains incapable of transferring their own nuclear DNA can be used(Fig. 2).

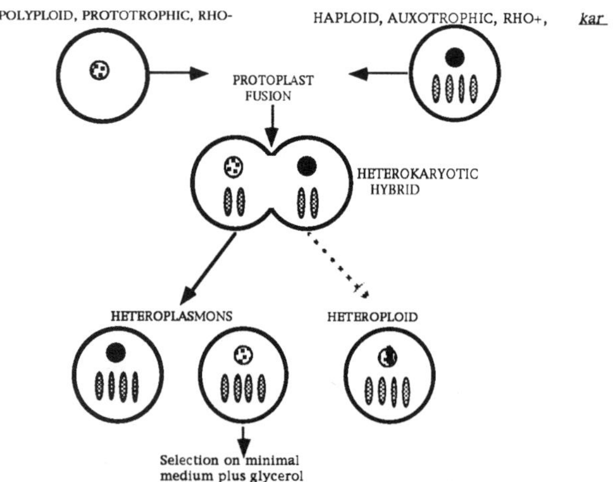

Fig.2. Formation of heteroplasmons which possess the nucleus of the bakers' yeast strain and the mitochondria of the laboratory strain.

4. OPTIMIZATION OF YEAST PRODUCTION

4.1. Yield

In a typical bakers' yeast propagation, the fermentable sugar for yeast growth is derived from molasses (1, 3). In addition to sucrose, which is predominant, beet molasses contain the sugar raffinose in the 0.5 to 8 % range. Only one third of the molecule is utilized by bakers' yeasts (1). The ß-fructosidase (invertase) enzyme present in the bakers' yeasts yields fructose and melibiose from raffinose (2). The fructose formed is readily assimilated by the yeasts. Bakers' yeasts are unable to utilize the dissacharide melibiose due to the lack of the enzyme α–galactosidase (melibiase) (1). Clearly, the productivity and the process economics could be markedly improved by the use of bakers' yeast strains that have the added capability of utilizing melibiose.

Protoplast fusion technique has been used to develop MEL$^+$ industrial hybrids capable of improving the molasses yield in yeast propagations (2). However, fusants between a laboratory and a industrial bakers' yeast consistently showed higher molasses yield but performed poorly in baking applications. The use of rDNA technology for the construction of MEL$^+$ bakers' yeasts has been reported (1, 2), by cloning the MEL1 gene from *Saccharomyces uvarum* into a shuttle vector. However, marketing considerations have resulted in a reluctance to introduce bacterial genes into industrial strains. A one-step dominant selection procedure has been developed (5) to construct recombinant bakers' yeast devoid of plasmid sequences. It consisted in a region of nonessential DNA from chromosome XIII in which SMR1 (sulfometuron resistance) and MEL1 dominant selection markers replaced internal sequences (Fig.3). Transformed MEL$^+$ bakers' yeast grow with growth rates similar to those of untransformed strains both in laboratory minimal and semiindustrial media (beet molasses) (J.M.Gasent-Ramírez, unpublished). When using melibiose or raffinose supplemented minimal media, the biomass/substrate yield obtained (Yxs) is considerably higher in the transformed strains as compared to the melibiose negative bakers' yeast. Such stable integration of MEL1 devoid of bacterial plasmid DNA facilitates the commercial application of recombinant DNA technology in the food and beverage industries.

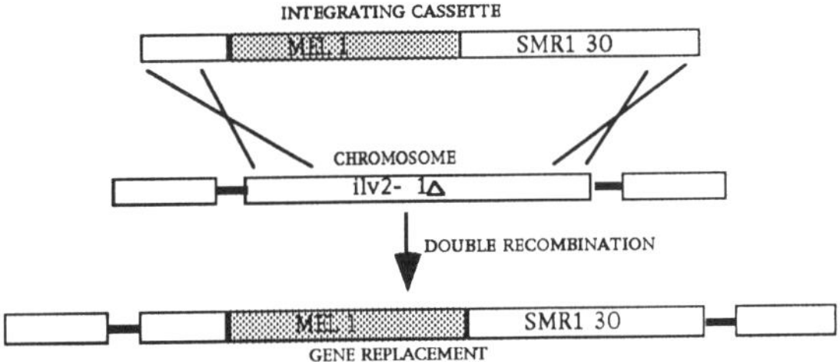

Fig.3. Gene replacement leading to the insertion of MEL1 gene without the addition of bacterial sequences.

4.2.Productivity

Bakers' yeast production is always carried out as fed-batch fermentions, in which the growth medium is fed incrementally to a starter culture over a given period of time, until the fermenter is filled. Under anaerobic conditions the yield of yeast biomass is low (about 7.5 g yeast weight per 100 g sugar) but under strictly aerobic conditions the yield reachs 54 g yeast solids per 100 g sugar (1, 2).

The purpose of bakers' yeast production should be to maximize growth and minimize alcohol fementation. For this reason and to maximize yeast yield, sugar levels have to be maintained very low to avoid catabolite repression. If sufficient oxygen is not supplied by aeration and the growth rate is not kept low (below 0.1 to 0.2 h^{-1}), ethanol is formed regardless of the aerobicity of the process and yeast yield is drastically reduced (Fig. 4). Bakers' yeasts are sensitive to glucose concentrations from 0.02 to 0.85 mM glucose (2). In consequence sugar must be fed continuously at a slow rate to avoid repression of the respiratory pathway enzymes (2). It would be most desirable to obtain both a high productivity (biomass produced per unit of time) and yield. However, no bakers' strains able to maximize yeast yield, maintaining full respiratory metabolim, and at a high growth rates (bakers' yeast can grow up to 0.4 to 0.5 h^{-1} growth rates) have been isolated.

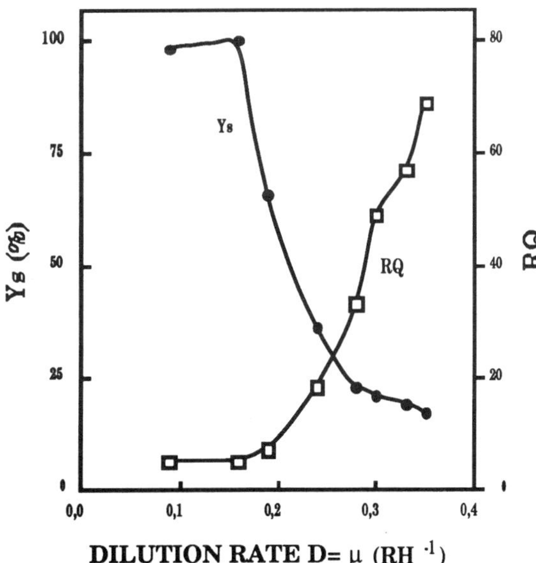

Fig. 4. Effect of the growth rate, $\mu(h^{-1})$, on biomass yield, Ys (g. cells per 100 g. sucrose) (%). RQ= respiratory quotient

5.STABILITY OF BAKERS' YEAST

Bakers' yeast are sold under different commercial types: desiccated (dried), frozen and fresh (compressed) (2, 3). Compressed yeasts are perishable (3-4 weeks) and have to be maintained at 0-4ºC. Dried yeasts can be maintained at room-temperature for at least one year and before use they have to be rehydrated with water at 35-45° C (5). During the rehydration process yeasts lose up to 31 % of their weight. Substances such as glutathion, a reducing agent, are lost so that the dough is not active. These compounds weaken the structure of the gluten, which does not retain CO_2 and in consequence the dough does not raise (1, 2, 3).

5.1.Role of mitochondria in viability

It seems that the loss of substances during the rehydration process is due to changes in the structure of the cell membranes. The phospholipid content decreases, the ergosterol content, as well as the degree of saturation of the fatty acids also change and as a consequence there is a decrease in viability (2). Cell membrane composition also influences the leavening of the dough and the viability during bakers' yeast storage (3). Airation and the presence of good functional mitochondria are essential to synthesize unsaturated fatty acids and ergosterol. Unsaturated fatty acids are in their turn essential for the mitochondria membranes to become functional, whereas ergosterol stabilizes cell membrane phospholipids (2).

On the other hand, the effects of reduced temperature and freezing on yeast have received little attention. Yeasts in general survive freezing in high percentages when they are quickly frozen (close to at least -80ºC). However, for food preservation and in commercial practice, relatively higher temperatures (-15 to -20ºC) are applied, so that the survival ratio is less and the freezing injury is higher (2). This is attributed to the formation of ice cristals causing intracellular dehydration and membrane damage. The effect of thawing seems also to be recrystallization of ice, causing further injury. Baked breads or cakes are often frozen to preserve their freshness for prolonged storage periods. It is believed that the loss of fermenting activity of the yeast occurs mainly during frozen storage, rather than during freezing of the dough (2, 3). Yeasts which possess a particular membrane structure are either more sensitive (osmotolerant yeasts) or more tolerant (alcohol-tolerant yeasts) to the lethal effect of freezing than are other yeasts (1, 2).

Some yeasts are highly tolerant to organic solvents (ethanol) and high temperature and possess cell membranes particularly resistant to these agents. Transference to bakers' yeast of mitochondria from these yeast strains might give rise to heteroplasmon bakers' yeast (Fig. 2) more tolerant to the described agents and to autolysis during storage at different temperatures. There are references in the literature describing both increase in viability of yeast strains which have received mitochondria from thermotolerant strains acting as donors (6), and correlation in yeast strains which simultaneously become resistant to organic solvents and to freezing. Preliminary results (A.C.Codón, unpublished) indicate the viability of the transference of mitochondria from laboratory donors to baker yeasts, after protoplast fussion and selection of cells with the appropriate nucleus.

5.2.Role of the carbohydrate content

The main carbon source present in molasses is sucrose, which is hydrolyzed into glucose and fructose. Part of this sugar is converted into storage material (trehalose and glucogen) which is degraded during long stationary periods (1). Threalose plays an important role in osmoregulation, acting as a protecting agent which maintains the structure of the molecules under stress conditions. Thermotolerance and resistance to dryness increase when yeasts have higher concentrations of trehalose. High fermentative capacity has also been associated with a high trehalose content so that, increasing trehalose levels has resulted in a proportional increase in fermentative capacity (2, 3). Viability decreases rapidly when the trehalose content diminishes to 20-5% of the cell dry weight (Fig.5).

The trehalose content depends on culture conditions. Possible correlations between trehalose content and viability (autolysis resistance) and fermentative capacity, both in wild type strains and in mutants which have maltose activity in the presence of glucose have been established (1, 2, 3). There are also references in the literature about strains which increase their trehalose content in the presence of glucose because they possess a constitutive MAL locus (3).

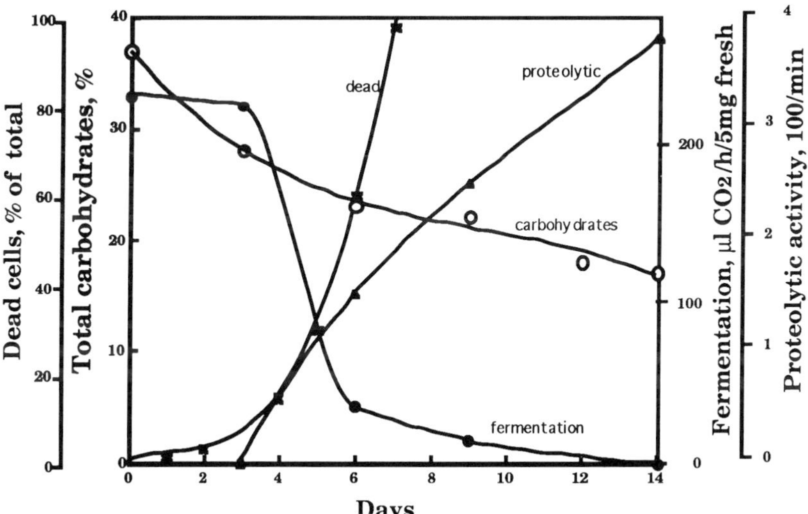

Fig.5. Decrease of the carbohydrate content and the fermentation activity and increase of dead cells and proteolytic activity of bakers' compressed yeast during storage at 35ºC. (reprinted from (7) with permission)

5.3. Role of the proteolytic activity

A simultaneous increase in cell death and proteolytic activity (Fig.5) during bakers' yeast storage has been observed (7). Viability might therefore be improved by the isolation of mutants with defective proteases (2). In order to do so it is necessary to check first whether the lack of one or several of these proteases makes a strain more resistant to autolysis, and if so, to get bakers' yeast defective in these proteases.

6. IMPROVEMENT OF DOUGH FERMENTATION

During the fermentation of the dough, bakers' yeasts produce rapid and efficiently CO_2 from fermentable sugars, mainly maltose, which results from the hydrolysis of starch carried out by endogenous amylases present in the dough. There are data in the literature which strongly support a relationship between capacity to hydrolyze maltose and fermentative capacity, so that a good fermentative capacity depends on a high potential to ferment maltose (1). Maltose hydrolysis requires the presence of this sugar and the process is repressed by glucose, which represses maltase and maltose-permease synthesis and inactivate maltase enzyme. The best fermentative strains are those which possess constitutive MAL genes. Bakers' yeast with these characteristics have been constructed by transformation with MAL genes possessing constitutive promoters (2). In addition strains insensitive to catabolite repression can also be isolated and are also able to ferment maltose in the presence of glucose. This can be favoured by selection of mutants resistant to the toxic analogue 2-deoxy-glucose, by isolating such mutants on 2-deoxy-glucose and raffinose. These mutants could be affected: in genes related to invertase or α–galactosidase; in unspecific regulatory genes related to carbon catabolism, which affect most hydrolytic enzymes; in genes specifically related to maltose metabolism, affecting for instance maltase stability in relation to its inactivation by glucose, etc. In fact, some isolated mutants, (A.C.Codón, unpublished) derepressed for glucose repression of maltose uptake, ferment the disaccharide without first requiring a 50-60% drop in glucose levels and result in faster fermentation rates. In the presence of glucose and absence of maltose some of the mutants lack maltase activity. However in the presence of both, maltose and glucose, they show a much higher maltase activity than that of the wild type. Experiments are currently being carried out to determine the number of genes involved in 2-deoxy-glucose resistance in the mutants.

7. IMPROVEMENT OF NUTRITION QUALITIES

Nutrients may appropriately be added to a bread to correct a dietary insufficiency or to restore its content to a level representative of the food prior to processing (proteins, vitamins, iron, manganese, etc.). For instance, protein malnutrition is a problem affecting some groups subsisting on diets of cereals. It may be desirable to improve protein intake by adding protein fortifiers to the dough (1). Alternatively, specific amino acids can be added to improve the bread quality. Lysine, and to a lesser extent methionine and threonine, are the limiting amino acids in cereal products. However, cost becomes a very important factor, especially when the dietary improvement is intended for large number of impoverished persons. In addition, effects of the supplement on the physiological properties of doughs and on organoleptic properties of the finished products are often limiting factors (1). Amino acid-enriched *Saccharomyces* yeasts can be added directly to a cereal to improve its nutritive value, indicating that amino acids need not go through a costly purification procedure, but might form an integral part of a food stuff. The resulting cell composition can vary depending on growing conditions, but yeast aminograms have in general a fairly high level of lysine and relatively poor levels of threonine and sulfur-containing amino acids. These levels can be increased isolating amino acid overproducer mutants (8, 9).

Slow-growing amino acid overproducer yeast mutants can be selected in batch cultures as resistant to the corresponding amino acid analog. However, in continuous culture (chemostat) only mutants able to grow at the highest growth rate are isolated so that their growth rates are in all cases similar to that of the wild type. This method provides a useful long-term selection system based on the fact that *Saccharomyces* bakers' yeasts acidify the medium during growth on fermentable carbon sources. Therefore the nutrient pump of the chemostat is switched on and off in response to the pH of the culture (Fig.6).

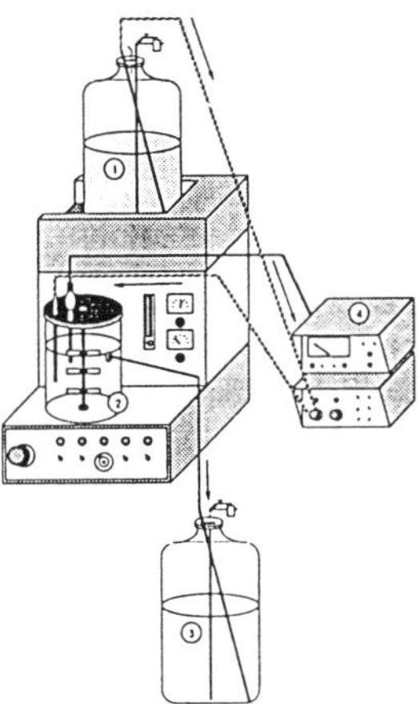

Fig.6. Chemostat that automatically controls culture density with a pH-meter

By this procedure amino acid-analog resistant mutants have been isolated in continuous culture able to produce almost 40 times more threonine, 150 times more methionine and 10 times more lysine than the wild type.(7,8). This is particularly interesting since, as indicated, the amino acid balance of cereal protein for human and animal feed is markedly improved by the addition of the nutritionally-limiting amino acids methionine, threonine and lysine.(1)

The aroma of bread is created during baking by thermal reactions between some of its individual components formed by yeast fermentation and between these compounds and other dough constituents such as sugars and amino acids. There is a tendency to reduce the time of dough fermentation, with the consequent decrease in flavour and aroma.(2). When shortening the fermentation time cysteine as a reducing compound is included in the dough composition. There have been attempts to produce fermentation flavours that could be used in leavened products. Amino acids can also be reacted with products of the yeast fermentation.(2). The abovementioned selection procedure would also make it possible to obtain mutants able to overproduce other amino acids such as cysteine or tyrosine, which are essential in the contribution of flavours and aroma to doughs and bread.

CONCLUSIONS

Techniques for the genetic manipulation of bakers' yeast are somehow established and are being applied to commercial targets, some of which have been briefly outlined here. It is becoming increasingly apparent that some bakers' yeasts are more amenable to recombinant DNA technology than classical genetic techniques. However, the use of the former methods will depend upon the perceived adaptability of the strains derived. And, it must not be forgotten that the yeast must be capable of producing the desired product but must also be compatible with the manufacture of a food product.

REFERENCES

1. S.A. Matz, in: Bakery Technology: Packaging, Nutrition, Product Development, QA. pp. 169-204. Elsevier Science Publ. 1989.
2. G. Reed and T.W. Nagodawithana, in: Yeast Technology, 2nd. Ed. Van Nostrand Reinhold, ed. 1991.
3. J.P. van der Walt, in: The Yeast, vol. I, pp. 95-117. A.H. Rose et al , eds. AC Press, 1987.
4. S. Fogel and J.W. Welch, in: Biological Research on Industrial Yeast, vol.I, pp. 99-110. G.G. Stewart et al , eds. CRC Press, 1987.
5. W. Xiao and G.H. Rank. Gene 76 : 99-107. 1989.
6. J. Jiménez and T. Benítez. Curr. Genet. 13 : 461-469. 1988.
7. H. Suomalainen. Curr. J. Appl. Microbiol. 1 : 1-12. 1975.
8. E. Martínez-Force and T. Benítez. Curr. Genet. 21 : 191-196. 1992.
9. E. Martínez-Force and T. Benítez. Biotechnol. Bioeng. 40 : 643-649. 1992.

ECB6: Proceedings of the 6th European Congress on Biotechnology
edited by L. Alberghina, L. Frontali, P. Sensi

Effect of HeNe and CO_2 lasers on yeast and dough

L. Bolognani[a], F. Casacci[b], N. Haegens[b], C. Magnani[a] and T. Venturelli[a]

[a] Department of "Biologia animale", Univ. of Modena. Via Berengario 14, 41100 Modena Italy

[b] Barilla G.R. F.lli s.p.a. Via Mantova 136, 43100 Parma Italy

1. INTRODUCTION

Laser effects on yeast (*Saccharomyces cerevisiae*) were studied by Karu and Lethokov (1)

In their pioniering work increment of yeast growth and protein synthesis were demonstrated.

This led us to suppose the involvment of laser radiation on redox processes coupled to ATP synthesis in as much as research on integrated biosystems in vertebrate at the mitochondrial level demonstrated the increase of ATP synthesis in the HeNe irradiated samples (2)

Our group focused the attention to ATP concentration and CO_2 production in yeast exposed to low power (HeNe) laser irradiation and to CO_2 laser irradiation.

2. MATERIAL AND METHODS

2.1. Test on yeast

100mg of commercial pressed yeast stored at + 4°C were suspended in 20 ml of 0,05 mol/l phosphate buffer pH 7,0.

1 g of sucrose was added and the volume brought up to 50 ml with buffer (final molarity 5,8 per 10^{-2} mol/l).

The volumetric flask was then immersed in a water bath at +36°C for 10 min. CO_2 production was measured by a manometric device (modified from Hayduck instrument (3) and adapted to microscale) connected with the volumetric flask. CO_2 measurements were taken at 10 min. intervals for 1 hour. The values were expressed as µl of CO_2 produced by 100 mg yeast in 1 hour (4).

Non irradiated samples were compared with HeNe irradiated samples.

2.2. Test on dough

15 min irradiated (Space-Valfivre HeNe laser at 632 nm,30 min at 10 cm, power density 8,25 mW/cm^2 energy density 14,84 joule/cm^2) yeast was used to prepare dough in duplicate samples:10 g of flour, 5,50 g water, 0,24 g yeast, 0,15 g NaCl were mixed in automatic equipment (National MFG Co. Lincoln Nebraska-USA) and pressed four times by a Morse Industrial Corporation,(Denver, Colorado, USA).

After 10 min. at +22°C CO_2 production by dough was measured in glass closed bottles connected with manometric device as in the previous section. CO_2 was measured after 10,20 and 30 min. at +37°C. Values were expressed as µl CO_2 per g^{-1} per h^{-1}.

HeNe irradiated samples were compared with non irradiated samples.

CO_2 laser (Polaris-Team Laser) was also tested for stimulating CO_2 production in dough. The samples were irradiated at 10600 nm for 5 min (pulsed at 0,05 seconds, setting at 6 Watt) on rotating device (15 rpm) for 5 min.

2.3. ATP determination on dough

ATP concentration was tested by luminometric procedure (luminometer LKB 1250 connected with a Linseis recorder). Reagent was lyophylized luciferase-luciferin reagent (from *Photinus pyralis*) in 0.02 M pH 7.0 phosphate buffer solution (Boehringer CLS Kit).

3. RESULTS

3.1. CO_2 production by HeNe irradiated yeast

Preirradiated yeast increases the CO_2 production respect to non irradiated samples as reported in Figure 1.

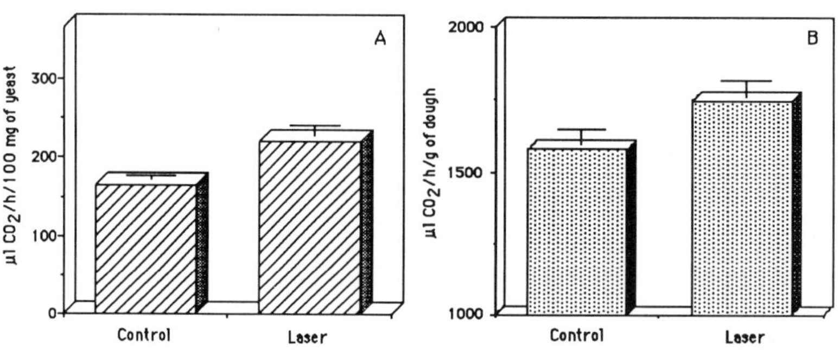

Figure 1.CO_2 production in yeast (A) and dough (B) treated with HeNe laser

The CO_2 production is enhanced by adding bicarboxylic acids or tricarboxylic acids

This means that redox decarboxylative processes are related to Krebs cycle and to the redox phosphorylative production. In fact ATP synthesis is increased in irradiated samples respect to controls (see 3.3 paragraph).

3.2. CO_2 production in dough

CO_2 production is enhanced in dough prepared with preirradiated yeast as reported in Figure 1B.

Infrared (10600 nm) CO_2 Laser (POLARIS TEAM LASER) is also effective in stimulating dough fermentation; the CO_2 produced was 2651 μl CO_2/g/h in treated versus 2207 μl CO_2/g/h in controls.

3.3 ATP production in dough

ATP production is increased when irradiated yeast is used to prepared dough. Values in ATP concentration (pmoles ATP/100 mg yeast) at increasing time on reported in the table 1.

Table 1.ATP concentration in dough at different times of fermentation

	To		10 min at 22°C		10 min at 37°C		20 min at 37°C		30 min at 37°C	
	CONTROL	+ LASER	CONTROL	+ LASER	CONTROL	+ LASER	CONTROL	+ LASER	CONTROL	+ LASER
AVERAGE	728	1286	1117	1764	1209	1964	1507	2229	1838	2745
STANDARD DEVIATION	± 340	± 617	± 199	± 782	± 232	± 163	± 281	± 686	± 283	± 1160
PERCENTAGE OF DIFFERENCE		+ 56,6		+ 36,7		+ 38,4		+ 32,4		+ 33,0
SIGNIFICANCE		p = 0,075		p = 0,075		p < 0,01		p < 0,05		p = 0,09

4.DISCUSSION

Laser energy in the range of red (632 nm) and far infrared are both effective in stimulating redox processes and coupled ATP production by enzymatic systems in yeast.

These experiments are related to the investigation of Salet *et al.*(5) and Passarella *et al.*(2) who demonstrated ATP increment in heart cell cultures and in purified mitochondria after HeNe irradiation. Pleiotropic mechanisms are likely to be involved in laser enzyme interaction and laser effects on membrane of eucariotic cells. Lubart (6) studied the effect of HeNe and 760 laser on Ca^{2+} transport trough cell membrane in bull sperm cells demonstrating changes in permeability.

Possible laser interaction with ATPase systems are suggested on the basis of experimental investigation on ATPase Na^{+}-K^{+} and myosin ATPase. Certainly mitochondrial ATPase are not the unique involved and specific photoabsorpting molecules could be predicted as candidate in explaining biological effect of laser according to Rochkind *et al.* (7). Independently from the hypothesis of laser mechanisms involved our data outlines the stimulating effect of laser energy on yeast and its possible application in technological fields.

REFERENCES

1. I. Karu and V.F. Letokhov, in: S. Martellucci and A.N Chester (eds.), Laser photobiology and photomedicine. Proceeding of course on laser application in biology and medicine. New York: Plenum Press, 1985, 57.

2.S. Passarella, L. Roncali, R. Cicero and E. Quagliariello, Laser in Life Sciences, 2 (1988) 161.

3. Hayduck Berl-Lunge D'anse, in: F. Villavecchia and G. Fabris (eds.), Metodos de analysis quimico industrial ed labor. Madrid, 1948, Vol. 5.

4. L. Bolognani, M. Costato, C. Magnani and N. Volpi, Laser & Technology, 2 (1992) in the press.

5.C. Salet, G. Moreno and F. Vinzens, Exp. Cell Res., 120 (1979) 25.

6.H. Friedmann, R. Lubart, I. Laulicht and S. Rochkind, J. of Photochem. and Photobiol., 11(1) (1991) 87.

7. L. Rochkind, N. Barrnea Razon, A. Bartal and M. Schwartz, Neurosurgery, 20 (1987) 843.

ECB6: Proceedings of the 6th European Congress on Biotechnology
edited by L. Alberghina, L. Frontali, P. Sensi

MIXED CULTURE PRE-FERMENT OF LACTIC AND PROPIONIC ACID BACTERIA FOR IMPROVED SHELF-LIFE OF SOUR RYE BREAD

P. Javanainen and Y.-Y. Linko

Helsinki University of Technology, Department of Chemical Engineering, Laboratory of Biotechnology and Food Engineering, SF-02150 Espoo, Finland

1. INTRODUCTION

For proper crumb, aroma and flavor formation acidification is necessary for most rye bread doughs containing more than 20% of rye flour[1]. Conventionally, a commercial starter culture, a spontaneous sour, or a portion of previous sour dough are used as seeding in rye sour dough fermentations. In order to prevent bread molding, we have previously shown that a pre-ferment based on controlled formation of lactic, acetic and propionic acid by a mixed culture of *Lactobacillus* sp. and *Propionibacterium* sp. is a good starter for improved wheat bread shelf-life through *in situ* biotechnical preservative formation[2-5]. A similar improvement with a mixed culture pre-ferment of *L. brevis* together with *P. jensenii* in the absence of yeast was suggested by the preliminary experiments with rye sour dough fermentation[6]. In the present work, the effect of factors such as dough yield, temperature and quantity of pre-ferment on rye sour dough fermentation based on pre-ferments of *L. brevis* together with *P. acidipropionici* or *P. jensenii* was examined with the aid of statistical experimental design and empirical modelling.

2. MATERIALS AND METHODS

2.1. Micro-organisms

Lactic and propionic acid bacteria used were *L. brevis* TKK 105-6-1, *P. acidipropionici* TKK 114-3-1 and *P. jensenii* TKK 114-4-1. *L. brevis* was maintained on Lactobacilli MRS broth (Difco). Propionic acid bacteria were maintained on a medium (pH 7.2) containing tryptone, 1%, yeast extract, 1% and tomato juice, 20% (*P. acidipropionici*), or a medium (pH 7.0) containing sodium lactate, 1%, tryptone, 1%, yeast extract, 1%, KH_2PO_4, 0.025%, and $MgSO_4{\cdot}7H_2O$, 0.0005% (*P. jensenii*). The inoculated stabs were incubated at 30°C for 20 h (lactic acid bacteria) and for 72 h (propionic acid bacteria), and stored at +5°C for up to six weeks.

2.2. Pre-fermentation

Sprout-damaged wheat flour, SDWF, (Falling Number 62, moisture 9.9%, protein 10.4%, ash 1.9%) used as substrate has been previously shown by Javanainen and Linko[4] to enhance lactic, acetic and propionic acid formation in mixed culture pre-fermentation. The substrate for culturing the microorganisms consisted of malt α-amylase-treated SDWF, 10%, wheat germ, 0.5%, $(NH_4)_2SO_4$, 1.34%, KH_2PO_4, 0.22%, $MgSO_4{\cdot}7H_2O$, 0.96%, $CaCl_2{\cdot}2H_20$,

0.0336%, and yeast extract, 0.1%. The pH was adjusted to 6.5, and before autoclaving (120°C, 20 min) 2% $CaCO_3$ was added to prevent the pH from falling below 5.2. The pre-fermentation was carried out for 5 d at 30°C by using a 5% inoculum each of the 20-h *L. brevis* culture and 72-h *P. acidipropionici* or *P. jensenii* cultures.

2.3. Optimization of the acidification process

A 2^3-factorial experimental design[7] with α-points (α = 1.353) and three replicates at the center point and a total number of 17 experiments (Table 1) was used in optimization. Dough yield (the weight of dough divided by the weight of flour, expressed in %), temperature (°C) and the amount of pre-ferment (% of flour weight) were chosen as the independent process variables in rye sour dough fermentations with mixed culture pre-ferments of lactic and propionic acid bacteria. Second degree polynomials were used to estimate the responses of pH (Y_{pH}), titratable acidity (Y_{TA}, mL 0.1 M NaOH/5 g flour), lactic acid concentration (Y_{LA}, g/100 g sour dough), acetic acid concentration (Y_{AA}, g/100 g sour dough), molar ratio of lactic to acetic acid ($Y_{L/A}$) and propionic acid concentration (Y_{PA}, g/100 g sour dough) according to Thompson[8]. The levels of significance were given as: *** $P < 0.001$; ** $P < 0.01$; * $P < 0.05$. Response surfaces were drawn as the function of two selected input process variables with the third process variable at the center point level.

2.4. Analytical methods

Titratable acidity and pH per 10 g (as is) of the sample were measured using standard procedures[9]. Titratable acidity was expressed as the amount (mL) of 0.1 M sodium hydroxyde consumed per 5 g flour. Acetic, lactic and propionic acids were determined by HPLC as described previously[2-5].

Table 1
The experimental design with the independent process variables of dough yield (X_1, %), temperature (X_2, °C) and the amount of pre-ferment (X_3, % flour weight); α = 1.353.

Experi-ment	Coded values x_1	x_2	x_3	Real values X_1 (%)	X_2 (°C)	X_3 (%)
1	-1	-1	-1	176	27	18
2	1	-1	-1	324	27	18
3	-1	1	-1	176	35	18
4	-1	-1	1	176	27	32
5	1	1	-1	324	35	18
6	-1	1	1	176	35	32
7	1	-1	1	324	27	32
8	1	1	1	324	35	32
9	-1.353	0	0	150	31	25
10	+1.353	0	0	350	31	25
11	0	-1.353	0	250	25	25
12	0	+1.353	0	250	37	25
13	0	0	-1.353	250	31	15
14	0	0	+1.353	250	31	35
15	0	0	0	250	31	25
16	0	0	0	250	31	25
17	0	0	0	250	31	25

3. RESULTS AND DISCUSSION

Dough yield and temperature were the dominating factors in the second degree polynomial models (Fig.1, eq. (1); Fig.2, eq. (2), eq. (3)) to describe pH, titratable acidity, lactic, acetic and propionic acid concentration, and molar ratio of lactic to acetic acid at 17, 24, 40 and 48 h acidifications. Under the experimental conditions for rye sour dough fermentation a pH less than 4, and a titratable acidity of about 16, a level typical of a fermented sour sponge[10], was usually obtained in 17 to 24 h with both pre-ferment combinations investigated. The strongest acidification shown by the lowest pH values (Fig. 1) and highest titratable acidities (Fig. 2) was obtained at a high dough yield of 350% (soft dough).

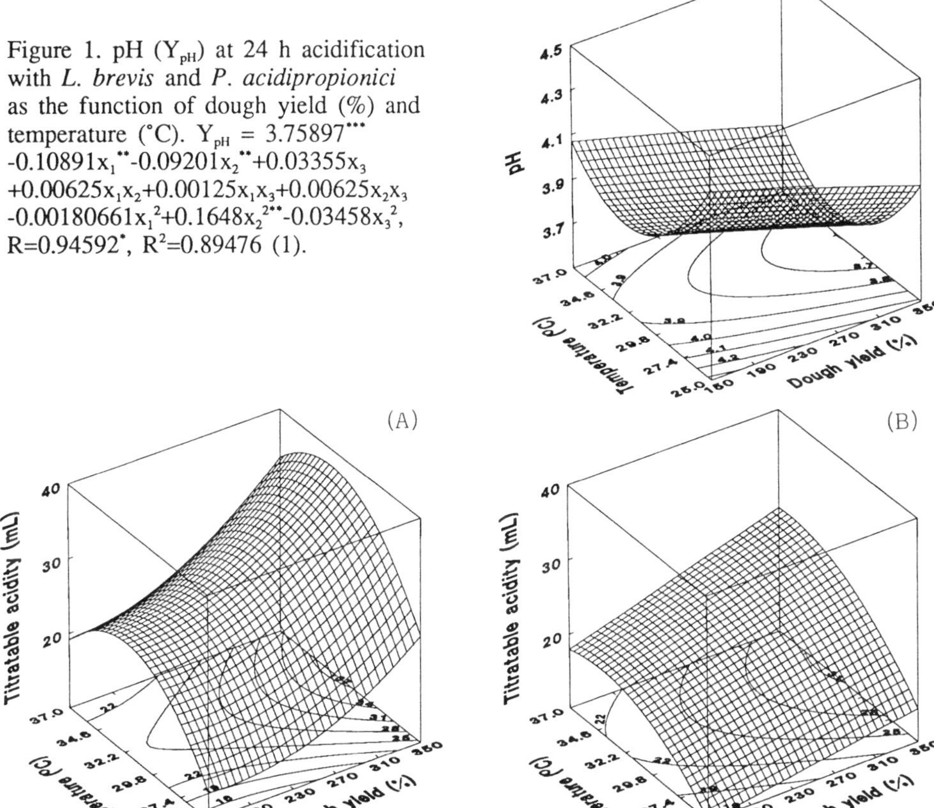

Figure 1. pH (Y_{pH}) at 24 h acidification with *L. brevis* and *P. acidipropionici* as the function of dough yield (%) and temperature (°C). $Y_{pH} = 3.75897^{***} -0.10891x_1^{**}-0.09201x_2^{**}+0.03355x_3 +0.00625x_1x_2+0.00125x_1x_3+0.00625x_2x_3 -0.00180661x_1^2+0.1648x_2^{2**}-0.03458x_3^2$, $R=0.94592^{*}$, $R^2=0.89476$ (1).

Figure 2. Titratable acidity (mL 0.1 M NaOH/5 g flour) (Y_{TA}) at 24 h acidification with *L. brevis* and *P. acidipropionici* (A) or *P. jensenii* (B) as the function of dough yield (%) and temperature (°C). A: $Y_{TA(A)} = 27.63323^{***}+4.61421x_1^{**}+2.72099x_2+0.75919x_3 +0.525x_1x_2 +0.35x_1x_3 +0.275x_2x_3+1.79009x_1^2-4.51929x_2^{2*}-0.88662x_3^2$, $R=0.90061$, $R^2=0.8111$ (2).
B: $Y_{TA(B)} = 25.89248^{***}+3.36313x_1^{***}+2.06773x_2^{***}+0.45753x_3 +0.1125x_1x_2 +0.2625x_1x_3 +0.0625x_2x_3 -0.22711x_1^2-2.39496x_2^{2***}+0.6095x_3^2$, $R=0.98659^{***}$, $R^2=0.97335$ (3).

Further, under the process conditions resulting in the highest measured propionic acid concentrations of 0.55 g/100 g sour dough with *P. acidipropionici* at 48 h or 0.4 g/100 g sour dough with *P. jensenii* at 24 h, titratable acidity values of 27.3 to 34.8 (*P. acidipropionici*) and 14.3 to 24.8 (*P. jensenii*), were obtained (Table 2). This indicated medium to strong sour rye dough. The lactic to acetic acid molar ratio of 1.9 to 4.0 was within the optimal range reported for sour dough bread[1]. Consequently, the mixed culture pre-ferment of lactic and propionic acid bacteria appeared to be a potential alternative to the traditional rye sour dough fermentation starters for improved mold-free bread shelf-life through *in situ* increased propionate level.

Table 2
Acidification conditions for the highest propionic acid concentrations.

	L. brevis and *P. acidipropionici*[a]			*L. brevis* and *P. jensenii*[b]		
Fermentation time (h)	pH	Titratable acidity (mL)	Lactic to acetic acid ratio	pH	Titratable acidity (mL)	Lactic to acetic acid ratio
17	3.72	27.3	4.00	4.40	14.3	1.86
24	3.79	30.3	4.00	3.96	19.7	2.62
40	3.69	34.5	3.87	-	-	-
48	3.68	34.8	1.93	3.83	24.8	1.91

[a] X_1 = 320%, X_2 = 35°C, X_3 = 32%; propionic acid concentration 0.55 g/100 g sour (48 h)
[b] X_1 = 176%, X_2 = 27°C, X_3 = 32%; propionic acid concentration 0.40 g/100 g sour (24 h)

REFERENCES

1. G. Spicher, In Biotechnology, Vol 5 (G. Reed, ed.), Verlag Chemie GmbH, Weinheim (1983) 1.
2. P. Javanainen, Y.-Y. Linko and P. Linko, Proc. 4th Eur. Congr. Biotechnol. Amsterdam 3 (1987a) 313.
3. P. Javanainen, Y.-Y. Linko and P. Linko, Proc. 23. Nordic Cereal Congr., The Danish Cereal Society, Copenhagen (1987b) 197.
4. P. Javanainen and Y.-Y. Linko, Food Biotechnol. (in press).
5. P. Javanainen and Y.-Y. Linko, J. Cereal Sci. (in press).
6. P. Javanainen and Y.-Y. Linko, 5th Eur. Congr. Biotechnol. Abstract Book, 1990, 487.
7. G.E.P. Box and J.S. Hunter, Ann. Math. Stat., 28 (1957) 195.
8. D. Thompson, J. Food Process Preserv., 6 (1982) 155.
9. Standardmethoden für Getreide, Mehl und Brot, Arbeitsgemeinschaft Getreideforschung, Verlag Moriz Schäfer, Detmold (1978).
10. E. Oura, H. Suomalainen and R. Viskari, In Economic Microbiology, Vol 7 (A.H. Rose, ed.), Academic Press Inc., Oxford (1982) 87.

ECB6: Proceedings of the 6th European Congress on Biotechnology
edited by L. Alberghina, L. Frontali, P. Sensi

MOULD INTERACTIONS BETWEEN CONTAMINANTS AND STARTER CULTURES USED FOR SAUSAGE FERMENTATION

Susanne Juul Andersen

Department of Biotechnology, Building 221, The Technical University of Denmark, DK-2800 Lyngby / Chr. Hansen´s Laboratorium, Denmark A/S. The experiments were carried out at The Department of Food Science and Technology, University of Reading, England, supervised by J.D. Owens and P.C. Cook.

1. INTRODUCTION

Traditionally, mould fermented sausages are produced by spontaneous colonization by the houseflora. At present, the application of well-characterized starter cultures for surface treatment of meat products increase. The purpose being to obtain a more uniform product in respect of taste, flavour and colour. It is therefore important to use a starter culture that is a strong competitor and capable of inhibiting growth of contaminants. Growth of contaminants on the sausages is undesirable because it may result in off-flavour or cause discolouration. Further, the contaminants may produce mycotoxins and thereby represent a potential health hazard. *Penicillium nalgiovense* and to a lesser degree *P. camemberti* species are frequently used as starter cultures in the ripening process of mould fermented sausages. In the present investigation, isolates belonging to the species *P. nalgiovense* and *P. camemberti* were examined to determine their antagonistic ability and their tolerance of antagonism. Colony interactions were studied by growing the potential starter cultures in dual culture with mould species often isolated from the surface of mould fermented sausages and generally considered contaminants. Reaction types and interaction index based on percentage inhibition of growth were determined.

2. MATERIALS AND METHODS

2.1 Fungal isolates

Five isolates belonging to the species *P. nalgiovense* (IBT12004, IBT12005, IBT13477, IBT3800 (type) and IBT3793) and one *P. camemberti* (IBT12012) were selected as starter cultures. The contaminants belonged to the species *P. commune*, *P. verrucosum*, *Aspergillus flavus* and *Cladosporium sphaerospermum*. The contaminants have all, except for *A. flavus*, been isolated from mould fermented sausages. *P. verrucosum* and *P. commune* respectively produce the mycotoxins ochratoxin A and cyclopiazonic acid. *A. flavus* is a potential producer of the mycotoxin aflatoxin. *C. sphaerospermum* produces black spots.

2.2 Media and growth conditions

The fungi were inoculated on Czapek yeast autolysate agar (CYA) and meat extract agar (ME). ME consists of 19.45% Lab–Lemco powder, 1.4% $NaNO_2$ and 1.5% Oxoid agar (techn. no.3) plus trace elements (5 ppm $CuSO_4$,$5H_2O$ and 10 ppm $ZnSO_4$,$7H_2O$). The plates were incubated 10 days at 25°C.

2.3 Inoculation

Plates were inoculated in triplicates at one and two points for examination of growth in single and dual culture. Conidia were transferred from 7–day–old cultures on CYA. The distance between inoculation points, in dual cultures, were : 25 mm in combination with *P.verrucosum* and *C. sphaerospermum*, 30 mm in combination with *P. commune* and 40 mm in combination with *A. flavus*. The fast growing *A. flavus* were inoculated two days later than the starter cultures in order to allow establishment of the starter cultures.

2.4 Examination of colony interaction in dual culture

Interactions between starter cultures and contaminants grown in dual culture were classified according to table 1.

Table 1 *Reaction types*

Reaction types	Classification description
A	Mutual intermingling (both fungi grow into one another without any macroscopic signs of interaction)
B	Inhibition of one organism on contact (the inhibitor species grows, unchanged or at reduced rate, up to and around or overgrows the inhibited colony)
C	Mutual inhibition on contact, or small space between colonies (< 2mm)
D	Mutual inhibition at a distance (≥ 2mm)

Adapted from Porter (1924) and Johnson & Curl (1978)

The percentage inhibition of radial growth was calculated from the following expression.

$$\text{Inhibition} = \frac{\text{Radius, single culture} - \text{Radius, dual culture}}{\text{Radius, single culture}} \times 100\ \%$$

Radial growth in both single culture and dual culture (in the direction where the two species were facing each other) were measured daily. The inhibition was determined on the day the two cultures growing in the dual culture would have met (dependent on the inoculation distance) if unaffected by each other. Based on the sum of percentage the starter cultures inhibits the four contaminants and the sum of percentage the starter cultures are inhibited growing in dual culture with the contaminants, an interaction index is calculated (Shearer & Zare–Maivan, 1988). In other words the interaction index (I) is the sum of rankings of ability to inhibit (R_1) and to resist inhibition (R_2). Therefore the smaller the interaction index, the better the competitive ability of the starter culture.

3. RESULTS AND DISCUSSION

The percentage of interactions classified according to the different reaction types (table 1) is shown in table 2. The reaction types of the interactions observed on CYA and ME showed inhibition of either one (CYA: 75%, ME: 17%) or both cultures (CYA: 25%, ME: 54%) on contact. Strong antagonistic interactions, i.e. the cultures inhibit each other at a distance, were observed on ME (29%) only. When the interactions were classified as reaction type B, the starter culture was the inhibiting culture in most cases on CYA. The exceptions were *P. commune* in combination with *P. nalgiovense* IBT3800 and *A. flavus* in combination with each of the starter cultures. The starter culture was always the inhibiting culture when the interactions were classified as reaction type B on ME. The classification in reaction types showed no correlation with the percentage inhibition determined.

Table 2 *Percentage of interactions classified according to the different reaction types*

Reaction types	A	B	C	D	B_S *	B_C *
CYA	0	75	25	0	46	29
ME	0	17	54	29	17	0

*) B_S = reaction type B: the starter culture is the inhibiting culture. B_C = reaction type B: the contaminant is the inhibiting culture.

The determined interaction index and growth rates are shown in table 3. The competitive abilities of the starter cultures are better, the smaller the value of the interaction index. When grown on CYA *P. nalgiovense* IBT3793 is the best competitor, but not the fastest growing starter culture. ME is rich in proteins and amino acids and, therefore, it is of considerable interest with reference to sausage production. Rated by the interaction index the best competitors are *P. nalgiovense* IBT12005, *P. camemberti* IBT 12012 and *P. nalgiovense* IBT12004 in respective order, when grown on ME. There was no correlation between growth rates of the cultures applied in single culture and the interaction index. When grown in single culture on ME, *P. nalgiovense* IBT12004 and IBT12005 were the fastest growing starter cultures and *P. camemberti* IBT12012 was the slowest growing starter culture. Hence the ability to compete is not determined by the growth rate. The ability to compete is dependent upon the growth medium as the starter cultures that compete best on CYA and ME are not the same. On ME the sum of percentage *P. camemberti* (IBT12012) is inhibited growing in dual culture with the contaminants is negative (table 3). The reason being that the growth rate of *P. camemberti* is higher (66,7%) when grown in combination with *C. sphaerospermum* compared to growth in single culture.

The contaminants that are inhibited the least by the starter cultures are *P. verrucosum* and *A. flavus* on CYA and *P. commune* on ME. *C. sphaerospermum* is the most inhibited contaminant on both CYA and ME.

Table 3 *Interaktionsindex (I) and growth rate*

CYA	Inhibition		Resistance			
Isolate	Σ% inhibition	Rank, R_1	Σ% inhibition	Rank, R_2	$I = R_1 + R_2$	Growth rate (mm/day)
IBT12004	28.4	6	58.4	6	12	2.7 ± 0.1
IBT12005	53.7	3	33.0	1	4	3.1 ± 0.1
IBT3793	68.8	1	34.4	2	3	2.8 ± 0.1
IBT3800	38.3	5	36.8	5	10	2.2 ± 0.1
IBT13477	43.7	4	35.7	4	8	3.1 ± 0.1
IBT12012	61.1	2	34.8	3	5	1.9 ± 0.1

ME	Inhibition		Resistance			
Isolate	Σ% inhibition	Rank, R_1	Σ% inhibition	Rank, R_2	$I = R_1 + R_2$	Growth rate (mm/day)
IBT12004	67.7	2	36.4	4	6	2.0 ± 0.1
IBT12005	81.4	1	29.9	2	3	2.0 ± 0.1
IBT3793	31.0	6	33.0	3	9	1.4 ± 0.1
IBT3800	53.7	4	45.3	5	9	1.6 ± 0.1
IBT13477	66.8	3	62.8	6	9	1.8 ± 0.1
IBT12012	53.7	4	−38.8	1	5	0.6 ± 0.1

REFERENCES

Johnson, L.F. & Curl, E.A. (1978). *Methods for research on the ecology of soil–borne plant pathogens*. Burgess publishing company, USA.

Porter, C.L. (1924). Concerning the characters of certain fungi as exhibited by their growth in the presence of other fungi. American journal of botany *II*, 168–188.

Shearer, C.A. & Zare–Maivan, H. (1988). *In vitro* hyphal interactions among wood– and leaf–inhabiting Ascomycetes and Fungi Imperfecti from freshwater habitats. Mycologia *80* (1), 31–37.

ECB6: Proceedings of the 6th European Congress on Biotechnology
edited by L. Alberghina, L. Frontali, P. Sensi

PARTIAL CHARACTERISATION OF PECTIN METHYLESTERASE INHIBITOR FROM KIWI FRUIT (ACTINIDIA CHINENSIS) PURIFIED BY AFFINITY CHROMATOGRAPHY

L. Camardella[1], G. di Prisco[1], A. Giovane[2], L. Quagliuolo[2] and C. Balestrieri[2]

[1]Institute of Protein Biochemistry and Enzymology, C.N.R. Via Marconi 10, 80125 Napoli, Italy
[2]Department of Biochemistry and Biophysics, University of Naples, Via Costantinopoli 16, 80138 Napoli, Italy

INTRODUCTION

One of the problems encountered in the production and storage of fruit juices and purees is the undesirable change of organoleptic properties, due to the enzymatic degradation of the fruit cell wall components.

Fruit juice is a system composed of two phases: the liquid phase, generally termed "serum", and the solid phase, which is called "cloud". This solid phase is composed of partially disrupted cells and cell wall debris and forms a thin suspension which contributes to the appearance and the characteristics of the juice. Soluble pectin is a component of fruit juice which stabilises the cloud.

Pectin is a major component of the plant cell wall; it is a polymer of galacturonic acid units linked by α, 1->4 glycosidic linkages; the carboxyl groups in position 6 are methylated to a various extent.

Pectin methylesterase (PME) is the first enzyme involved in the degradation of pectin; it produces a pectin with a lower methylation degree, which can then undergo further enzymatic cleavage by polygalacturonase. It is well known that PME activity changes the texture of fruit products (1, 2); therefore, control of PME activity is of primary importance in fruit product preparation and storage.

Soluble pectin in fruit juice can undergo deesterification of methyl groups by action of PME, with a consequent increase of free carboxyl groups which can then interact with calcium ions present in the juice; the resulting calcium polypectate is

----

This work is partially supported by the "Progetto Finalizzato Biotecnologia e Biostrumentazione" of the C.N.R.

insoluble and precipitates, destabilising the cloud and inducing undesired clarification of the juice. The most common technology used to overcome this problem is thermal inactivation (105-115°C) of PME by means of pasteurisation. However this procedure can alter the color and the flavour of the juice. Alternative procedures have therefore been investigated.

Kiwi fruit (Actinidia chinensis) have been reported to contain a glycoprotein which is a powerful inhibitor of PME (3); the effectiveness of partially purified PME inhibitor (PMEI) from kiwi fruit in the stabilisation of orange juice and concentrate has been evaluated (4).

An improved purification method and a preliminary molecular characterisation of the glycoprotein is herewith reported.

RESULTS AND DISCUSSION

1) Protein purification

The inhibitor activity was determined by detecting the residual PME activity after incubation of the enzyme for 5 min at 25°C in the presence of the inhibitor, by means of a standard spectrophotometric assay at pH 7.5 (5).

Purification of PMEI was achieved by means of affinity chromatography on PME-bound Sepharose resin prepared according to the following procedure. PME was purified from tomato by chromatography on heparin-Sepharose as described (6). The enzyme was dialysed against 0.1 M sodium bicarbonate pH 8.3, containing 0.5 M NaCl; CNBr-activated Sepharose 4B (Pharmacia) was allowed to swell in the same buffer. The enzyme was then mixed with the gel at the ratio of 2 mg of protein/ml of wet gel and was allowed to react for 4 h at room temperature under gentle stirring. The gel was then washed on a buchner funnel with the same buffer used for the coupling reaction, followed by several washes with 0.5 M NaCl. The gel was then treated with 0.1 M Tris-HCl pH 7.5 at 4°C overnight and washed with 10 mM Tris-HCl pH 7.5, containing 0.1 M NaCl.

Fully ripened kiwi fruit were peeled and homogenised in water; PMEI was found in the aqueous extract, whereas the PME activity, which is associated with the cell wall, remained in the precipitate. The aqueous extract containing the inhibitor was partially purified by precipitation with ammonium sulfate at 80% saturation, the precipitate was resuspended in 10 mM Tris-HCl pH 7.5, containing 0.1 M NaCl, dialysed against the same buffer and directly loaded onto the affinity column. After exhaustive washing with 10 mM Tris-HCl M pH 7.5, containing 0.5 M NaCl, the inhibitor was eluted in one step with 3 M NaCl in the same buffer.

The purified protein showed a single band on SDS-PAGE having an apparent molecular mass of 15 kDa; the isoelectric point as determined by isofocusing was relatively low, less than 3.5.

2) N-terminal amino acid sequence

The purified protein was reduced and alkylated with 4-vinylpyridine "in situ" and subjected to automatic Edman degradation on an Applied Biosystems Model 477 A protein sequencer, according to the procedure suggested by the manufacturer (7).

In the first cycle, two amino acid residues were detected in comparable amounts; the following ten cycles yielded clearly identifiable residues:

1 10
NH_2Glu/Ala-Asn-His-Leu-Ile-Ser-Glu-Ile-Cys-Pro-Lys

This result indicates that the protein contains a single polypeptide chain with unblocked amino terminus, showing heterogeneity at the first position.

3) Trypsin digestion

The purified protein was reduced in 6 M guanidine-HCl, 0.5 M Tris-HCl pH 8.0, by addition of dithiothreitol (75-fold molar excess to the protein) and incubation for 2 h at room temperature; then alkylated with 4-vinylpyridine (2000-fold molar excess to the protein) for 1 h at room temperature. The

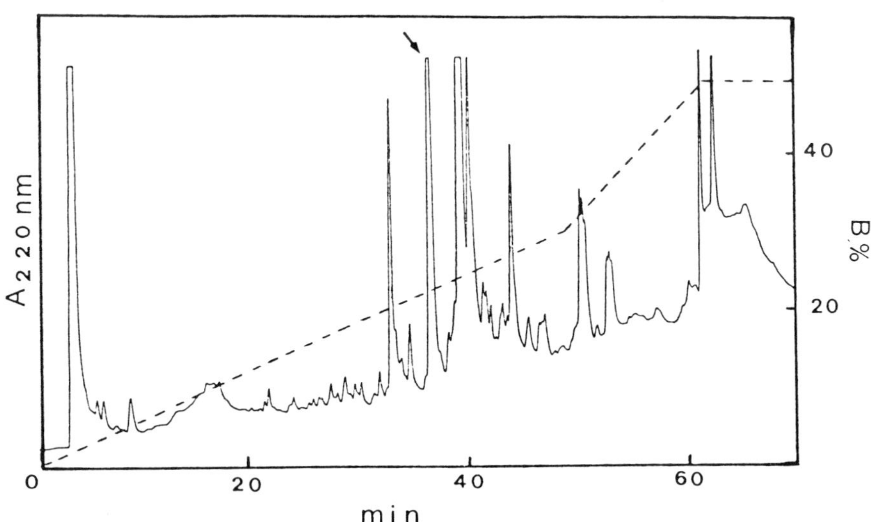

Figure 1. HPLC elution profile of peptides derived from PMEI trypsin digestion. The dashed line indicates solvent gradient composition.

protein was then freed from excess reagents by desalting on a prepacked Sephadex G-25 column (Pharmacia) equilibrated with 0.1% trifluoracetic acid (TFA). The eluate was lyophilised, the protein was resuspended in 1% ammonium bicarbonate at a concentration of 0.5 mg/ml, and trypsin was added (trypsin to protein ratio, 1:50 w/w). Digestion was carried out for 3 h at 37°C. The digest was then lyophilised, redissolved in 0.1% TFA and analysed by reverse-phase HPLC on an Aquapore RP 300 column (solvent A: 0.1% aqueous TFA; solvent B: 0.08% TFA in acetonitrile). Fig. 1 shows the separation pattern; the peak indicated by the arrow was subjected to amino acid sequence analysis and contains the peptide located at the amino terminal end. Analysis of amino acid composition and sequencing of the remaining peptides are under way.

The affinity chromatography procedure for the purification of PMEI is highly effective, since it allows to obtain the pure protein in good yield and in a single step. Amino acid sequence data indicate that PMEI consists of a single polypeptide chain; the finding of two residues at the first position can be explained by the presence of two isoforms with at least a single amino acid substitution.

In order to have information on the control of the synthesis and the expression of this protein in kiwi fruit, extensive sequencing of the protein is in progress; the aim is to design oligonucleotides on the basis of the amino acid sequence and to use them as probes for screening a DNA library from kiwi fruit and to isolate the corresponding gene.

REFERENCES

1) K. Pozsar-Hajnal and M. Polacsek-Racz
Acta Alimentaria 4 (1975) 271
2) C. Versteg, F. M. Rombouts, C. H. Spansen and W. Pilnik
J. Food Sci. 45 (1980) 969
3) C. Balestrieri, D. Castaldo, A. Giovane, l. Quagliuolo and
L. Servillo Eur. J. Biochem. 193 (1990) 183
4) D. Castaldo, A. Lovoi, L. Quagliuolo, L. Servillo,
C. Balestrieri and A. Giovane J. Food Sci. 56 (1990) 1632
5) A. E. Hagerman and P. J. Austin
J. Agric. Food Chem. 34 (1986) 440
6) D. Hawke and P. Yuan ABI User Bulletin no. 28 (1987)

Bioinformatics for Biotechnology

ECB6: Proceedings of the 6th European Congress on Biotechnology
edited by L. Alberghina, L. Frontali, P. Sensi

BIOINFORMATICS FOR BIOTECHNOLOGY

C. Saccone

Department of Biochemistry and Molecular Biology, University of Bari, 70125 Bari, Italy

1. INTRODUCTION

The rapid sequencing of nucleic acids and, in particular, the sequencing projects of entire genomes have generated a strong interest towards the methodologies for the treatment and analysis of sequence information.
Since the genetic material can be considered as a text written in a given language, the linguistic approach is becoming a very useful tool for deciphering genetic information.
To understand the language of the genes and then manipulate it, represents one of the challenges of modern Biology.
The evolutionary theory of the living matter, however, clearly indicates that the same information can be written in a large variety of forms, despite the adherence to a number of universal rules. In this context Molecular Evolution, which studies and compares the texts, becomes a powerful way for discovering such rules.
We report here some contributions given by our laboratory to the analysis of the information contained in the sequences of nucleic acids.
In the first part, we describe two algorithms, CORRELA [1] and WORDUP [2] which analyze known and unknown sequences following a linguistic approach.
In the second part, we present the Stationary Markov Clock [3] we have proposed for the quantitative measurement of the genetic distance between multialigned sequences and CONSTRAINT [4], a stochastic method for detecting hyper-variable or hyper-constrained regions in homologous nucleotide sequences which can supply useful information on the structure- function relationship of biological macromolecules.

2. LINGUISTIC ANALYSIS

2.1 CORRELA

CORRELA introduces a global method for pairwise sequence comparison involving a correlation analysis between the 4^n tetranucleotide frequencies present in two compared sequences. The correlation index between sequences X and Y, respectively L_X and L_Y nucleotides long, is defined as:

$$r = \frac{\sum_k (X_k - X_m)(Y_k - Y_m)}{\sqrt{\sum_k (X_k - X_m)^2 \sum_k (Y_k - Y_m)^2}}$$

where X_k and Y_k are the occurrence numbers for the k-th tetramer in sequence X and Y respectively with $X_m=(L_x-3)/256$ and $Y_m=(L_y-3)/256$, the average occurrences of each tetramer in sequence X and Y. r is an adimensional parameter, ranging -1 to +1. Tetramers revealed to be the optimal choice for avoiding false positives, i.e. high correlations without sequence similarity. If two sequences have a high similarity their correlation is necessarily near 1.

This method is based on observed quantities and does not depend upon the assumption of a theoretical model.

CORRELA may be used as a reliable test of overall sequence similarity. It does not involve sequence alignment procedures and permits the rapid screening of large amounts of sequence data. It can be used also in the important task of databases cleaning from redundancies, which constitute a bias in any kind of statistical analysis.

CORRELA has been successfully used to obtain purified databases for 5' and 3' untranslated human mRNA regions whose analysis, performed with WORDUP algorithm, is shown in the following paragraph.

2.2. WORDUP

This algorithm is based on a first order Markov analysis. It allows the analysis, in sets of unaligned sequences, known to be functionally related (e.g. promoter regions, introns, untranslated regions, etc.) but not evolutionary homologous, of the statistical pattern of DNA oligomers which are assumed to be Poisson distributed.

By using a very fast string matching algorithm, modified for application to the nucleotide four letter alphabet [5], we were able to single out words of length ranging from six to ten nucleotides, the typical size of the DNA sequence signals (e.g. promoters, DNA binding sites, ..) which due to their non random statistical properties may be biologically active.

The statistical significance of the k-th word can be established according to the elementary χ^2 statistics:

$$\chi_k^2 = \frac{(P(s_k) - \Pi(s_k))^2}{\Pi(s_k)}$$

where $P(s_k)$ and $\Pi(s_k)$ are the observed and expected number of occurrences for the w-mer k in the sequence sample under investigation. $\Pi(s_k)$ is calculated by

assuming a first order stationary Markov chain which takes into account dinucleotides and mononucleotides frequencies observed in each sequence.
This method has been applied to 521 eukaryotic promoter sequences extracted from release 24 of EPD database [6] and two datasets containing 1446 and 1702 sequences corresponding to 5' and 3' untranslated regions of human mRNAs.
The cis-acting sequences most commonly found in the region surrounding the transcription initiation site, such as TATA-box, CAAT-box and GC-box, were clearly identified in the promoter sequences. The AAUAAA hexamer typically found 15-25 nucleotides upstream the mRNA poly-A was also successfully identified. All this confirms the reliability of the analysis carried out with WORDUP. In addition, several other words whose biological role is still unknown were isolated (Table I).
Some interesting perspectives of WORDUP analysis consist in directing experimental work aimed at demonstrating the functional role of unknown sequence motifs and at assessing the biological role of anonymous sequences produced in large amounts, e.g. in megasequencing projects.

3. MOLECULAR EVOLUTIONARY ANALYSIS

3.1 The Stationary Markov Clock

It is our opinion that any molecular evolution model having a chance to be realistic must be stochastic. The necessity to follow a stochastic approach in the study of gene evolution is also dictated by purely biological reasons. When the genetic distance among organisms is measured at the level of nucleic acids instead of proteins, one is faced with a number of problems due to wide fluctuation in nucleotide composition among different organisms and to well-established properties of the genetic code , such as its triplet structure and degeneracy. This leaves us with the stochastical approach as the only appropriate means to describe molecular evolution at the nucleotide level.
A few years ago, we proposed a simple stochastic model of molecular evolution, the Stationary Markov Clock (SMC), and we demonstrated that it is at work in a large variety of evolutionary dynamics operating at the gene level.
The SMC allows us:

a) to determine the four-nucleotide substitution rate matrix, R_{ij}, which describes the propensity of nucleotide j to be substituted by nucleotide i independently from time T.

b) to calculate the effective average evolutionary rate of sequence sites (i.e. the rate of silent codon position in mRNA genes).

c) to construct phylogenetic trees.

Thanks to our method we have been able to determine the genetic distances between Primates [7], several orders of mammals [8,9] and, very recently , the cladistic relationship between human populations and thus the coalescence time of mitochondrial DNA of our species [10]. The latter have been calculated by using the sequences of the D-loop region and applying a correction factor to our method, which takes into account the number of variable sites in that particular region.

Table I. Most significant oligonucleotides obtained by analyzing words seven nucleotides long

	Occurrences		χ^2
	Observed	Expected	
EPD motifs			
GTATAAA	62	3.81	886.59
CTATAAA	61	4.35	735.87
TATATAA	63	5.92	550.07
TGACGTC	11	1.04	94.58
ATAAAAG	65	5.88	68.67
GGGGCGG	40	13.07	55.44
GATTGGC	13	2.33	48.56
TATATAT	27	6.72	44.38
TATTTAA	20	5.13	43.01
ATAAATA	59	6.90	37.49
Human 5'-UTR motifs			
CCGCCGC	138	66.59	76.58
AGTGGAT	26	6.76	54.76
CTGCAGC	88	42.29	49.41
CGGGACC	45	16.98	46.24
CGCCGCC	122	66.59	46.11
CCGGAGC	71	32.76	44.64
CTCCAGC	87	43.15	44.56
TGGATTT	30	9.74	42.14
TTCTGTG	32	11.09	39.43
CATCCTG	33	12.17	35.65
Human 3'-UTR motifs			
AAATAAA	476	159.16	630.73
CAATAAA	294	93.31	431.64
TAATAAA	287	104.68	317.55
TTATTTT	358	169.46	209.77
TATTTAT	211	87.34	175.08
TGTATTT	215	90.92	169.33
GAATAAA	192	81.02	152.02
TTTTTAA	385	207.57	151.67
ATATTTT	243	121.87	120.39
CAGCCTC	178	83.61	106.56

3.2 CONSTRAINT: a Stochastic method for detecting regions with different evolutionary dynamics in multialigned sequences

By applying SMC, we have devised a stochastic method able to trace the pattern of gene evolution in multialigned sequences. It allows us to detect gene regions with distinct evolutionary dynamics, e.g. regions which significantly deviate from an expected "average". An accurate detection of the hyper-variable or hyper-constrained regions may provide useful information on the structure-function relationship of biosequences. In addition, the selection of distinct evolutionary dynamics may help in the correct use of biosequences as reliable molecular clocks.

The identification of hyper-variable or hyper-constrained regions within a gene can be achieved comparatively starting from the multiple alignment of a set of homologous sequences. Consider n aligned homologous sequences of s sites. The multialignment can be written in a matrix form X, with $x_{rc} = \{A, C, G, T\}$. The matrix can be rewritten as a sequence of column vectors: $X_1,...,X_s$ with $\mathbf{X}_c=[x_{1c}, x_{2c}, x_{nc}]$, for the c-th site of the multialignment. For each site c we calculate a value p_c:

$p_c = 1$ if $x_{1c} = x_{2c} = ... = x_{nc}$
$= 0$ otherwise

In such a way, given a set of n homologous genes, we identify only two classes of sites: variable ($p_c = 0$) and invariable ($p_c = 1$). The permanence density of a gene in a window of w sites is defined as:

$$P_z(w) = \sum_{h=z}^{z+w+1} p_z / w, \qquad z = 1,....,s-w+1$$

and the average permanence density of the entire gene as:

$$\overline{P} = \frac{1}{s} \sum_{c=1}^{s} p_c$$

To appreciate the statistical meaning of the observed values of the permanence density, their expectation and statistical fluctuations, a suitable simulation procedure is performed in such a way that expected quantities $\pi_z(w)$ of permanence density are also calculated.

To perform a correct simulation, a phylogenetic tree must be assumed. By defining a path along the tree, all other sequences can be obtained

starting from any real one. The simulated sequences for a given tree topology can be calculated according to the SMC.
To appreciate the significant deviations in the permanence density of the observed values with respect to expected ones, obtained through simulation, for each window z, χ^2 values are calculated:

$$\chi_z^2(w) = \left[w(P_z(w) - \pi_z(w))^2 \right] / \pi_z(w)$$

According to the elementary statistics, the standard $\chi 2$ distribution determines the probability $P(\chi^2(w))$ that the z-th interval follows the average expected behaviour. Thus the "constraint index" is defined as

$$R(\chi_z^2(w)) = [1 - P(\chi_z^2(w))] \cdot f_z$$

The absolute value of this quantity represents the probability that the z-th interval will be anomalously constrained or variable. The value f_z is set equal to +1 or -1, as $Pz(w) > \pi_z(w)$ or $Pz(w) < \pi_z(w)$ respectively.

Figure 1. Observed ($P_z(w)$) and expected ($\pi_z(w)$) permanence density values with their standard deviation plotted versus position z by using a window size of 20 nucleotides for the second codon positions of cytochrome b.

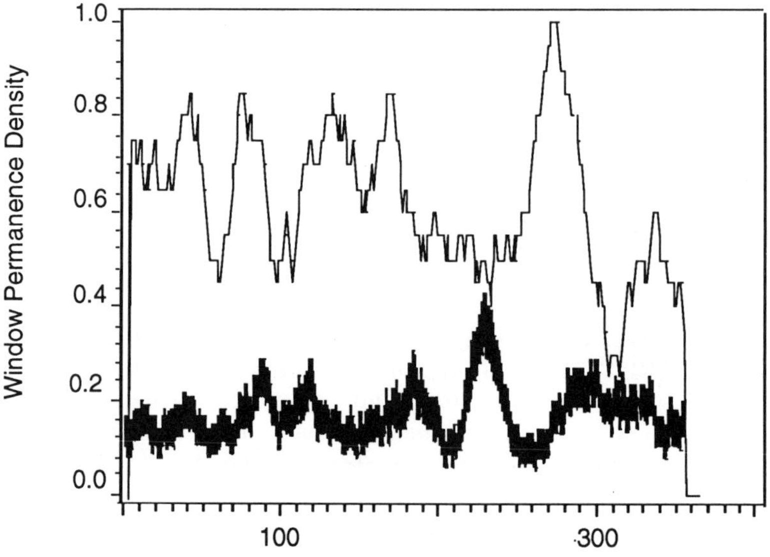

An example of application of CONSTRAINT method is reported for the second codon positions of some cytochrome b genes. This analysis clearly shows that the degree of base permanence is significantly higher than the expected value. This is due to the fact that substitution dynamics affects all the sites in the simulated process equally, while in reality only a fraction of sites change. The profile of the base permanence density reported in Figure 2 provides relevant indications on the pattern of functional constraints acting on the cytochrome b since the analysis of the second codon positions obviously reflects the evolutionary process at the level of the aminoacids. Our results, showing five principal peaks of base permanence corresponding to known functional regions of the protein, are in line with the structural model previously proposed for the mitochondrial subunit of cytochrome b [11].

REFERENCES

1. E. Pizzi, M. Attimonelli, S. Liuni, C. Frontali and C. Saccone. Nucleic Acid Res. 20 (1992) 131-136.

2. G. Pesole, N. Prunella, S. Liuni, M. Attimonelli and C. Saccone. Nucleic Acid Res. 20 (1992) 2871-2875.

3. C. Saccone, C. Lanave, G. Pesole, and G. Preparata. °In: Doolittle R.F. ed. Methods in Enzymology. Accademic Press, °Inc.1990; 183:570-583.

4. G. Pesole, M. Attimonelli, G. Preparata and C. Saccone. Molecular Phylogenetics and Evolution 1 (1992) 91-96.

5. R.S. Boyer and J.S. Moore. Comm. ACM 20(10) (1977) 62-72.

6. P. Bucher. EMBL Nucleotide Sequence Data Library Release 23, Postfach 10.2209, D-6900 Heidelberg, Germany.

7. E.C. Holmes, G. PESOLE and C. SACCONE. J. Hum. Evol. 18 (1989) 775-794

8. C. Saccone, G. Pesole and E. Sbis . J. Mol. Evol. 33 (1991) 83-91.

9. Pesole G., E. Sbis , F. Mignotte and C.Saccone. J. Mol. Evol. 33 (1991) 537-542.

10. Pesole G., E. Sbis , G. Preparata and C. Saccone. 9(4) (1992) 587-598.

11. N. Howell. J. Mol. Evol. 29 (1989) 157-169.

Acknowledgements

This work was financed by MURST (Italy), Progetto Finalizzato Biotecnologie e Biostrumentazioni (C.N.R., Italy) and Progetto Finalizzato Ingegneria Genetica (C.N.R., Italy).

ECB6: Proceedings of the 6th European Congress on Biotechnology
edited by L. Alberghina, L. Frontali, P. Sensi
1994 Elsevier Science B.V.

BIOREP - PERMANENT INVENTORY OF BIOTECHNOLOGY RESEARCH PROJECTS IN THE EUROPEAN COMMUNITIES

M.L.H. Lalieu

Library of the Royal Netherlands Academy of Arts and Sciences (KNAW), P.O. Box 41950, 1009 DD Amsterdam, The Netherlands

1. History

BIOREP comprises a directory of ongoing, publicly funded, biotechnology research in the EEC member states. It started as a pilot project in March 1987 with the aims of identifying trends in European biotechnology, furthering of scientific contacts and providing a basis for co-ordination and more efficient planning. This project was endorsed by the former Task Force for Biotechnology Information (TFBI) of Directorate-General XIII (Telecommunications, Information Industries and Innovation, Luxemburg).

In 1987-1989 the project was introduced into the member states and an organizational network of national focal points - ministries, funding bodies or national documentation centres - was established, which are responsible for data collection and input. In the first round 4469 project descriptions were obtained from 10 countries.

The database has been on-line available since July 1989 through ECHO, the European Commission Host Organisation in Luxemburg. A printed directory was issued January 1990 [1]*.

The outcome of the evaluation of the pilot project late 1990 has been that BIOREP has met its basic objectives, but that it needs reinforcement, particularly regarding database environment as well as product development and marketing, aimed at raising data quality and use on the one side and at cost recovery on the other, as offset for the limited possibilities of structural funding [2].

2. Present status and developments

Following a Limited Call for Proposals (issued April 1991) the project has been continued from January 1992 and was transferred to the Concertation Activities under the current biotechnology research programme BRIDGE of Directorate-General XII (Science, Research and Development, Brussels).

---

* The reports and newsletters mentioned belong to the category of grey literature; they are directly available from the author

Since the introduction of the project was 4-5 ago and in a number of countries functionaries had changed or even the focal point task was transferred to another organization, a new round of consultations with the national focal points has been accomplished, aimed at consolidation and with an outlook to the future.
In addition to the national focal points, the European Molecular Biology Laboratory (EMBL) contributes to the system, and the European biotechnology (BRIDGE) and biotechnology related agricultural and food research programmes (ECLAIR, FLAIR) are included as well.
The usage of the database is satisfactory and has increased from 1200 accessions in 1990 to 3214 accessions in 1992.
The developments in 1993, following the updating and addition of new information throughout the year and as a result of promotion, will be monitored carefully.

It is tried to improve quality and raise the status of BIOREP along the following lines:

■ Transformation of the organizational network of national focal points to a consortium structure (BIOREP Advisory Group).
This will be a basis for definite national commitments and for assessing the future level of funding, warrant maximum involvement of all parties and increase of coverage, and help BIOREP finding its future position in the European biotechnology information infrastructure.
This idea has met with a favourable reception. In consultation with the Commission of the European Communities it has been stipulated that the BIOREP Advisory Group will become a consultative group without official status. The constituent meeting will be held September 1993.

■ Transfer of BIOREP to a biotechnology user environment.
On ECHO, BIOREP is in a neutral environment between numerous other, non-biotechnology databases and thus has too low profile. Other hosts like DIMDI or DataStar are considered, to be accessed directly and/or via international biotechnology information networks like offered by MSDN (Microbial Strain Data Network) in Cambridge. Also the possibilities for BIOREP in the future Common Core Database (CCD), the future integrated information infrastructure for biotechnology in Europe, will be explored. The CCD pilot study has started March 1993.
The latest developments prompt to short-term decision making. BIOREP has clearly outgrown its pioneering stage, and would have to pay a considerable financial contribution if it is to stay on ECHO. This policy towards database producers has developed as a result of budgetary cutbacks, but also to prevent distortion of competition with commercial hosts.

■ Product development, marketing and distribution.
In addition to on-line access there is a need for printed products and CD-ROM. Ultimately a situation should be arrived at where people can consult the data at or in the vicinity of

their desks, preferably in conjunction with other biotechnology databases (CCD, "one-stop shopping").
In 1993 a feasibility study will be carried out into the implementation of additional profiles (national, organization and expertise), which will be relationally coupled to the project descriptions. This might add substantial value to the system and aims at catering for the needs of various user groups. However, duplication of already available biobusiness directories should be avoided.
The original idea behind the above-mentioned Limited Call was to position, market and distribute BIOREP as product package in collaboration with professionals in the field ("combination of skills"). But contacts with publishers did not lead to an offer. Therefore the solution should no longer been sought in a purely commercial partnership, but in the possibilities provided by the network and by the not-for-profit sector. As BIOREP can now be regarded as an infrastructural asset with a low threshold - and some national focal points have explicitly stressed the importance of easy accessibility and the public domain character of the information - this aspect should count heavily in the discussions on distribution and product development. From the BIOREP Advisory Group new initiatives will be taken.

■ Geographical extension and collaboration with non-EEC centres.
Since in the view of the Commission of the European Communities BIOREP is seen as a means not only of promoting greater mutual awareness within Europe, but of giving European biotechnology research a high profile in the global context, they would be happy to encourage the developments of wider links, at least as second priority.
For the next years there are three new types of contacts in the planning:

(a) Extension with the EFTA countries.
In the framework of the European Economic Space these countries will collaborate more closely with the EEC countries. In European research programs researchers from the EFTA countries even will become entitled to act as main contractors. Although their research projects and collaborations will be displayed in BIOREP through the files submitted by the Commission, the basis will be substantially broader if they would join as full participants (pan-European approach).
NORDINFO, the cooperative body for scientific and technical information and documentation of the Scandinavian countries, has already been contacted.

(b) Close collaboration with the COBIOTECH Information Centre (CIC) in Moscow, established in 1991.
This collaboration will aim at establishing a biotechnology information infrastructure for the Russian federation and at East-West biotechnology information exchange. Funding has to be sought for a joint project to be formulated later in 1993.

(c) Collaboration with the China-EC Biotechnology Centre (CEBC) in Beijing.
This centre was inaugurated 1 November 1991 as an international agency for scientific and technical cooperation and information exchange between Europe and China in sectors of biotechnological application. BIOREP has the potency to function as an important intermediary tool in the process of developing China-EC joint R&D activities.

All developments mentioned in this paper have been described recently in more detail [3,4].

Since its first hesitating steps BIOREP provides a number of new challenges. It is hoped that within a few years the system will have a high impact and that the products derived from it will become additional useful tools in the hands of the international biotechnology research community.

References

1. M.L.H. Lalieu (ed.), Permanent Inventory of Biotechnology Research Projects in the European Communities, Commission of the European Communities, Luxemburg, 1990 (EUR 12574)

2. M.L.H. Lalieu, BIOREP Newsletter, Issue 4, Amsterdam, February 1992 (ISSN 0923-4918)

3. M.L.H. Lalieu, BIOREP - Permanent Inventory of Biotechnology Research Projects in the European Communities, Amsterdam / Brussels, March 1993 (Progress Report 1992 + Annexes)

4. M.L.H. Lalieu, BIOREP Newsletter, Issue 5, Amsterdam, March 1993 (ISSN 0923-4918)

ECB6: Proceedings of the 6th European Congress on Biotechnology
edited by L. Alberghina, L. Frontali, P. Sensi

Hypervariable regions in B-type nuclear lamins

A. Lisa, G. Biamonti, G. Perini, F. Weighardt and C. Matessi

Istituto di Genetica Biochimica ed Evoluzionistica, Consiglio Nazionale delle Ricerche, Via Abbiategrasso 207, 27100 Pavia, Italy

Lamins are structural proteins of the nuclear lamina, a discrete layer lining the inner nuclear membrane [1]. Vertebrates express several distinct lamins which can be grouped in two types, A and B [2, 3]. A human cDNA, containing part of the gene for a B-type nuclear lamin, was recently isolated and sequenced [4]. The aminoacid sequence inferred from it has an overall similarity of 80% [4] with the carboxy end of the mouse lamin B_2 sequence, as published by [5]. However, an internal region of about 50 residues was found where the human and murine sequences diverge considerably. Very high levels of variability, roughly in the same region, also resulted from comparisons of the two mammalian sequences with the homologous proteins of chicken and *Xenopus laevis* [4]. In the attempt to clarify the evolutionary origin and the functional relevance of this localized variability of lamin B_2, we have done a detailed statistical analysis of the distribution of dissimilarity between the human and mouse sequences, both at the nucleotide and the amino acid level.

1. MATERIALS AND METHODS

The cDNA sequences of human and mouse lamin, as determined by [4] and [5] respectively, were obtained from release 33 of EMBL (accession no.s M94363 and X54098). The inferred amino acid sequences were aligned with the Needleman-Wunsch method [6], as implemented in the GAP program of the GCG package [7], using standard scores and gap penalties. The human sequence of 515 aa alignes with the 510 aa C-end of the mouse sequence, which has a total of 592 aa. In the aligned part of the mouse sequence two gaps, of 1 and 4 residues respectively, have been determined. The unaligned stretch of 82 aa at the N-end of the mouse lamin was ignored in subsequent analyses, since its human homologue has not been sequenced.

Local concentrations of dissimilarity between two aligned sequences, significantly higher than the level observed overall, have been identified with the help of a statistical procedure that can be summarized as follows.

Moving Average Dissimilarity (MAD). A window of size n ($n \approx 1/30$ of sequence length) is moved one position at the time along the aligned sequences and each time a dissimilarity value is computed as the proportion of positions in the window with a mismatch. The observed series of dissimilarities is then checked against what would be expected if mismatches were dispersed at random. For this purpose, we produce random permutations of the observed string of matches/mismatches and, for each of them, we compute the corresponding series of dissimilarities. The following 5 statistics are computed: maximum dissimilarity (d) in the series; maximal length (l_θ) of runs of dissimilarity values above the thresholds $t_\theta = p + \theta\sigma$ ($\theta = 0.5, 1, 1.5, 2$), where p is the observed overall dissimilarity and $\sigma = \sqrt{p(1-p)/n}$ is an estimate of the standard deviation of dissimilarity in a window of size n. From the sampling distributions of these statistics, resulting from 2000 random permutations, we estimated their corresponding (upper) critical values, at a 5% significance level, d° and $l°_\theta$. If in the observed series of dissimilarities we find any value greater than d°, or a run of values above t_θ longer than $l°_\theta$, we take this as evidence that the corresponding set of positions is a region of significantly high variability.

To evaluate the possibility that observed nucleotide differences result from neutral base substitutions, subject to functional constraints at the protein level, we estimated rates of non-synonymous and synonymous substitutions respectively [8], i.e. the rates at which substitutions, that respectively did or did not result in a change of amino acid, occured in the protein since the two lineages separated. This is more efficient than simply estimating rates of substitution for 1st and 2nd codon position, on one hand, and for 3rd position , on the other, since it accounts for silent substitutions at 1st position and non-silent substitutions at 3rd position. The proportions of synonymous and non-synonymous base changes were estimated with the unweighted pathway method of Nei and Gojobori [9]; substitution rates were then estimated from these, using the method of Jukes and Cantor [10]. For comparison with corresponding estimates available in the literature for other genes, rates were calculated on a per year basis, on the assumption that the man and mouse lineages separated 80 My ago [8].

2. RESULTS

With a MAD analysis of the whole human and mouse sequences we find two segments with significantly high variability, consisting respectively of nucleotides 919-1170 and 1378-1497 (counting from the first 5' site in the human sequence of 1545 nt). The union of these segments codes for 124 amino acids in man (119 in mouse) and is still very heterogeneous in dissimilarity, as shown by a second round of MAD and other statistical tests. Hence, we have further isolated from it two smaller segments (nt 955-1110 and 1414-1461), accounting for 68 aa in man (63 in mouse), the first of which was noted by [4] as conspicuously variable. These two segments jointly form a high variability (HV) region while the residual part, consisting of four segments flanking the two of HV, forms a region of medium variability (MV). What remains of the sequence, after detaching HV and MV, is a low variability region LV.

The different levels of dissimilarity in LV, MV and HV are summarized in Table 1. The proportion of amino acids that have changed between man and mouse in MV is intermediate to that of LV and HV. A more complex relation appears if changes that drastically modify the biochemical properties of amino acids are distinguished from those that do not. For this purpose amino acids are subdivided in 4 functional types: acidic, basic, polar and hydrophobic. Interestingly, most changes in MV preserve the functional type, while changes of type are as frequent here as in LV. By contrast, in HV a majority of substitutions introduce amino acids of a different functional type.

Table 1
Amino acid changes in different regions of lamin B_2

		% of amino acid changes		
	no. of aa	funct.lly different	funct.lly equal	total
LV region	391	5	6	11
MV region	56	4	17	21
HV region	63	44	31	75

From Tab. 1 it would seem that evolution in MV, although faster than in LV, was nevertheless much slower than in HV. The picture is different, however, if we consider evolution at the nucleotide level and distinguish synonymous from non-synonymous substitutions. As Table 2 shows, non-synonymous substitutions

Table 2
Rates of non-synonymous and synonymous substitutions per site per 10^9 years

	Non-synonymous	Synonymous
LV region	0.4 ± 0.06	3.4 ± 0.35
MV region	0.8 ± 0.21	6.3 ± 1.79
HV region*	–	6.3 ± 0.87
Proteins (average)†	0.9	4.7
Pseudogenes (average)†	–	4.9

(*) syn. and non-syn. rates are the same in this region; (†) data taken from [8]

in MV have occurred at a rate which is only slightly faster than in LV and quite comparable to non-synonymous rates typical of many proteins. However, synonymous substitutions apparently occurred in MV at the same very high rate as observed in HV for substitutions of both kinds. This rate is much greater in these regions than in LV and exceeds even the rates typical of pseudogenes.

3. DISCUSSION

The high substitution rate in HV and in synonymous sites of MV is not very consistent with fixation of neutral single base mutations, particularly in view of the much lower rate in synonymous sites of LV. Even if changes in HV were unchecked by constraints, they should occur at the same rate as in the equally unconstrained synonymous sites of LV. The extensive reshaping of HV can be easily explained, instead, by more drastic events such as insertion/deletions either of fragments of moderate size or of a few single bases which locally disrupted the codon reading frame. It seems then, that we should suppose that HV is essentially devoid of any relevant biological function. But the high synonymous rates in MV must also be explained. Since fixation of neutral mutations appears not a likely cause, the simplest hypothesis could be that, the disruptive events that acted on HV, initially also affected all sites of MV, which flanks HV. Later, base mutations and positive natural selection could have modified non-synonymous sites so as to restore the primitive functionality of this part of the protein. The signature of the primitive event, however, could be seen in the synonymous sites of the region. The high proportion of amino acids that, though changed, are of the same functional type in man and mouse can be easily viewed, under this hypothesis, as a partial but sufficient restoration of the ancestral activity. Consider, however, that if positive selection was strong enough to restore the function of MV, negative selection of equal strength should have prevented fixation of its initial loss, to begin with. Hence, under our hypothesis, since fixation appears to have indeed occured, the negative initial effects on MV must have been balanced by some positive effects of the changes in HV. We are thus lead to suggest that the high phylogenetic variability of the region HV might serve some useful biological purpose, perhaps associated to a species-specific activity of this particular domain of lamin B_2

ACKNOWLEDGMENTS

This reserch was financed by PF Ingegneria Genetica - CNR, which also provided a fellowship to A. Lisa. G. Perini is supported by a fellowship from AIRC.

REFERENCES

1. L. Gerace and B. Burke, Ann. Rev. Cell Biol., 4 (1988) 335.
2. J. W. Newport and D. J. Forbes, Ann. Rev. Bichem., 56 (1987) 535
3. R. Stick, Chromosoma, 101 (1992) 566.
4. G. Biamonti, M. Giacca, G. Perini, G. Contreas, L. Zentilin, F. Weighardt, M. Guerra, G. Della Valle, S. Saccone, S. Riva and A. Falaschi, Mol. Cell. Biol., 12 (1992) 3499.
5. T. H. Höger, K. Zatloukal, I. Waizenegger and G. Krohne, Chromosoma, 99 (1990) 379.
6. S. B. Needleman and C. D. Wunsch, J. Mol. Biol. 48 (1970) 443.
7. J. Devereux, P. Haeberli and O. Smithies, NAR, 12 (1984) 387.
8. M. Nei, Molecular Evolutionary Genetics, Columbia Univ. Press, New York, 1987.
9. M. Nei and T. Gojobori, Mol. Biol. Evol., 3 (1986) 418.
10. T. H. Jukes and C. R. Cantor, in H. N. Munro (ed), "Mammalian Protein Metabolism", Academic Press, New York, 1969.

ECB6: Proceedings of the 6th European Congress on Biotechnology
edited by L. Alberghina, L. Frontali, P. Sensi
1994 Elsevier Science B.V.

COMPUTER SYSTEM FOR GENES STRUCTURE PREDICTION

L. Milanesi[a], I.V. Ishenko, I.B. Rogozin, V. Kapitonov, N.A. Kolchanov, A. E. Kel, M.P. Ponomarenko

[a]Istituto di Tecnologie Biomediche Avanzate, CNR, via Ampere n. 56, 20131 Milano , Italy

Institute of Cytology and Genetics of the Russian Academy of Sciences, Siberian Department, Novosibirsk 630090, Russia.

1. ABSTRACT

In this paper we describe a computer system for the analysis of human genes structure. The recognition is made by using different techniques able to search for the potential splice sites, creating a set of potential coding regions along with the estimation of their coding potential and finally creating a gene with the maximal average coding potential. A specific program for promoter region prediction has also been developed. In order to obtain a more accurate prediction analysis a graphical interface has been provided for an easy use of the programs and for the visualisation of the results of the analysis.

2. INTRODUCTION

Recognition of gene coding regions is the important problem in computer analysis of human genome. Eukaryotic genes have a mosaic structure, they are composed of exons and introns and their boundaries are determined by the splice sites location. So, to ensure a successful search for genes coding regions in human genome a highly precision method to recognise the splice sites should be elaborated.

The characteristic of donor and acceptor splice sites are represented by very short consensus sequences containing just 2-3 conservative positions. These conservative sequences have quite a fixed location in genome, showing a considerable functional importance of these regions.

Due to the complexity of the structural and functional organisation of splice sites, it is very difficult to construct a highly precise method for recognition. For increasing the reliability in the prediction methodology, various authors suggest the use of combinations of methods and several methods have been developed for the coding regions prediction (1, 2, 3, 4).

3. SYSTEM AND METHODS

For the prediction of genes in new sequence, we have developed a computer system "GeneView" designed for the identification of the protein coding regions in human nucleotide sequences. GenView works in automated and an interactive mode. In the this mode the standard values are used during analysis, it is also possible to change the values and methods. An advanced graphical interface as been provided for an easy use of the programs and for visualizing the results.

GenView is organised in different modules that are able to perform a series of individual tasks.

3.1. Identification of the potential splice sites.

The first module of the program searches in the sequence for the donor or acceptor consensus, revealing the potential splice sites in nucleotide sequences of human genome.

In the second module these splice sites are verified by using another technique based on weight matrices (5).

3.2. Construction of a Potential Coding Fragment (PCF).

In the third module all Potential Coding Fragments (PCF) are detected and then an estimation of their protein coding potential is performed by using a dicodon statistic.

The PCF, with score of the coding potential $P_i > 0$ are selected for the further analysis as potential fragments of the gene coding regions.

3.3. Estimation of the protein coding potential of the revealed PCF.

We use two different levels of significance for the prediction of PCF.

Level 1 is able to detect "excellent" PCF. By using this level it is possible to predict with more accuracy the potential exones.

Level 2 is able to detect all "good" PCF. This level is used for reduce the possibility of missing some potential exones. In this case the over-prediction is increased.

3.4. Best genes prediction.

In the fourth module a set of potential genes from the set of revealed PCF is constructed. The aim is to select a limited set of genes that has a maximal coding potential. The coding potential Pg of gene consisting in m PCF separated by $m\text{-}1$ introns, is calculated as a sum of all PCF involved in this gene:

$$P_g = \sum_{i=1}^{m} P_i \tag{1}$$

Two PCF are considered present in one gene and separated by an intron only if they are not overlapping and the minimal intron length between two non-overlapped exons it is not less than 60 bp.

Let A and B be two consequent non-overlapped PCF (the PCF A being positioned upstream that B). After eliminating of the separating intron, these two PCF can form a longer fragment only at condition of saving the same translation frame.

Taking in consideration these rules, for the set of N revealed PCF it is possible to consider a matrix $Q = \{q(k,l)\}; k,l = 1,\ldots,N,$ which describes all possible variants for uniting two PCF.

$$q(k,l) = \begin{cases} 1, & \text{if k - th and l - th PCF satisfy the rules} \\ 0, & \text{otherwise} \end{cases} \tag{2}$$

The set of potential genes for a given region of the nucleotide sequence, is selected by using the path with a maximal weight through the non oriented graph set by the matrix Q. The weight is determined according to Eq. 1. The computations produce a number of paths with the maximal weights. The path with the greatest weight is assumed to correspond to a real gene.

The results of the computer analysis of the gene structure prediction are saved in a text file. It is also possible to visualise the results obtained by the automated analysis of the sequence by a graphic interface. In this case it is possible to analyse the obtained prediction in interactive mode.

The comparison between real data and the computational results showed that 16% of real exons ware under-prediction and 15% non real exons ware over-predicted.

In the correctly predicted exons set, 4% of total length were under-predicted and 13% was over-prediction. Control set includes 110 exons in 18 genes. Total length of exons is 16299 bp. In comparison with the mode of best exons prediction the over-prediction is large, but the large portion of real exons is predicted.

3.5. Promoter prediction.

The module for the prediction of promoter region is base on the algorithm that use uniform oligonucleotide distribution along Eukariotic promoters.

One of the revealed characteristic specific for promoter regions is the distribution of "TAWD" oligonucleotide in the region upstream the transcription start. In this case the oligonucleotides of "TAWD" types are concentrated within the narrow window from -35 to -20 position. In standard 4-th lettered alphabet the mentioned nucleotide "TAWD" may be represented as "TA A/T -C". It means that this oligonucleotide similar to "TATA" box characteristic of Eukaryotic promoters. Another characteristic revealed is represented by the high concentration of oligonucleotide "SANY" present in the same narrow window as for "TAWD" oligonucleotide. This "SANY" signal is less specific than mentioned TAWD oligonucleotide because it is also revealed in other promoter regions. In the revealed window (between -40 and -20 upstream from the transcription start) concentration of "SANY" signal is significantly higher than in other regions of promoter.

Another characteristic has been revealed in the promoter region with length of 400 bp, two groups of triplets monotonously increase or decrease concentration through all length of promoter region considered.

Table 1
Triplet combinations in promoter sequences (400 bp) with monotonously increasing or decreasing concentration upstream promoter regions

regularities	triplets combination
increasing concentration	ATA AGC TCG CTG CTC CGT CGC GTG GCG GGT GGC GGA
decreasing concentration	ATG ATT ACA AGA TTA TAC CAA CTT GAA

On the base of these characteristic we have developed a program "PROMOTER" for the prediction of promoter regions in human genes. A graphical interface is used for visualizing promoter signals such as: "CCAAT" box, "TATA" box and "Cap" signals.

ACKNOWLEDGEMENTS

This work was supported by "Genetic Engineering Project" CNR, Italy and Russian Human Genome project.

REFERENCES

1. Guigo R, Knudsen S, Drake N and Smith T (1992). Prediction of gene structure. J. Mol. Biol. 225.
2. Hutchinson GB and Hayden MR (1992). The predicition of exons through an analysis of spliceable open reading frames. Nucl. Acids Res. (in press).
3. Mural RJ, Einstein R, Guan X, Mann RC and Uberbacher EC (1992). An artificial intelligence approach to DNA sequence feature recognition. Trends Biotech. 10, 66-69.
4. Fields CA and Soderlund CA (1990). GM: a practical tool for automating DNA sequence analysis. CABIOS 6, 263-270.
5. Shapiro MB and Senapathy P (1987). RNA splice junctions of different classes of eukariotes: sequence statistics and functional